OKANAGAN COLLEGE LRC

P9-EKE-567

K
O

TD 427 .P35 D75 1989

Drinking water health
advisory.

DATE DUE

FEB 1 4 1991	
MAR 2 0 1991	DEC 2 3 2015
APR 4 1991	
APR 1 9 1991	
APR 1992 / NOV 4 1992	
SEP 1 7 1993	
NOV 3 0 1993	
APR - 5 1994	
APR 2 0 1994	
JAN 1 9 1995	
OCT 1 4 1997	
NOV 1 0 1998	
NOV 3 0 1999	
APR 2 2 2000	
DEC 1 3 2010	

BRODART, INC.   Cat. No. 23-221

OKANAGAN COLLEGE LIBRARY
BRITISH COLUMBIA

Drinking Water Health Advisory:

PESTICIDES

United States Environmental Protection Agency
Office of Drinking Water Health Advisories

 LEWIS PUBLISHERS

Library of Congress Cataloging-in-Publication Data

Drinking water health advisory. Pesticides.

  (United States Environmental Protection Agency,
Office of Drinking Water health advisories)
  Includes bibliographical references.
  1. Pesticides — Toxicology. 2. Drinking Water —
Health aspects — United States. I. Series.
RA1270.P4D75 1989   363.17'922   89–12175
ISBN 0–87371–235–8

COPYRIGHT © 1989 by LEWIS PUBLISHERS, INC.
ALL RIGHTS RESERVED

Neither this book nor any part may be reproduced or transmitted in
any form or by any means, electronic or mechanical, including photo-
copying, microfilming, and recording, or by any information storage
and retrieval system, without permission in writing from the publisher.

LEWIS PUBLISHERS, INC.
121 South Main Street, Chelsea, Michigan 48118

PRINTED IN THE UNITED STATES OF AMERICA

Preface

Scope and Purpose of the Health Advisory Program

The United States Environmental Protection Agency (USEPA) Office of Drinking Water (ODW) Health Advisory Program was initiated to provide information and guidance to individuals or agencies concerned with potential risk from drinking water contaminants for which no national regulations currently exist. The Health Advisories (HAs) discussed in this volume were developed in a cooperative effort with the Office of Pesticide Programs and in conjunction with the USEPA National Pesticide Program. HAs are prepared for contaminants that meet two criteria: (1) the contaminant has the potential to cause adverse health effects in exposed humans, and (2) the contaminant is either known to occur or might reasonably be expected to occur in drinking water supplies. Each HA contains information on the nature of the adverse health effects associated with the contaminant and the concentrations of the contaminant that would not be anticipated to cause an adverse effect following various periods of exposure. In addition, the HA summarizes information on available analytical methods and treatment techniques for the contaminant.

History and Present Status

The program was begun in 1978, and guidance was issued for the first 20 contaminants in 1979. At that time, the concentrations judged to be safe were termed "Suggested-No-Adverse-Response Levels" (SNARLs). These guidance values were retitled Health Advisories in 1981. To date, the USEPA has issued 98 HAs in final form, covering a wide variety of inorganic, pesticide and nonpesticide organic contaminants, and one microbial contaminant (*Legionella*). In addition, the USEPA is preparing additional HAs on various unregulated volatile organic chemicals, disinfectants and their by-products, and other inorganic contaminants.

Quality Assurance

Initial drafts of each HA undergo a series of thorough reviews before they are released to the public. The general technical content and the risk assessment values are reviewed by a group of independent expert scientists, an ODW Toxicological Review Panel, and any other USEPA offices with interest and expertise in the contaminant. The draft HAs are also distributed for review and comment by the public. Each HA is revised in response to criticisms and suggestions received during the review process before being released in final draft form. Each HA is periodically updated as significant new information becomes available that may impact the original conclusions or guidance values.

Acknowledgments

The development of each HA involves the participation of many individuals. The following members of the Health Advisory Program are acknowledged for their valuable contributions:

Michael B. Cook, Director, Office of Drinking Water
Susan Wayland, Deputy Director, Office of Pesticide Programs
Joseph A. Cotruvo, Ph.D., Director, Criteria and Standards Division, Office of Drinking Water
Edward V. Ohanian, Ph.D., Chief, Health Effects Branch, Office of Drinking Water
Jennifer Orme, M.S., Health Advisory Program Coordinator
Gerald F. Kotas, Director, National Pesticide Survey (1986-1988)
James Boland, Acting Director, National Pesticide Survey

Charles Abernathy, Ph.D.
Larry Anderson, Ph.D.
Ken Bailey, Ph.D.
Ambika Bathija, Ph.D.
Robert Cantilli, M.S.
Steve Clark, P.E.
Julie Du, Ph.D.
Penelope Fenner-Crisp, Ph.D.
Maria Gomez-Taylor, Ph.D.

William Hartley, Sc.D.
Krishan Khanna, Ph.D.
Peter Lassovsky, P.E.
Amal Mahfouz, Ph.D.
Bruce Mintz, B.S.
William Marcus, Ph.D.
Yogendra Patel, Ph.D.
James Murphy, Ph.D.
Robert Vanderslice, Ph.D.

The members of the Health Advisory Program would like to acknowledge the assistance provided by Life Systems, Inc. and Eastern Research Group, Inc.

Introduction

Health Advisories (HAs) are prepared by the Criteria and Standards Division, Office of Drinking Water (ODW) of the United States Environmental Protection Agency (USEPA) in Washington, DC. Documents summarized in this volume are part of the Health Advisory Program sponsored by ODW in response to the public need for guidance during emergency situations involving drinking water contamination. They provide technical guidance to public health officials on health effects, analytical methodologies, and treatment technologies associated with drinking water contamination.

The pesticide HAs were developed jointly by the Office of Pesticide Programs (OPP) and ODW in conjunction with the USEPA National Pesticide Survey. Each HA summarizes available data concerning the occurrence, pharmacokinetics, and health effects of a specific contaminant or mixture, as well as analytical methods and treatment technologies for the contaminant. The health effects data are used to estimate concentrations of the contaminant in drinking water that are not anticipated to cause any adverse noncarcinogenic health effects over specific exposure durations (see Table I). These HA concentrations include a margin of safety to protect sensitive members of the population (e.g., children, the elderly, pregnant women). Health Advisories are used only for guidance and are not legally enforceable in the United States. They are subject to change as new information becomes available.

The data for each HA were obtained through a comprehensive literature search covering available publications through September 1986. Some HAs may include additional information through July 1988 that became available after the initial literature search. Some HAs refer to data provided in studies submitted by private companies to the USEPA Office of Pesticide Programs under the Federal Insecticide, Fungicide and Rodenticide Act (FIFRA). These references are not readily available to the public, as they contain confidential business information.

Health Advisories were first made available to the public as drafts in August 1987. Comments were received until March 8, 1988. The 18 comments that were received were incorporated where appropriate.

For further information on the Health Advisories, contact the Safe Drinking Water Hotline (1–800–426–4791), or the Health Effects Branch, Criteria and Standards Division, Office of Drinking Water, USEPA, 401 M Street SW, Washington, DC 20460 (1–202–382–7571).

Table I. Health Advisories (HAs) Determined by the Office of Drinking Water

- *One-day HA* — The concentration of a chemical in drinking water that is not expected to cause any adverse noncarcinogenic effects for up to 5 consecutive days of exposure, with a margin of safety.

- *Ten-day HA* — The concentration of a chemical in drinking water that is not expected to cause any adverse noncarcinogenic effects up to 14 consecutive days of exposure, with a margin of safety.

- *Longer-term HA* — The concentration of a chemical in drinking water that is not expected to cause any adverse noncarcinogenic effects up to approximately 7 yr (10% of an individual's lifetime) of exposure, with a margin of safety.

- *Lifetime HA* — The concentration of a chemical in drinking water that is not expected to cause any adverse noncarcinogenic effects over a lifetime of exposure, with a margin of safety.

I. Assessment of Noncarcinogenic Risks

A. Selection of Data for Deriving Health Advisories

Health Advisories are based on data from animal or human studies of acceptable design. The first step in deriving HAs is a thorough review of the literature. For each study, the highest doses at which no adverse effects were observed in the test species (No-Observed-Adverse-Effect Levels — NOAELs) and the lowest doses at which adverse effects were observed (Lowest-Observed-Adverse-Effect Levels — LOAELs) are noted. For each HA, the most appropriate NOAEL (or LOAEL, if a NOAEL has not been identified) is selected from the available data based on the considerations described below.

A key factor in determining which NOAEL or LOAEL to use in calculating a particular HA is exposure duration. Ideally, the data will be taken from a study with an exposure duration comparable to the exposure duration for which the HA is being derived (see Table II). For example, a One-day HA is generally based on data from acute human or animal studies involving up to 7 days of exposure; a Ten-day HA is generally based on subacute animal studies involving 7 to 30 days of exposure.

Another factor that is considered in selecting the NOAEL or LOAEL is the route of exposure. An oral route (drinking water, gavage, or diet) is preferred.

Data from inhalation studies may also be used in deriving a HA when adequate ingestion data are not available. The relevance of data from subcutaneous or intraperitoneal studies is considered on a case-by-case basis.

Other factors that contribute to selection of the NOAEL or LOAEL are species sensitivity and the magnitude of the NOAEL/LOAEL relative to other NOAELs/LOAELs (generally the lowest concentration is used); the degree of confidence in the study; and whether the NOAEL or LOAEL is supported by other dose-response data.

B. Derivation of the One-day, Ten-day and Longer-term Health Advisories

Once the NOAEL or LOAEL has been selected, the One-day, Ten-day, and Longer-term HAs are derived using the following formula:

$$\text{HA} = \frac{\text{(NOAEL or LOAEL) (BW)}}{\text{(UF) (\_\_\_\_ L/d)}} = \text{\_\_\_\_ mg/L (or \_\_\_\_ } \mu\text{g/L)}$$

where:
- NOAEL = No-Observed-Adverse-Effect Level
  - or
- LOAEL = Lowest-Observed-Adverse-Effect Level
- BW = assumed body weight of protected individual
- UF = uncertainty factor (chosen in accordance with EPA or National Academy of Science [NAS] guidelines discussed below)
- ___ L/d = assumed daily water consumption of protected individual

The assumptions made concerning human body weights and water consumptions used in calculating each HA are given in Table II. For the One-day and Ten-day HAs, the protected individual is assumed to be a 10-kg child drinking 1 L/d of water. For the Lifetime HA, the protected individual is assumed to be a 70-kg adult consuming 2 L/d of water unless otherwise indicated. Longer-term HAs are calculated for both the child and the adult.

The uncertainty factor, chosen in accordance with NAS/ODW guidelines (Table III), ranges from 1 to 10,000 depending on the nature and quality of the data. Selection of the uncertainty factor is based principally upon scientific judgment and accounts for possible intra- and interspecies differences. Other considerations may necessitate the use of an additional uncertainty factor of 1 to 10. These considerations include the significance of the adverse health effect, pharmacokinetic factors, counterbalancing of beneficial effects, and the quality of the available data base for each contaminant.

Table II. Data Used to Develop Health Advisories (HAs) and Carcinogen Risk Estimates

	Assumed Weight of Protected Individual	Assumed Volume of Drinking Water Ingested/Day	Preferred Exposure Data for HA Development
One-day HA	10-kg child	1 L	Up to 7 days of exposure
Ten-day HA	10-kg child	1 L	7 to 30 days of exposure
Longer-term HA	10-kg child and 70-kg adult	1 L 2 L	Subchronic (90 days to 1 yr) (i.e., approximately 10% of the animal's lifetime)
Lifetime HA	70-kg adult	2 L	Chronic or subchronic
Cancer risk estimates	70-kg adult	2 L/day for 70 yrs	Chronic or subchronic

Table III. Guidelines Used in Selecting Uncertainty Factors for HAs[a]

- An uncertainty factor of 10 is generally used when good chronic or subchronic human exposure data identifying a NOAEL are available, and are supported by chronic or subchronic toxicity data in other species.

- An uncertainty factor of 100 is generally used when good chronic toxicity data identifying a NOAEL are available for one or more animal species (and human data are not available), or when good chronic or subchronic toxicity data identifying a LOAEL in humans are available.

- An uncertainty factor of 1,000 is generally used when limited or incomplete chronic or subchronic toxicity data are available, or when good chronic data that identify a LOAEL but not a NOAEL for one or more animal species are available.

- An uncertainty factor of 10,000 may be used when a subchronic study identifying a LOAEL but not a NOAEL is used.

[a]Source: NAS (1977, 1980) as modified by the USEPA Office of Drinking Water.

C. Derivation of the Lifetime Health Advisory

The One-day, Ten-day, and Longer-term HAs are based on the assumption that all exposure to the chemical comes from drinking water. Over a lifetime, however, other sources (e.g., food, air) may provide significant additional exposure, or may be the predominant exposure route to a chemical. An additional step is added to the calculation of the Lifetime HA to account for these sources.

The Lifetime HA is calculated in three steps. Together, the first two steps are identical to the calculation performed to derive the other HAs. In the first step, the NOAEL or LOAEL is divided by the uncertainty factor to determine a Reference Dose (RfD):

$$\text{Rfd} = \frac{\text{(NOAEL or LOAEL)}}{\text{(UF)}} = \underline{\hspace{1cm}} \text{ mg/kg bw/d}$$

The RfD is an estimate, with an uncertainty of perhaps an order of magnitude, of a daily exposure that is likely to be without appreciable risk of deleterious health effects in the human population (including sensitive subgroups) over a lifetime.

In the second step, the RfD is adjusted for an adult (with body weight assumed to be 70 kg) consuming 2 L water per day to produce the Drinking Water Equivalent Level (DWEL):

$$\text{DWEL} = \frac{\text{(RfD) (70 kg)}}{\text{(2 L/d)}} = \underline{\hspace{1cm}} \text{ mg/L } (\underline{\hspace{1cm}} \mu\text{g/L})$$

The DWEL represents the concentration of a substance in drinking water that is not expected to cause any adverse noncarcinogenic health effects in humans over a lifetime of exposure. The DWEL is calculated assuming that all exposure to the chemical comes from drinking water.

In the third step, the Lifetime HA is calculated by reducing the DWEL in proportion to the amount of exposure from drinking water relative to other sources (e.g., food, air). In the absence of actual exposure data, this relative source contribution (RSC) is generally assumed to be 20%. Thus:

$$\text{Lifetime HA} = \text{DWEL} \times \text{RSC} = \underline{\hspace{1cm}} \text{ mg/L } (\underline{\hspace{1cm}} \mu\text{g/L})$$

The value presented in μg/L is generally rounded to one significant figure.

Lifetime HAs are calculated for all noncarcinogenic chemicals (Groups D and E—see Table IV). For known (Group A) or probable (Group B) human carcinogens, carcinogenicity is usually considered the toxic effect of greatest concern. In general, a Lifetime HA is not recommended for Group A or B carcinogens. Instead, a mathematical model (usually the multistage) is used to determine theoretical upper-bound lifetime cancer risks based on the available

Table IV. EPA Scheme for Categorizing Chemicals According to Their
Carcinogenic Potential[a,b]

Group A: Human carcinogen

Sufficient evidence in epidemiologic studies to support causal association between
exposure and cancer

Group B: Probable human carcinogen

Limited evidence in epidemiologic studies (Group B1) *and/or* sufficient evidence
from animal studies (Group B2)

Group C: Possible human carcinogen

Limited or equivocal evidence from animal studies *and* inadequate or no data in
humans

Group D: Not classified

Inadequate or no human *and* animal evidence of carcinogenicity

Group E: No evidence of carcinogenicity for humans

No evidence of carcinogenicity in at least two adequate animal tests in different
species *or* in adequate epidemiologic and animal studies

[a]Source: USEPA 1986.
[b]Other factors such as genotoxicity, structure-activity relationships and benign versus
malignant tumors may influence classification.

cancer data (see Section II). For comparison purposes, a DWEL is calculated,
and the upper-bound cancer risk associated with lifetime exposure to the
DWEL is determined (see Section II).

For chemicals classified in Group C: Possible human carcinogen, ODW
applies an additional 10-fold uncertainty factor when deriving a Lifetime HA.
This extra uncertainty factor provides an additional margin of safety to
account for the possible carcinogenic effects of the chemical.

II. Assessment of Carcinogenic Risk

If toxicological evidence leads to the classification of a contaminant as a
human or probable human carcinogen (Groups A or B), mathematical models
are used to estimate an upper-bound excess cancer risk associated with lifetime
ingestion of drinking water. The data used in these estimates usually come
from lifetime exposure studies in animals. Upper-bound excess cancer risk
estimates may be calculated using models such as the one-hit, Weibull, logit,

probit, or multistage models (USEPA 1986). Since the mechanism of cancer is not well understood, there is no evidence to suggest that one model can predict risk more accurately than another. Therefore, the USEPA generally uses one of the more conservative models for its carcinogen risk assessment: the linearized multistage model (USEPA 1986). This model fits linear dose-response curves to low doses (NAS 1986). It is consistent with a no-threshold model of carcinogenesis, i.e., exposure to even a very small amount of the substance theoretically produces a finite increased risk of cancer.

The linearized multistage model uses dose-response data from the most appropriate carcinogenic study to calculate a carcinogenic potency factor (q_1*) for humans. The q_1* is then used to determine the concentrations of the chemical in drinking water that are associated with theoretical upper-bound excess lifetime cancer risks of 10^{-4}, 10^{-5}, and 10^{-6} (i.e., concentrations predicted to contribute an incremental risk of 1 in 10,000, 1 in 100,000, and 1 in 1,000,000 individuals over a lifetime of exposure). The following formula is used for this calculation:

$$\text{Concentration in drinking water} = \frac{(10^{-x})\,(70\text{ kg})}{(q_1{}^{*})\,(2\text{ L/d})} = \underline{\hspace{1cm}} \ \mu\text{g/L}$$

where: 10^{-x} = risk level (x = 4, 5, or 6)

  70 kg = assumed body weight of adult human

  q_1* = carcinogenic potency factor for humans as determined by the linearized multistage model in $(\mu\text{g/kg/d})^{-1}$

  2 L/d = assumed water consumption of adult human

The carcinogenic risk associated with lifetime exposure to the Drinking Water Equivalent Level (DWEL) is calculated using the following formula:

$$\text{Risk} = (\text{DWEL})\,\frac{(2\text{ L/d})\,(q_1{}^{*})}{(70\text{ kg})} = 10^{-x}$$

where: DWEL = Drinking Water Equivalent Level in μg/L

The theoretical upper-bound cancer risk associated with lifetime exposure to the DWEL is provided to assist the risk manager for comparison in assessing the overall risks.

III. Analytical Methods and Treatment Technologies

In addition to the health assessments, HAs also summarize information on analytical methods and treatment technologies for each contaminant. These methods and technologies include those validated as USEPA methods, as well as other methods that may be available. For further information on the analytical methods and treatment technologies for drinking water contaminants, contact the Safe Drinking Water Hotline (1–800–426–4791) or the Science and Technology Branch, Criteria and Standards Division, Office of Drinking Water, USEPA, 401 M Street SW, Washington, DC 20460 (1–202–382–3022).

References

NAS (1977) National Academy of Sciences. Drinking water and health, Vol I. NAS, Washington, DC.

NAS (1980) National Academy of Sciences. Drinking water and health, Vol II. NAS, Washington, DC.

NAS (1986) National Academy of Sciences. Drinking water and health, Vol VI. NAS, Washington, DC.

USEPA (1986) Guidelines for carcinogen risk assessment. Federal Register. 51(185):33992–34003.

Table of Contents

Drinking Water Health Advisory:

PESTICIDES

Acifluorfen

I. General Information and Properties

A. CAS No. 5094–66–6 (acid)
62476–59–9 (sodium salt)

B. Structural Formula

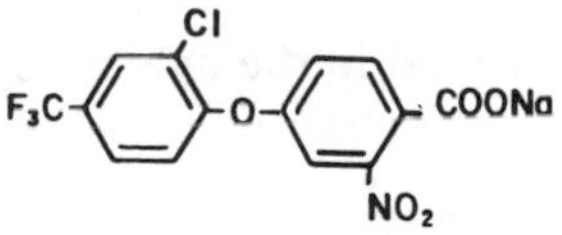

Sodium 5-(2-chloro-4-([trifluoromethyl]-phenoxy)-2-nitrobenzoate

C. Synonyms

- Blazer®; Carbofluorfen; RH-6201; Tackle®; Sodium acifluorfen (Meister, 1983).

D. Uses

- Acifluorfen is used as a selective pre- and post-emergence herbicide to control weeds and grasses in large-seeded legumes including soybeans and peanuts (Meister, 1983).

E. Properties (Windholz et al., 1983; Meister, 1983; CHEMLAB, 1985; WSSA, 1983)

Chemical Formula	$C_{14}H_7ClF_3NO_5$ (acid)
	$C_{14}H_6ClF_3NNaO_5$ (sodium salt)
Molecular Weight	361.66 (acid)
	383.65 (sodium salt)
Physical State (25°C)	Off-white solid (acid),
	brown crystalline powder/white powder
	(sodium salt)
Boiling Point	--
Melting Point	124–125°C (sodium salt)
	151.5–157°C (acid)

1

Density	--
Vapor Pressure (25°C)	24 mm Hg (45% H_2O solution)
Specific Gravity	--
Water Solubility (25°C)	> 25% (sodium salt)
	(dimensions not specified)
Log Octanol/Water Partition Coefficient	–4.85 (acid) (calculated)
Taste Threshold	--
Odor Threshold	--
Conversion Factor	--

F. Occurrence

- No information was found in the available literature on the occurrence of acifluorfen.

G. Environmental Fate

- Acifluorfen is stable to hydrolysis; no degradation was observed in solutions at pH 3, 6 or 9 within a 28-day interval. Varying temperatures (18 to 40°C) did not alter this stability. The half-life of the parent compound is 92 hours under continuous exposure to light approximating natural sunlight. The decarboxy derivative of acifluorfen was the primary degradate found in solution. It is suspected that a substantial percentage of the photodegradate parent is lost from solution (through volatilization or other mechanisms) (U.S. EPA, 1981).

- The half-life of acifluorfen in an aerobically incubated soil was found to be about 170 days; anaerobic degradation was more rapid (half-life about 1 month). The dominant residue compounds after 6 months' aerobic incubation were the parent compound and bound materials. After 2 months under anaerobic conditions, the acetamide of amino acifluorfen was the major degradate extracted from soil; the amino analog itself was also significant, and denitro acifluorfen was also formed (U.S. EPA, 1981).

- Acifluorfen applied at 0.75 lb active ingredient (a.i.)/A to a silt loam in Mississippi dissipated with a tentative half-life of 59 days. Leaching of the parent compound below 3 inches in the soil was negligible during the 179-day study. The dissipation of acifluorfen in two silt loam soils in Illinois receiving multiresidue treatments was somewhat slower; half-lives were 101 to 235 days (U.S. EPA, 1981).

- Acifluorfen applied to soil columns at highly excessive rates indicative of spills (682 lb a.i./A) is very mobile. Acifluorfen leached from the columns with 10 inches of water accounted for 79 to 93% of the acifluorfen applied. Aerobic aging of the residues in the column substantially reduced the mobility, and pesticide movement was inversely proportional to the soil CEC. Results from soil TLC (unaged residues only) predict mobility to be intermediate to mobile. Supplementary data from a batch adsorption study indicate that un-aged acifluorfen is weakly and reversibly adsorbed (U.S. EPA, 1981).

- Greenhouse studies have demonstrated that the uptake of acifluorfen by rotational crops decreases with aging of residues in soil (U.S. EPA, 1981).

II.   Pharmacokinetics

A. Absorption

- No information was found in the available literature on the absorption of acifluorfen.

B. Distribution

- No information was found in the available literature on the distribution of acifluorfen.

C. Metabolism

- No information was found in the available literature on the metabolism of acifluorfen.

D. Excretion

- No information was found in the available literature on the excretion of acifluorfen.

III.   Health Effects

A. Humans

- No information was found in the available literature on the health effects of acifluorfen in humans.

B. Animals

1. *Short-term Exposure*

- The Whittaker Corporation (no date, a) reported that the oral LD_{50} of Tackle 2S (a formulation containing 20.2% sodium acifluorfen) in the rat (strain not specified) was 2,025 mg/kg for males and 1,370 mg/kg for females.

- Meister (1983) reported that the acute dermal LD_{50} of Blazer® (technical grade, purity unspecified) in the rabbit is 450 mg/kg. The acute dermal LD_{50} of Tackle® (purity unspecified) in the rabbit is 2,000 mg/kg.

- Goldenthal et al. (1978a) presented the results of a 2-week range-finding study in which RH 6201 (a formulation containing 39.4% sodium aci-fluorfen) was administered to Charles River CD-1 mice (10/sex/dose) at dietary concentrations of 0, 625, 1,250, 2,500, 5,000 or 10,000 ppm. Assuming that 1 ppm in the diet of mice is equivalent to 0.15 mg/kg/day (Lehman, 1959), these doses correspond to about 0, 93.8, 187.5, 375.0, 750.0 or 1,500 mg/kg/day. No changes in general behavior or appearance were reported at any dose level. During the second week of the study, there was a decrease in body weight and food consumption in animals receiving 10,000 ppm (1,500 mg/kg/day). Gross pathological findings included pale kidneys, yellowish livers and reddish foci of hyperemia in the stomachs of several mice at the 5,000- and 10,000-ppm (750- and 1,500-mg/kg/day) dose levels. Absolute liver weight was increased in all test groups dosed at levels of 2,500 ppm (375 mg/kg/day) or greater. The increases were statistically significant (p < 0.01). A statistically significant (p < 0.01) increase in relative liver weight was reported at all dose levels. Based on the results of this study, a Lowest-Observed-Adverse-Effect Level (LOAEL) of 625 ppm (93.8 mg/kg/day) was identified.

- Piccirillo and Robbins (1976) administered RH 6201 (a formulation containing 39.8% sodium acifluorfen) to Wistar rats (5/sex/dose) for 4 weeks at dietary concentrations of 0, 5, 50, 500 or 5,000 ppm (reported to be equivalent to 0, 0.7, 7.6, 55.4 or 506.4 mg/kg/day for males and 0, 0.8, 8.3, 60.6 or 528.2 mg/kg/day for females). Assuming that these dietary levels reflect the concentration of the test compound and not the active ingredient, corresponding levels of sodium acifluorfen are 0, 0.3, 3.0, 22.1 and 201.6 mg/kg/day for males and 0, 0.3, 3.3, 24.0 and 210.2 mg/kg/day for females (Lehman, 1959). Results of the study indicated that body weight was decreased in males at 22.1 and 201.6 mg/kg/day, and food consumption was decreased in both males at 201.6 mg/kg/day and females at 210.2 mg/kg/day. Biochemical analyses revealed that serum glutamic pyruvic transaminase (SGPT) levels were increased in males at 22.1 and 201.6 mg/kg/day; in males that received 201.6 mg/kg/day, blood urea nitrogen (BUN) was increased and glucose levels were decreased. Changes in organ weights included increased absolute liver and kidney weights in males at 201.6 mg/kg/day, increased relative liver and kidney weights in males at 201.6 mg/kg/day and females at 210.2 mg/kg/day and increased relative liver weight in males only at 22.1 mg/kg/day. Based on the results of this study, a No-Observed-Adverse-Effect Level (NOAEL) of 3.0 mg/kg/day was identified.

2. *Dermal/Ocular Effects*

- In a dermal irritation study (Whittaker Corp., no date, b), Tackle 2S (a formulation containing 20.2% sodium acifluorfen) was applied occlusively (dose not specified) to the intact and abraded skin of rabbits. Effects observed included slight erythema, slight edema, blanching of the skin and eschar formation. Signs of dermal irritation at intact and abraded sites were absent by 8 days post-application. The test substance was considered to be a moderate dermal irritant at 72 hours.

- In a dermal irritation study, Weatherholtz et al. (1979b) applied RH 6201 (sodium acifluorfen) to the skin of New Zealand White rabbits (5/sex/dose; 10/sex/control). Three different formulations of RH 6201 were used in the study and each formulation was tested at 1.0 or 4.0 mL/kg/day. The authors indicated that for all RH 6201 formulations tested, the dose levels correspond to 50 or 200 mg/kg/day of the active ingredient. The test material was applied once daily for 5 days, followed by 2 days with no applications, over a 4-week period (total of 20 applications). At both dose levels, two of the formulations produced slight to well-defined irritation. At 200 mg/kg/day, central nervous system depression and a statistically significant decrease in body weight gain and food consumption were noted. The third formulation produced essentially the same effects, with the addition of "thinness," ataxia, slight tremors and mortality (2/5 males). Microscopic evaluations revealed chronic dermatitis, acanthosis and hyperkeratosis at both dose levels for all formulations.

- Madison et al. (1981) presented the results of the Buhler test for dermal sensitization in Hartley-derived albino guinea pigs. In this study, Tackle® (sodium acifluorfen; purity not specified) was not found to be a sensitizer when applied topically at a dose of 0.25 mL under occlusive binding.

- In an ocular irritation study (Whittaker Corp., no date, c), Tackle 2S (a formulation containing 20.3% sodium acifluorfen) was instilled into the eyes of rabbits. Signs of ocular irritation and lesions included opacities of the cornea, iritis, redness and chemosis of the conjunctiva and discharges from both washed and unwashed eyes. Four of six unwashed eyes and one of three washed eyes exhibited blistering of the conjunctiva. Three of six unwashed and one of three washed eyes exhibited pannus where corneal opacity had been.

- In an ocular irritation study (Weatherholtz et al., 1979a), 0.1 mL of Blazer 2S (purity not specified) was applied to the corneal surface of the eyes of Rhesus monkeys. Corneal opacity and conjunctival redness, swelling and discharge were observed in both washed and unwashed eyes. All treated eyes were free of signs of irritation by 14 days posttreatment.

3. *Long-term Exposure*

- Harris et al. (1978) administered RH 6201 (a formulation containing 39.4% sodium acifluorfen) in the diet to Sprague-Dawley rats (15/sex/dose) for 3 months at dose levels of 0, 75, 150 or 300 mg/kg/day. Assuming that these doses reflect levels of the test compound and not the active ingredient, corresponding levels of sodium acifluorfen would be 0, 29.6, 59.1 or 118.2 mg/kg/day. At the highest dose level (118.2 mg/kg/day), a number of effects were observed in male rats. These effects included decreased body weight (13%) and decreased food consumption (8%). Biochemical analyses of blood revealed increased alkaline phosphatase levels (32%), decreased total protein (8%) and decreased albumin (14%). No such effects were reported for female rats. (These biochemical analyses were performed on control and high-dose animals only.) Increased liver weight and microscopic liver changes (enlarged hepatocytes) were observed in male rats that received 59.1 or 118.2 mg/kg/day. In terms of the active ingredient, a NOAEL of 29.6 mg/kg/day was identified.

- Barnett (1982) administered Tackle 2S (a formulation containing 20.4 to 23.6% sodium acifluorfen) to Fischer 344 rats (30/sex/dose) for 90 days at dietary concentrations of 0, 20, 80, 320, 1,250, 2,500 or 5,000 ppm.

The author indicated that these dietary levels correspond to average compound intake levels of 0, 1.5, 6.1, 23.7, 92.5, 191.8 or 401.7 mg/kg/day for males and 0, 1.8, 7.4, 29.7, 116.0, 237.1 or 441.8 mg/kg/day for females. Assuming that these levels reflect test compound and not active ingredient intake, corresponding levels of sodium acifluorfen intake are approximately 0, 0.4, 1.4, 5.6, 21.8, 45.3 or 94.8 mg/kg/day for males and 0, 0.4, 1.8, 7.0, 27.4, 56.0 or 104.3 mg/kg/day for females (based on 23.6% active ingredient in test compound). At 5,000 ppm the following effects were observed: decreased body weight and food consumption in both sexes; decreased red blood cell (RBC) count, hemoglobin and hematocrit in both sexes; increased serum cholesterol and serum calcium, and decreased serum phosphorous in both sexes; increased alkaline phosphatase, SGPT and BUN levels in males; elevated urobilinogen in both sexes; increased liver size and discolored liver and kidneys in both sexes; and liver cell hypertrophy and increases in mitotic figures and individual cell deaths in both sexes. At 2,500 ppm the following effects were observed: decreased body weight in males; decreased RBC count, hemoglobin and hematocrit in both sexes; increased BUN levels in males; elevated urobilinogen in both sexes; increased liver size in both sexes; and liver cell hypertrophy and increases in mitotic figures and individual cell deaths in both sexes. At 1,250 ppm, the following effects were observed: increased liver size in males and liver cell hypertrophy in both sexes. The author identified 320 ppm as the NOAEL in this study. In terms of active ingredient concentration, this corresponds to a NOAEL of 5.6 mg/kg/day for males and 7.0 mg/kg/day for females.

• Mobil Environmental and Health Science Laboratory (1981) presented the 6-month interim results of a longer-term study in which Tackle 2S (a formulation containing approximately 75% sodium acifluorfen) was administered to beagle dogs (8/sex/dose) at dietary concentrations of 0, 20, 320 or 4,500 ppm. These dietary levels were reported to be equivalent to 0, 0.7, 9.0 or 160 mg/kg/day. Assuming that these levels reflect test compound and not active ingredient intake, corresponding levels of sodium acifluorfen intake are 0, 0.5, 6.8 or 120.0 mg/kg/day (based on 75% active ingredient in the test compound). Following 6 months of compound administration, two animals/sex/dose were sacrificed. The study reported a number of effects at the highest dose tested. These effects included decreased body weight and food consumption and increased liver weight in both sexes. Additionally, RBC count and hemoglobin concentration were decreased in both sexes. Clinical chemistry analyses revealed depressed serum cholesterol, increased alkaline phosphatase and transient elevation of BUN in both sexes. Males only showed increased levels of lactic dehydrogenase. No histopathological examinations were conducted. The NOAEL reported in this study was 320 ppm. In terms of the active ingredient, this corresponds to a NOAEL of 6.8 mg/kg/day.

- Barnett et al. (1982b) administered Tackle 2S (a formulation containing 19.1 to 25.6% sodium acifluorfen) to Fischer 344 rats (73/sex/dose) for 1 year at dietary levels of 0, 25, 150, 500, 2,500 or 5,000 ppm. Assuming that these dietary levels reflect the concentrations of the test compound and not the active ingredient, corresponding levels of sodium acifluorfen are 0, 6.4, 38.4, 128.0, 640.0 or 1,280 ppm (based on 25.6% active ingredient in the test compound). Assuming that 1 ppm in the diet of rats is equivalent to 0.05 mg/kg/day, these levels correspond approximately to 0, 0.3, 1.9, 6.4, 32.0 or 64.0 mg/kg/day (Lehman, 1959). No excess moribundity or mortality was associated with the ingestion of the test substance. At 5,000 ppm the following effects were observed: decreased mean body weight in both sexes; increased absolute and relative liver weight in both sexes; decreased protein production, decreased serum glucose, decreased triglyceride levels, increased alkaline phosphatase and creatine phosphokinase levels and sporadic increases in SGOT and SGPT in both sexes; a slight increase in the excretion of urobilinogen in both sexes; and the presence of acidophilic cells that were considered to be evidence of cytotoxic changes in the livers of both sexes. At 2,500 ppm, male rats showed increased absolute and relative liver weights. Based on the information presented in this study, a NOAEL of 500 ppm was identified for the test compound. In terms of the active ingredient, this corresponds to a NOAEL of 6.4 mg/kg/day.

- Spicer et al. (1983) administered Tackle 2S (a formulation containing 74.5 to 82.8% sodium acifluorfen) to beagle dogs (8/sex/dose) for 2 years at dietary concentrations of 0, 20, 300 or 4,500 ppm, reported to be equivalent to 0, 0.5, 7.3 or 121 mg/kg/day for males and 0, 0.5, 8.3 or 154 mg/kg/day for females. Assuming that these dietary levels reflect the concentration of the test compound and not the active ingredient, corresponding levels of sodium acifluorfen are 0, 0.4, 6.0 or 100.2 mg/kg/day for males and 0, 0.4, 6.9 or 127.5 mg/kg/day for females (based on 82.8% active ingredient in the test compound). At the highest dose, body weight was decreased (not statistically significant), and a corresponding (statistically significant) decrease in food consumption was also reported. Physical examination revealed heart anomalies in the high- and mid-dose groups. At the high dose, irregular heart rhythms and rapid or slow heart rates were reported in one male and four females. Also at this dose level, one male was found to have a systolic murmur.

At the mid-dose level, one animal of each sex had an irregular heart rhythm (accompanied by rapid heart rate in the male). At the highest dose tested, a number of changes were reported, including a statistically significant decrease in erythrocyte count, hemoglobin and hematocrit in both sexes; reductions in albumin and cholesterol; increased absolute and relative liver and kidney weights; and histopathological liver changes including centrilobular hepatocellular fatty vacuolation, bilirubin pigmentation and minimal foci of alteration. Renal tubules showed bilirubin pigmentation at all dose levels (most pronounced at the high dose). The authors concluded that this study showed clear evidence of target organ toxicity affecting the liver and possibly the kidney at the highest dose level. The authors identified 300 ppm (of test compound) as the NOAEL. In terms of the active ingredient, this corresponds to a NOAEL of 6.0 mg/kg/day.

• Goldenthal et al. (1979) administered RH 6201 (a formulation containing 39.4 to 40.5% sodium acifluorfen) to Charles River CD-1 mice (80/sex/dose) for 2 years in the diet at concentrations that provided dosage levels of 0, 1.25, 7.5 or 45.0 ppm of the active ingredient. After 16 weeks of administration, the 1.25 ppm dose was increased to 270 ppm. Assuming that 1 ppm in the diet of mice is equivalent to 0.15 mg/kg/day, these levels correspond to about 0, 0.19 (increased to 40.5), 1.13 and 6.8 mg/kg/day (Lehman, 1959). Two control groups were used in this study. One group received acetone in the diet (control 1), and the other received water in the diet (control 2). At the 40.5-mg/kg/day dose level, the following effects were observed: slight to marked elevations in alkaline phosphatase and SGPT levels, in both sexes, beginning after one year of exposure; increased absolute and relative liver weight in males; increased absolute liver weight in females; increased relative kidney weight in males; decreased absolute heart weight in males; cellular alterations in the livers of males consisting of focal pigmentation, focal hepatocytic necrosis, focal cellular alteration, nodular hepatocellular proliferation and hepatocellular carcinoma (the only statistically significant change was the focal cellular alteration); and focal pigmentation in the livers of females. At the 6.8 mg/kg/day dose level, the following effects were observed: occasional increases in alkaline phosphatase and SGPT levels in both sexes; decreased absolute heart weight in males; and focal pigmentation in the livers of females. The author indicated that changes with an apparent dose-related distribution included focal pigmentation, hepatocellular vacuolation, focal hepatocytic necrosis and nodular hepatocellular proliferation. The incidence of hepatocellular carcinoma in males of all treatment groups was approximately the same. A NOAEL of 7.5 ppm (1.13 mg/kg/day) was identified by the author.

4. *Reproductive Effects*

- In a three-generation reproduction study, Goldenthal et al. (1978b) administered RH 6201 (a formulation containing sodium acifluorfen) in the diet to Charles River CD rats. During the course of the study, the test compound was administered at various levels depending on the age of the animals. The F_1 generation received dose levels of 2.9, 17.3 or 104 ppm during the first 2 weeks of the study, and 5, 30 and 180 ppm for the remaining weeks of the generation (study weeks 3 to 17) [time-weighted average (TWA) dosage levels 4.8, 28.5 or 171.1 ppm]. The F_2 and F_3 generations received dosage levels of 180, 10 or 60 ppm during the first and second weeks of the generation; 312, 17.3 or 104 ppm during the third, fourth and fifth weeks of the generation; and 540, 30 or 180 ppm for the remaining weeks of the generation (TWAs for F_2 generation were 486.0, 27.0 or 162.0 ppm; TWAs for F_3 generation were 483.8, 26.7 or 161.3 ppm). The highest dietary TWA dose tested in this study was 486 ppm of active ingredient. Assuming that 1 ppm in the diet of rats is equivalent to 0.05 mg/kg/day, this corresponds to a dose of 24.3 mg/ kg/day (Lehman, 1959). No effects related to compound administration were observed in parents or pups in terms of general behavior, appearance or survival. Parental and pup body weights and food consumption were similar to controls. Fertility, gestation and viability indices were comparable for controls and treated groups. There were no biologically meaningful teratogenic effects in the second or third generation, based on mean number of viable fetuses, postimplantation losses, total implantations and corpora lutea per dam, mean fetal body weight, number of fetal anomalies and sex-ratio variations. No compound-related gross lesions were noted in third-generation pups necropsied. Based on the information presented, a NOAEL of 486 ppm (24.3 mg/kg/day) was identified. This NOAEL represents the highest dose tested.

- In a two-generation reproduction study, Lochry et al. (1986) administered technical-grade Tackle (sodium acifluorfen) of unspecified purity to rats at levels of 0, 25, 500 and 2,500 ppm. The compound was administered in the diet *ad libitum* to groups of 35 rats/sex/dose beginning at 47 days of age and continuing until sacrifice. In addition, the compound was also administered to groups of 40 rats/sex/dose from weaning until sacrifice. Reproductive parameters, mortality, body weight and a number of other end points were measured; in addition, both gross and histopathological examinations were conducted. The NOAEL for toxicity to both the parents and offspring was 25 ppm, based on mortality and kidney lesions at higher doses. Assuming that 1 ppm in the diet of rats is equivalent to 0.05 mg/kg/day, the NOAEL of 25 ppm in this study corresponds to 1.25 mg/kg/day (Lehman, 1959).

5. Developmental Effects

- Lightkep et al. (1980) administered Tackle 2S (a formulation containing 22.4% sodium acifluorfen) by oral intubation at doses of 0, 3, 12 or 36 mg/kg/day to New Zealand White rabbits (16/dose) on days 6 to 29 of gestation. The authors indicated that the administered doses were in terms of the active ingredient. At 36 mg/kg/day, there was a slight (nonsignificant) inhibition of maternal body weight gain and a marked (significant) inhibition of maternal food consumption. At this dose level, there was also possible interference with implantation and a slight decrease in average fetal body weight; neither of these changes was statistically significant. No gross, soft-tissue or skeletal malformations were observed in pups, fetuses or late resorptions at any dose level. Based on the information presented in this study, a NOAEL of 36 mg/kg/day was identified for maternal toxicity, fetal toxicity and teratogenicity. This NOAEL represents the highest dose tested.

- Florek et al. (1981) administered Tackle 2S (a formulation containing 22.4% sodium acifluorfen) by gavage at doses of 0, 20, 90 or 180 mg/kg/day to Sprague-Dawley rats (25/dose) on days 6 to 19 of gestation. The authors indicated that the administered doses were in terms of active ingredient. At 180 mg/kg/day, dams gained significantly less weight than controls. At 90 and 180 mg/kg/day, lower average fetal body weight and significantly delayed ossification of metacarpals and forepaw and hindpaw phalanges were noted. At 180 mg/kg/day, there was delayed ossification of caudal vertebrae, sternebrae and metatarsals. Additionally, at the highest dose level there was a significantly increased incidence of slight dilation of the lateral ventricle of the brain. The authors stated that the fetal effects were indicative of delayed fetal development. Based on the results of this study, a NOAEL of 90 mg/kg/day for maternal toxicity, a NOAEL of 20 mg/kg/day for fetotoxicity and a NOAEL of 180 mg/kg/day (the highest dose tested) for teratogenic effects were identified.

- Weatherholtz and Piccirillo (1979) administered RH 6201 (a formulation containing 39.8% sodium acifluorfen) by gavage at doses of 0, 20, 60 or 180 mg/kg/day to New Zealand White rabbits on days 7 to 19 of gestation. Maternal toxicity at 180 mg/kg/day included statistically significant weight loss and mortality. At 180 mg/kg/day, there was also evidence of fetal toxicity (mortality). Due to embryotoxicity and maternal toxicity at 180 mg/kg/day, teratogenic evaluations could not be performed at this dose level. At lower doses, no teratogenic effects were observed. Based on the results of this study, NOAELs of 60 mg/kg/day were identified for teratogenic effects, maternal toxicity and fetal toxicity.

6. *Mutagenicity*

- Schreiner et al. (1980) tested Tackle 2S (purity unspecified) in an Ames assay using *Salmonella typhimurium* strains TA98, TA100, TA1535, TA1537 and TA1538. The test compound was not found to be mutagenic, with or without metabolic activation, at concentrations up to 1.8 mg/plate.

- Brusick (1976) tested RH 6201 (purity not specified) in a mutagenicity assay using *Saccharomyces cerevisiae* strain D4 and *S. typhimurium* strains TA1535, TA1537, TA1538, TA98 and TA100. The compound was not found to be mutagenic, with or without metabolic activation, at concentrations up to 500 µg/plate.

- Putnam et al. (1981) tested Tackle 2S (purity not specified) in a dominant lethal assay using Sprague-Dawley rats. The compound was administered by gavage at doses of 0, 80, 360 or 800 mg/kg/day for 5 consecutive days. No detectable mutagenic activity, as defined by induction of fetal death, was reported.

- Myhr and McKeon (1981) conducted a primary rat (Fischer 344) hepatocyte unscheduled DNA synthesis (UDS) assay using Tackle 2S (purity not specified). The test compound did not induce a detectable level of UDS over a concentration range of 0.10 to 25 µg/mL. Treatment of hepatocytes with 50 µg/mL was almost completely lethal to the cells.

- Schreiner et al. (1981) tested Tackle 2S (purity not specified) in a bone marrow metaphase analysis using Sprague-Dawley rats. The animals were given the test compound by intubation at doses of 0, 0.37, 1.11 or 1.87 g/kg/day for 5 days. The test compound did not significantly increase clastogenic events in the bone marrow cells.

- Schreiner et al. (1980) tested Tackle 2S (purity not specified) in a murine lymphoma assay. The compound was tested without metabolic activation at 0.11 to 1.7 µg/mL, and with metabolic activation at 0.08 to 0.56 µg/mL. No detectable mutagenic activity was detected either with or without activation.

- Jagannath (1981) tested Tackle 2S (29.7% purity) in a mitotic recombination assay using *Saccharomyces cerevisiae* strain D5. The compound was tested at 0, 2.5, 5.0 or 7.5 μL/plate without metabolic activation, and at 7.5, 10.0 and 25.0 μL/plate with metabolic activation. In the absence of metabolic activation, the compound induced a dose-related increase in recombination events (significant at 5.0 μL/plate). With metabolic activation, a dose of 10.0 μL/plate induced an increase in recombination events. The authors reported that very few survivors were observed at 25.0 μL/plate.

- Bowman et al. (1981) tested Tackle 2S (purity not specified) in mutagenicity assays using *Drosophila melanogaster*. Assays included the Biothorax test of Lewis, a dominant lethal assay, an assay for Y-chromosome loss, and a White Ivory reversion assay. In all cases, the test compound was tested at concentrations of 15 mg/mL. Results of these assays were negative for somatic reversions of White Ivory and the Biothorax test of Lewis and positive for Y-chromosome loss and dominant lethal mutations.

7. *Carcinogenicity*

- Barnett et al. (1982b) administered Tackle 2S (a formulation containing 19.1 to 25.6% sodium acifluorfen) to Fischer 344 rats (73/sex/dose) for 1 year at dietary levels of 0, 25, 150, 500, 2,500 or 5,000 ppm. Assuming that these dietary levels reflect the concentrations of the test compound and not the active ingredient, corresponding levels of sodium acifluorfen are 0, 6.4, 38.4, 128.0, 640.0 or 1,280 ppm (based on 25.6% active ingredient in the test compound). Assuming that 1 ppm in the diet of rats is equivalent to 0.05 mg/kg/day, these doses correspond approximately to 0, 0.3, 1.9, 6.4, 32.0 or 64.0 mg/kg/day (Lehman, 1959). Histopathological examinations revealed no evidence of carcinogenicity at any dose level.

- Barnett et al. (1982a) administered Tackle® (a formulation containing 24% sodium acifluorfen) to B6C3F$_1$ mice (60/sex/dose) for 18 months at dietary concentrations of 0, 625, 1,250 or 2,500 ppm. (The high dose was reported to be the maximum tolerated dose.) The authors reported that the dietary levels corresponded to average compound intake values of 0, 118.96, 258.73 or 655.15 mg/kg/day for males, and 0, 142.50, 312.65 or 710.54 mg/kg/day for females. Assuming that these levels reflect test compound and not active ingredient intake, corresponding levels of sodium acifluorfen intake are 0, 28.55, 62.10 or 157.24 mg/kg/day for males and 0, 34.20, 75.04 or 170.53 mg/kg/day for females. An obvious dose-related depression of body weight was reported for all doses. Beginning in week 52 of the study and continuing with increasing frequency was the appearance of palpable abdominal masses. Gross necropsy revealed a dose-related increase in liver masses in both sexes. Histopathological examinations conducted at the 52-week interval revealed that the livers of six animals per sex of high-dose animals (157.24 mg/kg/day for males; 170.53 mg/kg/day for females) showed evidence of acidophilic cells. Males receiving this dose displayed a statistically significant increase in the frequency of hepatocellular adenomas. After 18 months of treatment, all 40 high-dose males and 27/47 high-dose females sacrificed were found to have a single benign hepatoma, multiple benign hepatomas or hepatocellular carcinomas. In the males, the incidence of single benign hepatoma and hepatocellular carcinomas was statistically significant. In the females, the incidence of single hepatomas was statistically significant.

- Goldenthal et al. (1979) administered RH 6201 (a formulation containing 39.4 to 40.5% sodium acifluorfen) to Charles River CD-l mice (80/sex/dose) for 2 years in the diet at concentrations that provided dose levels of 0, 1.25, 7.5 or 45.0 ppm of the active ingredient. After 16 weeks of administration, the 1.25 ppm dose was increased to 270 ppm. Assuming that 1 ppm in the diet of mice is equivalent to 0.15 mg/kg/day, these doses correspond to approximately 0, 0.19 (increased to 40.5), 1.13 or 6.8 mg/kg/day (Lehman, 1959). Two control groups were used in this study. One group received acetone in the diet (control 1) and the other received water in the diet (control 2). In males receiving the highest dose, there was a nonstatistically significant increase in the incidence of nodular hepatocellular proliferation and hepatocellular carcinoma, which indicated to the authors that these changes were dose-related.

- Coleman et al. (1978) administered RH 6201 (a formulation containing 39.8% sodium acifluorfen) to Charles River outbred albino CD COBS rats (approximately 75/sex/dose) for 2 years at changing dietary concentrations. Mean sodium acifluorfen intake values over the course of the study were 0, 1.25, 7.54 and 17.56 mg/kg/day for males and 0, 1.64, 9.84 and 25.03 mg/kg/day for females.

- Acifluorfen is structurally similar to nitrofen [2,4-dichloro-1-(4-nitrophenoxy) benzene; CAS No. 1836–75–7]. Nitrofen has been shown to be carcinogenic in Osborne-Mendel rats and B6C3F$_1$ mice (NCI, 1978, 1979; both as cited in NAS, 1985).

IV. Quantification of Toxicological Effects

A. One-day Health Advisory

No data were found in the available literature that were suitable for determination of a One-day HA value for acifluorfen. It is therefore recommended that the Ten-day HA value for a 10-kg child (2 mg/L, calculated below) be used at this time as a conservative estimate of the One-day HA value.

B. Ten-day Health Advisory

The study by Florek et al. (1981) has been selected to serve as the basis for determination of the Ten-day HA for a 10-kg child. In this study, Tackle 2S (a formulation containing 22.4% sodium acifluorfen) was administered by gavage at doses of 0, 20, 90 or 180 mg/kg/day to Sprague-Dawley rats (25/ dose) on days 6 to 19 of gestation. The authors indicated that the administered doses were in terms of active ingredient. At 180 mg/kg/day, dams reportedly gained significantly less weight than controls. At 90 and 180 mg/kg/day, lower average fetal body weight and significantly delayed ossification of metacarpals and forepaw and hindpaw phalanges were noted. At 180 mg/kg/day, there was delayed ossification of caudal vertebrae, sternebrae and metatarsals. Additionally, at the highest dose level there was a significantly increased incidence of slight dilation of the lateral ventricle of the brain. The authors stated that the fetal effects were indicative of delayed fetal development. No effects on implantations, litter size, fetal viability, resorption or fetal sex ratio were reported. Based on the results of this study, a NOAEL of 20 mg/kg/day for fetotoxicity was identified.

The Ten-day HA for the 10-kg child is calculated as follows:

$$\text{Ten-day HA} = \frac{(20 \text{ mg/kg/day}) (10 \text{ kg})}{(100) (1 \text{ L/day})} = 2 \text{ mg/L } (2{,}000 \text{ } \mu\text{g/L})$$

C. Longer-term Health Advisory

The study by Barnett (1982) had been selected to serve as the basis for determination of the Longer-term HA. In this study, the NOAEL was 5.6 mg/ kg/day based on an increase in the size of the liver in male rats. However, a lower NOAEL, 1.25 mg/kg/day, was recently identified in a two-generation rat reproduction study by Lochry et al. (1986). Since the NOAEL in the Lochry et al. (1986) study is numerically identical to the value on which the DWEL is based and since a two-generation reproduction study is suitable for calculating a Longer-term HA, it was determined that it is appropriate to base the Longer-term HA on the DWEL.

The Longer-term HA for a 10-kg child is calculated as follows:

$$\text{Longer-term HA} = \frac{(1.25 \text{ mg/kg/day}) (10 \text{ kg})}{(100) (1 \text{ L/day})} = 0.13 \text{ mg/L} (100 \text{ } \mu\text{g/L})$$

The Longer-term HA for the 70-kg adult is calculated as follows:

$$\text{Longer-term HA} = \frac{(1.25 \text{ mg/kg/day}) (70 \text{ kg})}{(100) (2 \text{ L/day})} = 0.44 \text{ mg/L} (400 \text{ } \mu\text{g/L})$$

D. Lifetime Health Advisory

A 2-year Charles River CD-1 mouse dietary study by Goldenthal et al. (1979) was originally selected to serve as the basis for determination of the DWEL for acifluorfen. In this study, a NOAEL of 1.13 mg/kg/day was identified. More recently, however, a two-generation rat reproduction study by Lochry et al. (1986) was identified that strongly supports the results of the Goldenthal (1979) study and identifies a NOAEL of 1.25 mg/kg/day.

Using the NOAEL of 1.25 mg/kg/day, the DWEL for acifluorfen is calculated as follows:

Step 1: Determination of the Reference Dose (RfD)

$$\text{RfD} = \frac{(1.25 \text{ mg/kg/day})}{(100)} = 0.013 \text{ mg/kg/day}$$

Step 2: Determination of the Drinking Water Equivalent Level (DWEL)

$$\text{DWEL} = \frac{(0.013 \text{ mg/kg/day}) (70 \text{ kg})}{(2 \text{ L/day})} = 0.437 \text{ mg/L} (400 \text{ } \mu\text{g/L})$$

Step 3: Determination of the Lifetime Health Advisory

Acifluorfen may be classified in Group B2: probable human carcinogen. A Lifetime HA is not recommended for acifluorfen. The estimated excess cancer risk associated with lifetime exposure to drinking water containing acifluorfen at 437 μg/L is approximately 4×10^{-4} (4.4×10^{-4}). This estimate represents the upper 95% confidence limit from extrapolations (U.S. EPA, 1988) prepared by using the Weibull 82 multistage model (time to death).

E. Evaluation of Carcinogenic Potential

- Four studies that evaluated the carcinogenic potential of sodium acifluorfen were identified. The results of one of these studies (Barnett et al., 1982a) indicated that sodium acifluorfen was carcinogenic in $B6C3F_1$ mice. The results of the other three studies (Goldenthal et al., 1979; Coleman et al., 1978; Barnett et al., 1982b) provided no evidence of carcinogenicity in two strains of rats and one strain of mice. However, due to deficiencies in the three negative studies, the results of these studies are not sufficient to contradict the results of the positive study. Each of these studies is discussed briefly below.

—   In the positive study (Barnett et al., 1982a), B6C3F$_1$ mice received sodium acifluorfen in the diet for 18 months. At the end of the study, the high-dose (157.24 mg/kg/day) male mice displayed a statistically significant increase in the incidence of single benign hepatomas and hepatocellular carcinomas. A statistically significant increase in the incidence of single hepatomas was observed in high-dose (170.53 mg/kg/day) females.

—   In one of the studies with negative results (Goldenthal et al., 1979) Charles River CD-l mice received sodium acifluorfen in the diet for 2 years at doses of 0, 0.19 (increased to 40.5 after 16 weeks), 1.13 or 6.8 mg/kg/day. Although no evidence of carcinogenicity was observed in this study, the dose levels tested were considerably lower than the level that produced positive results in the 18-month mouse feeding study (157.24 mg/kg/day) (Barnett et al., 1982a).

—   In the second study with negative results (Coleman et al., 1978), Charles River outbred albino CD COBS rats received sodium acifluorfen for 2 years at dietary levels up to 25.03 mg/kg/day (females) or 17.56 mg/kg/day (males). Although it is difficult to make cross-species comparisons, these levels are considerably lower than the level that produced positive results in the 18-month mouse feeding study (157.24 mg/kg/day) (Barnett et al., 1982a). In addition, no adverse effects were observed at any dose level used in this study, indicating that the maximum tolerated dose was not used.

—   In the third study with negative results (Barnett et al., 1982b), Fischer 344 rats received sodium acifluorfen for 1 year at dietary concentrations of 0, 0.3, 1.9, 6.4, 32.0 or 64.0 mg/kg/day. Although the results of this study were negative, a study duration of 1 year is not sufficient for assessing carcinogenic potential.

•   Acifluorfen is structurally similar to nitrofen [2,4-dichloro-1-(4-nitrophenoxy) benzene; CAS No. 1836–75–7]. Nitrofen has been shown to be carcinogenic in Osborne-Mendel rats and B6C3F$_1$ mice [NCI, 1978, 1979; both as cited in NAS (1985)]. Although data on nitrofen cannot be used to conclude that sodium acifluorfen is carcinogenic, these data do, to some extent, support the positive results of Barnett et al. (1982a).

- The International Agency for Research on Cancer has not evaluated the carcinogenic potential of acifluorfen.

- Evidence has been presented in one carcinogenicity study (Barnett et al., 1982a) showing that acifluorfen is carcinogenic to mice. Using the Weibull 82 increased multistage time-to-death with tumor model (since there is a survival disparity for male mice), U.S. EPA (1988) reported a unit risk, q_1^*, estimated in human equivalent (surface area conversion) as 3.55×10^{-2} $(mg/kg/day)^{-1}$ $[B_2]$. It was noted that q_1^* is an estimate of upper (95%) bound on risk and the true value of risk is unknown and may be as low as zero.

- Using this q_1^* value and assuming that a 70-kg human adult consumes 2 liters of water a day over a 70-year lifespan, the Weibull 82 multistage model estimates that concentrations of 100, 10 and 1 μg acifluorfen per liter may result in excess cancer risk of 10^{-4}, 10^{-5} and 10^{-6}, respectively (U.S. EPA, 1988).

- For comparison purposes drinking water concentrations associated with an excess risk of 10^{-6} were 0.2 μg/L, 0.7 μg/L, 7 μg/L, < 0.002 μg/L and < 0.002 μg/L for the multihit, one-hit, probit, logit and Weibull models, respectively.

- Applying the criteria described in EPA's guidelines for assessment of carcinogenic risk (U.S. EPA, 1986a), acifluorfen is classified in Group B2: probable human carcinogen.

V. Other Criteria, Guidance and Standards

- U.S. EPA has established residue tolerances for sodium acifluorfen in or on raw agricultural commodities that range from 0.01 to 0.1 ppm (CFR, 1985).

- The EPA RfD Workgroup has concluded that an RfD of 0.013 mg/kg/ day is appropriate for acifluorfen.

VI. Analytical Methods

- Analysis of acifluorfen is by a gas chromatographic (GC) method applicable to the determination of certain chlorinated acid pesticides in water samples (U.S. EPA, 1986b). In this method, approximately 1 liter of sample is acidified. The compounds are extracted with ethyl ether using a separatory funnel. The derivatives are hydrolyzed with potassium hydroxide, and extraneous organic material is removed by a solvent wash. After acidification, the acids are extracted and converted to their methyl esters using diazomethane as the derivatizing agent. Excess reagent is removed, and the esters are determined by electron capture GC. The method detection limit has not been determined for this compound, but it is estimated that the detection limits for analytes included in this method are in the range of 0.5 to 2 μg/L.

VII. Treatment Technologies

• Reverse osmosis (RO) is a promising treatment method for pesticide-contaminated water. As a general rule, organic compounds with molecular weights greater than 100 are candidates for removal by RO. Larson et al. (1982) report 99% removal efficiency of chlorinated pesticides by a thin-film composite polyamide membrane operating at a maximum pressure of 1,000 psi and at a maximum temperature of 113°F. More operational data are required, however, to specifically determine the effectiveness and feasibility of applying RO for the removal of acifluorfen from water. Also, membrane adsorption must be considered when evaluating RO performance in the treatment of acifluorfen-contaminated drinking water supplies.

VIII. References

Barnett, J.* 1982. Evaluation of ninety-day subchronic toxicity of Tackle® in Fischer 344 rats. GSRI Project No. 413–971–40. Rhone-Poulenc Agrochemie No. 372–80. Unpublished study. MRID 0122730.

Barnett, J., L. Jenkins and R. Parent.* 1982a. Evaluation of the potential oncogenic and toxicological effects of long-term dietary administration of Tackle® to B6C3F₁ mice. GSRI Project No. 413–984–41. Final Report. Unpublished study. MRID 00122732.

Barnett, J., L. Jenkins and R. Parent.* 1982b. Evaluation of the potential oncogenic and toxicological effects of long-term dietary administration of Tackle® to Fischer 344 rats. GSRI Project No. 413–985–41. Interim report. Unpublished study. MRID 00122735.

Bowman, J., C. Mackerer, S. Bowman, D.C. Jessup, R.C. Geil and B.W. Benson.* 1981. *Drosophila* mutagenicity assays of Mobil Chemical Company compound MC 10109 (MRI 533). Study No. 009–275–533–9. Unpublished study. MRID 00122737.

Brusick, D.* 1976. Mutagenicity evaluation of RH-6201. LBI Project No. 2547. Unpublished study. MRID 00083057.

CFR. 1985. Code of Federal Regulations. 40 CFR 180.383. July 1. p. 336.

CHEMLAB. 1985. The Chemical Information System, CIS, Inc. Baltimore, MD.

Coleman, M.E., T.E. Murchison, P.S. Sahota et al.* 1978. Three and twenty-four month oral safety evaluation study of RH-6201 in rats. DRC 5800. Final Report. Unpublished study. MRID 00087478.

Florek, M., M. Christian, G. Christian and E.M. Johnson.* 1981. Teratogenicity study of TACU 06238001 in pregnant rats. Argus Project 113–004. Unpublished study. MRID 00122743.

Goldenthal, E.I., D.C. Jessup, R.G. Geil and B.W. Benson.* 1978a. Two week range finding study in mice: 285–016. Unpublished study. MRID 00080568.

Goldenthal, E.I., D.C. Jessup and D. Rodwell.* 1978b. Three generation reproduction study in rats: RH-6201, 285–014a. Unpublished study. MRID 00107486.

Goldenthal, E.I., D.C. Jessup, R.G. Geil and B.W. Benson.* 1979. Lifetime dietary feeding study in mice: 285–013a. Unpublished study. MRID 00082897.

Harris, J.C., G. Cruzan and W.R. Brown.* 1978. Three month subchronic rat study. RH-6201. TRD-76P-30. Unpublished study. MRID 00080569.

Jagannath, D.* 1981. Mutagenicity of 06238001 lot LCM 266830-7 in the mitotic recombination assay with the yeast strain D5. Genetics Assay No. 5374. Final Report. Unpublished study. MRID 00122740.

Larson, R.E., P.S. Cartwright, P.K. Eriksson and R.J. Petersen. 1982. Applications of the FT-30 reverse osmosis membrane in metal finishing operations. Paper presented at Yokohama, Japan.

Lehman, A.J. 1959. Appraisal of the safety of chemicals in foods, drugs and cosmetics. Assoc. Food and Drug Off. U.S. Q. Bull.

Lightkep, G., G. Christian et al.* 1980. Teratogenic potential of TACU06238001 in New Zealand white rabbits (Segment II evaluation). Argus Project 113–003. Unpublished study. MRID 00122744.

Lochry, E.A., A. M. Hoberman and M. S. Christian.* 1986. Two-generation rat reproduction study. Argus Research Laboratories, Inc. Study No. 218–002.

Madison, P., R. Becci and R. Parent.* 1981. Guinea pig sensitization study. Buhler test for Mobil Corporation. Tackle 2S. FDRL Study No. 6738. Unpublished study. MRID 00122729.

Meister, R., ed. 1983. Farm chemicals handbook. Willoughby, OH: Meister Publishing Company.

Mobil Environmental and Health Science Laboratory.* 1981. A study of the oral toxicity of Tackle 2S in the dog. Mobil Study No. 1091–80. Six-month status report. Unpublished study. MRID 00122733.

Myhr, B. and M. McKeon.* 1981. Evaluation of 06238001 in the primary rat hepatocyte unscheduled DNA synthesis assay. MEHSL Study 1022-80. Final Report. Unpublished study. MRID 00122742.

NAS. 1985. National Academy of Sciences. Toxicity of selected contaminants. In: Drinking water and health, Vol. 6. Washington, DC: National Academy Press.

NCI. 1978. National Cancer Institute. Bioassay of nitrofen for possible carcinogenicity. Technical Report Series No. 26. DHEW Publication No. (NIH) 78-826. Washington, DC: U.S. Department of Health, Education and Welfare. 101 pp. Cited in: NAS. 1985.

NCI. 1979. National Cancer Institute. Bioassay of nitrofen for possible carcinogenicity. Technical Report Series No. 184. DHEW Publication (NIH) 79-1740. Washington, DC: U.S. Department of Health, Education and Welfare. 57 pp. Cited in: NAS. 1985.

Piccirillo, V.J. and T.L. Robbins.* 1976. Four week oral range findings study in rats. RII-6201. Unpublished study. MRID 00071892.

Putnam, D., L. Schechtman and W. Moore.* 1981. Activity of T1689 in the dominant lethal assay in rodents. MA Project No. T1689.116. Final Report. Unpublished study. MRID 00122738.

Schreiner, C.A., M.A. McKenzie and M.A. Mehlman.* 1980. An Ames Salmonella/mammalian microsome mutagenesis assay for determination of potential mutagenicity of Tackle 2S MCI0978. Study No. 511-80. Unpublished study. MRID 00061622.

Schreiner, C., M. Skinner and M. Mehlman.* 1981. Metaphase analysis of rat bone marrow cells treated in vivo with Tackle 2S. Study No. 1041-80. Unpublished study. MRID 00122741.

Spicer, E., L. Griggs, F. Marroquin, N.D. Jefferson and M. Blair.* 1983. Two year dietary toxicity study in dogs. (Tackle®) 450-0395. Unpublished study. MRID 00131162.

U.S. EPA. 1986a. U.S. Environmental Protection Agency. Guidelines for carcinogen risk assessment. Fed. Reg. 51(185):33992-34003. September 24.

U.S. EPA. 1986b. U.S. Environmental Protection Agency. U.S. EPA Method 3 — Determination of chlorinated acids in ground water by GC/ECD. Draft. January. Available from U.S. EPA's Environmental Monitoring and Support Laboratory, Cincinnati, OH.

U.S. EPA. 1988. U.S. Environmental Protection Agency. CRAVE carcinogenicity assessment for lifetime exposure. OPP. June 30.

Weatherholtz, W., S. Moore and G. Wolfe.* 1979a. Eye irritation study in monkeys. Blazer 2S. Project No. 417-396. Final Report. Unpublished study. MRID 00140887.

Weatherholtz, W., K. Peterson, M. Koka and R.W. Kapp.* 1979b. Four-week repeated dermal study in rabbits. RH-6201 formulations. Project No. 417–386. Final Report. Unpublished study. MRID 00140889.

Weatherholtz, W. and V. Piccirillo.* 1979. Teratology study in rabbits (RH-6201 LC). Final Report. Project No. 417–374. Unpublished study. MRID 00107485.

Whittaker Corporation.* No date, a. Acute oral LD_{50} rats. Study No. 410–0249. Unpublished study. MRID 00061625.

Whittaker Corporation.* No date, b. Primary dermal irritation—rabbit. Study No. 410–0286. Unpublished study. MRID 00061629.

Whittaker Corporation.* No date, c. Primary eye irritation—rabbits. Study No. 410–0252. Unpublished study. MRID 00061628.

Windholz, M., S. Budavari, R.F. Blumetti and E.S. Otterbein, eds. 1983. The Merck index, 10th ed. Rahway, NJ: Merck and Co., Inc.

WSSA. 1983. Weed Science Society of America. Herbicide handbook, 5th ed.

---

* Confidential Business Information submitted to the Office of Pesticide Programs.

Ametryn

I. General Information and Properties

A. CAS No. 834–12–8

B. Structural Formula

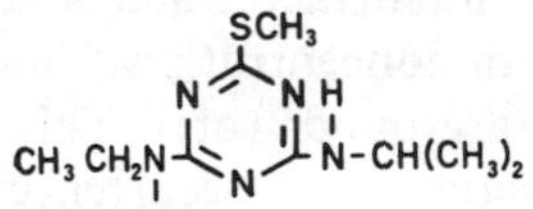

$$\text{CH}_3\text{CH}_2\text{N}(\text{H})-\!\!\!\overset{\overset{\displaystyle \text{SCH}_3}{|}}{\underset{\text{N}}{\text{N}}}\!\!\!-\text{N}(\text{H})-\text{CH(CH}_3)_2$$

2-(Ethylamino)-4-(isopropylamino)-6-(methylthio)-s-triazine

C. Synonyms

- N-ethyl-N'-(1-methylethyl)-6-(methylthio)-1,3,5-triazine-2,4-diamine; Ametrex; Ametryne; Cemerin; Crisatine; Evik 80W; Gesapax (WSSA, 1983; Meister, 1983).

D. Uses

- Ametryn is a selective herbicide for control of broadleaf and grass weeds in pineapple, sugarcane, bananas and plantains. It is also used as a postdirected spray in corn, as a potato vine desiccant and for total vegetation control (WSSA, 1983).

E. Properties (WSSA, 1983)

Chemical Formula	$C_9H_{17}N_5S$
Molecular Weight	227.35
Physical State	Colorless crystals
Boiling Point	--
Melting Point	84 to 85°C
Density	--
Vapor Pressure	8.4×10^{-7} mm Hg
Specific Gravity	--
Water Solubility	185 mg/L

Log Octanol/Water
   Partition Coefficient   –1.72 (calculated)
Taste Threshold   --
Odor Threshold   --
Conversion Factor   --

F. Occurrence

- Ametryn has been found in 2 of 1,190 surface water samples analyzed and in 24 of 560 ground-water samples (STORET, 1988). Samples were collected at 215 surface water locations and 513 ground-water locations, and ametryn was found in 6 states. The 85th percentile of all nonzero samples was 0.1 μg/L in surface water and 210 μg/L in ground-water sources. The maximum concentration found was 0.1 μg/L in surface water and 450 μg/L in ground water. This information is provided to give a general impression of the occurrence of this chemical in ground and surface waters as reported in the STORET database. The individual data points retrieved were used as they came from STORET and have not been confirmed as to their validity. STORET data are often not valid when individual numbers are used out of the context of the entire sampling regime, as they are here. Therefore, this information can only be used to form an impression of the intensity and location of sampling for a particular chemical.

G. Environmental Fate

- In aqueous solutions, ametryn is stable to natural sunlight, with a half-life of greater than 1 week. When exposed to artificial light for 6 hours, 75% of applied ametryn remained. One photolysis product was identified as 2-ethylamino-4-hydroxy-6-isopropylamino-s-triazine (Registrant CBI data).

- Ametryn is stable to photolysis on soil (Registrant CBI data).

- Soil metabolism of ametryn, under aerobic conditions, proceeds with a half-life of greater than 2 to 3 weeks. Metabolic products include 2-amino-4-isopropylamino-6-methylthio-s-triazine, 2-amino-4-ethyl-amino-6-methylthio-s-triazine and 2,4-diamino-6-methylthio-triazine. Under anaerobic conditions the rate of metabolism decreases ($t_{1/2}$ = 122 days) (Registrant CBI data).

- Under sterile conditions ametryn does not degrade appreciably. Therefore, microbial degradation is a major degradation pathway (Registrant CBI data).

- Neither ametryn nor its hydroxy metabolite leach past 6 inches in depth with normal rainfall. However, since both compounds are persistent, they may leach under exaggerated rainfall or flood and furrow irrigation. This behavior is seen with other triazines (Registrant CBI data).

- Ametryn's Freundlich soil-water partition coefficient values, Kd, range from 0.6 in sands to 5.0 in silty clay soils. Specifically, the Kd for a sandy loam is 4.8, and for two silty loams, 3.8 and 2.8, respectively.

- In the laboratory, Ametryn has a half-life of 36 days. In the field, Ametryn degraded with a half-life of 125 to 250 days (Registrant CBI data).

II.   Pharmacokinetics

A. Absorption

- Oliver et al. (1969) administered ^{14}C-labeled ametryn orally to Sprague-Dawley rats. Investigators stated that ametryn was administered by stomach tube to animals at dosage levels from 1 to 4 mg per animal. When the label was in the ring, 32.1% was excreted in the feces, indicating that over 70% had been absorbed. When the label was in the ethyl or isopropyl side chains, only 2 to 5% was excreted in the feces.

B. Distribution

- Oliver et al. (1969) administered ring-labeled ametryn orally to male and female Sprague-Dawley rats and measured distribution of label in tissues at 6, 48 and 72 hours after dosing. Tissue distribution at 6 hours was greatest in kidney, followed by liver, spleen, blood, lung, fat, carcass, brain and muscle. Blood levels remained relatively constant for 72 hours after dosing, while all other tissue levels dropped rapidly to < 0.1% of dose per gram of tissue.

C. Metabolism

- Oliver et al. (1969) administered ^{14}C-labeled ametryn orally to groups of six male and six female Sprague-Dawley rats. When the label was in the isopropyl side chain, 41.9% of the label appeared as CO_2. When the label was in the ethyl side chain, 18.1% of the label appeared as CO_2. This indicated that the side chains were extensively metabolized. When the ring was uniformly labeled with carbon-14 and the compound fed orally to rats, 58% was excreted in the urine but it was not determined whether excretion of the original compound or metabolites had occurred.

D. Excretion

- Oliver et al. (1969) studied the excretion of ametryn utilizing uniformly labeled compound with ^{14}C-ametryn in the ring or in the ethyl or isopropyl side chains. Forty-eight hours after oral dosing of six male and six female Sprague-Dawley rats, 57.6% of the ring-labeled activity had been excreted in the urine with 32.1% excreted in the feces (total 89.7% of dose). When the fed compound was labeled in the side chains, however, much of the ^{14}C was excreted in expired air as carbon dioxide. When the fed compound was labeled in the isopropyl side chain, rats excreted 41.9% of the label in expired air, 20% in the urine and 2% in the feces, and 7% remained in the carcass (total 70.9%) at 48 hours. When the ethyl side chain contained the label, 18.1% of the label was excreted as carbon dioxide, 45% in the urine, 5% in the feces and 9% remained in the carcass (total 77.1% of dose). After 72 hours, total recovery was approximately 88% for both of the side-chain labeled compounds.

III. Health Effects

A. Humans

- No information was found in the available literature on the health effects of ametryn in humans.

B. Animals

1. *Short-term Exposure*

- The following acute oral LD_{50} values for ametryn in rats were reported: Charles River CD rats, 1,207 mg/kg (males), 1,453 mg/kg (females) (Grunfeld, 1981); mixed male and female rats (strain not specified), 1,750 mg/kg (Stenger and Planta, 1961a); male and female Wistar rats, 1,750 mg/kg (Consultox Laboratories Limited, 1974).

- Piccirillo (1977) reported the results of a 28-day feeding study in male and female mice. Animals were 5 weeks of age and weighed 21 to 28 g at the beginning of the study. Animals (5/sex/dose) were fed diets containing 0, 100, 300, 600, 1,000, 3,000, 10,000 or 30,000 ppm of ametryn (technical). Based on the assumption that 1 ppm in the diet of mice is equivalent to 0.15 mg/kg/day (Lehman, 1959), these doses correspond to 0, 15, 45, 90, 150, 450, 1,500 or 4,500 mg/kg/day. At 30,000 ppm in the diet, all animals died within 2 weeks. At 10,000 ppm, 3 of the 10 died within 2 weeks. No other deaths occurred at any other dose level. Clinical signs in the two highest dose groups included hunched appearance, stained fur and labored respiration. At the 3,000-ppm dose level, only 1 of the 10 animals showed clinical signs of toxicity. Body weight gain was comparable in all survivors by the end of week 4. Gross pathology in animals that died showed a dark-red mucosal lining of the gastrointestinal tract and ulcerated areas of the gastric mucosa. There was no histopathological examination of tissues in this study.

* Stenger and Planta (1961b) reported a 28-day study of the toxicity of ametryn in rats. Dose levels of 100, 250 or 500 mg/kg/day were administered 6 days/week by gavage to groups of five male and five female rats. The study indicated that there was a control group but no data were given. At the 500-mg/kg/day dose level, animals became emaciated, weight gain was limited and 7 of 10 rats died. Histopathological examination of the animals that died indicated severe vascular congestion, centrilobular liver necrosis and fatty degeneration of individual liver cells. At 250 mg/kg/day, 1 of 10 rats died during the study and there was depressed growth rate in the survivors. Histological examination of liver, kidney, spleen, pancreas, heart, lung, intestine and gonads showed no major degenerative changes. No effects were reported in animals administered 100 mg/kg/day, which was identified as the No-Observed-Adverse-Effect-Level (NOAEL) in this study.

* Ceglowski et al. (1979) administered single oral doses of 88 or 880 mg/kg of ametryn to mice 5 days before, on the day of or 2 days after immunization with sheep erythrocytes (purity not specified). All mice receiving the highest dose (880 mg/kg) of ametryn had significant depression of splenic plaque-forming cell numbers when assayed 4 days later. Animals receiving the low dose showed no effect. Similarly, animals receiving 88 mg/kg for 8 or 28 consecutive days prior to immunization exhibited no significant reduction in antibody plaque formation.

2. Dermal/Ocular Effects

* Two of six rabbits showed mild skin irritation when ametryn was left in contact with intact or abraded skin (500 mg/2.5 cm^2) for 24 hours (Sachsse and Ullmann, 1977).

* In a sensitization study with Perbright White guinea pigs (Sachsse and Ullmann, 1977), 10 male and 10 female guinea pigs weighing 400 to 450 g received 10 daily intracutaneous 0.1-mL injections of 0.1% ametryn in polyethylene glycol:saline (70:30). Fourteen days after the last

dose, animals were challenged by an occlusive dermal application of ametryn or by an intradermal challenge. Animals showed no sensitization reaction following the dermal application of the challenge dose, but there was a positive response after the intradermal challenge.

- Kopp (1975) found that ametryn (technical grade) placed in the eyes of rabbits produced slight conjunctival redness at 24 hours. This cleared completely within 72 hours.

- Sachsse and Bathe (1976) applied 2,150 mg/kg or 3,170 mg/kg ametryn in suspension to the shaved backs of five male and five female rats weighing 180 to 200 g. The occlusive covering was removed at 24 hours, the skin was washed and animals were observed for 14 days. There was no local irritation or adverse reaction, and at necropsy there were no gross changes in the skin. The acute dermal LD_{50} in male and female rats was reported to be $> 3,170$ mg/kg.

- Ametryn (2,000 mg/kg) was applied daily to the skin of five male and five female rats weighing approximately 200 g (Consultox Laboratories Limited, 1974). After 14 days of treatment, no deaths had occurred and no other effects were reported. The 14-day dermal LD_{50} was reported to be $> 2,000$ mg/kg/day.

3. *Long-term Exposure*

- Domenjoz (1961) administered ametryn in water via stomach tube 6 days/week for 90 days to Meyer-Arendt rats (12/sex/dose). The initial material was 50% ametryn in a powder vehicle. Two dose levels of the material (20 or 200 mg/kg/day) provided dose levels of ametryn of 10 or 100 mg/kg/day. Two control groups were included; one group received water only and the other received the powder vehicle only suspended in water. Over the 90-day period, all animals gained weight at comparable rates and there was no visible effect on appearance or behavior. One control rat and one rat in the 100-mg/kg dosage group died. This death was not considered compound-related. At the 90-day necropsy, organ-to-body weight ratios were comparable to controls. Liver, kidney, spleen, heart, gonads, small intestine, colon, stomach, thyroid and lung were microscopically examined. The Lowest-Observed-Adverse-Effect-Level (LOAEL) was associated with fatty degeneration of the liver. Based on this study, a LOAEL of 100 mg/kg/day (the highest dose tested) was identified. All tissues were comparable to controls at the lowest dose (10 mg/kg/day), which was identified as the NOAEL.

4. *Reproductive Effects*

- No information was found in the available literature on the reproductive effects of ametryn.

5. *Developmental Effects*

- No information was found in the available literature on the developmental effects of ametryn.

6. *Mutagenicity*

- Anderson et al. (1972) reported that ametryn was not mutagenic in eight strains of *Salmonella typhimurium*. No metabolic activating system was utilized.

- Simmons and Poole (1977) also reported that ametryn was not mutagenic in five strains of *Salmonella typhimurium* (TA98, TA100, TA1535, TA1537 and TA1538), with or without metabolic activation provided by an S-9 fraction from rats pretreated with Aroclor 1254.

- Shirasu et al. (1976) reported ametryn was not mutagenic in the rec-assay system utilizing two strains of *Bacillus subtilis*, in reversion assays utilizing auxotrophic strains of *Escherichia coli* (WP2) and in *S. typhimurium* strains TA1535, TA1536, TA1537 and TA1538 (without metabolic activation).

7. *Carcinogenicity*

- No information was found in the available literature on the carcinogenic effects of ametryn.

IV.   Quantification of Toxicological Effects

A. One-day Health Advisory

No data were found in the available literature that were suitable for the determination of a One-day HA value for ametryn. It is therefore recommended that the Ten-day HA value for the 10-kg child (8.6 mg/L, calculated below) be used at this time as a conservative estimate of the One-day HA value.

OKANAGAN COLLEGE LIBRARY
BRITISH COLUMBIA

B. Ten-day Health Advisory

The study by Stenger and Planta (1961b) has been selected to serve as the basis for determination of the Ten-day HA value for the 10-kg child. This study identified a NOAEL of 100 mg/kg/day, based on normal weight gain and absence of histological evidence of injury in rats following 28 days of exposure by gavage. The study also identified a LOAEL of 250 mg/kg/day, based on reduced body weight gain, although no major histological changes were noted. One death occurred in the 250-mg/kg/day group, but it could not be determined if this was compound-related. The NOAEL identified in this study (100 mg/kg/day) is supported by the 28-day feeding study in rats by Piccirillo (1977), which identified a NOAEL of 150 mg/kg/day and a LOAEL of 450 mg/kg/day, and by the study of Ceglowski et al. (1979), which identified a NOAEL of 88 mg/kg/day and a LOAEL of 880 mg/kg/day.

Using the NOAEL of 100 mg/kg/day, the Ten-day HA for a 10-kg child is calculated as follows:

$$\text{Ten-day HA} = \frac{(100 \text{ mg/kg/day})(10 \text{ kg})(6/7)}{(100)\ (1 \text{ L/day})} = 8.6 \text{ mg/L } (9,000 \text{ } \mu g/L)$$

where: 6/7 = conversion from 6 to 7 days of exposure.

C. Longer-term Health Advisory

The 90-day oral dosing study in rats by Domenjoz (1961) has been selected to serve as the basis for determination of the Longer-term HA. At two dose levels (10 or 100 mg/kg/day), no deaths were reported and no other effects were noted during the 90-day period. Terminal necropsy findings and histological examination of tissues from treated animals were comparable to controls. At the highest dose tested, there was fatty degeneration in the livers examined. Based on these data, a NOAEL of 10 mg/kg/day (the lowest dose tested) was identified.

The Longer-term HA for a 10-kg child is calculated as follows:

$$\text{Longer-term HA} = \frac{(10 \text{ mg/kg/day})(10 \text{ kg})(6/7)}{(100)\ (1 \text{ L/day})} = 0.86 \text{ mg/L } (900 \text{ } \mu g/L)$$

where: 6/7 = conversion from 6 to 7 days of exposure.

The Longer-term HA for a 70-kg adult is calculated as follows:

$$\text{Longer-term HA} = \frac{(10 \text{ mg/kg/day})(70 \text{ kg})(6/7)}{(100)\ (2 \text{ L/day})} = 3 \text{ mg/L } (3,000 \text{ } \mu g/L)$$

where: 6/7 = conversion from 6 to 7 days of exposure.

D. Lifetime Health Advisory

Compound-specific, chronic ingestion data for ametryn are not available at this time. In the absence of appropriate ingestion studies, the Lifetime HA for ametryn is derived from the subchronic study in rats reported by Domenjoz (1961). At two dose levels (10 or 100 mg/kg/day), no deaths were reported during the 90-day period. Terminal necropsy findings and histological examination of tissues from treated animals were comparable to controls at the lowest dose level of 10 mg/kg/day. This study identified a NOAEL of 10 mg/kg/day (the lowest dose tested).

Using the NOAEL of 10 mg/kg/day, the Lifetime HA for ametryn is calculated as follows:

Step 1: Determination of the Reference Dose (RfD)

$$\text{RfD} = \frac{(10 \text{ mg/kg/day)} (6/7)}{(1,000)} = 0.009 \text{ mg/kg/d}$$

where: 6/7 = conversion from 6 to 7 days of exposure.

Step 2: Determination of the Drinking Water Equivalent Level (DWEL)

$$\text{DWEL} = \frac{(0.0086 \text{ mg/kg/day)} (70 \text{ kg})}{(2 \text{ L/day})} = 0.3 \text{ mg/L } (300 \text{ } \mu g/L)$$

Step 3: Determination of the Lifetime Health Advisory

Lifetime HA = (0.3 mg/L) (20%) = 0.06 mg/L (60 μg/L)

E. Evaluation of Carcinogenic Potential

- No carcinogenicity studies were found in the literature searched.

- The International Agency for Research on Cancer (IARC) has not evaluated the carcinogenic potential of ametryn.

- Applying the criteria described in EPA's guidelines for assessment of carcinogenic risk (U.S. EPA, 1986), ametryn may be classified in Group D: not classifiable. This category is for agents with inadequate or no animal evidence of carcinogenicity.

V.   Other Criteria, Guidance and Standards

- U.S. EPA has established residue tolerances for ametryn in or on raw agricultural commodities that range from 0.1 to 0.5 ppm (CFR, 1985).

VI. Analytical Methods

- Analysis of ametryn is by a gas chromatographic (GC) method applicable to the determination of certain nitrogen-phosphorus containing pesticides in water samples. In this method, approximately 1 liter of sample is extracted with methylene chloride. The extract is concentrated and the compounds are separated using capillary column GC. Measurement is made using a nitrogen-phosphorus detector. This method has been validated in a single laboratory, and estimated detection limits have been determined for the analytes in this method, including ametryn. The estimated detection limit is 2.0 μg/L (U.S. EPA, 1988).

VII. Treatment Technologies

- Available data indicate that granular-activated carbon (GAC) adsorption will remove ametryn from water.

- Whittaker (1980) experimentally determined adsorption isotherms for ametryn on GAC.

- Whittaker (1980) reported the results of GAC columns operating under bench-scale conditions. At a flow rate of 0.8 gpm/ft^2 and an empty bed contact time of 6 minutes, ametryn breakthrough (when effluent concentration equals 10% of influent concentration) occurred after 896 bed volumes (BV). When a bisolute ametryn-propham solution was passed over the same column, ametryn breakthrough occurred after 240 BV.

- In a laboratory study (Nye, 1984), GAC was employed as a possible means of removing ametryn from contaminated wastewater. The results show that the column exhaustion capacity was 111.2 mg ametryn adsorbed on 1 g of activated carbon.

- Treatment technologies for the removal of ametryn from water are available and have been reported to be effective. However, selection of individual or combinations of technologies to attempt ametryn removal from water must be based on a case-by-case technical evaluation, and an assessment of the economics involved.

VIII. References

Anderson, K.J., E.G. Leighty and M.T. Takahasi. 1972. Evaluation of herbicides for possible mutagenic activity. J. Agr. Food Chem. 20:649–656.

Ceglowski, W.S., D.D. Ercegrovich and N.S. Pearson. 1979. Effects of pesticides on the reticuloendothelial system. Adv. Exp. Med. Biol. 121:569–576.

CFR. 1985. Code of Federal Regulations. 40 CFR 180.258. July 1. pp. 300–301.

Consultox Laboratories Limited.* 1974. Ametryn: acute oral and dermal toxicity evaluation. Unpublished study. MRID 00060310.

Domenjoz, R.* 1961. Ametryn: Toxicity in long-term administration. Unpublished study. MRID 00034838.

Grunfeld, Y.* 1981. Ametryn 80 w.p.: acute oral toxicity in the rat. Unpublished study. MRID 00100573.

Kopp, R.W.* 1975. Acute eye irritation potential study in rabbits. Final Report. Project No. 915-104. Unpublished study. MRID 00060311.

Lehman, A.J. 1959. Appraisal of the safety of chemicals in foods, drugs and cosmetics. Assoc. Food Drug Off. U.S., Q. Bull.

Meister, R., ed. 1983. Farm chemicals handbook. Willoughby, OH: Meister Publishing Co.

Nye, J.C. 1984. Treating pesticide-contaminated wastewater. Development and evaluation of a system. American Chemical Society.

Oliver, W.H., G.S. Born and P.L. Zcimer. 1969. Retention, distribution, and excretion of ametryn. J. Agr. Food Chem. 17:1207–1209.

Piccirillo, V.J.* 1977. 28-day pilot feeding study in mice. Final Report. Project No. 483-126. Unpublished study. MRID 00068169.

Sachsse, K. and R. Bathe.* 1976. Acute dermal LD$_{50}$ in the rat of technical G34162. Project No. Siss. 5665. Unpublished study. MRID 00068172.

Sachsse, K. and L. Ullmann.* 1977. Skin irritation in the rabbit after single application of technical grade G34162. Unpublished study. MRID 00068174.

Shirasu, Y., M. Moriya, K. Kato, A. Furuhashi and T. Kada. 1976. Mutagenic screening of pesticides in the microbial system. Mutat. Res. 40:19–30.

Simmons, V.F. and D. Poole.* 1977. *In-vitro* and *in-vivo* microbiological assays of six Ciba-Geigy chemicals. SRI project LSC-5686. Final Report. Unpublished study. MRID 00060642.

Stenger, P. and V. Planta.* 1961a. Oral toxicity in rats. Unpublished study. MRID 00048226.

Stenger, P. and V. Planta.* 1961b. Subchronic toxicity test no. 257. Unpublished study. MRID 00048228.

STORET. 1988. STORET Water Quality File. Office of Water. U.S. Environmental Protection Agency (data file search conducted in May, 1988).

U.S. EPA. 1986. U.S. Environmental Protection Agency. Guidelines for carcinogen risk assessment. Fed. Reg. 51(185):33992–34003. September 24.

U.S. EPA. 1988. U.S. EPA Method 507 — Determination of nitrogen and phosphorus containing pesticides in water by GC/NPD. Draft. April. Available from U.S. EPA's Environmental Monitoring and Support Laboratory, Cincinnati, OH.

WSSA. 1983. Weed Science Society of America. Herbicide handbook, 5th ed. Champaign, IL: Weed Society of America. pp. 16–19.

Whittaker, K.F. 1980. Adsorption of selected pesticides by activated carbon using isotherm and continuous flow column systems. Ph.D. Thesis. Purdue University.

---

* Confidential Business Information submitted to the Office of Pesticide Programs.

Ammonium Sulfamate

I. General Information and Properties

A. CAS No. 7773–06–0

B. Structural Formula

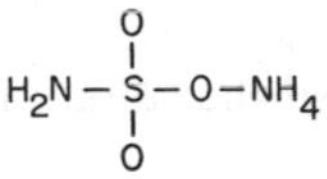

$$H_2N - \overset{\displaystyle O}{\underset{\displaystyle O}{\overset{|}{\underset{|}{S}}}} - O - NH_4$$

Ammonium sulfamate

C. Synonyms

- Amicide; Amidosulfate; Ammate; Ammonium amidosulfate; Ammonium amidosulfonate; Ammonium amidotrioxosulfate; AMS; Fyran 206k; Ikurin.

D. Uses

- Ammonium sulfamate is an herbicide used to control woody plant species. May be used for poison ivy control in apple and pear orchards (Meister, 1986).

E. Properties (Meister, 1986)

Chemical Formula	$H_6N_2O_3S$
Molecular Weight	114.14
Physical State (25°C)	Colorless crystals
Boiling Point	--
Melting Point	131 to 132°C
Density (20°C)	> 1
Vapor Pressure*	Negligible (< 10^{-7} mm Hg)
Water Solubility*	684 g/L
Specific Gravity	--
Log Octanol/Water Partition Coefficient	--
Taste Threshold	--

35

Odor Threshold   --
Conversion Factor   --

* (WSSA, 1983)

F. Occurrence

- No information was found in the available literature on the occurrence of ammonium sulfamate.

G. Environmental Fate

- Konnai et al. (1974) showed that ammonium sulfamate was very mobile in soil.

II.   Pharmacokinetics

A. Absorption

- No information was found in the available literature on the absorption of ammonium sulfamate.

B. Distribution

- No information was found in the available literature on the distribution of ammonium sulfamate.

C. Metabolism

- The metabolism of ammonium sulfamate in the urine of dogs was reported by Bergen and Wiley (1938); however, details of the study were not clearly defined for assessment.

D. Excretion

- Bergen and Wiley (1938) reported 80 to 84% excretion of sulfamic acid in the urine of two dogs following oral administration of ammonium sulfamate in capsules for 5 days.

III. Health Effects

A. Humans

- No information was found in the available literature on the health effects of ammonium sulfamate in humans.

B. Animals

1. Short-term Exposure

- An oral LD_{50} of 3,900 mg/kg in the rat is reported for ammonium sulfamate (Meister, 1986).

2. Dermal/Ocular Effects

- Five rats received a 20% aqueous solution of ammonium sulfamate (dose level not specified) on the shaved skin of the back. They were killed after 16 treatments on the 27th day of the period (Read and Hueber, 1938). Another five rats received a 50% aqueous solution of ammonium sulfamate on the shaved skin of the back. These animals were killed after 11 treatments on the 19th day of the study. It should be noted that the animals were not prevented from licking the chemical. Investigators reported that there were no gross pathological changes of importance in any of the animals. On microscopic pathological examination of the animals, the spleen of 9 of 10 animals had numerous macrophages with brown pigment. The stomach sections of seven animals revealed a brown, granular material in the surface capillaries of the mucosa.

- Read and Hueber (1938) orally administered 1 mL of a 50% aqueous solution of ammonium sulfamate (1.7 g/kg/day) to 10 rats on alternate days. Five rats were killed on the 27th day of the study after nine treatments, and the remaining five were killed on the 42nd day of the study after 15 treatments. Investigators reported that there were no gross pathological changes of importance in any of the animals. Microscopic pathology indicated the following: in one animal, superficial capillaries of the stomach mucosa occasionally contained yellow-brown granules; in three animals, there was slight vacuolation of the cytoplasm of liver cells about the central veins, but these changes were very mild; and in the spleen, three of the sections had moderate numbers of macrophages filled with hemosiderin. A fourth spleen section showed marked erythrophagia.

3. *Long-term Exposure*

- Gupta et al. (1979) reported the results of a 90-day study involving oral administration of 0, 100, 250 or 500 mg/kg of ammonium sulfamate to rats (20 animals to each dose group) 6 days a week. No adverse effects were observed with respect to appearance, behavior or survival of animals. No significant difference in the body weights of rats was observed, except in the case of rats receiving 500 mg/kg, where body weight was significantly less than controls after the end of 60 days. No significant changes in relative organ weights were noticed in any group of rats. Hematological examination conducted at 30, 60 and 90 days revealed nonsignificant increases in the numbers of neutrophils in the female adult and male weanling rats (500 mg/kg dose level) after 90 days. In the histological examination, organs in all the groups of animals appeared normal except that the liver of one adult rat (500 mg/kg) showed slight fatty degenerative changes after 90 days.

- Rosen et al. (1965) reported the findings of a study in female rats following administration of ammonium sulfamate at dietary levels of 1.1% (10 g/kg/day) or 2.1% (20 g/kg/day) for 105 days. No effect was detected at the 1% (10 g/kg/day) level of feeding, but growth retardation and a slight cathartic effect were observed at the 2% (20 g/kg/day) dietary level. No other information was provided by the authors.

- Sherman and Stula (1966) reported the results of a 19-month feeding study in 29-day-old CHR-CD male and female rats. Ammonium sulfamate was fed at dietary concentrations of 0, 350 or 500 ppm without any clinical or nutritional evidence of toxicity. There were no histopathological changes that could be attributed to the feeding of the test chemical. The observed pathologic lesions were interpreted as a result of spontaneous diseases.

4. *Reproductive Effects*

- Sherman and Stula (1966) reported the results of a three-generation reproduction study in rats. Rats receiving 0, 350 or 500 ppm ammonium sulfamate in the diet showed no evidence of toxicity as measured by histopathological evaluation and reproduction and lactation indices.

5. *Developmental Effects*

- No information was found in the available literature on the developmental effects of ammonium sulfamate.

6. *Mutagenicity*

- No information was found in the available literature on the mutagenic effects of ammonium sulfamate.

7. *Carcinogenicity*

- No information was found in the available literature on the carcinogenic effects of ammonium sulfamate.

IV.   Quantification of Toxicological Effects

A. One-day Health Advisory

No data were located in the available literature that were suitable for deriving a One-day HA value for ammonium sulfamate. It is recommended that the Longer-term HA value for the 10-kg child (20 mg/L, calculated below) be used at this time as a conservative estimate of the One-day HA value.

B. Ten-day Health Advisory

No data on ammonium sulfamate toxicity were located in the available literature that were suitable for calculation of a Ten-day HA value. It is recommended that the Longer-term HA value for the 10-kg child (20 mg/L, calculated below) be used at this time as a conservative estimate of the Ten-day HA value.

C. Longer-term Health Advisory

The subchronic oral toxicity study in rats by Gupta et al. (1979) is considered for the Longer-term HA. In this study, rats (female adults and male and female weanlings) received ammonium sulfamate orally at dose levels of 0, 100, 250 or 500 mg/kg/day for 90 days. Hematological and histological examinations at 30, 60 and 90 days revealed nonsignificant changes in hematological and histological measures. However, adult rats fed 500 mg/kg ammonium sulfamate showed lesser weight gain compared to other groups.

Using 250 mg/kg/day as a No-Observed-Adverse-Effect-Level (NOAEL), a Longer-term HA for the 10-kg child is calculated as follows:

$$\text{Longer-term HA} = \frac{(250 \text{ mg/kg/day}) (10 \text{ kg}) (6/7)}{(100) (1 \text{ L/day})} = 21.4 \text{ mg/L } (20,000 \text{ } \mu g/L)$$

where: 6/7 = conversion from 6 days to 7 days of exposure.

The Longer-term HA for a 70-kg adult is calculated as follows:

$$\text{Longer-term HA} = \frac{(250 \text{ mg/kg/day})(70 \text{ kg})(6/7)}{(100)\ (2 \text{ L/day})} = 75 \text{ mg/L } (80,000 \text{ } \mu\text{g/L})$$

where: 6/7 = conversion from 6 days to 7 days of exposure.

D. Lifetime Health Advisory

The study by Gupta et al. (1979) has been selected to serve as the basis for determination of the Lifetime HA even though the results of this subchronic study were based on 90 days' exposure. In this study, rats (female adults and weanling males and females) received ammonium sulfamate orally in drinking water at dose levels of 0, 100, 250 or 500 mg/kg/day for 90 days. The NOAEL was identified as 250 mg/kg/day, since the highest dose level of 500 mg/kg/day was associated with decreased body weight gain in rats over a 90-day exposure period. In a chronic feeding study reported by Sherman and Stula (1966), ammonium sulfamate was fed to rats at dietary levels of 0, 350 or 500 ppm over a 19-month period. The authors stated that these dose levels did not produce any significant clinical or histological changes in rats receiving the test compound, and any changes recorded were interpreted as being lesions of spontaneous diseases.

Using a NOAEL of 250 mg/kg/day, the Lifetime HA is calculated as follows:

Step 1: Determination of the Reference Dose (RfD)

$$\text{RfD} = \frac{(250 \text{ mg/kg/day})\ (6/7)}{(1,000)} = 0.214 \text{ mg/kg/day}$$

where: 6/7 = conversion from 6 days to 7 days of exposure.

Step 2: Determination of the Drinking Water Equivalent Level (DWEL)

$$\text{DWEL} = \frac{(0.214 \text{ mg/kg/day})\ (70 \text{ kg})}{(2 \text{ L/day})} = 7.5 \text{ mg/L } (8,000 \text{ } \mu\text{g/L})$$

Step 3: Determination of the Lifetime Health Advisory

$$\text{Lifetime HA} = (7.5 \text{ mg/L})\ (20\%) = 1.5 \text{ mg/L } (2,000 \text{ } \mu\text{g/L})$$

E. Evaluation of Carcinogenic Potential

- No studies were found in the available literature investigating the carcinogenic potential of ammonium sulfamate. Applying the criteria described in EPA's final guidelines for assessment of carcinogenic risk (U.S. EPA, 1986), ammonium sulfamate may be classified in Group D: not classifiable. This category is used for substances with inadequate or no animal evidence of carcinogenicity.

V.   Other Criteria, Guidance and Standards

- The American Conference of Government Industrial Hygienists (ACGIH) has adopted a Threshold Limit Value-Time-Weighted Average (TLV-TWA) of 10 mg/m^3 and a TLV Short-Term Exposure Limit (STEL) of 20 mg/m^3 for inhalation exposure (ACGIH, 1984).

VI.   Analytical Methods

- There is no standardized method for determination of ammonium sulfamate in water samples. A procedure has been reported for the estimation of ammonium sulfamate in certain foods, however (U.S. FDA, 1969). This procedure involves a colorimetric determination of ammonium sulfamate based on liberating SO$_4$ and reducing it to H$_2$S, which is measured after treating with zinc, p-aminodimethylaniline and ferric chloride to form methylene blue.

VII.   Treatment Technologies

- No information was found in the available literature on treatment technologies capable of effectively removing ammonium sulfamate from contaminated water.

VIII.   References

ACGIH. 1984. American Conference of Governmental Industrial Hygienists. Documentation of the threshold limit values for substances in workroom air, 3rd ed. Cincinnati, OH: ACGIH.

Bergen, D.S. and F.M. Wiley.* 1938. The metabolism of sulfamic acid and ammonium sulfamate. Unpublished report. Submitted to U.S. EPA, Office of Pesticide Programs, Washington, DC.

Gupta, B.N., R.N. Khanna and K.K. Datta. 1979. Toxicological studies of ammonium sulfamate in rats after repeated oral administration. Toxicology. 13:45–49.

Konnai, M., Y. Takeuchi and T. Takematsu. 1974. Basic studies on the residues and movements of forestry herbicides in soil. Bull. Coll. Agric. Utsunomiya Univ. 9(1):995–1012.

Meister, R., ed. 1986. Farm chemicals handbook. Willoughby, OH: Meister Publishing Co.

Read, W.T. and K.C. Hueber.* 1938. The pathology produced in rats following the administration of sulfamic acid and ammonium sulfamate. Unpublished report. MRID GS0016–0040.

Rosen, D.E., C.J. Krisher, H. Sherman and E.E. Stula. 1965. Toxicity studies on ammonium sulfamate. The Toxicologist. Fourth Annual Meeting, Williamsburg, VA. March 8–10.

Sherman, H. and E. Stula.* 1966. Toxicity studies on ammonium sulfamate. Unpublished report. MRID GS0016–0038.

U.S. EPA. 1986. U.S. Environmental Protection Agency. Guidelines for carcinogen risk assessment. Fed. Reg. 51(185):33992–34003. September 24.

U.S. FDA. 1969. U.S. Food and Drug Administration. Pesticide analytical manual, Vol. II. Washington, DC.

WSSA. 1983. Weed Science Society of America. Herbicide handbook, 5th ed. Champaign, IL.

---

* Confidential Business Information submitted to the Office of Pesticide Programs.

Atrazine

I. General Information and Properties

A. CAS No. 1912–24–9

B. Structural Formula

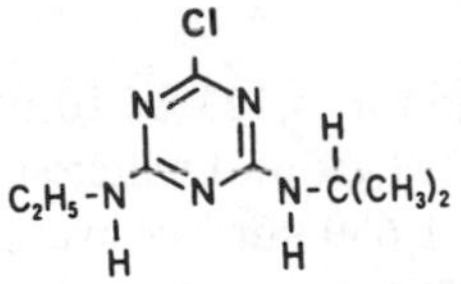

2-Chloro-4-ethylamino-6-isopropylamino-1,3,5-triazine

C. Synonyms

- AAtrex; Atranex; Crisatrina; Crisazine; Farmco Atrazine; Griffex; Shell Atrazine Herbicide; Vectal SC; Gesaprim; Primatol (Meister, 1987).

D. Uses

- Atrazine over the past 30 years has been the most heavily used herbicide in the U.S. It is used for nonselective weed control on industrial or noncropped land and selective weed control in corn, sorghum, sugar cane, pineapple and certain other plants (Meister, 1987).

E. Properties (Meister, 1987; Windholz, 1976)

Chemical Formula	$C_8H_{14}ClN_5$
Molecular Weight	215.72
Physical State	White, odorless, crystalline solid
Boiling Point	
(25 mm Hg)	--
Melting Point	175 to 177°C
Density (20°)	1.187
Vapor Pressure (20°C)	3.0×10^{-7} mm Hg
Specific Gravity	--
Water Solubility (22°C)	70 mg/L

43

Log Octanol/Water
   Partition Coefficient   2.33 to 2.71
Taste Threshold   --
Odor Threshold   --
Conversion Factor   --

F. Occurrence

- In a monitoring study of Mississippi River water, atrazine residues were found at a maximum level of 17 ppb; residues were detected throughout the year, with the highest concentrations found in June or July (Newby and Tweedy, 1976).

- Atrazine has been found in 4,123 of 10,942 surface water samples analyzed and in 343 of 3,208 ground water samples (STORET, 1988). Samples were collected at 1,659 surface water locations and 2,510 ground water locations. The 85th percentile of all nonzero samples was 2.3 μg/L in surface water and 1.9 μg/L in ground water sources. The maximum concentration found in surface water was 2,300 μg/L, and in ground water it was 700 μg/L. Atrazine was found in surface water of 31 states and in ground water in 13 states. This information is provided to give a general impression of the occurrence of this chemical in ground and surface waters as reported in the STORET database. The individual data points retrieved were used as they came from STORET and have not been confirmed as to their validity. STORET data are often not valid when individual numbers are used out of the context of the entire sampling regime, as they are here. Therefore, this information can only be used to form an impression of the intensity and location of sampling for a particular chemical.

- Atrazine has been found also in ground water in Pennsylvania, Iowa, Nebraska, Wisconsin and Maryland; typical positives were 0.3 to 3 ppb (Cohen et al., 1986).

G. Environmental Fate

- An aerobic soil metabolism study in Lakeland sandy loam, Hagerstown silty clay loam, and Wehadkee silt loam soils showed conversion of atrazine to hydroxyatrazine, after 8 weeks, to be 38%, 40% and 47% of the amount applied, respectively (Harris, 1967). Two additional degradates, deisopropylated atrazine and deethylated atrazine, were identified in a sandy loam study (Beynon et al., 1972).

- Hurle and Kibler (1976) studied the effect of water-holding capacity on the rate of degradation and found half-lives for atrazine of more than 125 days, 37 days and 36 days in sandy soil held at 4%, 35% and 70% water-holding capacity, respectively.

- In Oakley sandy loam and Nicollet clay loam, atrazine had a half-life of 101 and 167 days (Warnock and Leary, 1978).

- Carbon dioxide production was generally slow in several anaerobic soils: sandy loam, clay loam, loamy sand and silt loam (Wolf and Martin, 1975; Goswami and Green, 1971; Lavy et al., 1973).

- ^{14}C-Atrazine was stable in aerobic ground water samples incubated for 15 months at 10 or 25°C in the dark (Weidner, 1974).

- Atrazine is moderately to highly mobile in soils ranging in texture from clay to gravelly sand as determined by soil thin layer chromatography (TLC), column leaching and adsorption/desorption batch equilibrium studies. Atrazine on soil TLC plates was intermediately mobile in loam, sandy clay loam, clay loam, silt loam, silty clay loam, and silty clay soils, and was mobile in sandy loam soils. Hydroxyatrazine showed a low mobility in sandy loam and silty clay loam soils (Helling, 1971).

- Soil adsorption coefficients for atrazine in a variety of soils were: sandy loam (0.6), gravelly sand (1.8), silty clay (5.6), clay loam (7.9), sandy loam (8.7), silty clay loam (11.6) and peat (more than 21) (Weidner, 1974; Lavy, 1974; Talbert and Fletchall, 1965).

- Soil column studies indicated atrazine was mobile in sand, fine sandy loam, silt loam and loam; intermediately mobile in sand, silty clay loam and sandy loam; and low to intermediately mobile in clay loam (Weidner, 1974; Lavy, 1974; Ivey and Andrews, 1964; Ivey and Andrews, 1965).

- In a Mississippi field study, atrazine in silt loam soil had a half-life of less than 30 days (Portnoy, 1978). In a loam to silt loam soil in Minnesota, atrazine phytotoxic residues persisted for more than 1 year and were detected in the maximum-depth samples (30 to 42 inches) (Darwent and Behrens, 1968). In Nebraska, phytotoxic residues persisted in silty clay loam and loam soils 16 months after application of atrazine; they were found at depths of 12 to 24 inches. But atrazine phytotoxic residues had a half-life of about 20 days in Alabama fine sandy loam soil, although leaching may partially account for this value (Buchanan and Hiltbold, 1973).

- Under aquatic field conditions, dissipation of atrazine was due to leaching and to dilution by irrigation water, with residues persisting for 3 years in soil on the sides and bottoms of irrigation ditches to the maximum depth sampled, 67.5 to 90 cm (Smith et al., 1975).

- Ciba-Geigy (1988) recently submitted comments on the atrazine Health Advisory. These comments included a summary of the results of its studies on the environmental fate of atrazine. This summary indicated that laboratory degradation studies showed that atrazine is relatively stable in the aquatic medium under environmental pH conditions and indicated that atrazine degraded in soil by photolysis and microbial processes. The products of degradation are dealkylated metabolites, hydroxyatrazine and nonextractable (bound) residues. Atrazine and the dealkylated metabolites are relatively mobile whereas hydroxyatrazine is immobile.

- Ciba-Geigy (1988) also indicated that field dissipation studies conducted in California, Minnesota and Tennessee show no leaching of atrazine and metabolites below 6 to 12 inches of soil. The half-lives of atrazine in soil ranged between 20 to 101 days, except in Minnesota where degradation was slow. A forestry degradation study conducted in Oregon showed no adverse effects on either terrestrial or aquatic environments. Also, bioconcentration studies have shown low potential for bioaccummulation with a range of 15 to 77X.

II.   Pharmacokinetics

A. Absorption

- Atrazine appears to be readily absorbed from the gastrointestinal tract of animals. Bakke et al. (1972) administered single 0.53-mg doses of ^{14}C-ring-labeled atrazine to rats by gavage. Total fecal excretion after 72 hours was 20.3% of the administered dose; the remainder was excreted in urine (65.5%) or retained in tissues (15.8%). This indicates that at least 80% of the dose was absorbed.

B. Distribution

- Bakke et al. (1972) administered single 0.53-mg doses of ^{14}C-ring-labeled atrazine to rats by gavage. Liver, kidney and lung contained the largest amounts of radioactivity, while fat and muscle had lower residues than the other tissues examined.

- In a metabolism study by Ciba-Geigy (1983a), the radioactivity of ^{14}C-atrazine dermally applied to Harlan Sprague-Dawley rats at 0.25 mg/kg was distributed to a minor extent to body tissues. The highest levels were measured in liver and muscle at all time points examined; 2.1% of the applied dose was in muscle and 0.5% in liver at 8 hours.

- Khan and Foster (1976) observed that in chickens the hydroxy metabolites of atrazine accumulate in the liver, kidney, heart and lung. Residues of both 2-chloro and 2-hydroxy moieties were found in chicken gizzard, intestine, leg muscle, breast muscle and abdominal fat.

C. Metabolism

- The principal reactions involved in the metabolism of atrazine are dealkylation at the C-4 and C-6 positions of the molecule. There is also some evidence of dechlorination at the C-2 position. These data were reported by several researchers as demonstrated below.

- Bakke et al. (1972) administered single 0.53-mg doses of ^{14}C-ring-labeled atrazine to rats by gavage. Less than 0.1% of the label appeared in carbon dioxide in expired air. Most of the radioactivity was recovered in the urine (65.5% in 72 hours), including at least 19 radioactive compounds. More than 80% of the urinary radioactivity was identified as 2-hydroxyatrazine and its two mono-N-dealkylated metabolites. None of the metabolites identified contained the 2-chloro moiety [which may have been removed via hydrolysis during the isolation technique or by a dechlorinating enzyme as suggested by the *in vitro* studies of Foster et al. (1979), who found evidence for a dechlorinase in chicken liver homogenates incubated with atrazine].

- Bohme and Bar (1967) identified five urinary metabolites of atrazine in rats: the two monodealkylated metabolites of atrazine, their carboxy acid derivatives and the fully dealkylated derivative. All of these metabolites contained the 2-chloro group. The *in vitro* studies of Dauterman and Muecke (1974) also found no evidence for dechlorination of atrazine in the presence of rat liver homogenates.

- Similarly, Bradway and Moseman (1982) administered atrazine (50, 5, 0.5 or 0.005 mg/day) for 3 days to male Charles River rats and observed that the fully dealkylated derivative (2-chloro-4,6-diamino-s-triazine) was the major urinary metabolite, with lesser amounts of the two mono-N-dealkylated derivatives.

- Erickson et al. (1979) dosed Pittman-Moore miniature pigs by gavage with 0.1 g of atrazine (80W). The major compounds identified in the urine were the parent compound (atrazine) and deethylated atrazine (which contains the 2-chloro substituent).

- Hauswirth (1988) indicated that the rat metabolism studies taken together are sufficient to show that in the female rat dechlorination of the triazine ring and N-dealkylation are the major metabolic pathways. Oxidation of the alkyl substituents appears to be a minor and secondary metabolic route. The total body half-life is approximately one and one-half days. Atrazine and/or its metabolites appear to bind to red blood cells. Other tissue accumulation does not appear to occur.

D. Excretion

- Urine appears to be the principal route of atrazine excretion in animals. Following the administration of 0.5 mg doses of 14C-ring-labeled atrazine by gavage to rats, Bakke et al. (1972) reported that in 72 hours most of the radioactivity (65.5%) was excreted in the urine, 20.3% was excreted in the feces and less than 0.1% appeared as carbon dioxide in expired air. About 85 to 95% of the urinary radioactivity appeared within the first 24 hours after dosing, indicating rapid clearance.

- Dauterman and Muecke (1974) have reported that atrazine metabolites are conjugated with glutathione to yield a mercapturic acid in the urine. The studies of Foster et al. (1979) in chicken liver homogenates also indicate that atrazine metabolism involves glutathione.

- Ciba-Geigy (1983b) studied the excretion rate of ^{14}C-atrazine from Harlan Sprague-Dawley rats dermally dosed with atrazine dissolved in tetrahydrofuran at levels of 0.025, 0.25, 2.5 or 5 mg/kg. Urine and feces were collected from all animals at 24-hour intervals for 144 hours. Results indicated that atrazine was readily absorbed, and within 48 hours most of the absorbed dose was excreted, mainly in the urine and to a lesser extent in the feces. Cumulative excretion in urine and feces appeared to be directly proportional to the administered dose, ranging from 52% at the lowest dose to 80% at the highest dose.

III. Health Effects

A. Humans

1. *Short-term Exposure*

- A case of severe contact dermatitis was reported by Schlicher and Beat (1972) in a 40-year-old farm worker exposed to atrazine formulation. The clinical signs were red, swollen and blistered hands with hemorrhagic bullae between the fingers. Although it is noted that the exposure of this patient may have been inclusive to exposure to other chemicals in addition to atrazine, it is also noted that atrazine is a skin irritant in animal studies.

2. *Long-term Exposure*

- Yoder et al. (1973) examined chromosomes in lymphocyte cultures taken from agricultural workers exposed to herbicides including atrazine. There were more chromosomal aberrations in the workers during mid-season exposure to herbicides than during the off-season (no spraying). These aberrations included a fourfold increase in chromatid gaps and a 25-fold increase in chromatid breaks. During the off-season, the mean number of gaps and breaks was lower in this group than in controls who were in occupations unlikely to involve herbicide exposure. This observation led the authors to speculate that there is enhanced chromosomal repair during this period of time resulting in compensatory protection. However, these data may not be representative of the effect of atrazine since the exposed workers were also exposed to other herbicides.

B. Animals

1. *Short-term Exposure*

- Acute oral LD_{50} values of 3,000 mg/kg in rats and 1,750 mg/kg in mice have been reported for technical atrazine by Bashmurin (1974); the purity of the test compound was not specified.

- Acute oral studies conducted by Ciba-Geigy (1988) with atrazine [97% active ingredient (a.i.)] reflected the following LD_{50}s: 1,869 mg/kg in rats and > 3,000 mg/kg in mice.

- Molnar (1971) reported that when atrazine was administered by gavage to rats at 3,000 mg/kg, 6% of the rats died within 6 hours, and 25% of those remaining died within 24 hours. The rats that died during the first day exhibited pulmonary edema with extensive hemorrhagic foci, cardiac dilation and microscopic hemorrhages in the liver and spleen. Rats that died during the second day had hemorrhagic bronchopneumonia and dystrophic changes of the renal tubular mucosa. Rats sacrificed after 24 hours had cerebral edema and histochemical alterations in the lungs, liver and brain. It is noted that the dose used in this study was almost two times the LD_{50} (Ciba-Geigy, 1988).

- Gaines and Linder (1986) determined the oral LD_{50} for adult male and female rats to be 737 and 672 mg/kg respectively and 2,310 mg/kg for pups. It is, therefore, noted that young animals are more sensitive to atrazine than adults. This study also reflected that the dermal LD_{50} for adult rats was higher than 2,500 mg/kg.

- Palmer and Radeleff (1964) administered atrazine as a fluid dilution or in gelatin capsules to Delaine sheep and dairy cattle (one animal per dosage group). Two doses of 250 mg/kg atrazine caused death in both sheep and cattle. Sixteen doses of 100 mg/kg were lethal to the one sheep tested. At necropsy, degeneration and discoloration of the adrenal glands and congestion in lungs, liver and kidneys were observed.

- Palmer and Radeleff (1969) orally administered atrazine 80W (analysis of test material not provided) by capsule or by drench to sheep at 5, 10, 25, 50, 100, 250 or 400 mg/kg/day and to cows at 10, 25, 50, 100 or 250 mg/kg/day. The number of animals in each dosage group was not stated, and the use of controls was not indicated. Observed effects included muscular spasms, stilted gait and stance and anorexia at all dose levels in sheep and at 25 mg/kg in cattle. Necropsy revealed epicardial petechiae (small hemorrhagic spots on the lining of the heart) and congestion of the kidneys, liver and lungs. Effects appeared to be dose related. A Lowest-Observed-Adverse-Effect Level (LOAEL) of 5 mg/kg/day in sheep and a No-Observed-Adverse-Effect Level (NOAEL) of 10 mg/kg/day in cows can be identified from this study.

- Bashmurin (1974) reported that oral administration of 100 mg/kg of atrazine to cats had a hypotensive effect, and that a similar dose in dogs was antidiuretic and decreased serum cholinesterase (ChE) activity. No other details of this study were reported. Atrazine is not an organophosphate (OP), therefore, its effect on ChE may not be similar to the mechanism of ChE inhibition by OPs.

2. *Dermal/Ocular Effects*

- In a primary dermal irritation test in rats, atrazine at 2,800 mg/kg produced erythema but no systemic effects (Gzheyotskiy et al., 1977).

- Ciba-Geigy (1988) indicated that the studies it performed reflected dermal sensitization in rats but not irritation in rabbits' eyes.

3. *Long-term Exposure*

- Hazelton Laboratories (1961) fed atrazine to male and female rats for 2 years at dietary levels of 0, 1, 10 or 100 ppm. Based on the dietary assumptions of Lehman (1959), these levels correspond to doses of approximately 0, 0.05, 0.50 or 5.0 mg/kg/day. After 65 weeks, the 1.0-ppm dose was increased to 1,000 ppm (50 mg/kg/day) for the remainder of the study. No treatment-related pathology was found at 26 weeks, at 52 weeks, at 2 years, or in animals that died and were necropsied during the study. Results of blood and urine analyses were unremarkable. Atrazine had no effects on the general appearance or behavior of the rats. A transient roughness of the coat and piloerection were observed in some animals after 20 weeks of treatment at the 10- and 100-ppm levels but not at 52 weeks. Body weight gains, food consumption and survival were similar in all groups for 18 months, but from 18 to 24 months there was high mortality due to infections (not attributed to atrazine) in all groups, including controls, which limits the usefulness of this study in determining a NOAEL for the chronic toxicity of atrazine.

- In a 2-year study by Woodard Research Corporation (1964), atrazine (80W formulation) was fed to male and female beagle dogs for 105 weeks at dietary levels of 0, 15, 150 or 1,500 ppm. Based on the dietary assumptions of Lehman (1959), these levels correspond to doses of 0, 0.35, 3.5 or 35 mg/kg/day. Survival rates, body weight gain, food intake, behavior, appearance, hematologic findings, urinalyses, organ weights and histologic changes were noted. The 15-ppm dosage (0.35 mg/kg/day) produced no toxicity, but the 150-ppm dosage (3.5 mg/kg/day) caused a decrease in food intake as well as increased heart and liver weight in females. In the group receiving 1,500 ppm (35 mg/kg/day) atrazine, there were decreases in food intake and body weight gain, an increase in adrenal weight, a decrease in hematocrit and occasional tremors or stiffness in the rear limbs. There were no differences among the different groups in the histology of the organs studied. Based on these results, a NOAEL of 0.35 mg/kg/day can be identified for atrazine.

- In a study by Ciba-Geigy (1987b) using technical atrazine (97% a.i.), 6-month-old beagle dogs were assigned randomly to four dosage groups: 0, 15, 150 and 1,000 ppm. These doses correspond to actual average intake of 0, 0.48, 4.97 and 33.65/33.8 (male/female) mg/kg/day. Six animals/sex/group were assigned to the control and high-dose groups and four animals/sex/group were assigned to the low- and mid-dose groups. One mid-dose male, one high-dose male and one high-dose female had to be sacrificed moribund during the study period. Decreased body weight gains and food consumption were noted at the high-dose level. Statistically significant ($p < 0.05$) reductions in erythroid parameters (red cell count, hemoglobin and hematocrit) in high-dose males were noted throughout the study as well as mild increases in platelet counts in both sexes. Slight decreases in total protein and albumin ($p < 0.05$) were noted in high-dose males as well as decreased calcium and chloride in males and increased sodium and glucose in females. Decreases in absolute heart weight were noted in females and increased relative liver weight in males of the high-dose group. The mid-dose females reflected an increase in the absolute heart weight and heart/brain weight ratios. The most significant effect of atrazine in this study was reflected in the high-dose animals of both sexes as discrete myocardial degeneration. Clinical signs associated with cardiac pathology such as ascites, cachexia, labored/shallow breathing and abnormal EKG were observed in the group as early as 17 weeks into the study. Gross pathology reflected severe dilation of the right atrium and occasionally of the left atrium. These findings were also noted histopathologically as degenerative atrial myocardium (atrophy and myolysis). In the mid-dose group, two males and one female appeared to be affected with the cardiac syndrome but to a much lesser degree in the intensity of the noted responses. Therefore, the LOAEL in this study is 4.97 mg/kg/day and the NOAEL is 0.48 mg/kg/day.

• A 2-year chronic feeding/oncogenicity study (Ciba-Geigy, 1986) was recently evaluated by the Agency. In this study, technical atrazine (98.9% a.i.) was fed to 37- to 38-day-old Sprague-Dawley rats. The dosage levels used were 0, 10, 70, 500 or 1,000 ppm, equivalent to 0, 0.5, 3.5, 25 or 50 mg/kg/day (using Lehman's conversion factor, 1959). Twenty rats per sex per group were used to measure blood parameters and clinical chemistries and urinalysis. Fifty rats per sex per group were maintained on the treated and control diets for 24 months. An additional 10 rats per sex were placed on control and high-dose (1,000 ppm) diets for a 12-month interim sacrifice and another 10 per sex (control and high dose, 1,000 ppm) for a 13-month sacrifice (the 1,000 ppm group was placed on control diet for 1 month prior to sacrifice). The total numbers of animals/sex in the control and HDT groups were 90 and 70 for the 10-, 70- and 500-ppm groups. Histopathology was performed on all animals. At the mid- and high-dose, there was a decrease in mean body weights of males and females. Survival was decreased in high-dose females but increased in high-dose males. There were decreases in organ-to-body weight ratios in high-dose animals, which were probably the result of body weight decreases. Hyperplastic changes in high-dose males (mammary gland, bladder and prostate) and females (myeloid tissue of bone marrow and transitional epithelium of the kidney) were of questionable toxicologic importance. There was an increase in retinal degeneration and in centrolobular necrosis of the liver in high-dose females and an increase in degeneration of the rectus femoris muscle in high-dose males and females when compared to controls. Based on decreased body weight gain, the LOAEL for nononcogenic activities in both sexes is 25 mg/kg/day and the NOAEL is 3.5 mg/kg/day. However, oncogenic activities were noted at 3.5 mg/kg/day (70 ppm) and above as reflected in the increased incidence of mammary gland tumors in females.

- A recent 91-week oral feeding/oncogenicity study in mice by Ciba-Geigy (1987c) has been evaluated by the Agency. In this study, atrazine (97% a.i.) was fed to 5-week-old CD-1 mice, weighing 21.0/26.8 grams (female/male). The mice were randomly assigned to five experimental groups of approximately 60 animals/sex/ group. The dosages tested were 0, 10, 300, 1,500 and 3,000 ppm; these dosages correspond to actual mean daily intakes of 1.4, 38.4, 194.0 and 385.7 mg/kg/day for males, and 1.6, 47.9, 246.9 and 482.7 mg/kg/day for females. This study shows that there are dose-related effects at 1,500 ppm or 3,000 ppm atrazine: an increase in cardiac thrombi; a decrease in mean body weight gain at 12 and 91 weeks during the study; and decreases in erythrocyte count, hematocrit and hemoglobin concentration. Cardiac thrombi contributed to the deaths of the group of mice that did not survive to terminal sacrifice. The LOAEL is set at 1,500 ppm based upon decreases of 23.5% and 11.0% in mean body weight gain found at 91 weeks in male and female mice, respectively. Also, an increase in the incidence of cardiac thrombi is found in female mice in the 1,500 ppm exposure group. None of the above effects are found at 300 ppm, thus the NOAEL is set at 300 ppm (corresponding to 38.4 mg/kg/day in males and 47.9 mg/kg/day for females).

4. Reproductive Effects

- A three-generation study on the effects of atrazine on reproduction in rats was conducted by Woodard Research Corporation (1966). Groups of 10 males and 20 females received atrazine (80W) at dietary levels of 0, 50 or 100 ppm. Based on the dietary assumptions that 1 ppm in the diet of rats is equivalent to 0.05 mg/kg/day (Lehman, 1959), these levels correspond to doses of approximately 0, 2.5 or 5 mg/kg/day. Two litters were produced per generation but parental animals were chosen from the second litter after weaning for each generation. Young rats were maintained on the test diets for approximately 10 weeks in each generation. The third generation pups were sacrificed after weaning. It is noted that the parental animals of the first generation were fed only half of the dietary atrazine levels for the first 3 weeks of exposure. There were no adverse effects of atrazine on reproduction observed during the course of the three-generation study. A NOAEL of 100 ppm (5 mg/kg/day) was identified for this study. However, the usefulness of this study is limited due to the alteration of the atrazine content of the diet during important maturation periods of the neonates.

- A recent two-generation study in rats by Ciba-Geigy (1987a) was conducted using the 97% a.i. technical atrazine. Young rats, 47 to 48 days old, were maintained on the control and test diets for 10 weeks before mating. The concentrations used were 0, 10, 50 and 500 ppm (equivalent to 0, 0.5, 2.5 and 25 mg/kg/day using Lehman conversion factor, 1959). Thirty animals/sex/group were used in each generation; one litter was produced per generation. The level tested had no effect on mortality in either generation. Body weight and body weight gains were significantly depressed ($p < 0.05$) at the highest dose; however, food consumption was also decreased at this high-dose level in parental males and females during the premating period and for the first generation females (F_1) on days 0 to 7 of gestation. Neither histopathological effects nor other effects were noted during gross necropsy in either parental generation with the exception of increased testes' relative weight in both generations at the high dose. In pups of both generations, significant reduction ($p < 0.05$) in body weight was noted; however, this effect was only dose-related in the second generation (F_2) at both the mid- and high-dose levels on postnatal day 21. Therefore, maternal toxicity NOAEL is 2.5 mg/kg/day; the reproductive LOAEL is 2.5 mg/kg/day (reduced pup weight in F_2 generation on postnatal day 21) and the NOAEL is 0.5 mg/kg/day.

5. Developmental Effects

- In the three-generation reproduction study in rats conducted by Woodard Research Corporation (1966) (described above), atrazine at dietary levels of 50 or 100 ppm (2.5 or 5 mg/kg/day) resulted in no observed histologic changes in the weanlings and no effects on fetal resorption. No malformations were observed, and weanling organ weights were similar in controls and atrazine-treated animals. Therefore, a NOAEL of 100 ppm (5 mg/kg/day) was also identified for developmental effects in this study. However, the usefulness of this study is limited due to an alteration of the atrazine content of the diet during important maturation periods of the neonates.

- Atrazine was administered orally to pregnant rats on gestation days 6 to 15 at 0, 100, 500 or 1,000 mg/kg (Ciba-Geigy, 1971). The two higher doses increased the number of embryonic and fetal deaths, decreased the mean weights of the fetuses and retarded the skeletal development. No teratogenic effects were observed. The highest dose (1,000 mg/kg) resulted in 23% maternal mortality and various toxic symptoms. The 100 mg/kg dose had no effect on either dams or embryos and is therefore the maternal and fetotoxic NOAEL in this study.

- In a study by Ciba-Geigy (1984a), Charles River rats received atrazine (97%) by gavage on gestation days 6 to 15 at dose levels of 0, 10, 70, or 700 mg/kg/day. Excessive maternal mortality (21/27) was noted at 700 mg/kg/day, but no mortality was noted at the lower doses; also reduced weight gains and food consumption were noted at both 70 and 700 mg/kg/day. Developmental toxicity was also present at these dose levels. Fetal weights were severely reduced at 700 mg/kg/day; delays in skeletal development occurred at 70 mg/kg/day, and a dose-related runting was noted at 10 mg/kg/day and above. The NOAEL for maternal toxicity appears to be 10 mg/kg/day; however, this is also the LOAEL for developmental effects.

- New Zealand White rabbits received atrazine (96%) by gavage on gestation days 7 through 19 at dose levels of 0, 1, 5 or 75 mg/kg/day (Ciba-Geigy, 1984b). Maternal toxicity, evidenced by decreased body weight gains and food consumption, was present in the mid- and high-dose groups. Developmental toxicity was demonstrated only at 75 mg/kg/day by an increased resorption rate, reduced fetal weights, and delays in ossification. No teratogenic effects were indicated. The NOAEL appears to be 1 mg/kg/day.

- Peters and Cook (1973) fed atrazine to pregnant rats (four/group) at levels of 0, 50, 100, 200, 300, 400, 500 or 1,000 ppm in the diet throughout gestation. Based on an assumed body weight of 300 g and a daily food consumption of 12 g (Arrington, 1972), these levels correspond to approximately 0, 2, 4, 8, 12, 16, 20 or 40 mg/kg/day. The number of pups per litter was similar in all groups, and there were no differences in weanling weights. This study identified a NOAEL of 40 mg/kg/day for developmental effects. In another phase of this study, the authors demonstrated that subcutaneous (sc) injections of 50, 100 or 200 mg/kg atrazine on gestation days 3, 6 and 9 had no effect on the litter size, while doses of ≥ 800 mg/kg were embryotoxic. Therefore, a NOAEL of 200 mg/kg by the sc route was identified for embryotoxicity.

6. *Mutagenicity*

- Loprieno et al. (1980) reported that single doses of atrazine (1,000 mg/kg or 2,000 mg/kg, route not specified) produced bone marrow chromosomal aberrations in the mouse. No other details of this study were provided.

- Murnik and Nash (1977) reported that feeding 0.01% atrazine to male *Drosophila melanogaster* larvae significantly increased the rate of both dominant and sex-linked recessive lethal mutations. They stated, however, that dominant lethal induction and genetic damage may not be directly related.

- Adler (1980) reviewed unpublished work on atrazine mutagenicity carried out by the Environmental Research Programme of the Commission of the European Communities. Mutagenic activity was not induced even when mammalian liver enzymes (S-9) were used; however, the use of plant microsomes produced positive results. Also, in *in vivo* studies in mice, atrazine induced dominant lethal mutations and increased the frequency of chromatid breaks in bone marrow. Hence, the author suggested that activation of atrazine in mammals occurs independently of the liver, possibly in the acidic part of the stomach.

- As described previously, Yoder et al. (1973) studied chromosomal aberrations in the lymphocyte cultures of farm workers exposed to various pesticides including atrazine. During mid-season a 4-fold increase in chromatid gaps and a 25-fold increase in chromatid breaks was observed. During the off-season (no spraying), the number of gaps and breaks was lower than in controls, suggesting to the authors that there is enhanced chromosomal repair during the unexposed period.

- Recently, Spencer (1987) and Dearfield (1988) evaluated several *in vitro* and *in vivo* mutagenicity studies on atrazine that were recently submitted to U.S. EPA by Ciba-Geigy. They noted that most of these studies were inadequate with the exception of the following three tests: a *Salmonella* assay, an *E. coli* reversion assay, and a Host-mediated assay. The first two assays were negative for mutagenic effects; the results of the third assay were equivocal.

- Ciba-Geigy (1988) indicated that Brusick (1987) evaluated atrazine mutagenicity and that the weight-of-evidence analysis he used placed the chemical in a nonmutagenic status. The Agency (Dearfield, 1988) evaluated Brusick's analysis. It is noted that the use of the weight-of-evidence approach is not appropriate at the present time. The *in vivo* studies by Adler (1980) suggest a positive response. These findings have not been diminished by other atrazine studies. In addition, Dearfield (1988) indicated that the scheme used by Brusick in this analysis is flawed by the lack of calibration of the chemical test scores to an external standard and by the use of some studies that are considered inadequate by design to determine the mutagenic potential of atrazine.

7. *Carcinogenicity*

- Innes et al. (1969) investigated the tumorigenicity of 120 test compounds including atrazine in mice. Two F_1 hybrid stocks (C57BL/6 × Anf) F_1 and (C57BL/6 × AKR) F_1 were used. A dose of 21.5 mg/kg/day was administered by gavage to mice of both sexes from age 7 to 28 days. After weaning at 4 weeks, this dose level was maintained by feeding 82 ppm atrazine *ad libitum* in the diet for 18 months. The incidence of hepatomas, pulmonary tumors, lymphomas and total tumors in atrazine-treated mice was not significantly different from that in the negative controls.

- A 2-year feeding/oncogenicity study in rats by Ciba-Geigy (1986) has been evaluated recently by the Agency. Atrazine (98.9% a.i.) was fed to 37- to 38-day-old Sprague-Dawley rats. The dosage levels used were 0, 10, 70, 500 or 1,000 ppm, equivalent to 0, 0.5, 3.5, 25 or 50 mg/kg/day (using Lehman's conversion factor, 1959). The total number of animals/sex in the control and HDT groups was 90; and there were 70 animals/sex/group in the 10-, 70- and 500-ppm groups. Histopathology was performed on all animals. In females, atrazine was associated with a statistically significant increase in mammary gland fibroadenomas at 1,000 ppm; in mammary gland adenocarcinomas (including two carcinosarcomas at the HDT) at 70, 500 and 1,000 ppm; and in total mammary gland tumor bearing animals at 1,000 ppm. Each of these increases was associated with a statistically significant dose-related trend and was outside of the high end of the historical control range. In addition, U.S. EPA (1986a) indicated that there was evidence for decreased latency for mammary gland adenocarcinomas at the 12-month interim sacrifice that was already submitted by Ciba-Geigy in 1985. This study was also reported as positive in a briefing paper by Ciba-Geigy (1987a).

- A recent 91-week oral feeding/oncogenicity study in mice by Ciba-Geigy (1987d) has been evaluated by the Agency. In this study, atrazine (97% a.i.) was fed to 5-week-old CD-1 mice weighing 21.0/26.8 grams (female/male). The mice were randomly assigned to five experimental groups of approximately 60 animals/sex/group. The dosages tested were 0, 10, 300, 1,500 and 3,000 ppm; these dosages correspond to actual mean daily intake of 1.4, 38.4, 194.0 and 385.7 mg/kg/day for males, and 1.6, 47.9, 246.9 and 482.7 mg/kg/day for females. The following kinds of neoplasms were noted in this study: mammary adenocarcinomas, adrenal adenomas, pulmonary adenomas and malignant lymphomas. However, no dose-related or statistically significant increases were observed in the incidences of these neoplasms. Therefore, atrazine is not considered oncogenic in this strain of mice.

IV. Quantification of Toxicological Effects

A. One-day Health Advisory

No suitable information was found in the available literature for the determination of the One-day HA value for atrazine. It is therefore recommended that the Ten-day HA value calculated below for a 10-kg child of 0.1 mg/L (100 µg/L) be used at this time as a conservative estimate of the One-day HA value.

B. Ten-day Health Advisory

Two teratology studies by Ciba-Geigy, one in the rat (1984a) and one in the rabbit (1984b), were considered for the calculation of the Ten-day HA value. The rat study reflected a NOAEL of 10 mg/kg/day for maternal toxicity but this value was also the LOAEL for developmental toxicity, while the rabbit study reflected NOAELs of 5 mg/kg/day for developmental toxicity and 1 mg/kg/day for maternal toxicity. Thus, the rabbit appears to be a more sensitive species than the rat for maternal toxicity, hence, the rabbit study with a NOAEL of 1 mg/kg/day is used in the calculations below.

The Ten-day HA for a 10-kg child is calculated below as follows:

$$\frac{(1 \text{ mg/kg/day}) (10 \text{ kg})}{(100) (1 \text{ L/day})} = 0.1 \text{ mgL} (100 \text{ µg/L})$$

C. Longer-term Health Advisory

No suitable information was found in the available literature for the determination of the longer-term HA value for atrazine. It is, therefore, recommended that the adjusted DWEL for a 10-kg child of 0.05 mg/L (50 µg/L), calculated below, and the DWEL for a 70-kg adult of 0.2 mg/L (200 µg/L), calculated in the Lifetime Health Advisory section, be used at this time as conservative estimates of the Longer-term HA values.

For a 10-kg child, the adjusted DWEL is calculated as follows:

$$\text{DWEL} = \frac{(0.005 \text{ mg/kg/day}) \ (10 \text{ kg})}{1 \text{ L/day}} = 0.05 \text{ mg/L}$$

where: 0.005 mg/kg/day = RfD (see Lifetime Health Advisory section).

D. Lifetime Health Advisory

Three studies were considered for the development of the Lifetime HA. A 2-year dog feeding study (Woodard Research Corporation, 1964), a 1-year dog feeding study (Ciba-Geigy, 1987c) and a 2-year rat oral feeding/oncogenicity study (Ciba-Geigy, 1986).

The first study in dogs (Woodard Research Corporation, 1964) reflected a NOAEL of 0.35 mg/kg/day and a LOAEL of 3.5 mg/kg/day that was associated with increased heart and liver weights in females. The new 1-year dog study (1987c) reflected a NOAEL of 0.48 mg/kg/day and a LOAEL of 4.97 mg/kg/day based on mild cardiac pathology intensified at the higher dose tested 33.65/33.8 (male/female) mg/kg/day. The 2-year rat study (Ciba-Geigy, 1986) reflected a NOAEL at 3.5 mg/kg/day for systemic effect other than oncogenicity; however, this study indicated that atrazine caused mammary gland tumors at this dose level and above. No adverse effects were observed at the lowest dose tested, 0.5 mg/kg/day.

The 1964 dog study was initially used for the calculation of the RfD and the Lifetime HA. However, this study was partially flawed by the lack of information on the purity of the test material and by the inadequate documentation of the hematological data. Therefore, the recent 1-year dog study (Ciba-Geigy, 1987c), using technical atrazine (97% a.i.) is considered a more adequate study for the calculation of the RfD and the Lifetime HA. The NOAEL in this study, 0.48 mg/kg/day, is also supported by the NOAEL of 0.5 mg/kg/day in the two-generation reproduction study (Ciba-Geigy, 1987b) and by the fact that no systemic effects or tumors were noted at this dose level in the 2-year chronic feeding/oncogenicity study in rats (Ciba-Geigy, 1986). [Other studies: Woodard Research Corporation (1966) and Hazelton Laboratories (1961) identified long-term NOAEL values of 5 to 50 mg/kg/day and were not considered to be as protective as the dog studies for use in calculating the HA values for atrazine.]

Step 1: Determination of the Reference Dose (RfD)

$$\text{RfD} = \frac{0.48 \text{ mg/kg/day}}{(100)} = 0.005 \text{ mg/kg/day (rounded from 0.0048 mg/kg/day)}$$

Step 2: Determination of the Drinking Water Equivalent Level (DWEL)

$$\text{DWEL} = \frac{(0.0048 \text{ mg/kg/day}) \ (70 \text{ kg})}{(2 \text{ L/day})} = 0.168 \text{ mg/L} \ (200 \text{ μg/L})$$

Step 3: Determination of the Lifetime Health Advisory

$$\text{Lifetime HA} = \frac{(0.168 \text{ mg/L}) \ (20\%)}{10} = 0.003 \text{ mg/L} \ (3 \text{ μg/L})$$

E. Evaluation of Carcinogenic Potential

- A study submitted by Ciba-Geigy Corporation (1986) in support of the pesticide registration of atrazine indicated that atrazine induced an increased incidence of mammary tumors in female Sprague-Dawley rats. These findings have been further confirmed in a briefing by Ciba-Geigy (1987) on this study.

- Atrazine was not oncogenic in mice (Ciba-Geigy, 1987d).

- Three closely related analogs — propazine, terbutryn and simazine — are presently classified as Group C oncogens based on an increased incidence of tumors in the same target tissue (mammary gland) and animal species (rat) as was noted for atrazine.

- The International Agency for Research on Cancer has not evaluated the carcinogenic potential of atrazine.

- Applying the criteria described in EPA's guidelines for assessment of carcinogenic risk (U.S. EPA, 1986b), atrazine may be classified in Group C: possible human carcinogen. This category is used for substances with limited evidence of carcinogenicity in animals in the absence of human data.

V.   Other Criteria, Guidance and Standards

- Toxicity data on atrazine were reviewed by the National Academy of Sciences (NAS, 1977), and the study by Innes et al. (1969) was used to identify a chronic NOAEL of 21.5 mg/kg/day. Although at that time it was concluded that atrazine has low chronic toxicity, an uncertainty factor of 1,000 was employed in calculation of the ADI from that study, since only limited data were available. The resulting value (0.021 mg/kg/day) corresponds to an ADI of 0.73 mg/L in a 70-kg adult consuming 2 L of water per day.

• Tolerances for atrazine alone and the combined residues of atrazine and its metabolites in or on various raw agricultural commodities have been established (U.S. EPA, 1986c). These tolerances range from 0.02 ppm (negligible) in animal products (meat and meat by-products) to 15 ppm in various animal fodders.

VI. Analytical Methods

• Analysis of atrazine is by a gas chromatographic (GC) method, Method No. 507, applicable to the determination of certain nitrogen-phosphorus containing pesticides in water samples (U.S. EPA, 1988). In this method, approximately 1 L of sample is extracted with methylene chloride. The extract is concentrated and the compounds are separated using capillary column GC. Measurement is made using a nitrogen phosphorus detector. The method has been validated in a single laboratory. The estimated detection limit for the analytes in this method, including atrazine, is 0.13 μg/L.

VII. Treatment Technologies

• Treatment technologies which will remove atrazine from water include activated carbon adsorption, ion exchange, reverse osmosis, ozone oxidation and ultraviolet irradiation. Conventional treatment methods have been found to be ineffective for the removal of atrazine from drinking water (ESE, 1984; Miltner and Fronk, 1985a). Limited data suggest that aeration would not be effective in atrazine removal (ESE, 1984; Miltner and Fronk, 1985a).

• Baker (1983) reported that a 16.5-inch GAC filter cap using F-300, which was placed upon the rapid sand filters at the Fremont, Ohio, water treatment plant, reduced atrazine levels by 30 to 64% in the water from the Sandusky River. At Jefferson Parish, Louisiana, Lykins et al. (1984) reported that an adsorber containing 30 inches of Westvaco WV-G® 12 × 40 GAC removed atrazine to levels below detectable limits for over 190 days.

• At the Bowling Green, Ohio, water treatment plant, PAC in combination with conventional treatment achieved an average reduction of 41% of the atrazine in the water from the Maumee River (Baker, 1983). Miltner and Fronk (1985a) reported that in jar tests using spiked Ohio River water with the addition of 16.7 and 33.3 μg/L of PAC and 15 to 20 mg/L of alum, PAC removed 64 and 84%, respectively, of the atrazine.

Higher percent removals reflected higher PAC dosages. Miltner and Fronk (1985b) monitored atrazine levels at water treatment plants which utilized PAC in Bowling Green and Tiffin, Ohio. Applied at dosages ranging from 3.6 to 33 mg/L, the PAC achieved 31 to 91% removal of atrazine, with higher percent removals again reflecting higher PAC dosages.

- Harris and Warren (1964) reported that Amberlite IR-120 cation exchange resin removed atrazine from aqueous solution to less than detectable levels. Turner and Adams (1968) studied the effect of varying pH on different cation and anion exchange resins. At a pH of 7.2, 45% removal of atrazine was achieved with Dowex® 2 anion exchange resin and with $H_2PO_4^-$ as the exchangeable ion species.

- Chian et al. (1975) reported that reverse osmosis, utilizing cellulose acetate membrane and a cross-linked polyethelenimine (NS-100) membrane, successfully processed 40% of the test solution, removing 84 and 98%, respectively, of the atrazine in the solution.

- Miltner and Fronk (1985a) studied the oxidation of atrazine with ozone in both spiked distilled and ground water. Varying doses of ozone achieved a 70% removal of atrazine in distilled water and 49 to 76% removal of atrazine in ground water.

- Kahn and Schnitzer (1978) studied the effect of fulvic acid upon the photochemical stability of atrazine to ultraviolet irradiation. A 50% removal of atrazine was achieved much faster at higher pH conditions than at lower pH conditions. In the presence of fulvic acids, the time needed for ultraviolet irradiation to achieve 50% removal was almost triple the time required to achieve similar removals without the presence of fulvic acids. Since fulvic acids will be present in surface waters, ultraviolet irradiation may not be a cost-effective treatment alternative.

VIII. References

Ader, I.D. 1980. A review of the coordinated research effort on the comparison of test systems for the detection of mutagenic effects, sponsored by the E.E.C. Mutat. Res. 74:77–93.

Arrington, L.R. 1972. The laboratory animals. In: Introductory laboratory animal science. The breeding, care and management of experimental animals. Danville, IL: Interstate Printers and Publishers, Inc. pp. 9–11.

Baker, D. 1983. Herbicide contamination in municipal water supplies in northwestern Ohio. Final Draft Report 1983. Prepared for Great Lakes National Program Office, U.S. Environmental Protection Agency. Tiffin, OH.

Bakke, J.E., J.D. Larson and C.E. Price. 1972. Metabolism of atrazine and 2-hydroxyatrazine by the rat. J. Agric. Food Chem. 20:602–607.

Bashmurin, A.F. 1974. Toxicity of atrazine for animals. Sb. Rab. Leningrad Vet. Institute. 36:5–7. (English abstract only.)

Beynon, K.I., G. Stoydin and A.N. Wright. 1972. A comparison of the breakdown of the triazine herbicides cyanazine, atrazine and simazine in soils and in maize. Pestic. Biochem. Physiol. 2:153–161.

Bohme, E. and F. Bar. 1967. Uber den Abbau von Triazin-Herbiciden in tierischen Organismus. Food Cosmet. Toxicol. 5:23–28. (English abstract only.)

Bradway, D.E. and R.F. Moseman. 1982. Determination of urinary residue levels of the n-dealkyl metabolites of triazine herbicides. J. Agric. Food Chem. 30:244–247.

Brusick, D.J. 1987. An assessment of the genetic toxicity of atrazine: relevance to health and environmental effects. A document prepared for Ciba-Geigy Corporation (submitted to EPA in 1988 as a part of Ciba-Geigy comments on the HA). December.

Buchanan, G.A. and A.E. Hiltbold. 1973. Performance and persistence of atrazine. Weed Sci. 21:413–416.

Chian, E.S.K., W.N. Bruce and H.H.P. Fang. 1975. Removal of pesticides by reverse osmosis. Environmental Science and Technology. 9(1):52–59.

Ciba-Geigy. 1971. Rat reproduction study-test for teratogenic or embryotoxic effects. 10/1971.

Ciba-Geigy. 1983a. Dermal absorption of ^{14}C-atrazine by rats. Report No. ABR-83005. Accession No. 255815. May. Greensboro, NC: Ciba-Geigy Corporation.

Ciba-Geigy. 1983b. Excretion rate of ^{14}C-atrazine from dermally dosed rats. Report No. ABR-83081. Accession No. 255815. October. Greensboro, NC: Ciba-Geigy Corporation.

Ciba-Geigy Ltd. 1984a. A teratology study of atrazine technical in Charles River Rats. Toxicology/Pathology Report No. 60–84. MRID 00143008.

Ciba-Geigy Ltd. 1984b. Segment II. Teratology study in rabbits. Toxicology/Pathology Report No. 68–84. MRID 00143006.

Ciba-Geigy. 1985. Atrazine chronic feeding/oncogenicity study. One-year interim report. May 17. Greensboro, NC: Ciba-Geigy Corporation.

Ciba-Geigy. 1986. Twenty-four month combined chronic oral toxicity and oncogenicity in rats utilizing atrazine technical by American Biogenic Corp. Study No. 410–1102. Accession Nos. 262714–262727.

Ciba-Geigy. 1987a. Briefing paper on atrazine. December 1986. Analysis of chronic rat feeding study results. Greensboro, NC: Ciba-Geigy Corporation.

Ciba-Geigy. 1987b. Two-generation rat reproduction. Study No. 852063. MRID 404313–03.

Ciba-Geigy. 1987c. Atrazine technical 52-week oral feeding in dogs. Study No. 852008 and Pathology Report No. 7048. MRID 404313–01.

Ciba-Geigy. 1987d. Atrazine technical 91-week oral carcinogenicity study in mice. Study No. 842120. MRID 404313–02.

Ciba-Geigy. 1988. Comments on the atrazine draft health advisory. A letter from Thomas Parish to U.S. EPA/ODW. Greensboro, NC: Ciba-Geigy Corporation.

Cohen, S.Z., C. Eiden and M.N. Lorber. 1986. Monitoring ground water for pesticides in the U.S.A. In: Evaluation of pesticides in ground water. American Chemical Society Symposium Series. No. 315.

Cosmopolitan Laboratories.* 1979. Document No. 00541, EPA Accession No. 2–41725. CBI.

Darwent, A.L. and R. Behrens. 1968. Dissipation and leaching of atrazine in a Minnesota soil after repeated applications. In: Proc. North Cent. Weed Control Conf., December 3–5, 1968, Indiana. pp. 66–68.

Dauterman, W.C. and W. Muecke. 1974. *In vitro* metabolism of atrazine by rat liver. Pestic. Biochem. Physiol. 4:212–219.

Dearfield, K.L. 1988. An assessment of the genetic toxicity of atrazine; review of submitted studies and document prepared by D. Brusick for Ciba-Geigy. Memo (including an executive summary) from U.S. EPA, Office of Pesticide Programs. April 26.

ESE. 1984. Environmental Science and Engineering. Review of treatability data for removal of 25 synthetic organic chemicals from drinking water. Washington, DC: U.S. Environmental Protection Agency, Office of Drinking Water.

Erickson, M.D., C.W. Frank and D.P. Morgan. 1979. Determination of s-triazine herbicide residues in urine: studies of excretion and metabolism in swine as a model to human metabolism. J. Agric. Food Chem. 27:743–745.

Foster, T.S., S.U. Khan and M.H. Akhtar. 1979. Metabolism of atrazine by the soluble fraction (105,000 g) from chicken liver homogenates. J. Agric. Food Chem. 17:300–302.

Gaines, T.B. and R.E. Linder. 1986. Acute toxicity of pesticides in adult and weanling rats. Fundam. Appl. Toxicol. 7:299–308.

Goswami, K.P. and R.E. Green. 1971. Microbial degradation of the herbicide atrazine and its 2-hydroxy analog.

Gzhegotskiy, M.I., L.V. Shklraruk and L.A. Dychok. 1977. Toxicological characteristics of the herbicide zeazin. Vrach. Delo 5:133–136. In: Pesticides Abstract. 10:711–712, 1977.

Harris, C.I. and G.F. Warren. 1964. Adsorption and desorption of herbicides by soil. Weeds. 12:120–126.

Harris, C.I. 1967. Fate of 2-chloro-s-triazine herbicides in soil. J. Agric. Food Chem. 15:157–162.

Hauswirth, J.W. 1988. Summary on some atrazine toxicity studies submitted by Ciba-Geigy (including Metabolism Studies Nos. ABR-87116, 87048, 87087, 85104, 87115 and AG-520). Memo from U.S. EPA, Office of Pesticide Programs. May 3.

Hayes, W.J., Jr. 1982. Pesticides studied in man. Baltimore, MD: Williams and Wilkins.

Hazelton Laboratories.* 1961. Two-year chronic feeding study in rats. CBI. Document No. 000525. MRID 0059211.

Helling, C.S. 1971. Pesticide mobility in soils. II. Applications of soil thin-layer chromatography. Proc. Soil Sci. Soc. Am. 35:737–748.

Hurle, K. and E. Kibler. 1976. The effect of changing moisture conditions on the degradation of atrazine in soil. Proceedings of the British Crop Protection Conference. Weeds. 2:627–633.

Innes, J.R.M., B.M. Ulland and M.G. Valerio. 1969. Bioassay of pesticides and industrial chemicals for tumorigenicity in mice: a preliminary note. J. Natl. Cancer Inst. 42:1101–1114.

Ivey, M.J. and H. Andrews. 1964. Leaching of simazine, atrazine, diuron, and DCPA in soil columns. Unpublished study submitted by Ciba-Geigy. Greensboro, NC.

Ivey, M.J. and H. Andrews. 1965. Leaching of simazine, atrazine, diuron, and DCPA in soil columns. Unpublished study prepared by University of Tennessee, submitted by American Carbonyl, Inc. Tenafly, NJ.

Khan, S.U. and T.S. Foster. 1976. Residues of atrazine (2-chloro-4-ethyl-amino-6-isopropylamino-s-triazine) and its metabolites in chicken tissues. J. Agric. Food Chem. 24:768–771.

Khan, S.U. and M. Schnitzer. 1978. UV irradiation of atrazine in aqueous fulvic acid solution. Environmental Science and Health. B13:299–310.

Lavy, T.L. 1974. Mobility and deactivation of herbicides in soil-water systems. Project A-024-NEB. Available from National Technical Information Service, Springfield, VA. PB-238-632.

Lavy, T.L., F.W. Roeth and C.R. Fenster. 1973. Degradation of 2,4-D and atrazine at three soil depths in the field. J. Environ. Qual. 2:132–137.

Lehman, A.J. 1959. Appraisal of the safety of chemicals in foods, drugs and cosmetics. Assoc. Food and Drug Off. U.S., Q. Bull.

Loprieno, N., R. Barale, L. Mariani, S. Presciuttini, A.M. Rossi, I. Shrana, L. Zaccaro, A. Abbondandolo and S. Bonatti. 1980. Results of mutagenicity tests on the herbicide atrazine. Mutat. Res. 74:250. Abstract.

Lykins, B.W., Jr., E.E. Geldreich, J.Q. Adams, J.C. Ireland and R.M. Clark.

1984. Granular activated carbon for removing nontrihalomethane organics from drinking water. Cincinnati, OH: U.S. Environmental Protection Agency, Office of Research and Development, Municipal Environmental Research Laboratory.

Meister, R.G., ed. 1987. Farm chemicals handbook, 3rd ed. Willoughby, OH: Meister Publishing Co.

Miltner, R.J. and C.A. Fronk. 1985a. Treatment of synthetic organic contaminants for Phase II regulations. Progress report. U.S. Environmental Protection Agency, Drinking Water Research Division. July.

Miltner, R.J. and C.A. Fronk. 1985b. Treatment of synthetic organic contaminants for Phase II regulations. Internal report. U.S. Environmental Protection Agency, Drinking Water Research Division. December.

Molnar, V. 1971. Symptomatology and pathomorphology of experimental poisoning with atrazine. Rev. Med. 17:271-274. (English abstract only.)

Murnik, M.R., and C.L. Nash. 1977. Mutagenicity of the triazine herbicides atrazine, cyanazine, and simazine in *Drosophila melanogaster*. J. Toxicol. Environ. Health. 3:691-697.

NAS. 1977. National Academy of Sciences. Drinking Water and Health. Washington, DC: National Academy Press. pp. 533-539.

Newby, L. and B.G. Tweedy. 1976. Atrazine residues in major rivers and tributaries. Unpublished study submitted by Ciba-Geigy Corporation. Greensboro, NC.

Palmer, J.S. and R.D. Radeleff. 1964. The toxicological effects of certain fungicides and herbicides on sheep and cattle. Ann. N.Y. Acad. Sci. 111:729-736.

Palmer, J.S. and R.D. Radeleff. 1969. The toxicity of some organic herbicides to cattle, sheep and chickens. Production Research Report No. 106. U.S. Department of Agriculture, Agricultural Research Service. pp. 1-26.

Peters, J.W. and R.M. Cook. 1973. Effects of atrazine on reproduction in rats. Bull. Environ. Contam. Toxicol. 9:301-304.

Portnoy, C.E. 1978. Disappearance of bentazon and atrazine in silt loam soil. Unpublished study submitted by BASF Wyandotte Corporation. Parsippany, NJ.

Schlicher, J.E. and V.B. Beat. 1972. Dermatitis resulting from herbicide use — a case study. J. Iowa Med. Soc. 62:419-420.

Smith, A.E., R. Grover, G.S. Emmond and H.C. Korven. 1975. Persistence and movement of atrazine, bromocil, monuron, and simazine in intermittently tilled irrigation ditches. Can. J. Plant Sci. 55:809-816.

Spencer, H. 1987. Review of several mutagenicity studies on atrazine. U.S. EPA, Office of Pesticide Programs' review of a Ciba-Geigy data submission. Accession No. 284052. MRID 402466-01.

STORET. 1988. STORET Water Quality File. Office of Water. U.S. Environmental Protection Agency (data file search conducted in March, 1988).

Talbert, R.E. and O.H. Fletchall. 1965. The adsorption of some S-triazines in soils. Weeds. 13:46–52.

Turner, M.A. and R.S. Adams, Jr. 1968. The adsorption of atrazine and atratone by anion- and cation-exchange resins. Soil Sci. Amer. Proc. 32:62–63.

U.S. EPA. 1986a. U.S. Environmental Protection Agency. Atrazine chronic feeding/oncogenicity study preliminary incidence table of tumors regarding possible section 6(a)(2) effect. Washington, DC: U.S. EPA Office of Pesticide Programs.

U.S. EPA. 1986b. U.S. Environmental Protection Agency. Guidelines for carcinogen risk assessment. Fed. Reg. 51(185):33992–34003. September 24.

U.S. EPA. 1986c. U.S. Environmental Protection Agency. Code of Federal Regulations. Protection of the environment. Tolerances and exemptions from tolerances for pesticide chemicals in or on raw agricultural commodities. 40 CFR 180.220. p. 216.

U.S. EPA. 1988. U.S. Environmental Protection Agency. Method 507 — Determination of nitrogen and phosphorus containing pesticides in ground water by GC/NPD. Draft. April 14.

Warnock, R.E. and J.B. Leary. 1978. Paraquat, atrazine and Bladex dissipation in soils. Unpublished study prepared by Chevron Chemical Company, submitted by Shell Chemical Company. Washington, DC.

Weidner, C.W. 1974. Degradation in groundwater and mobility of herbicides. Master's thesis. University of Nebraska, Department of Agronomy.

Wolf, D.C. and J.P. Martin. 1975. Microbial decomposition of ring-^{14}C-atrazine, cyanuric acid, and 2-chloro-4,6-diamino-S-triazine. J. Environ. Qual. 4:134–139.

Woodard Research Corporation.* 1964. Two-year feeding study in dogs. CBI. Document No. 000525. MRID 00059213.

Woodard Research Corporation.* 1966. Three-generation reproduction study in rats. CBI. Document No. 000525. MRID 00024471.

Yoder, J., M. Watson and W.W. Benson. 1973. Lymphocyte chromosome analysis of agricultural workers during extensive occupational exposure to pesticides. Mutat. Res. 21:335–340.

Windholz, M., ed. 1976. The Merck index. 9th ed. Rahway, NJ: Merck and Co., Inc.

---

* Confidential Business Information submitted to the Office of Pesticide Programs.

Baygon (Propoxur)

I. General Information and Properties

A. CAS No. 114–26–1

B. Structural Formula

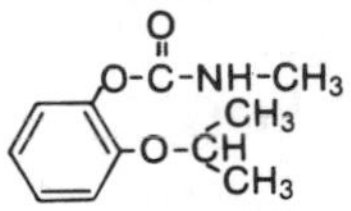

2-(1-Methylethoxy)-phenol methylcarbamate

C. Synonyms

- Baygon Propoxur (proposed common name); Aprocarb: Blattanex; BAY 39007; Bayer 39007; Pillargon; Propyon; Suncide; Tugon; OMS 33; Unden (Meister, 1984).

D. Uses

- Propoxur is a nonfood insecticide used on humans, animals and turf grass (Meister, 1984).

- Propoxur is an insecticide currently registered for use on a variety of sites: terrestrial nonfood (ornamentals, lawns and general outdoor areas), aquatic nonfood (sewage systems and stagnant water), forestry, domestic outdoor and indoor. Application rates are dependent on use. For use as a bait, applications include 1.81 to 18.14 g/1,000 ft^2. For use on turf, applications include 39.69 to 85.05 g/1,000 ft^2. For outdoor uses, applications include 0.046 to 0.175 lb/acre, 0.375 to 1.125 lb/mile with a 300-foot swath, 0.493 to 2% finished spray, 0.14 to 0.45/1,000 ft^2, 2.1% D applied directly to ant hills and 6 strips containing 10% Impr/gypsy moth monitoring trap. For treatment of premises, applications include 0.25 to 1% finished spray and 0.06 to 0.28 g/1,000 ft^2. For indoor treatments, applications include 0.25%, 0.25 to 2% P/T, 0.25 to 1.7% Impr, 1% RTU in traps/bait trays, one fly strip containing 0.4% Impr/1,000 ft^3, contact strips containing 4 to 10% Impr, shelf paper containing 0.6 to 1% Impr, and 0.493 to 2% finished spray. For treatment of animals, applications include 0.9% Impr (dab-on), 9.4% Impr (collars), 8.5 to 56.7 g/gal (dips), 0.125% RTU (shampoo) and 0.25 to 0.28% finished spray. Propoxur may be applied with ground or aerial equipment.

E. Properties (ACGIH, 1984; Meister, 1984; Worthing, 1983; CHEMLAB, 1985)

Chemical Formula	$C_{11}H_{15}O_3N$
Molecular Weight	209.24
Physical State (25°C)	White to tan crystalline solid
Boiling Point	--
Melting Point	91°C
Density (°C)	--
Vapor Pressure (120°C)	0.01 mm Hg
Specific Gravity	--
Water Solubility (20°C)	2000 mg/L
Log Octanol/Water Partition Coefficient	0.14
Taste Threshold	--
Odor Threshold	--
Conversion Factor	--

F. Occurrence

- Baygon has been found in 85 of 624 surface water samples analyzed and in 0 of 114 ground-water samples (STORET, 1988). Samples were collected at 21 surface water locations and 111 ground-water locations. The 85th percentile of all mean-zero samples was 0.96 μg/L in surface water. The maximum concentration found in surface water was also 0.96 μg/L. Baygon was found only in surface water and this finding was reported only in Michigan and Ohio. This information is provided to give a general impression of the occurrence of this chemical in ground and surface waters as reported in the STORET database. The individual data points retrieved were used as they came from STORET and have not been confirmed as to their validity. STORET data are often not valid when individual numbers are used out of the context of the entire sampling regime, as they are here. Therefore, this information can only be used to form an impression of the intensity and location of sampling for a particular chemical.

G. Environmental Fate

- Ring-labeled ^{14}C-propoxur (radiochemical purity 97%), at 5 ppm, degrades with a half-life of > 28 days in irradiated and nonirradiated dry sandy loam soil (McNamara and Moore, 1981). Propoxur comprised 75% of the applied amount in the irradiated soil at 28 days posttreatment. 2-(1-Methylethoxy)phenol comprised < 2% in both irradiated and nonirradiated soil at all sampling intervals. Approximately 1% of the applied radioactivity was isolated in trapping solutions from the irradiated samples. Radioactivity was not detected in trapping solutions from dark control samples.

- ^{14}C-propoxur (radiochemical purity > 95%), at 0.94 to 98.6 μg/10 mL, was very mobile (Freundlich K_{aas} 0.05 to 0.30) in sandy loam, silt loam and silty clay soil:distilled water slurries (1:5, soil:solution ratio) based on batch equilibrium studies (Lenz and Gronberg, 1980). Desorption was quite variable; 0 to 100% (0 to 5.9 μg/10 mL) of the adsorbed amount was desorbed from the soil.

- Aged (28 days) ring- and carbonyl-labeled ^{14}C-propoxur residues (78 to 89% propoxur) were very mobile in columns (12-inch length) of silt loam soil leached with 22.5 inches of water over a 45-day period (Atwell, 1976). Between 69 and 74% of the recovered ^{14}C-residues were leached through the columns. The remaining residues were distributed evenly throughout the columns. In the leachate, 100% of the carbonyl-labeled ^{14}C-residues and 77% of the ring-labeled ^{14}C-residues were identified as propoxur; 23% of the ring-labeled ^{14}C-residues were 2-(1-methylethoxy)phenol.

- Unaged ^{14}C-propoxur (radiochemical purity > 98%) was mobile in columns of sandy loam and silt loam soil; 70 to 79% of the applied radioactivity was leached from the columns (Moellhoff, 1983). Propoxur was the major ^{14}C-compound in both the leachates and the columns treated with unaged propoxur. Aged (30 days) ^{14}C-propoxur residues were slightly mobile in both soils: 72 to 75% of the applied radioactivity remained in the soil columns. Propoxur was the predominant compound in both the soil and leachate after leaching.

- Propoxur was very mobile on sand, sandy loam, sandy clay loam, silt loam and silty clay soil thin-layer chromatography (TLC) plates, with R_f values ranging from 0.70 to 0.89 (Thornton et al., 1976).

II. Pharmacokinetics

A. Absorption

- Vandekar et al. (1971) administered a single oral dose of 1.5 mg/kg of propoxur, 95% active ingredient (a.i.), to a 42-year-old male volunteer. About 45% of the dose was recovered in urine within 24 hours as o-isopropoxyphenol. Since vomiting occurred 23 minutes after ingestion, the authors assumed that much of the dose was expelled by this route, so the percent actually absorbed could not be calculated.

- Chemagro Corp. (no date) investigated the dermal absorption of ^{14}C-labeled Baygon in human subjects. Baygon (4 μg/cm^2, total dose less than 1 mg) was applied to the forearm of the subjects(s) in four tests: (1) application to the skin without preparation, (2) application after stripping of the skin with an adhesive tape, (3) application followed by occlusion and (4) application followed by induction of sweating. The amounts excreted (route not specified, but presumably in urine) after these treatments were 20, 51, 64 and 18%, respectively, indicating that Baygon is absorbed through the skin.

- Krishna and Casida (1965) administered single oral doses of ^{14}C-labeled Baygon (50 mg/kg) to Sprague-Dawley rats. After 48 hours, about 4% of the dose had been excreted in feces, and the remainder was detected in urine (64 to 72%), expired air (26%) or the body (4.2 to 7.9%). This indicated that Baygon had been well absorbed (at least 96%) from the gastrointestinal tract. Similar findings were reported by Foss and Krechniak (1980).

B. Distribution

- Foss and Krechniak (1980) investigated the fate of Baygon after oral administration of 50 mg/kg to male albino rats. Analysis of tissues indicated that Baygon levels were greatest in the kidneys, with somewhat lower levels in the liver, blood and brain.

C. Metabolism

- Dawson et al. (1964) administered single oral doses of 92.2 mg of Baygon (purity not specified) to six male volunteers, and single oral doses of 50 mg to three subjects. Urine samples were collected and analyzed for metabolites. A material identified as 2-isopropoxyphenol was observed in the urine of both groups. Similar results were reported by Vandekar et al. (1971).

• Foss and Krechniak (1980) investigated the metabolism of Baygon after both oral and intravenous administration of 50 mg/kg to male albino rats. Isopropoxyphenol was detected in tissues 10 minutes following administration, and the highest concentrations were attained between 30 and 60 minutes after dosing. This metabolite prevailed in the blood and liver, but in the kidney only unchanged Baygon could be detected. Eight hours postdosing, only traces of Baygon and its metabolites were detected in these tissues.

• Everett and Gronberg (1971) studied the metabolism of Baygon in Holtzman rats. Animals were dosed by gavage with Baygon (5 to 10 mg/kg) labeled with ^{14}C or ^{3}H in the carbonyl or the isopropyl groups. Pooled urine from eight rats (4/sex) dosed with 20 mg/kg/day of unlabeled Baygon for 4 days was used to isolate sufficient quantities of metabolites for identification of structure. Results indicated that the major pathway of Baygon metabolism involved depropylation to 2-hydroxyphenol-N-methyl carbamate and hydrolysis of the carbamate to isopropoxyphenol. Minor pathways involved ring hydroxylation at the five- or six-position, secondary hydroxylation of the 2'-carbon of the isopropoxy group and N-methyl hydroxylation. Metabolites that contained the 6-hydroxy group formed N-conjugates, while those that contained the 5-hydroxy group formed O-glucuronides.

D. Excretion

• Dawson et al. (1964) reported that in humans given single oral doses of 92.2 mg Baygon (purity not specified), 38% of the dose was excreted as phenols in urine over the next 24 hours; most was excreted in the first 8 to 10 hours.

• Krishna and Casida (1965) administered single oral doses of 50 mg/kg of ^{14}C-carbonyl-labeled Baygon to Sprague-Dawley rats. After 48 hours, recovery of label in excretory products was as follows: 64% (males) and 72% (females) in urine, 4% in feces (males and females), and 26% in expired carbon dioxide (males and females). Residual label in the body was 4.2% (males) and 7.9% (females). One-third of the excreted dose was hydrolyzed, with most of the remainder intact.

• Everett and Gronberg (1971) reported that 85% of orally administered ^{14}C-carbonyl-labeled Baygon (5 to 8 mg/kg) was recovered from Holtzman rats within 16 hours of dosing; 20 to 25% of the radioactivity appeared in the expired air, and 60% of the radioactivity appeared in the urine as conjugates. Also, Foss and Krechniak (1980) indicated that 85 to 95% of an oral dose (50 mg/kg) administered to male albino rats was excreted in urine with a half-life of 0.18 to 0.26 hour.

III.   Health Effects

A. Humans

1. *Short-term Exposure*

- Vandekar et al. (1971) studied the acute oral toxicity of Baygon in human volunteers. A 42-year-old man ingested a single oral dose of 1.5 mg/kg of propoxur (Baygon) (95% a.i., recrystallized). Cholinergic symptoms, including blurred vision, nausea, sweating, tachycardia and vomiting, began about 15 to 20 minutes after exposure. Effects were transient and disappeared within 2 hours. Cholinesterase (ChE) activity (measured spectrophotometrically) in red blood cells decreased to 27% of control values by 15 minutes after exposure, and returned to control levels by 2 hours. No effect was detected in plasma ChE activity. In a second test, a single dose of 0.36 mg/kg caused short-lasting stomach discomfort, blurred vision and moderate facial redness and sweating. Red blood cell ChE activity fell to 57% of control values within 10 minutes, then returned to control levels within 3 hours.

- Vandekar et al. (1971) administered five oral doses of 0.15 or 0.20 mg/kg to male volunteers at half-hour intervals (total dose of 0.75 or 1.0 mg/kg). In each subject, a symptomless depression of red blood cell ChE was observed; the lowest level, about 60% of control values, was reached between 1 and 2 hours following doses 3, 4 and 5. After the final dose, red blood cell ChE activity rose to control levels within about 2 hours. The authors noted that a dose of Baygon was tolerated better if it was divided into portions and given over time than if it was given as a single dose.

2. *Long-term Exposure*

- Davies et al. (1967) described the effects of a large-scale spraying operation in El Salvador in which Baygon (OMS-33, 100% a.i.) was used. The trial was planned so that medical assistance would be available, and appropriate clinical support could be provided to those affected by the spraying. The total amount of OMS-33 sprayed was 345 kg. Among the spraymen, exposure (expressed in person-days) was 70.5; 19 experienced symptoms (26% incidence). In the general population, the exposure was 3,340 person-days, and 35 experienced symptoms (1% incidence). The primary symptoms were headache, vomiting and nausea. In the spray-men, the symptoms occurred mostly in the first days, with no visible symptoms after that time. In severe cases, atropine was administered as an antidote. It was concluded that the acute toxicity symptoms were observed in a low incidence, and they were, in general, mild, evanescent, reversible, responsive to small doses of atropine, and tended to occur at the beginning of the spray program.

- Montazemi (1969) reported on the toxic effects of Baygon on the population of 26 villages in Iran that were sprayed with Baygon at the rate of 2 g/m^2 daily for 18 days. Selected inhabitants from six villages and sprayers were examined on days 2, 8 and 18 and after the completion of the spraying. Depression of ChE activity was found in the inhabitants and in the sprayers, but the sprayers generally had more severe symptoms. Atropine or belladonna was adequate to treat those exhibiting symptoms.

B. Animals

1. Short-term Exposure

- The acute oral LD_{50} value for technical Baygon (purity not specified) in male and female Sherman rats was reported to be 83 and 86 mg/kg, respectively (Gaines, 1969). The oral LD_{50} was reported to be 23.5 mg/kg in mice and 40 mg/kg in guinea pigs (NIOSH, 1987).

- Farbenfabriken Bayer (1961) determined an oral LD_{50} of 100 to 150 mg/kg (purity not specified) in male albino rats. Severe muscle spasms were observed, but no dose-response information was provided.

- Eben and Kimmerle (1973) studied the acute toxicity of Baygon in SPF-Wistar rats. Single oral doses of propoxur (98.7% a.i.), diluted with propylene glycol, were given by gavage to groups of three male rats at levels of 15, 20, 40 or 60 mg/kg; female rats were given doses of 10, 20, 40 or 60 mg/kg. Cholinesterase levels were measured in plasma, erythrocytes and brain at 10, 20 and 180 minutes after dosing. Maximum ChE depression was observed at 10 and 20 minutes in the plasma and erythrocytes, and at 180 minutes in the brain. The inhibition was dose-dependent and a no-effect level was not observed. In plasma, ChE was inhibited from 19% (low dose) to 63% (high dose) in males and from 0 to 32% in females. In erythrocytes, ChE was inhibited from 27% (low dose) to 63% (high dose) in males and from 15 to 45% in females. Based on ChE inhibition, this study identified a Lowest-Observed-Adverse-Effect Level (LOAEL) of 10 mg/kg/day.

- Farbenfabriken Bayer (1966) conducted a 9-week feeding study with Bay 39007 (purity not specified) in male and female rats (Elberfeld FB-30). Baygon was included in the diets of the male animals at dose levels of 0, 1,000, 2,000, 4,000 or 8,000 ppm. Based on the assumption that 1 ppm in the diet of rats is equivalent to 0.05 mg/kg/day (Lehman, 1959), this corresponds to doses of 0, 50, 100, 200 or 400 mg/kg/day, respectively. Females were given only one dose (4,000 ppm). The study was begun when the animals (15/dose level) were 4 weeks of age and weighed about 48 g. In males, food consumption and body weight were depressed in a dose-dependent manner. At the 4,000- and 8,000-ppm levels, the males were less lively and exhibited slightly shaggy coats. Gross pathologic examinations of all animals were conducted. Two males exposed to 4,000 ppm died during the study, one at 11 days (evidence of myocarditis) and one at 23 days. Two males also died at the 8,000-ppm level (at 23 and 25 days); one showed necrotic inflammation of the mucosa of the small intestine. Females (exposed to 4,000 ppm only) displayed decreased food consumption and reduced weight gain similar to that seen in exposed males. One of 15 female controls died at day 12 (death attributed to pneumonia), and two of 15 exposed females died, one at 7 days and one at 45 days (in this rat there was suppuration of the cerebellar bottom). There were apparently no measurements of ChE activity or other clinical tests performed during this study. It was concluded by the authors that the observed pathology could not be directly attributed to the presence of Baygon in the diet. Based on gross observations, the No-Observed-Adverse-Effect-Level (NOAEL) for male animals was identified as 2,000 ppm (100 mg/kg/day) and the LOAEL as 4,000 ppm (200 mg/kg/day). In females, 4,000 ppm (200 mg/kg/day, the only dose tested) was a toxic level.

- Eben and Kimmerle (1973) exposed SPF-Wistar rats (4/sex/dose) by gavage to doses of 3, 10 or 30 mg/kg/day of Baygon for 4 weeks. The high-dose animals (30 mg/kg/day) displayed cholinergic symptoms. Cholinesterase activity in plasma and red blood cells, measured 15 minutes after dosing on days 3, 8, 14, 21 and 28, was generally depressed in a dose-related manner at 10 and 30 mg/kg, but not at the 3-mg/kg dose. For example, on day 28, ChE activity in plasma was reduced by 0, 21 or 27% in males and by 14, 27 or 41% in females. In erythrocytes, ChE was inhibited by 9, 24 or 32% in males and by 11, 32 or 43% in females. No cumulative toxic effects were observed. Based on ChE inhibition, the NOAEL for this study was 3 mg/kg/day, and the LOAEL was 10 mg/kg/day.

2. *Dermal/Ocular Effects*

- The acute dermal LD_{50} of technical Baygon (purity not specified) was reported to be greater than 2,400 mg/kg for both male and female Sherman rats (Gaines, 1969).

- Crawford and Anderson (1971) indicated that 500 mg of technical Baygon (purity not specified, dissolved in acetone) did not cause any skin irritation within 72 hours of its application to the abraded or unabraded skin of mature New Zealand White rabbits (6/group).

- Heimann (1982) demonstrated that Baygon (98.8% pure) is not a skin sensitizer when tested in guinea pigs.

- Crawford and Anderson (1971) instilled 100 mg of technical Baygon (purity not specified) in the left eye of six rabbits. Examination at 48 and 72 hours revealed no evidence of ocular irritation or corneal damage.

3. *Long-term Exposure*

- Eben and Kimmerle (1973) fed propoxur (98.7% a.i.) to male SPF Wistar rats in the diet for 15 weeks. Doses were 0, 250, 750 or 2,000 ppm. Assuming that 1 ppm in the diet is equivalent to 0.05 mg/kg/day (Lehman, 1959), this corresponds to doses of about 0, 12.5, 37.5 or 100 mg/kg/day. Assays for ChE activity in plasma, erythrocytes and brain showed no constancy of inhibition and no dependence on the administered dose. No other details were given.

- Root et al. (1963) studied the effect of Bayer 39007 added to the diet of Sprague-Dawley rats for 16 weeks. The rats (12/sex/dose, weighing 72 to 145 g at the start of the feeding trial) were fed Baygon (technical, 95.1% pure) at dose levels of 0, 100, 200, 400 or 800 ppm. Assuming that 1 ppm in the diet of rats is equivalent to 0.05 mg/kg/day (Lehman, 1959), this corresponds to doses of 0, 5, 10, 20 or 40 mg/kg/day. Biweekly measurements revealed no changes in growth or food consumption. Cholinesterase was assayed in blood, brain and submaxillary glands of five animals of each sex at each dose level, and no inhibition was detected. Necropsies were performed on five animals of each sex at the termination of the study, and no significant pathology was found. It was concluded that the NOAEL for the rats was greater than 800 ppm (40 mg/kg/day, the highest dose tested).

- Suberg and Loeser (1984) conducted a chronic (106-week) feeding study of Baygon (99.4% a.i.) in rats (Elberfeld strain) at dose levels of 0, 200, 1,000 or 5,000 ppm for 106 weeks. Based on the assumption that 1 ppm in the diet of rats is equivalent to 0.05 mg/kg/day (Lehman, 1959), this corresponds to doses of about 0, 10, 50 or 250 mg/kg/day. There were 50 rats of each sex per dose level, plus an additional 10 of each sex for interim autopsies at the end of the first year. At the 200-ppm dose, there was no effect on food consumption or body weight; there were no cholinergic signs; and clinical chemistry, gross pathology, histopathology and organ weights showed no changes from control values. At 1,000 ppm, retarded weight gain was observed in males during the first 20 weeks. At 1,000 and 5,000 ppm, there were significant hyperplasia of urinary bladder epithelium (described in more detail in the Carcinogenicity section) and increased incidence of neuropathy. At the 5,000-ppm dose, both weight gain and food consumption were significantly retarded throughout the study; males showed increased thromboplastin time, and females had consistently lower mean plasma ChE activity than did controls or other test groups. No significant inhibition was reported in the examined dose group. Both sexes showed some degree of splenic atrophy, but there were no other significant changes in other organs. Based on body weight gain, the NOAEL for this study was identified as 200 ppm (10 mg/kg/day), and the LOAEL as 1,000 ppm (50 mg/kg/day).

- Loser (1968a) conducted a 2-year feeding study of Baygon in male and female SPF-Wistar rats (25/sex/dose). Starting at 1 month of age, the test material, BAY 39007 (99.8% a.i., technical), was included in the diet at levels of 0, 250, 750, 2,000 or 6,000 ppm. Based on the assumption that 1 ppm in the diet of rats is equivalent to 0.05 mg/kg/day (Lehman, 1959), this corresponds to doses of 0, 12.5, 37.5, 100 or 300 mg/kg/day. The control group consisted of 50 animals of each sex, while test groups contained 25 animals of each sex. Growth and behavior were observed, liver function and ChE activity were tested, and blood and urine were analyzed periodically. Necropsies on five animals of each sex were conducted at the termination of the experiment. The major adverse effects noted were low food consumption and low body weight in all animals at the 6,000-ppm dose level, and low body weight in the female (but not male) animals at the 2,000-ppm dose level. Cholinesterase determinations on blood (measured at the high dose only) revealed no changes; ChE activity was 9.8 and 9.9 units in control males and females, respectively, compared with 9.9 and 10.0 in exposed males and females. The author indicated that the methodology may have been too insensitive to detect small changes that may have occurred. No spasms or other symptoms of ChE inhibition were observed. No impairment of liver or kidney function was detected by clinical tests, but necropsy revealed increased liver weight at all doses greater than 250 ppm. Results of blood analysis were normal at all dose levels except at 6,000 ppm. Apart from increased liver weights, necropsy findings were unremarkable. Based on increased liver weights, this study identified a NOAEL of 250 ppm (12.5 mg/kg/ day) and a LOAEL of 750 ppm (37.5 mg/kg/day).

• Loser (1968b) conducted a 2-year study of Baygon toxicity in beagle dogs (4/sex/dose). The product, BAY 39007 (technical, 99.8% pure), was included in the diet at levels of 0, 100, 250, 750 or 2,000 ppm. Assuming that 1 ppm in the diet of dogs is equivalent to 0.025 mg/kg/ day (Lehman, 1959), this corresponds to doses of about 0, 2.5, 6.25, 18.7 or 50 mg/kg/day. The study was begun when the dogs (4/sex/dose) were 4 to 5 months old. Observations on the animals included weight and food consumption at periodic intervals, ChE determinations in blood at 16 weeks, clinical evaluations of blood and urine and tests for liver and kidney function. Necropsies were performed on animals that died during the study and at termination of the study. The appearance, behavior and food consumption of dogs at the 100-, 250- or 750-ppm levels were comparable to those of the controls. At the 2,000-ppm level, dogs of both sexes appeared to be weak and sick. One of the males and all four females at this dose died before completion of the study. During the first 6 months, dogs at this dose level exhibited quivering and spasms, particularly in the abdominal region, and food consumption was less than for the controls (especially in females); as expected, the dogs showed statistically significant depression in weight gain compared with the controls. Males, but not females, showed lower weights than did controls at the 750-ppm dose level, but the decrease was not statistically significant. Clinical analyses did not reveal any aberrations in the blood or any changes in liver or kidney function. However, increased liver weights were observed at necropsy, and serum electrophoresis performed at the time of sacrifice revealed decreased levels of some serum proteins, interpreted by the author as reflecting impaired protein synthesis. Cholinesterase determinations in whole blood at 16 weeks did not reveal any significant inhibition of activity. ChE inhibition at 100, 250, 750 and 2,000 ppm was 0, 11, 1 and 13%, respectively, in males, and 0, 10, 7 and 0%, respectively, in females. The author indicated that the assay method may have been too insensitive to detect small changes that may have occurred. Emaciation was the principal finding in dogs that died during the study; one female had abnormal liver parenchyma. The NOAEL for this study was 250 ppm (6.25 mg/kg/day), and the LOAEL (based on increased liver weight, decreased body weight and altered blood proteins) was 750 ppm (18.7 mg/kg/day).

- Bomhard and Loeser (1981) conducted a 2-year feeding study of propoxur (99.6% a.i.) in SPF CF_1/W74 mice at dose levels of 0, 700, 2,000 or 6,000 ppm. Assuming that 1 ppm in the diet of mice is equivalent to 0.15 mg/kg/day (Lehman, 1959), this corresponds to doses of about 0, 105, 300 or 900 mg/kg/day. Mice were 5 to 6 weeks of age, weighing 22 to 25 g at the beginning of the study; each group consisted of 50 animals of each sex, plus an additional 10/sex/group included for interim autopsy at 1 year. Body weight gain was slightly depressed in male mice at the 6,000-ppm level. Apart from this observation, all aspects of behavior, appearance, food intake, weight and mortality were comparable to control values. Clinical chemistry and blood studies, including glucose and cholesterol levels, were within the normal range for all groups, and there were no significant gross pathological or histopathological findings that could be attributed to the ingestion of Baygon. It was concluded that the male mice tolerated the pesticide at levels up to and including 2,000 ppm, while the female mice tolerated doses up to and including 6,000 ppm without adverse effects. Based on these conclusions, the NOAEL for this study was 2,000 ppm (300 mg/kg/day), and the LOAEL (based on depressed weight gain in males) was 6,000 ppm (900 mg/kg/day).

4. *Reproductive Effects*

- No multigeneration studies of the effects of Baygon on the reproductive function of animals were found in the available literature.

- In a developmental toxicity study in rabbits, Schlueter and Lorke (1981) observed no adverse effects on several reproductive end points. This study is described below.

5. *Developmental Effects*

- Schlueter and Lorke (1981) studied the effect of propoxur (99.6% a.i.) on Himalayan CHBB:HM rabbits during gestation. Propoxur was administered by gavage (in 0.5% cremophor) to 15 animals/dose at 0, 1, 3 or 10 mg/kg. No adverse effects were observed in the dams, and no changes were detected in implantation index, mean placental weight, resorption index or litter size. Embryos were examined for visceral and skeletal defects grossly, then were stained with Alizarin, and transverse sections were prepared using the Wilson technique. No adverse fetal effects were found at any dose level with respect to mean fetal weight, the percent of stunting, the percent of slight skeletal deviations or the malformation index. These results indicate that the NOAEL for maternal toxicity, teratogenicity and fetotoxicity is greater than 10 mg/kg/day (the highest dose tested).

- Lorke (1971) fed Baygon (technical, 98.4% a.i., 0.82% isopropoxyphenol) in the diet to female FB-30 rats on days 1 to 20 of gestation, at levels of 0, 1,000, 3,000 or 10,000 ppm (10/dose). Assuming that 1 ppm in the diet of rats is equivalent to 0.05 mg/kg/day (Lehman, 1959), this corresponds to doses of about 50, 150 or 500 mg/kg/day. The rats were 2.5 to 3.5 months of age, weighing 200 to 250 g at the time of the experiment. Cesarean sections were performed on day 20. External and internal examinations on fetuses were performed, and fetuses were subjected to skeletal staining. At the 3,000- and 10,000-ppm dose levels, average fetal weights were significantly lower than control values, but other fetal measurements were in the control range. No terata were observed at a higher incidence than in the control group. Data on fetal ossification were not adequately described for an adequate evaluation. Although this study appears to reflect a NOAEL of 1,000 ppm (50 mg/kg/day) based on fetotoxic effects, information obtained from this study is limited due to the small number of animals tested and an apparent dose-related decrease in maternal weight gain and fetal weight at the lowest dose tested (although these effects were not statistically significant).

6. Mutagenicity

- DeLorenzo et al. (1978) evaluated the mutagenic properties of Baygon and other carbamate pesticides by use of the *Salmonella* mutagenicity test of Ames. In assays using five strains of *Salmonella typhimurium*, no mutagenic activity was obtained with Baygon at 10 to 1,500 μg/plate (with or without microsomal activation).

- Moriya et al. (1983) tested Baygon at up to 5,000 μg/plate in five strains of *S. typhimurium* and one strain of *Escherichia coli* using the Ames technique (without metabolic activation) and observed no evidence of mutagenic activity.

- Blevins et al. (1977) used five mutants of *S. typhimurium* LT2 to examine the mutagenic properties of Baygon and other methyl carbamates and their nitroso derivatives. No mutagenic activity was observed with Baygon in this experiment using the Ames technique.

7. *Carcinogenicity*

- Suberg and Loeser (1984) conducted a chronic (106-week) feeding study of Baygon (99.4% a.i.) in rats (Elberfeld strain SPF strain, Bor. WISW) at dose levels of 0, 200, 1,000 or 5,000 ppm. Assuming that 1 ppm in the diet of rats is equivalent to 0.05 mg/kg/day (Lehman, 1959), this corresponds to doses of about 0, 10, 50 or 250 mg/kg/day. The study utilized 50 rats/sex/dose, plus an additional 10 of each sex included for interim necropsies at the end of the first year. At 5,000 ppm significant hyperplasia of the urinary bladder epithelium was noted. The incidence at this dose level after 2 years was 44/49 in males and 48/48 in females, as compared with 1/49 and 0/49 in control males and females, respectively. At 1,000 ppm, there was a smaller increased incidence (10/50 and 5/49 in males and females), respectively. No significant effect occurred at 200 ppm (1/50 and 0/49, males and females, respectively). Bladder papillomas were observed in both males (25/57) and females (28/48) at the highest dose after 2 years. In addition, at the 5,000-ppm level, carcinoma of the bladder was found in 8/57 males and 5/48 females, and carcinoma of the uterus was seen in 8/49 females, as compared with 3/49 for the control group. At the mid-dose level (1,000 ppm) only papillomas were noted in one male. The tumors of significance in this study are the uncommon bladder tumors (carcinoma and papillomas) with high incidences at the high-dose level. The combined tumor incidences were 34/57 males and 33/48 females at 5,000 ppm; 1/59 males and 0/48 females at 1,000 ppm, and none in the 200-ppm or control groups.

- Bomhard and Loeser (1981) conducted a 2-year feeding study of propoxur (99.5% a.i.) in SPF CF_1/W74 mice at dose levels of 0, 700, 2,000 or 6,000 ppm. Assuming that 1 ppm in the diet is equivalent to 0.15 mg/kg/day (Lehman, 1959), this corresponds to doses of about 0, 105, 300 or 900 mg/kg/day. Mice were 5 to 6 weeks of age, weighing 22 to 25 g at the beginning of the study; each group consisted of 50 animals of each sex, plus an additional 10/sex/group included for interim necropsy at 1 year. Gross and histological examination of tissues revealed no evidence of increased tumor frequency.

IV.   Quantification of Toxicological Effects

A. One-day Health Advisory

The study by Vandekar et al. (1971) has been selected to serve as the basis for determination of the One-day HA for Baygon. In this study, human volunteers who ingested single oral doses of 0.36 or 1.5 mg/kg displayed transient cholinergic signs accompanied by marked (43 and 75%, respectively) inhibition of red blood cell ChE (measured 10 to 15 minutes after exposure). Total doses of 0.75 or 1.0 mg/kg administered in five equal portions over 2 hours did not cause clinical signs, but inhibited red blood cell ChE by about 40%. A NOAEL was not identified; 0.36 mg/kg is taken as the LOAEL for bolus exposure, and 0.45 mg/kg (three-fifths of a 0.75-mg/kg/day total dose, administered in the first 3/5 doses) is the LOAEL when exposure to this dose is spread over several hours. It should be noted that both values are considerably lower than the NOAEL values for Baygon identified in subchronic and chronic feeding studies in animals, especially rodents. Possible reasons for this disparity are that humans may be more sensitive to this chemical than animals are; furthermore, single oral doses probably produce higher peak inhibitions than if the same total dose is ingested over a longer period of time. It is also likely that measurement of ChE activity 10 to 15 minutes after exposure (as in the case of human studies) detects peak inhibition, while sampling later reveals smaller effects (due to the reversible nature of ChE inhibition with carbamates). Since a child's exposure is more likely to occur in a manner similar to Vandekar's test, where doses were administered in five equal portions over time, and since the LOAEL of 0.45 mg/kg (three-fifths of a 0.75-mg/kg total dose) caused a similar level of ChE inhibition as the bolus dose of 0.36 mg/kg (ChE inhibition is in the 40% range), the LOAEL, 0.36 mg/kg, is used for the One-day HA calculation.

The One-day HA for a 10-kg child is calculated as follows:

$$\text{One-day HA} = \frac{(0.36 \text{ mg/kg/day}) (10 \text{ kg})}{(100) (1 \text{ L/day})} = 0.036 \text{ mg/L} (40 \text{ } \mu\text{g/L})$$

B. Ten-day Health Advisory

In addition to human studies by Vandekar et al. (1971) discussed above, two studies were considered for determination of the Ten-day HA. In a teratology study in rabbits by Schlueter and Lorke (1981), the NOAEL appeared to be higher than 10 mg/kg/day, the highest dose tested. In a teratology study in rats by Lorke (1971), the dietary administration of Baygon to animals during gestation was designed to assess both maternal and fetal effects. While sufficient data were obtained to derive a NOAEL of 50 mg/kg/day and a LOAEL of 150 mg/kg/day in rats, it is important to note that a dosage of 50 mg/kg/day was sufficient to kill all female animals in a chronic study in dogs by Loser (1968b); all deaths occurred before the end of the 2-year study period. Because humans appear to be more sensitive to Baygon than animals, the human study by Vandekar et al. (1971), used in the determination of the One-day HA value, is also the most suitable study for calculation of the Ten-day HA. The two LOAELs identified in this study, 0.36 mg/kg (bolus exposure) and 0.45 mg/kg/day (exposure to three-fifths of a 0.75-mg/kg total dose spread out over the day) resulted in the same level of red blood cell ChE inhibition. Therefore, the lower dose, 0.36 mg/kg/day, is used for calculation of the Ten-day HA.

The Ten-day HA for a 10-kg child is calculated as follows:

$$\text{Ten-day HA} = \frac{(0.36 \text{ mg/kg/day}) (10 \text{ kg})}{(100) (1 \text{ L/day})} = 0.036 \text{ mg/L } (40 \text{ } \mu\text{g/L})$$

C. Longer-term Health Advisory

No suitable information was found in the available literature for the determination of the Longer-term HA value for Baygon. It is, therefore, recommended that the modified Drinking Water Equivalent Level (DWEL) of 40 μg/L for a 10-kg child be used as a conservative estimate for a longer-term exposure. The DWEL of 100 μg/L, calculated below, should be used for the longer-term value for a 70-kg adult.

D. Lifetime Health Advisory

The 2-year feeding study in dogs by Loser (1968b) and the human study by Vandekar et al. (1971) have been considered for determination of the Lifetime HA. In the 2-year dog study by Loser (1968b), the chronic NOAEL was identified as 6.25 mg/kg/day and the LOAEL as 18.7 mg/kg/day. The dog NOAEL value is supported by the data of Loser (1968a) and of Suberg and Loeser (1984), which identified NOAEL values of 12.5 and 10 mg/kg/day, respectively, in chronic studies in rats. However, the dog appears to be far more sensitive at the higher doses than are rodents; all female dogs and some of the males in the high-dose group, 50 mg/kg/day, died before the end of the study period, while mild systemic toxicity was noted at this dose level in rats. Cholinesterase determinations were not performed in the dog study for use in comparison with human data. Due to the reversible nature of ChE inhibition by carbamates, a large difference is noted between the dosages that can cause biologically significant levels of ChE inhibition and the dosages that can produce cholinergic symptoms of toxicity (including death). Hence, in the absence of ChE data in the dog study, and because of the sensitivity of this end point in the determination of the toxicity of this chemical, the study by Vandekar et al.

(1971) in humans has been selected to serve as the basis for the Lifetime HA for Baygon. This study was discussed in the previous sections on the One-day and Ten-day HAs. The 2-year mouse study by Bomhard and Loeser (1981) was not considered, since the data suggest that the mouse is even less sensitive than the rat.

Using a human ChE LOAEL of 0.36 mg/kg/day, the Lifetime HA is calculated as follows:

Step 1: Determination of the Reference Dose (RfD)

$$RfD = \frac{(0.36 \text{ mg/kg/day})}{100} = \begin{array}{l} 0.004 \text{ mg/kg/day (after rounding} \\ \text{from } 0.0036 \text{ mg/kg/day)} \end{array}$$

Step 2: Determination of the Drinking Water Equivalent Level (DWEL)

$$DWEL = \frac{(0.0036 \text{ mg/kg/day}) (70 \text{ kg})}{(2 \text{ L/day})} = 0.126 \text{ mg/L} (100 \text{ } \mu g/L)$$

Step 3: Determination of the Lifetime Health Advisory

$$\text{Lifetime HA} = \frac{(0.126 \text{ mg/L}) (20\%)}{(10)} = 0.003 \text{ mg/L} (3 \text{ } \mu g/L)$$

E. Evaluation of Carcinogenic Potential

- Suberg and Loeser (1984) detected an increased frequency of urinary bladder epithelium hyperplasia, bladder papillomas and carcinomas, and carcinoma of the uterus in rats fed Baygon (250 mg/kg/day) for 2 years.

- Bomhard and Loeser (1981) did not detect an increased incidence of tumors in mice fed Baygon at doses up to 90 mg/kg/day for 2 years.

- The International Agency for Research on Cancer (IARC) has not evaluated the carcinogenic potential of Baygon.

- Applying the criteria described in EPA's guidelines for assessment of carcinogenic risk (U.S. EPA, 1986a), Baygon may be classified in Group C: possible human carcinogen. This category is for substances with limited evidence of carcinogenicity in animals in the absence of human data. However, this classification group may be considered preliminary at present (U.S. EPA, 1987b), since the U.S. EPA Office of Pesticide Programs (OPP) has classified this chemical in Group B2: probable human carcinogen (U.S. EPA, 1987a). A resolution will be reached between OPP and the Cancer Assessment Group (CAG) in the near future.

V.   Other Criteria, Guidance and Standards

- Residue tolerances have not been established for Baygon by the OPP.

- The American Conference of Governmental Industrial Hygienists (ACGIH, 1984) has proposed a threshold limit value of 0.5 mg/m^3.

- The World Health Organization (WHO) calculated an ADI of 0.02 mg/kg/day for Baygon (Vettorazzi and Van den Hurk, 1985).

VI.   Analytical Methods

- Analysis of Baygon is by a high-performance liquid chromatographic (HPLC) procedure used for the determination of N-methyl carbamoyloximes and N-methylcarbamates in water samples (U.S. EPA, 1986b). In Method 531.1, the water sample is filtered and a 400-μL aliquot is injected into a reverse-phase HPLC column. Separation of compounds is achieved using gradient elution chromatography. After elution from the HPLC column, the compounds are hydrolyzed with sodium hydroxide. The methyl amine formed during hydrolysis is reacted with o-phthalaldehyde (OPA) to form a fluorescent derivative that is detected with a fluorescence detector. This method has been validated in a single laboratory. The limit of detection for this method for Baygon is 1.0 μg/L.

VII.   Treatment Technologies

- Available data indicate granular activated carbon (GAC) adsorption to be a possible Baygon removal technique.

- Adsorption of Baygon on GAC proceeds in accordance with both Freundlich and Langmuir isotherms (El-Dib et al., 1974; Whittaker et al., 1982).

- One full-scale laboratory test was carried out on a commercially available system (Dennis et al., 1983; Kobylinski et al., 1984).

- Different levels of Baygon (20 mg/L, 60 mg/L and 100 mg/L) were added to tap water. At a flow rate of 67.4 ppm, the column removed 99% of the Baygon in 3.5, 8.5 and 21 hours, respectively, using only 45 lb of granular carbon.

VIII. References

ACGIH. 1984. American Conference of Governmental Industrial Hygienists. Documentation of the threshold limit values for substances in workroom air, 3rd ed. Cincinnati, OH: ACGIH.

Atwell, S.H. 1976. Leaching characteristics of Baygon on aged soil. Report No. 50718. Unpublished study received Oct. 21. 1981, under 3125-306. Submitted by Mobay Chemical Corp., Kansas City, MO. CDL:246088-L. MRID 00085769.

Blevins, R.D., M. Lee and J.D Regan. 1977. Mutagenicity screening of five methyl carbamate insecticides and their nitroso derivatives using mutants of *Salmonella typhimurium* LT2. Mutat. Res. 56:1-6.

Bomhard, E. and E. Loeser.* 1981. Propoxur, the active ingredient of Baygon: chronic toxicity study in mice (two-year feeding experiment). Bayer Report No. 9954;69686. Bayer A.G., Institut fur Toxicologie. Unpublished study. MRID 00100546.

Chemagro Corporation.* No date. Toxicity study on humans. Report No. 28374. Unpublished study. MRID 00045091.

CHEMLAB. 1985. The Chemical Information System, CIS, Inc. Baltimore, MD.

Crawford, C.R. and R.H. Anderson.* 1971. The skin and eye irritating properties of (R) Baygon technical and Baygon 70% WP to rabbits. Report No. 29706. Unpublished study. MRID 00045097.

Davies, J.E., J.J. Freal and R.W. Babione. 1967. Toxicity studies: field trial of OMS-33 insecticide in El Salvador. Report No. 23933. World Health Organization. CDL:091768-F. Unpublished study. MRID 00052281.

Dawson, J.A., D.F. Heath, J.A. Rose, F.M. Thain and J.B. Word. 1964. The excretion by humans of the phenol derived from 2-isopropoxyphenyl N-methylcarbamate. Bull. WHO. 30:127-134.

DeLorenzo, F., N. Staiano, L. Silengo and R. Cortese. 1978. Mutagenicity of Diallate, Sulfallate and Triallate and relationship between structure and mutagenic effects of carbamates used widely in agriculture. Cancer Res. 38:13-15.

Dennis, W.H., A.B. Rosencrance, T.M. Trybus, C.W.R. Wade and E.A. Kobylinski.1983. Treatment of pesticide-laden wastewaters from Army pest control facilities by activated carbon filtration using the carbolator treatment system. Frederick, MD: U.S. Army Bioengineering Research and Development Laboratory, Ft. Detrick.

Eben, A. and G. Kimmerle.* 1973. Propoxur: effect of acute and subacute oral doses on acetylcholinesterase activity in plasma, erythrocytes, and brain of rats. Report No. 4262. Report No. 39114. Unpublished study. MRID 00055148.

El-Dib, M.A., F.M. Ramadan and M. Ismail. 1974. Adsorption of Sevin and Baygon on granular activated carbon. Water Res. 9:795–798.

Everett, L.J. and R.R. Gronberg.* 1971. The metabolic fate of Baygon (o-isopropoxyphenylmethyl carbamate) in the rat. Chemagro Corp. Research and Development Department. Report No. 28797. Unpublished study. MRID 00057737.

Farbenfabriken Bayer.* 1961. Toxicity of Bayer 39007 (Dr. Bocker 5812315). Report No. 6686. Farbenfabriken Bayer Aktiengesellschaft. Unpublished study. MRID 00040433.

Farbenfabriken Bayer.* 1966. Two-month feeding test with Bayer 39007. Report No. 17466. Institut fur Toxikologie. Unpublished study. MRID 00035412.

Foss, W. and J. Krechniak. 1980. The fate of propoxur in rat. Arch. Toxicol. 4:346–349.

Gaines, T.B. 1969. Acute toxicity of pesticides. Toxicol. Appl. Pharmacol. 14:515–534.

Heimann, K. 1982. Propoxur (the active ingredient of Baygon and Unden): study of sensitization effects on guinea pigs. Bayer Report No. 11218. Mobay Report 82567. Prepared by Bayer AG. Institut fur Toxikologie. Unpublished study. MRID 00141139.

Kobylinski, F.A., W.H. Dennis and A.B. Rosencrance. 1984. Treatment of pesticide-laden wastewater by recirculation through activated carbon. American Chemical Society.

Krishna, J.G. and J.F. Casida.* 1965. Fate of the variously labeled methyl- and dimethyl-carbamate-^{14}C insecticide chemicals in rats. Report No. 16440. Unpublished study. MRID 00049234.

Lehman, A. J. 1959. Appraisal of the safety of chemicals in foods, drugs and cosmetics. Assoc. Food Drug Off. U.S., Q. Bull.

Lenz, M.F. and R.R. Gronberg. 1980. Soil adsorption and desorption of Baygon. Report No. 69016. Unpublished study received Oct. 21, 1981 under 3125–306. Submitted by Mobay Chemical Corp., Kansas City, MO. CDL:246088-N. MRID 00085770.

Lorke, D.* 1971. BAY 39007: examination for embryotoxic effects among rats. Report No. 2388. Report No. 29035. MRID 00045094.

Loser, F.* 1968a. BAY 39007: chronic toxicological studies on rats. Report No. 726. Report No. 22991. Unpublished study. MRID 00035425.

Loser, F.* 1968b. BAY 39007: chronic toxicological studies on dogs. Report No. 669. Report No. 22814. Unpublished study. MRID 00035423.

McNamara, F.T. and K.D. Moore. 1981. Photodecomposition of Baygon on soil. Report No. 69476. Unpublished study received Oct. 21, 1981 under 3125–306. Submitted by Mobay Chemical Corp., Kansas City, MO. CDL:246088-F. MRID 00085765.

Meister, R., ed. 1984. Farm chemicals handbook. Willoughby, OH: Meister Publishing Company.

Moellhoff, E. 1983. Leaching behavior of aged propoxur (Baygon) residues (laboratory study): RA-162-172B. Unpublished study. Prepared by Bayer AG. 12 p. MRID 00149025.

Montazemi, K. 1969. Toxicological studies of Baygon insecticide in Shabankareh area, Iran. Trop. Geogr. Med. 21:186-190.

Moriya, M., T. Ohta, K. Wantanabe, T. Mivazawa, K. Kato and Y. Shirasu. 1983. Further mutagenicity studies on pesticides in bacterial reversion assay systems. Mutat. Res. 116:185-216.

NIOSH. 1987. National Institute for Occupational Safety and Health. Registry of toxic effects of chemical substances. Sweet, D.U., ed. Cincinnati, OH: National Institute for Occupational Safety and Health. Microfiche edition.

Root, M., J. Cowan and J. Doull.* 1963. Subacute oral toxicity of Bayer 39007 to male and female female (sic) rats. Report No. 10685. Unpublished study. MRID 00040447.

Schlueter, G. and D. Lorke.* 1981. Propoxur, the active ingredient of Baygon: study of embryotoxic and teratogenic effects on rabbits after oral administration. Bayer Report No. 10183. Mobay ACD Report No. 80034. Bayer AG Institut fur Toxicologie. Unpublished study. MRID 00100547.

STORET. 1988. STORET Water Quality File. Office of Water. U.S. Environmental Protection Agency (data file search conducted in March, 1988).

Suberg, H. and H. Loeser.* 1984. Chronic toxicological study with rats (feeding study over 106 weeks). Report 12870. Unpublished Mobay study No. 88501. Prepared by Bayer Institute of Toxicology. Unpublished study. MRID 00142725.

Thornton, J.S., J.B. Hurley and J.J. Obrist. 1976. Soil thin-layer mobility of twenty-four pesticide chemicals. Mobay Report No. 51016. Prepared by Mobay Chemical Corp., Kansas City, MO. Submitted by Chevron Chemical Company, Richmond, CA. Acc. No. 259942. Reference 6.

U.S. EPA. 1986a. U.S. Environmental Protection Agency. Guidelines for carcinogen risk assessment. Fed. Reg. 51(185):33992-34003. September 24.

U.S. EPA. 1986b. U.S. Environmental Protection Agency. Method 531.1 – Measurement of N-methyl carbamoyloximes and N-methylcarbamates in ground water by direct aqueous injection HPLC with post column derivatization. Draft. January. Available from EPA's Environmental Monitoring and Support Laboratory, Cincinnati, OH.

U.S. EPA. 1987a. U.S. Environmental Protection Agency. Qualitative and quantitative risk assessment for Baygon. Memo from Bernice Fisher to Dennis Edwards. Office of Pesticide Programs. April 3.

U.S. EPA. 1987b. U.S. Environmental Protection Agency. Supplemental discussion of Baygon classification. Memo from Arthur Chiu to William H. Farland. Cancer Assessment Group. April 6.

Vandekar, M., R. Plestina and K. Wilhelm. 1971. Toxicity of carbamates for mammals. Bull. WHO. 44:241–249.

Vettorazzi, G. and G.W. Van den Hurk. 1985. Pesticides reference index. Joint Meeting on Pesticide Residues (JMPR) 1961–1984.

Whittaker, K.F., J.C. Nye, R.F. Wukash, R.J. Squires, A.C. York and H.A. Razimier. 1982. Collection and treatment of wastewater generated by pesticide application. EPA 600/2-82-028. Cincinnati, OH: U.S. Environmental Protection Agency.

Worthing, C.R. 1983. The pesticide manual: a world compendium, 7th ed. London: BCPC Publishers.

---

* Confidential Business Information submitted to the Office of Pesticide Programs.

Bentazon

I. General Information and Properties

A. CAS No. 25057–89–0

B. Structural Formula

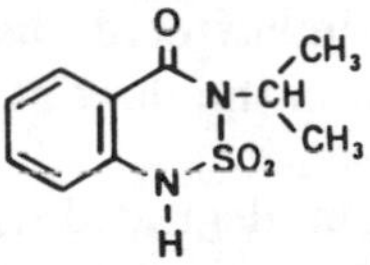

3-(1-Methylethyl)-1H-2,1,3-benzothiadiazin-4(3H)-one-2,2-dioxide

C. Synonyms

- Basagran; Bendioxide; Bentazone (Worthing, 1983).

D. Uses

- Bentazon is a selective postemergent herbicide used to control broadleaf weeds in soybeans, rice, corn, peanuts, dry beans, dry peas, snap beans for seed, green lima beans and mint (Meister, 1986).

E. Properties (Worthing, 1983)

Chemical Formula	$C_{10}H_{12}N_2O_3S$
Molecular Weight	240.3
Physical State	Colorless crystalline powder
Boiling Point	--
Melting Point	137 to 139°C
Density	--
Vapor Pressure (20°C)	$< 0.1 \times 10^{-7}$ mm Hg
Water Solubility	500 mg/L
Specific Gravity	--
Log Octanol/Water Partition Coefficient	--
Taste Threshold	--
Odor Threshold	--
Conversion Factor	--

F. Occurrence

- Bentazon was not found in sampling performed at two water supply stations in the STORET database (STORET, 1988). No information on the occurrence of bentazon was found in the available literature.

G. Environmental Fate

- Bentazon, at 1 ppm, was stable to hydrolysis for up to 122 days in unbuffered water (initial pH 5, 7, and 9) at 22°C (Drescher, 1972c). The bentazon degradate, 2-amino-N-isopropyl benzamide (AIBA) at 1 ppm, was stable to hydrolysis in unbuffered, distilled water at pH 5, 7, and 9, during 28 days' incubation in the dark at 22°C (Drescher, 1973b).

- ^{14}C-Bentazon at 2 to 10 ppm, degraded with a half-life of less than 2 to 14 weeks in a sandy clay loam, loam, and three loamy sand soils (Drescher and Otto, 1973a; Drescher and Otto, 1973b). The soils were incubated at 14 to 72% of field capacity and 23°C. The bentazon degradation rate was not affected by soil moisture content but was decreased by lowering the temperatures to 8 to 10°C. At pH 6.4, the degradation rate in a loamy sand soil was 2.5 times longer than at pH 4.6 and 5.5. The bentazon degradate, AIBA, was identified at less than 0.1 ppm. AIBA degraded in loamy sand soil with a half-life of 1 to 10 days (43% of field capacity). ^{14}C-Bentazon at 1.7 ppm did not degrade appreciably in a loamy sand soil during 8 weeks of incubation; AIBA was detected at a maximum concentration of 0.008 ppm.

- Bentazon did not adsorb to Drummer silty clay loam, adjusted to pH 5 and 7, and 11 other soils tested at pH 5 (Abernathy and Wax, 1973). In the same study, using soil TLC, ^{14}C-bentazon was very mobile in 12 soils, ranging in texture from sand to silty clay loam, with an R_f value of 1.0.

- Bentazon was very mobile in a variety of soils, ranging in texture from loamy sand to silty clay loam and muck, based on soil column tests (Drescher and Otto, 1972; Abernathy and Wax, 1973; Drescher, 1973a; Drescher, 1972a). Approximately 73 to 103% of the bentazon applied to the columns was recovered in the leachate.

- AIBA (100 μg applied to loamy sand soil) was very mobile (Drescher, 1972b). After leaching a 12-inch soil column with 500 mL (10 inches) of distilled water, 86.3% of the applied material was found in the leachate.

- Bentazon has the potential to contaminate surface waters as a result of its mobility in runoff water and application to rice fields (Devine, 1972; Wuerzer, 1972).

- In the field, bentazon at 0.75 to 10 lb a.i./acre dissipated with a half-life of less than or equal to 1 month in the upper 6 inches of soil, ranging in texture from sand to clay (Daniels et al., 1976; Devine and Hanes, 1973; Stoner and Hanes, 1974b; Stoner and Hanes, 1974a; BASF Wyandotte Corporation, 1974; Devine and Tietjens, 1973; Devine et al., 1973). In the majority of soils (6 of 9), bentazon had a half-life of less than 7 days. AIBA was detected at less than or equal to 0.09 ppm. Collectively, the available data indicated that geographic region (NC, TX, MS, AL, MN or ID) and crops treated (peanuts, soybeans, corn or fallow soil) had little or no effect on the dissipation rate of bentazon in soil.

II.   Pharmacokinetics

A. Absorption

- Male and female rats (200 to 250 g) given 0.8 mg ^{14}C-bentazon in 1 mL of 50% ethanol by stomach tube excreted 91% of the administered dose in the urine within 24 hours. This suggests that bentazon is almost completely absorbed when ingested (Chasseaud et al., 1972).

B. Distribution

- Whole-body autoradiography of rats indicated high levels of radioactivity in the stomach, liver, heart and kidneys after 1 hour of dosing with ^{14}C-bentazon. Radioactivity was not observed in the brain or spinal cord (Chasseaud et al., 1972).

C. Metabolism

- Bentazon is poorly metabolized. Two unidentified metabolites were detected (Chasseaud et al., 1972).

D. Excretion

- Rats given radiolabeled bentazon excreted 91% of the administered dose in the urine as parent compound. Feces contained 0.9% of the administered dose (Chasseaud et al., 1972).

III.   Health Effects

A. Humans

- No information on the health effects of bentazon in humans was found in the available literature.

B. Animals

1. *Short-term Exposure*

- The oral LD_{50} of bentazon in the rat was reported to be 2,063 mg/kg (Meister, 1986).

- LD_{50} values for bentazon in the rat, dog, cat and rabbit are reported to be 1,100, 900, 500 and 750 mg/kg, respectively (RTECS, 1985).

2. *Dermal/Ocular Effects*

- Previously available information was invalidated. Therefore, it was not possible to utilize that information for the development of the Health Advisory.

3. *Long-term Exposure*

- Leuschner et al. (1970) reported the effects of bentazon in beagle dogs. Beagle dogs (three dogs/sex/dose) were given 0 (control), 100, 300, 1,000 and 3,000 ppm (0, 2.5, 7.5, 25 and 75 mg/kg/day; Lehman, 1959) of bentazon for 13 weeks. At a dose level of 3,000 ppm, overt signs of toxicity, including weight loss and ill health, were observed; 1/3 males and 2/3 females died. At 3,000 ppm, all males showed signs of prostatitis. Similar signs were observed in one male each at the 300- and 1,000 ppm levels. This study suggests a NOAEL of 100 ppm (2.5 mg/kg/day).

4. *Reproductive Effects*

- Previously available information was invalidated. Therefore, it was not possible to utilize that information for the development of the Health Advisory.

5. *Developmental Effects*

- Previously available information was invalidated. Therefore, it was not possible to utilize that information for the development of the Health Advisory.

6. *Mutagenicity*

- Previously available information was invalidated. Therefore, it was not possible to utilize that information for the development of the Health Advisory.

7. *Carcinogenicity*

- Previously available information was invalidated. Therefore, it was not possible to utilize that information for the development of the Health Advisory.

IV. Quantification of Toxicological Effects

A. One-day Health Advisory

No data were found in the available literature that were suitable for determination of One-day HA values. It is therefore recommended that the Longer-term HA value for a 10-kg child (0.3 mg/L) be used at this time as a conservative estimate of the One-day HA.

B. Ten-day Health Advisory

No data were found in the available literature that were suitable for determination of Ten-day HA values. It is therefore recommended that the Longer-term HA value for a 10-kg child (0.3 mg/L) be used at this time as a conservative estimate of the Ten-day HA.

C. Longer-term Health Advisory

A 13-week study in beagle dogs has been selected for the calculation of a Longer-term HA (Leuschner et al., 1970). Beagle dogs (three dogs/sex/dose) were given 0 (control), 100, 300, 1,000 and 3,000 ppm (0, 2.5, 7.5, 25 and 75 mg/kg/day; Lehman, 1959) of bentazon for 13 weeks. At a dose level of 3,000 ppm, overt signs of toxicity, including weight loss and ill health, were observed; 1/3 males and 2/3 females died. At 3,000 ppm, all males showed signs of prostatitis. Similar signs were observed in one male each at the 300- and 1,000-ppm levels. This study suggests a NOAEL of 100 ppm (2.5 mg/kg/day).

Utilizing this NOAEL, a Longer-term HA for a 10-kg child is calculated as follows:

$$\text{Longer-term HA} = \frac{(2.5 \text{ mg/kg/day}) (10 \text{ kg})}{(100) (1 \text{ L/day})} - 0.25 \text{ mg/L} (300 \text{ } \mu g/L)$$

The Longer-term HA for the 70-kg adult is calculated as follows:

$$\text{Longer-term HA} = \frac{(2.5 \text{ mg/kg/day}) (70 \text{ kg})}{(100) (2 \text{ L/day})} = 0.875 \text{ mg/L } (900 \text{ } \mu g/L)$$

D. Lifetime Health Advisory

Lifetime studies were not available to calculate a Lifetime HA. However, with the addition of another safety factor of 10 for studies of less-than-lifetime duration, the Lifetime HA may be calculated from the 13-week feeding study in dogs (Leuschner et al., 1970).

Using the NOAEL of 2.5 mg/kg/day, the Lifetime HA for bentazon is calculated as follows:

Step 1: Determination of the Reference Dose (RfD)

$$\text{RfD} = \frac{(2.5 \text{ mg/kg/day})}{(1,000)} = 0.0025 \text{ mg/kg/day}$$

Step 2: Determination of the Drinking Water Equivalent Level (DWEL)

$$\text{DWEL} = \frac{(0.0025 \text{ mg/kg/day}) (70 \text{ kg})}{(2 \text{ L/day})} = 0.0875 \text{ mg/L } (90 \text{ } \mu g/L)$$

Step 3: Determination of the Lifetime Health Advisory

$$\text{Lifetime HA} = (0.0875 \text{ mg/L}) (20\%) = 0.0175 \text{ mg/L } (20 \text{ } \mu g/L)$$

E. Evaluation of Carcinogenic Potential

- No valid data are available to make a determination of the carcinogenic potential of bentazon.

- Applying the criteria described in EPA's guidelines for assessment of carcinogenic risk (U.S. EPA, 1986b), bentazon may be classified in Group D: not classifiable. This category is for agents with inadequate animal evidence of carcinogenicity.

V.   Other Criteria, Guidance and Standards

- In response to a bentazon-tolerance review petition, EPA's Office of Pesticide Programs has concluded that "a tolerance cannot be supported at this time."

VI. Analytical Methods

* Analysis of bentazon is by a gas chromatographic (GC) method applicable to the determination of bentazon in water samples (U.S. EPA, 1985). In this method, an aliquot of sample is acidified and extracted with ethyl acetate. The extract is dried, concentrated to 1 to 2 mL, and methylated with diazomethane. The methylated extracts are analyzed by gas chromatography with flame photometric detection. The method detection limit for bentazon has not been determined.

VII. Treatment Technologies

* There is no information available regarding treatment technologies used to remove bentazon from contaminated water.

VIII. References

Abernathy, J.R. and L.M. Wax. 1973. Bentazon mobility and adsorption in twelve Illinois soils. Weed Science. 21(3):224–226.

BASF Wyandotte Corporation. 1974. Analytical residue reports (soil and water): bentazon. Unpublished study.

Chasscaud, L.F., D.R. Hawkins, B.D. Cameron, B.J. Fry and V.H. Saggers. 1972. The metabolic fate of bentazon. Xenobiotica. 2(3):269–276.

Daniels, J., J. Gricher and T. Boswell. 1976. Determination of bentazon (BAS 351-H) residues in sand soil samples from Yoakum, Texas. Report No. IRDC-3; BWC Project No. I-2-G-73. Unpublished study prepared by International Research and Development Corporation. Submitted by BASF Wyandotte Corporation, Wyandotte, MI.

Devine, J.M. 1972. Determination of BAS 351-H (3-isopropyl-lH-2,1,3-benzothiadiazin-4[3H]-one-2,2-dioxide) residues in soil and runoff water. Report No. 133.

Devine, J.M. and R.E. Hanes. 1973. Determination of residues of BAS 351-H (3-isopropyl-lH-2,1,3-benzothiadiazin-4[3H]-one-2,2-dioxide) and its benzamide metabolite, AIBA (2-amino-N-isopropyl benzamide), in Sharkey silty clay soil from Greenville, Mississippi. Field Experiment No. 72–99. Unpublished study prepared by State University of New York–Oswego, Lake Ontario Environmental Laboratory and United States Testing Company, Inc. Submitted by BASF Wyandotte Corporation, Parsippany, NJ.

Devine, J.M. and F. Tietjens. 1973. Determination of BAS 351-H (3-isopropyl-1H-2,1,3-benzothiadiazin-4[3H]-one-2,2-dioxide) residues in Commerce silt loam soil from Greenville, Mississippi. Field Experiment No.

72–76. Unpublished study prepared by State University of New York–Oswego, Lake Ontario Environmental Laboratory and United States Testing Company, Inc. Submitted by BASF Wyandotte Corporation, Parsippany, NJ.

Devine, J.M., C. Carter, L.W. Hendrick et al. 1973. Determination of residues of BAS 351-H (3-isopropyl-1H-2,1,3-benzothiadiazin-4[3H]-one-2,2-dioxide) and its benzamide metabolite, AIBA (2-amino-N-isopropyl benzamide), in Webster Glencoe silty clay loam soil from Prior Lake, Minnesota. Field Experiment No. III-B-6–72. Unpublished study prepared by State University of New York–Oswego, Lake Ontario Environmental Laboratory and others. Submitted by BASF Wyandotte Corporation, Parsippany, NJ.

Drescher, N. 1972a. A comparison between the leaching of bentazon and 2,4-D through a soil in a model experiment. Laboratory Report No. 679.

Drescher, N. 1972b. Leaching of 2-amino-N-isopropyl benzamide (AIBA) from the soil. Laboratory Report No. 682.

Drescher, N. 1972c. The effect of pH on the rate of hydrolysis of bentazon (BAS 351-H) in water. Laboratory Report No. 1107. Translation. Unpublished study prepared by Badische Anilin- and Soda-Fabrik, AG. Submitted by BASF Wyandotte Corporation, Parsippany, NJ.

Drescher, N. 1973a. Leaching of bentazon in a muck soil. Laboratory Report No. 1138.

Drescher, N. 1973b. The influence of pH on the hydrolysis of the bentazon metabolite AIBA (2-amino-N-isopropyl benzamide) in water. Laboratory Report No. 1136.

Drescher, N. and S. Otto. 1972. Penetration and leaching of bentazon in soil. Laboratory Report No. 1099. Translation. Unpublished study prepared by BASF, AG. Submitted by BASF Wyandotte Corporation, Parsippany, NJ.

Drescher, N. and S. Otto. 1973a. Degradation of bentazon (BAS 351-H) in soil. Report No. 1140.

Drescher, N. and S. Otto. 1973b. Degradation of bentazon (BAS 351-H) in soil. Report No. 1149.

Lehman, A.J. 1959. Appraisal of the safety of chemicals in foods, drugs and cosmetics. Assoc. Food Drug Off. U.S.

Leuschner, F., A. Leuschner, W. Schwerdtfeger and H. Otto.* 1970. 13-Week toxicity study of 3-isopropyl-lH-2,1,3-benzothiadiazin-4(3H)-one-2,2-dioxide to beagles when administered with the food. Unpublished report prepared by Laboratory of Pharmacology and Toxicology, W. Germany. September 28. Acc. Nos. 112129, 092225.

Meister, R., ed. 1986. Farm chemicals handbook. Willoughby, OH: Meister Publishing Co.

RTECS. 1985. Registry of Toxic Effects of Chemical Substances. National

Institute for Occupational Safety and Health. National Library of Medicine Online File.

Stoner, J.H. and R.E. Hanes. 1974a. Determination of residues of bentazon and AIBA (2-amino-N-isopropyl benzamide) in Commerce silt loam soil from Greenville, MS. Field Experiment No. 73–41. Unpublished study prepared in cooperation with Stoner Laboratories, Inc. and United States Testing Company, Inc. Submitted by BASF Wyandotte Corporation, Parsippany, NJ.

Stoner, J.H. and R.E. Hanes. 1974b. Determination of residues of bentazon (BAS 351-H) and AIBA in Commerce silt loam soil from Greenville, MS. Field Experiment No. 73–43. Unpublished study prepared in cooperation with Stoner Laboratories, Inc. and United States Testing Company, Inc. Submitted by BASF Wyandotte Corporation, Parsippany, NJ.

STORET. 1988. STORET Water Quality File. Office of Water. U.S. Environmental Protection Agency (data file search conducted in May, 1988).

U.S. EPA. 1985. U.S. Environmental Protection Agency. U.S. EPA Method 107 – Revision A, bentazon. Fed. Reg. 50:40701. October 4.

U.S. EPA. 1986a. U.S. Environmental Protection Agency. RfD Work Group. Worksheet dated April 7.

U.S. EPA. 1986b. U.S. Environmental Protection Agency. Guidelines for carcinogen risk assessment. Fed. Reg. 51(185):33992–34003. September 24.

Worthing, C.R., ed. 1983. The pesticide manual. Great Britain: The Lavenham Press, Ltd. p. 39.

Wuerzer, B. 1972. Bentazon model box runoff study. Runoff Report 73–6. Unpublished study prepared in cooperation with United States Testing Company. Submitted by BASF Wyandotte Corporation, Parsippany, NJ.

---

* Confidential Business Information submitted to the Office of Pesticide Programs.

Bromacil

I. General Information and Properties

A. CAS No. 314-40-9

B. Structural Formula

$$\underset{\text{Br}}{\overset{\text{H}_3\text{C}}{}}\quad \overset{\text{H}}{\underset{\text{O}}{\text{N}}} \quad \text{O}$$

5-Bromo-6-methyl-3-(1-methylpropyl)-2,4(1H,3H)-pyrimidinedione

C. Synonyms

- Borea; Bromax; Hyvar; Uragan; Urox B (Meister, 1988).

D. Uses

- Bromacil is used as a herbicide for general weed or brush control in noncrop areas; particularly useful against perennial grasses (Meister, 1988).

E. Properties (Windholz et al., 1983)

Chemical Formula	$C_9H_{13}O_2N_2Br$
Molecular Weight	261.11
Physical State (25°C)	White crystalline solid
Boiling Point	--
Melting Point	158–160°C
Density	--
Vapor Pressure (100°C)	8×10^{-4} mm Hg
Specific Gravity	--
Water Solubility (20°C)	815 mg/L
Log Octanol/Water Partition Coefficient	--
Taste Threshold	--
Odor Threshold	--
Conversion Factor	--

F. Occurrence

- Bromacil has been found in Florida ground water; a typical positive was 300 ppb (Cohen et al., 1986).

- Bromacil has not been found in any of 3 surface water samples collected at 2 locations or in any of 841 ground-water samples collected at 834 locations (STORET, 1988). This information is provided to give a general impression of the occurrence of this chemical in ground and surface waters as reported in the STORET database. The individual data points retrieved were used as they came from STORET and have not been confirmed as to their validity. STORET data are often not valid when individual numbers are used out of the context of the entire sampling regime, as they are here. Therefore, this information can only be used to form an impression of the intensity and location of sampling for a particular chemical.

G. Environmental Fate

- Bromacil in aqueous solution was stable when exposed to simulated sunlight for 6 days (Moilanen and Crosby, 1974). Only one minor (<4%) photolysis product (5-bromo-6-methyluracil) was identified. An aqueous solution of bromacil at 1 ppm lost all herbicidal activity after exposure to UV light for 10 minutes, but at 10 ppm and 15 minutes' irradiation herbicidal activity was still present (Kearney et al., 1964). However, bromacil in an aqueous solution (pH 9.4) containing the photosensitizer methylene blue was rapidly degraded under direct sunlight with a half-life of < 1 hour (Acher and Dunkelblum, 1979).

- More than 26 soil fungi representative of several taxonomic groups, including Fungi Imperfecti, Ascomycetes and Zygomycetes, were capable of metabolizing bromacil as their sole carbon source (Wolf et al., 1975; Torgeson, 1969; Torgeson and Mee, 1967; Boyce Thompson Institute for Plant Research, 1971).

- Data from soil metabolism studies indicate that bromacil at 8 ppm had a half-life of about 6 months in aerobic loam soil incubated at 31°C (Zimdahl et al., 1970). However, 10% of applied bromacil at approximately 3 ppm was slowly degraded to CO_2 in an aerobic sandy loam soil after 330 days at 22°C (Wolf, 1974; Wolf and Martin, 1974). In anaerobic sandy loam soil, bromacil at approximately 3 ppm had a calculated half-life of approximately 144 days. No CO_2 evolved from the sterilized soil treated with bromacil within 145 days, indicating that degradation was microbial.

- Bromacil is mobile in soil. Phytotoxic residues of bromacil leached 19 cm in clay and silty clay loam soils eluted with the equivalent of 4.3 acre-inches of water (Signori et al., 1978). In mucky peat, loam and loamy sand soils eluted with the equivalent of 13 to 15 cm of water, bromacil leached to 10-, 25-, and >30-cm depths, respectively (Day, 1976). Utilizing soil thin-layer chromatographic techniques ^{14}C-bromacil was evaluated to be mobile (R_f 0.7) in a silty clay loam soil (Helling, 1971). Bromacil is not adsorbed by montmorillonite, illite or humic acid to any great extent [Freundlich K (adsorption coefficient) ≤ 10 at 25°C]; however, at 0°C bromacil was adsorbed (Freundlich K 126) to humic acid (Haque and Coshow, 1971; Volk, 1972). Adsorption appeared to increase with decreasing temperatures.

- Data from field dissipation studies showed that bromacil phytotoxic residues persisted in soils ranging in texture from sand to clay for > 2 years following a single application of bromacil at ≥ 2.6 lb a.i./A (active ingredient/acre) (Bunker et al., 1971; Stecko, 1971).

II.   Pharmacokinetics

A. Absorption

- Workers who were exposed to bromacil during production, formulation and packaging excreted unchanged bromacil and the 5-bromo-3-*sec*-butyl-6-hydroxymethyluracil metabolite in the urine (E. I. duPont de Nemours and Co., 1966b). Unchanged bromacil and the metabolite were also detected in the urine of rats fed bromacil in the diet (E. I. duPont de Nemours and Co., 1966a). Although these data indicate that bromacil is absorbed, sufficient information was not available to quantify the extent of absorption.

B. Distribution

- No information was found in the available literature on the distribution of bromacil.

C. Metabolism

- Workers at a bromacil production plant excreted unchanged bromacil and the 5-bromo-3-*sec*-butyl-6-hydroxymethyluracil metabolite, present as the glucuronide and/or sulfonate conjugate, in urine (E. I. duPont de Nemours and Co., 1966b).

- Gardiner et al. (1969) fed rats (age and strain not specified) food containing 1,250 ppm bromacil for 4 weeks. Assuming 1 ppm equals 0.05 mg/kg/day in the older rat (Lehman, 1959), this dietary level corresponds to about 62.5 mg/kg/day. Analysis of the urine of these rats revealed the presence of unchanged bromacil and the 5-bromo-3-*sec*-butyl-6-hydroxymethyluracil metabolite (primarily as the glucuronide and/or sulfonate conjugate). Five other minor metabolites were also detected: 5-bromo-3-(2-hydroxy-1-methylpropyl)-6-methyluracil; 5-bromo-3-(2-hydroxy-1-methylpropyl)-6-hydroxmethyluracil; 3-*sec*-butyl-6-hydroxymethyluracil; 5-bromo-3-(3-hydroxyl-1-methylpropyl)-6-methyluracil; and 3-*sec*-butyl-6-methyluracil. An unidentified bromine-containing compound with a molecular weight of 339 was also detected.

D. Excretion

- In humans exposed to bromacil during its formulation and packaging, urinary excretion products included 0.1 ppm parent compound and 6.3 ppm 5-bromo-3-*sec*-butyl-6-hydroxymethyluracil, present mostly as a conjugate (E. I. duPont de Nemours and Co., 1966b).

- Rats were fed bromacil (1,250 ppm in the diet) for 4 weeks; urine was collected daily during weeks 3 and 4 of the study. Analysis of the urine revealed the presence of 20 ppm unchanged bromacil and 146 ppm of the 5-bromo-3-*sec*-butyl-6-hydroxymethyluracil metabolite (conjugated and unconjugated form) (E. I. duPont de Nemours and Co., 1966a; Gardiner et al., 1969).

III. Health Effects

A. Humans

- No information was located in the available literature on the health effects of bromacil in humans.

B. Animals

- Most of the animal data available are from unpublished studies identified prior to the published report by Sherman and Kaplan (1975). These authors stated that an 80% wettable bromacil powder was used in all tests discussed in their report except for eye irritation studies in which a 50% wettable bromacil powder was used. All dosages and feeding levels, unless otherwise stated, were based on the active ingredient, bromacil.

1. *Short-term Exposure*

- The oral LD_{50} value for male ChR-CD rats was calculated to be 5,200 mg/kg (Sherman and Kaplan, 1975). Clinical signs of toxicity included rapid respiration, prostration and initial weight loss.

- In one male mongrel dog, a single oral dose in capsules of 5,000 mg/kg caused nausea, vomiting, fatigue, incoordination and diarrhea (Sherman and Kaplan, 1975). It was not possible to estimate a lethal oral dose for bromacil in dogs because vomiting occurred almost immediately in another dog at doses of 100 or 250 mg/kg given 5 days apart.

- Sherman and Kaplan (1975) repeated a study in which bromacil was administered to groups of six male ChR-CD rats by gastric intubation at dose levels of 650, 1,035 or 1,500 mg/kg/day, 5 days/week for 2 weeks (10 doses). Four of six animals died at the high dose. Five of six survived exposure to 1,035 mg/kg/day, but showed gastrointestinal and nervous system disturbances, and focal liver cell hypertrophy and hyperplasia. All animals survived the low dose with similar, but less severe, pathological changes. The 650-mg/kg/day dose is identified as the Lowest-Observed-Adverse-Effect Level (LOAEL) in this study.

- Palmer (1964) reported that sheep that received bromacil at oral doses of 250 mg/kg for 5 days developed weakness in the legs and incoordination. Recovery from these symptoms usually took several weeks. Administration of 100 mg/kg/day for 11 days induced an 11% weight loss but no observable clinical symptoms.

2. *Dermal/Ocular Effects*

- Bromacil (applied as a 50% aqueous solution of the 80% wettable powder) was only mildly irritating to the intact and abraded skin of young guinea pigs exposed for periods of up to 3 weeks. It was more irritating to the skin of older animals. Bromacil did not produce skin sensitization (E. I. duPont de Nemours and Co., 1962).

- Sherman and Kaplan (1975) reported that when bromacil was applied dermally to rabbits, the lethal dose was greater than 5,000 mg/kg, the maximum feasible dose. No clinical signs of toxicity and no gross pathological changes were observed.

- Bromacil, as a 50% aqueous suspension, was mildly irritating to the skin of young guinea pigs, but only slightly more irritating to the skin of older animals. It was not a skin sensitizer (Sherman and Kaplan, 1975).

- Sherman and Kaplan (1975) reported that bromacil (0.1 mL of a 10% suspension in mineral oil) resulted in only mild temporary conjunctivitis in both the washed and unwashed eyes of rabbits. No corneal injury was observed when a dose of 10-mg dry powder was applied directly to the eye.

3. *Long-term Exposure*

- Zapp (1965) discussed a study, also reported by Sherman and Kaplan (1975), in which 10 male and 10 female ChR-CD rats were fed dietary levels of 0, 50, 500 or 2,500 ppm bromacil for 90 days. This corresponds to doses of about 0, 2.5, 25 or 125 mg/kg/day, assuming 1 ppm equals 0.05 mg/kg/day in an older rat (Lehman, 1959). Because no signs of toxicity were observed at any dose, the high dose was increased to 5,000 ppm (about 250 mg/kg/day) after 6 weeks; to 6,000 ppm (about 300 mg/kg/day) after 10 weeks; and to 7,500 ppm (about 375 mg/kg/day) after 11 weeks. This dosing pattern resulted in reduced food intake and mild histological changes in thyroid and liver. No compound-related effects on weight gain, hematology, urinalysis or histology were detected at the two lowest doses; 25 mg/kg/day was identified as the No-Observed-Adverse-Effect Level (NOAEL) in this study.

- Sherman et al. (1966, also reported by Sherman and Kaplan, 1975) fed groups of 36 male and 36 female ChR-CD rats food containing 0, 50, 250 or 1,250 ppm bromacil for 2 years. This corresponds to doses of about 0, 2.5, 12.5 or 62.5 mg/kg/day, assuming 1 ppm equals 0.05 mg/kg/day in older rats (Lehman, 1959). Females at the highest dose showed decreased weight gain (p < 0.001). No other toxic effects were observed in a variety of parameters measured, including mortality, hematology, urinalysis, serum biochemistry, gross pathology, organ weight or histopathology, except for a slight thyroid hyperplasia at the high dose. This study identified a NOAEL of 12.5 mg/kg/day.

- Beagle dogs (3/sex/dose level) were fed a nutritionally complete diet containing 0, 50, 250 or 1,250 ppm bromacil for 2 years (Sherman et al., 1966; also reported by Sherman and Kaplan, 1975). This corresponds to doses of about 0, 1.25, 6.25 or 31.2 mg/kg/day, assuming 1 ppm equals 0.025 mg/kg/day in the dog (Lehman, 1959). No nutritional, clinical, hematological, urinary, blood chemistry or histopathologic evidence of toxicity was detected in any group. This study identified a NOAEL of 31.2 mg/kg/day.

- Kaplan et al. (1980) administered bromacil (approximately 95% pure) to
  CD-1 mice (80/sex/dose) for 78 weeks at dietary levels of 0, 250, 1,250
  or 5,000 ppm. Based on information presented by the authors, these
  dietary levels correspond to doses of 0, 39.6, 195 or 871 mg/kg/day for
  males and 0, 66.5, 329 or 1,310 mg/kg/day for females. During the first
  year of the study, a compound-related decrease in body weight gain was
  observed in male mice receiving 5,000 ppm and in female mice receiving
  1,250 ppm. The treatment and control groups exhibited no significant
  (p < 0.05) differences in food consumption. Mortality in the 5,000-ppm
  females was significantly (p < 0.05) greater than in the controls. Liver
  changes noted in treated mice consisted of increased mean and relative
  weights in the 1,250-ppm females and the 5,000-ppm males; an increased
  incidence of diffuse hepatocellular hypertrophy in the 1,250- and 5,000-
  ppm males and in the 5,000-ppm females; an increased incidence of
  centrilobular vacuolation in 250-ppm males; an increased incidence of
  scattered hepatocellular necrosis in 5,000-ppm males; and the presence
  of extravasated erythrocytes in the hypertrophied hepatocytes of the
  1,250- and 5,000-ppm males. The authors felt that centrilobular vacuolation and hypertrophy were probably related to enzyme induction. The
  toxicological significance of extravasated erythrocytes in the hypertrophied hepatocytes was unclear. Compound-related changes in the testes
  of mice consisted of an increased incidence of spermatocyte necrosis,
  sperm calculi and mild interstitial-cell hypertrophy/hyperplasia in the
  1,250- and 5,000-ppm males and a dose-related increase in the incidence
  of testicular tubule atrophy in *all* male treatment groups. Based on
  changes in testes, a LOAEL of 250 ppm (39.6 mg/kg/day) is identified
  for male mice. A NOAEL of 250 ppm (66.5 mg/kg/day) was identified
  for female mice.

4. Reproductive Effects

- Sherman et al. (1966; also reported by Sherman and Kaplan, 1975)
  reported the effects of bromacil on reproduction in a three-generation
  study in rats. Twelve male and twelve female weanling ChR-CD rats
  were fed bromacil in the diet at 0 or 250 ppm. This corresponds to doses
  of about 0 or 12.5 mg/kg/day, assuming 1 ppm in the diet equals 0.05
  mg/kg/day for older rats (Lehman, 1959). Animals were bred after 12
  weeks, and the F_{1b} and the F_{2b} generations were maintained on the same
  diets as their parents. No evidence of adverse effects on reproduction or
  lactation performance was observed. Examination of the F_{2b} generation
  revealed no evidence of gross or histopathological effects. This study
  identified a minimum NOAEL of 12.5 mg/kg/day.

5. *Developmental Effects*

- Paynter (1966; also reported by Sherman and Kaplan, 1975) administered bromacil to New Zealand White rabbits (8 or 9 per dosage) at dietary levels of 0, 50 or 250 ppm on days 8 through 16 of gestation. Assuming 1 ppm equals 0.03 mg/kg/day in the rabbit (Lehman, 1959), these dietary levels correspond to about 0, 1.5 or 7.5 mg/kg/day. No significant differences between the conception rates of the control and test groups were observed. Control and test group litters were comparable in terms of litter size, mean pup length, mean litter weight, number of stillbirths and number of resorption sites. No gross malformations were observed in any animals. Skeletal clearing revealed no abnormalities in bone structure in any animals. Based on reproductive and teratogenic end points, a NOAEL of 250 ppm (7.5 mg/kg/day) was identified.

- Pregnant rats (strain not specified) were exposed to aerosols of bromacil (165 mg/m³) on days 7 to 14 of gestation. No prenatal changes or teratogenic effects were observed (no further details were provided) (Dilley et al., 1977).

6. *Mutagenicity*

- In a sex-linked recessive lethal test (Valencia, 1981), *Drosophila melanogaster* (Canton-S wild-type stock) were exposed to bromacil in food at levels of 2, 3, 5 or 2,000 ppm. Bromacil was found to be weakly mutagenic at the 2,000-ppm dose level.

- Riccio et al. (1981) reported that bromacil (tested concentrations not specified) was not mutagenic with or without metabolic activation in assays conducted using *Saccharomyces cerevisiae* strains D3 and D7.

- Siebert and Lemperle (1974) reported that bromacil was not mutagenic when tested at a concentration of 1,000 ppm using *S. cerevisiae* strain D4.

- Simmon et al. (1977) reported that bromacil was not mutagenic in an *in vivo* mouse dominant-lethal assay and the following *in vitro* assays: unscheduled DNA synthesis in human fibroblasts (WI-38 cells); reverse mutation in *Salmonella typhimurium* strains TA1535, TA1537, TA1538 and TA100, and in *Escherichia coli* WP2; mitotic recombination in *S. cerevisiae*; and preferential toxicity assays in *E. coli* (strains W3110 and p3478) and *Bacillus subtilis* (strains H17 and M45).

- In a modified Ames assay (Rashid, 1974), bromacil was not mutagenic in *S. typhimurium* strains TA1535 and TA1538 when tested at concentrations up to 325 μg/plate.

- In an assay designed to test for thymine replacement in mouse DNA (McGahen and Hoffman, 1963), Swiss-Webster white mice received bromacil by oral intubation at 100 mg/kg twice daily for 2 days, followed by 50 mg/kg twice daily for 8 days. Under the conditions of the assay, bromacil was not recognized as a thymine analog by the mouse.

- Bromacil did not show any signs of mutagenicity in a variety of microbial test systems (Jorgenson et al., 1976; Woodruff et al., 1984).

- In the Ames test, bromacil (5% concentration) induced revertants in three of six *Salmonella* strains tested (Njage and Gopalan, 1980).

- Bromacil did not induce sex-linked recessive lethals in *D. melanogaster* (Gopalan and Njage, 1981).

7. *Carcinogenicity*

- Sherman et al. (1966) fed groups of 36 male and 36 female weanling ChR-CD rats bromacil in the diet for 2 years. Dietary levels were 0, 50, 250 or 1,250 ppm (about 0, 2.5, 12.5 or 62.5 mg/kg/day, based on Lehman, 1959). There was no effect on mortality, and the only treatment-related lesion detected by histological examination was a slight increase in the incidence of light-cell and follicular-cell hyperplasia in the thyroid at the high dose. One high-dose female was found to have follicular-cell adenoma.

- Kaplan et al. (1980) administered bromacil (approximately 95% pure) to CD-1 mice (80/sex/dose) for 78 weeks at dietary levels of 0, 250, 1,250 or 5,000 ppm. Based on information presented by the authors, these dietary levels correspond to compound intake levels of 0, 39.6, 195 or 871 mg/kg/day for males and 0, 66.5, 329 or 1,310 mg/kg/day for females. In males, the combined incidences of hepatocellular adenomas plus carcinomas/number of animals examined were 10/74, 11/71, 8/71 and 19/70 (p < 0.05) at 0, 250, 1,250 and 5,000 ppm, respectively. Hepatocellular carcinoma incidences were 5/74, 4/71, 4/71 and 9/70 (p > 0.05) at 0, 250, 1,250 and 5,000 ppm, respectively. These tumors were found predominantly in mice that survived to terminal sacrifice. No effect on liver tumor incidence was observed in females.

IV.   Quantification of Toxicological Effects

A. One-day Health Advisory

No studies were located which are suitable for derivation of a One-day HA for bromacil. The acute dosing study in dogs reported by Sherman and Kaplan (1975) is not considered because the bolus treatment with 100 mg/kg by capsule could inaccurately indicate the ability of this species to tolerate more divided doses as in diet or drinking water, as suggested by the 31.2 mg/kg/day NOAEL in the 2-year study in dogs by Sherman et al. (1966). The Ten-day HA, derived below, of 5 mg/L for a 10-kg child is proposed as a conservative One-day HA.

B. Ten-day Health Advisory

The 2-week oral study in rats by Sherman and Kaplan (1975) has been selected as the basis for the Ten-day HA for bromacil. Animals were dosed by gavage for 10 days over a period of 2 weeks. The lowest dose tested (650 mg/kg/day) produced mild pathological changes in the liver, and this value was identified as a LOAEL. Although a lower oral LOAEL of 100 mg/kg was evident in sheep (Palmer, 1964), this study was not selected for HA development because of uncertainty in using ruminants in estimating risk to humans from oral exposure.

Using a LOAEL of 650 mg/kg/day, the Ten-day HA for a 10-kg child is calculated as follows:

$$\text{Ten-day HA} = \frac{(650 \text{ mg/kg/day})(5/7)(10 \text{ kg})}{(1,000)\,(1 \text{ L/day})} = 4.6 \text{ mg/L } (5,000 \text{ } \mu\text{g/L})$$

where: 5/7 = correction for dosing 5 days per week.

C. Longer-term Health Advisory

The 90-day study by Zapp (1965) has been selected to serve as the basis for the Longer-term HA for bromacil. Rats were fed diets containing up to 500 ppm without any adverse effects. This study identified a NOAEL of 500 ppm (about 25 mg/kg/day).

Using a NOAEL of 25 mg/kg/day, the Longer-term HA for a 10-kg child is calculated as follows:

$$\text{Longer-term HA} = \frac{(25 \text{ mg/kg/day})\,(10 \text{ kg})}{(100)\,(1 \text{ L/day})} = 2.5 \text{ mg/L } (3,000 \text{ } \mu\text{g/L})$$

Using a NOAEL of 25 mg/kg/day, the Longer-term HA for a 70-kg adult is calculated as follows:

$$\text{Longer-term HA} = \frac{(25 \text{ mg/kg/day}) (70 \text{ kg})}{(100) (2 \text{ L/day})} = 8.8 \text{ mg/L } (9{,}000 \text{ } \mu\text{g/L})$$

D. Lifetime Health Advisory

The chronic feeding study in rats by Sherman et al. (1966) has been selected to serve as the basis for the Lifetime HA. This study identified a dietary LOAEL of 1,250 ppm and a NOAEL of 250 ppm, based on weight gain and mild thyroid hyperplasia. This NOAEL corresponds to 12.5 mg/kg/day. The same NOAEL is evident in a three-generation reproduction study in rats by Sherman et al. (1966). The long-term feeding studies in dogs by Sherman et al. (1966) and mice by Kaplan et al. (1980) were not selected, since the demonstrated NOAEL was the lowest in the rat study.

Using a NOAEL of 12.5 mg/kg/day, the Lifetime HA is derived as follows:

Step 1: Determination of the Reference Dose (RfD)

$$\text{RfD} = \frac{(12.5 \text{ mg/kg/day})}{(100)} = 0.125 \text{ mg/kg/day}$$
$$\text{(rounded to 0.13 mg/kg/day)}$$

Step 2: Determination of the Drinking Water Equivalent Level (DWEL)

$$\text{DWEL} = \frac{(0.13 \text{ mg/kg/day}) (70 \text{ kg})}{(2 \text{ L/day})} = 4.55 \text{ mg/L } (5{,}000 \text{ } \mu\text{g/L})$$

Step 3: Determination of the Lifetime Health Advisory

$$\text{Lifetime HA} = \frac{(4.6 \text{ mg/L}) (20\%)}{10} = 0.09 \text{ mg/L } (90 \text{ } \mu\text{g/L})$$

E. Evaluation of Carcinogenic Potential

- Bromacil has not been determined to be carcinogenic, although an increased incidence of hepatocellular adenomas plus carcinomas was observed in male but not female CD-1 mice fed bromacil in the diet at a dose level of 871 mg/kg/day for 78 weeks (Kaplan et al., 1980).

- The International Agency for Research on Cancer has not evaluated the carcinogenic potential of bromacil.

- Applying the criteria described in EPA's guidelines for assessment of carcinogenic risk (U.S. EPA, 1986), bromacil is classified in Group C: possible human carcinogen. This category is for substances with limited evidence of carcinogenicity in animals in the absence of human data.

- The U.S. EPA has not published excess lifetime cancer risks for this material.

V.   Other Criteria, Guidance and Standards

- The NAS (1977) has calculated an acceptable daily intake (ADI) of 0.0125 mg/kg/day, based on a chronic NOAEL of 12.5 mg/kg/day in rats and an uncertainty factor of 1,000. A Suggested-No-Adverse-Response Level (SNARL) of 0.088 mg/L was calculated based on an assumed water consumption of 2 L/day by a 70-kg adult, with 20% contribution from water.

- The U.S. EPA Office of Pesticide Programs (EPA/OPP) previously calculated an ADI of 62.5 μg/kg/day, based on a NOAEL of 6.25 mg/kg/day in a 2-year feeding study in dogs (Sherman et al., 1966) and an uncertainty factor of 100. This was updated to 130 μg/kg/day based on a 2-year rat feeding study using a NOAEL of 12.5 mg/kg/day and a 100-fold uncertainty factor.

- A tolerance of 0.1 ppm bromacil in or on citrus fruits and pineapples has been set by the EPA/OPP (CFR, 1985). A tolerance is a derived value based on residue levels, toxicity data, food consumption levels, hazard evaluation and scientific judgment, and it is the legal maximum concentration of a pesticide in or on a raw agricultural commodity or other human or animal food (Paynter et al., no date).

- The American Conference of Governmental Industrial Hygienists (ACGIH, 1984) has recommended a threshold limit value (TLV) of 1 ppm, and a short-term exposure limit (STEL) of 2 ppm.

VI.   Analytical Methods

- Analysis of bromacil is by a gas chromatographic (GC) method applicable to the determination of certain organonitrogen pesticides in water samples (U.S. EPA, 1988). This method requires a solvent extraction of approximately 1 liter of sample with methylene chloride using a separatory funnel. The methylene chloride extract is dried and exchanged to acetone during concentration to a volume of 10 mL or less. The compounds in the extract are separated by gas chromatography and measurement is made with a thermionic bead detector. This method has been validated in a single laboratory, and estimated detection limits have been determined for the analytes in the method, including bromacil. The estimated detection limit is 2.5 $\mu g/L$.

VII. Treatment Technologies

- No information was found in the available literature on treatment technologies used to remove bromacil from contaminated water.

VIII. References

ACGIH. 1984. American Conference of Governmental Industrial Hygienists. Documentation of the threshold limit values for substances in workroom air, 3rd ed. Cincinnati, OH: ACGIH. p. 11.

Acher, A.J. and E. Dunkelblum. 1979. Identification of sensitized photooxidation products of bromacil in water. J. Agric. Food Chem. 27(6):1184–1187.

Boyce Thompson Institute for Plant Research. 1971. Interaction of herbicides and soil microorganisms. Washington, D.C.: U.S. EPA, Office of Research and Monitoring.

Bunker, R.C., W.C. LeCroy, D. Katchur and T.C. Ellwanger, Jr. 1971. Preliminary evaluation of herbicides on native grassland in Florida. Frederick, MD: Department of the Army, Fort Detrick. Department of the Army Technical Memorandum No. 232. Available from NTIS, Springfield, VA.

CFR. 1985. Code of Federal Regulations. 40 CFR 180.210. p. 287. July 1.

Cohen, S.Z., C. Eiden and M. N. Lorber. 1986. Monitoring ground water for pesticides in the U.S.A. In: Evaluation of pesticides in ground water. American Chemical Society Symposium Series. In press.

Day, E.W.* 1976. Laboratory soil leaching studies with tebuthiuron. Unpublished studies received Feb. 18, 1977, under 1471–109. Submitted by Elanco Products Co., Div. of Eli Lilly and Co., Indianapolis, IN. CDL: 095854-I. MRID 00020782.

Dilley, J.V., N. Chernoff, D. Kay, N. Winslow and G.W. Newell. 1977. Inhalation teratology studies of five chemicals in rats. Toxicol. Appl. Pharmacol. 41:196.

E. I. duPont de Nemours and Co.* 1962. Toxicological information: 5-bromo-3-sec-butyl–6-methyl-uracil. Unpublished report. MRID 00013246.

E. I. duPont de Nemours and Co.* 1966a. Effect of enzymatic hydrolysis on the concentration of bromacil and the principal bromacil metabolite in rat urine. Unpublished report. MRID 00013274.

E. I. duPont de Nemours and Co.* 1966b. Analysis of urine from bromacil production workers. Unpublished report. MRID 00013273.

Gardiner, J.A., R.W. Reiser and H. Sherman. 1969. Identification of the metabolites of bromacil in rat urine. J. Agri. Food Chem. 17:967–973.

Gopalan, H.N.B. and G.D.E. Njage. 1981. Mutagenicity testing of pesticides. Genetics. 97:544.

Haque, R. and W.R. Coshow. 1971. Adsorption of isocil and bromacil from aqueous solution onto some mineral surfaces. Environ. Sci. Tech. 5:139–141.

Helling, C.S. 1971. Pesticide mobility in soils. I. Parameters of thin-layer chromatography. Proc. Soil Sci. Soc. Am. 35:732–737.

Jorgenson, T.A., C.J. Rushbrook and G.W. Newell. 1976. *In vivo* mutagenesis investigation of ten commercial pesticides. Toxicol. Appl. Pharmacol. 37:109.

Kaplan, A.M., H. Sherman, J.C. Summers, P.W. Schneider, Jr. and C.K. Wood.* 1980. Long-term feeding study in mice with 5-bromo-3-*sec*-butyl-6-methyluracil (INN-976; Bromacil). Haskell Laboratory Report No. 893–80. Final Report. Unpublished study. MRID 00072782.

Kearney, P.C., E.A. Woolson, J.R. Plimmer and A.R. Isensee. 1964. Decontamination of pesticides in soils. Residue Rev. 29:137–149.

Lehman, A.J. 1959. Appraisal of the safety of chemicals in foods, drugs and cosmetics. Assoc. Food Drug Off. U.S., Q. Bull.

McGahen, J.W. and C.E. Hoffman. 1963. Action of 5-bromo-3-*sec*-butyl-6-methyluracil as regards replacement of thymine on mouse DNA. Nature. 199: 810–811.

Meister, R., ed. 1988. Farm chemicals handbook. Willoughby, OH: Meister Publishing Company.

Moilanen, K.W., and D.G. Crosby. 1974. The photodecomposition of bromacil. Arch. Environ. Contam. Toxicol. 2(1):3–8.

NAS. 1977. National Academy of Sciences. Drinking water and health, Vol. 1. Washington, DC: National Academy Press.

Njage, G.D.E. and H.N.B. Gopalan. 1980. Mutagenicity testing of some selected food preservatives, herbicides and insecticides: II Ames Test. Bangladesh J. Bot. 9(2):141–146.

Palmer, J.S. 1964. Toxicity of methyluracil and substituted urea and phenol compounds to sheep. J. Am. Vet. Med. Assoc. 145:787–789.

Paynter, O.E.* 1966. Reproduction study—rabbits. Project No. 201–163. Unpublished study including letter dated May 27, 1966 from O.E. Paynter to Wesley Clayton, Jr. MRID 00013275.

Paynter, O.E., J.G. Cummings and M.H. Rogoff. No date. United States pesticide tolerance system. Unpublished. Washington, DC: U.S. EPA Office of Pesticide Programs.

Rashid, K.A.* 1974. Mutagenesis induced in two mutant strains of *Salmonella typhimurium* by pesticides and pesticide degradation products. Master's

Thesis. Pennsylvania State Univ., Dept. of Entomology. Unpublished study. MRID 00079923.

Riccio, E., G. Shepherd, A. Pomery, K. Mortelmans and M.D. Waters.* 1981. Comparative studies between the *S. cerevisiae* D3 and D7 assays of eleven pesticides. Environ. Mutagen. 3:327. Abstract P63.

Sherman, H. and A.M. Kaplan. 1975. Toxicity studies with 5-bromo-3-sec-butyl-6-methyluracil. Toxicol. Appl. Pharmacol. 34:189–196.

Sherman, H., J.R. Barnes and E.F. Stula.* 1966. Long-term feeding tests with 5-bromo-3-secondary butyl-6-methyluracil (INN-976; Hyvar(R)X; Bromacil). Report No. 21–66. Unpublished study. MRID 00076371.

Siebert, D. and E. Lemperle. 1974. Genetic effects of herbicides: induction of mitotic gene conversion in *Saccharomyces cerevisiae*. Mutat. Res. 22:111–120.

Signori, L.H., R. Deuber and R. Forster. 1978. Leaching of trifluralin, atrazine, and bromacil in three different soils. Noxious Plants. I(1):39–43.

Simmon, V.F., A.D. Mitchell and T.A. Jorgenson.* 1977. Evaluation of selected pesticides as chemical mutagens: *in vitro* and *in vivo* studies. Unpublished study. MRID 05009139.

Stecko, V. 1971. Comparison of the persistence and the vertical movement of the soil-applied herbicides simazine and bromacil. In: Proceedings of the 10th British Weed Control Conference, Vol. 1. Droitwich, England: British Weed Control Conference. pp. 303–306.

STORET. 1988. STORET Water Quality File. Office of Water. U.S. Environmental Protection Agency (data file search conducted in May, 1988).

Torgeson, D.C. 1969. Microbial degradation of pesticides in soil. In: Gunckel, J.E., ed. Current topics in plant science. New York: Academic Press. pp. 58–59.

Torgeson, D.C. and H. Mee. 1967. Microbial degradation of bromacil. In: Proceedings of the Northeastern Weed Control Conference, Vol. 21. Farmingdale, NY: Northeastern Weed Control Conference. p. 584.

U.S. EPA. 1985. U.S. Environmental Protection Agency. U.S. EPA Method 633 – Organonitrogen pesticides. Fed. Reg. 50:40701. October 4.

U.S. EPA. 1986. U.S. Environmental Protection Agency. Guidelines for carcinogen risk assessment. Fed. Reg. 51(185):33992–34003. September 24.

U.S. EPA. 1988. U.S. Environmental Protection Agency. U.S. EPA Method 507 – Determination of nitrogen- and phosphorus-containing pesticides in water by gas chromatography with a nitrogen-phosphorus detector (GC/NPD). Draft. April 15. Available from U.S. EPA's Environmental Monitoring and Support Laboratory, Cincinnati, OH.

Valencia, R.* 1981. Mutagenesis screening of pesticides: "Drosophila." Prepared by Warf Institutes, Inc., for the Environmental Protection Agency.

Available from the National Technical Information Service. EPA 600/1/81/017. Unpublished study. MRID 00143567.

Volk, V.V. 1972. Physico-chemical relationships of soil-pesticide interactions. In: Progress Report, Oregon State University Environmental Health Science Centre. Corvallis, OR. pp. 186–199.

Windholz, J., S. Budaveri, R.F. Blumetti and E.S. Otterbein, eds. 1983. The Merck index, 10th ed. Rahway, NJ: Merck and Company, Inc.

Wolf, D.C. 1974. Degradation of bromacil, terbacil, 2,4-D and atrazine in soil and pure culture and their effect on microbial activity. Diss. Abstr. Int. B. 34(10):4783–4784.

Wolf, D.C. and J.P. Martin. 1974. Microbial degradation of 2-carbon-14 bromacil and terbacil. Proc. Soil Sci. Soc. Am. 38:921–925.

Wolf, D.C., D.I. Bakalivanov and J.P. Martin. 1975. Reactions of bromacil in soil and fungus cultures. Soil Sci. Ann. XXVI(2):35–48.

Woodruff, R.C., J.P. Phillips and D. Irwin. 1984. Pesticide-induced complete and partial chromosome loss in screens with repair-defective females of *Drosophila melanogaster*. Environ. Mutagen. 5:835–846.

Zapp, J.A., Jr.* 1965. Toxicological information: bromacil: 5-bromo-3-*sec*-butyl-6-methyluracil. Unpublished study. MRID 00013243.

Zimdahl, R.L., V.H. Freed, M.L. Montgomery and W.R. Furtick. 1970. The degradation of triazine and uracil herbicides in soil. Weed Res. 10:18–26.

---

* Confidential Business Information submitted to the Office of Pesticide Programs.

Butylate

I. General Information and Properties

A. CAS No. 2008-41-5

B. Structural Formula

$$C_2H_5-S-\overset{\overset{\displaystyle O}{\|}}{C}-N[CH_2CH(CH_3)_2]_2$$

Carbamothioic acid, bis(2-methylpropyl)-, S-ethyl ester

C. Synonyms

- S-ethyl diisobutylthiocarbamate; S-ethyl bis(2-methylpropyl) carbamothioate; ethyl N,N-diisobutyl thiocarbamate; S-ethyl-diisobutyl thiocarbamate; ethyl-N,N-diisobutyl thiolcarbamate; R-1910; Sutan® + ; Anelda Plus; Genate Plus (Meister, 1988).

D. Uses

- Butylate is used as a selective preplant herbicide (Meister, 1988).

E. Properties (BCPC, 1977)

Chemical Formula	$C_{11}H_{23}NOS$
Molecular Weight	217.41
Physical State (25°C)	Clear liquid, aromatic odor
Boiling Point	138°C
Melting Point	--
Density (25°C)	0.9417
Vapor Pressure (25°C)	1.3×10^{-3} mm Hg
Specific Gravity	--
Water Solubility (20°C)	45 mg/L
Log Octanol/Water Partition Coefficient	--
Taste Threshold	--
Odor Threshold	--
Conversion Factor	--

F. Occurrence

- Butylate has been found in 91 of 836 surface water samples analyzed and in 2 of 152 ground-water samples (STORET, 1988). Samples were collected at 80 surface water locations and 61 ground-water locations, and butylate was found in 5 states. The 85th percentile of all nonzero samples was 0.17 μg/L in surface water and 138,000 μg/L in ground-water sources. The maximum concentration found was 4.70 μg/L in surface water and 138,000 μg/L in ground water. This information is provided to give a general impression of the occurrence of this chemical in ground and surface waters as reported in the STORET database. The individual data points retrieved were used as they came from STORET and have not been confirmed as to their validity. STORET data are often not valid when individual numbers are used out of the context of the entire sampling regime, as they are here. Therefore, this information can only be used to form an impression of the intensity and location of sampling for a particular chemical.

G. Environmental Fate

- Butylate degrades fairly rapidly in moist soils under aerobic conditions; half-lives were 3 to 10 weeks (Thomas and Holt, 1979; Shell Development Company, 1975; Stauffer Chemical Company, 1975a). Under anaerobic conditions, butylate degrades with a half-life of 13 weeks (Thomas et al., 1978). Butylate sulfoxide is the major degradate, but S-ethyl-2,2-dimethyl-2-hydroxyethylisobutyl thiocarbamate, diisobutylformamide, diisobutylamine, diisobutylthiocarbamate and isobutylamine were also identified as degradates (Thomas and Holt, 1979; Thomas et al., 1978; Shell Development Company, 1975; Stauffer Chemical Company, 1975a).

- Butylate is slightly mobile to highly mobile in soils ranging in texture from silty clay loam to gravelly sand (Gray and Weierich, 1966; Lavy, 1974; Thomas and Holt, 1979; Weidner, 1974).

- Butylate is fairly volatile; 45 to 50% of ^{14}C-butylate applied to moist (20% moisture) Sorrento clay loam was recovered as volatile radioactivity over 3 weeks following treatment. Volatile radioactivity was characterized as butylate (Thomas and Holt, 1979).

- In the field, butylate dissipated more readily in a soil in Florida than in a silty clay loam in California, probably leaching beyond the 6-inch sampling depth. The estimated half-lives in the upper 6 inches of the sand were 28 and 18 days when a 4 lb/gal Mcap and a 6.7 lb/gal EC formulation, respectively, were applied at 8 lb a.i./A (active ingredient/acre). For the silty clay loam, estimated half-lives were more than 64 days for both the Mcap and a 7 lb/gal EC formulation applied at 8 lb a.i./A (Stauffer Chemical Company, 1975b,c).

• Butylate has a low bioaccumulation potential in bluegill sunfish. A bioconcentration factor of 33 was found in the edible portion of fish dosed with ^{14}C-butylate at 0.01 or 1 ppm for 28 days. The nonedible portion of fish dosed at 0.01 and 1 ppm exhibited bioconcentration factors of 174 and 122, respectively. After 10 days of depuration, 50 to 67% of the day-28 residues was lost (Sleight, 1973).

II. Pharmacokinetics

A. Absorption

• Data relating specifically to the absorption of butylate were not located in the available literature; however, some information was obtained from a metabolism study by Hubbell and Casida (1977). Doses of 12.3 or 156.0 mg/kg ^{14}CO-labeled butylate were administered by gavage to male albino Sprague-Dawley rats weighing 190 to 210 g. Within 48 hours, 27.3 and 31.5% of the administered radioactivity was recovered in the urine, and 60.9 and 64.0% was expired as $^{14}CO_2$ in the low- and high-dose groups, respectively. These results indicate that butylate is appreciably absorbed from the gastrointestinal tract of rats.

B. Distribution

• Hubbell and Casida (1977) measured the tissue radioactivity 48 hours after the administration by gavage of 12.3 or 156.0 mg ^{14}CO-labeled butylate/kg to male Sprague-Dawley rats. At the low dose, 2.4% of the administered radioactivity was retained in the body, with levels of radioactivity equivalent to 276 ppb in the blood, 524 ppb in the kidney, 710 ppb in the liver and a range of 182 to 545 ppb in other tissues (brain, fat, heart, lung, muscle, spleen and testes). At the high dose, 2.2% of the radioactivity was retained in the body with 2,076 ppb in the blood, 5,320 ppb in the kidney, 7,720 ppb in the liver and 1,720 to 5,560 ppb in other tissues.

C. Metabolism

- Hubbell and Casida (1977) followed the metabolism of butylate in male Sprague-Dawley rats based upon identification of the 48-hour urinary metabolites of ^{14}CO-labeled preparations of butylate (12.3 or 156 mg/kg). Degradation of administered butylate metabolites was also assessed. Approximately 40% of the administered ^{14}CO-butylate was metabolized by ester cleavage and ^{14}CO$_2$ liberation without going through the sulfoxide (the major metabolite) as an intermediate. The metabolites from all compounds were essentially the same qualitatively and quantitatively. The metabolites for ^{14}CO-butylate included, as percent of urinary radioactivity, 4.3% as the N,N-diisobutyl mercapturic acid, 17.1% as the N-isobutylmercapturic acid, 0.8% as the mercaptoacetic acid derivative, 11.7% as the glycine conjugate of the mercaptoacetic acid derivative and about 66% as at least 15 other metabolites.

- S-(1-^{14}C)ethyl-Sutan®, orally administered at about 110 mg Sutan®/kg, was readily degraded and excreted by male and female Sprague-Dawley rats (Thomas et al., 1980). Cleavage of the S-ethyl moiety and the incorporation of the two-carbon fragment into intermediary metabolic pathways accounted for > 70% of the total administered radiocarbon. Urinary excretion of ^{14}C-hippuric acid, ethyl methyl sulfoxide and ethyl methyl sulfone was evident.

D. Excretion

- Hubbell and Casida (1977) administered 12.3 or 156 mg/kg of ^{14}CO-labeled butylate by gavage to adult male Sprague-Dawley rats. Within 24 hours, 60.9 and 64.0% of the administered radioactivity was expired as CO$_2$, 27.3 and 31.5% was excreted in the urine and 3.3 and 4.7% were excreted in the feces in the low- and high-dose groups, respectively.

- A study by Bova et al. (1978) indicates that biotransformation of S-(1-^{14}C) ethyl-Sutan® in male and female Sprague-Dawley rats given oral doses of 83.5 to 133.5 mg Sutan®/kg involves rapid cleavage of the S-ethyl moiety. Degradation of this fragment of the molecule results in the release of ^{14}CO$_2$ as the major product of metabolism, accounting for 69% of the total administered dose. This rapid production of ^{14}CO$_2$ may account for the relatively high levels (7.8%) of ^{14}C found in the tissues after 8 days. Urine and feces accounted for 13.9 and 3.2% of the ^{14}C dose, respectively.

• Data obtained from a 3-day balance and tissue residue study by Thomas et al. (1979) show that $(1\text{-}^{14}C\text{-isobutyl})$Sutan® is readily eliminated by male and female Sprague-Dawley rats after a single oral dose (about 100 mg Sutan®/kg). More than 99% of the administered radiocarbon was recovered from the animals within 72 hours after dosing. Most of the dose (94%) was recovered within 24 hours after treatment. Less than 0.5% of the radiocarbon remained in the tissues after 72 hours, and the Sutan® equivalents in organ and tissue samples were all less than 2 ppm. Urine, feces and expired $^{14}CO_2$ accounted for 93.7, 4.0 and 2.0% of the dose, respectively.

III.   Health Effects

A. Humans

• No information was found in the available literature on the health effects of butylate in humans.

B. Animals

1. *Short-term Exposure*

• The acute oral LD_{50} value in male and female rats given butylate technical (85.71% pure) was 3.34 and 3.0 g/kg, respectively (Raltech, 1979).

2. *Dermal/Ocular Effects*

• Skin irritation was observed in rabbits topically exposed to 2 g butylate technical (85.71% pure) for 24 hours (Raltech, 1979).

• Topical application of R-1910 6E technical (97.5% pure) at doses of 20 and 40 mg a.i./kg, 5 days per week for a total of 21 applications, was without observed effect except for local skin irritation (Woodard Research Corp., 1967a).

• Application of butylate technical (85.71% pure) to the eyes of rabbits resulted in irritation and corneal opacity. No corneal opacity was in eyes washed after treatment (Raltech, 1979).

3. *Long-term Exposure*

• Dietary feeding of R-1910 Technical (97.5% pure) to male and female Charles River rats at dose levels of 32, 16 and 8 mg/kg/day for 13 weeks was without observable adverse effect. The high dose (32 mg/kg/day) was identified as the No-Observed-Adverse-Effect Level (NOAEL) for this study (Woodard Research Corp., 1967b).

- Dietary feeding of Sutan® Technical and Sutan® Analytical (purities not specified) to male Sprague-Dawley rats at dose levels as high as 180 mg/kg/day (the NOAEL) for 15 weeks was without observable adverse effect (Scholler, 1976).

- Results of a toxicity study in which male and female beagle dogs were fed R-1910 Technical (97.6% pure) at dietary levels of 450, 900 and 1,800 ppm (corresponding to doses of 11, 23 and 45 mg/kg/day, assuming 1 ppm equals 0.025 mg/kg/day from Lehman, 1959) for 16 weeks were unremarkable (Woodard Research Corp., 1967c). Hence, 45 mg/kg/day is identified as a NOAEL.

- Sutan® Technical (purity not given) was given orally by capsule to male and female beagle dogs (5/dose/sex) at doses of 5, 25, or 100 mg/kg/day for 12 months. Liver-to-body weight ratios were increased (p < 0.05) in males given 25 or 100 mg/kg/day. The effect at the mid dose mainly reflected an 18% decrease in body weight compared to controls. Decreased body weights, increased liver weights and liver lesions (males only) were observed in males and females treated with the high dose. The NOAEL is 5 mg/kg/day (Biodynamics, Inc., 1987).

- Sutan® Technical (98% pure) was fed in the diet to male and female Sprague-Dawley rats at dose levels of 10, 30 and 90 mg/kg/day for 56 weeks. One group of rats was given 90 mg/kg/day for 15 weeks followed by 180 mg/kg/day for 41 weeks. No systemic effects were found at 10 mg/kg/day. Testes/body weight ratios were significantly (p < 0.05) lower in terminally sacrificed males given 30 and 90 mg/kg/day. Slight (8 to 15%) nonsignificant (p > 0.05) mean body weight decreases were found in 30 and 90 mg/kg males and 90 mg/kg females. Liver to body weight increases and testicular lesions were found with the highest doses. Blood clotting parameters were affected at all doses, with the effects at 10 mg/kg/day being significant (p < 0.05) decreases in factor II times in males and activated partial thromboplastin times in females. The 10 mg/kg/day dose is considered a NOAEL as explained below in the Longer-term Health Advisory section (Hazelton Laboratories, Inc., 1978).

- R-1910 Technical (purity not specified) was fed in the diet to male and female Sprague-Dawley CD rats at dose levels of 50, 100, 200 and 400 mg/kg/day for 2 years (Biodynamics, 1982). Although significantly (p < 0.05) elevated liver-to-body weight ratios occurred in terminally sacrificed males given 50 mg/kg/day, this effect was not observed in animals from this dose group sacrificed at 12 and 18 months. Hence, 50 mg/kg/day was identified as a NOAEL. In males and females, body weights were significantly ($p < 0.05$) reduced, and liver-to-body weight ratios were significantly ($p < 0.05$) increased with doses above 50 mg/kg/day. Neoplastic nodules and periportal hypertrophy in the liver were significantly ($p < 0.05$) increased in males given 400 mg/kg/day (Biodynamics, Inc., 1982).

- Male and female Charles River CD-1 mice were given Sutan® Technical (98% pure) in the diet at dose levels of 20, 80 and 120 mg/kg/day for 2 years. No effects were found at 20 mg/kg/day (the NOAEL). Kidney and liver lesions were noted with higher doses (International Research and Development Corporation; IRDC, 1979).

4. *Reproductive Effects*

- Twenty-five male and 25 female CrlCD(SD)BR rats were fed diets containing 0, 200, 1,000, or 4,000 ppm (0, 10, 50, or 200 mg/kg/day) Sutan® Technical (98.2% pure) for 63 days before the animals were mated (P_0 group) (Stauffer Chemical Company, 1986). Two matings were done to produce the F_{1a} and F_{1b} litters of the first generation. Parental rats obtained from the first generation (P_1 group) were mated three times to produce the F_{2a}, F_{2b} and F_{2c} litters of the second generation. Assessments included survival, body and organ weights, reproductive success and pathology. Significant ($p < 0.05$) effects at 50 mg/kg/day were increased liver-to-body weight ratios in P_0 females, decreases in body weights of P_1 females and in food consumption by P_0 males, decreased body weights in F_{2a} pups and decreased brain weights in F_{1b} male weanlings. At 200 mg/kg/day, increased ($p < 0.05$) incidences of dilated renal pelvis and retinal folds in F_{1b} rats were evident. The NOAEL is 10 mg/kg/day.

5. *Developmental Effects*

- Sutan® Technical (98.2% pure) was administered by gavage to pregnant rats at doses of 40, 400 and 1,000 mg/kg/day on days 6 through 20 of gestation. The 40 mg/kg/day dose was without observable effect (the NOAEL). Higher doses decreased body weight gain in dams, increased liver-to-body weight ratios in dams, decreased fetal body weights, increased incidences of misaligned sternebrae and delayed ossification and increased early resorptions. Sutan® was not teratogenic in this study (Stauffer Chemical Company, 1983).

- Administration of R-1910 Technical (97.6% pure) in the diet to pregnant Charles River mice at dose levels of 4, 8 and 24 mg/kg/day either on days 6 through 18 of gestation or from day 6 until natural delivery was without observable effect (NOAEL) on dams and fetuses (Woodard Research Corp., 1967d).

- Administration of Sutan® Technical (99% pure) by gavage to pregnant rabbits at doses of 0, 10, 100 or 500 mg/kg/day on gestation days 6 through 18 resulted in a NOAEL of 500 mg/kg/day for developmental effects and 100 mg/kg/day for maternal toxicity. Maternal body weight was decreased (p < 0.05) with 500 mg/kg/day (Stauffer Chemical Company, 1987).

6. *Mutagenicity*

- Butylate was not mutagenic in *Salmonella typhimurium* strains TA1535, TA1537, TA1538 and TA100 with or without the S-9 activating fraction (Eisenbeis et al., 1981).

- In *Drosophila melanogaster*, butylate treatment increased the frequency of sex-linked recessive lethals but had no effect on the frequency of dominant lethals (Murnik, 1976).

7. *Carcinogenicity*

- R-1910 Technical was not determined to be carcinogenic in the 2-year rat study by Biodynamics, Inc. (1982), but a significant (p < 0.05) increase in neoplastic nodules in liver in high-dose males was evident. Neoplastic nodules were found in 2/69, 6/69, 1/69, 1/70 and 9/70 males given 0 ppm (control), 50 ppm, 100 ppm, 200 ppm and 400 ppm, respectively. Hepatocellular carcinomas were found in 2/69, 3/69, 4/69, 3/70 and 2/70 males given 0 ppm (control), 50 ppm, 100 ppm, 200 ppm and 400 ppm, respectively.

- Sutan® Technical was not carcinogenic in the 2-year mouse study by IRDC (1979).

IV.   Quantification of Toxicological Effects

A. One-day Health Advisory

No information was found in the available literature that was suitable for determination of the One-day HA value for butylate. It is, therefore, recommended that the Ten-day HA value (2 mg/L, calculated below) be used at this time as a conservative estimate of the One-day HA value.

B. Ten-day Health Advisory

The teratology study in mice by Woodard Research Corporation (1967d) has been selected to serve as the basis for determination of the Ten-day HA value for butylate because it provides a short-term NOAEL (24 mg/kg/day for 13 days) for both maternal and fetal toxicity. The teratology study in rats by Stauffer Chemical Company (1983), which identified a NOAEL of 40 mg/kg/day (for 15 days) for maternal and fetal effects, could also be considered; however, because doses higher than the 24 mg/kg/day NOAEL were not included in the Woodard study (1967d), the effect levels in this study are uncertain. Furthermore, the agent was given in the diet in the Woodard study (1967d) and by gavage in the Stauffer (1983) study. Therefore, dose-response comparisons in terms of both effect and no-effect levels between the Woodard (1967d) and Stauffer (1983) studies cannot be made.

Using a NOAEL of 24 mg/kg/day, the Ten-day HA for a 10-kg child is calculated as follows:

$$\text{Ten-Day HA} = \frac{(24 \text{ mg/kg/day}) \ (10 \text{ kg})}{(1 \text{ L/day}) \ (100)} = 2.4 \text{ mg/L } (2,000 \ \mu\text{g/L})$$

C. Longer-term Health Advisory

The two-generation reproduction study in rats by Stauffer Chemical Company (1986) has been selected as the basis for the Longer-term Health Advisory.

The 56-week feeding study with Sutan® Technical in rats by Hazelton Laboratories (1978) is a possible basis for a Longer-term HA. However, effects observed in this study were not evident with higher doses in the 2-year feeding study with R-1910 Technical in rats by Biodynamics, Inc. (1982). Effects on blood clotting parameters (decrease in factor II times in males and activated partial thromboplastin times in females) at the 10 mg/kg/day dose and higher in the Hazelton (1978) study are considered to be of questionable toxicological significance because it is not certain whether they actually represent adverse effects and because these effects were not found in the 2-year rat study by Biodynamics (1982).

The 16-week and 13-week feeding studies with R-1910 Technical in dogs and rats, respectively, by Woodard Research Corp. (1967b,c) can also be proposed for calculation of the Longer-term HA. However, the highest estimated dose of 45 mg/kg/day was the NOAEL in the dog study, and the highest dose of 32 mg/kg/day was the NOAEL in the rat study. Each generation in the two-generation study in rats by Stauffer (1986) can, in effect, also be considered an assessment of subchronic toxicity. Because this study does provide a NOAEL

(10 mg/kg/day) and a LOAEL (50 mg/kg/day), it is preferred over the studies by Woodard Research Corp. (1967b,c) in which the data suggest that higher doses in these studies could have been NOAELs approximating the dose that is a LOAEL in the Stauffer (1986) study.

Using a NOAEL of 10 mg/kg/day, the Longer-term HA for a 10-kg child is calculated as follows:

$$\text{Longer-term HA} = \frac{(10 \text{ mg/kg/day}) (10 \text{ kg})}{(100) (1 \text{ L/day})} = 1.0 \text{ mg/L} (1{,}000 \text{ } \mu g/L)$$

The Longer-term HA for a 70-kg adult is calculated as follows:

$$\text{Longer-term HA} = \frac{(10 \text{ mg/kg/day}) (70 \text{ kg})}{(100) (2 \text{ L/day})} = 3.5 \text{ mg/L} (4{,}000 \text{ } \mu g/L)$$

D. Lifetime Health Advisory

The 12-month chronic toxicity study with Sutan® in dogs by Biodynamics, Inc. (1987) is selected to serve as the basis for the Lifetime HA value for butylate. Of the available data, this study is considered to represent the most sensitive endpoint of toxicity in terms of the lowest NOAEL.

The Lifetime HA is calculated as follows:

Step 1: Determination of the Reference Dose (RfD)

$$\text{RfD} = \frac{(5 \text{ mg/kg/day})}{(100)} = 0.05 \text{ mg/kg/day} (50 \text{ } \mu g/kg/day)$$

Step 2. Determination of the Drinking Water Equivalent Level (DWEL)

$$\text{DWEL} = \frac{(0.05 \text{ mg/kg/day}) (70 \text{ kg})}{(2 \text{ L/day})} = 1.8 \text{ mg/L} (1{,}800 \text{ } \mu g/L)$$

Step 3: Determination of the Lifetime Health Advisory

$$\text{Lifetime HA} = (1.8 \text{ mg/L})(20\%) = 0.36 \text{ mg/L} (360 \text{ } \mu g/L)$$

E. Evaluation of Carcinogenic Potential

- Available toxicity data do not determine butylate to be carcinogenic. Although a significant (p < 0.05) increase in neoplastic nodules in controls in the 2-year study by Biodynamics (1982) was found, concurrent control incidence (2.89%) was low compared to laboratory historical control data (average of 7.07%).

- Applying the criteria described in EPA's guidelines for assessment of carcinogenic risk (U.S. EPA, 1986), butylate may be placed in Group D: not classifiable. This category is for substances that show inadequate evidence of carcinogenicity in animals and humans.

- The U.S. EPA has not calculated excess lifetime cancer risks for this material.

V.   Other Criteria, Guidance and Standards

- Residue tolerances for butylate have been established by the U.S. EPA (1985) and include 0.1 ppm in or on corn grain, fresh corn, corn forage and fodder, sweet corn and popcorn. A tolerance is a derived value based on residue levels, toxicity data, food consumption levels, hazard evaluation and scientific judgment, and it is the legal maximum concentration of a pesticide in or on a raw agricultural commodity or other human or animal food (Paynter et al., no date).

- The U.S. EPA Office of Pesticide Programs has calculated a provisional ADI of 70 μg/kg/day based on the 20-mg/kg/day NOAEL in the 2-year mouse study by IRDC (1979) and a 300-fold uncertainty factor (used because of data gaps, including a chronic feeding study in dogs, a reproduction study in rats and a teratology study in rabbits, in the total data package).

VI.   Analytical Methods

- Analysis of butylate is by a gas chromatographic (GC) method applicable to the determination of certain nitrogen-phosphorus–containing pesticides in water samples (U.S. EPA, 1988). In this method, approximately 1 liter of sample is extracted with methylene chloride. The extract is concentrated and the compounds are separated using capillary column GC. Measurement is made using a nitrogen-phosphorus detector. The method detection limit has not been determined for butylate, but it is estimated that the detection limits for analytes included in this method are in the range of 0.1 to 2 μg/L. This method has been validated in a single laboratory and estimated detection limits have been determined for the analytes, including butylate. The estimated detection limit is 0.15 μg/L.

VII.   Treatment Technologies

• No information was found in the available literature on treatment technologies capable of effectively removing butylate from contaminated water.

VIII.   References

BCPC. 1977. British Crop Protection Council. Pesticide manual, 5th ed. Nottingham, England: Boots Company, Ltd. p. 593.

Biodynamics, Inc.* 1982. A two-year oral toxicity/carcinogenicity study of R-1910 in rats. Project No. 78–2169. Submitted to Stauffer Chemical Company, Richmond, CA. Unpublished final report. MRID 00125678.

Biodynamics, Inc.* 1987. A twelve-month oral toxicity study of Sutan® Technical in dogs. Study No. T-12651. Submitted to Stauffer Chemical Company, Richmond, CA. Unpublished final report. MRID 40389101.

Bova, D.L., J.R. DeBaun, J.C. Petersen and J.J. Menn.* 1978. Metabolism of [ethyl-^{14}C] Sutan® in the rat: balance and tissue residue. Stauffer Chemical Company, Richmond, CA. Unpublished final report. MRID 00043681.

Casida, J.E., R.A. Gray and H. Tilles. 1974. Thiocarbamate sulfoxides: potent, selective and biodegradable herbicides. Science. 184:573–574.

Eisenbeis, S.J., D.L. Lynch and A.E. Hampel. 1981. The Ames mutagen assay tested against herbicides and herbicide combinations. Soil Sci. 131(1):44–47.

Gray, R.A. and A.J. Weierich.* 1966. Behavior and persistence of S-ethyldiisobutylthiocarbamate (Sutan®) in soils. Stauffer Chemical Company, Richmond, CA. Unpublished study.

Hazelton Laboratories, Inc.* 1978. Fifty-six-week feeding study in rats: Sutan® Technical. Project No. 132–135. Submitted to Stauffer Chemical Company, Richmond, CA. Unpublished final report. MRID 00035843.

Hubbell, J.P. and J.E. Casida. 1977. Metabolic fate of the N,N'-dialkylcarbamoyl moiety of thiocarbamate herbicides in rats and corn. J. Agric. Food Chem. 25(2):404–413.

IRDC.* 1979. International Research and Development Corporation. Sutan® Technical: Lifetime oral study in mice. Submitted to Stauffer Chemical Company, Richmond, CA. Unpublished final report. MRID 00035844.

Lavy, T.L. 1974. Mobility and deactivation of herbicides in soil-water systems. Project A-024-NEB. University of Nebraska, Water Resources Research Institute. Submitted by Shell Chemical Company, Washington DC. Available from National Technical Information Service (NTIS), Springfield, VA. PB-238-632.

Lehman, A.J. 1959. Appraisal of the safety of chemicals in foods, drugs and cosmetics. Assoc. Food Drug Off. U.S., Q. Bull.

Meister, R., ed. 1988. Farm chemicals handbook. Willoughby, OH: Meister Publishing Company.

Murnik, M.R. 1976. Mutagenicity of widely used herbicides. Genetics. 83:S54.

Paynter, O.E., J.G. Cummings and M.H. Rogoff. No date. United States Pesticide Tolerance System. U.S. EPA, Office of Pesticide Programs. Unpublished draft report.

Raltech.* 1979. Project Nos. 74489 and 733422. Submitted to Stauffer Chemical Company, Richmond, CA. Unpublished final report.

Scholler, J.* 1976. Fifteen-week oral (diet) toxicity study with Sutan® Technical and Analytical in male rats. Experiment 7. Unpublished final report. MRID 00021844.

Shell Development Company.* 1975. Dissipation of Bladex® herbicide and Sutan® in soil following application of Bladex®, Sutan®, or a tank mix of Bladex® and Sutan®. TIR-24–134–74. Unpublished study.

Sleight, B.H., III.* 1973. Exposure of fish to ^{14}C-labeled Sutan®: accumulation, distribution, and elimination of ^{14}C residues. Unpublished study prepared by Bionomics, Inc. Submitted by Stauffer Chemical Company. Richmond, CA.

Stauffer Chemical Company.* 1975a. Dissipation of Bladex® herbicide and Sutan® in soil following application of Bladex®, Sutan®, or a tank mix of Bladex® and Sutan®: TIR-24–134–74. Unpublished study submitted by Stauffer Chemical Company. Richmond, CA.

Stauffer Chemical Company.* 1975b. Residues from Sutan® on soil: FSDS Nos. A-9229, A-9229–1, A-9229–2, A-10366. Unpublished study by Stauffer Chemical Company. Richmond, CA.

Stauffer Chemical Company.* 1975c. Soil residue data of Sutan® combinations and R-25788: FSDS Nos. A-9229, A-9229–1, A-9229–2, A-10366. Unpublished study by Stauffer Chemical Company. Richmond, CA.

Stauffer Chemical Company.* 1983. A teratology study in CD rats with Sutan® Technical. Project No. T-11713. Unpublished final report by Stauffer Chemical Company. Richmond, CA. MRID 000131032.

Stauffer Chemical Company.* 1986. A 2-generation reproduction study in rats with Sutan®. Study No. T-11940. Unpublished final report by Stauffer Chemical Company. Richmond, CA. EPA Accession No. 263612–263614.

Stauffer Chemical Company.* 1987. A teratology study in rabbits with Sutan® Technical. Study No. T-12999. Unpublished final report by Stauffer Chemical Company. Richmond, CA. MRID 40389102.

STORET. 1988. STORET Water Quality File. Office of Water. U.S. Environmental Protection Agency (data file search conducted in May, 1988).

Thomas, D.B., J.B. Miaullis, A.R. Vispetto and J. Osuna.* 1979. Metabolism of [isobutyl-^{14}C] Sutan® in the rat: balance and tissue residue study. Unpublished final report by Stauffer Chemical Company. Richmond, CA. MRID 00043680.

Thomas, D.L.B., J.C. Petersen and J.R. DeBaun.* 1980. Metabolism of [1-^{14}C-ethyl] Sutan® in the rat: urinary metabolite identification. Unpublished final report by Stauffer Chemical Company. Richmond, CA. MRID 00043682.

Thomas, V.M. and C.L. Holt.* 1979. Behavior of Sutan® in the environment: MRC-B-76; MRC-78-02. Unpublished study, submitted by Stauffer Chemical Company, Richmond, CA.

Thomas, V.M., C.L. Holt and P.A. Bussi.* 1978. Anaerobic soil metabolism of Sutan® selective herbicide: MRC-B-98; MRC-79-13. Unpublished study. Submitted by Stauffer Chemical Company. Richmond, CA.

U.S. EPA. 1985. U.S. Environmental Protection Agency. Residue tolerances for S-ethyl-diisobutyl thiocarbamate. CFR 180.232. July 1. p. 294.

U.S. EPA. 1986. U.S. Environmental Protection Agency. Guidelines for carcinogen risk assessment. Fed. Reg. 51(185):33992–34003. September 24.

U.S. EPA. 1988. U.S. Environmental Protection Agency. U.S. EPA Method 507 — Determination of nitrogen- and phosphorus-containing pesticides in water by GC/NPD. Draft. April 15. Available from U.S. EPA's Environmental Monitoring and Support Laboratory, Cincinnati, OH.

Weidner, C.W.* 1974. Degradation in groundwater and mobility of herbicides. Master's thesis. University of Nebraska, Department of Agronomy. Unpublished study submitted by Shell Chemical Company. Washington, DC.

Woodard Research Corporation.* 1967a. R-1910 6-E: subacute dermal toxicity; 21-day experiment with rabbits. Submitted to Stauffer Chemical Company. Richmond, CA. Unpublished final report. MRID 00026312.

Woodard Research Corporation.* 1967b. R-1910: safety evaluation by dietary feeding to rats for 13 weeks. Submitted to Stauffer Chemical Company, Richmond, CA. Unpublished final report. MRID 00026313.

Woodard Research Corporation.* 1967c. R-1910: safety evaluation by dietary feeding to dogs for 16 weeks. Submitted to Stauffer Chemical Company, Richmond, CA. Unpublished final report. MRID 00026314.

Woodard Research Corporation.* 1967d. R-1910: safety evaluation by teratological study in the mouse. Submitted to Stauffer Chemical Co., Richmond, CA. Unpublished final report. MRID 000129544.

---

* Confidential Business Information submitted to the Office of Pesticide Programs.

Carbaryl

I. General Information and Properties

A. CAS No. 63–25–2

B. Structural Formula

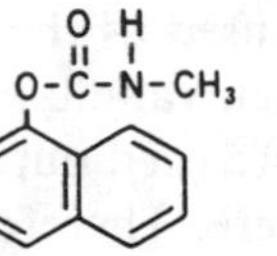

$$O-\overset{\overset{O}{\|}}{C}-\overset{\overset{H}{|}}{N}-CH_3$$

1-Naphthalenol methylcarbamate

C. Synonyms

- Arilate; Bercema NMC50; Caprolin; Sevin; Vioxan (Meister, 1983).

D. Uses

- Carbaryl is a contact insecticide used for the control of pests on more than 100 different crops, forests, lawns, ornamentals, shade trees and rangeland (Meister, 1983).

E. Properties (Windholz et al., 1983; CHEMLAB, 1985)

Chemical Formula	$C_{12}H_{11}O_2N$
Molecular Weight	201.22
Physical State (25°C)	White crystals
Boiling Point	--
Melting Point	145°C
Density (20°C)	1.232
Vapor Pressure (25°C)	$< 4 \times 10^{-5}$ mm Hg
Specific Gravity	--
Water Solubility (30°C)	120 mg/L
Log Octanol/Water Partition Coefficient	0.14
Taste Threshold	--
Odor Threshold	--
Conversion Factor	--

F. Occurrence

- Carbaryl has been found in 58 of 640 surface water samples analyzed and in none of 1,541 ground-water samples (STORET, 1988). Samples were collected at 185 surface water locations and 1,409 ground-water locations, and carbaryl was found in six states. The 85th percentile of all nonzero samples was 260 μg/L in surface water and 0 μg/L in ground-water sources. The maximum concentration found was 180,000 μg/L in surface water and 0 μg/L in ground water. This information is provided to give a general impression of the occurrence of this chemical in ground and surface waters as reported in the STORET database. The individual data points retrieved were used as they came from STORET and have not been confirmed as to their validity. STORET data are often not valid when individual numbers are used out of the context of the entire sampling regime, as they are here. Therefore, this information can only be used to form an impression of the intensity and location of sampling for a particular chemical.

G. Environmental Fate

- ^{14}C-Carbaryl (purity unspecified) at 10 ppm was relatively stable to hydrolysis in buffered solutions at pH 3 and 6. It hydrolyzed at pH 9 with a half-life of 3 to 5 hours when incubated at 25°C (Khasawinah and Holsing, 1977a). At 35°C, ^{14}C-carbaryl was stable at pH 3, and hydrolyzed with a half-life of > 28 days and 30 to 60 minutes at pH 6 and 9, respectively. l-Naphthol was the major degradate formed.

- ^{14}C-Carbaryl (purity unspecified) at 5 ppm photodegraded slowly in 0.1 M phosphate buffer solutions, with 4.39 to 4.49 ppm remaining as parent compound after 18 days of irradiation (Khasawinah and Holsing, 1977b). In a 2% acetone solution, ^{14}C-carbaryl accounted for 3.63 to 3.65 ppm after 18 days. l-Naphthol and several unidentified compounds were found at < 0.07 ppm.

- Under aerobic conditions, ^{14}C-carbaryl (> 99% pure) at 1 ppm degraded with a half-life of 7 to 14 days in a sandy loam soil maintained at 15 or 23° to 25°C, and 14 to 28 days in a clay loam soil maintained at 23° to 25°C (Khasawinah and Holsing, 1978). Degradation was slightly slower in sterile soils (half-lives of 14 to 56 days). The majority of the applied radioactivity was bound to the soil or had been evolved as ^{14}CO$_2$ by the end of the test period (112 days). No degradates were found.

- Under aerobic conditions, ^{14}C-carbaryl (> 99% pure) at 1 ppm degraded with a half-life of 84 to 112 days in a flooded sandy loam soil (Khasawinah and Holsing, 1978). At 168 days after treatment, ^{14}C-carbaryl accounted for 42% of the applied radioactivity in the soil and water layer. 4-Hydroxy carbaryl was found at < 0.3% of the applied radioactivity in soil samples taken after 112 days. Approximately 20% of the total radioactivity was soil-bound at 112 days.

II.   Pharmacokinetics

A. Absorption

- Comer et al. (1975) reported the results of tests conducted in factory workers exposed to carbaryl during the formulation of 4 and 5% carbaryl dust. Carbaryl exposure via the skin was measured by attachment of a special gauze pad to various parts of the body, and inhaled carbaryl was measured by the use of special filter pads in face masks. Calculated exposures were 73.90 and 1.10 mg/hour for the dermal and respiratory routes, respectively. The total exposure was 75 mg/hour, or 600 mg/day. Absorption levels were determined by estimation of the carbaryl metabolite 1-naphthol in urine. It was determined that during an 8-hour workday the total absorption of carbaryl would be 5.6 mg. This is about 0.9% of the total exposure, and the authors interpreted this to mean that dermal absorption was not complete.

- Feldmann and Maibach (1974) applied 4 μg/cm^2 of ^{14}C-labeled carbaryl (position of label not specified) dissolved in acetone to one or both forearms of apparently healthy male volunteers. The area of application was left unwashed and unprotected for 24 hours. Based on the excretion rate, it was determined that 73.9% of the applied carbaryl was absorbed through the skin.

- Houston et al. (1974) reported that ^{14}C-carbamyl-labeled carbaryl administered by gavage to male rats at doses of 0.5 mg/kg (given as 0.5 mL of 0.5% propylene glycol in water) rapidly appeared in the systemic circulation. Within a few minutes, the plasma level was 50 ng/mL. A maximum level of 150 ng/mL was reached in less than 10 minutes and steadily declined to 20 ng/mL at 120 minutes. Only 4.6% of the dose was excreted in the feces, indicating that at least 95.3% had been absorbed.

- Falzon et al. (1983) administered single doses of 20 mg/kg of ^{14}C-carbaryl (in olive oil) to six female rats by gavage. After 24 hours, 5.8% of the label was recovered in the feces, indicating that about 94.2% had been absorbed.

B. Distribution

- The distribution of ^{14}C-carbonyl-labeled carbaryl in male and female rats after administration of 1.5 mg/kg by stomach tube was examined in eight body tissues (Krishna and Casida, 1965). The amounts detected (μM/kg) in males and females, respectively, were: cecum, 0.17 and 0.60; esophagus, 0.05 and 0.05; large intestine, 0.20 and 0.30; small intestine, 0.06 and 0.08; kidney, 0.06 and 0.07; liver, 0.11 and 0.112; spleen, 0.05 and 0.08; and stomach, 0.07 and 0.14.

- Falzon et al. (1983) administered single oral doses of 20 mg/kg of ^{14}C-carbaryl to female Wistar rats by gavage. The amounts detected 24 hours after administration were 0.11% in the brain, 3.87% in the digestive tract and 13.31% in the carcass.

C. Metabolism

- Human tissues obtained by either biopsy or autopsy were incubated using an *in vitro* organ-maintenance technique with ^{14}C-(N-methyl)-labeled carbaryl (Chin et al., 1974). The following tissues were examined: for males — lung, liver and kidney; for females — liver, placenta, vaginal mucosa, uterus and uterine tumor (leiomyoma). Hepatic tissues metabolized carbaryl by hydrolysis and/or demethylation, hydroxylation and oxidation followed by conjugation. The primary hydrolytic product was 1-naphthol (42% by 24 hours at pH 7.8). The kidney produced naphthyl glucuronide; the uterus, lung and placenta produced naphthyl sulfate from carbaryl. The vaginal mucosa produced glucuronide and sulfate conjugates, but only a slight amount of conjugating activity (naphthol sulfate) was found in the uterine leiomyoma.

- Houston et al. (1974) administered ^{14}C-carbamyl-labeled carbaryl (0.5 mg/kg) to male rats by gavage. Within 48 hours, 54.5% of the label had been excreted in the urine as metabolites (not identified). In addition, 32.9% was excreted as CO_2. This indicated that carbaryl was extensively metabolized in rats.

D. Excretion

- Comer et al. (1975) studied the excretion of 1-naphthol in the urine of workers who were exposed to carbaryl in a pesticide formulation plant. The workers were exposed to carbaryl both dermally (73.9 mg/hour) and by inhalation (1.1 mg/hour). Analyses of urine samples indicated that the excretion rate of 1-naphthol varied from 0.004 to 3.4 mg/hour, with a mean value of 0.5 mg/hour. This corresponds to an excretion rate of 0.7 mg carbaryl/hour. Following exposure to carbaryl at the start of the workday, the urinary level of 1-naphthol increased, reached its maximum level during the late afternoon and evening hours, and then dropped to a lower level before the start of the next day's workday.

- Urinary excretion of topically applied radiolabeled carbaryl in healthy male volunteers was measured by Feldman and Maibach (1974). A total of 26.1% of the dose was recovered in the urine over a 5-day period.

- Krishna and Casida (1965) administered single doses of 1.5 mg/kg of ^{14}C-carbonyl-labeled carbaryl orally to rats. Excretion of the label for male and female animals, respectively, was as follows: expired carbon dioxide, 26% and 26%; urine, 64.0% and 72.0%; and feces, 4.0% and 4.0%.

- Houston et al. (1974) administered ^{14}C-carbamyl-labeled carbaryl (0.5 mg/kg) by gavage to male rats. The label was almost completely excreted within 48 hours, with the following distribution: expired carbon dioxide, 32.9%; urine, 54.5%; and feces, 4.6%. Less than 1% of the label in urine was unchanged carbaryl. About 6.0% of the label remained in the body. Biliary excretion was examined by bile-duct cannulation. Within 6 hours, 30 to 33% of the administered dose was present in the bile; after 6 hours, the amount in the bile leveled off.

III.   Health Effects

A. Humans

1. *Short-term Exposure*

- Vandekar (1965) investigated the effects of large-scale carbaryl spraying in a village in Nigeria. No quantitative estimates of exposure were obtained, but plasma cholinesterase (ChE) activity was decreased by about 15% in eight applicators (spraymen) and by an average of 8% in 63 villagers.

2. *Long-term Exposure*

- Wills et al. (1968) studied the subchronic toxicity of carbaryl in human volunteers. Groups of five or six men were given daily oral doses of 0, 0.06 or 0.13 mg/kg/day for 6 weeks. At the lower dose, no significant effects were detected on kidney function, electroencephalogram, hematology, blood chemistry, urinalysis or plasma and red blood cell ChE activity. At the higher dose, the only detectable effect was a slight increase in the urinary ratio of amino acid nitrogen to creatinine. This was interpreted to suggest a slight decrease in resorption of amino acids in the kidney. This effect was fully reversible. Based on these observations, a No-Observed-Adverse-Effect Level (NOAEL) of 0.06 mg/kg/day was identified.

B. Animals

1. *Short-term Exposure*

- Carpenter et al. (1961) investigated the acute oral toxicity of carbaryl in several species. Cats were found to be most sensitive (2/2 deaths at 250 mg/kg). Guinea pigs, rats and rabbits were less sensitive, with calculated LD_{50} values of 280, 510 and 710 mg/kg, respectively. No deaths were reported in dogs administered doses of up to 795 mg/kg/day.

- The acute oral toxicity of carbaryl in male Sprague-Dawley rats was studied by Rittenhouse et al. (1972). Carbaryl (99.9% active) dissolved in acetone and propylene glycol (10% v/v) was administered in a single dose at four dose levels to six animals per level. Animals were observed for 14 days following treatment. Dose levels were 439, 658, 986 or 1,481 mg/kg. Mortalities observed at these levels were 0/6, 0/6, 4/6 and 5/6 rats, respectively. Most deaths occurred in the first 24 hours. The LD_{50} was calculated to be 988 mg/kg. Animals at all dose levels exhibited symptoms of ChE inhibition, but ChE activity was not measured. No other parameters were reported.

- Carpenter et al. (1961) fed single oral doses of carbaryl in capsules to female mongrel dogs as follows: 250 mg/kg (one animal), 375 mg/kg (four animals) or 500 mg/kg (one animal). Signs of overstimulation of the parasympathetic nervous system were observed at the two higher doses, but not at 250 mg/kg. These signs included increased respiration, lacrimation, salivation, urination, defecation, muscular twitching, constriction of pupils, poor coordination and vomiting. Plasma ChE was not affected at 375 mg/kg, but a transient decrease (24 to 33%) was observed in erythrocyte ChE at this dose. After 1 day, the appearance of the animals was normal and no adverse CNS effects were noted. Based on the absence of visible external effects or inhibition of ChE, this study identified a NOAEL of 250 mg/kg.

- Carpenter et al. (1961) also administered single oral doses of carbaryl (560 mg/kg, by gavage in corn oil) to three groups of rats (seven to nine per group). Groups were sacrificed after 0.5, 4 or 24 hours, and ChE activity was measured in plasma, erythrocytes and brain. Plasma ChE was slightly lower (7 to 14%) than control, but this was not statistically significant. In erythrocytes, ChE was inhibited 42% after 0.5 hours, but this returned to near normal (86% of control) within 24 hours. Brain ChE activity was inhibited 30% after 0.5 hours, and this returned toward normal (91% of control) by 24 hours.

- Weil et al. (1968) fed carbaryl in the diet for 1 week to Harlan-Wistar albino rats (42 days old) at concentrations yielding ingested doses of 0, 10, 50, 250 or 500 mg/kg/day. Body weight gain was decreased in animals exposed to 50 mg/kg/day or higher. At 10 mg/kg/day, ChE activity was not significantly affected in plasma, red blood cells or brain. At 50 mg/kg/day, plasma ChE was decreased 15 and 18% and red blood cell ChE was decreased 26 and 47% in females and males, respectively. At higher doses, larger decreases in plasma and red blood cell ChE were seen, brain ChE was also decreased (26 and 23% at 250 mg/kg/day and 28 and 33% at 500 mg/kg/day in males and females, respectively). After 1 day on control diet, these effects on ChE were entirely reversed. Based on these data, a NOAEL of 10 mg/kg/day and a Lowest-Observed-Adverse-Effect Level (LOAEL) of 50 mg/kg/day were identified in rats.

2. Dermal/Ocular Exposure

- Carpenter et al. (1961) applied 0.01 mL of 10% carbaryl in acetone (a dose of 1 mg) to the clipped skin of the belly of five rabbits. No irritation was detected.

- Gaines (1960) applied a series of doses of carbaryl dissolved in xylene to the skin of Sherman rats. The dermal LD_{50} value was greater than 4,000 mg/kg for both males and females.

- Carpenter et al. (1961) detected a weak skin sensitization reaction in 4 of 16 male albino guinea pigs given eight intracutaneous injections of 0.1 mL of 0.1% carbaryl (0.1 mg/dose). The challenge dose (not specified) was given 3 weeks later, and examinations for sensitization reaction were performed 24 and 48 hours thereafter.

- Carpenter et al. (1961) applied carbaryl to the eyes of rabbits and evaluated corneal injury. Technical carbaryl (98% pure) applied as a 10% suspension in propylene glycol caused mild injury in 1/5 eyes. A 25% aqueous suspension caused no injury, and 50 mg of powder caused only traces of corneal necrosis.

3. Long-term Exposure

- Wistar rats (5/sex, 45 days old) were fed carbaryl (as Compound 7744; purity not specified) in the diet for 90 days at levels of 0.0037, 0.011, 0.033 or 0.10% (Weil, 1956). Assuming that 1 ppm in the diet of young rats is equivalent to approximately 0.10 mg/kg/day (Lehman, 1959), this corresponds to doses of about 3.7, 11, 33 or 100 mg/kg/day. The author stated that there were no significant changes in appetite or weight gain when compared to the controls; micropathology revealed no changes in lung, liver or kidney tissue at any dose level. It was concluded that for these end points the effect level for toxicity is higher than 0.10%, which is equivalent to a NOAEL of about 100 mg/kg/day (the highest dose tested).

- Carbaryl was administered to male rats by gavage at a level of 200 mg/kg, 3 days a week for 90 days (Dikshith et al., 1976). This corresponds to an average dose of 86 mg/kg/day. The control animals received vehicle (peanut oil) on a similar schedule. There were no overt toxicological signs in these rats, and no marked biological changes were seen in testes, liver and brain (enzymatic determinations) except for ChE activity, which was inhibited 34% in blood ($p < 0.001$) and 12% in brain ($p < 0.05$). No significant histological changes were noted in testes, epididymis, liver or kidney. Based on ChE inhibition, the LOAEL for this study was identified as 86 mg/kg/day.

- Carpenter et al. (1961) fed carbaryl to male and female basenji-cocker dogs (four or five per dose) for 1 year. Dietary levels were about 0, 24, 95 or 414 ppm, which were adjusted to supply ingested doses of 0, 0.45, 1.8 or 7.2 mg/kg/day. No compound-related effects were detected on mortality, body weight, hematocrit, hemoglobin, leukocyte count, blood chemistry, plasma or erythrocyte ChE activity, or liver and kidney weights. Microscopic examination of tissues revealed diffuse cloudy swelling of renal nephrons and focal debris in glomeruli of dogs fed the higher dose. These conditions were also observed in controls, but less frequently, and the authors judged they were not early stages of toxic degeneration. One dog at the low dose displayed a transient hind leg weakness after 189 days. This disappeared within 3 weeks, although dosing was continued throughout. Subsequent microscopic examination revealed no differences between this dog and others. A NOAEL of 7.2 mg/kg/day (the highest dose tested) was identified.

- Schering (1963) administered carbaryl (5.0 mg/kg/day) by gavage to 25 male and 25 female rats, 5 days per week for 18 months. No effects were observed on weight gain, organ weights, urinalysis, hematology or histologic appearance of tissues. The authors concluded that 5.0 mg/kg/day was a NOAEL in rats.

- Carpenter et al. (1961) studied the toxicity of carbaryl in a 2-year feeding study in rats. Groups of 20 male and 20 female CF-N rats (60 days old) were maintained on a diet containing 0, 50, 100, 200 or 400 ppm dry Sevin. Based on measured food consumption and body weights, the authors reported the doses to be equivalent to 0, 2.0, 4.0, 7.9 or 15.6 mg/kg/day in males, and 0, 2.4, 4.6, 9.6 or 19.8 mg/kg/day in females. No adverse effects were detected on life span, food consumption, body weight gain, liver and kidney weights, cataract formation or hematocrit. Histological examination after 1 year revealed mild changes in the kidney, characterized by cloudy swelling of the nephrons. This was statistically significant ($p < 0.004$) at the high dose. Cloudy swelling of hepatic chords was also observed at the high dose, and this was significant after 2 years ($p < 0.002$). No histological changes were detectable at the lower doses. Based on these observations, a NOAEL of 7.9 mg/kg/day for males and 9.6 mg/kg/day for females was identified.

4. Reproductive Effects

- Weil (1972) investigated the reproductive effects of carbaryl in female rats exposed either by gavage or by feeding. Doses of 0, 2.5 and 10 mg/kg/day ingested from the diet for three generations resulted in no statistically significant, dose-related effects on fertility, gestation, lactation or pup viability. Doses of 100 mg/kg/day given by gavage (5 days/week, beginning at 5 weeks of age) resulted in maternal mortality, reduced fertility and signs of ChE inhibition. These signs were not seen in animals ingesting doses of up to 200 mg/kg/day from the diet.

- Murray et al. (1979) studied the reproductive effects of carbaryl (99% active ingredient) in female CF$_1$ mice. Carbaryl was administered by gavage at 100 or 150 mg/kg/day, or by feeding in the diet at 5,660 ppm (calculated by the authors to be equivalent to 1,166 mg/kg/day). At the gavage dose of 150 mg/kg/day, the mice gained less weight and exhibited significant maternal toxicity, including salivation, ataxia and lethargy, and 10/37 females died during the experimental period. At 100 mg/kg/day by gavage, a single maternal death occurred, but weight gain was normal and no other evidence of maternal toxicity was observed. No maternal deaths or signs of ChE inhibition were seen among the mice supplied carbaryl in the diet (1,166 mg/kg/day), although there was a significant (p < 0.05) decrease in body weight gain on days 10 through 15. The incidence of pregnancy, the average number of live fetuses/litter and the incidence of resorptions were not altered by carbaryl for either route of administration. Mean fetal body weight and length were significantly (p < 0.05) lower than control values among litters given carbaryl in the diet, but were not affected among those given carbaryl by gavage. Based on maternal reproductive effects, the NOAEL in mice was identified as 100 mg/kg/day.

- In an investigation using Sprague-Dawley rats, carbaryl was administered by gavage at levels of 1, 10 or 100 mg/kg/day for 3 months prior to and throughout gestation (Lechner and Abdel-Rahman, 1984). Carbaryl of formulation grade (purity not specified) was administered in corn oil. Dams were sacrificed on day 20 for examination. Animals receiving 100 mg/kg/day showed a significant decrease in weight gain during the gestational period, occurring primarily in the third week (days 15 to 20). There was also a slight decrease in the number of implantation sites and live fetuses per dam after treatment at this dose level. Fetal weights and body length for all three doses were within the range of control values. There were no overt signs of maternal toxicity that suggested ChE inhibition. Based on maternal weight gain and number of implantations, the NOAEL in this study was identified as 10 mg/kg/day.

- Collins et al. (1971) reported the effects of carbaryl in the diet on various reproductive parameters over three generations of rats. Osborne-Mendel rats were fed 0, 2,000, 5,000 or 10,000 ppm carbaryl in the diet. Assuming that 1 ppm in the diet of rats is equivalent to 0.05 mg/kg/day (Lehman, 1959), these levels correspond to doses of about 0, 100, 250 or 500 mg/kg/day. At 10,000 ppm, no litters were produced after the first litter of the second generation; decreases were observed in the fertility, viability, survival and lactation indices in all litters at this dose. The survival index also showed a decrease at the 5,000-ppm level. Dose-related decreases were observed in the ratio of average number of animals weaned per number of litters at both 5,000 and 10,000 ppm. At all three dose levels there was a decrease in weanling weights. In rats, the LOAEL was identified as 2,000 ppm (100 mg/kg/day).

- Collins et al. (1971) reported the effects of carbaryl in a three-generation study in gerbils. Carbaryl was fed at dose levels of 0, 2,000, 4,000, 6,000 or 10,000 ppm. Assuming that 1 ppm in the diet of rats is equivalent to

0.05 mg/kg/day (Lehman, 1959), and that gerbils are similar to rats in terms of body weight and food consumption, this corresponds to doses of about 0, 100, 200, 300 or 500 mg/kg/day. No second litters were produced in the third generation at 10,000 ppm. Decreases in the viability index were observed at 6,000 and 10,000 ppm. Dose-related decreases in the survival index were also observed. The average number of animals weaned per litter was also decreased. Based on these findings, a LOAEL of 6,000 ppm (300 mg/kg/day) and a NOAEL of 4,000 ppm (200 mg/kg/day) were identified.

5. Developmental Effects

- Weil et al. (1971) exposed pregnant Harlan-Wistar rats to carbaryl in the diet on days 5 to 15 of gestation. Ingested doses were 0, 20, 100 or 500 mg/kg/day. Animals were sacrificed on days 19 to 21, and fetuses were examined for soft-tissue and skeletal abnormalities. No increased incidence of teratogenic anomalies was detected at any dose level. Based on this information, a NOAEL of 500 mg/kg/day (the highest dose tested) was identified.

- Murray et al. (1979) administered 200 mg/kg/day carbaryl to female rabbits by gavage on days 6 to 18 of gestation. Fetuses were removed and examined for developmental defects. There was a significantly ($p < 0.05$) higher incidence of omphalocele in fetuses from exposed animals than in the controls. The anomalies occurred in litters from does that showed the greatest weight losses during the experimental period. No other anomalies were seen at this dose level. At 150 mg/kg/day, there were single cases of omphalocele, hemivertebrae and conjoined nostrils with missing nasal septum, but no fetal alterations occurred at an incidence significantly different from that of the control group. Based on fetal defects, the LOAEL for the rabbit was identified as 150 mg/kg/day.

- Golbs et al. (1975) orally administered carbaryl to Wistar rats at doses of 200 or 350 mg/kg on days 5, 7 and 9, or on days 11, 13 and 15 of the gestation period. In one group of rats, 200 mg/kg was administered on days 5, 7, 9, 11, 13 and 15. Doses of 350 mg/kg given during late gestation (days 11 to 15) delayed fetal development, whereas the same dose given at the earlier interval (days 5 to 9) resulted in loss of fertilized ova and more pronounced retardation in development of individual fetuses. Similar results were produced by the 200-mg/kg dose given on alternate days from day 5 through day 15. It was concluded that carbaryl produces dose-dependent effects on intrauterine development in rats.

Based on this study, a LOAEL of 200 mg/kg (100 mg/kg/day) was identified.

- Murray et al. (1979) studied the teratogenic effects of carbaryl in CF_1 mice. Carbaryl was administered by gavage at 100 or 150 mg/kg/day, or by feeding in the diet at 5,660 ppm (calculated by the authors to be equivalent to 1,166 mg/kg/day). No major malformations were detected among the offspring of dams given carbaryl by either route at incidences significantly different than concurrent or historical controls. Delayed ossification of skull bones and of sternebrae occurred significantly more often among litters from dams given carbaryl in the diet, but not in litters from gavage-administered dams. Based on developmental observations in fetuses, the NOAEL in this study was identified as 150 mg/kg/day.

- Lechner and Abdel-Rahman (1984) administered carbaryl to Sprague-Dawley rats by gavage for 3 months prior to and throughout gestation at doses of 0, 1, 10 or 100 mg/kg/day. Dams were sacrificed on day 20, and fetuses were examined for external, skeletal and visceral malformations. There were no statistically significant increases of serious anomalies at any dose level. The authors concluded that in the rats tested, carbaryl displayed no evidence of teratogenicity. On this basis, a NOAEL of 100 mg/kg/day (the highest dose tested) was identified.

- Benson et al. (1967) fed mice carbaryl in their diet (intake levels of 10 or 30 mg/kg/day) on day 6 through termination of gestation. Some dams were allowed to deliver naturally, and others were delivered by Cesarean section. There were no differences between the offspring of the two treated groups and the controls in sex ratio, incidence of anomalies or in ossification. Based on this information, a NOAEL of 30 mg/kg/day (the highest dose tested) was identified.

6. *Mutagenicity*

- The effects of pesticides on scheduled and unscheduled DNA synthesis of rat thymocytes and human lymphocytes were studied by Rocchi et al. (1980). Carbaryl (99.2% pure) in the rat thymocyte culture inhibited thymidine uptake 15, 22 and 99% at levels of 1, 10 and 100 μg/mL, respectively. In the human lymphocytes, a dose of 50 μg/mL produced 62% inhibition on scheduled DNA synthesis, but had no effect on unscheduled DNA synthesis.

7. *Carcinogenicity*

- Carpenter et al. (1961) fed carbaryl to groups of CF-N rats (20/sex/ dose) for 2 years. Concentrations in the diet were 0, 50, 100, 200 or 400 ppm, reported by the authors to be equal to doses of 0, 2.0, 4.0, 7.9 or 15.6 mg/kg/day in males and 0, 2.4, 4.6, 9.6 or 19.8 mg/kg/day in females. Based on gross and histological examination of tissues, no increased frequency of any tumor type was detected. The total number of tumors seen at each of the five concentrations tested was 10, 12, 8, 9 and 12, occurring in 9, 11, 7, 6 and 11 rats, respectively.

- Schering (1963) dosed 25 male and 25 female rats by gavage with 5.0 mg/kg/day carbaryl for 18 months. Based on histological examination of tissues, no effects of carbaryl on tumor frequency were detected.

- Carbaryl (30 mg/kg/day) was administered by gavage to mongrel rats daily for 22 months (Andrianova and Alekseev, 1969). At the termination of the study, 46 of the original 48 controls survived and one animal had a malignant tumor. In the treated rats, 12 of the original 60 survived to 22 months, and 4 of these had malignancies (25%). It was concluded that carbaryl was carcinogenic in this investigation. Statistical analyses of the results were not presented.

IV.   Quantification of Toxicological Effects

A. One-day Health Advisory

No data were found in the available literature that were suitable for determination of the One-day HA value. It is recommended that the Ten-day HA value for a 10-kg child (1.0 mg/L, calculated below) be used at this time as a conservative estimate of the One-day HA value.

B. Ten-day Health Advisory

The study by Weil et al. (1968) has been selected to serve as the basis for determination of the Ten-day HA for the 10-kg child. This study identified a NOAEL of 10 mg/kg/day in rats fed carbaryl in the diet for 7 days, based on inhibition of ChE in plasma and red blood cells.

The Ten-day HA for a 10-kg child is calculated as follows:

$$\text{Ten-day HA} = \frac{(10 \text{ mg/kg/day}) (10 \text{ kg})}{(100)(1 \text{ L/day})} = 1.0 \text{ mg/L } (1,000 \text{ } \mu\text{g/L})$$

C. Longer-term Health Advisory

No data were found in the available literature that were suitable for the determination of a Longer-term HA value. It is therefore recommended that the DWEL, adjusted for a 10-kg child (1.0 mg/L, calculated below) be used as a conservative estimate of the Longer-term HA value.

For a 10-kg child, the adjusted DWEL is calculated as follows:

$$DWEL = \frac{(0.1 \text{ mg/kg/day})(10 \text{ kg})}{1 L/\text{day}} = 1.0 \text{ mg/L}$$

where: 0.1 mg/kg/day = RfD (see Lifetime Health Advisory Section, below).

D. Lifetime Health Advisory

The 2-year feeding study in rats by Carpenter et al. (1961) has been selected to serve as the basis for determination of the Lifetime HA for carbaryl. This study identified a NOAEL of 9.6 mg/kg/day, based on absence of effects on mortality, body weight, organ weight, hematology, cataract frequency or histopathology. This value is supported by a 1-year feeding study in dogs, which identified a NOAEL of 7.2 mg/kg/day (Carpenter et al., 1961), and an 18-month oral study in rats, which identified a NOAEL of 5.0 mg/kg/day (Schering, 1963); however, these latter studies were not selected because exposure was less-than-lifetime.

Using the NOAEL of 9.6 mg/kg/day, the Lifetime HA for carbaryl is calculated as follows:

Step 1: Determination of the Reference Dose (RfD)

$$RfD = \frac{(9.6 \text{ mg/kg/day})}{(100)} = 0.1 \text{ mg/kg/day}$$

Step 2: Determination of the Drinking Water Equivalent Level (DWEL)

$$DWEL = \frac{(0.1 \text{ mg/kg/day}) (70 \text{ kg})}{(2 \text{ L/day})} = 3.5 \text{ mg/L} (4,000 \text{ } \mu g/L)$$

Step 3: Determination of the Lifetime Health Advisory

Lifetime HA = (3.5 mg/L) (20%) = 0.70 mg/L (700 μg/L)

E. Evaluation of Carcinogenic Potential

- The International Agency for Research on Cancer (IARC) (1976) has classified carbaryl in Group 3; i.e., this chemical cannot be classified as to its carcinogenicity for humans.

- Applying the criteria described in EPA's guidelines for assessment of carcinogenic risk (U.S. EPA, 1986), carbaryl may be classified in Group D: not classifiable. This category is for substances with inadequate animal evidence of carcinogenicity.

V. Other Criteria, Guidance and Standards

- The U.S. EPA Office of Research and Development determined an Acceptable Daily Intake (ADI) of 0.096 mg/kg/day based on a rat chronic oral NOAEL of 9.6 mg/kg/day (Carpenter et al., 1961) with an uncertainty factor of 100.

- The National Academy of Sciences (NAS) determined an ADI of 0.082 mg/kg/day based on a rat chronic oral NOAEL of 8.2 mg/kg/day (Union Carbide, 1958) and an uncertainty factor of 100.

- The NAS has also determined a Suggested-No-Adverse-Response Level (SNARL) of 0.574 mg/L, based on an ADI of 0.082 mg/kg/day (70-kg adult consuming 2 L/day and a 20% source contribution factor) (NAS, 1977).

- The U.S. EPA has established residue tolerances for carbaryl in or on raw agricultural commodities that range from 0.1 to 100 ppm (CFR, 1985).

VI. Analytical Methods

- Analysis of carbaryl is by a high performance liquid chromatographic (HPLC) procedure used for the determination of N-methyl-carbamoyloximes and N-methylcarbamates in drinking water (U.S. EPA, 1988). In this method the water sample is filtered and a 400-μL aliquot is injected into a reverse phase HPLC column. Separation of compounds is achieved using gradient elution chromatography. After elution from the HPLC column, the compounds are hydrolyzed with sodium hydroxide. The methyl amine formed during hydrolysis is reacted with o-phthalaldehyde (OPA) to form a fluorescent derivative which is detected using a fluorescence detector. The method detection limit has been estimated to be 2.0 μg/L for carbaryl.

VII.   Treatment Technologies

- Available data indicate that granular-activated carbon (GAC) adsorption, ozonation and conventional treatment will remove carbaryl from water. The percentage removal efficiency ranged from 43 to 99%.

- Whittaker (1980) determined adsorption isotherms using GAC on laboratory prepared carbaryl in water solutions.

- Pilot studies proved that GAC is 99% effective for carbaryl removal (Whittaker et al., 1980 and 1982). Two columns, each packed with 37 kg (80 lbs) of two different GAC, were studied at an empty bed contact time of 8 minutes and an optimum flow rate of 1 gpm.

- Laboratory studies for both batch and flow-through columns were used to examine carbaryl adsorption on two different GAC particle sizes (Whittaker et al., 1982). Data were fitted to both Langmuir and Freundlich isotherms; the monolayer capacity was calculated to be 800 moles carbaryl/g and 1,250 moles carbaryl/g for the 1.2 mm and 0.6 mm GAC, respectively.

- Ozonation has been 99% effective in removing carbaryl and its hydrolysis product, naphthol, from aqueous solution (Shevchenko et al., 1982). Carbaryl and naphthol were not detected in the treated effluent after the addition of 24.8 mg/L and 4.8 mg/L of ozone, respectively. Before ozonation can be used to treat carbaryl contaminated drinking water, however, the identity and toxicity of the resulting degradates must be established.

- Conventional water treatment by alum coagulation, 30-minute settling period and filtration removed 56% of the carbaryl present (Whittaker et al., 1982). Alum dosage of 100 mg/L plus the addition of 1 mg/L of anionic polymer achieved this degree of removal of carbaryl from wastewater.

- A 3-day settling period without any chemical treatment yielded a 50% carbaryl concentration reduction (Holiday and Hardin, 1981).

- Treatment technologies for the removal of carbaryl from water are available and have been reported to be effective. However, selection of individual or combinations of technologies to attempt carbaryl removal from water must be based on a case-by-case technical evaluation, and an assessment of the economics involved.

VIII.   References

Andrianova, M.M. and I.V. Alekseev.* 1969. Carcinogenic properties of Sevin, Maneb, Ciram and Cineb. Vopr. Pitan. 29:71–74. Unpublished report. MRID 00080671.

Benson, B., W. Scott and R. Beliles.* 1967. Sevin: safety evaluation by teratological study in the mouse. Unpublished report. MRID 00118363.

Carpenter, C.P., C.S. Weil, P.E. Palm, M.W. Woodside, J.H. Nair and H.F. Smyth. 1961. Mammalian toxicity of 1-napthyl-N-methylcarbamate (Sevin insecticide). J. Agr. Food Chem. 9:30–39.

CFR. 1985. Code of Federal Regulations. 40 CFR 180.169. July 1. pp. 274–276.

CHEMLAB. 1985. The Chemical Information System, CIS, Inc. In: U.S. EPA. 1985. U.S. Environmental Protection Agency. Pesticide survey chemical profile. Final Report. Contract No. 68–01–6750. Office of Drinking Water.

Chin, B.H., J.M. Eldridge and L.J. Sullivan. 1974. Metabolism of carbaryl by selected human tissues using an organ-maintenance technique. Clin. Toxicol. 7(1):37–56.

Collins, T.F.X., W.H. Hansen and H.V. Keeler. 1971. The effect of carbaryl on reproduction of the rat and the gerbil. Toxicol. Appl. Pharmacol. 19:202–216.

Comer, S.W., D.C. Staiff, J.F. Armstrong and H.R. Wolfe. 1975. Exposure of workers to carbaryl. Bull. Environ. Contam. Toxicol. 13(4):385–391.

Dikshith, T.S.S., P.K. Gupta, J.S. Gaur, K.K. Datta and A.K. Mathur. 1976. Ninety-day toxicity of carbaryl in male rats. Environ. Res. 12:161–170.

Falzon, M., Y. Fernandez, C. Cambon-Gros and S. Mitjavila. 1983. Influence of experimental hepatic impairment on the toxicokinetics and the anticholinesterase activity of carbaryl in the rat. J. Appl. Toxicol. 3(2):87–89.

Feldmann, R.J. and H.I. Maibach.* 1974. Percutaneous penetration of some herbicides in man. Toxicol. Appl. Pharmacol. 28:126–132. Unpublished report. MRID 00031050.

Gaines, V.B.* 1960. The acute toxicity of pesticides to rats. Toxicol. Appl. Pharmacol. 2:88–99. MRID 00005467.

Golbs, S., M. Kuehnert and F. Leue. 1975. Prenatal toxicity of Sevin (carbaryl) for Wistar rats. Arch. Exp. Veterinaermed. 29(4):607–614.

Holiday, A.D. and D.P. Hardin. 1981. Activated carbon removes pesticides from wastewater. Chem. Eng. 88(6):88–89.

Houston, J.B., D.G. Upshall and J.W. Bridges. 1974. Pharmacokinetics and metabolism of two carbamate insecticides, carbaryl and landrin, in the rat. Xenobiotica. 5(10):637–648.

IARC. 1976. International Agency for Research on Cancer. IARC monographs on the evaluation of carcinogenic risk of chemicals to man. Lyon, France: IARC. 12:37–48.

Khasawinah, A.M. and G.C. Holsing.* 1977a. Hydrolysis of carbaryl in aqueous buffer solutions. In: Metabolism and environmental fate, Carbaryl Registration Standard. Unpublished study received Nov. 30, 1984 under 264–327. Submitted by Union Carbide Corporation, Research Triangle Park, N.C. Accession No. 255799.

Khasawinah, A.M. and G.C. Holsing.* 1977b. Photodegradation of carbaryl in aqueous buffer solutions. In: Metabolism and environmental fate, Carbaryl Registration Standard. Unpublished study received Nov. 30, 1984 under 264–327. Submitted by Union Carbide Corporation, Research Triangle Park, N.C. Accession No. 255799.

Khasawinah, A.M. and G.C. Holsing.* 1978. Fate of carbaryl in soil. In: Metabolism and environmental fate, Carbaryl Registration Standard. Unpublished study received Nov. 30, 1984 under 264–327. Submitted by Union Carbide Corporation, Research Triangle Park, N.C. Accession No. 255799.

Krishna, J.G. and J.E. Casida.* 1965. Fate of ten variously labeled methyl- and dimethyl-carbamate-C_{14} insecticide chemicals in rats. Unpublished report. MRID 00049134.

Lechner, D.M.W. and M.S. Abdel-Rahman. 1984. A teratology study of carbaryl and malathion mixtures in rat. J. Toxicol. Environ. Health. 14:267–278.

Lehman, A.J. 1959. Appraisal of the safety of chemicals in foods, drugs and cosmetics. Assoc. Food Drug Off. U.S.

Meister, R., ed. 1983. Farm chemicals handbook. Willoughby, OH: Meister Publishing Company.

Murray, F.J., R.E. Staples and B.A. Schwetz. 1979. Teratogenic potential of carbaryl given to rabbits and mice by gavage or by dietary inclusion. Toxicol. Appl. Pharmacol. 51(1):81–89.

NAS. 1977. National Academy of Sciences. Drinking water and health. Washington, DC: National Academy Press.

Rittenhouse, J.R., J.K. Narcisse and R.D. Cavalli.* 1972. Acute oral toxicity to rats of Orthene in combination with five other cholinesterase-inhibiting materials. Unpublished report. MRID 00014933.

Rocchi, P., P. Perocco, W. Alberghini, A. Fini and G. Prodi. 1980. Effect of pesticides on scheduled and unscheduled DNA synthesis of rat thymocytes and human lymphocytes. Arch. Toxicol. 45:101–108.

Schering, A.G.* 1963. Promecarb (SN 34615): long-term feeding study in rats (ZK No. 3858). Unpublished report. MRID 00081723.

Shevchenko, M.A., P.N. Taran and P.V. Marchenko. 1982. Modern methods for purifying water from pesticides. Soviet Journal of Water Chemistry and Technology. 4(4):53–71.

STORET. 1988. STORET Water Quality File. Office of Water. U.S. Environmental Protection Agency (data file search conducted in May, 1988).

Union Carbide. 1958. Chronic oral feeding of Sevin to rats. Internal Report No. 21–88. Cited in: NAS. 1977. National Academy of Sciences. Drinking water and health. Washington, DC: National Academy Press.

U.S. EPA. 1984. U.S. Environmental Protection Agency. Method 531 — Measurement of N-methyl carbamoyloximes and N-methylcarbamates in drinking water by direct aqueous injection HPLC with post-column derivatization. Cincinnati, OH: Environmental Monitoring and Support Laboratory.

U.S. EPA. 1986. U.S. Environmental Protection Agency. Guidelines for carcinogen risk assessment. Fed. Reg. 51(185):33992–34003. September 24.

U.S. EPA. 1988. U.S. Environmental Protection Agency. Method 531.1 — Measurement of N-methyl carbamoyloximes and N-methylcarbamates in drinking water by direct aqueous injection HPLC with post-column derivatization. April 15, 1988. Cincinnati, OH: Environmental Monitoring and Support Laboratory.

Vandekar, M. 1965. Observations of the toxicity of carbaryl, folithion and 3-isopropylphenyl-N-methylcarbamate in a village-scale trial in southern Nigeria. Bull. W.H.O. 33:107–115. MRID 000365173.

Weil, C.J.* 1956. Special report on subacute oral toxicity studies on Compound 7744. Unpublished report. MRID 00076124.

Weil, C., M.W. Woodside, J. Bernard, D. Crawford and P. Baker.* 1968. Sevin: results of feeding in the diet of rats for one week and for one week plus one day on control diets. Unpublished report. MRID 00118383.

Weil, C.S., M.W. Woodside, C.P. Carpenter and H.F. Smyth. 1971. Current status of tests of carbaryl for reproductive and teratogenic effects. Unpublished report. MRID 0008064.

Weil, C.S. 1972. Comparative study of dietary inclusion versus stomach intubation on three generations on reproduction, on teratology and on mutagenesis. Unpublished report. MRID 00125161.

Whittaker, K.F. 1980. Adsorption of selected pesticides by activated carbon using isotherm and continuous flow column systems. Ph.D. Thesis, Purdue University.

Whittaker, K.F., J.C. Nye, R.F. Wukasch and H.A. Kazimier. 1980. Cleanup and collection of wastewater generated during the cleanup of pesticide application equipment. In: Control of Hazardous Material Spills, Proceedings of a National Conference. pp. 141–144.

Whittaker, K.F., J.C. Nye, R.F. Wukasch, R.J. Squires, A.C. York and

H.A.Kazimier. 1982. Collection and treatment of wastewater generated by pesticide application. EPA Report No. 600/2–82–028.

Wills, J.H., E. Jameson and F. Coulston. 1968. Effects of oral doses of carbaryl on man. Clin. Toxicol. 1:265–271.

Windholz, M., S. Budavari, R.F. Blumetti and E.S. Otterbein, eds. 1983. The Merck Index, 10th ed. Rahway, NJ: Merck and Co., Inc. pp. 246–247.

Zabezhinski, M.A.* 1970. Possible carcinogenic effect of β-Sevin. Voprosy Onkoologii. 16:106–107. In Russian: translation. Unpublished report. MRID 00086672.

---

* Confidential Business Information submitted to the Office of Pesticide Programs.

Carboxin

I. General Information and Properties

A. CAS No. 5234–68–4

B. Structural Formula

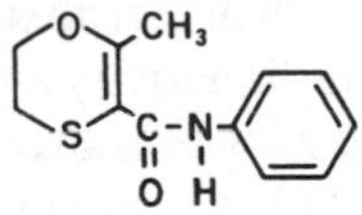

5,6-Dihydro-2-methyl-N-phenyl-1,4–Oxathiin-3-carboxamide

C. Synonyms

- Carbathiin; Carboxine; D-735; DCMO; DMOC; F735; Vitavax (Meister, 1983).

D. Uses

- Carboxin is a systemic fungicide, a seed protectant, and a wood preservative (Mcister, 1983).

E. Properties (Meister, 1983; Windholz et al., 1983; Wo and Shapiro, 1983; Worthing, 1983; TDB, 1985)

Chemical Formula	$C_{12}H_{13}O_2NS$
Molecular Weight	235.31
Physical State (25°C)	Crystals
Boiling Point	--
Melting Point	93 to 95°C
Density	--
Vapor Pressure (20°C)	< 1 mm Hg
Specific Gravity	--
Water Solubility (25°C)	170 mg/L
Log Octanol/Water Partition Coefficient	--
Taste Threshold	--

Odor Threshold   --
Conversion Factor   --

F. Occurrence

- No information was found in the available literature on the occurrence of carboxin.

G. Environmental Fate

- Carboxin is rapidly metabolized (oxidized by flavin enzymes found in fungi mitochondria) in aerobic soil. When applied to soil (aerobic conditions), more than 95% of the carboxin was degraded within 7 days. The major degradation product was carboxin sulfoxide, which represented 31 to 54% of the applied radioactivity at 7 days after treatment. Several minor degradation products were also formed (carboxin sulfone, p-hydroxy carboxin and $^{14}CO_2$). Carboxin was degraded in sterile soil but at a much slower rate than in nonsterile soil (46 to 72% degraded in 7 days). This would indicate that soil metabolism of carboxin under aerobic conditions is primarily by microbial processes. Carboxin sulfoxide is stable in anaerobic soil (Chin et al., 1969, 1970a,b, 1972; Dzialo and Lacadie, 1978; Dzialo et al., 1978; Spare, 1979).

- Carboxin sulfoxide, a major metabolite of carboxin, photodegrades to unknown compounds. After 7 days of incubation, 49% of the applied radioactivity was present as unknown compounds (Smilo et al., 1977).

- Carboxin does not readily adsorb to soil [K value (adsorption coefficient) < 1] and both carboxin and carboxin sulfoxide are very mobile in soil with about half of the applied radioactivity leaching through 12-inch columns of clay loam soils (Lacadie et al., 1978; Dannals et al., 1976).

- In aqueous solution, carboxin was oxidized to carboxin sulfoxide and carboxin sulfone within 7 days (Chin et al., 1970a).

II.   Pharmacokinetics

A. Absorption

- Waring (1973) administered carboxin (Vitavax) by gavage to groups of four to six female New Zealand White rabbits (age not specified; 2.5 to 3 kg) and Wistar rats (age not specified; 200 to 250 g) at 1 mMol/kg (235 mg/kg). In the rats, an average of 40% of the dose was excreted in the feces, mostly as unchanged carboxin. In the rabbits, an average of 10%

was recovered in the feces. These data suggest that carboxin is not completely absorbed from the gut, especially in rats.

B. Distribution

- Waring (1973) administered single oral doses of carboxin (Vitavax, 6.3 μCi/rat) to female Wistar rats (age not specified; 200 to 250 g). Carboxin was labeled either in the heterocyclic or aromatic ring and distribution of label was assessed by autoradiography of whole-body sections. After 2 hours, label was localized in the liver, intestinal tract and salivary gland. After 6 hours, label was also present in the kidney. Only trace levels remained in any tissue after 48 hours. There were no differences in the distribution of the two labeled compounds.

- Nandan and Wagle (1980) fed carboxin to male albino rats (age not specified) for 28 days at dietary levels of 0, 100, 1,000 or 10,000 ppm. Based on the dietary assumptions of Lehman (1959), 1 ppm in the diet of rats equals approximately 0.05 mg/kg/day. Therefore, these levels correspond to 0, 5, 50 and 500 mg/kg/day. In animals fed the highest dose, maximum levels were detected in the liver (140 μg/g), with lower levels in the kidney (123 μg/g), heart (58 μg/g) and muscle (22 μg/g).

C. Metabolism

- In the study by Waring (1973), as described previously, female New Zealand White rabbits (age not specified; 2.5 to 3 kg) and Wistar rats (age not specified; 200 to 250 g) were given single oral doses of carboxin by gavage at 1 mMol/kg (235 mg/kg). The principal metabolic pathway was found to be ortho- or parahydroxylation, followed by glucuronidation. In the rats, 32% of the dose was excreted in urine as glucuronides and 7% as unconjugated phenols. In the rabbits, 85% of the dose was excreted in urine as glucuronides and 3% as free phenols. The pattern of phenolic metabolites was the same for carboxin labeled in either the heterocyclic or the aromatic rings, indicating that cleavage of the compound did not occur.

D. Excretion

- In the study by Waring (1973), as described previously, female New Zealand White rabbits (age not specified; 2.5 to 3 kg) and Wistar rats (age not specified; 200 to 250 g) were given single oral doses of carboxin by gavage at 1 mMol/kg (235 mg/kg). In the rats, 41% was excreted in the feces (largely unchanged carboxin) and 54% was excreted in the urine (15% parent compound, 32% glucuronides, 7% free phenols). In the rabbits, 10% was excreted in the feces and 90% was excreted in the urine (2% parent compound, 85% glucuronides, 3% free phenols).

III.   Health Effects

A. Humans

1. *Short-term Exposure*

- A 7-year-old boy developed headaches and vomiting within 1 hour after ingesting several handfuls of wheat seed treated with carboxin. He was administered ipecac (an emetic) and was asymptomatic 2 hours later. No estimate of the ingested dose was provided (PIMS, 1980).

2. *Long-term Exposure*

- No information was found in the available literature on the long-term exposure health effects of carboxin in humans.

B. Animals

1. *Short-term Exposure*

- Reagan and Becci (1983) reported that the acute oral LD_{50} for technical carboxin (purity not specified) in young CD–1 mice (age not specified) was 4,150 mg/kg for males and 2,800 mg/kg for females. The average LD_{50} was reported to be 3,550 mg/kg.

- RTECS (1985) reported that the acute oral LD_{50} for carboxin (purity not specified) in the rat (age not specified) was 430 mg/kg.

- Nandan and Wagle (1980) fed carboxin to male albino rats (age not specified) for 28 days at dietary levels of 0, 100, 1,000 or 10,000 ppm. Based on the authors' measurements of food consumption and assuming average body weights of 0.1 kg, these levels corresponded to doses of about 0, 5.5, 59.0 or 311 mg/kg/day. A Lowest-Observed-Adverse-Effect Level (LOAEL) of 100 ppm (5.5 mg/kg/day) was tentatively identified in this study based on fluid accumulation in the liver. However, due to a number of deficiencies in this study, it is not possible to accurately evaluate its validity. These deficiencies include a lack of information on the test animals (e.g., condition at study initiation, numbers used) and the absence of statistical analyses.

2. *Dermal/Ocular Effects*

- Holsing (1968a) applied carboxin (D-735; purity and vehicle not specified) to the intact or abraded abdominal skin of rabbits (10/sex/dose; age not specified) at concentrations of 1,500 or 3,000 mg/kg. Five animals of each sex served as controls. Test animals were exposed occlusively for 6 to 8 hours, 5 days per week, for 3 weeks (15 applications). No signs of dermal irritation were observed. The test material stained the skin and precluded readings for erythema.

3. *Long-term Exposure*

- Ozer (1966) administered carboxin (D-735; purity not specified) to weanling FDRL (Wistar-derived) rats (10/sex/dose; controls: 15/sex) for 90 days at dietary concentrations of 0, 200, 600, 2,000, 6,000 or 20,000 ppm, intended by the author to correspond to approximate dosage levels of 0, 10, 30, 100, 300 or 1,000 mg/kg/day. All animals survived the 90-day treatment period. Growth, food efficiency, hematology, blood chemistry and urinalysis were reported to be similar in all groups with the exception of increased blood urea nitrogen and decreased hemoglobin at the 12-week interval in females that received 20,000 ppm (1,000 mg/kg/day). No significant dose-related gross pathological changes were observed. Microscopically, a significant number of inflammatory degenerative renal changes were found in animals that received doses of 600 ppm (30 mg/kg/day) or higher. These changes included focal chronic inflammation, protein casts and cortical tubular degeneration. In two animals that received 2,000 ppm (100 mg/kg/day), some fibrosis in the medulla was observed. Based on renal changes, a LOAEL of 600 ppm (30 mg/kg/day) and a No-Observed-Adverse-Effect Level (NOAEL) of 200 ppm (10 mg/kg/day) can be identified.

- Jessup et al. (1982) administered carboxin (technical Vitavax; purity not specified) to 6-week-old Charles River CD-1 mice (50/sex/dose; controls: 75/sex) for approximately 84 weeks at dietary concentrations of 0, 50, 2,500 or 5,000 ppm. The authors indicated that these dietary levels corresponded to doses of about 0, 8, 385 or 751 mg/kg/day for males and 0, 9, 451 or 912 mg/kg/day for females. No compound-related effects on general behavior or appearance were reported. Survival rates of females receiving 5,000 ppm (912 mg/kg/day) were significantly (p < 0.01) lower than controls. No compound-related effects on body weight gain, food consumption or various hematological parameters were reported. No gross pathologic lesions that were considered to be related to compound administration were observed at necropsy in any mice in any treatment group. Microscopically, compound-related effects on the liver, consisting of hypertrophy of the centrilobular parenchymal cells, were observed in mice in the 2,500- or 5,000-ppm dose groups (385 and 751 mg/kg/day for males; 451 and 912 mg/kg/day for females). No other nonneoplastic lesions that could be attributed to compound administration were observed. The NOAEL in this study is 50 ppm (8 mg/kg/day for males; 9 mg/kg/day for females) based on hepatic effects.

- Holsing (1969a) administered carboxin (technical D-735; considered to be 100% active ingredient) to Charles River rats (30/sex/dose; controls: 60/sex) for 2 years at dietary concentrations of 0, 100, 200 or 600 ppm. Based on the dietary assumptions of Lehman (1959), 1 ppm in the diet of rats equals approximately 0.05 mg/kg/day. Therefore, these dietary levels correspond to dose levels of approximately 0, 5, 10 or 30 mg/kg/day. While the age of the animals was not specified, the weights of the male rats at initiation ranged from 65 to 88 g and the weight of the female rats ranged from 59 to 85 g. No compound-related effects in terms of physical appearance, behavior, hematology, blood chemistry or urinalysis were reported at any dose level. Observations at terminal necropsy did not reveal any compound-related gross or microscopic changes in the organs of animals at any dose level. At the 600-ppm level (30 mg/kg/day), body weight gain was significantly depressed in both sexes, and food consumption by males was lower than that of controls throughout most of the study (significantly lower during the first 26 weeks). Food consumption by females at all dose levels was generally comparable to controls. Compound-related effects included an increase in mortality at 18 months in males that received 600 ppm (30 mg/kg/day), and changes in absolute and relative organ weights at all dose levels, including increases in thyroid weight and decreases in kidney, heart and spleen weight and histopathological changes in the kidneys at the 12-month interval in both sexes at 200 and 600 ppm. Most of these effects were inconsistent and were not observed at the end of the study period. At the end of the 2-year study, decreased kidney weights were observed in males at 600 ppm (30 mg/kg/day). Therefore, based on the information presented in this study, a NOAEL of 200 ppm (10 mg/kg/day) was identified.

- Holsing (1969b) administered carboxin (technical D-735; considered to be 100% active ingredient) to young adult beagle dogs (4/sex/dose; controls: 6/sex) for 2 years at dietary concentrations of 0, 100, 200 or 600 ppm. Based on the dietary assumptions of Lehman (1959), 1 ppm in the diet of rats equals approximately 0.05 mg/kg/day. Therefore, these dietary levels have been calculated to correspond approximately to 0,

2.5, 5.0 or 15.0 mg/kg/day. No treatment-related effects were reported on survival; body weight gain; food consumption; organ weights; organ-to-body weight ratios; hematological, blood chemistry or urinary parameters; liver and kidney function tests; or gross and histopathological observations. Based on this information, a NOAEL of 600 ppm (15 mg/kg/day; the highest dose tested) was identified.

4. Reproductive Effects

- In a three-generation reproduction study, Holsing (1968b) administered carboxin (technical D-735; 97% active ingredient) to Charles River rats (10 males/dose, 20 females/dose; controls [age not specified]: 15 males, 30 females) at dietary concentrations of 0, 100, 200 or 600 ppm. Based on the dietary assumptions of Lehman (1959), these dietary levels have been calculated to correspond to dose levels of approximately 0, 5, 10 or 30 mg/kg/day. Criteria evaluated included fertility, gestation, live birth and lactation indices, litter size and the physical appearance and growth of the pups. No compound-related effects on reproductive performance were reported at any dose level. A compound-related effect on the progeny (moderate growth suppression in the nursing male and female pups of all three generations) was observed at the 600-ppm (30 mg/kg/day) dose level. Based on the information presented in this study, a NOAEL of 200 ppm (10 mg/kg/day) was identified.

5. Developmental Effects

- Schardein and Laughlin (1981) administered technical Vitavax (carboxin; 99% active ingredient) by gavage at doses of 0, 75, 375 or 750 mg/kg/day to 7- to 8-month-old Dutch Belted rabbits (10/dose) on days 6 through 27 of gestation. The compound was administered in a 0.5% carboxymethyl cellulose vehicle. No treatment-related effects on maternal mortality, appearance, behavior or body weight were reported. Four females aborted on days 27 and 28 of gestation (one at 375 mg/kg/day, three at 750 mg/kg/day). Examination for fetal malformations revealed no compound-related differences between the control and treatment groups. Based on the frequency of abortion, a NOAEL of 75 mg/kg/day and a LOAEL of 375 mg/kg/day were identified.

- Knickerbocker (1977) administered carboxin (technical Vitavax; purity not specified) in corn oil by gavage at doses of 0, 4, 20 or 40 mg/kg/day to sexually mature (age not specified) Sprague-Dawley rats (20/dose) on days 6 through 15 of gestation. No compound-related effects were observed on reproduction, gestation or in skeletal or soft-tissue development. Based on the information presented, a NOAEL of 40 mg/kg/day (the highest dose tested) was identified.

6. *Mutagenicity*

- Brusick and Weir (1977) conducted a mutagenicity assay using *Salmonella typhimurium* strains TA1535, TA1537, TA1538, TA98 and TA100, and *Saccharomyces cerevisiae* strain D4. Carboxin (purity not specified) was tested without activation at concentrations up to 500 μg/plate and with activation at concentrations up to 100 μg/plate. No mutagenic activity was detected in this assay.

- Byeon et al. (1978) reported that carboxin (Vitavax; purity not specified) tested at concentrations up to 1 mg/plate was not found to be mutagenic in an Ames assay using *S. typhimurium* strains TA1535, TA1538, TA98 and TA100.

- Brusick and Rabenold (1982) conducted an Ames assay using technical carboxin (Vitavax; 98% active ingredient) at concentrations up to 5,000 μg/plate. No mutagenic activity was detected, with or without activation, in *S. typhimurium* strains TA1535, TA1537, TA1538, TA98 and TA100.

- Myhr and McKeon (1982) reported the results of a primary rat hepatocyte unscheduled DNA synthesis assay using carboxin (technical Vitavax; 98% active ingredient). The test compound produced significant increases in the nuclear labeling of primary rat hepatocytes over a concentration range of 5.13 to 103 μg/mL.

7. *Carcinogenicity*

- Holsing (1969a) administered carboxin (technical D-735; considered to be 100% active ingredient) to Charles River rats (30/sex/dose; controls: 60/sex) for 2 years at dietary concentrations of 0, 100, 200 or 600 ppm. Based on the dietary assumptions of Lehman (1959), 1 ppm in the diet of rats equals approximately 0.05 mg/kg/day. While the age of the animals was not specified, the weights of the male rats at initiation ranged from 65 to 88 g and the weights of the female rats ranged from 59 to 85 g. Therefore, dietary levels correspond to approximately 0, 5, 10 or 30 mg/kg/day. No evidence of increased tumor frequency was detected by either gross or histological examination of tissues.

- Jessup et al. (1982) administered carboxin (technical Vitavax; purity not specified) to 6-week-old Charles River CD-l mice (50/sex/dose; controls: 75/sex) for approximately 84 weeks at dietary concentrations of 0, 50, 2,500 or 5,000 ppm. The authors indicated that these dietary levels corresponded to dosage levels of approximately 0, 8, 385 or 751 mg/kg/day for males and 0, 9, 451 or 912 mg/kg/day for females. Survival rates of females receiving 5,000 ppm (912 mg/kg/day) were significantly (p < 0.01) lower than those of controls. No compound-related gross pathologic lesions were observed at necropsy in any treatment group. Microscopically, compound-related effects on the liver, consisting of hypertrophy of the centrilobular parenchymal cells, were observed in mice in the 2,500- or 5,000-ppm dose groups (385 and 751 mg/kg/day for males; 451 and 912 mg/kg/day for females). In males, the incidence of pulmonary adenoma/alveolar-bronchiolar adenoma was 13/75, 7/49, 7/50 and 17/50 at 0, 50, 2,500 and 5,000 ppm, respectively. The incidence at the high dose (34%) may have been compound-related based on comparison with the incidence in controls (17%). The difference was statistically significant (p < 0.01) using Cox's test for adjusted trend and the Kruskall Wallis tests for life-table data and adjusted incidence. However, based on the opinions of pathologists who reviewed the data and on historical data on tumor incidence in control Charles River CD-1 mice, the authors concluded that the increased incidence was not compound-related. Historical data indicate that in six 18-month studies, the incidence of lung adenomas ranged from 6.3 to 16.7%; in seven 20- to 22-month studies, the incidence of lung adenomas ranged from 4.0 to 31.1%.

IV.   Quantification of Toxicological Effects

A. One-day Health Advisory

Appropriate data for calculating a One-day HA value are not available. It is recommended that the Longer-term HA value for the 10-kg child (1.0 mg/L, calculated below) be used as the One-day HA value.

B. Ten-day Health Advisory

Appropriate data for calculating a Ten-day HA value are not available. The 22-day rabbit teratogenicity study by Schardein and Laughlin (1981) was considered for the development of the Ten-day HA. However, the NOAEL (75 mg/kg/day) identified in this study is far in excess of the NOAEL (10 mg/kg/day) identified in the 90-day rat feeding study reported by Ozer (1966), suggesting that the rat is the more sensitive species. It is therefore recommended that the Longer-Term HA value for the 10-kg child (1.0 mg/L, calculated below) be used as the Ten-day value.

C. Longer-term Health Advisory

The study by Ozer (1966) has been selected to serve as the basis for calculating the Longer-term HA for carboxin. In this study, weanling rats were exposed to carboxin in the diet for 90 days. At 30 mg/kg/day, there was histological evidence of renal injury. At 10 mg/kg/day, no effects were detected on any parameter measured, including growth, hematology, blood chemistry, urinalysis, gross pathology and histopathology. Based on these data, a NOAEL of 10 mg/kg/day was identified. This value is supported by the subchronic (84-week) feeding study in mice by Jessup et al. (1982), which identified a NOAEL of 8 to 9 mg/kg/day based on the absence of effects on appearance, behavior, mortality, weight gain, hematology, gross pathology and histopathology.

The Longer-term HA for the 10-kg child is calculated as follows:

$$\text{Longer-term HA} = \frac{(10 \text{ mg/kg/day}) (10 \text{ kg})}{(100) (1 \text{ L/day})} = 1.0 \text{ mg/L} (1,000 \text{ } \mu g/L)$$

The Longer-term HA for the 70-kg adult is calculated as follows:

$$\text{Longer-term HA} = \frac{(10 \text{ mg/kg/day}) (70 \text{ kg})}{(100) (2 \text{ L/day})} = 3.5 \text{ mg/L} (4,000 \text{ } \mu g/L)$$

D. Lifetime Health Advisory

The study by Holsing (1969a) has been selected to serve as the basis for calculation of the Lifetime HA for carboxin. In this study, rats were exposed to carboxin in the diet for 2 years. At 10 mg/kg/day, no significant effects were detected on appearance, behavior, body weight, mortality, hematology, blood chemistry, urinalysis, gross pathology or histopathology. Based on these data, a NOAEL of 10 mg/kg/day was identified. This value is supported by a 90-day rat study (Ozer, 1966), which also identified a NOAEL of 10 mg/kg/day; a 2-year feeding study in dogs by Holsing (1969b), which identified a NOAEL of 15 mg/kg/day; and an 84-week mouse study (Jessup et al., 1982), which identified a NOAEL of 8 mg/kg/day for males and 9 mg/kg/day for females.

Using the NOAEL of 10 mg/kg/day, the Lifetime HA for carboxin is calculated as follows:

Step 1: Determination of the Reference Dose (RfD)

$$\text{RfD} = \frac{(10 \text{ mg/kg/day})}{(100)} = 0.1 \text{ mg/kg/day}$$

Step 2: Determination of the Drinking Water Equivalent Level (DWEL)

$$DWEL = \frac{(0.1 \text{ mg/kg/day}) (70 \text{ kg})}{(2 \text{ L/day})} = 3.5 \text{ mg/L } (4{,}000 \text{ } \mu g/L)$$

Step 3: Determination of the Lifetime Health Advisory

$$Lifetime \text{ HA} = (3.5 \text{ mg/L}) (20\%) = 0.7 \text{ mg/L } (700 \text{ } \mu g/L)$$

E. Evaluation of Carcinogenic Potential

- Jessup et al. (1982) reported a possible compound-related increase in pulmonary adenoma/alveolar-bronchiolar adenoma frequency in male CD-1 mice that received carboxin in the diet at 751 mg/kg/day.

- Holsing (1969a) fed Charles River rats carboxin at dietary levels up to 30 mg/kg/day for 2 years, and detected no compound-related histopathologic changes. This study is limited, however, by the following factors: inadequate numbers of animals were used; survival was generally poor and, therefore, late-developing lesions may not have been detected; all tissues from all animals were not examined microscopically; and there was no adjustment in dietary levels of carboxin to account for growth of the test animals.

- The International Agency for Research on Cancer has not evaluated the carcinogenic potential of carboxin.

- Applying the criteria described in EPA's guidelines for assessment of carcinogenic risk (U.S. EPA, 1986a), carboxin is classified in Group D: not classifiable. This category is for substances with inadequate human and animal evidence of carcinogenicity or for which no data are available.

V.   Other Criteria, Guidance and Standards

- No existing criteria or standards for oral exposure to carboxin were located.

- The U.S. EPA Office of Pesticide Programs (OPP) has proposed an Acceptable Daily Intake (ADI) of 0.4 mg/kg/day, based on a NOAEL of 200 ppm established in a 2-year rat feeding study and an uncertainty factor of 100 (U.S. EPA, 1981).

- The U.S. EPA has established residue tolerances for carboxin in or on raw agricultural commodities that range from 0.01 to 0.5 ppm (CFR, 1979).

VI.   Analytical Methods

• Analysis of carboxin is by a gas chromatographic (GC) method applicable to the determination of certain nitrogen-phosphorus–containing pesticides in water samples (U.S. EPA, 1986b). In this method, approximately 1 liter of sample is extracted with methylene chloride. The extract is concentrated and the compounds are separated using capillary column GC. Measurement is made using a nitrogen-phosphorus detector. The method detection limit has not been determined for carboxin but it is estimated that the detection limits for analytes included in this method are in the range of 0.1 to 2 $\mu g/L$.

VII.   Treatment Technologies

• No information regarding treatment techniques to remove carboxin from contaminated waters is currently available.

VIII.   References

Brusick, D.J. and R.J. Weir.* 1977. Mutagenicity evaluation of D-735. CBI Project No. 1683. Final Report 53727. Unpublished study. MRID 00053727.

Brusick, D. and C. Rabenold.* 1982. Mutagenicity evaluation of technical grade Vitavax in the Ames *Salmonella* microsome plate test. CBI Project No. 20988. Final Report. Unpublished study. MRID 00132453.

Byeon, W., H.H. Hyun and S.Y. Lee.* 1978. Mutagenicity of pesticides in the *Salmonella*/microsomal enzyme activation system. Korean J. of Microbiol. 14:128–134. MRID 00061590.

Chin, W.T., L.E. Dannals and N. Kucharczyk.* 1972. Environmental fate studies on Vitavax. Unpublished study. Submitted by Uniroyal Chemical, Bethany, CT. CDL:093515-A. MRID 00002935.

Chin, W.T., G.M. Stone and A.E. Smith.* 1969. Fate of D735 in soil. Unpublished study. Submitted by Uniroyal Chemical, Bethany, CT. CDL:091420. MRID 00003041.

Chin, W.T., G.M. Stone and A.E. Smith.* 1970a. Degradation of carboxin (Vitavax) in water and soil. J. Agric. Food Chem. 18(4):731–732. MRID 05002176.

Chin, W.T., G.M. Stone, A.E. Smith and B. von Schmeling.* 1970b. Fate of carboxin in soil, plants, and animals. In: Proc. Fifth British Insecticide and Fungicide Conf., Nov. 17–20, 1969, Brighton, England. Vol. 2. pp. 322–327. MRID 05004996.

CFR. 1979. Code of Federal Regulations. 40 CFR 180.301. July 1. p. 527.

Dannals, L.E., C.R. Campbell and R.A. Cardona.* 1976. Environmental fate studies on Vitavax. Status report II on PR 70–15, including three updated methods. Unpublished study. Submitted by Uniroyal Chemical, Bethany, CT. CDL:223866-A. MRID 00003114.

Dzialo, D.G. and J.A. Lacadie.* 1978. Aerobic soil study of ^{14}C-Vitavax in sandy soil: Project no. 7746–1. Unpublished study. Submitted by Uniroyal Chemical, Bethany, CT. CDL:236662-F. MRID 00003225.

Dzialo, D.G., J.A. Lacadie and R.A. Cardona.* 1978. Anaerobic soil metabolism of ^{14}C-Vitavax in sandy soil. Unpublished study. Submitted by Uniroyal Chemical, Bethany, CT. CDL:236662-G. MRID 00003226.

Holsing, G.C.* 1968a. Summary: repeated dermal (Leary design)—rabbits. Project No. 798–148. Unpublished study. MRID 00021626.

Holsing, G.C.* 1968b. Three-generation reproduction study—rats. Final Report. Project No. 798–104. Unpublished study. MRID 00003032.

Holsing, G.C.* 1969a. 24-Month dietary administration—albino rats. Final Report. Project No. 798–102. Unpublished study. MRID 00003031.

Holsing, G.C.* 1969b. Two-year dietary administration—dogs. Final Report. Project No. 798–103. Unpublished study. MRID 00003030.

Jessup, D., G. Gunderson and R. Gail.* 1982. Lifetime carcinogenicity study in mice (Vitavax): 399–002a. Unpublished study. MRID 00114139.

Knickerbocker, M.* 1977. Teratologic evaluation of Vitavax technical in Sprague-Dawley rats. Unpublished study. MRID 00003102.

Lacadie, J.A., D.R. Gerecke and R.A. Cardona.* 1978. Vitavax ^{14}C laboratory column leaching study in clay loam. Project No. 7758. Unpublished study. Submitted by Uniroyal Chemical, Bethany, CT. CDL:236662-H. MRID 00003227.

Lehman, A.J. 1959. Appraisal of the safety of chemicals in foods, drugs and cosmetics. Assoc. Food Drug Off. U.S., Q. Bull.

Matthews, R.J.* 1973. Acute LD_{50} rats, oral. Final Report. Unpublished study. MRID 00003012.

Meister, R., ed. 1983. Farm chemicals handbook. Willoughby, OH: Meister Publishing Co.

Myhr, B. and M. McKeon.* 1982. Evaluation of Vitavax technical grade in the primary rat hepatocyte unscheduled DNA synthesis assay. CBI Project No. 20991. Unpublished study. MRID 00132454.

Nandan, D. and D.S. Wagle. 1980. Metabolic effects of carboxin in rats. Symp. Environ. Pollut. Toxicol. pp. 305–312.

Ozer, B.L.* 1966. Report: Subacute (90 day) feeding studies with D-735 in rats. Unpublished study. MRID 00003063.

PIMS. 1980. Pesticide Incident Monitoring System. Summary of reported incidents involving carboxin. Report No. 383. Washington, DC: Health

Effects Branch, Hazard Evaluation Division, Office of Pesticide Programs, U.S. Environmental Protection Agency. October.

Reagan, E. and P. Becci.* 1983. Acute oral LD$_{50}$ assay in mice: (Vitavax Technical). FDRL Study No. 7581A. Unpublished study. MRID 00128469.

RTECS. 1985. Registry of Toxic Effects of Chemical Substances. National Institute for Occupational Safety and Health. National Library of Medicine Online File.

Schardein, J.L. and K.A. Laughlin.* 1981. Teratology study in rabbits: 399–042. Unpublished study. MRID 00086054.

Smilo, A.R., J.A. Lacadie and B. Cardona.* 1977. Photochemical fate of Vitavax in solution. Unpublished study. Submitted by Uniroyal Chemical, Bethany, CT. CDL:231932-C. MRID 00003088.

Spare, W.* 1979. Report: Vitavax microbial metabolism in soil and its effect on microbes. Unpublished study. Prepared by Biospherics, Inc., in cooperation with United States Testing Co., Inc. Submitted by Uniroyal Chemical, Bethany, CT. CDL:098029-A. MRID 00005540.

TDB. 1985. Toxicology Data Bank. MEDLARS II. National Library of Medicine's National Interactive Retrieval Service.

U.S. EPA. 1981. U.S. Environmental Protection Agency. Carboxin. Pesticide Registration Standard. Washington, D.C.: Office of Pesticides and Toxic Substances.

U.S. EPA. 1986a. U.S. Environmental Protection Agency. Guidelines for carcinogen risk assessment. Fed. Reg. 51(185):33992–34003. September 24.

U.S. EPA. 1986b. U.S. Environmental Protection Agency. U.S. EPA Method 1 — Determination of Nitrogen and Phosphorus Containing Pesticides in Ground Water by GC/NPD. Draft. January. Available from U.S. EPA's Environmental Monitoring and Support Laboratory, Cincinnati, OH.

Waring, R.N. 1973. The metabolism of Vitavax by rats and rabbits. Xenobiotica. 3:65–71.

Windholz, M., S. Budavari, R.F. Blumetti and E.S. Otterbein, eds. 1983. The Merck index, 10th ed. Rahway, NJ: Merck and Co., Inc.

Wo, C. and R. Shapiro.* 1983. EPA acute oral toxicity. Report No. T-3449. Unpublished study. MRID 00143944.

Worthing, C. R. 1983. The pesticide manual. British Crop Protection Council.

---

* Confidential Business Information submitted to the Office of Pesticide Programs.

Chloramben

I. General Information and Properties

A. CAS No. 133-90-4

B. Structural Formula:

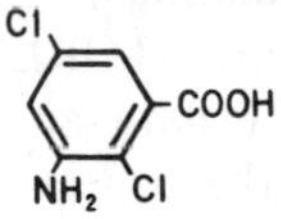

3-Amino-2-5-dichlorobenzoic acid

C. Synonyms

- Acp-m-728, Ambiben; Abiben; Amibin; Amoben; Chlorambed; Chlorambene; NCI-C00055 ornamental weeder; Ornamental weeder; Vegaben; Vegiven (U.S. EPA, 1985).

D. Uses

- Chloramben is used as a preemergent herbicide for weed control (Meister, 1983).

E. Properties (U.S. EPA, 1985; CHEMLAB, 1985)

Chemical Formula	$C_7H_5O_2NCl_2$
Molecular Weight	206.02
Physical State (25°C)	Crystals
Boiling Point	--
Melting Point	200–201°C
Density	--
Vapor Pressure	7×10^{-3} mm Hg (100°C)
Specific Gravity	--
Water Solubility (25°C)	700 mg/L
Log Octanol/Water Partition Coefficient	2.32
Taste Threshold	--

165

Odor Threshold   --
Conversion Factor   --

F. Occurrence

- Chloramben has been found in 13 of 34 surface water samples analyzed and in 1 of 566 ground-water samples (STORET, 1988). Samples were collected at 11 surface water locations and 322 ground-water locations, and chloramben was found in only 1 state. The 85th percentile of all nonzero samples was 2.1 μg/L in surface water and 1.7 μg/L in ground-water sources. This information is provided to give a general impression of the occurrence of this chemical in ground and surface waters as reported in the STORET database. The individual data points retrieved were used as they came from STORET and have not been confirmed as to their validity. STORET data are often not valid when individual numbers are used out of the context of the entire sampling regime, as they are here. Therefore, this information can only be used to form an impression of the intensity and location of sampling for a particular chemical.

G. Environmental Fate

- Sodium chloramben appears to be resistant to hydrolysis. Limited studies indicate that there is no loss of phytotoxicity when aqueous solutions of chloramben are kept in the dark (U.S. EPA, 1981a).

- Photodegradation of aqueous solutions of sodium chloramben appears to occur readily in sunlight. Total loss of phytotoxicity occurs in 2 days. Loss of phytotoxicity on dry soil is somewhat slower, about 30% in 48 hours (U.S. EPA, 1981a).

- Soil bacteria bring about a loss of phytotoxicity in sodium chloramben after several weeks. It appears that this is due to a decarboxylation. The rate of reaction appears to be independent of soil pH within the range of 4.3 to 7.5 (U.S. EPA, 1981a).

- The mobility of sodium chloramben is governed principally by its high solubility in water and its apparent limited strength of adsorption to soil particles. It appears to easily leach down in most soil types by rainfall (U.S. EPA, 1981a).

- Probably all plants grown in contact with sodium chloramben take up the compound. In some plants the subsequent movement of compound away from the roots is very slow, whereas in others it readily spreads throughout the plant. The fate of chloramben in plants includes decomposition, a detoxifying conjugation which proceeds fairly rapidly, or a detoxifying conjugation which goes slowly, if at all (U.S. EPA, 1981a).

- The methyl ester of chloramben acid appears to have the expected properties of a carboxylic acid ester. It is apparently not hydrolyzed after a short period in contact with water at slightly acid pH values (5 to 6). Bacteria-mediated hydrolysis appears to be quick: approximately 50% of the ester is converted to the free acid in about 1 week when in contact with wet soil. A subsequent and slower bacterial reaction, shown by a loss of phytotoxicity, is probably a decarboxylation, as with sodium chloramben (U.S. EPA, 1981a).

- The leaching behavior of the methyl ester is governed by its aqueous solubility, which is much lower than that of the sodium salt (120 ppm and 250,000 ppm, respectively). For a given rainfall, the ester seems to leach down about 15% of the distance traveled by the sodium salt (U.S. EPA, 1981a).

II.   Pharmacokinetics

A. Absorption

- Chloramben is rapidly absorbed from the gastrointestinal tract of Sprague-Dawley female rats (Andrawes, 1984). Based on radioactivity recovered in urine (96.7%) and expired air (0.2%), about 97% of an oral dose (5 μCi/rat) of chloramben is absorbed.

B. Distribution

- Andrawes (1984) reported low levels (up to 0.5% of the administered dose) of chloramben in liver, kidney, lung, muscle, plasma and red blood cells of rats 96 hours after a single oral dose (by gavage).

C. Metabolism

- In rats dosed by gavage, Andrawes (1984) reported that the parent compound accounted for 70% of the applied dose in 24-hour urine.

- Andrawes (1984) identified 5 of 24 urinary metabolites: 3-amino-5-chlorobenzoic acid; 3-aminobenzoic acid; 2,5-dihydroxybenzoic acid; 3,5-dihydroxybenzoic acid; and 2,5-dichloroaniline. Together, these constituted 1.4% of the administered dose.

- Metabolism of chloramben in rats proceeded through dechlorination, deamination, decarboxylation and hydroxylation. Metabolism through oxidative ring cleavage was negligible (Andrawes, 1984).

D. Excretion

- Rats administered chloramben (5 μCi/rat) by gastric intubation excreted over 99% of the dose within 3 to 4 days, mostly within the first 24 hours (Andrawes, 1984). Approximately 96.7% was eliminated in the urine, with lesser amounts in the feces (4.1%) and respiratory gases (0.2%). Only 0.6% remained in the carcass after 3 to 4 days.

III.   Health Effects

A. Humans

- No information was found in the available literature on the human health effects of chloramben.

B. Animals

1. *Short-term Exposure*

- Acute oral LD_{50} values for chloramben range from 2,101 mg/kg (Field, 1980) to 5,000 mg/kg (Field and Carter, 1978) in rats; the acute dermal LD_{50} in rabbits has been reported to be > 2,000 (Field and Field, 1980) or > 5,000 mg/kg (Field, 1978).

- Rees and Re (1978) reported an acute (1 hr) LC_{50} of > 200 mg/L in rat inhalation studies.

- Keller (1959) fed male Holtzman Sprague-Dawley rats (10/dose) chloramben [100% active ingredient (a.i.)] for 28 days in the diet at dose levels of 0, 1,000, 3,000 or 10,000 ppm. Assuming that 1 ppm in the diet of rats is equivalent to 0.05 mg/kg/day (Lehman, 1959), this corresponds to doses of 0, 50, 150 or 500 mg/kg/day. Body weights, food consumption, general appearance and behavior and histopathology were evaluated. There were no statistically significant differences between the treated rats and untreated controls in any parameter measured. Based on this information, a No-Observed-Adverse-Effect Level (NOAEL) of 10,000 ppm (500 mg/kg/day), the highest dose tested, was identified.

2. *Dermal/Ocular Effects*

- Gabriel (1969) applied chloramben (4 to 8 g/kg) to intact and abraded skin of 16 male albino rabbits (8/dose). Test animals were observed for 14 days. No evidence of skin irritation was observed under conditions of the study.

- In a study by Myers et al. (1982), a 1.0% (w/w) chloramben sodium salt suspension produced little or no sensitization reactions in male albino Hartley guinea pigs.

3. *Long-term Exposure*

- In studies by Beliles (1976), weanling Golden Syrian hamsters (12/sex/ dose) were administered technical chloramben (purity not specified) at dose levels of 0, 100, 1,000 or 10,000 ppm (reported to be equivalent to 0, 11, 115 and 1,070 mg/kg/day) in the diet for 90 days. Food consumption, body and organ weights and histopathology were evaluated. No treatment-related adverse effects were reported for any parameter evaluated. Based on this information, a NOAEL of 10,000 ppm (1,070 mg/ kg/day), the highest dose tested, was identified.

- In an 18-month feeding study (U.S. EPA, 1981b), Crl:COBS CD-1 mice (50/sex/dose) were administered technical chloramben (purity not specified) at dietary levels of 0, 100, 1,000 or 10,000 ppm. Assuming that 1 ppm in the diet of mice is equivalent to 0.15 mg/kg/day (Lehman, 1959), this corresponds to doses of about 0, 15, 150 and 1,500 mg/kg/ day. No compound-related effects were observed in terms of survival, general appearance, behavior or changes in body weight. Statistically significant (p < 0.05) changes in organ weights included decreased liver weight in males at 100 ppm, decreased kidney weight in males at 10,000 ppm and decreased kidney weight in females at 10,000 ppm. Since the values for these observations were within normal ranges for this species and no trends were established, the organ weight changes were not attributed to compound administration. Histopathological examinations revealed alterations in the livers of all treated mice. The primary hepatocellular reaction was a histomorphological hepatocellular alteration compatible with that observed in enzyme induction. The typical cellular changes included hepatocyte hypertrophy, increased nuclear size and chromatin content and dense granular eosinophilic cytoplasm. Other changes included scattered foci of individual or small groups of degenerating hepatocytes, hepatocyte vacuolation, cytoplasmic eosinophilic inclusions, and multiple focal small granulomas. Based on the reported hepatic effects, this study identifies a Lowest-Observed-Adverse-Effect Level (LOAEL) of 100 ppm (15 mg/kg/day).

- NCI (1977) administered technical-grade chloramben (90 to 95% active ingredient) to Osborne-Mendel rats (50/sex/dose) and B6C3F$_1$ mice (50/sex/dose) for 80 weeks at dietary levels of 10,000 or 20,000 ppm. Assuming that 1 ppm in the diet of rats is equivalent to 0.05 mg/kg/day and 1 ppm in the diet of mice is equivalent to 0.15 mg/kg/day (Lehman, 1959), this corresponds to doses of 500 or 1,000 mg/kg/day for rats and 1,500 or 3,000 mg/kg/day for mice. Matched controls consisted of 10 animals per sex for each species. Pooled controls consisted of the matched controls plus 75 rats/sex and 70 mice/sex from similarly performed bioassays. Body weights and mortality did not differ between control and treatment groups for both species, and the various (unspecified) clinical signs observed were similar in the control and treatment groups for both species. Based on this information, a NOAEL of 20,000 ppm (1,000 mg/kg/day for rats and 3,000 mg/kg/day for mice), the highest dose tested, was identified for each species.

- In studies conducted by Paynter et al. (1963), albino rats (35/sex/dose) were administered chloramben (97% pure) in the diet for 2 years at dose levels of 0, 100, 1,000 or 10,000 ppm. Assuming that 1 ppm in the diet of rats is equivalent to 0.05 mg/kg/day (Lehman, 1959), this corresponds to doses of 0, 5, 50 or 500 mg/kg/day. Untreated rats (70/sex/dose) were observed concurrently. The general appearance and behavior, growth, food consumption, clinical chemistry, hematology and histopathology in the treated rats did not differ significantly from the untreated controls. Based on this information, a NOAEL of 10,000 ppm (500 mg/kg/day), the highest dose tested, was identified.

- Hazleton and Farmer (1963) administered technical chloramben (97% pure) in the feed to 16 young adult beagle dogs (4/sex/dose) for 2 years at dietary levels of 0, 100, 1,000 or 10,000 ppm. Assuming that 1 ppm in the diet of dogs is equivalent to 0.025 mg/kg/day (Lehman, 1959), this corresponds to doses of 0, 2.5, 25 or 250 mg/kg/day. General appearance and behavior, food consumption, body weight, hematology, biochemistry, urinalysis and histopathology of the treated dogs did not differ significantly from the untreated controls. Based on this information, a NOAEL of 10,000 ppm (250 mg/kg/day), the highest dose tested, was identified.

- Johnston and Seibold (1979) administered technical chloramben to Sprague-Dawley rats for 2 years at dietary concentrations of 0, 100, 1,000 or 10,000 ppm. Assuming that 1 ppm in the diet of rats is equivalent to 0.05 mg/kg/day (Lehman, 1959), this corresponds to doses of 0, 5, 50 and 500 mg/kg/day. No compound-related effects were observed on any parameters measured including body weight, food consumption, hematology, clinical chemistry, urinalysis, gross pathology and histopathology. Based on this information, a NOAEL of 10,000 ppm (500 mg/kg/day), the highest dose tested, was identified.

4. Reproductive Effects

- In a three-generation study (Gabriel, 1966), three groups of albino rats (8 females and 16 males/dose) were administered 0, 500, 1,500 or 4,500 ppm chloramben (purity not specified) in the diet for 9 weeks prior to breeding, during breeding and during weaning periods. Assuming that 1 ppm in the diet of rats is equivalent to 0.05 mg/kg/day (Lehman, 1959), these dietary levels correspond to doses of about 0, 25, 75 or 225 mg/kg/day. Untreated animals served as controls. Following treatment, various parameters were measured, including indices of fertility, gestation, viability and lactation. No adverse effects were reported in any parameter measured. Based on this information, a NOAEL of 4,500 ppm (225 mg/kg/day), the highest dose tested, was identified for reproductive effects.

5. Developmental Effects

- Beliles and Mueller (1976) administered technical chloramben (purity not specified) to pregnant CFE rats (20/dose) by incorporation into the diets on days 6 through 15 of gestation. No compound-related changes were seen among dams treated at levels of 0, 500, 1,500 and 4,500 ppm. Assuming that 1 ppm in the diet of rats is equivalent to 0.05 mg/kg/day (Lehman, 1959), this corresponds to doses of about 0, 25, 75 or 225 mg/kg/day. Fetal mortality was increased, and data suggestive of decreased fetal skeletal development were observed in fetuses from dams treated at 4,500 ppm (225 mg/kg/day). At 1,500 ppm (75 mg/kg/day), there was no significant increase in embryo mortality; however, there was a generalized reduction in skeletal development. Fetuses of dams treated with 500 ppm (25 mg/kg/day) were similar in all respects to those of untreated control dams. Based on this information, a NOAEL of 4,500 ppm (225 mg/kg/day), the highest dose tested, was identified for maternal toxicity and teratogenicity. The NOAEL for fetotoxicity was identified as 500 ppm (25 mg/kg/day).

- Holson (1984) conducted studies in which New Zealand White rabbits (24/dose) were administered chloramben (sodium salt, 83% a.i. by weight) by gavage at dose levels of 0, 250, 500 or 1,000 mg/kg during days 6 through 18 of gestation. A NOAEL of 1,000 mg/kg/day, the highest dose tested, was identified, since the test compound did not produce maternal or fetal toxicity or teratogenic effects at any dose level tested. Other end points were not monitored.

6. *Mutagenicity*

- Chloramben was found to be negative in several indicator systems for potential mutagenic activity, including several microbial assays (Anderson et al., 1967; Eisenbeis et al., 1981; Jagannath, 1982), an *in vivo* bone marrow cytogenetic assay (Ivett, 1985) and a primary rat hepatocytes unscheduled DNA synthesis test (Myhr and McKeon, 1982).

- Results were positive for the *in vitro* cytogenic test using Chinese hamster ovary cells (Galloway and Lebowitz, 1982).

7. *Carcinogenicity*

- In an 18-month feeding study (U.S. EPA, 1981b), Crl:COBS CD-1 mice (50/sex/dose) were administered technical chloramben (purity not specified) at dietary levels of 0, 100, 1,000 or 10,000 ppm. Assuming that 1 ppm in the diet of mice is equivalent to 0.15 mg/kg/day (Lehman, 1959), this corresponds to doses of about 0, 15, 150 and 1,500 mg/kg/day. Hepatocellular carcinomas (trabecular type) were present in 1/50 low-dose and 1/50 high-dose males. In no case was vascular invasion or secondary spread of the nodular carcinoma masses observed. Hepatocellular adenomas were present only in males as follows: 5/50 control, 2/50 low-dose, 2/48 intermediate-dose and 5/50 high-dose. However, due to a number of deficiencies in this study (e.g., missing data, significant tissue autolysis), no conclusion can be made regarding the oncogenic potential of the test material.

- NCI (1977) administered 10,000 or 20,000 ppm technical chloramben (90 to 95% active ingredient) in the feed to Osborne-Mendel rats (50/sex/dose) and B6C3F$_1$ mice (50/sex/dose) for 80 weeks followed by up to 33 weeks of postexposure observation. Assuming that 1 ppm in the diet of rats is equivalent to 0.05 mg/kg/day and 1 ppm in the diet of mice is equivalent to 0.15 mg/kg/day (Lehman, 1959), this corresponds to doses of 500 or 1,000 mg/kg/day for rats and 1,500 or 3,000 mg/kg/day for mice. Under conditions of the study, no compound-related tumors were reported in male or female rats or male mice. Hepatocellular carcinomas were reported in female mice, but in a retrospective audit of this bioassay by Drill et al. (1982), it was reported that the incidence of hepatocellular carcinomas in both the low-dose and high-dose female mice was lower than the maximal incidence of corresponding tumors in historical groups. It was concluded that there was no association between chloramben and the occurrence of hepatocellular carcinomas under conditions of the assay. However, since exposure was for only 80 weeks, this study may not have been adequate to detect late-occurring tumors.

- Paynter et al. (1963) reported no evidence of carcinogenic activity in albino rats (35/sex/dose) that received chloramben (97% pure) in the diet for 2 years at dose levels of 0, 100, 1,000 or 10,000 ppm. Assuming that 1 ppm in the diet of rats is equivalent to 0.05 mg/kg/day (Lehman, 1959), this corresponds to doses of 0, 5, 50 or 500 mg/kg/day.

- Johnston and Seibold (1979) reported no evidence of carcinogenic activity in Sprague-Dawley rats administered 0, 100, 1,000 or 10,000 ppm technical chloramben in the diet for 2 years. Assuming that 1 ppm in the diet of rats is equivalent to 0.05 mg/kg/day (Lehman, 1959), this corresponds to doses of 0, 5, 50 or 500 mg/kg/day. No compound-related effects were observed on any other parameters measured, including body weight, food consumption, hematology, clinical chemistry, urinalysis, gross pathology and histopathology.

IV.   Quantification of Toxicological Effects

A. One-day Health Advisory

No data were found in the available literature that were suitable for determination of the One-day HA value. It is therefore recommended that the Ten-day HA value for a 10-kg child (3 mg/L, calculated below) be used at this time as a conservative estimate of the One-day HA value.

B. Ten-Day Health Advisory

The rat teratology study by Beliles and Mueller (1976) has been selected to serve as the basis for determination of the Ten-day HA value for a 10-kg child for chloramben. In this study, a NOAEL of 225 mg/kg/day, the highest dose tested, was identified for maternal toxicity and teratogenicity while a NOAEL of 25 mg/kg/day was identified for fetotoxicity (skeletal development) in rats exposed on days 6 to 15 of gestation. There is some question as to whether it is appropriate to base a Ten-day HA for the 10-kg child on fetotoxicity observed in a teratology study. However, this study is of appropriate duration and the fetus may be more sensitive than the 10-kg child.

The studies by Keller (1959) and Holson (1984) have not been selected, since the NOAEL values identified in these studies (500 and 1,000 mg/kg/day, respectively) are much higher than the NOAEL identified by Beliles and Mueller (1976).

Using the NOAEL of 25 mg/kg/day, the Ten-day HA for the 10-kg child is calculated as follows:

$$\text{Ten-day HA} = \frac{(25 \text{ mg/kg/day}) (10 \text{ kg})}{(100) (1 \text{ L/day})} = 2.5 \text{ mg/L} (3{,}000 \ \mu g/L)$$

C. Longer-term Health Advisory

No data were found in the available literature that were suitable for the determination of the Longer-term HA. It is, therefore, recommended that an adjusted DWEL for a 10-kg child (0.15 mg/L = 200 μg/L) and the DWEL for a 70-kg adult (0.525 mg/L = 500 μg/L) be used at this time for the Longer-term HA values.

For a 10-kg child, the adjusted DWEL is calculated as follows:

$$\text{DWEL} = \frac{(0.015 \text{ mg/kg/day}) (10 \text{ kg})}{(1 \text{ L/day})} = 0.15 \text{ mg/L} (200 \ \mu g/L)$$

where: 0.015 mg/kg/day = RfD (see Lifetime Health Advisory section).

D. Lifetime Health Advisory

The 18-month feeding study by the Huntingdon Research Center (U.S. EPA, 1981b) has been selected to serve as the basis for determination of the Lifetime HA for chloramben. In this study, Crl:COBS CD-1 mice were administered technical chloramben at dietary levels of 0, 100, 1,000 or 10,000 ppm (0, 15, 150 or 1,500 mg/kg/day). Hepatocellular alterations were observed in mice in all treatment groups, and a LOAEL of 100 ppm (15 mg/kg/day) was identified. Other studies of appropriate duration identify NOAELs that are higher than the LOAEL of 15 mg/kg/day. For example, Hazleton and Farmer (1963) identified a NOAEL of 250 mg/kg/day in a 2-year study in dogs, and both Paynter et al. (1963) and Johnston and Seibold (1979) identified a NOAEL of 500 mg/kg/day in 2-year rat studies.

Using the LOAEL of 15 mg/kg/day, the Lifetime HA for chloramben is calculated as follows:

Step 1: Determination of the Reference Dose (RfD)

$$\text{RfD} = \frac{(15 \text{ mg/kg/day})}{(1{,}000)} = 0.015 \text{ mg/kg/day}$$

Step 2: Determination of the Drinking Water Equivalent Level (DWEL)

$$DWEL = \frac{(0.015 \text{ mg/kg/day}) (70 \text{ kg})}{(2 \text{ L/day})} = 0.525 \text{ mg/L } (500 \text{ μg/L})$$

Step 3: Determination of the Lifetime Health Advisory

$$\text{Lifetime HA} = (0.525 \text{ mg/L}) (20\%) = 0.105 \text{ mg/L } (100 \text{ μg/L})$$

E. Evaluation of Carcinogenic Potential

- NCI (1977) evaluated the carcinogenic potential of orally administered chloramben (10,000 or 20,000 ppm, equivalent to 500 or 1,000 mg/kg/ day) to Osborne-Mendel rats (50/sex/dose) and B6C3F$_1$ mice (20/sex/ dose) for 80 weeks. It was concluded in a retrospective audit of this assay (Drill et al., 1982) that under conditions of this study, chloramben is not carcinogenic. Since exposure was for only 80 weeks, this experiment may not have been adequate to detect late-occurring tumors. Johnston and Seibold (1979) reported no evidence of carcinogenic activity in Sprague-Dawley rats that received chloramben in the diet for 2 years at concentrations up to 500 mg/kg/day. The Huntingdon Research Center (1978; cited in U.S. EPA, 1981b) reported no evidence of carcinogenicity in Crl:COBS CD-1 mice that received chloramben in the diet for 18 months at concentrations up to 1,500 mg/kg/day. However, due to a number of deficiencies in this study, no conclusion can be made regarding the oncogenic potential of the test material. Paynter et al. (1963) reported no evidence of carcinogenicity in albino rats that received chloramben in the diet for 2 years at concentrations up to 500 mg/kg/day.

- The International Agency for Research on Cancer has not evaluated the carcinogenicity of chloramben.

- Applying the criteria described in EPA's guidelines for assessment of carcinogenic risk (U.S. EPA, 1986a), chloramben may be classified in Group D: not classifiable. This category is for agents with inadequate human and animal evidence of carcinogenicity.

V.   Other Criteria, Guidance and Standards

- NAS has determined an Acceptable Daily Intake of 0.25 mg/kg/day with a Suggested-No-Adverse-Effect Level of 1.75 mg/L (U.S. EPA, 1985).

- U.S. EPA has established a residue tolerance for chloramben in or on raw agricultural commodities of 0.1 ppm (CFR, 1985).

VI.   Analytical Methods

- Chloramben may be analyzed using a gas chromatographic (GC) method applicable to the determination of chlorinated acids in water samples (U.S. EPA, 1986b). In this method, approximately 1 liter of sample is acidified. The compounds are extracted with ethyl ether using a separatory funnel. The derivatives are hydrolyzed with potassium hydroxide, and extraneous organic material is removed by a solvent wash. After acidification, the acids are extracted and converted to their methyl esters using diazomethane as the derivatizing agent. Excess reagent is removed, and the esters are determined by electron-capture (EC) gas chromatography. The method detection limit has not been determined for this compound.

VII.   Treatment Technologies

- No data were found for the removal of chloramben from drinking water by conventional treatment.

- No data were found for the removal of chloramben from drinking water by activated carbon treatment. However, due to its low solubility and its high molecular weight, chloramben probably would be amenable to activated carbon adsorption.

- No data were found for the removal of chloramben from drinking water by ion exchange. However, chloramben is an acidic pesticide and these compounds have been readily adsorbed in large amounts by ion exchange resins. Therefore, chloramben probably would be amenable to an ion exchange.

- No data were found for the removal of chloramben from drinking water by aeration. However, the Henry's Coefficient can be estimated from available data on solubility (700 mg/L at 25°C) and vapor pressure (7×10^{-3} mm Hg at 100°C). Due to its estimated Henry's Coefficient of

0.15 atm, chloramben probably would not be amenable to aeration or air stripping.

VIII.   References

Anderson, K.J., E.G. Leighty and M.T. Takahashi.* 1967. Evaluation of herbicides for possible mutagenic properties. Unpublished study. MRID 00025376.

Andrawes, N.* 1984. Amiben: metabolism of ^{14}C-chloramben in the rat. Project No. 852R10. Unpublished study. Union Carbide. MRID 00141157.

Beliles, R.P.* 1976. Ninety-day toxicity study in hamsters: technical chloramben. LBI Project No. 2595. Final Report. Unpublished study. MRID 00131187.

Beliles, R.P. and S. Mueller.* 1976. Teratology study in rats: technical chloramben. LBI Project No. 2577. Final Report. Unpublished study. MRID 0096618.

CFR. 1985. Code of Federal Regulations. 40 CFR 180.226. July 1. p. 298.

CHEMLAB. 1985. The Chemical Information System, CIS, Inc. Cited in U.S. EPA, 1985. (U.S. Environmental Protection Agency. Pesticide survey chemical profile. Final Report. Contract No. 68-01-6750. Office of Drinking Water).

Drill, V., S. Friess, H. Hayes et al.* 1982. Retrospective audit of the bioassay of chloramben for possible carcinogenicity. Unpublished study. MRID 00126379.

Eisenbeis, S.J., D.L. Lynch and A.E. Hampel. 1981. The Ames mutagen assay tested against herbicides and herbicide combination. Soil Sci. 131(1):44–47.

Field, W.E.* 1978. Acute dermal application (LD_{50}) – rabbit. Study No. CDC-AM-012-78. Unpublished study. MRID 00100319.

Field, W.* 1980. Oral LD_{50} in rats: chloramben 10G. Study No. CDC-UC-158. MRID 00128640.

Field, W.E. and W. Carter.* 1978. Oral LD_{50} in rats. Study No. CDC-AM-015-78. MRID 00100318.

Field, W. and G. Field.* 1980. Acute dermal toxicity in rabbits: (AXF-1107). Study No. CDC UC-16-180. Unpublished study. MRID 00128644.

Gabriel, K.L.* 1966. Reproduction study in albino rats with AmChem Products, Inc. Amiben (3-amino-2,5-dichlorobenzoic acid). Project No. 20-064. Unpublished study. MRID 00100202.

Gabriel, K.L.* 1969. Acute dermal toxicity – rabbits. Unpublished study. MRID 00023483.

Galloway, S. and H. Lebowitz.* 1982. Mutagenicity evaluation of chloramben (sodium salt), in an *in vitro* cytogenetic assay measuring chromosome aberration frequencies in Chinese hamster ovary (CHO) cells. Project No. 20990. Final Report. Unpublished study. MRID 00112855.

Hazleton, L.W. and K. Farmer.* 1963. Two year dietary feeding—dog. Final Report. Unpublished study. MRID 00100201.

Holson, J.* 1984. Teratology study of chloramben sodium salt in New Zealand White rabbits. Science Applications (1282018). MRID 00144930.

Ivett, J. 1985.* Clastogenic evaluation of chloramben in the mouse bone marrow cytogenetic assay. Final Report. LBI Project No. 22202. Unpublished study. MRID 00144363.

Jagannath, D.* 1982. Mutagenicity evaluation of chloramben sodium salt in Ames *Salmonella*/microsome plate test. Project No. 20988. Final Report. Unpublished study. MRID 00112853.

Johnston, C.D. and H.R. Seibold.* 1979. Two-year carcinogenesis study in rats: technical chloramben. LBI Project No. 20576. Final Report. Unpublished study. MRID 00029806.

Keller, J.G.* 1959. Twenty-eight day dietary feeding—rats. Unpublished study. MRID 00100199.

Lehman, A.J. 1959. Appraisal of the safety of chemicals in foods, drugs and cosmetics. Assoc. Food Drug Off. U. S., Q. Bull.

Meister, R., ed. 1983. Farm chemicals handbook. Willoughby, OH: Meister Publishing Co.

Myers, R., S. Christopher, H. Zimmer-Weaver et al.* 1982. Chloramben sodium salt: dermal sensitization study in the guinea pig. Project No. 45–162. Unpublished study. MRID 00130275.

Myhr, B. and M. McKeon.* 1982. Evaluation of chloramben sodium salt in the primary rat hepatocyte unscheduled DNA synthesis assay. Project No. 20991. Final report. Unpublished study. MRID 00112854.

NCI. 1977. National Cancer Institute. Bioassay of chloramben for possible carcinogenicity. Technical Report Series No. 25.

Paynter, O.E., M. Kundzin and T. Kundzin.* 1963. Two-year dietary feeding—rats. Final Report. Unpublished study. MRID 00100200.

Rees, D.C. and T.A. Re.* 1978. Inhalation toxicity of amiben sodium salt 3599 in adult Sprague-Dawley rats. Laboratory No. 5764b. Unpublished study. MRID 00100322.

STORET. 1988. STORET Water Quality File. Office of Water. U.S. Environmental Protection Agency (data file search conducted in May, 1988).

U.S. EPA.* 1981. U.S. Environmental Protection Agency. EPA Reg. No. 264–138, chloramben; 18-month oncogenic study in mice. Washington, DC: U.S. EPA, Office of Pesticide Programs. Memorandum from William Dykstra to Robert Taylor. Accession No. 242821–2. January 15.

U.S. EPA. 1985. U.S. Environmental Protection Agency. Pesticide survey chemical profile. Final Report. Contract No. 68–01–6750. Office of Drinking Water.

U.S. EPA. 1986a. U.S. Environmental Protection Agency. Guidelines for carcinogen risk assessment. Fed. Reg. 51(185):33992–34003. September 24.

U.S. EPA. 1986b. U.S. Environmental Protection Agency. Method #3 — Determination of chlorinated acids in ground water by GC/ECD. Draft. January. Available from U.S. EPA's Environmental Monitoring and Support Laboratory, Cincinnati, OH.

---

* Confidential Business Information submitted to the Office of Pesticide Programs.

Chlorothalonil

I. General Information and Properties

A. CAS No. 1897–45–6

B. Structural Formula

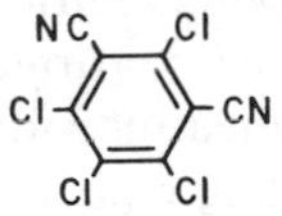

2,4,5,6-Tetrachloro-1,3-benzenedicarbonitrile

C. Synonyms

- Tetrachloroisophthalonitrile; Bravo; Chloroalonil; Chlorthalonil; Daconil; Exothern; Forturf; Nopcocide N96; Sweep; Termil; TPN; DAC-2787; T-117–11; DTX-77–0033; DTX-77–0034; DTX-77–0035.

D. Uses (Meister, 1986)

- Chlorothalonil is used as a broad-spectrum fungicide.

E. Properties (Meister, 1986; CHEMLAB, 1985; Meister, 1983; Windholz et al., 1983)

Chemical Formula	$C_8N_2Cl_4$
Molecular Weight	265.89
Physical State (25°C)	White, crystalline solid
Boiling Point	350°C
Melting Point	250 to 251°C
Density	--
Vapor Pressure (40°C)	2.0×10^{-6} mm Hg at 25°C
Specific Gravity	--
Water Solubility (25°C)	1.2 mg/L at 25°C
Log Octanol/Water Partition Coefficient	7.62×102
Taste Threshold	--

Odor Threshold   --
Conversion Factor   --

F. Occurrence

- Chlorothalonil has been found in 1 of 4 surface water samples analyzed from 3 surface water stations and in none of the 633 ground-water samples (STORET, 1988). Samples were collected at 3 surface water locations and 627 ground-water locations; and at the 1 location where it was found in Missouri, the concentration was 6,500 $\mu g/L$. This information is provided to give a general impression of the occurrence of this chemical in ground and surface waters as reported in the STORET database. The individual data points retrieved were used as they came from STORET and have not been confirmed as to their validity. STORET data are often not valid when individual numbers are used out of the context of the entire sampling regime, as they are here. Therefore, this information can only be used to form an impression of the intensity and location of sampling for a particular chemical.

G. Environmental Fate

- Ring-labeled ^{14}C-chlorothalonil, at 0.5 to 1.5 ppm, was stable to hydrolysis for up to 72 days in aqueous solutions buffered at pH 5 and 7 (Szalkowski, 1976b). At pH 9, chlorothalonil hydrolyzed with half-lives of 33 to 43 days and 28 to 72 days in solutions to which ring-labeled ^{14}C-chlorothalonil was added at 0.52 and 1.5 ppm, respectively. After 72 days of incubation, pH 9-buffered solutions treated with chlorothalonil at 1.5 ppm contained 36.4% chlorothalonil, 48.9% 3-cyano-2,4,5,6-tetrachlorobenzamide (DS-19211) and 11.3% 4-hydroxy-2,5,6-trichloroisophthalonitrile (DAC-3701).

- At 1,000 ppm, the chlorothalonil degradation product, ^{14}C-DAC-3701, was not hydrolyzed in aqueous solutions buffered at pH 5, 7, and 9 after 72 days of incubation (Szalkowski, 1976b).

- Ring-labeled ^{14}C-chlorothalonil and its major degradate, ring-labeled ^{14}C-DAC-3701, were stable to photolysis on two silt loam and three silty clay loam soils, after UV irradiation for the equivalent of 168 12-hour days of sunlight (Szalkowski, no date).

- ^{14}C-Chlorothalonil is degraded with half-lives of 1 to 16, 8 to 31 and 7 to 16 days in nonsterile aerobic sandy loam, silt loam and peat loam soils, respectively, at 77 to 95°F and 80% of field moisture capacity (Szalkowski, 1976a). When chlorothalonil wettable powder (WP) was applied to nonsterile soils ranging in texture from sand to silty clay loam, at 76 to 100°F and 6% soil moisture, it was degraded with half-lives of 4 to more than 40 days; increasing either soil moisture content (0.6 to 8.9%) or incubation temperature (76 to 100°F) enhanced chlorothalonil degradation (Stallard and Wolfe, 1967). Soil pH (6.5 to 8) does not appear to influence or only negligibly influences the degradation rate of chlorothalonil; however, soil sterilization greatly reduced the degradation rate. The major degradate identified in nonsterile aerobic soil was DAC-3701, representing up to 69% of the applied radioactivity. Other identified degradates included DS-19221, trichloro-3-carboxybenzamide, 3-cyanotrichlorohydroxybenzamide and 3-cyanotrichlorobenzamide (Stallard and Wolfe, 1967; Szalkowski, 1976a; Szalkowski et al., 1979).

- ^{14}C-Chlorothalonil was immobile (R_f 0.0) and the degradate ^{14}C-DAC-3701 was found to have low to intermediate mobility (R_f 0.25 to 0.43) in two silt loam and three silty clay loam soils, as evaluated using soil thin-layer chromatography (TLC) (Szalkowski, no date). Based on batch equilibrium tests, chlorothalonil has a relatively low mobility (high adsorption) in silty clay loam (K = 26), silt (K = 29) and sandy loam (K = 20) soils but is intermediately mobile (low adsorption) in a sand (K = 3) (Capps et al., 1982). Soil organic matter content did not appear to influence the mobility of chlorothalonil in soil.

- The chlorothalonil degradate DAC-3701 is mobile in sand, loam, silty clay loam and clay soils (Wolfe and Stallard, 1968a). After eluting a 6-inch soil column with the equivalent of 5 inches of water, approximately 57, 84, 10 and 84% of the applied DAC-3701 was recovered in the leachate of the sand, loam, silty clay loam and clay soil columns, respectively.

- Chlorothalonil (4.17 lb/gal FlC) was degraded with a half-life of 1 to 3 months in sandy loam and silt loam soils when applied alone at 8.34 lb active ingredient (a.i.)/A or in combination with benomyl (50% wettable powder) at 1.35 lb a.i./A (Johnston, 1981). The treated soils were maintained at 80% of moisture capacity in a greenhouse.

- Under field conditions, the half-life of chlorothalonil (75% wettable powder) in a sandy loam soil was between 1 and 2 months following the last of five consecutive weekly applications totaling 15 lb a.i./A (Stallard et al., 1972). Little movement of chlorothalonil (0.01 to 0.17 ppm) below the 0- to 3-inch depth occurred throughout the 8-month study. Small amounts (0.01 to 0.21 ppm) of the degradate DAC-3701 were found in soil samples collected up to 5 months posttreatment. No chlorothalonil or DAC-3701 was detected (less than 1 ppb) in a nearby stream up to 7 months posttreatment, or in ground-water samples (10-foot depth) up to 8 months posttreatment. Cumulative rainfall over the study period was 26.22 inches.

II. Pharmacokinetics

A. Absorption

- Ryer (1966) administered ^{14}C-chlorothalonil (dose not specified) orally to albino rats (3/sex, strain not specified). In 48 hours posttreatment, 60% of the radioactivity was detected in the feces, suggesting that at least 40% of the oral dose was absorbed.

- Skinner and Stallard (1967) reported that rats receiving 1.54 mg of ^{14}C-chlorothalonil (containing 2.8 μCi) in a 500 mg/kg dose (route not specified) eliminated 88% of the administered dose unchanged in the feces over 264 hours, suggesting that 12% was absorbed.

- Skinner and Stallard (1967) reported that mongrel dogs receiving a single oral dose (by capsule) of 500 mg/kg of chlorothalonil, eliminated 85% of the administered dose as the parent compound within 24 hours posttreatment, suggesting that 15% was absorbed.

B. Distribution

- Ryer (1966) administered ^{14}C-chlorothalonil (dose not specified) to albino rats (3/sex, strain not specified) by oral intubation. After 11 days, the carcasses retained 0.44% of the dose while 0.05% of the dose remained in the gastrointestinal tract. The highest residues occurred in the kidneys, which averaged 0.01% of the dose for the six rats. Lesser amounts were detected in the eyes, brain, heart, lungs, liver, thyroid and spleen.

- Ribovich et al. (1983) administered single doses of ^{14}C-chlorothalonil by oral intubation to CD-1 mice at levels of 0, 1.5, 15 or 105 mg/kg. Twenty-four hours posttreatment, the stomach, liver, kidneys, fat, small intestine, large intestine, lungs and heart accounted for less than 3% of the administered dose. The stomach and kidneys had the highest concentration at all doses tested. The compound was eliminated from the stomach and kidneys by 168 hours posttreatment.

- Wolfe and Stallard (1968b) reported a study in which dogs and rats received chlorothalonil in the diet for 2 years at 1,500 to 30,000 ppm. The amount of the 4-hydroxy-2,5,6-trichloroisophthalonitrile metabolite that was detected in the kidney tissue of dogs was less than 1.5 ppm; less than 3.0 ppm was detected in liver tissue from dogs and rats. The authors concluded that the metabolite was not stored in animal tissue.

C. Metabolism

- In the Wolfe and Stallard (1968b) study, only a small amount of the metabolite, 4-hydroxy-2,5,6-trichloroisophthalonitrile, was detected in the kidney tissue of dogs (< 1.5 ppm) and in liver tissue from dogs and rats (< 3 ppm).

- Marciniszyn et al. (1983b) reported that when Osborne-Mendel rats were administered single oral doses of ^{14}C-chlorothalonil by intubation at levels of 0, 5, 50, 200 or 500 mg/kg, no metabolites of chlorothalonil were unequivocally identified in urine.

D. Excretion

- The Ryer study (1966) revealed that, at the end of 11 days, an average of 88.45% of the administered dose was excreted in the feces, 5.14% in the urine and 0.32% in expired gases as CO_2.

- Skinner and Stallard (1967) presented results which demonstrated that 88% of a dose (route unspecified) of chlorothalonil was eliminated unchanged in the feces. Only 5.2% was eliminated via the urine and negligible amounts were detected in expired air.

- Ribovich et al. (1983) administered single doses of ^{14}C-chlorothalonil by oral intubation to CD-1 mice at levels of 0, 1.5, 15 or 105 mg/kg. The total recoveries of radioactivity 24 hours posttreatment were 93% for the low dose, 81% for the mid-dose and 62% for the high dose. The major route of elimination was the feces and was complete at 24 hours posttreatment for the low- and mid-dose animals, and by 96 hours for the high-dose animals.

- Marciniszyn et al. (1981) reported a study in which single doses of ^{14}C-chlorothalonil were administered intraduodenally to male Sprague-Dawley rats at 0.5, 5, 10, 50, 100 or 200 mg/kg. Biliary excretion of radioactivity was monitored for 24 hours. Percent recovery of radioactivity was 27.8, 20.7, 16.8, 6.4, 7.8 and 6% for each dose level, respectively.

- Marciniszyn et al. (1983a) administered ^{14}C-chlorothalonil intra-duodenally to male Sprague-Dawley rats (donor animals) at a dose of 5 mg/kg. Bile was collected for 24 hours following administration. Some of the collected bile was administered intraduodenally to recipient rats; bile was also collected from these animals for 24 hours. Data from the donor rats indicated that 1 to 6% of the administered radioactivity was excreted in the bile within 24 hours after dosing. Approximately 19% of the radioactivity in bile administered to recipient rats was excreted within 24 hours after dosing. These data suggest that enterohepatic recirculation plays a role in the metabolism of chlorothalonil in rats.

- Pollock et al. (1983) administered ^{14}C-chlorothalonil by gavage to male Sprague-Dawley rats at dose levels of 5, 50 or 200 mg/kg. They subsequently determined blood concentrations of radioactivity. The authors hypothesized that, at 200 mg/kg, an elimination mechanism (urinary, biliary and/or metabolism) was saturated, since the kinetics were nonlinear at this dose.

III.   Health Effects

- The purity of the administered chlorothalonil is assumed to be $\geq$ 90% for all studies described below, unless otherwise noted.

A. Humans

- Johnsson et al. (1983) reported that chlorothalonil exposure resulted in contact dermatitis in 14 of 20 workers involved in woodenware preservation. The wood preservative used by the workers consisted mainly of "white spirit," with 0.5% chlorothalonil as a fungicide. Workers exhibited erythema and edema of the eyelids, especially the upper eyelids, and eruptions on the neck, wrist and forearms. Results of a patch test conducted with 0.1% chlorothalonil in acetone were positive in 7 of 14 subjects. Reactions ranged from a few erythematous papules to marked papular erythema with a brownish hue without infiltration.

B. Animals

1. *Short-term Exposure*

- Powers (1965) reported that the acute oral LD_{50} of chlorothalonil (75% wettable powder) in Sprague-Dawley rats was > 10 g/kg.

- Doyle and Elsea (1963) reported that the acute oral LD_{50} of DAC-2787 (chlorothalonil), assumed by authors to be 100%, in male Dublin SD rats was > 10 g/kg when the compound was administered in corn oil.

2. Dermal/Ocular Effects

- Doyle and Elsea (1963) reported that the dermal LD_{50} of DAC-2787 chlorothalonil (assumed by the authors to be 100%), applied as a paste made of approximately equal parts of DAC-2787 and corn oil, in albino rabbits was > 10 g/kg. At dermal concentrations of 1, 2.15, 4.64 or 10 g/kg (24-hour exposure on abdominal skin), the compound produced mild to moderate skin irritation characterized by erythema, edema, atonia and desquamation.

- Doyle and Elsea (1963) reported that when 3 mg of DAC-2787 chlorothalonil (assumed by authors to be 100%, application method unknown) was applied to the eyes of albino rabbits, eye irritation was limited to mild conjunctivitis that subsided largely or completely within 7 days.

- Auletta and Rubin (1981) reported the results of eye irritation studies in cynomologus monkeys and New Zealand White rabbits using BRAVO 500, a 40% formulation containing 96% pure chlorothalonil. In both species, 0.1 mL of the test substance was instilled into the conjunctival sac of one eye. Each species displayed mild and transient ocular irritation as evidenced by corneal opacities that were reversed by 4 days postinstillation. The animals also showed slight to moderate iridial and conjunctival effects which were also reversible. Rinsing reduced conjunctival and iridial effects and prevented formation of corneal opacities.

3. Long-term Exposure

- Blackmore and Shott (1968) administered technical DAC 2787 (chlorothalonil), assumed by authors to be 100%, to Charles River rats (50/sex/dose) for 90 days at dietary levels of 0, 4, 10, 20, 30, 40 or 60 ppm (approximately 0, 0.2, 0.5, 1.0, 1.5, 2.0 or 3.0 mg/kg/day; Lehman, 1959). No compound-related effects were reported regarding physical appearance, growth, survival, terminal clinical values, organ weights or organ-to-body weight ratios. Microscopically, the kidneys exhibited occasional vacuolation and swelling of the epithelial cells lining the deeper proximal convoluted tubules. These changes were more numerous and more severe in the two highest dose groups. The authors stated that the difference between the two highest dose groups (2.0 and 3.0 mg/kg/day) and the controls was distinct, but the difference between the lower dose groups and controls was not clear. Based on this information, a NOAEL of 30 ppm (1.5 mg/kg/day) is identified.

- Wilson et al. (1981) administered chlorothalonil (98%) in the diet to Charles River CD rats (20/sex/dose) for 90 days at doses of 0, 40, 80, 175, 375, 750 or 1,500 mg/kg/day. At doses of 375 mg/kg/day or higher, significant decreases in body weight were reported. Decreases in glucose levels, blood urea nitrogen and serum thyroxine were attributed by the investigators to body weight effects. A dose-related decrease in serum glutamic-pyruvic transaminase (SGPT) was noted in all test groups. Significant increases in kidney weights were also noted in males at 40, 80, 175 and 375 mg/kg, while in females increased kidney weights were noted at 80, 175 and 750 mg/kg. There were dose-related increases in kidney-to-body weight ratios in both sexes at all doses. Focal acute gastritis occurred in some rats of both sexes at all doses and this effect was inversely related to dose. A LOAEL of 40 mg/kg/day (the lowest dose tested) is identified in this study.

- Colley et al. (1983) administered technical-grade chlorothalonil in the diet to Charles River rats (27 males and 28 females per dose) for 13 weeks at concentrations of 0, 1.5, 3.0, 10 or 40 mg/kg/day. Histopathological examination revealed that at a dose of 3.0 mg/kg/day or greater, all males displayed an increased number of irregular intracytoplasmic inclusion bodies in the renal proximal convoluted tubules. A NOAEL of 1.5 mg/kg/day is identified in this study.

- Shults et al. (1983) administered technical chlorothalonil (98.4%) to Charles River CD-1 mice (15/sex/dose) for 90 days at dietary concentrations of 0, 7.5, 15, 50, 275 or 750 ppm (males: 0, 1.2, 2.5, 8.5, 47.7 or 123.6 mg/kg/day; and females: 0, 1.4, 3.0, 9.8, 51.4 or 141.2 mg/kg/day). No treatment-related effects were noted on survival, physical condition, body weight, food consumption or gross pathology. At 750 ppm (141 mg/kg/day), an increase in alkaline phosphatase levels was observed in females only. Increased kidney weight was reported in males dosed at 750 ppm (124 mg/kg/day) and in females dosed at 275 and 750 ppm (51.0 and 141 mg/kg/day). Histopathologically, dose-related changes in the forestomach of mice were characterized by hyperplasia and hyperkeratosis of squamous epithelial cells. These changes were observed in the 50-, 275- and 750-ppm dose groups. No other treatment-related histopathological changes were reported. A NOAEL of 15 ppm (2.5 mg/kg/day) is identified for this study.

- Paynter and Murphy (1967) administered DAC 2787 (chlorothalonil) to beagle dogs (4/sex/dose) for 16 weeks at dietary concentrations of 0, 250, 500 or 750 ppm (approximately 0, 6.3. 12.5 or 18.8 mg/kg/day; Lehman, 1959). No effects attributable to chlorothalonil were noted in terms of appearance, behavior, appetite, elimination, body weight changes, gross pathology or organ weights. Hematological, biochemical and urinalysis values were generally within accepted limits in treated and control animals, except for slightly elevated protein-bound iodine values in treated dogs (especially high-dose females). No compound-related histopathology was noted. Based on this, a NOAEL of 750 ppm (18.8 mg/kg/day) is identified.

- Hastings et al. (1975) administered chlorothalonil to Wistar albino rats (15/sex/dose for treatment groups, 30/sex for controls) for four months at dietary concentrations of 0, 1, 2, 4, 15, 30, 60 or 120 ppm (approximately 0, 0.05, 0.1, 0.2, 0.8, 1.5, 3 or 6 mg/kg/day; Lehman, 1959). No significant differences between treated and control groups were seen in body weight, food consumption, mortality or gross pathological changes. Histopathological examination of the kidneys revealed no demonstrable effects at any dose level. A NOAEL of 120 ppm (6 mg/kg/day) is identified.

- Blackmore et al. (1968) administered DAC 2787 (chlorothalonil) to Charles River rats (35/sex/dose) for 22 weeks at dietary concentrations of 0, 250, 500, 750 or 1,500 ppm (approximately 0, 12.5, 25, 37.5 or 75 mg/kg/day; Lehman, 1959). At all dose levels, male rats gained less weight from weeks 11 to 22. Females gained less weight from weeks 9 to 22 at 750 and 1,500 ppm (37.5 or 75 mg/kg/day). Food consumption values were similar for all groups. No differences between control and test animals were reported for various hematological parameters, urinalysis and plasma and urine electrolytes. Results of gross necropsy revealed that livers and kidneys of males treated at 750 or 1,500 ppm (37.5 or 75 mg/kg/day) were larger than controls. Microscopic examinations demonstrated dose-related compound-induced alterations in the kidneys of both sexes at all doses. These changes were characterized by irregular swelling of the tubular epithelium, epithelial degeneration and tubular dilatation. There was a significant increase in renal tubular diameter in males at all dose levels. Accordingly, a LOAEL of 250 ppm (12.5 mg/kg/day) is identified.

- Blackmore and Kundzin (1969) administered technical-grade DAC 2787 (chlorothalonil), assumed by authors to be 100%, to rats (strain not specified) (35/sex/dose) for 1 year at dietary concentrations of 0, 4, 10,

20, 30, 40 or 60 ppm. The authors indicated that these dietary levels correspond to 0, 0.2, 0.5, 1.0, 1.5, 2.0 or 3.0 mg/kg/day. No compound-related effects on physical appearance, behavior, growth, food consumption, survival, clinical laboratory values, organ weights or gross pathology were noted. Microscopically, there were kidney alterations in both sexes at 40 and 60 ppm (2.0 and 3.0 mg/kg/day). These alterations occurred primarily in the deeper cortical tubules and consisted of increased vacuolation of epithelial cells accompanied by swelling or hypertrophy of the affected cells, often with the deposition of an eosinophilic droplet material in the cytoplasm of the vacuole. Statistical significance was not addressed. A NOAEL of 30 ppm (1.5 mg/kg/day) is identified.

- Holsing and Voelker (1970) administered technical chlorothalonil, assumed by authors to be 100%, to beagle dogs (8/sex/dose) for 104 weeks at dietary concentrations of 0, 60 or 120 ppm (approximately 0, 1.5 or 3 mg/kg/day; Lehman, 1959). After 2 years of administration, compound-related histopathological changes were observed in the kidneys of males fed 120 ppm (3 mg/kg/day). Males fed 60 ppm (1.5 mg/kg/day) and females fed both dose levels were comparable to controls. The observed changes included increased vacuolation of the epithelium in both the convoluted and collecting tubules and increased pigment in the convoluted tubular epithelium. Clinical findings, terminal body weight, organ-to-body weight ratios and gross pathology revealed no conclusive compound-related trends. A NOAEL of 60 ppm (1.5 mg/kg/day) is identified.

- Tierney et al. (1983) administered technical grade chlorothalonil (97.7% pure) to Charles River CD-1 mice (60/sex/dose) for 2 years at dietary concentrations of 0, 750, 1,500 or 3,000 ppm. The authors indicated that these dietary levels were approximately 0, 119.4, 251.1 or 517.4 mg/kg/day for males and 0, 133.6, 278.5 or 585.0 mg/kg/day for females. No treatment-related effects on body weight, food consumption, physical condition or hematological parameters were noted. A slightly increased mortality rate was noted in males receiving 3,000 ppm (517.4 mg/kg/day). Also, kidney-to-body weight ratios and kidney-to-brain weight ratios were increased significantly in all test groups. Gross necropsy revealed a number of renal effects including kidney enlargement, discoloration, surface irregularities, pelvic dilation, cysts, nodules and masses. Effects on the stomach included an increased incidence in masses or nodules. In the stomach and esophagus, nonneoplastic histopathological effects were noted at all dose levels, and included hyperplasia and hyperkeratosis of the squamous mucosa. This was considered to be indicative of mucosal irritation. Other changes in the stomach included mucosal and submucosal inflammation, focal necrosis or ulcers of mucosa and hyperplasia of glandular mucosa. Reported histopathological effects on the kidney included an increase in the incidence and severity of glomerulonephritis, cortical tubular degeneration and cortical cysts. These changes were not dose-related, but they did occur at higher incidences in treated animals. Based on the information presented in this study, a LOAEL of 750 ppm (119.4 mg/kg/day—males; 133.6 mg/kg/day—females) is identified.

4. Reproductive Effects

- In a three-generation reproduction study, Paynter and Kundzin (1967) administered a mixture containing 93.6% chlorothalonil to Charles River rats (10 males and 20 females per dose) at dietary concentrations of 0 or 5,000 ppm (approximately 0 or 250 mg/kg/day; Lehman, 1959). At the dose tested, the test material produced significant growth suppression in the nursing litters of each generation. Reproductive performance was not affected and pups showed no malformations attributable to the test substance. Body weight gains for exposed male and female rats of each generation were lower than controls.

5. Developmental Effects

- Rodwell et al. (1983) administered technical chlorothalonil (98% pure) by gavage, in aqueous methylcellulose, at doses of 0, 25, 100 or 400 mg/kg/day to Sprague-Dawley Cobs CD rats (25/dose level) on days 6 to 15 of gestation. No compound-related external, internal or skeletal malformations were observed in fetuses. At 400 mg/kg/day, maternal toxicity was noted (as evidenced by changes in appearance, three deaths, and decreased body weight gain and food consumption). A slight increase in the number of early embryonic deaths was associated with this maternal toxicity. This study identifies a NOAEL of 400 mg/kg/day for teratogenic effects and a NOAEL of 100 mg/kg/day for maternal toxicity.

- Shirasu and Teramoto (1975) administered chlorothalonil (99.3% pure) by gavage to Japanese white rabbits (eight controls, nine per dose) at doses of 0, 5 or 50 mg/kg/day on days 6 to 18 of gestation. At 50 mg/kg/day, four of the nine does aborted. No compound-related growth retardation or malformations were noted in offspring in any test group. This study identifies a NOAEL of 50 mg/kg/day for teratogenic effects and a NOAEL of 5 mg/kg/day for maternal toxicity.

6. *Mutagenicity*

- Quinto et al. (1981) reported that chlorothalonil (purity and concentrations not specified) was not mutagenic, with or without metabolic activation, in five tester strains of *Salmonella typhimurium*.

- Wei (1982) reported that chlorothalonil (90% pure), at concentrations up to 764 μg/plate, was not mutagenic in *S. typhimurium* strains TA1535, TA1537, TA1538, TA100 or TA98, with or without liver or kidney activation systems.

- Kouri et al. (1977c) reported that DTX-77–0035 (97.8% chlorothalonil) at concentrations up to 6.6 μg/plate did not induce point mutations in *S. typhimurium* strains TA1535, TA100, TA1537, TA1538 or TA98, with or without S-9 activation.

- Shirasu et al. (1975) reported the results of a reverse mutation test using *S. typhimurium* strains TA1535, TA1537, TA1538, TA98 and TA100 and *Escherichia coli* WP2 hcr+ and WP2 hcr–. Chlorothalonil (99.3%) failed to produce an effect without activation at concentrations up to 500 pg/plate; negative results also were obtained with activation at chlorothalonil concentrations up to 100 pg/plate.

- Kouri et al. (1977b) reported the effects of DTX-77–0033 (97.8% chlorothalonil) in a DNA repair assay using *S. typhimurium* strains TA1978 and TA1538. Chlorothalonil, dissolved in dimethylsulfoxide at 1 mg/mL and tested at 2, 10 and 20 μL of the stock solution per plate, was found to be active in both strains with or without metabolic activation.

- DeBertoldi et al. (1980) reported that a commercial preparation of chlorothalonil (2,500 ppm of active ingredient) did not induce mitotic gene conversions in *Saccharomyces cerevisiae* in the presence or absence of metabolic activation systems. In tests on *Aspergillus nidulans* using both resting and germinating conidia, chlorothalonil (up to 200 ppm) did not induce mitotic gene conversions.

- Shirasu et al. (1975) reported that, at concentrations up to 200 μg/disk, chlorothalonil (99.3%) was negative in a rec-assay using *Bacillus subtilis* strains H17 and M45.

- Kouri et al. (1977a) exposed Chinese hamster cells (V-79) and mouse fibroblast cells (BALB/3T3) *in vitro* to DTX-77–0034 (chlorothalonil, 97.8% pure) at concentrations of 0.3 μg/mL (for V-79 cells) or 0.03 μg/mL (for mouse fibroblast cells). The V-79 cells were tested without metabolic activation; the BALB/3T3 cells were tested with and without metabolic activation. Chlorothalonil was not mutagenic in either cell type.

- Mizens et al. (1983a) reported the results of a micronucleus test in Wistar rats, Swiss CFLP mice and Chinese hamsters. Rats were dosed at 0, 8, 40, 200, 1,000 or 5,000 mg/kg; mice and hamsters received 0, 4, 20, 100, 500 or 2,500 mg/kg. All animals were dosed by gavage and all received two doses, 24 hours apart. Technical chlorothalonil (98.2% a.i.) did not induce bone marrow erythrocyte micronuclei in any of the species tested.

- Legator (1974) reported the results of an *in vivo* cytogenetic test on DAC-2787 (chlorothalonil, more than 99% pure) in mice (strain not specified) using the micronuclei procedure. The test compound was administered by gavage for 5 days at a concentration of 6.5 mg/kg/day. At this concentration, chlorothalonil did not increase the number of cells with micronuclei.

- Legator (1974) presented the results of a host-mediated assay using male Swiss albino mice and *S. typhimurium* strains G-46, TA1530, C-207, TA1531, C-3076, TA1700, D-3056 and TA1724. Mice (10/dose) received DAC-2787 (chlorothalonil, 99% pure) by gavage for 5 days at 6.5 mg/kg/day. The compound did not produce any measurable mutagenic response when initially evaluated *in vitro* against the eight tester strains of *S. typhimurium*. When the tester strains were inoculated into treated mice, no increase in mutation frequency was observed.

- Legator (1974) presented the results of a dominant lethal assay in which male mice (strain not specified) were dosed with DAC-2787 (chlorothalonil, 99% pure) for 5 days at 6.5 mg/kg/day. These mice were mated with untreated females, and the number of early fetal deaths and preimplantation losses were measured. There was no significant difference in the fertility rates between test and control animals during weeks 1 to 7. At week 8, there was a significant decrease in fertility in the test group.

- Mizens et al. (1983b) presented the results of a chromosomal aberration test in Chinese hamsters. The test animals received two doses of technical chlorothalonil, 24 hours apart, by gavage at concentrations of 0, 8, 40, 200, 1,000 or 5,000 mg/kg. At 5,000 mg/kg, a statistically significant increase in bone marrow chromosomal abnormalities was observed. However, the authors concluded that this effect could not be attributed to chlorothalonil (98.2% a.i.) because the animals exhibited toxic responses to dosing.

7. *Carcinogenicity*

- NCI (1980) reported the results of a study in which technical-grade chlorothalonil (98.5%) was administered to Osborne-Mendel rats (50/sex/dose) for 80 weeks at time-weighted average (TWA) dietary doses for both males and females of 5,063 or 10,126 ppm, respectively. These dietary doses have been calculated to correspond to approximately 253 and 506 mg/kg/day (Lehman, 1959). Matched controls consisted of groups of 10 untreated rats of each sex; pooled controls consisted of the matched controls combined with 55 untreated male or female rats from other bioassays. An observation period of 30 to 31 weeks followed dosing. Clinical signs that appeared with increased frequency in dosed rats included hematuria and, from week 72 on, bright yellow urine. Adenomas and carcinomas of renal tubular epithelium occurred with a significant (p = 0.03, males; p = 0.007, females) dose-related trend. The frequency of renal tumors was statistically greater in the high-dose males (p = 0.035) and high-dose females (p = 0.016) than in corresponding controls (males: pooled controls, 0/62; low-dose, 3/46; high-dose, 4/49; females: pooled controls, 0/62; low-dose, 1/48; high-dose, 5/50). The observed adenomas and carcinomas were considered to be histogenically related. Results of this study were interpreted as sufficient evidence of carcinogenicity in Osborne-Mendel rats.

- NCI (1980) also reported a study in which technical-grade chlorothalonil (98.5%) was administered to B6C3F$_1$ mice (50/sex/dose) for 80 weeks at TWA dietary doses of 2,688 or 5,375 ppm for males and 3,000 or 6,000 ppm for females. These dietary doses have been calculated to correspond to approximately 403.2 or 806.3 mg/kg for males and 450 or 900 mg/kg for females (Lehman, 1959). Matched controls consisted of 10 untreated mice of each sex; pooled controls consisted of the matched controls combined with 50 untreated male or female mice from other bioassays. An observation period of 11 to 12 weeks followed dosing. Since the dosed female mice did not show depression in mean body weights or decreased survival compared with the controls, they may have been able to tolerate a higher dose. No tumors were found to occur at a greater incidence among dosed animals than among controls. It was concluded that, under the conditions of this bioassay, chlorothalonil was not carcinogenic in B6C3F$_1$ mice.

- Tierney et al. (1983) administered technical-grade chlorothalonil (97.7% pure) to Charles River CD-1 mice (60/sex/control and dose groups) for 2 years at dietary concentrations of 0, 750, 1,500 or 3,000 ppm. The authors indicated that these dietary levels were equivalent to 0, 119, 251 or 517 mg/kg/day for males and 0, 133, 278 or 585 mg/kg/day for females. Increased incidences of squamous cell tumors of the foresto-mach were noted in all treatment groups. These tumors consisted princi-pally of carcinomas, although papillomas were also seen. This increased incidence was statistically significant in females dosed at 1,500 ppm (279 mg/kg/day). No clear dose-related trend in the incidence of these tumors was observed. A slight increase in the incidence of tumors of the glandular epithelium of the fundic stomach was observed in dosed ani-mals; this increase was neither statistically significant nor dose-related. When the numbers of animals with epithelial tumors of the fundic or forestomach were combined, the incidence of these tumors showed a statistically significant increase in the 1,500- and 3,000-ppm female dose groups (279 and 585 mg/kg/day). No treatment-related renal neoplasms were seen in any female dose group. Increased incidences of adenomas and carcinomas in renal cortical tubules were noted in all treated groups of male mice. These changes did not show a dose-response relationship; the increased incidence was statistically significant only in the 750 ppm (251 mg/kg/day) group. The authors concluded that the administration of chlorothalonil caused an increase in the incidence of primary gastric tumors and, in male mice only, caused an increase in the incidence of renal tubular neoplasms.

• Wilson et al. (1985) gave technical chlorothalonil (98.1% pure with less than 0.03% hexachlorobenzene) to Fischer 344 rats (60/sex/dose) in their diet at dose levels of 0, 40, 80 or 175 mg/kg/day. Males were treated for 116 weeks, while females received the chemical for 129 weeks. Survival among the various groups was comparable. In both sexes, at the high-dose level, there were significant decreases in body weights. In addition, there were also significant increases in blood urea nitrogen and creatinine, while there were decreases in serum glucose and albumin levels. In both sexes, there were dose-dependent increases in kidney carcinomas and adenomas at doses above 40 mg/kg/day. In the high-dose females, there was also a significant increase in stomach papil-lomas. The data show that, in the Fischer 344 rat, chlorothalonil is a carcinogen.

IV. Quantification of Toxicological Effects

A. One-day Health Advisory

No information was found in the available literature that was suitable for determination of a One-day HA for chlorothalonil. Accordingly, it is recommended that the Longer-term HA value (200 μg/L, calculated below) for a 10-kg child be used at this time as a conservative estimate of the One-day HA value.

B. Ten-day Health Advisory

No information was found in the available literature that was suitable for determination of a Ten-day HA for chlorothalonil. Accordingly, it is recommended that the Longer-term HA value (200 μg/L, calculated below) for a 10-kg child be used at this time as a conservative estimate of the Ten-day HA value.

C. Longer-term Health Advisory

The studies by Colley et al. (1983), Blackmore and Kundzin (1969) and Blackmore and Shott (1968) have been selected to serve as the basis for the Longer-term HA for chlorothalonil. In the study by Colley et al., technical-grade chlorothalonil was administered in the diet to Charles River rats for 13 weeks at concentrations of 0, 1.5, 3.0, 10 or 40 mg/kg/day. Histopathological examinations revealed that at doses of 3.0 mg/kg/day or greater, male rats displayed an increased number of intracytoplasmic inclusion bodies in the proximal convoluted renal tubules. Blackmore and Shott (1968), gave technical-grade chlorothalonil in the diet to Charles River rats for 90 days at doses of 0, 0.2, 0.5, 1.0, 1.5, 2.0 or 3.0 mg/kg/day. At the two highest dose levels, the kidneys exhibited occasional vacuolation and swelling of the epithelial cells lining the deeper proximal convoluted tubules. In the Blackmore and Kundzin (1969) study, technical-grade chlorothalonil was administered in the diet to rats for 1 year at doses of 0, 0.2, 0.5, 1.0, 1.5, 2.0 or 3.0 mg/kg/day. At the 2 higher doses, there were alterations in the deeper convoluted renal tubules in both sexes. Each of the studies identified a NOAEL of 1.5 mg/kg/day.

The Longer-term HA for a 10-kg child is calculated as follows:

$$\text{Longer-term HA} = \frac{(1.5 \text{ mg/kg/day}) (10 \text{ kg})}{(100) (1 \text{ L/day})} = 0.15 \text{ mg/L} (200 \text{ μg/L})$$

The Longer-term HA for a 70-kg adult is calculated as follows:

$$\text{Longer-term HA} = \frac{(1.5 \text{ mg/kg/day}) (70 \text{ kg})}{(100) (2 \text{ L/day})} = 0.525 \text{ mg/L} (500 \text{ }\mu\text{g/L})$$

D. Lifetime Health Advisory

The study by Holsing and Voelker (1970) has been selected to serve as the basis for the Lifetime HA for chlorothalonil. In this study, technical-grade chlorothalonil was administered to beagle dogs (8/sex/dose) for 104 weeks at dietary concentrations of 0, 60 or 120 ppm (0, 1.5 or 3.0 mg/kg/day). The results following 2 years of administration revealed compound-related histopathological changes in the kidneys of males fed 120 ppm (3 mg/kg/day). Males fed 60 ppm (1.5 mg/kg/day) and females fed both dose levels were comparable to controls. The observed changes included increased vacuolation of the epithelium in both the convoluted and collecting tubules and increased pigment in the convoluted tubule epithelium. From these results, a NOAEL of 1.5 mg/kg was identified.

Using this NOAEL, the Lifetime HA is derived as follows:

Step 1: Determination of the Reference Dose (RfD)

$$\text{RfD} = \frac{(1.5 \text{ mg/kg/day})}{(100)} = 0.015 \text{ mg/kg/day}$$

Step 2: Determination of the Drinking Water Equivalent Level (DWEL)

$$\text{DWEL} = \frac{(0.015 \text{ mg/kg/day}) (70 \text{ kg})}{2 \text{ L/day}} = 0.525 \text{ mg/L} (500 \text{ }\mu\text{g/L})$$

Step 3: Determination of the Lifetime Health Advisory

Chlorothalonil is classified in Group B2: probable human carcinogen. Accordingly, a Lifetime HA is not recommended for chlorothalonil.

The estimated excess cancer risk associated with lifetime exposure to drinking water containing chlorothalonil at 525 μg/L (the DWEL) is 3.5×10^{-4}. This estimate represents the upper 95% confidence limit from extrapolations prepared by Office of Pesticide Programs (OPP) and Office of Drinking Water (ODW) using the linearized, multistage model. The actual risk is unlikely to exceed this value, but there is considerable uncertainty as to the accuracy of risks calculated by this methodology.

E. Evaluation of Carcinogenic Potential

- In an NCI bioassay (1980), technical-grade chlorothalonil was administered in the diet at 253 or 506 mg/kg/day to Osborne-Mendel rats for 80 weeks. A statistically significant increase in the frequency of renal tumors was observed in high-dose males and females.

- NCI (1980) reported that chorothalonil was not carcinogenic in $B6C3F_1$ mice when administered in the diet, at 403 or 806 mg/kg and 450 or 900 mg/kg for males and females, respectively, for 80 weeks. However, Tierney et al. (1983) concluded that chlorothalonil was carcinogenic in Charles River CD-1 rats which received the compound (0, 119, 251 or 517 mg/kg/day for males and 0, 134, 279 or 585 mg/kg/day for females) in the diet for 2 years. Increased incidences of squamous cell papilloma and carcinoma of the forestomach were noted in all treatment groups. This increase was statistically significant only in the mid-dose females. Increased incidences of adenoma and carcinoma of the renal cortical tubules were observed in all treatment groups. Again, no dose-response was noted, since these increases were statistically significant only in the mid-dose males.

- The International Agency for Research on Cancer (IARC, 1983) has evaluated the carcinogenic potential of chlorothalonil and concluded that there is limited evidence of carcinogenicity in experimental animals.

- Applying the criteria described in EPA's guidelines for assessment of carcinogenic risk (U.S. EPA, 1986), chlorothalonil is classified in Group B2: probable human carcinogen. This category is for chemicals for which there is inadequate evidence of carcinogenicity from human studies and sufficient evidence from animal studies.

- From the Wilson et al. (1985) data, OPP calculated a q_1^* of 2.4×10^{-2} $(mg/kg/day)^{-1}$. The 95% upper limit lifetime dose in drinking water associated with a 10^{-6} excess risk level is 1.5 $\mu g/L$. Corresponding levels for 10^{-5} and 10^{-4} are 15 and 150 $\mu g/L$, respectively. While recognized as statistically alternative approaches, the range of risks described by using any of these modeling approaches has little biological significance unless data can be used to support the selection of one model over another. In the interest of consistency of approach and in providing an upper bound on the potential cancer risk, the Agency has recommended use of the linearized multistage approach. However, for completeness, the 10^{-6} risk numbers for other models will be given. These values, at the 10^{-6} level, are: multihit -9 μg/L; one hit -2 μg/L; probit -51 μg/L; logit -0.8 μg/L; and Weibel -0.6 μg/L.

V.   Other Criteria, Guidance and Standards

- The WHO Temporary Acceptable Daily Intake (TADI) is 0.005 mg/kg/day (Vettorazzi and Van den Hurk, 1985).

- EPA/OPP has calculated a PADI of 0.015 mg/kg/day based on the NOAEL of 1.5 mg/kg/day identified in the 2-year dog study (Holsing and Voelker, 1970) and an uncertainty factor of 100 (U.S. EPA, 1984a).

- U.S. EPA established tolerances for residue levels in or on raw agricultural commodities of 0.1 to 5 ppm (CFR, 1985).

VI.   Analytical Methods

- Analysis of chlorothalonil is by a gas chromatographic (GC) method applicable to the determination of certain chlorinated pesticides in water samples (U.S. EPA, 1988). In this method, approximately 1 liter of sample is extracted with methylene chloride. The extract is concentrated and the compounds are separated using capillary column GC. Measurement is made using an electron capture detector. This method has been validated in a single laboratory, and estimated detection limits have been determined for analytes in this method, including chlorothalonil. The estimated detection limit for chlorothalonil is 0.025 μg/L.

VII.   Treatment Technologies

- Reverse osmosis (RO) is a promising treatment method for pesticide-contaminated water. As a general rule, organic compounds with molecular weights greater than 100 are candidates for removal by RO. Larson et al. (1982) reported 99% removal efficiency of chlorinated pesticides by a thin-film composite polyamide membrane operating at a maximum pressure of 1,000 psi and a maximum temperature of 113°F. More operational data are required, however, to specifically determine the effectiveness and feasibility of applying RO for the removal of chlorothalonil from water. Also, membrane adsorption must be considered when evaluating RO performance in the treatment of chlorothalonil-contaminated drinking water supplies.

VIII.   References

Auletta, C.S. and L.F. Rubin.* 1981. Eye irritation studies in monkeys and rabbits with Bravo 500. Report DS-2787. Unpublished study. MRID 00077176.

Blackmore, R.H, and L.D. Shott.* 1968. Final report: three-month feeding study—rats. Project No. 200–205. Unpublished study. MRID 00087316.

Blackmore, R.H., L.D. Shott, M. Kundzin et al.* 1968. Final report: four-month feeding study—rats (22 weeks). Project No. 200–198. Unpublished study. MRID 00057701.

Blackmore, R.H. and M. Kundzin.* 1969. Final report: 12-month feeding study—rats. Project No. 200–205. Unpublished study. MRID 00087358.

Capps, T.M., J.P. Marciniszyn, A.F. Markes and J.A. Ignatoski.* 1982. Document No. 555–4EF-81–0261–001, Section J, Vol. VI. Submitted by Diamond Shamrock Corporation.

CFR. 1985. Code of Federal Regulations. 40 CFR 180.275. July 1.

CHEMLAB. 1985. The Chemical Information System, CIS, Inc. Baltimore, MD.

Colley, J., L. Syred, R. Heywood et al.* 1983. A 13-week subchronic toxicity study of T-117–11 in rats (followed by a 13-week withdrawal period). Unpublished study. MRID 00127852.

DeBertoldi, M., M. Griselli, M. Giovannetti and R. Barale. 1980. Mutagenicity of pesticides evaluated by means of gene conversion in *Saccharomyces cerevisiae* and *Aspergillus nidulans*. Environ. Mut. 2:359–370.

Doyle, R.L. and J.R. Elsea.* 1963. Acute oral, dermal and eye toxicity and irritation studies on DAC-2787: N-107. Unpublished study. MRID 00038909.

Hastings, T.F., M. Dickson, W.M. Busey et al.* 1975. Four-month dietary toxicity study—rats chlorothalonil. Project No. 24–201. Unpublished study. MRID 00040463.

Holsing, G. and R. Voelker.* 1970. 104-week dietary administration—dogs: Daconil 2787 (Technical). Project No. 200–206. Unpublished study. MRID 00114304.

IARC. 1983. International Agency for Research on Cancer. IARC Monographs on the evaluation of carcinogenic risks of chemicals to humans. Miscellaneous pesticides. IARC Monographs, Vol. 30. pp. 319–328.

Johnsson, M., M. Buhagen, H.L. Leira and S. Solvang. 1983. Fungicide-induced contact dermatitis. Contact Dermat. 9:285–288.

Johnston, E.F.* 1981. Soil disappearance studies with Benlate fungicide and Bravo 500 fungicide, alone and in combination. Document No. AMR-06–81. Unpublished study. Submitted by E.I. duPont de Nemours and Co., Wilmington, DE.

Kouri, R.E., R. Joglekar and D.P.A. Fabrizio.* 1977a. Activity of DTX-77-0034 in an *in vitro* mammalian cell point mutation assay. Unpublished study. MRID 00030289.

Kouri, R.E., A.S. Parmar, J.M. Kuzava et al.* 1977b. Activity of DTX-77-0033 in a test for differential inhibition of repair deficient and repair competent strains of *Salmonella typhimurium*: repair test. Final report. Unpublished study. MRID 00030288.

Kouri, R.E., A.S. Parmar, J.M. Kuzava et al.* 1977c. Activity of DTX-77-0035 in the *Salmonella*/microsomal assay for bacterial mutagenicity. Unpublished study. MRID 00030290.

Larson, R.E., P.S. Cartwright, P.K. Eriksson and R.J. Petersen. 1982. Applications of the FT-30 reverse osmosis membrane in the metal finishing operations. Paper presented at Yokohama, Japan.

Legator, M.S.* 1974. Report on mutagenic testing with DAC 2787. Unpublished study. MRID 00040464.

Lehman, A.J. 1959. Appraisal of the safety of chemicals in foods, drugs and cosmetics. Assoc. Food Drug Off. U.S., Q. Bull.

Marciniszyn, J., J. Killeen and J. Ignatoski.* 1981. Dose-response determination of the excretion of radioactivity in rat bile following intraduodenal administration of ¹⁴C-chlorothalonil (¹⁴C-DS-2787). Unpublished study. MRID 00137132.

Marciniszyn, J., J. Killeen and J. Ignatoski.* 1983a. Recirculation of radioactivity in rat bile following intraduodenal administration of bile containing ¹⁴C-chlorothalonil label. Unpublished study. MRID 00137130.

Marciniszyn, J., J. Killeen and J. Ignatoski.* 1983b. Identification of major chlorothalonil metabolites in rat urine. Unpublished study. MRID 00137129.

Meister, R., ed. 1986. Farm chemicals handbook. Willoughby, OH: Meister Publishing Company.

Mizens, M., J. Killeen and J. Ignatoski.* 1983a. The micronucleus test in the rat, mouse and hamster using chlorothalonil. Unpublished study. MRID 00127853.

Mizens, M., J. Killeen and J. Ignatoski.* 1983b. The chromosomal aberration test in the rat, mouse and hamster using chlorothalonil. Unpublished study. MRID 00127854.

NCI. 1980. National Cancer Institute. Bioassay of chlorothalonil for possible carcinogenicity (NTP #TR-041). U.S. Public Health Service. U.S. Department of Health, Education and Welfare.

Paynter, O.E. and M. Kundzin.* 1967. Final report: three-generation reproduction study—rats. Project No. 200-155. Unpublished study. MRID 00091289.

Paynter, O.E. and J.C. Murphy.* 1967. Final report: 16-week dietary feeding—dogs. Project No. 200–200. Unpublished study. MRID 00057698.

Pollock, G., J. Marciniszyn, J. Killeen et al.* 1983. Levels of radioactivity in blood following oral administration of ^{14}C-chlorothalonil (^{14}C-DS-2787) to male rats. Unpublished study. MRID 00137127.

Powers, M.B.* 1965. Acute oral administration—rats. Project No. 200–167. Unpublished study. MRID 00038910.

Quinto, I., G. Martire, G. Vricella, F. Riccardi, A. Perfumo, R. Giulivo and F. DeLorenzo. 1981. Screening of 24 pesticides by *Salmonella*/microsome assay: mutagenicity of benazolin, metoxuron and paraoxon. Mutat. Res. 85:265.

Ribovich, M., Pollock G., J. Marciniszyn et al.* 1983. Balance study of the distribution of radioactivity following oral administration of ^{14}C–chlorothalonil (^{14}C-DS-2787) to male mice. Unpublished study. MRID 00137125.

Rodwell, D., M. Mizens, N. Wilson et al.* 1983. A teratology study in rats with technical chlorothalonil. Unpublished study. MRID 00130733.

Ryer, F.H.* 1966. Radiotracer metabolism study. Unpublished study. MRID 00038918.

Shirasu, Y., M. Moriya and K. Watanabe.* 1975. Mutagenicity testing on Daconil in microbial systems. Unpublished study. MRID 00052947.

Shirasu, Y. and S. Teramoto.* 1975. Teratogenicity study of Daconil in rabbits. Unpublished study. MRID 00127855.

Shults, S., J. Laveglia, J. Killeen et al.* 1983. A 90-day feeding study in mice with technical chlorothalonil. Unpublished study. MRID 00138148.

Skinner, W.A. and D.E. Stallard.* 1967. Daconil 2787 animal metabolism studies. Unpublished study. MRID 00038917.

Stallard, D.E. and A.L. Wolfe.* 1967. The fate of 2,4,5,6-tetrachloro-isophthalonitrile (Daconil 2787) in soil. Unpublished study. Submitted by Diamond Alkali Company, Cleveland, OH.

Stallard, D.E., A.L. Wolfe and W.C. Duane.* 1972. Evaluation of the leaching of chlorothalonil under field conditions and its potential to contaminate underground water supplies. Unpublished study. Submitted by Diamond Shamrock Agricultural Chemicals, Cleveland, OH.

STORET. 1988. STORET Water Quality File. Office of Water. U.S. Environmental Protection Agency (data file search conducted in May, 1988).

Szalkowski, M.B.* No date. Photodegradation and mobility of Daconil and its major metabolite on soil thin films. Unpublished study. Submitted by Diamond Shamrock Agricultural Chemicals, Cleveland, OH.

Szalkowski, M.B.* 1976a. Effect of microorganisms upon the soil metabolism of Daconil and 4-hydroxy-2,5,6-trichloroisophthalonitrile. Unpublished study. Submitted by Diamond Shamrock Agricultural Chemicals, Cleveland, OH.

Szalkowski, M.B.* 1976b. Hydrolysis of Daconil and its metabolite, 4-hydroxy-2,5,6-trichloroisophthalonitrile, in the absence of light at pH levels of 5, 7, and 9. Updated method. Unpublished study. Submitted by Diamond Shamrock Agricultural Chemicals, Cleveland, OH.

Szalkowski, M.B., J.J. Mannion, D.E. Stallard et al.* 1979. Quantification and characterization of the biotransformation products of 2,4,5,6-tetrachloroisophthalonitrile (chlorothalonil, DS-2787) in soil. Unpublished study. Submitted by Diamond Shamrock Agricultural Chemicals, Cleveland, OH.

Tierney, W., N. Wilson, J. Killeen et al.* 1983. A chronic dietary study in mice with technical chlorothalonil. Unpublished study. MRID 00127858.

U.S. EPA. 1984a. U.S. Environmental Protection Agency. Proposed guidelines for carcinogen risk assessment; request for comments. Fed. Reg. 49(227):46294–46301. November 23.

U.S. EPA. 1984b. U.S. Environmental Protection Agency. Chlorothalonil (case GS0097) Pesticide Registration Standard. Washington, DC: Office of Pesticide Programs.

U.S. EPA. 1986. U.S. Environmental Protection Agency. Guidelines for carcinogen risk assessment. Fed. Reg. 51(185):33992–34003. September 24.

U.S. EPA. 1988. U.S. Environmental Protection Agency. U.S. EPA Method 508 – Determination of chlorinated pesticides in water by GC/ECD. Draft. April 15. Available from the U.S. EPA's Environmental Monitoring and Support Laboratory, Cincinnati, OH.

Vettorazzi, G. and G.W. Van den Hurk. 1985. The pesticide reference index, JMPR 1961–1984. Geneva, Switzerland: World Health Organization.

Wei, C. 1982. Lack of mutagenicity of the fungicide 2,4,5,6-tetrachloro-isophthalonitrile in the Ames *Salmonella*/microsome test. Appl. Environ. Microbiol. 43:252–4.

Wilson, N., J. Killeen, J. Ignatoski et al.* 1981. A 90-day toxicity study of technical chlorothalonil in rats. Unpublished study. MRID 00127850.

Wilson, N., J. Killeen, and J. Ignatoski.* 1985. A tumorigenicity study of technical chlorothalonil in rats. Document No. 099–5TX–80–0234–008. Unpublished study. Prepared by ADS Biotech Corp. p. 2269. MRID 00146945.

Windholz, M., S. Budavari, R.F. Blumetti and E.S. Otterbein, eds. 1983. The Merck index — an encyclopedia of chemicals and drugs, 10th ed. Rahway, NJ: Merck and Company, Inc.

Wolfe, A.L. and D.E. Stallard.* 1968a. The fate of DAC-3701 (4-hydroxy-2,5,6-trichloroisophthalonitrile) in soil. Unpublished study. Submitted by Diamond Shamrock Agricultural Chemicals, Cleveland, OH.

Wolfe, A.L. and D.E. Stallard.* 1968b. Analysis of tissues and organs for storage of the Daconil metabolite 4-hydroxy-2,5,6-trichloro-isophthalonitrile. Unpublished study. MRID 00087254.

---

* Confidential Business Information submitted to the Office of Pesticide Programs.

Cyanazine

I. General Information and Properties

A. CAS No. 21725-46-2

B. Structural Formula

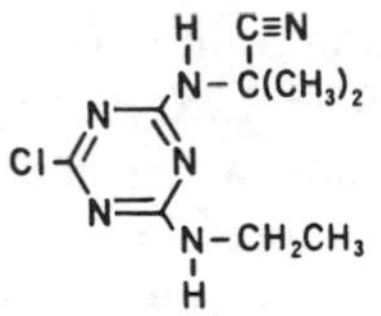

2-[[4-Chloro-6-(ethylamino)-1,3,5-triazin-2-yl]amino]-2-methylpropanenitrile

C. Synonyms

- Cyanazine (common name); Bladex; Fortrol; Payze; SD1518; VL19804; DW3418; and WL19805 (Meister, 1983).

D. Uses

- Cyanazine is used as a pre- and postemergence herbicide for the control of annual grasses and broad leaf weeds (U.S. EPA, 1984a).

E. Properties (U.S. EPA, 1984a; Meister, 1983; CHEMLAB, 1985)

Chemical Formula	$C_9H_{13}ClN_6$
Molecular Weight	240.7
Physical State (25°C)	White crystalline solid
Boiling Point	--
Melting Point	167.5 to 169°C
Density	0.35 (fluffed) to 0.45 (packed) g/cc
Vapor Pressure (20°C)	1.6×10^{-9} to 7.5×10^{-9} mm Hg
Water Solubility (25°C)	171 mg/L
Log Octanol/Water Partition Coefficient	2.24

Taste Threshold   --
Odor Threshold   --
Conversion Factor   --

F. Occurrence

- Cyanazine has been found in 1,708 of 5,297 surface water samples analyzed and in 21 of 1,821 ground-water samples (STORET, 1988). Samples were collected at 392 surface water locations and 1,314 ground-water locations. The 85th percentile of all nonzero samples was 4.11 μg/L in surface water and 0.20 μg/L in ground-water sources.

  The maximum concentration found in surface water was 1,300 μg/L and in ground water was 3,500 μg/L. Cyanazine was found in surface water in seven states and in ground-water in five states. This information is provided to give a general impression of the occurrence of this chemical in ground and surface waters as reported in the STORET database. The individual data points retrieved were used as they came from STORET and have not been confirmed as to their validity. STORET data are often not valid when individual numbers are used out of the context of the entire sampling regime, as they are here. Therefore, this information can only be used to form an impression of the intensity and location of sampling for a particular chemical.

- Cyanazine was identified in drinking water in New Orleans, Louisiana, in concentrations ranging from 0.01 to 0.35 μg/L (NAS, 1977).

- Cyanazine was monitored in a newly-built reservoir on the Des Moines River in Iowa from September 1977 through November 1978. Agricultural runoff (from corn and soybeans) was a major source of pollution in the river: levels of 71 to 457 ng/L were detected during the active months of May through August; levels of 2 to 151 ng/L were detected during September through December; and zero levels were found from January through April (U.S. EPA, 1984a; NAS, 1977).

- Cyanazine has been found in surface water in Ohio river basins (Datta, 1984).

- Cyanazine has also been found in ground-water in Iowa and Pennsylvania; typical positives found were 0.1 to 1.0 ppb (Cohen et al., 1986).

G. Environmental Fate

- ^{14}C-Cyanazine, at 5 to 10 ppm, degraded with a half-life of 2 to 4 weeks in an air-dried sandy clay loam soil, 7 to 10 weeks in a sandy loam soil, 10 to 14 weeks in a clay soil, and 9 weeks in a fresh sandy clay soil incubated in the dark at 22°C and field capacity (Osgerby et al., 1968). Three degradation products, the amide and two acids, were identified in all four soils; a fourth degradate, the amine, was found only in the air-dried sandy clay loam soil.

- Freundlich K values were 0.72 for a sandy loam soil (2.0% organic matter), 2.0 for a sandy clay soil (5.4% organic matter), 1.25 for a sandy clay loam soil (6.8% organic matter) and 6.8 for a clay soil (16% organic matter) treated with unaged ^{14}C-cyanazine (Osgerby et al., 1968). No linear correlation was found between organic matter content and adsorption.

- ^{14}C-Cyanazine readily moved through columns of sandy clay loam (52% of applied compound) and loamy sand (18% of applied) soil leached with 78 cm of water over a 13-day period; unaged ^{14}C-cyanazine was intermediately mobile on sandy clay loam and of low mobility on loamy sand soil thin-layer chromatography (TLC) plates (R_f 0.36 and 0.20, respectively) (McMinn and Standen, 1981). Aerobically and anaerobically aged ^{14}C-cyanazine residues, primarily the amide degradate (SD 20258), were intermediately mobile to mobile on sandy clay loam soil TLC plates.

- Aged ^{14}C-cyanizine residues readily leached through columns containing sand (47.8% of applied compound), loamy sand (69.7% of applied compound) and sandy loam (26.9% of applied compound) soils eluted with 20 cm of water (Eadsforth, 1984). The amide degradation product (SD 20258) was predominant in the leachate from the sandy soil (45% of radioactivity in leachate); the acid degradate (SD 20196) was predominant in leachate from the loamy sand (84%) and sandy loam (47%) soils. Unaltered cyanazine and SD 31222 were also identified in leachate from all three soils ($\leq$ 6% of recovered residues).

II. Pharmacokinetics

A. Absorption

- Studies by Shell Chemical Company (1969) and Hutson et al. (1970) indicated that cyanazine is rapidly absorbed from the gastrointestinal tract when administered orally at low dosage levels to three different animal species: rat, dog and cow. Measurements of urinary, fecal and biliary excretion indicated that 80 to 88% of 2,4,6-^{14}C-labeled cyanazine was eliminated within 4 days from the rat and dog, and within 21 days from the cow. The initial dosages were 1 to 4 mg/kg for the rat, 0.8 mg/animal for the dog and 5 ppm in the total ration of the cow. The dosages were administered by gavage in the rat studies and in gelatin capsules in the dog study.

B. Distribution

- In rats treated with a single oral dose of 4 mg/kg cyanazine, samples of the carcass, skin and gut reflected 2.02, 0.62 and 2.73% residual radioactivity, respectively, 4 days after exposure (Shell, 1969).

- In cows, samples of brain, liver, kidney, muscle and fat reflected concentrations of 0.55, 0.27, 0.24, 0.14 to 0.06 and less than 0.06 ppm cyanazine, respectively, after 21 days of continuous exposure to feed that contained 5 ppm cyanazine; however, when a lower dosage (0.2 ppm) was used in the feed, the detectable residues in each of these tissues were less than 0.05 ppm (Shell, 1969).

C. Metabolism

- Based on the analyses of metabolites in urine, the major metabolic pathways of cyanazine in the rat and cow involved: (1) conversion of the cyano group to an amide to form 2-chloro-4-ethylamino-6-(1-amido-methylethylamino)-s-thiazine; (2) conversion of N-deethylation to form 2-chloro-4-amino-6-(1-cyanol-methyl-ethylamino)-s-triazine; (3) conversion of the cyano group of deethylate cyanazine to form the amide of deethylated cyanazine, 2-chloro-4-amino-6-(1-amino-1-methylethyl-amino)-s-triazine; (4) dechlorination via glutathione, partial hydrolysis of glutathione conjugate and N-acetylation to form mercapturic acid, N-acetyl-S[4-amino-6-(1-cyano-1-methylethylamino)] L-cysteine and (5) dechlorination via hydrolysis (occurs only in the cow) to form 2-hydroxy-4-ethylamino-6-(1-carboxy-1-methylethylamino)-s-triazine and 2-hydroxy-4-amino-6-(1,carboxy-1-methylamino)-s-triazine, respectively (Shell, 1969).

- Studies by Shell Chemical Company (1969) and Hutson et al. (1970) in rats with ring-labeled and side-chain-labeled cyanazine (cyano-^{14}C, isopropyl-^{14}C and ethylamino-^{14}C) indicated that only the ethylamino-^{14}C side chain underwent extensive degradation, since 47% of the initial radioactivity was detected in the exhaled carbon dioxide. Thus, N-deethylation was found to be a major route of degradation of cyanazine.

- Crayford and Hutson (1972) identified 5 metabolites in urine of rats, an additional 2 (total 7) in feces and 4 metabolites in bile.

- Crayford et al. (1970) studied the metabolism of two major plant metabolites, DW4385 and DW4394, in rats. These two compounds were identified in the rat metabolism studies by Crayford and Hutson (1972) as 2-hydroxy-4-ethylamino-6-(1-carboxy-1-methylamino)-s-triazine (DW4385) and as 2-hydroxy-4-amino-6-(1-carboxy-1-methylethyl-amino)-s-triazine (DW4394). Approximately 91% of compound DW4385 and 84% of compound DW4394 were recovered unchanged from urine and feces.

D. Excretion

- Orally administered low doses of cyanazine (described above) were rapidly excreted in the urine and feces of rats and dogs (Shell, 1969; Hutson et al., 1970; Crayford and Hutson, 1972).

- In rats treated with 1 to 4 mg/kg cyanazine by gavage, a total of 88% of cyanazine was eliminated in 4 days. Elimination via urine was almost equal to elimination via feces; about 5.37% of the administered cyanazine remained in the body; and approximately 21% of the 1 mg/kg dose appeared in the bile within the first 20 hours (Shell, 1969).

- Hutson et al. (1970) reported that 33% of an oral dose of cyanazine was excreted in the urine of rats within 24 hours.

- A study in rats with ^{14}C-labeled 4-ethyl-amino cyanazine indicated that 47% of the radioactivity was eliminated in carbon dioxide (Shell, 1969).

- In dogs administered 0.8 mg of cyanazine in gelatin capsules, 51.67 and 36.29% of the dose were eliminated in the urine and feces, respectively, over a 4-day period (Shell, 1969).

- In cows exposed to treated feed (5 ppm cyanazine) for 21 consecutive days, the amount of daily excretion of radioactivity in urine and feces was constant throughout the study period. The total cyanazine equivalents in urine and feces were 53.7 and 26.8% of the dose, respectively. The concentration in milk was reported as 0.022 ppm (Shell, 1969).

III.   Health Effects

A. Humans

- No information was found in the available literature on the health effects of cyanazine in humans.

B. Animals

1. *Short-term Exposure*

- Acute oral LD_{50} values reported for rats range from 149 to 835 mg/kg (SRI, 1967b; NIOSH, 1987; Young and Adamik, 1979b; Meister, 1983). In these studies, the percentage of active ingredient (a.i.) in the tested product(s) was not clearly identified. However, studies by Walker et al. (1974) with technical cyanazine (97% a.i.) in three different animal species reflected LD_{50}s of 182, 380 and 141 mg/kg for the rat, mouse and rabbit, respectively.

- The acute dermal LD_{50} in rabbits treated with technical cyanazine (purity unspecified) was > 2,000 mg/kg (SRI, 1967a; Young and Adamik, 1979c); in rats, the LD_{50} was > 1,200 mg/kg (97% a.i.) (Walker et al., 1974).

- The acute inhalation LC_{50} for cyanazine dust (% a.i. not specified) in rats was > 2.28 mg/L/hr (Bishop, 1976) (thus cyanazine would be classified in toxicity category III).

- In a study by Walker et al. (1968), groups of 10 female CFE rats, 5 months old, were treated by gavage with single oral doses of 1, 5 or 25 mg/kg of a wettable powder formulation (75% a.i.); the control group received water. No diuretic effects were produced in the rats receiving the formulation; however, serum protein and potassium concentrations increased at the high dose, and serum osmolality increased at 5 mg/kg, the Lowest-Observed-Adverse-Effect Level (LOAEL). The No-Observed-Adverse-Effect Level (NOAEL) in this study appeared to be 1 mg/kg; however, this study did not provide enough information to determine the presence or absence of more significant effects at this dosage level.

- A 4-week oral toxicity study by Walker et al. (1968) was performed using groups of 10 male and 10 female CFE rats, 5 weeks of age, receiving diets containing 1, 10 or 100 ppm cyanazine (75% or 97% a.i.) for 4 weeks; these doses are equivalent to 0.05, 0.5 or 5 mg/kg/day (Lehman, 1959). A control group of 20 animals/sex was used. After 4 weeks, urine samples were collected for 16 hours (overnight), and blood samples were used to determine the kidney function. Reductions in body weight and food intake were noted at the high-dose level. Osmolal clearance decreased in males, and this change was associated with a decrease in free water clearance in both the low- and mid-dose groups. In females, decreased urine and increased serum osmolality were observed in the mid-dose group, and both creatinine clearance and urine potassium concentrations increased in the low-dose group. The LOAEL in this study appeared to be 0.05 mg/kg/day (lowest dose tested) based on kidney function tests, although additional information was not available to determine if any other significant adverse effects were noted at this level.

2. Dermal/Ocular Effects

- Cyanazine caused mild eye irritation at 100 mg (Young and Adamik, 1979a) and slight skin irritation at 2,000 mg (Young and Adamik, 1979c) in rabbits. A skin sensitization test in guinea pigs was negative (Walker et al., 1974; Young and Adamik, 1979d).

3. Long-term Exposure

- In a 13-week oral study in dogs (Walker and Stevenson, 1968a; Walker et al., 1974), groups of 5- to 7-month old beagle dogs, 4 animals/sex/treatment group, were given daily doses of 1.5, 5 or 15 mg/kg/day cyanazine in gelatin capsules. A control group of 5 animals/sex was given empty capsules. The test material caused emesis within the first hour of dosing in all of the high-dose males. Reduced body weight gain was also noted in the high-dose group during the second half of the study period, as well as increased kidney and liver weights in the females of this group. Thus, the LOAEL was 15 mg/kg/day and the NOAEL was 5 mg/kg/day.

- In a 13-week mouse feeding study (Fish et al., 1979), groups of 12 animals/sex/dose were fed diets containing 10, 50, 500, 1,000 or 1,500 ppm, equivalent to 1.5, 7.5, 75, 150 or 225 mg/kg/day (Lehman, 1959). The control group consisted of 24 animals/sex. Body weight gain reduction was observed in both sexes at 75 mg/kg/day and above. Statistically significant increases in liver weights were observed in both sexes at 75 mg/kg/day and above. Thus, the LOAEL was 75 mg/kg/day and the NOAEL was 7.5 mg/kg/day.

- An initial 13-week rat feeding study by Walker et al. (1968) was performed using 0.1, 1.0 or 100 ppm (equivalent to 0.005, 0.05 or 0.5 mg/kg/day; Lehman, 1959) of technical cyanazine (purity not specified: 97% or 75% a.i.) in feed. Each dosage group had 20 animals/sex; the control group had 40 animals/sex. Body weight gain decreased in all dosage groups in males and in the high-dose female group. A NOAEL was not reflected in this study for males, although it appeared to be 0.05 mg/kg/day for females.

- Walker and Stevenson (1968b) repeated the above study in rats at dose levels of 1.5, 3, 6, 12, 25, 50 or 100 ppm; these levels are equivalent to 0.075, 0.15, 0.30, 1.25, 2.5 or 5 mg/kg/day (Lehman, 1959). Similar effects were noted; however, a NOAEL of 25 ppm (1.25 mg/kg/day) was identified.

- In a 2-year study in dogs (Walker et al., 1970a), groups of 4- to 6-month-old beagle dogs were treated with technical cyanazine (97% a.i., in gelatin capsules) at dose levels of 0.625, 1.25 or 5 mg/kg/day. Each group consisted of 4 animals/sex. The control group consisted of 6 animals/sex and received empty gelatin capsules. Frequent emesis within 1 hour of dosing was observed throughout the study period in the high-dose group; this effect was associated with reduction of growth rate and serum protein. The NOAEL appeared to be 1.25 mg/kg/day; however, this NOAEL should be considered with reservations because the study did not provide adequate explanation relative to missing histological data on one of four female dogs in the 1.25 mg/kg/day dosage group. In addition, the reported data were limited to a summary report.

- In a 2-year study in mice (Shell, 1981), cyanazine technical (purity not specified) was given in feed to CD mice at 10, 25, 50, 250 or 1,000 ppm, equivalent to 1.5, 3.75, 7.5, 37.5 or 150 mg/kg/day (Lehman, 1959); 50 animals/sex were used in the treatment groups, and 100 animals/sex were used as controls. Toxic effects reported at the two high-dose levels, 37.5 and 150 mg/kg/day, included poor appearance and skin sores, increased mortality in the female animals in both groups, increased relative brain weight in both sexes, increased relative liver weight in the two female groups, and decreased absolute and differential leukocyte values in both sexes. Anemia was noted at 150 mg/kg/day in the females, as well as increased blood protein and increased relative kidney weight. Cyanazine did not demonstrate an oncogenic potential in this study. The NOAEL for systemic toxicity in mice appeared to be 50 ppm (7.5 mg/kg/day).

- Two chronic feeding studies in rats were available for review. In one study (Walker et al., 1970b; also cited in Walker et al., 1974), groups of 24 CFE rats/sex/dose received diets containing 6, 12, 25 or 50 ppm, equivalent to 0.3, 0.6, 1.25 or 2.5 mg/kg/day (Lehman, 1959) cyanazine (97% a.i.); 45 rats/sex were used as controls. The authors indicated that no effects due to cyanazine were noted in this study, although reduction in growth rate was noted in both sexes at 2.5 mg/kg/day and in females at 1.25 mg/kg/day. A review of this study (U.S. EPA, 1984b) indicated that cyanazine appeared to be tumorigenic in both male and female rats based on the increased incidences of thyroid tumors in all treatment groups as compared to the study's control group; increased incidences of adrenal tumors also were noted in all male treatment groups. However, this study was considered unacceptable because of several deficiencies: a limited number of tissues per animal were examined microscopically; the tumor incidences were calculated based on the number of animals tested rather than on the number of specific tissues histologically examined; gross examination and histologic findings for nonneoplastic lesions were not adequately reported; and only limited hematology, clinical blood chemistry and urinalyses data were presented.

- Simpson and Dix (1973) repeated the above 2-year study using 1, 3 or 25 ppm, equivalent to 0.05, 0.15 or 1.25 mg/kg/day in the diet of rats (Lehman, 1959); however, convulsions were noted in the rats 3 months after the study initiation and throughout the remainder of the study period. Approximately 42% of the animals were affected, and the incidence was not considered to be dose-related. The incidence of animals with convulsions was similar in both the control and high-dose male groups (21/48 and 11/24, respectively).

- A one-year feeding study in dogs using atrazine (98% a.i.) by Dickie (1986) has been evaluated by the Agency. Five experimental groups of 6 animals/sex/group were exposed to the following doses: 0, 10, 25, 100 or 200 ppm. These doses were equivalent to actual consumption of 0, 0.27, 0.68, 3.20 or 6.11 mg/kg/day for males and 0, 0.28, 0.72, 3.02 or 6.39 mg/kg/day for females, respectively. No systemic toxicity was noted at 10 or 25 ppm. However, dose-related decreases in body weight and body weight gains were noted at 100 and 200 ppm, as well as elevated platelet counts, and reduced levels of total protein, albumin and calcium in both sexes. At these two high-dose levels, there were also slight but not statistically significant decreases in spleen weights and increases in relative liver weights in females, and increases in liver weights and decreases in testes weights in males. Other noted changes in organ weights (i.e., heart, lung and kidneys) were not considered significant at these dose levels since they were not consistent with changes in the absolute and relative organ weight values. No gross or microscopic findings related to treatment were noted. Thus, in this study, the LOAEL is 3.1 mg/kg/day and the NOAEL is 0.7 mg/kg/day.

4. *Reproductive Effects*

- A three-generation reproduction study in Long-Evans rats (Eisenlord et al., 1969) using technical cyanazine (unknown percentage a.i.) at dietary levels of 3, 9, 27 or 81 ppm (0.15, 0.45, 1.35 or 4.05 mg/kg/day based on the dietary assumptions of Lehman, 1959) did not reflect a significant effect on reproduction parameters. The NOAEL in this study appeared to be 1.35 mg/kg/day; the LOAEL was 4.05 mg/kg/day (highest dose tested) based on findings related to reduced body weight gain in parental animals, and increased relative brain weight and decreased relative kidney weight in F_{3b} female weanlings.

- In a repeat two-generation reproduction study in Sprague-Dawley rats (WIL Research Laboratory, 1987), atrazine (100% a.i.) was administered in feed at 0, 25, 75, 150 or 250 ppm. These doses are equivalent to actual food consumption of 0, 1.8, 5.3, 11.1 or 18.5 mg/kg/day, respectively; however, these values changed during lactation to 0, 3.8, 11.2, 23.0 or 37.1 mg/kg/day, respectively. Dose-related decreases in pups' viability and body weights were noted at 75 ppm and above; therefore, the NOAEL for reproduction may be 25 ppm (3.8 mg/kg/day). However, this level, 25 ppm, (equivalent to 1.8 mg/kg/day during nonlactating periods of the study) may be considered as the LOAEL for parental animals due to the noted decreases in body weight and food consumption at this level.

5. *Developmental Effects*

- Cyanazine appeared to cause developmental toxicity in two animal species, the rabbit and the rat, and teratogenic effects in the rat (Bui, 1985b).

- In the rabbit study (Shell Toxicology Laboratory, 1982), 7- to 11-month-old New Zealand White rabbits were orally dosed with cyanazine (98% a.i.) in gelatin capsules at levels of 0, 1, 2 or 4 mg/kg/day on gestation days 6 through 18 (22 dams/dose/group). At 2 and 4 mg/kg/day, maternal toxic effects included anorexia, weight loss, death and abortion. Alterations in skeletal ossification sites, decreased litter size, and increased postimplantation loss were observed at 2 and 4 mg/kg/day. Malformations were also noted at 4 mg/kg/day as demonstrated by anophthalmia/microphthalmia, dilated brain ventricles, domed cranium and thoracoschisis; however, these responses were observed at levels in excess of maternal toxicity. The maternal and developmental toxicity NOAELs were 1 mg/kg/day.

- In a rat study by Lu et al. (1981, 1982), 122-day-old Fischer 344 rats (30 dams/group) were administered cyanazine (98.5% a.i.) by gavage at dose levels of 0, 1.0, 2.5, 10.0 or 25.0 mg/kg/day on gestation days 6 through 15; the dosages were suspended in a 0.2% Methocel emulsion vehicle. Maternal body weight reductions during dosing were noted at the 10- and 25-mg/kg/day levels. Diaphragmatic hernia associated with liver protrusion, microphthalmia and anophthalmia were observed at the 25 mg/kg/day dose level. A teratogenic NOAEL could not be determined from this study at 10 mg/kg/day and a maternal toxicity NOAEL at 2.5 mg/kg/day.

- The above study was repeated in the same strain of rats, Fischer 344, by Lochry et al. (1985) in order to further examine the malformations reported in the study by Lu et al. (1981). In this study, the dams (70/dosage group) were 86 days old. Cyanazine (98% a.i.) was administered by gavage in an aqueous suspension of 0.25% (w/v) methyl cellulose at dose levels of 0, 5, 25 or 75 mg/kg/day on days 6 through 15 of gestation. One-half of the dams in each group were selected for Cesarean delivery on day 20 of gestation. The remaining half of the dams in each group were allowed to deliver, and both they and their pups were observed for 21 days before sacrifice. Maternal body weight reductions during dosing were noted in all dosage groups and appeared to be partly associated with lower food intake during the dosing period. Alteration in skeletal ossification sites were also observed in the fetuses at all dose levels. Teratogenic effects were demonstrated at 25 and 75 mg/kg/day as anophthalmia/microphthalmia, dilated brain ventricles and cleft palate in the fetuses, and abnormalities of the diaphragm (associated with liver protrusion) in pups sacrificed at time of weaning. The maternal and developmental toxicity NOAELs were lower than 5 mg/kg/day (lowest dose tested), and the teratogenic NOAEL was 5 mg/kg/day (Bui, 1985a).

- An additional study in Sprague-Dawley rats (Shell, 1983) did not reflect any maternal or developmental toxicity at the highest dose tested, 30 mg/kg/day.

6. *Mutagenicity*

- The mutagenic potential of cyanazine has not been investigated adequately, and only limited information was available for evaluation.

- A study by Dean et al. (1974a) using technical cyanazine (80% a.i.) in mice of both sexes did not reflect any increase in chromosomal aberrations in the bone marrow cells. The animals were examined at 8- and 24-hour intervals after oral dosing with 50 or 100 mg/kg cyanazine. However, the sensitivity of this test was potentially compromised because the positive control data did not reflect a significant number of aberrations: the percent of cells showing chromatid gaps in the positive control (cyclophosphamide) was not statistically significant at the $p < 0.05$ level (U.S. EPA, 1984b).

- Dean et al. (1974b) used technical cyanazine (purity not specified) to induce dominant lethal effects in male CF_1 mice. The test was negative at the dose levels tested (80, 160 and 320 mg/kg). However, this study appeared to be invalid because there was no positive control for comparison of data, and a range-finding test was not performed to select the appropriate dosages used in this study (U.S. EPA, 1984b).

- Cyanazine is a member of the triazine family of herbicides. It is known that the triazines follow similar metabolic pathways (i.e., N-dealkylation, S-dealkylation or O-dealkylation and conjugation with glutathion) that result in common or closely related metabolites. Waters et al. (1980) noted that a triazine herbicide (atrazine) gave a positive mutagenic response in the *Drosophila* sex-linked recessive lethal test (DRL), although this chemical gave a negative response in an *in vitro* test battery with microorganisms. Hence, the potential for cyanazine to give a positive response in a similar test exists (U.S. EPA, 1984b).

7. *Carcinogenicity*

- Cyanazine was not determined to have a carcinogenic potential in a 2-year mouse study (Shell, 1981).

- Cyanazine was not oncogenic in 2-year rat studies by Walker et al. (1970b) or by Simpson and Dix (1973); however, these studies were deficient (see description of these studies under the section entitled Long-term Exposure) and are considered to be inadequate by design to determine the oncogenic potential of cyanazine.

IV.   Quantification of Toxicological Effects

A. One-day Health Advisory

No information was found in the available literature for determination of the One-day HA for cyanazine. It is therefore recommended that the Ten-day HA value for a 10-kg child, calculated below as 0.10 mg/L (100 μg/L), be used at this time as a conservative estimate of the One-day HA value.

B. Ten-day Health Advisory

The teratology study in rabbits by Shell Toxicology Laboratory (1982) has been selected as the basis for determination of the Ten-day HA for cyanazine because it provides a short-term NOAEL (1 mg/kg/day for 13 days) for both maternal and fetal toxicity. This study also reflects the lowest NOAEL when compared with the teratology studies in rats described earlier, two in Fischer 344 rats (Lu et al., 1981; Lochry et al., 1985) and one in Sprague-Dawley rats (Shell, 1983).

Using a NOAEL of 1 mg/kg/day, the Ten-day HA for a 10-kg child is calculated as follows:

$$\text{Ten-day HA} = \frac{(1 \text{ mg/kg/day}) (10 \text{ kg})}{(100) (1 \text{ L/day})} = 0.10 \text{ mg/L} (100 \ \mu\text{g/L})$$

C. Longer-term Health Advisory

No information was suitable for the determination of the Longer-term HA for cyanazine. It is therefore recommended that the adjusted Drinking Water Equivalent Level (DWEL) of 0.02 mg/L (20 μg/L) be used for a 10-kg child as a conservative estimate for the Longer-term HA value and the DWEL of 0.07 μg/L (70 μg/L), calculated in the Lifetime Health Advisory section, be used for a 70-kg adult.

For a 10-kg child, the adjusted DWEL is calculated as follows:

$$\text{DWEL} = \frac{(0.002 \text{ mg/kg/day}) (10 \text{ kg})}{1 \text{ L/day}} = 0.02 \text{ mg/L} (20 \ \mu\text{g/L})$$

where: 0.002 mg/kg/day = RfD (see Lifetime Health Advisory section)

D. Lifetime Health Advisory

Five chronic studies were available for evaluation: (1) a 2-year oncogenic study in mice (Shell, 1981) with a potential NOAEL of 50 ppm (approximately 7.5 mg/kg/day when using a conversion factor for food consumption of 15% of the body weight); (2) a 2-year feeding study in dogs (Walker et al., 1970a) with a NOAEL of 1.25 mg/kg/day; (3) a 2-year feeding/oncogenic study in rats (Walker et al., 1970b, also cited in Walker et al., 1974) with a NOAEL of 12 ppm (approximately 0.6 mg/kg/day when using a conversion factor for food consumption of 5% of the body weight); however, this study was considered unacceptable (U.S. EPA, 1984b) due to several deficiencies in the study report (see Long-term Exposure section); (4) a second 2-year feeding study in rats (Simpson and Dix, 1973), which was also considered inadequate because the control group reflected an effect, i.e., convulsions, that was suggestive of cross-dosing; and (5) a 1-year feeding study in dogs (Dickie, 1986) with a NOAEL of 25 ppm (approximately 0.7 mg/kg/day based on actual mean food consumption of males and females).

The NOAEL in the mouse study (7.5 mg/kg/day) can be considered for this calculation; however, this NOAEL is higher than the NOAEL in the Walker et al. (1970a) dog study (1.25 mg/kg/day) or in the Walker et al. (1970b) rat study (0.6 mg/kg/day). Since this rat study is considered unacceptable and since the second rat study (Simpson and Dix, 1973) appeared to be flawed by the invalidity of the control group, the 2-year dog study (Walker et al., 1970a) was used previously for the Lifetime HA calculations, using a NOAEL of 1.25 mg/kg/day. This study was of marginal acceptability because only a summary report was available for evaluation and histopathological data were missing for 1/4 females at 1.25 mg/kg/day. However, this NOAEL was supported by a similar NOAEL from a 13-week rat subchronic feeding study by Walker and Stevenson (1968b). Thus, using this NOAEL from both studies (i.e., the 2-year dog study and 13-week rat study) and applying a large uncertainty factor of 1,000-fold were appropriate for the calculation of the RfD and the Lifetime HA (in the absence of more adequate chronic studies at that time). At the present time, the 1-year dog feeding study by Dickie (1986) is a more adequate study for these calculations.

The NOAEL of 25 ppm (equivalent to 0.7 mg/kg/day based on mean actual food consumption in male and female dogs) from the Dickie study (1986) is used in the calculation of the RfD. In this 1-year study, 30 male and female dogs (6 animals/sex/dose) were fed diets containing 0, 10, 25, 100 or 200 ppm cyanazine (98% a.i.). These doses were equivalent to actual mean intakes of 0, 0.27, 0.68, 3.20 or 6.11 mg/kg/day in males and 0, 0.28, 0.72, 3.02 or 6.39 mg/kg/day in females, respectively. No systemic toxicity was noted at 25 ppm (0.7 mg/kg/day) in both sexes. The LOAEL was 100 ppm (3.1 mg/kg/day) based on reduced body weights and body weight gains, elevated platelet counts, and reduced levels of total protein, albumin and calcium in males and females. There were also slight, not statistically significant, decreases in spleen weights and increases in liver weights in the females and increases in liver weights and decreases in testes weights in the males. No gross or microscopic findings related to treatment were noted.

Using a NOAEL of 0.7 mg/kg/day, the Lifetime HA is calculated as follows:

Step 1: Determination of the Reference Dose (RfD)

$$RfD = \frac{(0.7 \text{ mg/kg/day})}{(100)(3)} = 0.002 \text{ mg/kg/day}$$

where: 3 = modifying factor used to compensate for the lack of a chronic rat study as required by the Office of Pesticide Programs.

Step 2: Determination of the Drinking Water Equivalent Level (DWEL)

$$DWEL = \frac{(0.002 \text{ mg/kg/day}) (70 \text{ kg})}{(2 \text{ L/day})} = 0.07 \text{ mg/L} (70 \text{ } \mu g/L)$$

Step 3: Determination of the Lifetime Health Advisory

Lifetime HA = (0.07 mg/L) (20%) = 0.014 mg/L (10 μg/L)

E. Evaluation of Carcinogenic Potential

- Available toxicity data indicate that cyanazine was not carcinogenic in mice (Shell, 1981) or rats (Walker et al., 1970b, 1974; Simpson and Dix, 1973). In the rat, some increases were noted in the incidences of both thyroid tumors (male and female rats) and adrenal tumors (male rats); however, these increases were not statistically significant.

- Cyanazine is a chloro-s-triazine derivative that has a chemical structure analogous to atrazine, propazine and simazine. These three analogs were found to significantly ($p < 0.05$) increase the incidence of mammary tumors in the Sprague-Dawley rat and they are classified as group C oncogens. Based on structure-activity relationship, cyanazine may reflect a similar pattern of toxicity in this strain of rats. A new 2-year oncogenic study is required from the manufacturer of this chemical to fill this data gap in the toxicity profile of cyanazine.

- Applying the criteria described in EPA's guidelines for assessment of carcinogenic risk (U.S. EPA, 1986a), cyanazine may be classified in Group D: not classifiable. This category is used for substances with inadequate animal evidence of carcinogenicity.

V.   Other Criteria, Guidance and Standards

- U.S. EPA Office of Pesticide Programs (OPP) has established residue tolerances for cyanazine ranging from 0.05 to 0.10 ppm in or on raw agricultural commodities (U.S. EPA, 1985) based on a Provisional ADI (PADI) of 0.0013 mg/kg/day.

VI.   Analytical Methods

- Analysis of cyanazine is by a high-performance liquid chromatographic (HPLC) method applicable to the determination of cyanazine in water samples, Method #4 (U.S. EPA, 1986b). In this method, 1 L of sample is extracted with methylene chloride using a separatory funnel. The methylene chloride extract is dried and concentrated to a volume of 10 mL or less. HPLC is used to permit the separation of compounds and measurement is conducted with an ultraviolet (UV) detector. Using this method, the estimated detection limit for cyanazine is 0.3 μg/L.

VII.   Treatment Technologies

- Whittaker (1980) experimentally determined adsorption isotherms for cyanazine on GAC.

- Available data indicate that granular-activated carbon (GAC) adsorption appears to be an effective method of cyanazine removal from water. However, selection of individual or combinations of technologies to attempt cyanazine removal from water must be based on a case-by-case technical evaluation, and an assessment of the economics involved.

VIII.   References

Bishop, A.L.* 1976. Report to Shell Chemical Company: acute dust inhalation toxicity study in rats. Unpublished study received July 18, 1979 under 201–279. Prepared by Industrial Bio-Test Laboratories, Inc., submitted by Shell Chemical Co., Washington, D.C. CDL:098395-A. MRID #00022789. (Cited in U.S. EPA, 1984b.)

Bui, Q.Q.* 1985a. Review of a developmental toxicity study (teratology and post-natal study). U.S. EPA internal memo from author to Robert Taylor reviewing study cited in Shell Development Company (1985), Report No. 619–002, Accession No. 257867.

Bui, Q.Q.* 1985b. Overview of the teratogenic potential of Bladex (cyanazine). U.S. EPA internal memo from author to Herb Harrison, dated June 5, 1985.

CHEMLAB. 1985. The Chemical Information System. CIS Inc., Baltimore, MD.

Cohen, S.Z., C. Eiden and M.N. Lorber. 1986. Monitoring ground water for pesticides in the U.S.A. In: Evaluation of pesticides in ground water. American Chemical Society Symposium Series (in press).

Crayford, J.V., E.C. Hoadley, B.A. Pikering et al.* 1970. The metabolism of the major plant metabolites of Bladex (DW 4385 and DW 4394) in the rat. Group research report TLGR. 0081.70. Unpublished study prepared by Shell Research, Ltd. MRID #000223871. (Cited in U.S. EPA, 1984b.)

Crayford, J.V. and D.H. Hutson.* 1972. Metabolism of the herbicide 2-chloro-4-(ethylamino)-6-(1-cyano-1-methylethylamino)-S-triazine in the rat. MRID #00022856. Pesticide Biochem. Physiol. 2:295-307. (Cited in U.S. EPA, 1985a; U.S. EPA, 1984b.)

Datta, P.R. 1984. Internal memorandum: review of six documents regarding monitoring of pesticides in northwestern Ohio rivers. Washington, DC: U.S. Environmental Protection Agency.

Dean, B.J., K.R. Senner, B.D. Perquin and S.M.A. Doak.* 1974a. Toxicity studies with Bladex chromosome studies on bone marrow cells of mice after two daily oral doses of Bladex. Unpublished study Report No. TLGR.0032074 received August 13, 1976 under 6F1729. Prepared by Shell Research, Ltd., submitted by Shell Chemical Co., Washington, D.C. CDL:095245-B. MRID #00023836. (Cited in U.S. EPA, 1984b.)

Dean, B.J., E. Thorpe and D.E. Stevenson.* 1974b. Toxicity studies on Bladex: dominant-lethal assay in male mice after single dose of Bladex. Unpublished study received August 13, 1976. Prepared by Shell Research, Ltd. for Shell Chemical Co., Washington, D.C. CDL:095245-C. MRID #00023837. (Cited in U.S. EPA, 1984b.)

Dickie, B.C. 1986. One-year oral feeding study in dogs with the triazine herbicide – cyanazine. Study #6160-104 and addendum #107F. Unpublished study performed by Hazelton Laboratories for duPont deNemours & Co. MRID 40081901 and 40229001.

Eadsforth, C.V. 1984. The leaching behavior of Bladex and its degradation products in German soils under laboratory conditions. Expt. No. 2994. Unpublished study submitted by Shell Chemical Company, Washington, DC.

Eisenlord, G., G.S. Loquvam and S. Leung.* 1969. Results of reproduction study of rats fed diets containing SD 15418 over three generations. Report No. 47. Unpublished study received on unknown date under 9G0844. Prepared by Hine Laboratories, Inc., submitted by Shell Chemical Co., Washington, DC. CDL:095023-D. MRID #00032346. (Cited in U.S. EPA, 1985b.)

Fish, A., R.W. Hend and C.E. Clay.* 1979. Toxicity studies on the herbicide Bladex: a three-month feeding study in mice: TLGR.0021.79. Unpublished study received July 19, 1979 under 201–279. Submitted by Shell Chemical Co., Washington, DC. CDL:09835-C. (Cited in U.S. EPA, 1984b.)

Hutson, D.H., E.C. Hoadley, M.H. Griffiths and C. Donninger. 1970. Mercapturic acid formation in the metabolism of 2-chloro-4-ethylamino-6-(l-methyl-1-cyanoethylamino)-s-triazine in the rat. J. Agric. Food. Chem. 18:507–512. (Data also available in U.S. EPA, 1984b, MRID #00032348, Shell Chemical Co., 1969.)

Lehman, A.J. 1959. Appraisal of the safety of chemicals in foods, drugs and cosmetics. Assoc. Food Drug Off. U.S.

Lochry, E.A., A.M. Hoberman and M.S. Christian.* 1985. Study of the developmental toxicity of technical Bladex herbicide (SD-15418) in Fischer-344 rats. Unpublished report. Submitted by Shell Oil Company, prepared by Argus Research Laboratory, Inc., Horsham, PA. Report No. 619–002. Dated April 18, 1985.

Lu, C.C., B.S. Tang, E.Y. Chai et al.* 1981. Technical Bladex (R) (SD 15418) teratology study in rats. Project No. 61230. Unpublished study received January 4, 1982 under 201–179. Submitted by Shell Chemical Co., Washington, DC. CDL:070584-A. MRID #00091020. (Cited in Lu et al., 1982, and in U.S. EPA, 1984b.)

Lu, C.C., B.S. Tang and E.Y. Chai. 1982. Teratogenicity evaluations of technical Bladex in Fischer-344 rats. Teratology. 25(2):59A-60A.

McMinn, A.L. and M.E. Standen. 1981. The mobility of Bladex and its degradation products in soil under laboratory conditions. Unpublished study submitted by Shell Chemical Company, Washington, DC.

Meister, R., ed. 1983. Farm chemicals handbook. Willoughby, OH: Meister Publishing Company.

Mirvish, S.S. 1975. Formation of N-nitroso compounds: chemistry, kinetics, and *in vivo* occurrence. Submitted by Shell Oil Co., Washington, DC. CDL:070584-A.

NAS. 1977. National Academy of Sciences. Drinking water and health. Washington, DC: National Academy Press.

NIOSH. 1987. National Institute for Occupational Safety and Health. Registry of toxic effects of chemical substances. Rockville, MD: U.S. DHEW, PHS, CDC. (Cited in U.S. EPA, 1984a.)

Osgerby, J.M., D.F. Clarke and A.T. Woodburn. 1968. The decomposition and adsorption of DW 3418 (WL 19,805) in soils. Unpublished study submitted by Shell Chemical Company, Washington, DC.

Plewa, M.J. and J.M. Gentile. 1976. Mutagenicity of atrazine: a maize-microbe bioassay. Mutat. Res. 38:287–292.

Shell.* 1969. Metabolism of cyanazone. Unpublished study submitted by Shell Chemical Company. MRID #00032348. (Cited in U.S. EPA, 1984b.)

Shell.* 1981. Two-year oncogenicity study in the mouse. Unpublished report submitted under pesticide petition number 9F2232. EPA accession number 247295 to -298.

Shell.* 1983. Teratogenic evaluation of Bladex in SD CD rats. Unpublished report submitted by Shell Development Company. Prepared by Research Triangle Institute. Project No. 31T-2564. Report dated May 16, 1983, submitted to the EPA on July 6, 1983. EPA Accession No. 071738. (Cited in U.S. EPA, 1984b.)

Shell Toxicology Laboratory (Tunstall).* 1982. A teratology study in New Zealand White rabbits given Bladex orally. A report prepared by Sittingbourne Research Center, England. Project No. 221/81, Experiment No. AHB-2321, November, 1982. Submitted on February 1, 1983 as document SBGR.82.357 by Shell Oil Co., Washington, DC. Accession No. 071382. (Cited in U.S. EPA, 1984b.)

Simpson, B.J. and K.M. Dix.* 1973. Toxicity studies on the s-triazine herbicide Bladex: second 2-year oral experiment in Research Limited, London. Dated July 1973. EPA Accession No. 251954, -955 and -956.

SRI.* 1967a. Stanford Research Institute. Acute dermal toxicity of SD-15418 (technical cyanazine). Project 868–1. Report No. 39, January 4, 1967. Submitted by Shell Chemical Co., Washington, DC. Pesticide Petition No. 9G0844. Accession No. 91460. (Cited in U.S. EPA, 1984b.)

SRI.* 1967b. Stanford Research Institute. Acute oral toxicity of SD-15418 (technical cyanazine). Project 55 868, Report No. 43, May 26, 1967. Submitted by Shell Chemical Co., Washington, DC. Pesticide Petition No. 9G0844. Accession No. 91460. (Cited in U.S. EPA, 1984b.)

STORET. 1988. STORET Water Quality File. Office of Water. U.S. Environmental Protection Agency (data file search conducted in May, 1988).

U.S. EPA. 1984a. U.S. Environmental Protection Agency. Draft health and environmental effects profile for cyanazine. Cincinnati, OH: Environmental Criteria and Assessment Office.

U.S. EPA.* 1984b. U.S. Environmental Protection Agency. Cyanazine toxicology data review for registration standard. Washington, DC: Office of Pesticide Programs.

U.S. EPA. 1985. U.S. Environmental Protection Agency. 40 CFR. 180.307.

U.S. EPA. 1986a. U.S. Environmental Protection Agency. Guidelines for carcinogen risk assessment. Fed. Reg. 51(185):33992–34003. September 24.

U.S. EPA. 1986b. U.S. Environmental Protection Agency. U.S. EPA Method 4-Determination of pesticides in ground water by HPLC/UV. January,

1986 draft. Available from U.S. EPA's Environmental Protection Monitoring and Support Laboratory, Cincinnati, OH.

Walker, A.I.T., R. Kampjes and G.G. Hunter.* 1968. Toxicity studies in rats on the s-triazine herbicide (DW 3418): (a) 13-week oral experiments; (b) the effect on kidney function. Group Research Report TLGR.0007.69. Unpublished study received Oct. 17, 1969 under 9G0844. Prepared by Shell Research, Ltd., England, submitted by Shell Chemical Co., Washington, DC. CDL:091460-H. MRID #00093200. (Cited in U.S. EPA, 1984b; Walker, et al., 1974.)

Walker, A.I.T. and D.E. Stevenson.* 1968a. The toxicity of the s-triazine herbicide (DW 3418): 13-week oral toxicity experiment in dogs. Group Research Report TLGR.0016.68. Unpublished study received Oct. 17, 1969 under 9G0844. Prepared by Shell Research, Ltd., England, submitted by Shell Chemical Co., Washington, DC. CDL:091460-G. MRID #00093199. (Cited in U.S. EPA, 1984b; Walker et al., 1974.)

Walker, A.I.T. and D.E. Stevenson.* 1968b. The toxicity of the s-triazine herbicide (DW 3418): 13-week oral experiment in rats. Group Research Report TLGR.0017.68. Unpublished study received Oct. 17, 1969 under 9G0844. Prepared by Shell Research, Ltd., England, submitted by Shell Chemical Co., Washington, DC. MRID #00093198. (Cited in U.S. EPA, 1984b; Walker et al., 1974.)

Walker, A.I.T., E. Thorpe and C.G. Hunter.* 1970a. Toxicity studies on the s-triazine herbicide Bladex (DW 3418): two-year oral experiment with dogs. Group Research Report TLGR.0065.70. Unpublished study received December 4, 1970 under OF0998. Prepared by Shell Research, Ltd., England, submitted by Shell Chemical Co., Washington, DC. CDL:091724-R. MRID #00065483.

Walker, A.I.T., E. Thorpe and C.G. Hunter.* 1970b. Toxicity studies on the s-triazine herbicide Bladex (DW 3418): two-year oral experiment with rats. An unpublished report prepared by Tunstall Laboratory. Submitted by Shell Research, Ltd., London. (TLGR.0063.70). EPA Accession Nos. 251, 949–251, 953; Pesticide Petition No. OF0998 CDL:091724-Q. MRID #00064482.

Walker, A.I.T., V.K. Brown, J.R. Kodama, E. Thorpe and A.B. Wilson. 1974. Toxicological studies with the 1,3,5-triazine herbicide cyanazine. Pestic. Sci. 5(2):153–159.

Waters, M.D., V.F. Simmon, A.D. Mitchell, T.A. Jorgenson and R. Valencia. 1980. An overview of short-term tests for the mutagenic and carcinogenic potential of pesticides. J. Environ. Sci. Health. 6:867–906.

Whittaker, K.F., 1980. Adsorption of selected pesticides by activated carbon using isotherm and continuous flow column systems. Ph.D. Thesis. Lafayette, IN: Purdue University.

WIL Research Laboratories. 1987. Two-generation rat reproduction study. Project No. WIL 93001, August 12. Unpublished study submitted by duPont. MRID #403600–01.

Wolfe, N.L., R.G. Zapp, J.A. Gordon and R.C. Fincher. 1975. N-Nitrosoatrazine: formation and degradation. 170th Amer. Chem. Soc. Meeting. Abstracts. American Chemical Society. p. 23.

Young, S.M. and E.R. Adamik.* 1979a. Acute eye irritation study in rabbits with SD 15418 (technical Bladex (R) herbicide). Code 16–8–0–0. Project No. WIL-1223–78. Unpublished study received Jan. 10, 1980 under 201281. Submitted by Shell Chemical Co., Washington, DC. CDL:099198-E. MRID #00026427. (Cited in U.S. EPA, 1984b.)

Young, S.M. and E.R. Adamik.* 1979b. Acute oral toxicity study in rats with SD 15418 (technical Bladex (R) herbicide). Code 16–8–0–0. Project No. WIL-1223–78. Unpublished study received Jan. 10, 1980 under 201281. Submitted by Shell Chemical Co., Washington, DC. CDL:099198-C. MRID #00026424. (Cited in U.S. EPA, 1984b.)

Young, S.M. and E.R. Adamik.* 1979c. Acute dermal toxicity study in rabbits with SD 15418 (technical Bladex (R) herbicide). Code 16–8–0–0. Project No. WIL-1223–78. Unpublished study received Jan. 10, 1980 under 201281. Submitted by Shell Chemical Co., Washington, DC. CDL:099198-C. MRID #00026425. (Cited in U.S. EPA, 1984b.)

Young, S.M. and E.R. Adamik.* 1979d. Delayed contact in hypersensitivity study in guinea pigs with SD 15418 (technical Bladex (R) herbicide). Code 16–8–0–0. Project No. WIL-1223–78. Unpublished study received Jan. 10, 1980 under 201–281. Submitted by Shell Chemical Co., Washington, DC. CDL:099198-F. MRID #00026428. (Cited in U.S. EPA, 1984b.)

Zendzian, R.P. 1985. Review of a study on Bladex dermal absorption. U.S. EPA internal memo to G. Werdig reviewing study by Jeffcoat, A.R. (Research Triangle Institute, RTI/3134/01F, Dec. 1984). Accession No. 256324. Dated February 20.

---

* Confidential Business Information submitted to the Office of Pesticide Programs.

Dalapon

I. General Information and Properties

A. CAS No. 75–99–0

B. Structural Formula

$$CH_3CCl_2COOH$$

(2,2-Dichloropropionic acid)

C. Synonyms

- Dalapon (ANSI, BSI, WSSA), DPA, Basfapon and Basfapon B (discontinued by BASF Wyandotte); Basfapon/Basfapon N, BH Dalapon and Crisapon (Crystal Chemical Inter-America); Dalapon 85, Dalapon-Na, Ded-Weed and Devipon (Devidayal); Dowpon, Dowpon M, Gramevin and Radapon (discontinued by Dow); Revenge (Hopkins); Unipon (Meister, 1984).

D. Uses

- Dalapon (2,2-dichloropropionic acid) is used as a herbicide in the form of its sodium and/or magnesium salts to control grasses in crops, drainage ditches, along railroads and in industrial areas (U.S. EPA, 1984).

E. Properties (U.S. EPA, 1984; Reinert and Rogers, 1987)

Chemical Formula	$C_3H_4Cl_2O_2$
Molecular Weight	143 (acid form)
Physical State (room temp.)	liquid
Boiling Point	185 to 190°C
Melting Point	20°C
Density (°C)	--
Vapor Pressure	--
Specific Gravity	--
Water Solubility (25°C)	50,000 mg/L

Log Octanol/Water
   Partition Coefficient   1.47
Taste Threshold   --
Odor Threshold   --
Conversion Factor   --

F. Occurrence

- Dalapon has been found in none of the surface water or ground-water samples analyzed from 14 samples taken at 14 locations (STORET, 1988).

G. Environmental Fate

- The sodium salt of dalapon has been shown to hydrolyze slowly in water to produce pyruvic acid, and the rate of hydrolysis increases with increasing temperature. After 175 hours, the extent of hydrolysis at 25°C for 1%, 5% and 18% dalapon solutions was 0.41%, 0.61% and 0.8%, respectively (Brust, 1953).

- Hydrolysis of solutions of either dalapon or dalapon sodium salt is accelerated at alkaline pH values. For example, hydrolysis of dalapon sodium salt at 60°C was 20% complete in 30 hours at which time the equilibrium pH was 2.3. In contrast, hydrolysis was 50% complete in 30 hours when the pH was maintained at 12 during the experiment (Tracey and Bellinger, 1958).

- Based on reaction rate studies, Kenaga (1974) concluded that both dalapon salt and dalapon would have chemical hydrolysis half-lives of several months at temperatures less than 25°C and at initial solution concentrations of less than 1%. Considering the more rapid rate of microbial degradation, the author concluded that it does not appear that chemical hydrolysis of dalapon is a particularly significant degradative pathway in soils.

- Because of its high water solubility and lack of affinity for soil particles, appreciable adsorption of dalapon on suspended or bottom sediments is not expected in natural waters. Chemical degradation and volatilization probably occur too slowly to account for substantial loss of dalapon from water. Aquarium studies conducted by Smith et al. (1972) provide evidence that volatility is not a route for significant loss of dalapon from water.

- Microbial degradation is by far the most important process affecting the fate of dalapon in soil. Other processes which are of lesser importance are adsorption, leaching and runoff, chemical degradation and volatilization. Based on the light absorption characteristics of aqueous solutions of sodium salts of dalapon, it has been concluded that photodecomposition of dalapon in field applications is improbable (Kearney et al., 1965).

- Although dalapon is subject to hydrolysis under field conditions, chemical degradation is considered to be very slow and is unlikely to be an important factor in the dissipation of dalapon from soil. Smith et al. (1957) and Brust (1953) demonstrated that dalapon and its sodium salt can undergo hydrolysis to pyruvate and HCl.

- Although the laboratory studies indicate that dalapon is a highly mobile compound (Warren, 1954; Helling, 1971; Kenaga, 1974) and should be readily leachable from soils, field data show that under many practical conditions dalapon does not move beyond the first 6-inch depth of soil. This is probably because microbial action proceeds at a faster rate than leaching under favorable conditions (Kenaga, 1974).

- The microbial degradation of dalapon in soil has been well established. Thiegs (1955) compared the rates of degradation of dalapon in autoclaved and non-autoclaved soils. The concentration of dalapon (59 ppm) in the autoclaved soil did not change after incubation at 100°F for 1 week, while in the unsterilized soil, dalapon disappeared in 4 to 5 weeks after one application and in 1 week after the second application of 50 ppm. Based on the observations that dalapon decomposition is adversely affected by low soil moisture, low pH, temperatures below 20° to 25°C, and large additions of organic matter, Holstun and Loomis (1956) concluded that dalapon degradation was a function of microbiological activity.

II.   Pharmacokinetics

A. Absorption

- In both dogs and humans, orally administered dalapon is quickly excreted in the urine. Dogs administered a single oral dose of 500 mg/kg dalapon sodium salt excreted 65 to 70% of the administered dose in 48 hours (Hoerger, 1969). In a 60-day feeding study, dogs receiving 50 and 100 mg/kg of dalapon sodium salt excreted 25 to 53% of the administered dose in the urine (Hoerger, 1969). Human subjects consuming five successive daily oral doses of 0.5 mg of dalapon sodium salt excreted approximately 50% of the administered dose over an 18-day period (Hoerger, 1969). These data suggest that dalapon is well absorbed from the gastrointestinal tract.

B. Distribution

- Chronic oral administration of dalapon did not result in significant bioaccumulation in either rats or dogs (Paynter et al., 1960). In both rats and dogs, the highest levels of dalapon were found in the kidneys, followed by the muscle and the fat (Paynter et al., 1960).

C. Metabolism

- Although inadequate data are available to characterize dalapon metabolism in humans, data in cattle (Redemann and Hanaker, 1959) suggest that dechlorination may be involved in the metabolism of dalapon.

D. Excretion

- Available information suggests that at least 50% of orally administered dalapon is eliminated via the kidneys in dogs and humans (Hoerger, 1969).

III.   Health Effects

A. Humans

1. *Short-term Exposure*

- No information on the short-term health effects of dalapon in humans was found in the available literature.

2. *Long-term Exposure*

- No information on the long-term health effects of dalapon in humans was found in the available literature.

B. Animals

1. *Short-term Exposure*

- The sodium salt of dalapon is relatively nontoxic, with an oral LD_{50} ranging from 3,860 mg/kg in the female rabbit to 7,570 mg/kg in the female rat (Paynter et al., 1960).

2. *Dermal/Ocular Effects*

- Concentrated sodium dalapon solutions have been found to be irritating to the skin and eyes of rabbits (Paynter et al., 1960).

3. *Long-term Exposure*

- In a 90-day dietary study by Paynter et al. (1960), male and female rats were exposed to sodium dalapon (65% pure) at levels of 0, 11.5, 34.6, 115, 346 or 1,150 mg/kg/day. Increases in kidney and liver weight were observed in both sexes at 346 and 1,150 mg/kg/day. The No-Observed-Adverse-Effect Level (NOAEL) in this study was identified as 11.5 mg/kg/day based on increases in kidney weight at higher doses. (See discussion under Longer-term Health Advisory below.)

- In a 1-year study, sodium dalapon (65% pure) was administered to dogs by capsule at levels of 0, 15, 50 or 100 mg/kg/day. Based on increases in kidney weight at 100 mg/kg/day, the NOAEL was identified as 50 mg/kg/day (Paynter et al., 1960).

- With the exception of an increase in kidney weight in male rats, sodium dalapon (65% pure) was without effect in a 2-year dietary study (Paynter et al., 1960); the NOAEL in this study was 15 mg/kg/day. (See discussion under Longer-term Health Advisory below.)

4. *Reproductive Effects*

- Administered in the diet, sodium dalapon (65% pure) had no effects on reproduction in the rat at dose levels of approximately 30, 100 or 300 mg/kg/day (Paynter et al., 1960).

5. *Developmental Effects*

- Sodium dalapon (purity not specified) was not teratogenic in the rat at doses as high as 2,000 mg/kg/day (Emerson et al., 1971; Thompson et al., 1971). In the study by Emerson et al., sodium dalapon was administered orally to pregnant rats over a 10-day period (days 6 through 15 of gestation) at doses of 0, 500, 1,000 or 1,500 mg/kg/day. Although no compound-related teratogenic response was seen, there was a decrease in weight gain in the dams at the lowest level tested, 500 mg/kg/day. Decreased weight gain was also observed in the pups, but only at higher levels (1,000 and 1,500 mg/kg/day). This study identified a LOAEL of 500 mg/kg/day.

- Dalapon (sodium/magnesium salt, purity 99.3%) was administered orally (gavage) to New Zealand white female rabbits on days 6 through 18 of gestation at doses of 30, 100, and 300 mg/kg/day. Nineteen animals were treated at the high dose (300 mg/kg/day) and 16 animals per group were used as controls as well as the low-, mid- and high-dose groups (BASF, 1987). Administration of dalapon at the mid and high dose elicited maternal toxicity: significant decreases in body weight gain and food consumption. A LOAEL of 100 mg/kg/day and a NOAEL of 30 mg/kg/day were identified.

4. *Mutagenicity*

- Dalapon was not mutagenic in a variety of organisms including *Salmonella typhimurium, Escherichia coli,* T4 bacteriophage, *Streptomyces coelicolor* and *Aspergillus nidulans* (U.S. EPA, 1984). Although Kurinnyi et al. (1982) reported that dalapon increased chromosome aberrations in mice, the inadequate technical detail presented precluded an evaluation of this study.

5. *Carcinogenicity*

- No evidence of a carcinogenic response was observed in a 2-year chronic feeding study in which sodium dalapon (65% pure) was administered to rats at levels as high as 50 mg/kg/day for a period of 2 years (Paynter et al., 1960).

IV.   Quantification of Toxicological Effects

A. One-day Health Advisory

No data were found in the available literature that were suitable for determination of the One-day HA value for dalapon. It is therefore recommended that the Ten-day HA value for a 10-kg child (3 mg/L, calculated below) be used at this time as a conservative estimate of the One-day HA value.

B. Ten-day Health Advisory

The rat teratology study by Emerson et al. (1971) had been considered to serve as the basis for determination of the Ten-day HA for a 10-kg child. However, in this study, the chemical purity was not specified, and the lowest dose administered (500 mg/kg/day, LOAEL) had maternal toxicity. A recent teratology study with rabbits (BASF, 1987) was selected to serve as the basis for the Ten-day HA. This study specified the purity of the chemical and provided a lower LOAEL (100 mg/kg/day), and a NOAEL of 30 mg/kg/day. In this study, groups of inseminated New Zealand white rabbits were given oral doses of 0, 30, 100 and 300 mg/kg/day dalapon (Dowpon M: Na/Mg salt, purity 99.3%) on days 6 through 18 of gestation. Significant decreases in maternal body weight, and in food and water consumption were noted in the mid- and high-dose groups. Fetal body weight from dams given the high dose was also decreased. However, no adverse effects were noted in the low-dose group, and a NOAEL of 30 mg/kg/day was identified. Standards for dalapon are commonly expressed in terms of the acid rather than the salt. Thus, it is necessary to convert the NOAEL for the Na/Mg salt, 30 mg/kg/day to the equivalent value for the acid. It is assumed that Dowpon M was an approximately 5:1 mixture of sodium and magnesium salts.

$$\text{The NOAEL for}\atop\text{dalapon as acid} = \frac{(30 \text{ mg/kg/day})(143)(7)}{(165)(5) + (308.5)(1)} = 26.5 \text{ mg/kg/day}$$

where:

$$30 \text{ mg/kg/day} = \text{NOAEL for the Na/Mg salt.}$$
$$143 = \text{formula weight of dalapon as acid.}$$
$$165 = \text{molecular weight of sodium dalapon.}$$
$$308.5 = \text{molecular weight of magnesium dalapon.}$$
$$7 = \text{total number of dalapon acid moiety in Dowpon M.}$$
$$5 = \text{number of sodium dalapon per molecule of Dowpon M.}$$
$$1 = \text{number of magnesium dalapon per molecule of Dowpon M. (Each magnesium binds two acid moieties.)}$$

The Ten-day HA for a 10-kg child is calculated as follows:

$$\text{Ten-day HA} = \frac{(26.5 \text{ mg/kg/day})(10 \text{ kg})}{(100)(1 \text{ L/day})} = 2.7 \text{ mg/L } (3,000 \text{ } \mu g/L)$$

C. Longer-term Health Advisory

The results of Paynter et al. (1960) suggest that the subchronic and chronic toxicity of dalapon are much the same. Specifically, in a 97-day rat subchronic dietary study, sodium dalapon (65% sodium dalapon; 16% sodium salts of related chloropropionic acids; 2% sodium pyruvate; 5% sodium chloride; 5% water; 7% undetermined) produced an increase in kidney weight in female rats at 34.6 mg/kg/day and higher exposure levels but not at 11.5 mg/kg/day (NOAEL). Similarly, in a 2-year rat chronic dietary study, sodium dalapon exposure (65% pure) resulted in an increase in male kidney weight at 50 mg/kg/day but not at 15 mg/kg/day (NOAEL). Considering both Paynter et al. (1960) rat dietary studies together, the 15 mg/kg/day NOAEL for sodium dalapon is appropriate to calculate both a Longer term HA and a Lifetime HA.

It is customary to express dalapon standards in terms of the acid rather than the salt. The NOAEL used to derive the Longer-term HA is based on studies (Paynter et al., 1960) in which rats were exposed to sodium dalapon that was 65% pure. Thus, a NOAEL for dalapon as the pure acid must be calculated:

$$\text{The NOAEL for dalapon as pure acid} = \frac{(15 \text{ mg/kg/day}) \ (0.65) \ (143)}{165} = 8 \text{ mg/kg/day}$$

where:

15 mg/kg/day = NOAEL for 65% pure sodium dalapon.
0.65 = purity of sodium dalapon used in determining NOAEL.
143 = molecular weight of dalapon as acid.
165 = molecular weight of sodium dalapon.

The Longer-term HA for a 10-kg child is calculated as follows:

$$\text{Longer-term HA} = \frac{(8 \text{ mg/kg/day}) \ (10 \text{ kg})}{(100) \ (3) \ (1 \text{ L/day})} = 0.26 \text{ mg/L } (300 \ \mu g/L)$$

where: 3 = additional uncertainty factor to allow for deficiencies in the quality of the data base.

The Longer-term HA for a 70-kg adult is calculated as follows:

$$\text{Longer-term HA} = \frac{(8 \text{ mg/kg/day}) \ (70 \text{ kg})}{(100) \ (3) \ (2 \text{ L/day})} = 0.9 \text{ mg/L } (900 \ \mu g/L)$$

where: 3 = additional uncertainty factor to allow for deficiencies in the quality of the data base.

D. Lifetime Health Advisory

As discussed under Longer-term HA above, the data used to determine the Lifetime HA are identical to those used to determine the Longer-term HA. Using the NOAEL of 8 mg/kg/day from the 2-year rat study by Paynter et al. (1960), the Lifetime HA for the 70-kg adult is calculated as follows:

Step 1: Determination of the Reference Dose (RfD)

$$\text{RfD} = \frac{(8 \text{ mg/kg/day})}{(100) \ (3)} = 0.03 \text{ mg/kg/day}$$

where: 3 = additional uncertainty factor to allow for deficiencies in the quality of the data base.

Step 2: Determination of the Drinking Water Equivalent Level (DWEL)

$$\text{DWEL} = \frac{(0.03 \text{ mg/kg/day}) \ (70 \text{ kg})}{(2 \text{ L/day})} = 0.9 \text{ mg/L } (900 \ \mu g/L)$$

Step 3: Determination of the Lifetime Health Advisory

Lifetime HA = (0.9 mg/L) (20%) = 0.18 mg/L (200 μg/L)

E. Evaluation of Carcinogenic Potential

• No evidence of carcinogenicity was found in a 2-year dietary study in which sodium dalapon was administered to rats at levels as high as 50 mg/kg/day (Paynter et al., 1960).

• Applying the criteria described in EPA's guidelines for assessment of carcinogenic risk (U.S. EPA, 1986a), dalapon may be classified in Group D: not classifiable. This group is for substances with inadequate human and animal evidence of carcinogenicity.

V.   Other Criteria, Guidance and Standards

• The American Conference of Governmental Industrial Hygienists suggests a Threshold Limit Value (TLV) of 1 ppm (6 mg/m^3) as a time-weighted average for an 8-hour workday.

• Tolerances have been established for dalapon in a wide variety of agricultural commodities (CFR, 1985) ranging from 0.1 ppm in milk to 75 ppm in flaxseed.

VI.   Analytical Methods

• Analysis of dalapon is by a gas chromatographic (GC) method applicable to the determination of certain chlorinated acid pesticides in water samples (U.S. EPA, 1986b). In this method, approximately 1 liter of sample is acidified. The compounds are extracted with ethyl ether using a separatory funnel. The derivatives are hydrolyzed with potassium hydroxide, and extraneous organic material is removed by a solvent wash. After acidification, the acids are extracted and converted to their methyl esters using diazomethane as the derivatizing agent. Excess reagent is removed, and the esters are determined by electron-capture GC. The method detection limit has not been determined for this compound.

VII.   Treatment Technologies

- No information on treatment technologies capable of effectively removing dalapon from contaminated water was found in the available literature.

VIII.   References

BASF Corporation. 1987. Dalapon: Oral (gavage) teratogenicity study in the rabbit. BASF Corporation, Chemicals Division, Germany. MRID No. 40125701.

Brust, H. 1953. Hydrolysis of dalapon sodium salt solutions. E.C. Britton Research Laboratory, The Dow Chemical Co., Midland, MI. November 4, 1953. Cited in Kenaga, 1974.

CFR. 1985. Code of Federal Regulations. 40 CFR 180.150.

Emerson, J.L., D.J. Thompson and C.G. Gerbig. 1971. Results of teratological studies in rats treated orally with 2,2-dichloropropionic acid (dalapon) during organogenesis. Report HH-417, Human Health Research and Development Laboratories, The Dow Chemical Co., Zionsville, IN (cited in Kenaga, 1974).

Helling, C.S. 1971. Pesticide mobility in soils, I, II, III. Proc. Soil Sci. Soc. Amer. 35:732–748.

Hoerger, F. 1969. The metabolism of dalapon. Blood absorption and urinary excretion patterns in dogs and human subjects. Unpublished report. The Dow Chemical Co. (cited in Kenaga, 1974).

Holston, J.T. and W.E. Loomis. 1956. Leaching and decomposition of 2,2-dichloropropionic acid in several Iowa soils. Weeds. 4:205–217.

Kearney, P.C., et al. 1965. Behavior and fate of chlorinated aliphatic acids in soils. Adv. Pest. Control Res. 6:1–30.

Kenaga, E.E. 1974. Toxicological and residue data useful in the environmental safety evaluation of dalapon. Residue Rev. 53:109–151.

Kurinnyi, A.I., M.A. Pilinskaya, I.V. German and T.S. L'vova. 1982. Implementation of a program of cytogenic study of pesticides: Preliminary evaluation of cytogenic activity and potential mutagenic hazard of 24 pesticides. Tsitologiya i Genetika. 16:45–49.

Meister, R., ed. 1984. Farm chemicals handbook. Willoughby, OH: Meister Publishing Co.

Paynter, O.E., T.W. Tusing, D.D. McCollister and V.K. Rowe. 1960. Toxicology of dalapon sodium (2,2-dichloropropionic acid, sodium salt). Agr. Food Chem. 8:47–51.

Redemann, C.T. and J.W. Hanaker. 1959. The lactic secretion of metabolic products of ingested sodium 2,2-dichloropropionate by the dairy cow. Agricultural Research, The Dow Chemical Co. Seal Beach, CA (cited in Kenaga, 1974).

Reinert, K.H. and J.H. Rodgers. 1987. Fate and persistence of aquatic herbicides. Rev. Environ. Contamination and Tox. 98:61–98.

Smith, G.N., M.E. Getzendaner and A.H. Kutschinski. 1957. Determination of 2,2-dichloropropionic acid (dalapon) in sugar cane. J. Agr. Food Chem. 5:675. Cited in Kenaga, 1974.

Smith, G.N., Y.S. Taylor and B.S. Watson. 1972. Ecological studies on dalapon (2,2-dichloropropionic acid). Unpublished report NBE-16. Chemical Biology Res., The Dow Chemical Co., Midland, MI (cited in Kenaga, 1974).

STORET. 1988. STORET Water Quality File. Office of Water. U.S. Environmental Protection Agency (data file search conducted in May, 1988).

Thiegs, B.J. 1955. The stability of dalapon in soils. Down to Earth. Fall. Cited in Kenaga, 1974.

Thompson, D.J., C.G. Gerbig and J.L. Emerson. 1971. Results of tolerance study of 2,2-dichloropropionic acid (dalapon) in pregnant rats. Unpublished report HH-393. Human Research and Development Center, The Dow Chemical Co. (cited in Kenaga, 1974).

Tracey, W.J. and R.R. Bellinger, Jr. 1958. Hydrolysis of sodium 2,2-dichloropropionate in water solution. Midland Division, The Dow Chemical Co., Midland, MI (cited in Kenaga, 1974).

U.S. EPA. 1984. U.S. Environmental Protection Agency. Draft health and environmental effects profile for dalapon. Environmental Criteria and Assessment Office, Cincinnati, OH.

U.S. EPA. 1986a. U.S. Environmental Protection Agency. Guidelines for carcinogen risk assessment. Fed. Reg. 51(185):33992–34003. September 24.

U.S. EPA. 1986b. U.S. Environmental Protection Agency. U.S. EPA Method #3 — Determination of chlorinated acids in ground water by GC/ECD. January 1986 draft. Available from U.S. EPA's Environmental Monitoring and Support Laboratory, Cincinnati, OH.

Warren, G.F. 1954. Rate of leaching and breakdown of several herbicides in different soils. NC Weed Control Conf. Proc., 11th Ann. Meeting, Fargo, ND (cited in Kenaga, 1974).

---

* Confidential Business Information submitted to the Office of Pesticide Programs.

DCPA (Dacthal)

I. General Information and Properties

A. CAS No. 1861–32–1

B. Structural Formula

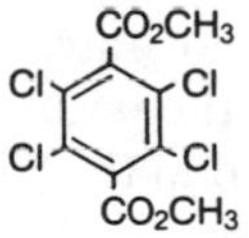

Dimethyl tetrachloroterephthalate

C. Synonyms

- 2,3,5,6-Tetrachlorodimethyl-1,4-benzenedicarboxylic acid; DCPA; Chlorothal; Dacthalor; DAC; DAC-4; DAC-893; DCP (Meister, 1983).

D. Uses

- DCPA is used as a selective preemergence herbicide used to control various annual grasses in turf, ornamentals, strawberries, certain vegetable transplants, seeded vegetables, cotton, soybeans and field beans (Meister, 1983).

E. Properties (Meister, 1983; Windholz et al., 1983; CHEMLAB, 1985)

Chemical Formula	$C_{10}H_6O_4Cl_4$
Molecular Weight	331.99
Physical State (25°C)	Crystals
Boiling Point	--
Melting Point	156°C
Density (°C)	--
Vapor Pressure (25°C)	2.5×10^{-6} mm Hg at 25°C
Specific Gravity	--
Water Solubility (25°C)	0.5 mg/L at 25°C
Log Octanol/Water Partition Coefficient	6.8×10^3

239

Taste Threshold   --
Odor Threshold   --
Conversion Factor   --

F. Occurrence

- Dacthal has been found in 386 of 1,995 surface water samples analyzed and in 12 of 982 ground-water samples (STORET, 1988). Samples were collected at 584 surface water locations and 844 ground-water locations, and dacthal was found in eight states. The 85th percentile of all nonzero samples was 0.39 μg/L in surface water and 0.05 μg/L in ground-water sources. The maximum concentration found was 8.74 μg/L in surface water and 0.05 μg/L in ground water. This information is provided to give a general impression of the occurrence of this chemical in ground and surface waters as reported in the STORET database. The individual data points retrieved were used as they came from STORET and have not been confirmed as to their validity. STORET data are often not valid when individual numbers are used out of the context of the entire sampling regime, as they are here. Therefore, this information can only be used to form an impression of the intensity and location of sampling for a particular chemical.

G. Environmental Fate

- In aqueous solutions, dacthal is not very photolabile with a half-life of greater than 1 week. Dacthal is also stable versus soil photolysis (Registrant Standard Science chapter for dacthal).

- Soil metabolism of dacthal proceeds with a half-life of greater than 2 to 3 weeks. Degradation rate is affected by temperature. No degradation of dacthal has been observed in sterile soils (half-life of 1,590 days) (Registrant Standard Science chapter for dacthal).

- Degradation products of dacthal include monomethyltetrachloroterephthalate (MTP) and tetrachloroterephthalic acid (TTA) (Registrant Standard Science chapter for dacthal).

- TTA has been shown to be very mobile in soils, whereas dacthal is not (Registrant Standard Science chapter for dacthal).

II.   Pharmacokinetics

A. Absorption

- Tusing (1963) reported that three humans receiving single oral doses of pure dacthal (25 mg) excreted up to 6% of the doses in the urine as metabolites over a 3-day period. When three other humans were administered 50-mg doses of dacthal, approximately 12% was excreted in the urine over a similar time period, indicating that at least 12% of a 50-mg dose was absorbed in humans.

- Skinner and Stallard (1963) reported that following administration of single oral doses of dacthal (100 or 1,000 mg/kg) by capsule to dogs, about 97% of the administered doses were eliminated as the parent compound in the feces by 96 hours. Approximately 3% of dacthal was converted to the monomethylester of tetrachloroterephthalic acid (DAC 1449). Two percent was eliminated in the urine and 1% in the feces. Less than 1% (0.07%) of DAC 1449 was converted to tetrachloroterephthalic acid (DAC 954), which was also excreted in the urine. The results indicated that dacthal was absorbed poorly (about 3%) from the gastrointestinal tract of dogs.

B. Distribution

- Skinner and Stallard (1963) reported that following a single oral dose of dacthal (100 or 1,000 mg/kg) to dogs, there was no storage of dacthal in the kidneys, liver or fat. However, DAC 954 was found in the kidneys. The authors also reported that no dacthal was found in the kidneys or liver of dogs that had been administered dacthal-T (dacthal containing 1.1% of the monomethyl ester and 1.7% of the tetrachlorophthalate) at 10,000 ppm (250 mg/kg/day) in the diet for 2 years. The kidneys, liver and fat contained DAC 1449, while the kidneys contained DAC 954 only. Both dacthal and DAC 1449 were found in the fat of dogs treated with 10,000 ppm.

C. Metabolism

- Tusing (1963) reported that humans who took single oral doses of pure dacthal (25 or 50 mg) converted 3 to 4% of the dose to DAC 1449 within 24 hours. After 3 days, approximately 6% of the 25-mg dose and 11% of the 50-mg dose were converted to DAC 1449. At either dose, less than 1% was converted to DAC 954 in the 1- or 3-day time period.

- Skinner and Stallard (1963) reported that in dogs administered single oral doses of dacthal, small amounts were converted to DAC 1449 (3%) or DAC 954 (0.07%).

- Hazleton and Dieterich (1963) reported similar results when dogs were administered dacthal (10,000 ppm; 250 mg/kg bw) in the diet for 2 years.

D. Excretion

- In human studies (Tusing, 1963), 6% of a single 25-mg oral dose was excreted in urine as DAC 1449 and 0.5% as DAC 954 over a 3-day period. Approximately 11% of the 50-mg dose was converted to DAC 1449 and 0.6% was converted to DAC 954. The parent compound was not found in the urine at either dose.

- Skinner and Stallard (1963) reported that following the administration of a single oral dose (100 or 1,000 mg/kg) to dogs, 90 and 97% were eliminated unchanged in the feces at 24 hours and 96 hours, respectively. Approximately 3% was converted to DAC 1449; of this, 2% was eliminated in the urine and 1% in the feces.

III. Health Effects

A. Humans

- Tusing (1963) reported that pure dacthal, administered as single 25-mg or 50-mg oral doses to volunteer subjects (three at each dose), did not cause any observable effects. Assuming 70-kg body weight, these amounts correspond to doses of 0.36 or 0.71 mg/kg. Hemograms, liver, kidney and urine analyses from the six human volunteers were normal.

B. Animals

1. Short-term Exposure

- The acute oral LD_{50} for male and female albino rats (Spartan strain) was reported to be greater than 12,500 mg/kg (Wazeter et al., 1974a).

- The acute oral LD_{50} for male and female beagle dogs was reported to be greater than 10,000 mg/kg (Wazeter et al., 1974b).

- Keller and Kundzin (1960) administered pure dacthal to weanling male Sprague-Dawley rats (10/dose) in the diet for 28 days at levels of 0, 0.0082, 0.0824 or 0.824%. Based upon body weight and compound consumption data provided by the investigators, these dietary levels correspond to approximately 0, 7.6, 78.6 or 758 mg/kg/day. Following treatment, no effects on growth, food consumption, survival, body weights, organ weights, gross pathology and histopathology were observed. This study identifies a NOAEL of 758 mg/kg/day (the highest dose tested).

- Keller (1961) reported that oral administration (by capsule) of 800 mg/kg/day of DCPA (88.5% active ingredient) to beagle dogs (2/sex) for 28 days resulted in loss of body weight, reduced appetite, increased liver weight and liver-to-body weight ratio and centrilobular liver congestion and degeneration.

2. *Dermal/Ocular Effects*

- The acute dermal LD_{50} value for albino rabbits was reported to be greater than 10,000 mg/kg (Elsea, 1958). Elsea also reported that dacthal, when applied to rabbit skin, did not cause irritation or sensitization.

- Johnson et al. (1981) applied dacthal (2,000 mg/kg) for 24 hours to shaved intact or abraded back or flank skin of New Zealand White rabbits (5/sex) in a paste form. Desquamation (which ranged from very slight to slight) and very slight erythema were observed. There was no pathology noted, and dacthal caused only slight irritation in some animals.

- A single application of 3.0 mg of dacthal to the eyes of albino rabbits produced a mild degree of irritation that subsided completely within 24 hours following treatment (Elsea, 1958).

3. *Long-term Exposure*

- Goldenthal et al. (1977) fed CD rats (15/sex/dose) disodium dacthal in the diet for 90 days at dose levels of 0, 50, 500, 1,000 or 10,000 ppm. Based upon compound consumption and body weight data provided by the authors, these dietary levels are approximately 0, 3.6, 36.4, 74 or 732 mg/kg/day for males and 0, 4.2, 43.2, 82.3 or 856 mg/kg/day for females. General behavior, appearance, body weight, food consumption, ophthalmoscopic evaluation, hematology, clinical chemistry, urinalysis, gross pathology and histopathology were comparable for treated and control groups. A NOAEL of 10,000 ppm (732 mg/kg/day for males and 856 mg/kg/day for females, the highest dose tested) was identified for this study.

- Hazleton and Dieterich (1963) fed beagle dogs (4/sex/dose) dacthal in the diet at 0, 100, 1,000 or 10,000 ppm for 2 years. Based upon body weight and food consumption data provided in the report, these dietary levels are approximately 0, 2.6, 17.7 or 199 mg/kg/day for males and 0, 3, 20.7 or 238 mg/kg/day for females. Physical appearance, behavior, food consumption, hematology, biochemistry, urinalysis, organ weight, organ-to-body weight ratio, gross pathology and histopathology were comparable in treated and control groups at all dose levels. A NOAEL of 10,000 ppm (199 mg/kg/day for males and 238 mg/kg/day for females, the highest dose tested) was identified for this study.

- Paynter and Kundzin (1963b) fed albino rats (35/sex/dose; 70/sex for controls) dacthal in the diet for 2 years at 0, 100, 1,000 or 10,000 ppm. Based on food consumption and body weight data provided in the report, these dietary levels correspond approximately to 0, 5, 50 or 500 mg/kg/day. Physical appearance, behavior, hematology, biochemistry, organ weights, body weights, gross pathology and histopathology of treated and control animals were monitored. After 3 months at 10,000 ppm, slight hyperplasia of the thyroid was reported in both sexes. After 1 year, increased hemosiderosis of the spleen of females occurred at 10,000 ppm and there were slight alterations in the centrilobular cells of the liver of both sexes. Kidney weights were increased significantly in males fed 10,000 ppm at the end of the 2-year study. Based on these data, a NOAEL of 1,000 ppm (50 mg/kg/day) was identified.

4. *Reproductive Effects*

- Paynter and Kundzin (1964) conducted a two-generation study using albino rats. Animals (8 males/16 females) were fed dacthal in the diet at dose levels of 0, 0.1 or 1.0% for 24 weeks, prior to mating. Assuming that 1 ppm in the diet of rats is equivalent to 0.05 mg/kg/day (Lehman, 1959), this corresponds to doses of 0, 50 or 500 mg/kg/day. This study reported an evaluation of data collected on the second parental generation (P_2) and through weaning of the first litter (F_{2a}). The authors reported that a second litter (F_{2b}) was not obtained. Following treatment, the following indices were evaluated: fertility, gestation, live births and lactation. Since the fertility index was 37% (6/16) at the 1% dose, 75% (12/16) at the 0.1% dose and only 19% (3/16) in controls, no conclusions could be reached. The lactation index for the 0.1% group was significantly lower than controls. No other adverse reproductive effects were observed.

- Paynter and Kundzin (1963a) performed a one-generation reproduction study in albino rats. Animals were given dacthal in the diet at 0, 1,000 or 10,000 ppm. Assuming that 1 ppm in the diet of rats is equivalent to 0.05 mg/kg/day (Lehman, 1959), this corresponds to doses of about 0, 50 or 500 mg/kg/day. No effects were detected on fertility, gestation, number of live births or lactation. Based on this information a NOAEL of 10,000 ppm (500 mg/kg/day, the highest dose tested) was identified.

5. Developmental Effects

- Powers (1964) fed pregnant New Zealand White rabbits (6/dose) dietary levels of dacthal-T (0, 1,000 or 10,000 ppm) on days 8 to 16 of gestation. Assuming that 1 ppm in the diet of rabbits is equivalent to 0.03 mg/kg/day (Lehman, 1959), this corresponds to about 0, 30 or 300 mg/kg/day. Following treatment, fetal toxicity (number of live/dead or resorptions), maternal effects (appearance, behavior, body weight) and visceral and skeletal anomalies were evaluated. No adverse effects were observed at any dose level tested. This study identified a developmental NOAEL of 300 mg/kg/day (the highest dose tested).

6. Mutagenicity

- No significant increase in mutation frequency was observed in *Drosophila melanogaster* larvae that had been fed media containing 0.1 to 10 mM dacthal (Paradi and Lovenyak, 1981).

- Dacthal had no mutagenic activity, with or without activation, in *Salmonella* assays (Auletta et al., 1977), in *in vivo* cytogenetic tests (Kouri et al., 1977b), in DNA repair tests (Auletta and Kuzava, 1977) or in dominant lethal tests (Kouri et al., 1977a).

7. Carcinogenicity

- Paynter and Kundzin (1963b) fed albino rats (35/sex/dose; 70/sex for controls) dacthal-T for 2 years at dose levels of 0, 100, 1,000 or 10,000 ppm. Based upon compound consumption and body weight provided in the report, these dietary levels correspond approximately to 0, 5, 50 or 500 mg/kg/day. Based on gross and histologic examination, neoplasms of various tissues and organs were similar in type, localization, time of occurrence, and incidence in control and treated animals.

IV.   Quantification of Toxicological Effects

A. One-day Health Advisory

No information was found in the available literature that was suitable for deriving a One-day HA. The study in humans by Tusing (1963) was not selected since only low doses (0.36 or 0.71 mg/kg) were tested, and longer-term studies in animals suggest the no-effect level may be much higher. It is therefore recommended that the Ten-day HA value for the 10-kg child (80 mg/L, calculated below) be used at this time as a conservative estimate of the One-day HA.

B. Ten-day Health Advisory

The 28-day feeding study in rats by Keller and Kundzin (1960) has been selected to serve as the basis for determination of the Ten-day HA. In this study, no adverse effects on growth, organ weight, food consumption, gross pathology or histopathology were detected at 758 mg/kg/day.

The Ten-day HA for the 10-kg child is calculated as follows:

$$\text{Ten-day HA} = \frac{(758 \text{ mg/kg/day}) (10 \text{ kg})}{(100) (1 \text{ L/day})} = 75 \text{ mg/L } (80{,}000 \text{ } \mu g/L)$$

C. Longer-term Health Advisory

No studies were found in the available literature that were suitable for deriving the Longer-term HA value for dacthal. It is therefore recommended that the Drinking Water Equivalent Level (DWEL) of 20 mg/L (20,000 μg/L), calculated in the Lifetime Health Advisory section, be used for the Longer-term HA value for an adult, and that the DWEL adjusted for a 10-kg child, 5.0 mg/L (5,000 μg/L), calculated below, be used for the Longer-term HA value for a child.

For a 10-kg child, the adjusted DWEL is calculated as follows:

$$\text{DWEL} = \frac{(0.5 \text{ mg/kg/day}) (10 \text{ kg})}{1 \text{ L/day}} = 5.0 \text{ mg/L } (5{,}000 \text{ } \mu g/L)$$

where: 0.5 mg/kg/day = RfD (see Lifetime Health Advisory section)

D. Lifetime Health Advisory

The 2-year study in rats by Paynter and Kundzin (1963b) has been selected to serve as the basis for determination of the Lifetime HA value for dacthal. This study identified a NOAEL of 50 mg/kg/day, based on absence of effects on appearance, behavior, hematology, blood chemistry, organ weight, body weight, gross pathology and histopathology in male rats. The LOAEL was 500 mg/kg/day, based on thyroid hyperplasia, histological changes in the liver and increased kidney weights.

Using this study, the Lifetime HA is derived as follows:

Step 1: Determination of the Reference Dose (RfD)

$$RfD = \frac{(50 \text{ mg/kg/day})}{(100)} = 0.5 \text{ mg/kg/day}$$

Step 2: Determination of the Drinking Water Equivalent Level (DWEL)

$$DWEL = \frac{(0.5 \text{ mg/kg/day}) (70 \text{ kg})}{(2 \text{ L/day})} = 17.5 \text{ mg/L } (20,000 \text{ } \mu g/L)$$

Step 3: Determination of the Lifetime Health Advisory

Lifetime HA = (17.5 mg/L) (20%) = 3.5 mg/L (4,000 μg/L)

E. Evaluation of Carcinogenic Potential

- Paynter and Kundzin (1963b) fed dacthal to rats for 2 years and reported no evidence of carcinogenic effects at dose levels up to 10,000 ppm (450 mg/kg/day for males and 555 mg/kg/day for females). This study is limited in that the relatively small numbers of animals used (35/sex/dose; 70/sex for controls) and the removal of animals (10/sex/dose; 20/sex for controls) for interim sacrifice may have resulted in there being too few animals available for observation of late-developing tumors.

- The International Agency for Research on Cancer has not evaluated the carcinogenic potential of dacthal.

- Applying the criteria described in EPA's guidelines for assessment of carcinogenic risk (U.S. EPA, 1986), dacthal may be classified in Group D: not classifiable. This category is for substances with inadequate animal evidence of carcinogenicity.

V.   Other Criteria, Guidance and Standards

- U.S. EPA has established residue tolerances for dacthal in or on raw agricultural commodities that range from 0.5 ppm to 15.0 ppm (U.S. EPA, 1985).

VI. Analytical Methods

- Analysis of dacthal is by a gas chromatographic (GC) method applicable to the determination of certain chlorinated pesticides in water samples (U.S. EPA, 1988). In this method, approximately 1 liter of sample is extracted with methylene chloride. The extract is concentrated and the compounds are separated using capillary column GC. Measurement is made using an electron capture detector. This method has been validated in a single laboratory and estimated detection limits have been determined for the analytes in the method, including dacthal. The estimated detection limit is 0.02 μg/L.

VII. Treatment Technologies

- Reverse osmosis (RO) is a promising treatment method for pesticide-contaminated waters. As a general rule, organic compounds with molecular weights greater than 100 are candidates for removal by RO. Larson et al. (1982) report 99% removal efficiency of chlorinated pesticides by a thin-film composite polyamide membrane operating at a maximum pressure of 1,000 psi and a maximum temperature of 113°F. More operational data are required, however, to specifically determine the effectiveness and feasibility of applying RO for the removal of dacthal from water. Also, membrane adsorption must be considered when evaluating RO performance in the treatment of dacthal-contaminated drinking water supplies.

VIII. References

Auletta, A. and J. Kuzava.* 1977. Activity of DTX-77-0005 in a test for differential inhibition of repair deficient and repair competent strains of *Salmonella typhimurium*. Microb. Assoc. Rpt. DS-0001. Unpublished study. MRID 00100776.

Auletta, A., A. Parmar and J. Kuzava.* 1977. Activity of DTX-0003 in the *Salmonella*/microsomal assay for bacterial mutagenicity. Microb. Assoc. Rpt. DS-002. Unpublished study. MRID 00100774.

CHEMLAB. 1985. The Chemical Information System, CIS, Inc., Baltimore, MD.

Elsea, J.R.* 1958. Acute oral administration; acute dermal application; acute eye application. Unpublished study. MRID 00045823.

Goldenthal, E.I., F.X. Wazeter, D. Jessup et al.* 1977. 90-day toxicity study in rats. MRID 00100773.

Hazleton, L.N. and W.H. Dieterich.* 1963. Final report: two-year dietary feeding—dogs. Unpublished study. MRID 00083584.

Johnson, D., J. Myer and A. Olafsson.* 1981. Acute dermal toxicity (LD_{50}) study in albino rats. Unpublished study. MRID 00110553.

Keller, J.G. 1961.* 28-day oral administration—dogs. Unpublished study. MRID 00083573.

Keller, J.G. and M. Kundzin.* 1960. 28-Day dietary feeding study—rats. Unpublished study. MRID 00083571.

Kouri, R., A. Parmar, J. Kuzava et al.* 1977a. Activity of DTX-77-0004 in the dominant lethal assay in rodents for mutagenicity. Microb. Assoc. Proj. No. T1077. Final Report. Unpublished study. MRID 00100775.

Kouri, R., A. Parmar, J. Kuzava et al.* 1977b. The activity of DTX-77-0006 in the *in vivo* cytogenetic assay in rodents for mutagenicity. Microb. Assoc. Proj. No. T1083. Unpublished study. MRID 00107907.

Larson, R.E., P.S. Cartwright, P.K. Eriksson and R.J. Petersen. 1982. Applications of the FT-30 reverse osmosis membrane in metal finishing operations. Paper presented in Yokohama, Japan.

Lehman, A.J. 1959. Appraisal of the safety of chemicals in foods, drugs and cosmetics. Assoc. Food Drug Off. U.S., Q. Bull.

Meister, R., ed. 1983. Farm chemicals handbook. Willoughby, OH: Meister Publishing Company.

Paradi, E. and M. Lovenyak. 1981. Studies on genetical effect of pesticides in *Drosophila melanogaster*. Acta. Biol. Sci. Hung. 32:119–122.

Paynter, O.E. and M. Kundzin.* 1963a. Reproduction study—albino rats. MRID 00083578.

Paynter, O.E. and M. Kundzin.* 1963b. Two year dietary administration—rats. Final Report. MRID 00083577.

Paynter, O.E. and M. Kundzin.* 1964. Reproductive study—rats. Second phase: Project No. 200. Unpublished study. MRID 00053082.

Powers, M.B. 1964.* Reproductive study—rabbits. Unpublished study. MRID 00053088.

Skinner, W.A. and D.E. Stallard.* 1963. Dacthal animal metabolism studies. MRID 00083579.

STORET. 1988. STORET Water Quality File. Office of Water. U.S. Environmental Protection Agency (data file search conducted in May, 1988).

Tusing, T.W.* 1963. Oral administration—humans. MRID 00083583.

U.S. EPA. 1985. U.S. Environmental Protection Agency. Code of Federal Regulations. 40 CFR 180.185. July 1. pp. 280–281.

U.S. EPA. 1986. U.S. Environmental Protection Agency. Guidelines for carcinogen risk assessment. Fed. Reg. 51(185):33992–34003. September 24.

U.S. EPA. 1988. U.S. Environmental Protection Agency. Method 515.1 - Determination of chlorinated pesticides in water by GC/ECD. Draft. April

15. Available from U.S. EPA's Environmental Monitoring and Support Laboratory, Cincinnati, OH.

Wazeter, F.X., E.I. Goldenthal and W.P. Dean.* 1974a. Acute oral toxicity (LD_{50}) in male and female albino rats. Unpublished study. MRID 00031872.

Wazeter, F.X., E.I. Goldenthal and W.P. Dean.* 1974b. Acute oral toxicity (LD_{50}) in beagle dogs. Unpublished study. MRID 00031873.

Windholz, M., S. Budavari, R.F. Blumetti and E.S. Otterbein, eds. 1983. The Merck index — an encyclopedia of chemicals and drugs, 10th ed. Rahway, NJ: Merck and Company, Inc.

---

* Confidential Business Information submitted to the Office of Pesticide Programs.

Diazinon

I. General Information and Properties

A. CAS No. 333–41–5

B. Structural Formula

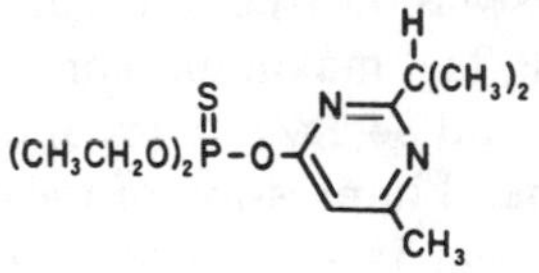

0,0-Diethyl-0-(2-isopropyl-4-methyl-6-pyrimidinyl)phosphorothiote

C. Synonyms

- Antigal; AG-500; Basudin; Bazudin; Ciazinon; Ducutox; Dassitox; Dazzel; Dianon; Diater; Diaterr-Fos; Diazajet; Diazide; Diazitol; Diazol; Dicid; Dimpylat; Dizinon; Dyzol; Exodin; Flytrol; Galesan; Kayazinon; Necidol/Nucidol; R-Fos; Spectacide; Spectracide (Meister, 1985).

D. Uses

- Diazinon is used for soil insecticide; insect control in fruit, vegetables, tobacco, forage, field crops, range, pasture, grasslands and ornamentals; nematocide in turf; seed treatment and fly control (Meister, 1985).

E. Properties (Meister, 1983; Windholz et al., 1983)

Chemical Formula	$C_{12}H_{21}O_3N_2SP$
Molecular Weight	304.36
Physical State (25°C)	Colorless oil
Boiling Point	83 to 84°C (0.002 mm Hg)
Melting Point	--
Density	--
Vapor Pressure (20°C)	1.4×10^{-4}
Specific Gravity	--
Water Solubility (20°C)	40 mg/L

251

Log Octanol/Water
Partition Coefficient   --
Taste Threshold   --
Odor Threshold   --
Conversion Factor   --

F. Occurrence

- Diazinon has been found in 6,026 of 22,291 surface water samples analyzed and in 74 of 3,633 ground-water samples (STORET, 1988). Samples were collected at 3,555 surface water locations and 2,835 ground-water locations, and diazinon was found in 46 states. The 85th percentile of all nonzero samples was 0.01 μg/L in surface water and 0.25 μg/L in ground-water sources. The maximum concentration found was 33,400 μg/L in surface water and 84 μg/L in ground-water. This information is provided to give a general impression of the occurrence of this chemical in ground and surface waters as reported in the STORET database. The individual data points retrieved were used as they came from STORET and have not been confirmed as to their validity. STORET data are often not valid when individual numbers are used out of the context of the entire sampling regime, as they are here. Therefore, this information can only be used to form an impression of the intensity and location of sampling for a particular chemical.

G. Environmental Fate

- ^{14}C-Diazinon (99% pure), at 7 or 51 ppm on sandy loam soil, degraded with a half-life of 37.4 hours after exposure to natural light (Blair, 1985). The degradate, oxypyrimidine, was detected at a maximum concentration of 19.60% (after 13.5 hours) of applied material when exposed to natural sunlight. After 35.5 hours (37.4 hours is the half-life) of sunlight exposure, 20.7% of the radiolabeled material was in soil-bound residues (some of which contained oxypyrimidine), 24.4% was oxypyrimidine and 39.7% was diazinon. Losses of 7% were attributed to volatilization of diazinon and degradates (of which 0.5% was carbon dioxide). The total ^{14}C-radioactive material balance was 87 to 89% at the 0 hour and 84% at all other experimental points.

- ^{14}C-Diazinon (99% pure) degraded in sandy loam soil with a half-life of 17.3 hours when exposed to natural sunlight (Martinson, 1985). The degradate, oxypyrimidine, was detected at maximum concentrations of 23.72% (after 32.6 hours) of applied material after exposure to natural sunlight. The degradate 2-(1'-hydroxy-1'-methyl)ethyl-4-methyl-6-

hydroxypyrimidine was present after 8 hours of natural sunlight exposure in 3.6% of the applied material but was not present in the unexposed samples. An unidentified degradate resulting from nonphotolytic degradation (since it was also present in nonexposed samples) accounted for about 7% of the applied material under sunlight.

- In a Swiss sandy loam soil at 75% of field capacity and 25°C, ring-labeled ^{14}C-diazinon (97% pure) applied at 10 ppm rapidly degraded to 2-isopropyl-4-methyl-6-hydroxypyrimidine (IMHP) with a half-life of less than 1 month. Within 14 days, only 12.3% of the activity was found as the parent; 72.9% was identified as IMHP. Breakdown of IMHP was slower than that of diazinon, and 49% of the applied radioactive material was in the form of IMHP after 84 days. After 166 days, the amount of IMHP decreased to 4.7% of the applied material. Increased recoveries of $^{14}CO_2$ (55.6% after 166 days) and unextracted ^{14}C residues (15.1% after 166 days) corresponded to IMHP breakdown. No other major metabolites were found. Radioactivity in the H_2SO_4 and ethylene glycol traps was $< 1\%$ of the applied ^{14}C throughout the study and material balance was generally above 80% of the applied material (Keller, 1981).

II.   Pharmacokinetics

A. Absorption

- Mucke et al. (1970) reported that in both male and female rats, 69 to 80% of orally administered diazinon is excreted in the urine within 12 hours. This indicates that diazinon is well absorbed from the gastrointestinal tract.

B. Distribution

- The retention of diazinon labeled with ^{14}C in the pyrimidine ring and in the ethoxy groups was investigated in Wistar rats (Mucke et al., 1970). Doses of 0.1 mg/rat were administered by stomach tube daily for 10 days. Tissue levels 8 hours after the final dose were as follows: stomach and esophagus, 0.25%; small intestine, 0.65%; cecum/colon, 0.76%; liver, 0.16%; spleen, 0.01%; pancreas, 0.01%; kidney, 0.04%; lung, 0.02%; testes, 0.02%; muscle, 0.77%; and fat, 0.23%.

- Chickens were fed diazinon at levels of 2, 20 or 200 ppm in their food for a period of 7 weeks (Mattson and Solga, 1965). Assuming that 1 ppm in the diet of chickens is equivalent to 0.125 mg/kg/day, this corresponds to doses of about 0.25, 2.5 or 25 mg/kg/day (Lehman, 1959). At the end of the feeding period, tissues from the animals fed 200 ppm (25 mg/kg/day) in the diet were analyzed for diazinon. There was no diazinon detected in fat, white or dark muscle, heart, kidney, liver, gizzard or eggs. The limit of sensitivity of the method was 0.05 ppm. There appeared to be no accumulation of diazinon in the body at 200 ppm (25 mg/kg/day) in the diet.

C. Metabolism

- The metabolism of diazinon ^{14}C-labeled in the pyrimidine ring was investigated in Wistar rats (200 g) after administration by stomach tube (Mucke et al., 1970). In addition to some unchanged diazinon, three major metabolites, all with the pyrimidine ring intact, were identified in the urine, and to a lesser degree in the feces. A fourth fraction containing polar materials was also found. The three main metabolites were the result of a split at the oxygen-phosphorus bond, with subsequent hydroxylation of the isopropyl side chain. There was no significant expiration of labeled carbon dioxide, further indicating that the pyrimidine nucleus remained intact.

- The metabolism of diazinon was investigated *in vitro* in rat liver microsomes obtained from adult male rats (Nakatsugawa et al., 1969). It was found that diazinon underwent a dual oxidative metabolism consisting of activation to diazoxon and degradation to diethyl phosphorothioic acid. The authors noted that they had observed similar pathways in studies with parathion and malathion, and these results emphasized the importance of microsomal oxidation in the degradation of organophosphate esters, indicating that many of the so-called phosphatase products or hydrolysis products may actually be oxidative metabolites.

D. Excretion

- The excretion of diazinon labeled with ^{14}C in the pyrimidine ring and in the ethoxy groups was investigated after administration by stomach tube to Wistar rats (Mucke et al., 1970). The diazinon was excreted rapidly by both male and female animals, and 50% of the administered dose was recovered within 12 hours. Of this, 69 to 80% was excreted in the urine, and 18 to 25% in the feces. There was negligible expiration of labeled carbon dioxide. There was no evidence of accumulation of diazinon in any tissue.

III.   Health Effects

- Diazinon is a reactive organophosphorus compound, and many of its toxic effects are similar to those produced by other substances of this class. Characteristic effects include inhibition of acetyl cholinesterase (ChE) and central nervous system (CNS) depression.

A. Humans

1. *Short-term Exposure*

- Wedin et al. (1984) described a case report of diazinon poisoning in a 26-year-old man who deliberately ingested a preparation containing an unknown concentration of diazinon in an apparent suicide attempt. Upon admission to the hospital, the patient exhibited most of the usual symptoms of organophosphate poisoning, including muscarinic, nicotinic and CNS manifestations. During treatment and monitoring, it was noted that the urine output was very low and was dark and cloudy in appearance. By the second day, the urine was found to contain moderate amorphous crystals that could not be identified. With increased intravenous fluids, the urine output increased, but the crystaluria persisted and increased up to the 4th day, with a gradual decrease for the next 5 days, at which time the patient was discharged. Serum creatinine and urea nitrogen levels remained normal throughout this period. It was noted that this phenomenon may have been related to the specific pesticide formulation that had been ingested, but the authors suggested that renal function should be monitored more closely in persons with organophosphate poisoning.

- Two men reportedly developed "marked" inhibition of plasma ChE following the administration (route not specified) of five doses of 0.025 mg/kg/day. A dose of 0.05 mg/kg/day for 28 days reduced plasma ChE in three men by 35 to 40%. In other tests, each involving three to four men, doses ranging from 0.02 to 0.03 mg/kg/day produced reductions in plasma cholinesterase activity of 0, 15 to 20 and 14%. In no case was there any effect on red blood cell cholinesterase activity or on hematology, serum chemistry or urinalysis. Thus, 0.02 mg/kg/day was identified as a No-Observed-Adverse-Effect Level (NOAEL) in humans (FAO/WHO, 1967; cited in Hayes, 1982).

2. *Long-term Exposure*

- No information was found in the available literature on the long-term health effects of diazinon in humans.

B. Animals

1. *Short-term Exposure*

- The acute oral toxicity of diazinon MG8 (a yellow oily liquid, 1,200 mg/mL) was studied in male albino rats (238 to 321 g) by DeProspo (1972). Four groups of six rats each were given a single dose of diazinon by gavage and then observed for 7 days. Dose levels administered were 157, 313, 625 or 1,250 mg/kg. Within 4 hours of administration, animals at the three higher levels displayed symptoms of lethargy, tremors, convulsions and runny noses. Mortality in the four groups was 0/6, 2/6, 5/6 and 6/6, respectively, with death occurring between 8 and 24 hours after exposure. At 2 days, the remaining animals at the two intermediate levels had recovered. There was no mention of adverse symptoms at the lowest dose level. Gross necropsy (performed only on animals that died) did not reveal abnormal findings. The acute oral LD_{50} value was calculated to be 395.6 mg/kg.

- Hazelette (1984) investigated the effects of dietary hypercholesteremia (HCHOL) on sensitivity to diazinon in inbred male C56BL/6J mice. The LD_{50} of diazinon in HCHOL mice was nearly half that of diazinon administered to normal mice (45 versus 84 mg/kg). Cholesterol feeding increased ChE activity in both blood and liver, and these increases were negated by diazinon. Hepatic diazinon levels were also higher in the HCHOL animals. It was concluded that HCHOL resulted in an increase in susceptibility to, and toxicity of, diazinon.

- Adult mongrel dogs (1/sex/dose) were fed diazinon (0 or 1.0 ppm in the diet) for a period of 6 weeks (Doull and Anido, 1957). Assuming that 1 ppm in the diet of dogs is equivalent to 0.025 mg/kg/day, this corresponds to doses of about 0 or 0.025 mg/kg/day (Lehman, 1959). Serum and erythrocyte ChE determinations were made on a weekly basis before and during exposure. Neither plasma nor red blood cell ChE varied by more than ±15% from control in exposed animals of either sex, and there were no observed changes in body weight for the test period. The apparent NOAEL for this study, based on blood chemistry parameters, is 0.025 mg/kg.

- The effect of diazinon (tech. grade) on blood cell ChE activity was investigated in sheep after the administration of single oral doses by gavage of 50, 65, 100, 200 or 250 mg/kg (Anderson et al., 1969). Twenty-six sheep were used in the study groups. Prior to dosing, 245 untreated sheep were used to determine the normal range of erythrocyte ChE values. A typical severe clinical response consisted of profuse salivation, ataxia, dyspnea, dullness, anorexia and muscle twitching. In mild cases, only dullness and anorexia were seen, but were sufficiently pronounced to enable differentiation between normal and affected animals. Sheep that were clinically affected by diazinon suffered a depression of ChE of more than 75%. However, there were five animals (at the 50-mg/kg dose level) that tolerated depressions of 80 to 90% without clinical effect. The ChE values fell to minimum values within 1 to 4 hours, and remained close to this level until about 8 hours after dosing, during which time symptoms were observed. In those showing maximum depressions of 80% or more, the ChE activity returned to about half its normal value by the 5th day, and thereafter recovered only very slowly during a period of several weeks.

• Davies and Holub (1980a) compared the subacute toxicity of diazinon (approximately 99%) in male and female Wistar rats. The diazinon was incorporated into a semipurified diet at levels of 2 or 25 ppm. Assuming that 1 ppm in the diet of rats is equivalent to 0.05 mg/kg/day, this corresponds to doses of about 0.1 or 1.2 mg/kg/day (Lehman, 1959). Effects on ChE activity were periodically assessed during a 28- to 30-day feeding period. Levels of 25 ppm (1.2 mg/kg/day) diazinon in the diet for 30 days produced more significant reduction of ChE activity in plasma (22 to 30%) and brain (5 to 9%) among treated females than among treated males. Erythrocyte ChE activity was significantly more depressed (13 to 17%) in treated females relative to males at days 21 to 28 of the feeding period. At no time was ChE activity in any tissue more reduced among treated males than females. At the 2-ppm (0.1 mg/kg/day) dose level for 7 days, diazinon failed to affect erythrocyte ChE activity in either sex relative to controls. Plasma ChE activities of treated males were not significantly different from control values, but treated females showed significant depression (29%) of plasma ChE activity. This investigation indicated that the female rat is more sensitive to the toxicity of dietary diazinon than the male. Based on the inhibition of ChE in the female animals observed at 2 ppm, the Lowest-Observed-Adverse-Effect Level (LOAEL) for this study was identified as 0.1 mg/kg/day.

2. Dermal/Ocular Effects

• Nitka and Palanker (1980) investigated the primary dermal irritation and primary ocular irritation characteristics of a commercial formulation of diazinon in New Zealand White rabbits. The percentage of diazinon in the formulation was not given. After administration of a single application of 0.5 mL to abraded and intact skin of six rabbits, the formulation was judged not to be a primary dermal irritant. Nine rabbits were used to examine the effect of administration of a single dose of 0.1 mL of the formulation in one eye, and the results indicated that it was not an ocular irritant.

3. *Long-term Exposure*

- Female Wistar rats were fed a semipurified diet containing 0 or 0.1 to 15 ppm diazinon (99%) for up to 92 days with no visible toxic effects (Davies and Holub, 1980b). Weight gain and food consumption were comparable to controls. Feeding studies up to 90 days revealed that rats were highly sensitive to diazinon after 31 to 35 days of exposure, as judged by reduction in plasma and erythrocyte cholinesterase (ChE) activities. Plasma ChE was judged most sensitive. A NOAEL of 0.1 ppm, which the authors translated to an equivalent daily intake of 9 μg/kg/day, is based on plasma ChE inhibition noted for up to 35 days of feeding.

- Woodard et al. (1965) exposed monkeys (3/sex/dose) to diazinon orally for 52 weeks. The animals were started at doses of 0.1, 1.0 or 10 mg/kg/day for the first 35 days, but these doses were lowered to 0.05, 0.5 or 5.0 mg/kg/day for the remainder of the study, apparently because of poor food consumption and decreased weight gain. During the 52 weeks, body weight gain was slightly depressed in all treated groups, and soft stools were observed in all animals, with diarrhea in three animals (dose not specified). One female at the 0.5-mg/kg dose level had significant weight loss and signs of dehydration, emaciation, pale skin coloration and an unthrifty hair coat. One female at this level (it is not clear whether it is the same animal just mentioned) exhibited decreased hemoglobin and a rapid sedimentation rate at 39 and 53 weeks. Plasma ChE was inhibited 93% at the high-dose and 23% at the mid-dose levels, but no inhibition was noted at 0.05 mg/kg (the low dose). Red blood cell ChE was inhibited 90%, 0% and 0% at the high, mid and low doses, respectively. Other biochemical parameters were normal. Based on inhibition of ChE, a NOAEL of 0.05 mg/kg/day and a LOAEL of 0.5 mg/kg/day were identified in this study.

- Barnett and Kung (1980) fed Charles River CD-1 mice diazinon (87.6%) in the diet at levels of 0, 4, 20 or 100 ppm for 18 months (males) or 19 months (females). Assuming that 1 ppm in the diet of mice is equivalent to 0.15 mg/kg/day, this corresponds to doses of about 0, 0.6, 3 or 15 mg/kg/day (Lehman, 1959). Groups of 60 animals of each sex were used at each treatment level, and a similar group served as controls. In males, there was a significant reduction in weight gain at the highest dose.

Weight reduction was significant in all female groups, although it did not appear to be dose or treatment related. There were no significant trends in mortality. Animals showed skin irritation, loss of hair, skin lesions and piloerection. Gross and microscopic examinations showed no inflammatory, degenerative, proliferative or neoplastic lesions due to the administration of diazinon. A LOAEL of 4 ppm (0.6 mg/kg/day) was identified for the mouse in this study.

• Horn (1955) fed diazinon (25%) to groups of 20 male and 20 female rats at 0, 10, 100 or 1,000 ppm in the diet for 104 weeks. Assuming that 1 ppm in the diet of rats is equivalent to 0.05 mg/kg/day, this corresponds to dose levels of about 0, 0.5, 5 or 50 mg/kg/day (Lehman, 1959). The rats were started on the diet as weanlings weighing 62 to 63 g. In preliminary studies, the highest dose caused significant growth retardation. The animals for this group were initially given 100 ppm diazinon, which was increased gradually over a period of 11 weeks to the 1,000-ppm level. Mortality occurred in all groups, including the controls, and pneumonia was common. In all groups, body weight and food consumption were comparable to the controls. Hematocrit values for males at 1,000 ppm were significantly depressed when compared to controls. At 10 ppm, plasma ChE was inhibited by 60 to 67%, red blood cell ChE was inhibited 24 to 42% and brain ChE was inhibited 8 to 10%. At 100 or 1,000 ppm, there was 95 to 100% inhibition of ChE in plasma and blood cells. At 100 ppm, brain ChE was inhibited 19 (males) to 53% (females), and this increased to 41 (males) to 59% (females) at 1,000 ppm. There were no significant gross pathological findings. Based on inhibition of blood and plasma ChE, the LOAEL for this study was identified as 10 ppm (0.5 mg/kg/day).

4. Reproductive Effects

• Johnson and Cronin (1965) conducted a three-generation reproduction study in Charles River rats. Beginning 70 days before mating, groups of 20 females were fed diazinon (as 50% wettable powder) in the diet at 4 or 8 ppm. Assuming that 1 ppm in the diet of rats is equivalent to 0.05 mg/kg/day, this corresponds to doses of about 0.2 or 0.4 mg/kg/day (Lehman, 1959). The end points monitored included general maternal condition; number of live and dead fetuses; number of pups per litter; mean pup and litter weights; gross pathology of F_{1a}, F_{2a} and F_{3a} animals; and histopathology of F_{3b} animals. All findings were reported to be normal, but there were no detailed data provided. A NOAEL of 8 ppm (0.4 mg/kg/day), the highest dose tested, was identified in this study.

- Diazinon was administered by gavage at dose levels of 0, 7, 25 or 100 mg/kg to groups of 18 to 22 New Zealand White rabbits on days 6 to 18 of gestation (Harris et al., 1981). At the 100-mg/kg level, 9/22 animals died. This was not statistically significant (p < 0.07) using the Fisher Exact Test, although it was thought to be biologically significant by the authors. Of these nine animals, seven showed lesions indicative of gastrointestinal toxicity. At this dose, animals also were observed to have tremors and convulsions and were anorexic and hypoactive. These symptoms were not observed in animals at the 7- and 25-mg/kg levels. One rabbit at the 25-mg/kg level aborted on day 27, and all fetuses were dead. At this dose there were no significant changes in weight gain compared to the control, and no changes in the corpora lutea. There were also no statistically significant changes in implantation sites; proportion of live, dead or resorbed fetuses per litter; fetal weights; or sex ratios. Based on these data, the NOAEL for reproductive effects for the rabbit was identified as 7 mg/kg/day.

5. Developmental Effects

- Diazinon at dose levels of 7, 25 or 100 mg/kg was administered by gavage to New Zealand White rabbits on days 6 to 18 of gestation (Harris et al., 1981). Groups of 18 to 22 rabbits, 4 to 5 months of age and weighing 3.0 to 4.1 kg, were given diazinon in 0.2% sodium carboxymethyl cellulose (CMC), and a group of controls was given 0.2% CMC only. At the 100-mg/kg level, 9/22 animals died, and although this mortality was not quite significant (p < 0.07) using the Fisher Exact Test, it was thought to be biologically significant by the authors. There were no significant differences in abnormalities between the control and treated groups, and it was concluded that diazinon was neither fetotoxic nor teratogenic in the rabbit at these dose levels. With respect to fetal effects, a NOAEL of 100 mg/kg/day, the highest dose tested, was identified. Based on maternal toxicity, a NOAEL of 25 mg/kg/day is identified.

- Tauchi et al. (1979) administered diazinon by gavage to groups of 30 pregnant rats for 11 days (days 7 to 17 of gestation), at dose levels of 0, 0.53, 1.45 or 4.0 mg/kg/day. In each group, 20 animals were delivered by Cesarean section on day 17, while the remaining 10 were allowed to deliver normally. There were no effects on behavior or learning ability, and no pathological lesions were detected at 10 weeks. It was concluded that diazinon was not teratogenic at the doses tested. The NOAEL for fetal effects in this study was 4.0 mg/kg/day, the highest dose tested.

6. *Mutagenicity*

- Four strains of *Salmonella typhimurium* were used to assay the mutagenic potential of diazinon (Marshall et al., 1976). Negative results were found by these investigators as well.

- The mutagenicity of diazinon was tested in bacterial reversion-assay systems with five strains of *Salmonella typhimurium* and one strain of *Escherichia coli* (Moriya et al., 1983). No evidence of mutagenic activity was noted in any of the test systems.

- Fritz (1975) conducted a dominant lethal study in NMRI-derived albino mice. Single doses of diazinon were administered orally to males at levels of 15 or 45 mg/kg. After exposure, the males were mated to untreated females several times over a period of 6 weeks. There were no significant differences in mating ratios, the number of implantations or embryonic deaths (resorptions), and no adverse effects were observed in the progeny at either dose level. It was concluded that diazinon did not produce dominant lethal mutations in this test at the doses used.

7. *Carcinogenicity*

- A chronic bioassay for possible carcinogenicity of diazinon was conducted in Fischer 344 rats and $B6C3F_1$ mice (NCI, 1979). Groups of 50 animals were fed diazinon in the diet at the following levels: rats, 400 or 800 ppm; mice, 100 or 200 ppm. Assuming that 1 ppm in the diet of rats and mice is equivalent to 0.05 and 0.15 mg/kg/day, respectively, this corresponds to doses of about 20 or 40 mg/kg/day in rats and about 15 or 30 mg/kg/day in mice (Lehman, 1959). There was some hyperactivity noted in animals of both species, but there was no significant effect on either weight gain or mortality. There was no incidence of tumors that could be clearly related to diazinon, and it was concluded that diazinon was not carcinogenic in either species.

- Charles River CD-1 mice were fed diazinon (87.6%) in the diet at levels of 4, 20 or 100 ppm for 18 months (males) or 19 months (females) (Barnett and Kung, 1980). Assuming that 1 ppm in the diet of mice is equivalent to 0.15 mg/kg/day, this corresponds to doses of about 0.6, 3 or 15 mg/kg/day (Lehman, 1959). Groups of 60 animals of each sex were used at each treatment level, and a similar group served as controls. In males at the highest dose level there was a significant difference in weight gain from the controls. Weight reduction was significant in all female treatment groups, but it did not appear to be dose or treatment related. There were no significant trends in mortality. Gross and microscopic examinations showed no inflammatory, degenerative, proliferative or neoplastic lesions due to the administration of diazinon, and the study was judged to be negative with respect to carcinogenicity.

IV.   Quantification of Toxicological Effects

A. One-day Health Advisory

No information was found in the available literature that was suitable for determination of the One-day HA value. It is therefore recommended that the Ten-day HA value for a 10-kg child (0.02 mg/L, calculated below) be used at this time as a conservative estimate of the One-day HA value.

B. Ten-day Health Advisory

The most sensitive indicator of the effects of diazinon is inhibition of ChE. However, this effect is reversible, and significant inhibition of this enzyme often occurs without production of clinically significant effects. Consequently, selection of a NOAEL or LOAEL value based only on inhibition of ChE, in the absence of any other toxic signs, is a highly conservative approach.

The study in humans described by Hayes (1982) has been selected to serve as the basis for determination of the Ten-day HA value for diazinon. Although this study is a secondary source, it establishes a NOAEL in humans based on the most sensitive end point, i.e., ChE. Hayes reported that in human volunteers, short-term exposure to doses of 0.02 mg/kg/day did not result in decreased ChE levels, while doses of 0.025 to 0.05 mg/kg/day caused ChE reductions of 15 to 40%. This NOAEL (0.02 mg/kg/day) is supported by studies in animals; e.g., based on blood and serum ChE, Doull and Anido (1957) reported a NOAEL of 0.05 mg/kg/day in a 6-week study in dogs.

Using a NOAEL of 0.02 mg/kg/day, the Ten-day HA for a 10-kg child is calculated as follows:

$$\text{Ten-day HA} = \frac{(0.02 \text{ mg/kg/day}) (10 \text{ kg})}{(10) (1 \text{ L/day})} = 0.02 \text{ mg/L } (20 \text{ } \mu\text{g/L})$$

C. Longer-term Health Advisory

The study by Woodard et al. (1965) has been selected to serve as the basis for the Longer-term HA. Based on inhibition of plasma ChE in monkeys exposed for 52 weeks, this study identified a NOAEL of 0.05 and a LOAEL of 0.5 mg/kg/day. These values are supported by the NOAEL for ChE inhibition of 0.025 mg/kg/day identified in a 6-week feeding study in dogs (Doull and

Anido, 1957) and by the LOAEL of 0.5 mg/kg/day identified by Horn (1955), based on ChE inhibition in rats exposed for 2 years.

Using a NOAEL of 0.05 mg/kg/day, the Longer-term HA for a 10-kg child is calculated as follows:

$$\text{Longer-term HA} = \frac{(0.05 \text{ mg/kg/day}) (10 \text{ kg})}{(100) (1 \text{ L/day})} = 0.005 \text{ mg/L } (5.0 \text{ }\mu\text{g/L})$$

Using a NOAEL of 0.05 mg/kg/day, the Longer-term HA for a 70-kg adult is calculated as follows:

$$\text{Longer-term HA} = \frac{(0.05 \text{ mg/kg/day}) (70 \text{ kg})}{(100) (2 \text{ L/day})} = 0.0175 \text{ mg/L } (20 \text{ }\mu\text{g/L})$$

D. Lifetime Health Advisory

Available lifetime studies were not judged adequate for use in the determination of the Lifetime HAs since toxicological end points and numbers of animals tested were limited. Therefore, the 13-week study of Davies and Holub (1980b) has been selected to serve as the basis for determination of the Lifetime HA, with an additional safety factor of 10 for studies of less than a lifetime. This study identified a NOAEL of 0.009 mg/kg/day.

Using a NOAEL of 0.009 mg/kg/day, the Lifetime HA is derived as follows:

Step 1: Determination of the Reference Dose (RfD)

$$\text{RfD} = \frac{(0.009 \text{ mg/kg/day})}{(100)} = 0.00009 \text{ mg/kg/day}$$

Step 2: Determination of the Drinking Water Equivalent Level (DWEL)

$$\text{DWEL} = \frac{(0.00009 \text{ mg/kg/day}) (70 \text{ kg})}{(2 \text{ L/day})} = 0.00315 \text{ mg/L } (3 \text{ }\mu\text{g/L})$$

Step 3: Determination of the Lifetime Health Advisory

$$\text{Lifetime HA} = (0.00315 \text{ mg/L}) (20\%) = 0.00063 \text{ mg/L } (0.6 \text{ }\mu\text{g/L})$$

E. Evaluation of Carcinogenic Potential

- Two studies on the carcinogenicity of diazinon in mice have been reported (NCI, 1979; Barnett and Kung, 1980). Neither study revealed any evidence of carcinogenicity.

- The International Agency for Research on Cancer has not evaluated the carcinogenic potential of diazinon.

- Applying the criteria described in EPA's guidelines for assessment of carcinogenic risk (U.S. EPA, 1986), diazinon may be classified in Group E: evidence of noncarcinogenicity for humans. This category is for substances that show no evidence of carcinogenicity in at least two adequate animal tests or in both epidemiologic and animal studies.

V.   Other Criteria, Guidance and Standards

- The NAS (1977) has calculated an ADI of 0.002 mg/kg/day, based on a NOAEL in humans of 0.02 mg/kg/day and an uncertainty factor of 10. Assuming average body weight of a human adult of 70 kg, daily consumption of 2 liters of water and a 20% contribution from water, NAS (1977) calculated a Suggested-No-Adverse-Effect Level of 0.014 mg/L.

VI.   Analytical Methods

- Analysis of diazinon is by a gas chromatographic (GC) method applicable to the determination of certain nitrogen-phosphorus containing pesticides in water samples (U.S. EPA, 1988). In this method, approximately 1 liter of sample is extracted with methylene chloride. The extract is concentrated and the compounds are separated using a capillary column GC. Measurement is made using a nitrogen phosphorus detector. This method has been validated in a single laboratory, and estimated detection limits have been determined for the analytes in the method, including diazinon. The estimated detection limit is 0.25 μg/L.

VII.   Treatment Technologies

- Available data indicate that reverse osmosis (RO), granular-activated carbon (GAC) adsorption and ozonation will remove diazinon from water. The percent removal efficiency ranged from 75 to 100%.

- Laboratory studies indicate that RO is a promising treatment method for diazinon-contaminated waters. Chian (1975) reported 100% removal efficiency using a cross-linked polyethylenimine (NS-100) membrane and 99.88% removal efficiency with a cellulose acetate (CA) membrane. Both membranes operated separately at 600 psi and a flux rate of 8 to 12 gal/ft^2/day. Membrane adsorption, however, is a major concern and must be considered, as breakthrough of diazinon would probably occur once the adsorption potential of the membrane was exhausted.

- GAC is effective for diazinon removal. Dennis and Kobylinski (1983) and Dennis et al. (1983) reported 94.5%, 90.5% and 76% diazinon removal efficiency from wastewater in 6-hour treatment periods with 45 lbs of GAC. Also, 95% diazinon removal efficiency was achieved in an 8-hour treatment period with 40 lbs of GAC.

- Whittaker (1980) experimentally determined that GAC adsorption isotherms for diazinon and diazinon-methyl parathion solutions in distilled water indicate that treatment with GAC can be used to remove diazinon.

- UV/03 oxidation treatment appears to be an effective diazinon removal method. UV/03 oxidized 75% of diazinon at 3.4 gm/L ozone dosage and a retention time of 204 minutes. When lime pretreatment was used, UV/03 oxidized 99 + % of diazinon at 4.1 gm/L ozone dosage and 240 minutes retention time (Zeff et al., 1984).

- Some treatment technologies for the removal of diazinon from water are available and have been reported to be effective. However, selection of individual or combinations of technologies to attempt diazinon removal from water must be based on a case-by-case technical evaluation, and an assessment of the economics involved.

VIII. References

Anderson, P.H., A.F. Machin and C.N. Hebert. 1969. Blood cholinesterase activity as an index of acute toxicity of organophosphorus pesticides in sheep and cattle. Res. Vet. Sci. 10:29–33.

Barnett, J.W. and A.H.C. Kung. 1980. Carcinogenicity evaluation with diazinon technical in albino mice. Chicago, IL: Industrial Bio-Test Laboratories, Inc.

Blair, J.* 1985. Photodegradation of diazinon on soil: Study No. 6015–208. Unpublished study prepared by Hazelton Laboratories America, Inc. 130 pp. MRID 00153230.

Chian, E.S.K., W.N. Bruce and H.H.P. Fang. 1975. Removal of pesticides by reverse osmosis. Environ. Sci. Technol. 9(1):52–59.

Davies, D.B. and B.J. Holub. 1980a. Comparative subacute toxicity of dietary diazinon in the male and female rat. Toxicol. Appl. Pharmacol. 54:359–367.

Davies, D.B. and B.J. Holub. 1980b. Toxicological evaluation of dietary diazinon in the rat. Arch. Environ. Contam. Toxicol. 9:637–650.

Dennis, W.H. and E.A. Kobylinski. 1983. Pesticide-laden wastewater treatment for small waste generators. J. Environ. Sci. Health. B18(13):317–331.

Dennis, W.H., A.B. Rosencrance, T.M. Trybus, C.W.R. Wade and E.A. Kobylinski. 1983. Treatment of pesticide-laden wastewaters from Army pest control facilities by activated carbon filtration using the carbolator treatment system. Technical Report 8203. Frederick, MD: U.S. Army Medical Bioengineering Research and Development Laboratory.

DeProspo, J.R.* 1972. Acute oral toxicity in rats: diazinon MG8. Prepared for Geigy Agricultural Chemicals. Princeton, NJ: Affiliated Medical Research. MRID 00034096.

Doull, J. and P. Anido.* 1957. Effects of diets containing guthion and/or diazinon on dogs. Chicago, IL: Department of Pharmacology, University of Chicago. MRID 00046789.

FAO/WHO. 1967. Food and Agricultural Organization of the United Nations/World Health Organization. Evaluation of some pesticide residues in food. Geneva, Switzerland: FAO PL:CP/15, WHO/Food Add/67.32.

Fritz, H. 1975.* Mouse: dominant lethal study of diazinon technical. Basle, Switzerland: Ciba-Geigy Ltd. MRID 00109037.

Harris, S.B., J.F. Holson and K.R. Fite.* 1981. A teratology study of diazinon in New Zealand White rabbits. Science Applications, Inc., La Jolla, CA, for Ciba-Geigy Corporation, Greensboro, NC. MRID 00079017.

Hayes, W.J. 1982. Pesticides studied in man. Baltimore, MD: Williams and Wilkins.

Hazelette, J.R. 1984. Dietary hypercholesteremia and susceptibility to the pesticide diazinon. Diss. Abstr. Int. B. 44:2116.

Horn, H.J.* 1955. Diazinon 25W: chronic feeding-104 weeks. Prepared for Geigy Agricultural Chemicals Division of Ciba-Geigy Corp. Falls Church, VA: Hazelton Laboratories. MRID 00075932.

Johnson, C.D. and M.T.I. Cronin.* 1965. Diazinon: three–generation reproduction study in the rat. Woodard Research Institute for Geigy Research Laboratory. MRID 00055407.

Keller, A.* 1981. Degradation of Basudin in aerobic soil. Project Report 37/81. Accession No. 251777. Report 7. Unpublished study received Nov. 5, 1982 under 4581–351. Prepared by Ciba-Geigy, Ltd., Switzerland, submitted by Agchem Div., Pennwalt Corp., Philadelphia, PA. CDL:248818-L. MRID 00118031.

Lehman, A.J. 1959. Appraisal of the safety of chemicals in foods, drugs and cosmetics. Assoc. Food Drug Off. U.S. Topeka, KS.

Marshall, T.C., H.W. Dorough and H.E. Swim. 1976. Screening of pesticides for mutagenic potential using Salmonella typhimurium mutants. J. Agric. Food Chem. 24(3):560–563.

Martinson, J.* 1985. Photolysis of diazinon on soil. Final Report. Biospherics Project No. 85-E-044 SP. Unpublished study prepared by Biospherics Inc. 135 pp. MRID 00153229.

Mattson, A.J. and J. Solga.* 1965. Analysis of chicken tissues for diazinon after feeding diazinon for seven weeks. Geigy Research Laboratories. MRID 00135229.

Meister, R., ed. 1983. Farm chemicals handbook. Willoughby, OH: Meister Publishing Company.

Meister, R., ed. 1985. Farm chemicals handbook. Willoughby, OH: Meister Publishing Company.

Moriya, M., T. Ohta, K. Watanabe, T. Miyazawa, K. Kato and Y. Shirasu. 1983. Further mutagenicity studies on pesticides in bacterial reversion assay systems. Mutat. Res. 116:185–216.

Mucke, W., K.O. Alt and H.O. Esser. 1970. Degradation of ^{14}C-labeled diazinon in the rat. J. Agr. Food Chem. 18(2):208–212.

Nakatsugawa, T., N.M. Tolman and P.A. Dahm. 1969. Oxidative degradation of diazinon by rat liver microsomes. Biochem. Pharmacol. 18:685–688.

NAS. 1977. National Academy of Sciences. Drinking water and health. Washington, DC: National Academy Press.

NCI. 1979.* National Cancer Institute. Bioassay of diazinon for possible carcinogenicity. Carcinogenicity Testing Program. DHEW Publication No. NIH 79-1392. Bethesda, MD: NCI-NIH. MRID 00073372.

Nitka, S. and A.L. Palanker.* 1980. Primary dermal irritation in rabbits; primary ocular irritation in rabbits. Final report. Study No. 80147 for Boyle-Midway, Cranford, NJ. MRID 00050966.

Tauchi, K., N. Igarashi, H. Kawanishi and K. Suzuki.* 1979. Teratological study of diazinon in the rat. Japan: Institute for Animal Reproduction. MRID 00131150.

STORET. 1988. STORET Water Quality File. Office of Water. U.S. Environmental Protection Agency (data file search conducted in May, 1988).

U.S. EPA. 1986. U.S. Environmental Protection Agency. Guidelines for carcinogen risk assessment. Fed. Reg. 51(185):33992–34003. September 24.

U.S. EPA. 1988. U.S. Environmental Protection Agency. U.S. EPA Method 507—determination of nitrogen- and phosphorus-containing pesticides in water by GC/NPD. Draft. Available from U.S. EPA's Environmental Monitoring and Support Laboratory, Cincinnati, OH. April 15.

Wedin, G.P., C.M. Pennente and S.S. Sachdev. 1984. Renal involvement in organo-phosphate poisoning. J. Am. Med. Assoc. 252:1408.

Whittaker, K.F. 1980. Adsorption of selected pesticides by activated carbon using isotherm and continuous flow column systems. Ph.D. Thesis. Purdue University.

Windholz, M., S. Budavari, R.F. Blumetti and E.S. Otterbein, eds. 1983. The Merck Index, 10th ed. Rahway, NJ: Merck and Co., Inc.

Woodard, M.W., K.O. Cockrell and B.J. Lobdell.* 1965. Diazinon 50W:

safety evaluation by oral administration for 104 weeks; 52-week report. Woodard Research Corporation. MRID 00064320.

Zeff, J.D., E. Leitis and J.A. Harris. 1984. Chemistry and application of ozone and ultraviolet light for water reuse—Pilot plant demonstration. Proceedings of Industrial Waste Conference. Vol. 38, pp. 105–116.

---

* Confidential Business Information submitted to the Office of Pesticide Programs.

Dicamba

I. General Information and Properties

A. CAS No. 1918-00-9

B. Structural Formula

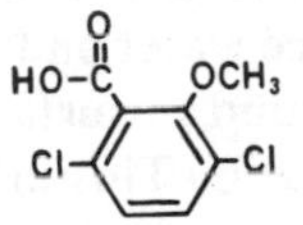

2-Methoxy-3,6-dichlorobenzoic acid

C. Synonyms

- Banes; Banex; Banlen; Banvel D; Banvel; Brush buster; Dianat; Dianate; Dicambe; Mediben; Mondak; MDBA; Velsicol Compound R.

D. Uses

- Dicamba is an herbicide used to control broadleaf weeds in field and silage corn; grain sorghum; small grains; asparagus; grass seed crops; turf; pasture; rangeland; and noncropland areas such as fence rows, roadways and wastelands. It is also used for control of brush and vines in noncropland, pasture and rangeland areas (Berg, 1986).

E. Properties (Berg, 1986; CHEMLAB, 1985; Windholz et al., 1983; Worthing, 1983; WSSA, 1983)

Chemical Formula	$C_8H_6Cl_2O_3$
Molecular Weight	221.04
Physical State (at 25°C)	Crystals
Boiling Point	--
Melting Point	114 to 116°C
Density	--
Vapor Pressure (20°C)	3.41×10^{-5} mm Hg
Specific Gravity	--
Water Solubility (20°C)	4,500 mg/L at 25°C

Log Octanol/Water
   Partition Coefficient   3.67 (calculated)
Taste Threshold   --
Odor Threshold   --
Conversion Factor   --

F. Occurrence

- Dicamba has been found in 262 of 806 surface water samples analyzed and in 2 of 230 ground-water samples (STORET, 1988). Samples were collected at 151 surface water locations and 192 ground-water locations; dicamba was found in 9 states. The 85th percentile of all nonzero samples was 0.15 µg/L in surface water and 0.07 µg/L in ground water. The maximum concentration found in surface water was 3.3 µg/L, while in ground water it was 0.07 µg/L. This information is provided to give a general impression of the occurrence of this chemical in ground and surface waters as reported in the STORET database. The individual data points retrieved were used as they came from STORET and have not been confirmed as to their validity. STORET data are often not valid when individual numbers are used out of the context of the entire sampling regime, as they are here. Therefore, this information can only be used to form an impression of the intensity and location of sampling for a particular chemical.

G. Environmental Fate

- In several aerobic soil metabolism studies, dicamba (acid or salt form not specified) had half-lives of 1 to 6 weeks in sandy loam, heavy clay, silty clay, clay loam, sand and silt loam soils at 18 to 38°C and 40 to 100% of field capacity. Degradation rates decreased with decreasing temperature and soil moisture (Smith, 1973a,b;1974; Smith and Cullimore, 1975; Suzuki, 1978;1979).

- For the dimethylamine salt, half-lives in sandy loam and loam soils ranged from 17 to 32 days (Altom and Stritzke, 1973). Phytotoxic residues, detected by a nonspecific bioassay method, have persisted in aerobic soil for almost 2 years (Sheets, 1964; Sheets et al., 1968).

- Based on soil thin-layer chromatography (TLC), dicamba (acid or salt form not specified) is highly mobile in sandy loam, silt loam, sandy clay loam, clay loam, loam, silty clay loam and silty clay soils (Helling, 1971; Helling and Turner, 1968).

- The free acid of dicamba and the dimethylamine salt were not appreciably adsorbed to any of five soils ranging from heavy clay to loamy sand (Grover and Smith, 1974). The dicamba degradation product, 3,6-dichlorosalicylic acid, adsorbed to sandy loam (30%), clay and silty clay (55%) (Smith, 1973a,b; Smith and Cullimore, 1975).

- Losses of 12 to 19% of the applied radioactivity from nonsterile soils indicated that metabolism contributes substantially more to ^{14}C-dicamba losses than does volatilization (Burnside and Levy, 1965; 1966).

- Under field conditions, dicamba (acid or salt form not specified) had half-lives of 1 to 2 weeks in a clay and a sandy loam soil when applied at 0.27 and 0.53 lb/acre (A). At either application rate, less than 30 ppb of dicamba remained after 4 weeks (Scifres and Allen, 1973). In another study, using a nonspecific bioassay method of analysis, dicamba phytotoxic residues dissipated within 2 years in loam and silty clay loam (Burnside et al., 1971).

- Ditchbank field studies indicated vertical movement of dicamba in soil; the soil layers at 6 to 12 inches contained a maximum of 0.07 ppm and 0.28 ppm in canals treated at 0.66 and 1.25 lb/A, respectively (Salman et al., 1972).

II. Pharmacokinetics

A. Absorption

- Atallah and Yu (1980) reported that mice, rats, rabbits and dogs administered single oral doses of ^{14}C-dicamba (99% purity, approximately 100 mg/kg) excreted an average of 85% of the administered dose in urine, as the parent compound, in 48 hours after dosing.

- Similar findings were reported for rats by Tye and Engel (1967) (96% excreted in 24 hours) and by Whitacre et al. (1976) (83% excreted unchanged in 24 hours). The data indicate that dicamba is rapidly absorbed from the gastrointestinal tract.

B. Distribution

- The retention of dicamba (99% purity, approximately 100 mg/kg) was investigated in rats, mice, rabbits and dogs following single doses by oral intubation (Atallah and Yu, 1980). They found that tissue levels were low and that dicamba did not accumulate in mammalian tissues.

- Tye and Engel (1967) also found low levels of dicamba in kidneys, liver and blood. These data also indicate that dicamba does not accumulate.

C. Metabolism

- The metabolism of ^{14}C-dicamba (99% purity) was investigated in mice, rats, rabbits and dogs after administration of approximately 100 mg/kg *per os* (Atallah and Yu, 1980). Between 90 to 99% of the dicamba was recovered unchanged in the urine of all four species. 3,6-Dichloro-2-hydroxybenzoic acid (DCHBA, a metabolite) was not detected in any urine sample at a level greater than 1% of the dose. There was also a small amount of unknown metabolites totaling about 1%.

D. Excretion

- Atallah and Yu (1980) investigated the excretion of ^{14}C-dicamba (99% purity) after a single oral dose (approximately 100 mg/kg) in mice, rats, dogs and rabbits, and reported that 67 to 93% of the administered dose was excreted in urine of the four species within 16 hours. The compound was found to a lesser degree in feces (0.5 to 9%) and various tissues (0.17 to 0.5%) 16 hours postdosing.

III. Health Effects

A. Humans

- The Pesticide Incident Monitoring System database revealed 10 incident reports involving humans from 1966 to March 1981 for dicamba alone (U.S. EPA, 1981). Six of the 10 reported incidents involved spraying operations. No concentrations were specified. Exposed workers developed muscle cramps, dyspnea, nausea, vomiting, skin rashes, loss of voice or swelling of cervical glands. Four additional incidences involved children. One resulted in coughing and dizziness in a child involved in an undescribed agricultural incident. Three other children who sucked mint leaves from a ditch bank previously sprayed with dicamba were asymptomatic.

B. Animals

1. *Short-term Exposure*

- Reported acute oral LD_{50} values for technical dicamba [85.8% active ingredient (a.i.)] range from 757 to 1,414 mg/kg (Witherup et al., 1962) in rats. The acute oral LD_{50} in mice has been reported to be > 4,640 mg/kg (Witherup et al., 1962) and 316 mg/kg in hens (Roberts et al., 1983).

- An acute inhalation LC_{50} of > 200 mg/L was reported in Spartan strain rats (Wazeter et al., 1973).

- The neurotoxic effects of dicamba in hens were studied by Roberts et al. (1983). Technical dicamba (86.82% a.i.) was administered *per os* (10 hens/dose) in doses of 0, 79, 158 or 316 mg/kg. Two groups of ten hens each were dosed at 316 mg/kg. The various groups were observed for 21 days following treatment. No signs of ataxia were observed at any dose level tested. Histopathological evaluation of nervous tissue from 13 hens treated at 316 mg/kg demonstrated sciatic nerve damage in 6 hens (46%). The authors attributed this alteration to prolonged recumbency (inability to stand) rather than a direct effect of dicamba. Based on the absence of delayed neurotoxicity and sciatic nerve damage, a NOAEL of 158 mg/kg is identified for this study.

- Rats (2/sex/dose) of the Charles River CD strain were fed diets containing 658 to 23,500 ppm of technical dicamba (85.8% a.i.) for up to 3 weeks (Witherup et al., 1962). Assuming that 1 ppm in the diet of rats is equivalent to 0.05 mg/kg/day (Lehman, 1959), these levels correspond to about 32.9 to 1,175 mg/kg/day. No adverse effects on physical appearance, behavior, food consumption, body or organ weights, gross pathology or histopathology were reported. Based on this information, a NOAEL of 1,175 mg/kg/day (the highest dose tested) is identified.

2. *Dermal/Ocular Effects*

- Wazeter et al. (1974) reported an acute LD_{50} for technical Banvel (85.8% a.i.) of > 2,000 mg/kg in dermal studies on New Zealand White rabbits (administration vehicle not stated).

- Heenehan et al. (1978) studied the sensitization potential of technical dicamba (86.8% a.i.) in Hartley albino guinea pigs. The compound was applied as a 10% suspension to the shaved backs of guinea pigs (5/sex) for 6 hours three times per week for 3 weeks. Following nine sensitizing doses, two challenge doses were applied. Dicamba was judged to cause moderate dermal sensitization.

- Technical dicamba (86.8% a.i.) was applied to the shaved backs of New Zealand White rabbits (4/sex/dose) in doses of 0, 100, 500 or 2,500 mg/kg/day, 5 days per week for 3 weeks (Dean et al., 1979). Slight skin irritation was observed at 100 mg/kg, and moderate irritation at 500 mg/kg/day and above. No changes were observed in general appearance, behavior, body weight, organ weight, biochemistry, hematology or urinalysis.

- Thompson (1984) instilled single doses (0.1 g) of technical dicamba (purity not specified) into the conjunctival sacs of nine New Zealand rabbits; three eyes were washed and six were not washed. Dicamba was severely irritating and corrosive to both washed and unwashed eyes.

3. *Long-term Exposure*

- Laveglia et al. (1981) fed CD rats (20/sex/dose) technical dicamba (86.8% a.i.) in the diet for 13 weeks in doses of 0, 1,000, 5,000 or 10,000 ppm. Assuming that 1 ppm in the diet of rats is equivalent to 0.05 mg/kg/day (Lehman, 1959), this corresponds to doses of about 0, 50, 250 or 500 mg/kg/day. No compound-related effects were observed in general appearance, hematology and biochemistry or in urinalysis values, survival and gross pathology at any dose levels tested. However, there was a decrease in mean body weight for both sexes (6.3% in females and 7.5% in males) at 10,000 ppm (500 mg/kg/day). The decrease in body weight was lower ($p < 0.05$) at week 13 compared to controls. At this dose, there was also a decrease ($p < 0.05$) in the wet weight of the kidney. In addition, an increase ($p < 0.01$) in liver weight, when expressed as % of body weight, was reported. A NOAEL of 5,000 ppm (250 mg/kg/day) can be identified for this study.

- Male Wistar rats (20/dose) were fed diets containing technical dicamba at 0, 31.6, 100, 316, 1,000 or 3,162 ppm (corresponding to doses of 0, 3.8, 12, 37.3, 119 or 364 mg/kg/day) for 15 weeks (Edson and Sanderson, 1965). Following treatment, general behavior, physical appearance, food consumption, organ weights, gross pathology and histopathology were evaluated. However, the authors presented data only for the evaluation of body and organ weights. Hematological, urinalysis or clinical chemistry studies were not reported. No adverse effects were observed in the parameters measured at 316 ppm (37 mg/kg/day) or less. Relative liver-to-body weight ratios increased (p value not specified) at 1,000 and 3,162 ppm (119 and 364 mg/kg/day). Based on these data, the authors identified a NOAEL of 316 ppm (37 mg/kg/day).

- Davis et al. (1962) fed beagle dogs (3/sex/dose) technical dicamba (90% a.i.) in the diet in doses of 0, 5, 25 or 50 ppm for 2 years. Assuming that 1 ppm in the diet of dogs is equivalent to 0.025 mg/kg/day, (Lehman, 1959), this corresponds to doses of about 0, 0.125, 0.625 or 1.25 mg/kg/day. No compound-related effects were observed on survival, food consumption, hematology, urinalysis and organ weights. Although a decrease in body weight was observed in males at 25 and 50 ppm and in females at 50 ppm, no individual data except for body weight were reported, and no statistical evaluations were made. The authors did not present data on gross pathology, and histopathology was done only on the heart, lung, liver and kidney. Based on this marginal information, a NOAEL or LOAEL will not be identified.

- Sprague-Dawley rats (32/sex/dose) were fed technical dicamba (90% a.i.) in the diet for 2 years in doses of 0, 5, 50, 100, 250 or 500 ppm (Davis et al., 1962). Assuming that 1 ppm in the diet of rats is equivalent to 0.05 mg/kg/day (Lehman, 1959), this corresponds to doses of about 0, 0.25, 2.5, 5, 12.5 or 25 mg/kg/day. The authors reported no adverse effects upon survival, body weight, food consumption, organ weight, hematologic values or histology at the dose levels tested. Since no data were presented for evaluation of pharmacologic effects, gross pathology, urinalysis or clinical chemistry and incomplete histological data were presented, a NOAEL could not be determined for this study.

- Dicamba (Technical Reference Standard, 86.8% a.i.) was administered for 1 year to male and female beagle dogs (4/sex/dose) in their diet at mean dosage levels of 0, 2, 11 or 52 mg/kg/day (Blair, 1986). There were no chemical-related changes in behavior, food consumption, body weight, clinical hematology, biochemistry or urinalysis. In addition, no compound-specific effects on gross or microscopic pathology were observed. Based upon these data, a NOAEL of 52 mg/kg/day can be identified.

4. *Reproductive Effects*

- Charles River CD rats (20 females and 10 males/dose) were fed diets containing technical dicamba (87.2% a.i.) in doses of 0, 5, 50, 100, 250 or 500 ppm through three generations (Witherup et al., 1966). Assuming that 1 ppm in the diet of rats is equivalent to 0.05 mg/kg/day (Lehman, 1959), this corresponds to doses of about 0, 0.25, 2.5, 5, 12.5 or 25 mg/kg/day. Fertility index, gestation index, viability index, lactation index and pup development were comparable in treated and control rats. A NOAEL of 500 ppm (25 mg/kg/day) was identified.

5. *Developmental Effects*

- Technical dicamba (87.7% a.i.) was administered *per os* to pregnant New Zealand White rabbits (23 to 27/dose) at doses of 0, 0.5, 1, 3, 10 or 20 mg/kg/day from days 6 through 18 of gestation (Wazeter et al., 1977). No maternal toxicity, fetotoxicity or teratogenic effects were observed at 1 and 3 mg/kg/day. There were slightly reduced fetal and maternal body weights and increased postimplantation losses in the 10 mg/kg/day dose group when compared to untreated controls. The author did not consider these differences to be significant and identified a developmental toxicity NOAEL of 10 mg/kg/day. However, the EPA Office of Pesticide Programs (OPP) identified a maternal and fetotoxic NOAEL of 3 mg/kg/day, based on a reduction in body weights and increased postimplantation losses at the highest dose.

• Pregnant albino rats (20 to 24/dose) were administered technical-grade dicamba by gavage at dose levels of 0, 64, 160 or 400 mg/kg/day on days 6 through 19 of gestation (Smith et al., 1981). No maternal toxicity was observed up to 160 mg/kg/day. Dicamba-treated dams in the 400 mg/ kg/day dosage group exhibited ataxia and reduced body weight gain; they consumed less food during the dosing period when compared with controls given vehicle alone (p < 0.05). No fetotoxicity or developmental effects were observed at the dose levels tested. Based on these findings, a NOAEL for maternal toxicity of 160 mg/kg/day is identified. The NOAEL for fetotoxic and developmental effects is 400 mg/kg/day (the highest dose tested).

6. *Mutagenicity*

• Moriya et al. (1983) reported that dicamba (up to 5,000 μg/plate) exhibited no mutagenic activity against *Salmonella typhimurium* (TA98, TA100, TA1535, TA1537 and TA1538) or *Escherichia coli* (WP2 hcr) either with or without metabolic activation.

• An increased number of chromosomal aberrations (p < 0.01) were reported in mouse bone marrow cells exposed to 500 mg/kg dicamba (Kurinnyi et al., 1982). No other details were presented and the data were not presented in the English summary. Accordingly, the significance of this report is unknown.

7. *Carcinogenicity*

• Sprague-Dawley rats (32/sex/dose) were administered dicamba (90% a.i.) in the diet for 2 years at doses of 0, 5, 50, 100, 250 or 500 ppm (Davis et al., 1962). Assuming that 1 ppm in the diet of rats is equivalent to 0.05 mg/kg/day (Lehman, 1959), this corresponds to doses of about 0, 0.25, 2.5, 5, 12.5 or 25 mg/kg/day. The treated rats did not differ from the untreated control animals with respect to the incidence, types and time of appearance of tumors.

IV.   Quantification of Toxicological Effects

A. One-day Health Advisory

No information was found in the available literature that was suitable for determination of the One-day HA value for dicamba. Accordingly, it is recommended that the Ten-day HA value of 0.3 mg/L (calculated below) for a 10-kg child be used at this time as a conservative estimate of the One-day HA.

B. Ten-day Health Advisory

The developmental toxicity study by Wazeter et al. (1977) has been selected to serve as the basis for the Ten-day HA value for dicamba. In this study,. pregnant rabbits administered technical dicamba (87.7% a.i) by gastric intubation at dosage levels of 0, 0.5, 1, 3, 10 or 20 mg/kg/day from days 6 through 18 of gestation showed slightly reduced maternal body weights at 10 mg/kg/day. Similarly, fetal body weights were slightly reduced, and postimplantation losses were increased in the 10 mg/kg/day dose group. Based on these data, a maternal and fetal toxicity NOAEL of 3 mg/kg/day is identified.

The Ten-day HA for a 10-kg child is calculated as follows:

$$\text{Ten-day HA} = \frac{(3 \text{ mg/kg/day}) (10 \text{ kg})}{(1 \text{ L/day}) (100)} = 0.3 \text{ mg/L} (300 \text{ } \mu\text{g/L})$$

C. Longer-term Health Advisory

No studies found in the available literature were suitable for determining a Longer-term HA value for dicamba. One 13-week rat study (Laveglia et al., 1981) and one 15-week rat study (Edson and Sanderson, 1965) reported NOAELs (250 mg/kg/day and 37 mg/kg/day, respectively) that were higher than the NOAEL (3 mg/kg/day) of the rabbit study (Wazeter et al., 1977) selected to derive the Ten-day HA value. It is therefore recommended that the Reference Dose (RfD) derived below in the calculation of the Lifetime HA (0.03 mg/kg/day) be used at this time as the basis for the Longer-term HA. As a result, the Longer-term HA for the 10-kg child is 0.3 mg/L (300 μg/L) and for the 70-kg adult is 1.0 mg/L (1,000 μg/L).

D. Lifetime Health Advisory

Although the l-year dog study by Blair (1986) identifies a NOAEL of 52 mg/kg/day, this NOAEL will not be used to calculate the RfD. This decision is based on the facts that the IRDC study did not address reproductive and developmental toxicity and that the study of Wazeter et al. (1977) identifies, according to EPA/OPP, a maternal and fetotoxic NOAEL of 3 mg/kg/day.

Since the reproductive and developmental effects appear to be seen at the lower exposure levels and to protect against potential fetotoxic effects, the RfD will be calculated using the NOAEL of 3 mg/kg/day.

The Lifetime HA is derived from this NOAEL as follows:

Step 1: Determination of the Reference Dose (RfD)

$$RfD = \frac{3 \text{ mg/kg/day}}{(100)} = 0.03 \text{ mg/kg/day}$$

Step 2: Determination of the Drinking Water Equivalent Level (DWEL)

$$DWEL = \frac{(0.03 \text{ mg/kg/day}) (70 \text{ kg})}{(2 \text{ L/day})} = 1.0 \text{ mg/L } (1{,}000 \text{ } \mu g/L)$$

Step 3: Determination of the Lifetime Health Advisory

$$\text{Lifetime HA} = (1.05 \text{ mg/L}) (20\%) = 0.2 \text{ mg/L } (200 \text{ } \mu g/L)$$

E. Evaluation of Carcinogenic Potential

- One study on the carcinogenicity of dicamba in rats has been reported (Davis et al., 1962). Although it revealed no evidence of carcinogenicity, it is limited in scope.

- The International Agency for Research on Cancer has not evaluated the carcinogenicity of dicamba.

- Applying the criteria described in EPA's guidelines for assessment of carcinogenic risk (U.S. EPA, 1986), dicamba is classified in Group D: not classifiable. This category is used for substances with inadequate evidence of carcinogenicity in animal studies.

V.   Other Criteria, Guidance and Standards

- The NAS (1977) has calculated an Acceptable Daily Intake (ADI) of 0.00125 mg/kg/day based on a NOAEL of 1.25 mg/kg/day from a 2-year feeding study in dogs and an uncertainty factor of 1,000. Assuming a body weight of 70 kg and a 20% source contribution factor, NAS calculated a Suggested-No-Adverse-Response Level (SNARL) of 0.009 mg/L.

- Residue tolerances from 0.05 to 40 ppm have been established for a variety of agricultural products (U.S. EPA, 1985).

VI.   Analytical Methods

- Analysis of dicamba is by a gas chromatographic (GC) method applicable to the determination of certain chlorinated acid pesticides in water samples (U.S. EPA, 1988). In this method, approximately 1 liter of sample is acidified and the compounds are extracted with ethyl ether using a separatory funnel. The derivatives are hydrolyzed with potassium hydroxide and extraneous organic material is removed by a solvent wash. After acidification, the acids are extracted and converted to their methyl esters using diazoxethane as the derivatizing agent. Excess reagent is removed and the esters are determined by electron capture GC. This method has been validated in a single laboratory and estimated detection limits have been determined for the analytes in this method. The estimated detection limit for dicamba is 0.081 $\mu g/L$.

VII.   Treatment Technologies

- Available data indicate granular-activated carbon (GAC) adsorption to be a possible removal technique for dicamba.

- Whittaker et al. (1982) report that a reduction of pH from 7 to 3 increased the extent of dicamba GAC adsorption. No system performance was reported.

VIII.   References

Atallah, Y.H. and C.C. Yu.* 1980. Comparative pharmacokinetics and metabolism of dicamba in mice, rats, rabbits and dogs. Unpublished study. MRID 00128088.

Altom, J.D. and J.R. Stritzke. 1973. Degradation of dicamba, picloram, and four phenoxy herbicides in soils. Weed Sci. 21:556-560.

Berg, G.L. 1986. Farm chemicals handbook. Willoughby, OH: Meister Publishing Co.

Blair, M.* 1986. Dicamba — one-year dietary toxicity study in dogs. IRDC Report No. 163-696. Unpublished study. EPA Accession No. 55947-3. December 19.

Burnside, O.C. and T.L. Levy. 1965. Dissipation of dicamba. Unpublished study. Prepared by the University of Nebraska, Department of Agronomy. Submitted by Velsicol Chemical Corporation, Chicago, IL.

Burnside, O.C. and T.L. Levy. 1966. Dissipation of dicamba. Weeds. 14:211-214.

Burnside, O.C., G.A. Wicks and C.R. Fenster. 1971. Dissipation of dicamba, picloram, and 2,3,6-TBA across Nebraska. Weed Sci. 19:323–325.

CHEMLAB. 1985. The Chemical Information System, CIS, Inc. Baltimore, MD.

Davis, R.K., W.P. Jolley, K.L. Stemmer et al.* 1962. The feeding for two years of the herbicide 2-methoxy-3,6-dichlorobenzoic acid to rats and dogs. Unpublished study. MRID 00028248.

Dean, W.P., E.I. Goldenthal, D.C. Jessup et al.* 1979. Three-week dermal toxicity study in rabbits. IRDC No. 163–620. MRID 00128090.

Edson, E.F. and D.M. Sanderson. 1965. Toxicity of the herbicides 2-methoxy-3,6-dichlorobenzoic acid (dicamba) and 2-methoxy-3,5,6-trichlorobenzoic acid (tricamba). Food Cosmet. Toxicol. 3:299–304.

Grover, R. and A.E. Smith. 1974. Adsorption studies with the acid and dimethylamine forms of 2,4-D and dicamba. Can. J. Soil Sci. 54:179–186.

Heenehan, P.R., W.E. Rinehart and W.G. Braun.* 1978. A dermal sensitization study in guinea pigs. Compound: Banvel 45, Banvel technical. Project No. 5055–78. Unpublished study. MRID 00023691.

Helling, C.S. 1971. Pesticide mobility in soils. II. Applications of soil thin-layer chromatography. Soil Sci. Soc. Amer. Proc. 35:737–748.

Helling, C.S. and B.C. Turner. 1968. Pesticide mobility: determination of soil thin-layer chromatography. Science. 162:562–563.

Kurinnyi, A.I., M.A. Pilinskaya, I.V. German and T.S. L'vova. 1982. Implementation of a program of cytogenetic study of pesticides: preliminary evaluation of cytogenetic activity and potential mutagenic hazard of 24 pesticides. Tsitol. Genet. 16:45–49.

Laveglia, J., D. Rajasekaran and L. Brewer.* 1981. Thirteen-week dietary toxicity study in rats with dicamba. IRDC No. 163–671. Unpublished study. MRID 00128093.

Lehman, A.J. 1959. Appraisal of the safety of chemicals in foods, drugs and cosmetics. Assoc. Food Drug Off. U.S., Q. Bull.

Moriya, M., T. Ohta, K. Watanabe, T. Miyazawa, K. Kato and Y. Shirasu. 1983. Further mutagenicity studies on pesticides in bacterial reversion assay systems. Mutat. Res. 116:185–216.

NAS. 1977. National Academy of Sciences. Drinking water and health, Vol. 1. Washington, DC: National Academy of Science Press.

Roberts, N., C. Fairley, C. Fish et al.* 1983. The acute oral toxicity (LD_{50}) and neurotoxic effects of dicamba in the domestic hen. HRC Report No. 24/8355. Unpublished study. MRID 00131290.

Salman, H.A., T.R. Bartley and A.R. Hattrup. 1972. Progress report of residue studies on dicamba for ditchbank weed control. Report No. REC-ERC-72–6. U.S. Department of the Interior, Bureau of Reclamation, Applied Sciences Branch, Division of General Research, Engineering and Research

Center. USDI, Br. Available from National Technical Information Center, Springfield, VA.

Scifres, C.F. and T.J. Allen. 1973. Dissipation of dicamba from grassland soils of Texas. Weed Sci. 21:393–396.

Sheets, T.J. 1964. Greenhouse persistence study with dicamba and tricamba. Letter sent to Warren H. Zick. U.S. Agricultural Research Service, Crops Research Division, Crops Protection Research Branch, Pesticide Investigations — Behavior in Soils. Unpublished study. January 3.

Sheets, T.J., J.W. Smith and D.D. Kaufman. 1968. Persistence of benzoic and phenylacetic acids in soils. Weed Sci. 16:217–222.

Smith, A.E. 1973a. Degradation of dicamba in prairie soils. Weed Res. 13:373–378.

Smith, A.E. 1973b. Transformation of dicamba in Regina heavy clay. J. Agric. Food Chem. 21:708–710.

Smith, A.E. 1974. Breakdown of the herbicide dicamba and its degradation products 3,6-dichlorosalicylic acid in prairie soils. J. Agric. Food Chem. 22:601–605.

Smith, A.E. and D.R. Cullimore. 1975. Microbiological degradation of the herbicide dicamba in moist soils at different temperatures. Weed Res. 15:59–62.

Smith, S.H., C.K. O'Loughlin, C.M. Salamon et al.* 1981. Teratology study in albino rats with technical dicamba. Toxigenetics Study No. 450-0460. Unpublished study. MRID 00084024.

STORET. 1988. STORET Water Quality File. Office of Water, U.S. Environmental Protection Agency. (Data file search conducted in May, 1988.)

Suzuki, H.K. 1978. Dissipation of Banvel and in combination with other herbicides in two soil types. Report No. 196. Unpublished study prepared in cooperation with IRDC. Submitted by Velsicol Chemical Corporation, Chicago, IL.

Suzuki, H.K. 1979. Dissipation of Banvel or Banvel in combination with other herbicides: two soil types. Report No. 197. Unpublished study. Prepared in cooperation with Craven Laboratories, Inc. Submitted by Velsicol Chemical Corporation, Chicago, IL.

Thompson, G.* 1984. Primary eye irritation study in albino rabbits with technical dicamba. Study No. Will 15134. Unpublished study. Will Research Laboratories, Inc. MRID 00144232.

Tye, R. and D. Engel. 1967. Distribution and excretion of dicamba by rats as determined by radiotracer technique. J. Agric. Food Chem. 15:837–840.

U.S. EPA. 1981. U.S. Environmental Protection Agency. Summary of reported incidents involving dicamba. Pesticide incident monitoring system. Report No. 432. Washington, DC: Office of Pesticide Programs.

U.S. EPA. 1985. U.S. Environmental Protection Agency. Code of Federal Regulations. 40 CFR 180.227. July 1.

U.S. EPA. 1986. U.S. Environmental Protection Agency. Guidelines for carcinogen risk assessment, Fed. Reg. 51(185):33992–34003. September 24.

U.S. EPA. 1988. U.S. Environmental Protection Agency. U.S. EPA Method 515.1 — Determination of chlorinated acids in water by gas chromatography with an electron capture detector. Draft. April 15. Available from U.S. EPA's Environmental Monitoring and Support Laboratory, Cincinnati, OH.

Wazeter, F.X., E.I. Goldenthal, W.P. Dean et al.* 1973. Acute inhalation exposure in male albino rats. IRDC No. 163–191. Unpublished study. MRID 00028234.

Wazeter, F.X., E.I. Goldenthal, W.P. Dean et al.* 1974. I. Acute toxicity studies in rats and rabbits. IRDC No. 163–295. Unpublished study. MRID 00025372.

Wazeter, F.X., E.I. Goldenthal, D.C. Jessup et al.* 1977. Pilot teratology study in rabbits. IRDC No. 163–436. Unpublished study. MRID 00025373.

Whitacre, D.M., L.I. Diaz, P. Schnur et al.* 1976. Metabolism of ^{14}C-dicamba in female rats. Unpublished study. MRID 00025363.

Whittaker, K.F., J.C. Nye, R.F. Weekash, R.J. Squires, A.C. York and H.A. Razemier. 1982. Collection and treatment of wastewater generated by pesticide application. EPA600/2-82-028. Cincinnati, OH: U.S. Environmental Protection Agency, Office of Environmental Criteria and Assessment.

Windholz, M., S. Budavari, R.F. Blumetti, and E.S. Otterbein, eds. 1983. The Merck index — An encyclopedia of chemicals and drugs, 10th ed. Rahway, NJ: Merck and Company, Inc.

Witherup, S., K.L. Stemmer and H. Schlecht.* 1962. The cumulative toxicity of 2-methoxy-3,6-dichlorobenzoic acid (Banvel D) and 2-methoxy-3,5,6-trichlorobenzoic acid (Banvel T) when fed to rats. Unpublished study. MRID 00022503.

Witherup, S., K.L. Stemmer, M. Roell et al.* 1966. The effects exerted upon the fertility of rats and upon the viability of their offspring by the introduction of Banvel D into their diets. Unpublished study. MRID 00028249.

Worthing, C.R., ed. 1983. The pesticide manual: a world compendium, 7th ed. London: BCPC Publishers.

WSSA. 1983. Weed Science Society of America. Herbicide handbook, 5th ed. Champaign, IL: WSSA.

---

* Confidential Business Information submitted to the Office of Pesticide Programs.

1,3-Dichloropropene

I. General Information and Properties

A. CAS No. 542–75–6

B. Structural Formula

$ClCH_2$, H $ClCH_2$, Cl
  $\diagdown$ $\diagup$ $\diagdown$ $\diagup$
   C = C   C = C
  $\diagup$ $\diagdown$ $\diagup$ $\diagdown$
  H   Cl   H   H

1,3-Dichloropropene (approximately 46% trans/42% cis)

C. Synonyms

- Dichloro-1,3-propene; 1,3-dichloro-1-propene; cis/trans-1,3-dichloro-propene; 1,3-D; DCP; D-D (approximately 28% cis/27% trans).

D. Uses

- DCP is the active ingredient in Telone®, a registered trademark of the Dow Chemical Company.

- The pesticide 1,3-dichloropropene (DCP) is a broad spectrum soil fumigant to control plant pests. Its major use is for nematode control on crops grown in sandy soils of the eastern, southern and western United States.

- The usage of DCP has increased due to cancellation of the once widely used product containing ethylene dibromide (EDB) and dibromochloropropane (DBCP) (U.S. EPA, 1986a).

- Estimated usage of DCP containing products in 1984 to 1985 ranged from about 34 to 40 million pounds (U.S. EPA, 1986a).

E. Properties (Dow Chemical U.S.A., 1977, 1982; Clayton and Clayton, 1981)

Chemical Formula	$C_3H_4Cl_2$
Molecular Weight	110.98 (pure isomers)
Physical State (25°C)	Pale yellow to yellow liquid
Boiling Point	About 104°C (104.3°C, cis; 112°C, trans)
Melting Point	--
Density (25°C)	1.21 g/mL
Vapor Pressure (25°C)	27.3 mm Hg
Specific Gravity	About 1.2 (20/20°C)
Water Solubility (25°C)	0.1 to about 0.25% (1 to 2.5 g/L) reported; miscible with most organic solvents
Log Octanol/Water Partition Coefficient	25
Taste Threshold	--
Odor Threshold	--
Conversion Factor (25°C)(air)	1 mg/L = 220 ppm; 1 ppm = 4.54 mg/m^3

F. Occurrence

- In California (Maddy et al., 1982), 54 wells were examined in areas where Telone or D-D were used for several years. The well water did not have measurable amounts of DCP (< 0.1 ppb).

- Monitoring data from New York have shown positive results for DCP in ground-water (U.S. EPA, 1986b).

- In deep well sampling in southern California (65 to 1,200 foot depths), no DCP was detected. In shallow wells (3 to 4 meters) around potato fields in Suffolk County, NY, DCP was detected up to 138 days after application (OPP, 1988).

- DCP has been found in 41 of 1,088 surface water samples analyzed and in 10 of 3,949 ground-water samples (STORET, 1988). Samples were collected in 800 surface water locations and 2,506 ground-water locations; DCP was found in 13 states. The range of concentrations found in ground-water was 0.2 µg/L to 90 µg/L. The 85th percentile of all non-zero samples was 1.3 µg/L in surface water and 3.4 µg/L in ground-water. This information is provided to give a general impression of the occurrence of this chemical in ground and surface waters as reported in the STORET database. The individual data points retrieved were used as they came from STORET and have not been confirmed as to their validity. STORET data are often not valid when individual numbers are used out of the context of the entire sampling regime, as they are here. Therefore, this information can only be used to form an impression of the intensity and location of sampling for a particular chemical.

G. Environmental Fate

- Available data indicate that DCP does leach to ground water. However, the relative hydrolytic instability of the parent compound would mitigate the potential for extensive contamination (U.S. EPA, 1986b; U.S. EPA, 1986c).

- The half-life of 1,3-DCP in soil was reported by Laskowski et al. (1982) to be approximately 10 days, while Van Dijk (1974) reported it to be 3 to 37 days depending on soil conditions and analytical methods.

- DCP hydrolyzes as a function of temperature not as a function of pH. At 10°C, the half-life is 51 days, while at 20°C it is 10 to 13 days. Chloroallyl alcohol is the main hydrolytic degradate. Some photolysis of DCP does occur (OPP, 1988).

- In laboratory aerobic soil metabolism studies, DCP degrades to chloroallyl alcohol in 20 to 30 days where soil pH is between 5.0 and 7.0, the temperature is between 15 and 20°C and the organic matter content is from 1.5 to 11.6% in sandy loam or clay soils. In anaerobic soil metabolism studies, DCP degrades to chloroallyl alcohol to less than 8% in 30 days. For anaerobic aquatic metabolism studies, the half-life was reported to be about 20 days at pHs of 6.9 to 7.5 (OPP, 1988).

- In a field dissipation study done in the Netherlands, DCP (220 to 250 lb/A) injected into the soil at 9- to 19-cm depths was found to move rapidly downward over a 2-week period. In a similar study in Delano, CA, DCP was injected to 81 cm at 1,310 L/ha to 1,638 L/ha (= lb/acre). Samples at 14 days noted the presence of DCP (up to 0.5 ppm) at all depths to 8 feet (OPP, 1988).

II.   Pharmacokinetics

A. Absorption

- Toxicity studies indicate that DCP is absorbed from skin, respiratory and gastrointestinal systems (Clayton and Clayton, 1981).

- Oral administration of DCP in rats resulted in approximately 90% absorption of the administered dose (Hutson et al., 1971).

B. Distribution

- Radiolabeled ^{14}C D-D (55% DCP) was administered orally in arachis oil in rats. After 4 days, most of the administered dose, based on measured radioactivity, was recovered, primarily in urine, and there were insignificant amounts (less than 5%) remaining in the gut, feces, skin and carcass (Hutson et al., 1971).

C. Metabolism

- cis-Dichloropropene in corn oil was given as a single oral dose (20 mg/kg bw) to two female Wistar rats. Urine and feces were collected separately. The main urinary metabolite (92%) was N-acetyl-S-[(cis)-3-chloroprop-2-enyl] cysteine. The cis-DCP has also been shown to react with glutathione in the presence of rat liver cystol to produce S[(cis)-3-chloroprop-2-enyl] glutathione. The cis-DCP is probably biotransformed to an intermediate glutathione conjugate and then follows the mercapturic acid pathway and is excreted in the urine as a cysteine derivative (Climie and Morrison, 1978).

- In a study conducted by Dietz et al. (1984), rats and mice administered (via gavage) up to 50 and 100 mg DCP/kg bw, respectively, demonstrated no evidence of metabolic saturation.

D. Excretion

- In two studies (Hutson et al., 1971; Climie and Morrison, 1978) ^{14}C cis- and/or trans-DCP, administered orally in rats, were excreted primarily in the urine in 24 to 48 hours. When pulmonary excretion was evaluated (Hutson et al., 1971), the cis and trans isomers were 3.9% and 23.6% of the administered dose, respectively. Most of the cis-DCP was excreted in the urine.

III.   Health Effects

A. Humans

1. *Short-term Exposure*

- The only known human fatality occurred a few hours after accidental ingestion of D-D mixture. The dosage was unknown. Symptoms were abdominal pain, vomiting, muscle twitching and pulmonary edema. Treatment by gastric lavage failed (Gosselin et al., 1976).

- Inhalation of DCP at high-vapor concentrations resulted in gasping, refusal to breathe, coughing, substernal pain and extreme respiratory distress at vapor concentrations over 1,500 ppm (Gosselin et al., 1976).

2. *Long-term Exposure*

- Venable et al. (1980) studied 64 male workers exposed to three carbon compounds, including DCP, to determine if fertility was adversely affected. The exposed study population was divided into $\leq$ 5 years exposure and > 5 years exposure. Sperm counts and percent normal sperm forms were the major variables evaluated. Although the study participation rate for the exposed group was only 64%, no adverse effects on fertility were observed.

B. Animals

1. *Short-term Exposure*

- DCP is moderately toxic via single-dose oral administration. A technical product containing 92% cis-/trans-DCP was administered by gavage as a 10% solution in corn oil to rats. The oral LD_{50}s in male and female rats were 713 and 740 mg/kg, respectively (Torkelson and Oyen, 1977). In another study, the oral LD_{50} in the mouse for both males and females was 640 mg/kg (Toyoshima et al., 1978).

2. *Dermal/Ocular Effects*

- The percutaneous LD_{50}s for male and female mice dosed with DCP were greater than 1,211 mg/kg (Toyoshima et al., 1978).

- The percutaneous administration of DCP in rabbits (3 g/kg) resulted in mucous nasal discharge, depressed respiration and decreased body movements. The LD_{50} by this route was 2.1 g/kg (Torkelson and Oyen, 1977).

- Primary eye irritation and primary dermal irritation studies in rabbits indicated that DCP causes severe conjunctival irritation, moderate transient corneal injury and slight skin erythema/edema. Eye irritation was reversible 8 days postinstillation. The dermal LD_{50} in rabbits was 504 mg/kg (Dow, 1978).

3. *Long-term Exposure*

- Rats, guinea pigs, rabbits and dogs were exposed to 4.5 or 13.6 mg/m^3 DCP in air for 7 hours per day and 5 days per week for 6 months. The only effect noted was slight cloudy swelling of renal tubular epithelium in male rats exposed to the high dose (Torkelson and Oyen, 1977).

- Fischer 344 rats and CD-l albino mice were exposed to Telone II® (Production Grade) by inhalation exposure, 6 hours per day for 13 weeks at concentrations of 11.98, 32.14 or 93.02 ppm. Gross pathology revealed an increased incidence of kidney discoloration in the treated male rats relative to the control group. The significance of this lesion is unknown (Coate et al., 1979).

- Solutions of Telone (78.5% DCP) in propylene glycol were administered by gavage to 10 rats/sex/dose for 6 days per week for a period of 13 weeks. The dose levels were 1, 3, 10 or 30 mg/kg/day. The control groups were given propylene glycol. The daily administration of DCP to rats by stomach intubation up to a dosage of 30 mg/kg/day did not result in any major adverse effects. No significant effects on body weight, food consumption, hematology and histopathology were noted. However, at the 10 and 30 mg/kg/day doses, the relative weight of the kidney of males was higher than controls. The authors conclude that the no-toxic-effect level for DCP was between 3 and 10 mg/kg/day. The actual No-Observed-Adverse-Effect Level (NOAEL) was 3 mg/kg/day (Til et al, 1973). This is the only study that can be used to develop a reference dose. However, because the design does not ideally address drinking water, a modifying factor will be used.

- The National Toxicology Program (NTP, 1985) evaluated the chronic toxicity and carcinogenicity of Telone II® in rats and mice. These studies utilized Telone II® fumigant containing approximately 89% cis- and trans-DCP. Groups of 52 male and female Fischer 344/N rats (doses 0, 25 or 50 mg/kg) and 50 male and female B6C3F$_1$ mice (doses 0, 50 or 100 mg/kg) were gavaged with Telone II® in corn oil, 3 days per week up to 104 weeks. Ancillary studies were conducted in which dose groups containing five male and female rats were killed after receiving Telone II® for 9, 16, 21, 24 or 27 months. Toxic effects (noncarcinogenic) included basal cell or epithelial hyperplasia of the forestomach of rats and mice at all treatment levels of DCP. Epithelial hyperplasia of the urinary bladder of mice occurred at both treatment levels in males and females. Kidney hydronephrosis also occurred in mice. The study in male mice was considered inadequate due to the deaths of vehicle control animals. Many chronic toxicity parameters (hematology/clinical chemistry) were not determined. The DCP used in the NTP study had a different stabilizer from the current Telone II®.

• Stott et al. (1987) exposed groups of male and female B6C3F$_1$ mice (70 animals/sex/exposure concentration) to vapors of Telone II® soil fumigant for 6 hours/day, 5 days/week for up to 24 months at 0, 5, 20 or 60 ppm. Urinary bladder effects, including hyperplasia of bladder epithelium, were noted in both sexes at 20 and 60 ppm. Hypertrophy and hyperplasia of the nasal respiratory mucosa were observed in most 60-ppm exposed mice of both sexes and in 20-ppm exposed females. Hyperplasia of the epithelial lining of the nonglandular portion of the stomach was observed in 60-ppm exposed males.

• Lomax et al. (1987) exposed groups of 70 male and female Fischer 344 rats to vapors of Telone II® soil fumigant for 6 hours/day, 5 days/week for up to 24 months at targeted concentrations of 0, 5, 20 or 60 ppm. The NOAEL was 20 ppm. The highest dose caused histopathological changes in nasal tissue as well as a decrease in body weight gain during the first year of this study. Males and females exposed to 60 ppm showed decreased thickness and erosions of the nasal epithelium as well as minimal submucosa fibrosis.

4. *Reproductive Effects*

• Groups of male and female Wistar rats were exposed to technical D-D at 0, 64, 145 and 443 mg/m^3 (0, 14, 32 or 94 ppm) for 5 days per week over 10 weeks. Male mating indices, fertility indices and reproductive indices were not affected by D-D exposure. No gross morphological changes were seen in sperm. Female mating, fertility and other reproductive indices were normal. Litter sizes and weights were normal and pup survival over 4 days was not influenced by exposure (Clark et al., 1980).

• Breslin et al. (1987) exposed by inhalation groups (F$_0$) of 30 male and 40 female rats for 10 weeks, 6 hours/day, 5 days/week to Telone II® at concentrations of 0, 10, 30 and 90 ppm prior to breeding. Exposure was increased to 7 days/week during breeding at weeks 11 to 13. Exposure of F$_1$ male and female parents to Telone II® began after weaning (approximately week 32 of the study) and continued for 12 weeks (5 days/week and 6 hours/day). The NOAEL for reproductive effects in the study was ≥ 90 ppm, the highest dose tested. Conception indices of females were somewhat reduced in the F_1 and F_2 generations. At 90 ppm, both males and females developed hyperplasia of respiratory epithelium and focal degeneration of olefactory tissue. Decreased body weight was observed in males and females exposed to 90 ppm.

5. *Developmental Effects*

- Hanley et al. (1987) investigated the effects of inhalation exposure to DCP on fetal development in rats. Pregnant Fischer 344 rats were exposed to 0, 20, 60 and 120 ppm DCP for 6 hours/day during gestation days 6 to 15. Maternal body weight gain was depressed in all of the DCP-exposed rats in a dose-related manner. Therefore, the Lowest-Observed-Adverse-Effect Level (LOAEL) for this effect was 20 ppm DCP. There was also significant depression of feed consumption in all exposed rats, along with decreases in water consumption in rats exposed to 120 ppm DCP. At 120 ppm there were significant increases in relative kidney weights and decreases in absolute liver weights in all exposed rats. There was a statistical increase in the incidence of delayed ossification of the vertebral centra of rats exposed to 120 ppm DCP. This effect is of little toxicological significance due to maternal toxicity observed at 120 ppm DCP.

- Hanley et al. (1987) also studied the effects of inhalation exposure to DCP on fetal development in rabbits. Pregnant New Zealand White rabbits were exposed to 0, 20, 60 or 120 ppm DCP for 6 hours/day during gestation days 6 through 18. In rabbits, evaluation of maternal weight gain over the entire exposure period indicated significant exposure-related decreases in both the 60- and 120-ppm groups. Therefore, the NOAEL was 20 ppm DCP. Statistically significant decreases in the incidence of delayed ossification of the hyoid and presence of cervical spurs among the exposed group were considered within normal variability in rabbits.

6. *Mutagenicity*

- Tests of commercial formulations containing DCP (DeLorenzo et al., 1975; Flessel, 1977; Neudecker et al., 1977; Brooks et al., 1978; Sudo et al., 1978; Stolzenberg and Hine, 1980), a mixture of pure cis-DCP and trans-DCP (DeLorenzo et al., 1975), and pure cis-DCP (Brooks et al., 1978) were positive in the *Salmonella typhimurium* strains TA1535 and TA100 with and without metabolic activation. These results indicate that DCP acts by base-pair substitution and is a direct acting mutagen.

- DCP may be a mutagen that acts via frame shift mutation since studies by DeLorenzo et al. (1975) reported positive results in TA1978 (with and without metabolic activation) for a commercial mixture of DCP and a mixture of pure cis- and trans-DCP.

- A commercial mixture of DCP and pure cis-DCP were also positive with and without metabolic activation in *Salmonella typhimurium* strain TA98 (Flessel, 1977; Sudo et al., 1978; Brooks et al., 1978).

- Sudo et al. (1978) tested DCP in a reverse mutation assay with *E. coli* B/r Wp2 with negative results.

- DCP was negative for reverse mutation in the mouse host-mediated test with *S. typhimurium* G46 in studies by Shirasu et al. (1976) and Sudo et al. (1978).

7. *Carcinogenicity*

- Fischer 344 rats of each sex were gavaged with Telone II® in corn oil at doses of 0, 25 and 50 mg/kg/day for 3 days per week. A total of 77 rats/sex were used for each dose group (52 animals/sex/group were dosed for 104 weeks in the main oncogenicity study; and in an ancillary study, 5 animals/sex/group were sacrificed after 9, 16, 21, 24 and 27 months' exposure to DCP). No increased mortality occurred in treated animals. Neoplastic lesions associated with Telone II® included squamous cell papillomas of the forestomach (male rats: 1/52; 1/52; 9/52; female rats: 0/52; 2/52; 3/52), squamous cell carcinomas of the forestomach (male rats: 0/52; 0/52; 4/52) and neoplastic nodules of the liver (male rats: 1/52; 6/52; 7/52). The increased incidence of forestomach tumors was accompanied by a positive trend of forestomach basal cell hyperplasia in male and female rats of both treated groups (25 and 50 mg/kg/day). The highest–dose level tested in rats (50 mg/kg/day) approximated a maximum tolerated dose level (NTP, 1985).

- B6C3F$_1$ mice of each sex were gavaged with Telone II® in corn oil at doses of 0, 50 and 100 mg/kg/day for 104 weeks. A total of 50 mice/sex were used for each dose group. Due to excessive mortality in control male mice from myocardial inflammation approximately 1 year after the initiation of the study, conclusions pertaining to oncogenicity were based on concurrent control data and NTP historical control data. Neoplastic lesions associated with the administration of Telone II® included squamous cell papillomas of the forestomach (female mice: 0/50; 1/50; 2/50), squamous cell carcinomas of the forestomach (female mice: 0/50; 0/50; 2/50), transitional cell carcinomas of the urinary bladder (female mice: 0/50; 8/50; 21/48) and alveolar/bronchiolar adenomas (female mice: 0/50; 3/50; 8/50). The increased incidence of forestomach tumors was accompanied by an increased incidence of stomach epithelial cell hyperplasia in males and females at the highest dose level tested (100 mg/kg/day); and the increased incidence of urinary bladder transitional cell carcinoma was accompanied by a positive trend of bladder hyperplasia in male and female mice of both treated groups (50 and 100 mg/kg/day). Incidences of neoplasms were not significantly increased in male mice (NTP, 1985).

- Thirty female Ha:ICR Swiss mice received weekly subcutaneous injections of cis-DCP. The dose was 3 mg DCP/mouse in 0.05 mL trioctanoin delivered to the left flank. After 77 weeks, there was an increased incidence of fibrosarcomas at the site of injection. Six of the 30 exposed mice developed the tumors. There were no similar lesions in the controls (Van Duuren, 1979).

- Stott et al. (1987) exposed groups of male and female B6C3F$_1$ mice (70 animals/sex/dose) to vapors of Telone II® for 6 hours/day, 5 days/week for up to 24 months at 0, 5, 20 or 60 ppm. The only tumorigenic effect was an increased incidence in benign lung tumors (bronchiolalveolar adenomas) in the 60-ppm exposed males. There were no tumorigenic effects in the lower-dose males or at any of the doses in females.

- Lomax et al. (1987) exposed Fisher 344 rats (70 rats/sex/dose) to vapors of Telone II® (0, 5, 20 and 60 ppm). The 2-year exposure by inhalation did not result in increases in tumor incidence.

IV. Quantification of Toxicological Effects

A. One-day Health Advisory

There are not sufficient data to derive a One-day Health Advisory value for DCP. It is recommended that the Longer-term HA value for a 10-kg child (30 µg/L, calculated below) be used at this time as a conservative estimate of the One-day HA value.

B. Ten-day Health Advisory

There are not sufficient data to derive a Ten-day HA value for DCP. It is recommended that the Longer-term HA value for a 10-kg child (30 µg/L, calculated below) be used as a conservative estimate of the Ten-day HA value.

C. Longer-term Health Advisory

The Til et al. (1973) 13-week subchronic gavage study in rats has been selected to serve as the basis for calculating the Longer-term HA for DCP. This study resulted in a LOAEL of 10.0 mg/kg/day based on increased relative kidney weight in males. No adverse effects were noted at the next lowest dose (3.0 mg/kg/day). Therefore, the NOAEL is 3.0 mg/kg/day.

Based on the NOAEL of 3.0 mg/kg/day determined in this study, the Longer-term HAs are calculated as follows:

For a 10-kg child:

$$\text{Longer-term HA} = \frac{(3.0 \text{ mg/kg/day}) (10 \text{ kg})}{(100) (10) (1 \text{ L/day})} = 0.03 \text{ mg/L} (30 \text{ } \mu g/L)$$

where: 10 = modifying factor, selected since this was the only useful gavage study available and classified as supplementary data. Also there were considerable toxicological data gaps.

For a 70-kg adult:

$$\text{Longer-term HA} = \frac{(3.0 \text{ mg/kg/day}) (70 \text{ kg})}{(100) (10) (2 \text{ L/day})} = 0.105 \text{ mg/L} (100 \text{ } \mu g/L)$$

where: 10 = modifying factor, selected since this was the only useful gavage study available and classified as supplementary data. Also there were considerable toxicological data gaps.

D. Lifetime Health Advisory

The Lifetime HA for a 70-kg adult has been determined on the basis of the study in rats by Til et al. (1973), as described above.

Using the NOAEL of 3.0 mg/kg/day, as determined in that study, the DWEL is calculated as follows:

Step 1: Determination of the Reference Dose (RfD)

$$\text{RfD} = \frac{(3.0 \text{ mg/kg/day})}{(1,000) (10)} = 0.0003 \text{ mg/kg/day}$$

where: 10 = modifying factor selected since this was the only useful gavage study available and classified as supplementary data. Also there were considerable toxicological data gaps.

Step 2: Determination of the Drinking Water Equivalent Level (DWEL)

$$\text{DWEL} = \frac{(0.0003 \text{ mg/kg/day}) (70 \text{ kg})}{(2 \text{ L/day})} = .011 \text{ mg/L} (10 \text{ } \mu g/L)$$

Step 3: Determination of the Lifetime Health Advisory

Lifetime HAs are not recommended for Group A or B carcinogens. DCP is a Group B2, probable human carcinogen. The estimated cancer risk associated with lifetime exposure to drinking water containing DCP at 10 μg/L is approximately 5.0 x 10^{-5}. This estimate represents the upper 95% confidence limit using the linearized multistage model. The actual risk is unlikely to exceed this value.

E. Evaluation of Carcinogenic Potential

- DCP may be classified as a B2, probable human carcinogen, based on sufficient evidence of tumor production in two rodent species and by two routes of administration.

- Data on an increased incidence of squamous cell papilloma or carcinoma of the forestomach in rats exposed to DCP (NTP, 1985) were used for a quantitative assessment of cancer risk due to DCP. Based on the data from this study and using the linearized multistage model, a carcinogenic potency factor (q_1*) for humans of 1.75×10^{-1} (mg/kg/day)$^{-1}$ was calculated.

- The drinking water concentrations corresponding to increased lifetime cancer risks of 10^{-4}, 10^{-5} and 10^{-6} (one excess cancer per one million population) for a 70-kg adult consuming 2 L/day are 20 μg/L, 2 μg/L and 0.2 μg/L, respectively.

- The forestomach tumor data in male rats used to calculate the q_1* value (NTP, 1985) consisted of the 2-year study data excluding the ancillary studies data. The ancillary studies involved serial sacrifice of animals (at 9, 16, 21, 24 and 27 months). It is not appropriate to include these data in the lifetime predictive model used (multistage).

- For comparison purposes, drinking water concentrations associated with an excess risk of 10^{-6} were 0.2 μg/L, 3.6 mg/L, 0.03 μg/L and 0.004 μg/L for the one-hit, Weibull, probit and logit models, respectively.

V.   Other Criteria, Guidance and Standards

- The ACGIH recommended 1 ppm (5 mg/m^3) as a Threshold Limit Value for DCP (Clayton and Clayton, 1981).

VI. Analytical Methods

- No specific methods have been published by U.S. EPA for analysis of DCP in water. However, EPA Method 524.2 (U.S. EPA, 1986d) and EPA Method 502.2 (U.S. EPA, 1986e), both for volatile organic compounds in water, should be suitable for analysis of DCP. Both are standard purge and trap capillary column gas chromatographic techniques. While an estimated detection limit has not been calculated for the two isomers of 1,3-dichloropropene, work done with 1,1-dichloropropene would indicate a range for 1,3-DCP of 0.02 to 0.05 μg/L.

VII. Treatment Technologies

- There are no specific publications on treatment of 1,3-DCP. However, adequate treatment by granular activated carbon (GAC) should be possible. Freundlich carbon absorption isotherms for DCP indicate reasonably high adsorption capacity (U.S. EPA, 1980).

VIII. References

Breslin, W.J., H.D. Kirk, C.M. Streeter, J.F. Quast and J.R. Szabo.* 1987. Telone II® soil fumigant: two-generation inhalation reproduction study in Fischer 344 rats. Prepared by Health and Environmental Sciences USA. Submitted by Dow Chemical U.S.A., Midland, MI. MRID 403124.

Brooks, T.M., B.J. Dean and A.S. Wright.* 1978. Toxicity studies with dichloropropenes: mutation studies with 1,3-D and cis-1,3-dichloropropene and the influence of glutathione on the mutagenicity of cis-1,3-dichloropropene in *Salmonella typhimurium*. Group research report (Shell Research, Ltd.) TLGR.0081 78. Unpublished study by Shell Chemical Co., Washington, DC. MRID 61059.

Clark, D., D. Blair and S. Cassidy.* 1980. A 10-week inhalation study of mating behavior, fertility and toxicity in male and female rats. Group research report (Shell Research, Ltd.) TLGR.80.023. Unpublished study by Dow Chemical U.S.A., Midland, MI. MRIDs 117055, 103280, 39691.

Clayton, G.D. and F.E. Clayton. 1981. Patty's industrial hygiene and toxicology, 3rd ed., Vol. 2B, New York, NY: John Wiley and Sons, Inc. pp. 3573–3577.

Climie, I.J.G., and B.J. Morrison.* 1978. Metabolism studies on (Z)1,3-dichloropropene in the rat. Group research report (Shell Research, Ltd.) TLGR.0101. 78. Unpublished study by Dow Chemical U.S.A., Midland, MI. MRID 32984.

Coate, W.B., D.L. Keenan, R.J. Hardy and R.W. Voelker.* 1979. 90-day inhalation-toxicity study in rats and mice: Telone II®. Project No. 174-127. Final report. Unpublished study by Hazleton Laboratories America, Inc., for Dow Chemical U.S.A., Midland, MI. MRID 119191.

DeLorenzo, F., S. Degl Innocenti and A. Ruocco.* 1975. Mutagenicity of pesticides containing 1,3-dichloropropene. University of Naples, Italy. Submitted by Dow Chemical U.S.A., Midland, MI. MRID 119179.

Dietz, F.K., E.A. Hermann and J.C. Ramsey. 1984. The pharmacokinetics of ^{14}C-1,3-dichloropropene in rats and mice following oral administration. Toxicologist. 4:585. Abstract.

Dow Chemical U.S.A.* 1977. Telone II® soil fumigant: product chemistry. MRID 00119178.

Dow Chemical U.S.A.* 1978. Summary of human safety data: summary of studies 099515-I and 099515-J. Unpublished study by Dow Chemical U.S.A., Midland, MI. MRID 39676.

Dow Chemical U.S.A. 1982. A data sheet giving the chemical and physical properties of the chemical. A complete statement of the names and percentages by weight of each active inert ingredient in the formulation to be shipped. Midland, MI: Dow Chemical U.S.A. MRID 115213.

Flessel, P.* 1977. Mutagen testing program, mutagenic activity of Telone II® in the Ames Salmonella assay. Letter dated Apr. 8. Prepared by Calif. Dept. Health, submitted by Dow Chemical U.S.A., Midland, MI. MRIDs 120906, 67534.

Gosselin, R.E., H.C. Hodge, R.P. Smith and M.N. Gleason. 1976. Clinical toxicology of commercial products, 4th ed. Baltimore, MD: The Williams and Wilkins Co. p. 120.

Hanley, T.R., J.A. John-Greene, J.T. Young, L.L. Calhoun and K.S. Rao. 1987. Evaluation of the effects of inhalation exposure to 1,3-dichloropropene on fetal development in rats and rabbits. Fundam. Appl. Toxicol. 8:562-570.

Hutson, D.H., J.A. Moss and B.A. Pickering.* 1971. The excretion and retention of components of the soil fumigant D-D and their metabolites in the rat. Food Cosmet. Toxicol. 9:677-680. Midland, MI: Dow Chemical U.S.A. MRID 39690.

Laskowski, D., C. Goring, P. McCall and R. Swan. 1982. Terrestrial environment. Environ. Risk Anal. Chem. 25:198-240.

Lomax, L.G., L.L. Calhoun, W.T. Stott and L.E. Frauson.* 1987. Telone II® soil fumigant: 2-year inhalation chronic toxicity-oncogenicity study in rats. Prepared by Health and Environmental Sciences, USA. Submitted by Dow Chemical Company, Midland, MI. MRID 403122.

Maddy, K., H. Fong, J. Lowe, D. Conrad and A. Fredrickson. 1982. A study of well water in selected California communities for residues of 1,3 dichloropropene, chloroallyl alcohol, and 49 organophosphate or chlorinated hydrocarbon pesticides. Bull. Environ. Contam. Toxicol. 29:354–359.

Neudecker, T., A. Stefani and D. Heschler. 1977. In vivo mutagenicity of soil nematocide 1,3-dichloropropene. Experientia. 33:1084–1085.

NTP. 1985. National Toxicology Program. NTP technical report on the toxicology and carcinogenesis studies of Telone II® in F344/N rats and B6C3F₁ mice (gavage studies). NTP TR 269, NIH Pub. No. 85-2525. May 1985.

OPP. 1988. Office of Pesticide Programs. Washington, DC: Exposure Assessment Branch, U.S. EPA.

Shirasu, Y., M. Moriga and K. Kato.* 1976. Mutagenicity testing on D-D in microbial systems. Prepared by Institute of Environmental Toxicology. Submitted by Shell Chemical Co., Washington, DC. MRID 61050.

Stolzenberg, S. and C. Hine. 1980. Mutagenicity of 2- and 3-carbon halogenated compounds in Salmonella/mammalian microsome test. Environmental Mutagenesis. 2:59–66.

STORET. 1988. STORET Water Quality File. Office of Water. U.S. Environmental Protection Agency (data file search conducted in May, 1988).

Stott, W.T., K.A. Johnson, L.L. Calhoun, S.K. Weiss and L.E. Frauson.* 1987. Telone II® soil fumigant: 2-year inhalation chronic toxicity-oncogenicity study in mice. Prepared by Health and Environmental Sciences, USA. Submitted by Dow Chemical Company, Midland, MI. MRID 403123.

Sudo, S., M. Nakazawa and M. Nakazono.* 1978. The mutagenicity test on 1,3-dichloropropene in bacteria test systems. Prepared by Nomura Sogo Research Institute. Submitted by Dow Chemical U.S.A., Midland, MI. MRID 39688.

Til, H.P., M.T. Spanjers, V.J. Feron and P.J. Reuzel.* 1973. Sub-chronic (90-day) toxicity study with Telone in albino rats. Report No. R4002. Final report. Unpublished study (Central Institute for Nutrition and Food Research) submitted by Dow Chemical U.S.A., Midland, MI. MRIDs 39684, 67977.

Torkelson, T.R. and F. Oyen.* 1977. The toxicity of 1,3-dichloropropene as determined by repeated exposure of laboratory animals. American Industrial Hygiene Association Journal. 38:217–223. Midland, MI: Dow Chemical U.S.A. MRID 39686.

Toyoshima, S., R. Sato and S. Sato. 1978. The acute toxicity test on Telone II® in mice. Unpublished study by Dow Chemical U.S.A., Midland, MI. MRID 39683.

U.S. EPA. 1980. U.S. Environmental Protection Agency. Carbon adsorption isotherms for toxic organics. EPA 60018-80-023. April.

U.S. EPA. 1986a. U.S. Environmental Protection Agency. 1,3-Dichloropropene, a digest of biological and economic benefits and regulatory implications. Benefits and Use Division, Office of Pesticide Programs.

U.S. EPA. 1986b. U.S. Environmental Protection Agency. 1,3-Dichloropropene: initiation of special review; availability of registration standard notice. Fed. Reg. 51(195):36161. October 8, 1986.

U.S. EPA. 1986c. U.S. Environmental Protection Agency. Guidance for the reregistration of pesticide products containing 1,3-dichloropropene as the active ingredient. Washington, DC: Office of Pesticides and Toxic Substances. September. 111 pp.

U.S. EPA. 1986d. U.S. Environmental Protection Agency. Volatile organic compounds in water by purge and trap capillary gas chromatography/mass spectrometry. Washington, DC: Office of Drinking Water. Aug.

U.S. EPA. 1986e. U.S. Environmental Protection Agency. Volatile organic compounds in water by purge and trap capillary column gas chromatography with photoionization and electrolytic conductivity detectors in series. Washington, DC: Office of Drinking Water.

Van Dijk, H. 1974. Degradation of 1,3-dichloropropenes in soil. Agro-Ecosystems. 1:193–204.

Van Duuren, B.L., B.M. Goldschmidt and G. Loewengart.* 1979. Carcinogenicity of halogenated olefinic and aliphatic hydrocarbons in mice. Journal of the National Cancer Institute. 63(6):1433–1439. MRID 94723.

Venable, J.R., C.D McClimans, R.E. Flake and D.B. Dimick.* 1980. A fertility study of male employees engaged in the manufacture of glycerine. Journal of Occupational Medicine. 22(2):87–91. Midland, MI: Dow Chemical U.S.A. MRID 117052.

---

* Confidential Business Information submitted to the Office of Pesticide Programs.

Dieldrin

I. General Information and Properties

A. CAS No. 60–57–1

B. Structural Formula

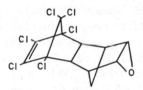

3,4,5,6,9,9-hexachloro-1a,2,2a,3,6,6a,7,7a-octahydro-
2,7:3,6-dimethanonaphth[2,3-b]oxirene (Windholz, 1983)

C. Synonyms

- HEOD; Alvit; Quintox; Octalox (IPCS, 1987).

D. Uses

- Dieldrin was formerly used for control of soil insects, public health insects, termites and many other pests. These uses have been cancelled and manufacture discontinued in the United States (Meister, 1983).

E. Properties (NAS, 1977; Weast and Astle, 1982; Windholz, 1983)

Chemical Formula	$C_{12}H_8Cl_6O$
Molecular Weight	380.93
Physical State	Crystals
Boiling Point	--
Melting Point	175 to 176°C
Density	--
Vapor pressure (20°C)	3.1×10^{-6} mm Hg
Water Solubility (25°C)	0.25 mg/L
Log Octanol/Water Partition Coefficient	--
Taste Threshold	--

Odor Threshold (water)   0.04 mg/L
Conversion Factor   --

F. Occurrence

- Dieldrin has been found in 7,320 of 50,473 surface water samples analyzed and in 223 of 5,443 ground-water samples (STORET, 1988). Samples were collected at 9,021 surface water locations and 4,131 ground-water locations, and Dieldrin was found in 48 states, Canada and Puerto Rico. The 85th percentile of all nonzero samples was 0.01 μg/L in surface water and 0.10 μg/L in ground-water sources. The maximum concentration found was 301 μg/L in surface water and 10.08 μg/L in ground water. This information is provided to give a general impression of the occurrence of this chemical in ground and surface waters as reported in the STORET database. The individual data points retrieved were used as they came from STORET and have not been confirmed as to their validity. STORET data are often not valid when individual numbers are used out of the context of the entire sampling regime, as they are here. Therefore, this information can only be used to form an impression of the intensity and location of sampling for a particular chemical.

G. Environmental Fate

- Dieldrin is stable and highly persistent in the environment.

- Dieldrin has the longest half-life of the chlorinated hydrocarbons in water 1-meter deep (half-life = 723 days) (MacKay and Wolkoff, 1973).

II.   Pharmacokinetics

A. Absorption

- A single oral dose of dieldrin at 10 mg/kg body weight (bw) administered in corn oil to male Sprague-Dawley rats produced consistent concentrations of dieldrin in plasma, muscle, brain, kidney and liver for periods up to 48 hours (Hayes, 1974); hence, it was absorbed.

B. Distribution

- Rats given a single oral dose of dieldrin at 10 mg/kg showed concentrations of dieldrin in fat, muscle, liver, blood, brain and kidney. The highest concentration of dieldrin was in fat. The lowest concentration was in the kidney (Hayes, 1974).

C. Metabolism

- Both the CFE rat and CF_1 mouse, following a single oral dose of dieldrin (not less than 85% HEOD) at 3 and 10 mg/kg in olive oil, respectively, metabolized dieldrin to 9-hydroxydieldrin, 6,7-trans-dihydroaldrindiol and some unidentified metabolites. The rat, but not the mouse, also metabolized dieldrin to pentachloroketone (Baldwin and Robinson, 1972).

D. Excretion

- Female rats infused with total doses of 8 to 16 mg ^{36}Cl-dieldrin/kg bw excreted approximately 70% of the infused dose in the feces over a period of 42 days, while only about 10% of the dose was recovered in the urine. Excretion was markedly increased by restriction of the diet indicating that the concentration of dieldrin in the blood increased as fat was mobilized (Heath and Vandekar, 1964).

III.   Health Effects

A. Humans

- Dieldrin has been reported to cause hypersensitivity and muscular fasciculations that may be followed by convulsive seizures and respective changes in the EEG pattern. Acute symptoms of intoxication include hyperirritability, convulsions and/or coma sometimes accompanied by nausea, vomiting and headache, while chronic intoxication may result in fainting, muscle spasms, tremors and loss of weight. The lethal dose for humans is estimated to be about 5 g (ACGIH, 1984).

B. Animals

1. *Short-term Exposure*

- RTECS (1985) reported the acute oral LD_{50} values of dieldrin in the rat, mouse, dog, monkey, rabbit, pig, guinea pig and hamster as 38.3, 38, 65, 3, 45, 38, 49 and 60 mg/kg, respectively.

2. *Dermal/Ocular Effects*

- Aldrin or Dieldrin (dry powder) applied to rabbit skin for 2 hours/day, 5 days/week for 10 weeks had no discernible effects (IPCS, 1987).

3. *Long-term Exposure*

- Groups of Osborne-Mendel rats, 12/sex/level, were fed 0, 0.5, 2, 10, 50, 100 or 150 ppm dieldrin (recrystallized, 100% active ingredient) in their diet for 2 years. These doses correspond to approximately 0, 0.025, 0.1, 0.5, 2.5, 5.0 or 7.5 mg/kg/day, respectively (Lehman, 1959). Survival was markedly decreased at levels of 50 ppm and above. Liver-to-body weight ratios were significantly increased at all treatment levels, with females showing the effect at 0.5 ppm and males at 10 ppm and greater. Microscopic lesions were described as being characteristic of chlorinated hydrocarbon exposure. These changes were minimal at the 0.5 ppm level. Male rats, at the two highest dose levels (100 and 150 ppm), developed hemorrhagic and/or distended urinary bladders usually associated with considerable nephritis (Fitzhugh et al., 1964). A Lowest-Observed-Adverse-Effect Level (LOAEL) of 0.025 mg/kg/day, the lowest dose tested, was identified in this study.

- Mongrel dogs, 1/sex/dose level (2/sex at 0.5 mg/kg/day), fed dieldrin (recrystallized, 100% active ingredient) at 0.2 to 10 mg/kg/day, 6 days/week for up to 25 months, showed toxic effects including weight loss and convulsions at dosages of 0.5 mg/kg/day or more. Survival was inversely proportional to dose level. No toxic effects, gross or microscopic, were seen at a dose level of 0.2 mg/kg/day (Fitzhugh et al., 1964).

- Groups of Carworth Farm "E" strain rats, 25/sex/dose level, were fed dieldrin (> 99% purity) in the diet at 0.0, 0.1, 1.0 or 10.0 ppm for 2 years. These doses correspond to approximately 0, 0.005, 0.05 or 0.5 mg/kg/day, respectively (Lehman, 1959). At 7 months, the 1-ppm intake level was equivalent to approximately 0.05 and 0.06 mg/kg/day for males and females, respectively. No effects on mortality, body weight, food intake, hematology and blood or urine chemistries were seen. At the 10-ppm level, all animals became irritable after 8 to 13 weeks of treatment and developed tremors and occasional convulsions. Liver weight and liver-to-body weight ratios were significantly increased in females receiving both 1.0 and 10 ppm. Pathological findings described as organochlorine-insecticide changes of the liver were found in one male and six females at the 10-ppm level. No evidence of tumorigenesis was found (Walker et al., 1969).

- Groups of beagle dogs (5/sex/dose) were treated daily by capsule with dieldrin (> 99% purity) at 0.0, 0.005 or 0.05 mg/kg in olive oil for 2 years. No treatment-related effects were seen in general health, behavior, body weight or urine chemistry. A significant increase in plasma alkaline phosphatase activity in both sexes and a significant decrease in serum protein concentration in males receiving the high dose were not associated with any clinical or pathological change. Liver weight and liver-to-body weight ratios were significantly increased in females receiving the high dose, 0.05 mg/kg/day, but no gross or microscopic lesions were found. There was no evidence of tumorigenic activity (Walker et al., 1969).

- Dieldrin (> 99% pure) was administered to CF_1 mice of both sexes (30/ sex/dose) in the diet for 128 weeks. Dosages were 1.25, 2.5, 5, 10 or 20 ppm dieldrin. These doses are equivalent to 0.19, 0.38, 0.75, 1.5 or 3 mg/kg body weight (Lehman, 1959). At the 20-ppm dose level, approximately 25% of the males and nearly 50% of the females died during the first 3 months of the experiment. Palpable intra-abdominal masses were detected after 40, 75 or 100 weeks in the 10, 5 and 2.5-ppm-treated groups, respectively. At 1.25 ppm, liver enlargement was not palpable and morbidity was similar to that of controls. A No-Observed-Adverse-Effect Level (NOAEL) cannot be established because clinical chemistry parameters were not determined (Walker et al., 1972).

4. Reproductive Effects

- Coulston et al. (1980) studied the reproductive effects of dieldrin in Long-Evans rats. Pregnant rats (18–20/dose) were administered 0 or 4 mg/kg bw dieldrin by gavage daily from day 15 of gestation through 21 days postpartum. The treated group did not differ from the control group when examined for fecundity, number of stillbirths, perinatal mortality and total litter weights.

5. Developmental Effects

- Pregnant Syrian golden hamsters given 30 mg/kg bw dieldrin (≥ 99% pure) in corn oil on days 7, 8 or 9 of gestation manifested an embryocidal and teratogenic response as evidenced by a statistically significant increase in fetal deaths, a decrease in live fetal weight and an increased incidence of webbed foot, cleft palate and open eye (Ottolenghi et al., 1974). Similar anomalies were observed in CD_1 mice administered 15 mg/kg bw dieldrin on day 9 of gestation, but no effect was seen on fetal survival or weight.

- Dieldrin (87% pure) was not found to be teratogenic in the CD rats and CD₁ mice administered doses of 1.5, 3.0 or 6.0 mg/kg/day by gastric intubation on days 7 through 16 of gestation. Fetal toxicity, as indicated by a significant decrease in numbers of caudal ossification centers at the 6.0-mg/kg/day dose level and a significant increase in the number of supernumerary ribs in one study group at both the 3.0- and 6.0-mg/kg/day dose level, was reported in the experiments in mice. Maternal toxicity in the high-dose rats was indicated by a 41% mortality and a significant decrease in weight gain; similarly, mice receiving 6.0 mg/kg/day showed a significant decrease in maternal weight gain. A significant increase in liver-to-body weight ratio in one group of maternal mice was reported at both 3.0 and 6.0 mg/kg/day (Chernoff et al., 1975).

6. *Mutagenicity*

- Dieldrin was not mutagenic in the *Salmonella*/microsome test with and without S-9 mix (McCann et al., 1975).

- Dieldrin significantly decreased the mitotic index and increased chromosome abnormalities in STS mice bone marrow cells in an *in vivo* study. Similar observations were made in human WI-38 embryonic lung cells in an *in vitro* test that also gave evidence of cytotoxicity, as indicated by degree of cell degeneration (Majumdar et al., 1976).

7. *Carcinogenicity*

- A dose-related increase in the incidence of hepatocellular carcinomas was observed in B6C3F₁ mice, with the incidence in the high-dose males being significantly ($p = 0.025$) higher when compared to pooled controls (NCI, 1978). Mice were given dieldrin (technical grade, > 85% purity) in the diet at concentrations of 2.5 or 5 ppm for 80 weeks. These doses correspond to approximately 0.375 or 0.75 mg/kg/day, respectively (Lehman, 1959).

- Osborne-Mendel rats treated with dieldrin at Time-weighted Average (TWA) doses of 29 or 65 ppm in the diet (approximately 1.45 or 3.25 mg/kg/day, respectively, based on Lehman, 1959) for 80 weeks, did not elicit treatment-related tumors (NCI, 1978).

- Diets containing 0.1, 1.0 or 10 ppm dieldrin (> 99% purity), when given to mice of both sexes for 132 weeks, were associated with an increased incidence of liver tumors at all dose levels tested (Walker et al., 1972). These doses are equivalent to approximately 0.015, 0.15 or 1.5 mg/kg/day, respectively (Lehman, 1959).

IV. Quantification of Toxicological Effects

A. One-day Health Advisory

No data were found in the available literature that were suitable for determination of a One-day HA value for dieldrin. It is therefore recommended that the modified DWEL for a 10-kg child (0.0005 mg/L, calculated below) be used as a conservative estimate for the One-day HA value.

For a 10-kg child, the adjusted DWEL is calculated as follows:

$$DWEL = \frac{(0.00005 \text{ mg/kg/day}) (10 \text{ kg})}{1 \text{L/day}} = 0.0005 \text{ mg/L}$$

where: 0.00005 mg/kg/day = RfD (see Lifetime Health Advisory section).

B. Ten-day Health Advisory

No data were found in the available literature that were suitable for determination of a Ten-day HA value for dieldrin. It is therefore recommended that the modified DWEL for a 10-kg child (0.0005 mg/L) be used as a conservative estimate for the Ten-day HA value.

C. Longer-term Health Advisory

No data were found in the available literature that were suitable for determination of a Longer-term HA value for dieldrin. It is therefore recommended that the modified DWEL for a 10-kg child (0.0005 mg/L) be used as a conservative estimate for the Longer-term HA value.

D. Lifetime Health Advisory

The study of Walker et al. (1969), in which rats were fed dieldrin in the diet at 0.0, 0.1, 1.0 or 10 ppm for 2 years (approximately 0, 0.005, 0.05 or 0.5 mg/kg/day based on Lehman, 1959), has been selected as the basis for calculating the DWEL. In this study, liver weight and liver-to-body weight ratios were significantly increased in females receiving 1 and 10 ppm, while pathological changes consistent with exposure to organochlorides were evident at the 10-ppm level. This study established a NOAEL of 0.1 ppm (equivalent to 0.005 mg/kg/day).

Using a NOAEL of 0.005 mg/kg/day, the Lifetime HA is calculated as follows:

Step 1: Determination of the Reference Dose (RfD)

$$RfD = \frac{0.005 \text{ mg/kg/day}}{100} = 0.00005 \text{ mg/kg/day}$$

Step 2: Determination of the Drinking Water Equivalent (DWEL)

$$DWEL = \frac{(0.00005 \text{ mg/kg/day})(70 \text{ kg})}{2 \text{ L/day}} = 0.00175 \text{ mg/L } (2 \text{ } \mu g/L)$$

Step 3: Determination of the Lifetime Health Advisory

Dieldrin may be classified in Group B2: probable human carcinogen. A Lifetime HA is not recommended for dieldrin.

The estimated excess cancer risk associated with lifetime exposure to drinking water containing dieldrin at 1.75 $\mu g/L$ is approximately 8.05×10^{-4}. This estimate represents the upper 95% confidence limit from extrapolations prepared by EPA's Carcinogen Assessment Group (U.S. EPA, 1987) using the linearized multistage model. The actual risk is unlikely to exceed this value.

E. Evaluation of Carcinogenic Potential

- Applying the criteria described in EPA's final guidelines for assessment of carcinogenic risk (U.S. EPA, 1986), dieldrin may be classified in Group B2: probable human carcinogen.

- Evidence has been presented in several carcinogenicity studies showing that dieldrin is carcinogenic to mice. Thirteen data sets from these studies are adequate for quantitative risk estimation. Utilizing the linearized multistage model, the U.S. EPA performed potency estimates for each of these data sets. The geometric mean of the potency estimates, $q_1^* = 16$ $(mg/kg/day)^{-1}$, was estimated as the potency for the general population (U.S. EPA, 1987).

- Using this q_1^* value and assuming that a 70-kg human adult consumes 2 liters of water a day over a 70-year lifespan, the linearized multistage model estimates that concentrations of 0.219, 0.0219 and 0.00219 μg dieldrin per liter may result in excess cancer risk of 10^{-4}, 10^{-5} and 10^{-6}, respectively.

- The linearized multistage model is only one method of estimating carcinogenic risk. From the data contained in U.S. EPA (1987), it was determined that five of the thirteen data sets were suitable for determining slope estimates for the probit, logit, Weibull and gamma multihit models. Using the geometric mean of these slope estimates (13 for multistage, 5 for other models) at their upper 95% confidence limits, the following comparisons of unit risk (i.e., a 70-kg man consuming 2 liters of water per day containing 1 μg/L of dieldrin over a lifetime) can be made: multistage, 4.78 x 10^{-4}; probit, 7.7×10^{-12}; logit, 5.09×10^{-6}; Weibull, 1.13×10^{-4}; multihit, 5.68×10^{-4}. Each model is based on different assumptions. No current understanding of the biological mechanisms of carcinogenesis is able to predict which of these models is more accurate than another.

- IARC (1982) concluded that there is limited evidence that dieldrin is carcinogenic in laboratory animals.

V. Other Criteria, Guidance and Standards

- ACGIH (1984) has established a short-term exposure limit (STEL) of 0.75 mg/m^3 and an 8-hour Threshold Limit Value (TLV)-TWA exposure of 0.25 mg/m^3 for dieldrin.

- U.S. EPA (1980) has recommended ambient water quality criteria of 0.71 ng/L for dieldrin. It is based on a carcinogenic potency factor (q_1*) of 30.37 (mg/kg/day)$^{-1}$ derived from the incidence of hepatocellular carcinoma in a mouse feeding study conducted by Walker et al. (1972).

- Residue tolerances ranging from 0.02 to 0.1 ppm have been established for dieldrin in or on agricultural commodities (U.S. EPA, 1985).

- WHO (1982) established guidance of 0.03 μg dieldrin/L in drinking water.

VI. Analytical Methods

- Determination of dieldrin is by Method 508, Determination of Chlorinated Pesticides in Water by Gas Chromatography with an Electron Capture Detector (U.S. EPA, 1988). In this procedure, a measured volume of sample of approximately 1 liter is solvent extracted with methylene chloride by mechanical shaking in a separatory funnel or mechanical tumbling in a bottle. The methylene chloride extract is isolated, dried and concentrated to a volume of 5 mL after solvent substitution with methyl tert-butyl ether (MTBE). Chromatographic conditions are described which permit the separation and measurement of the analytes in the extract by GC with an electron capture detector (ECD). This method has been validated in a single laboratory, and estimated detection limits have been determined for the analytes in the method, including dieldrin, for which the estimated detection limit is 0.02 μg/L.

VII. Treatment Technologies

- Available data indicate that reverse osmosis (RO), granular-activated carbon (GAC) adsorption, ozonation and conventional treatment will remove dieldrin from water. The percent removal efficiency ranges from 50 to 99 + %.

- Laboratory studies indicate that RO is a promising treatment method for dieldrin-contaminated waters. Chian et al. (1975) reported 99 + % removal efficiency for two types of membranes operating at 600 psig and a flux rate of 8 to 12 gal/ft^2/day. Membrane adsorption, however, is a major concern and must be considered, since breakthrough of dieldrin would probably occur once the adsorption potential of the membrane was exhausted.

- GAC is effective for dieldrin removal. Pirbazari and Weber (1983) reported 99 + % dieldrin removal efficiency of a GAC column operating at an empty bed contact time (EBCT) of 15 minutes and a hydraulic loading of 1.4 gal/ft^2/min, for the entire test period (approximately 7.5 months).

- Pirbazari and Weber (1983) determined adsorption isotherms using GAC on dieldrin in water solutions. Resin adsorption was also found to remove dieldrin from water. The Freundlich values determined by the authors indicate that the tested resins are not quite as effective as GAC in the removal of dieldrin from water.

- Ozonation treatment appears to be an effective dieldrin removal method. Treatment with 36 mg/L ozone (O_3) removed 50% of dieldrin while 11 mg/L O_3 removed only 15% of dieldrin (Robeck et al., 1965).

- Conventional water-treatment techniques using alum coagulation, sedimentation and filtration proved to be 55% effective in removing dieldrin from contaminated potable water supplies (Robeck et al., 1965). Lime- and soda-ash softening with ferric chloride as a coagulant did not improve upon the removal efficiency achieved with alum alone.

- Oxidation with chlorine and potassium permanganate is ineffective in degrading dieldrin (Robeck et al., 1965).

- Treatment technologies for the removal of dieldrin from water are available and have been reported to be effective. However, selection of individual or combinations of technologies to attempt dieldrin removal from water must be based on a case-by-case technical evaluation and an assessment of the economics involved.

VIII. References

ACGIH. 1984. American Conference of Governmental Industrial Hygienists. Documentation of the threshold limit values for substances in workroom air. 3rd ed. Cincinnati, OH: ACGIH. p. 139.

Baldwin, M.K. and J. Robinson. 1972. A comparison of the metabolism of HEOD (Dieldrin) in CF_1 mouse with that in the CFE rat. Food Cosmet. Toxicol. 10:333–351.

Chernoff, N., R.J. Kavlock, J.R. Kathrein, J.M. Dunn and J.K. Haseman. 1975. Prenatal effects of dieldrin and photodieldrin in mice and rats. Toxicol. Appl. Pharmacol. 31:302–308.

Chian, E.S., W.N. Bruce and H.H.P. Fang. 1975. Removal of pesticides by reverse osmosis. Environ. Sci. Technol. 9(1):52–59.

Coulston, F., R. Abraham and R. Mankes. 1980. Reproductive study in female rats given dieldrin, alcohol or aspirin orally. Albany, NY: Albany Medical College of Union University. Institute of Comparative and Human Toxicology. Cited in IPCS, 1987.

Fitzhugh, O.G., A.A. Nelson and M.L. Quaife. 1964. Chronic oral toxicity of aldrin and dieldrin in rats and dogs. Food Cosmet. Toxicol. 2:551–562.

Hayes, W.J., Jr. 1974. Distribution of dieldrin following a single oral dose. Toxicol. Appl. Pharmacol. 28:485–492.

Heath, D.F. and M. Vandekar. 1964. Toxicity and metabolism of dieldrin in rats. Br. J. Ind. Med. 21:269–279.

IARC. 1982. International Agency for Research on Cancer. IARC monographs on the evaluation of the carcinogenic risk of chemicals to humans: chemicals, industry process and industries associated with cancer in humans. IARC Monographs Vols. 1–29, Supplement 4. Geneva, Switzerland: World Health Organization.

IPCS. 1987. International Programme on Chemical Safety. Environmental Health Criteria for Aldrin and Dieldrin. United Nations Environment Programme. International Labour Organization. Geneva: World Health Organization.

Lehman, A. 1959. Appraisal of the safety of chemicals in foods, drugs and cosmetics. Association of Food and Drug Officials of the United States.

MacKay, D. and A.W. Wolkoff. 1973. Rate of evaporation of low-solubility contaminants from water bodies to atmosphere. Environ. Sci. Technol. 7:611.

Majumdar, S.K., H.A. Kopelman and M.J. Schnitman. 1976. Dieldrin-induced chromosome damage in mouse bone marrow and WI-38 human lung cells. J. Hered. 67:303–307.

McCann, J., E. Choi, E. Yamasaki and B.N. Ames. 1975. Detection of carcinogens as mutagens in the Salmonella/microsome test: Assay of 300 chemicals. Proc. Natl. Acad. Sci. 72(12):5135–5139.

Meister, R., ed. 1983. Farm chemicals handbook. Willoughby, OH: Meister Publishing Company.

NAS. 1977. National Academy of Sciences. Drinking water and health. Vol. 1. Washington, DC: National Academy Press. pp. 556–571.

NCI. 1978. National Cancer Institute. Bioassay of aldrin and dieldrin for possible carcinogenicity. Technical Report Series No. 21.

Ottolenghi, A.D., J.K. Haseman and F. Suggs. 1974. Teratogenic effects of aldrin, dieldrin, and endrin in hamsters and mice. Teratology. 9:11–16.

Pirbazari, M. and W.J. Weber. 1983. Removal of dieldrin from water by activated carbon. J. Environ. Eng. 110(3):656–669.

Robeck, G.G., K.A. Dostal, J.M. Cohen and J.F. Kreessl. 1965. Effectiveness of water treatment processes in pesticide removal. J. AWWA. (Feb):181–199.

RTECS. 1985. Registry of toxic effects of chemical substances. National Institute for Occupational Safety and Health. National Library of Medicine Online File.

STORET. 1988. STORET Water Quality File. Office of Water. U.S. Environmental Protection Agency (data file search conducted in May, 1988).

U.S. EPA. 1980. U.S. Environmental Protection Agency. Ambient water quality criteria for aldrin/dieldrin. EPA 440/5–80–019. NTIS Acc. No. PB 81–117301. Washington, DC: U.S. EPA.

U.S. EPA. 1984a. U.S. Environmental Protection Agency. Method 608, organo-chlorine pesticides and PCBs. Fed. Reg. 49(209):43234–43443. October 26.

U.S. EPA. 1984b. U.S. Environmental Protection Agency. Method 625, base/neutrals and acids. Fed. Reg. 49(209):43234–43443. October 26.

U.S. EPA. 1985. U.S. Environmental Protection Agency. Code of Federal Regulations. 40 CFR 180.137. July 1.

U.S. EPA. 1986. U.S. Environmental Protection Agency. Guidelines for carcinogen risk assessment. Fed. Reg. 51(185):33992–34003. September 24.

U.S. EPA. 1987. U.S. Environmental Protection Agency. Carcinogenicity assessment of aldrin and dieldrin. Washington, DC: Carcinogen Assessment Group, Office of Research and Development, U.S. EPA.

U.S. EPA. 1988. U.S. Environmental Protection Agency. Method 508 - Determination of chlorinated pesticides in water by gas chromatography with an electron capture detector. Available from EPA's Environmental Monitoring and Support Laboratory, Cincinnati, Ohio.

Walker, A.I.T., D.E. Stevenson, J. Robinson, E. Thorpe and M. Roberts. 1969. The toxicology and pharmacodynamics of dieldrin (HEOD). Two-year oral exposures of rats and dogs. Toxicol. Appl. Pharmacol. 15:345–373.

Walker, A.I.T., E. Thorpe and D.E. Stevenson. 1972. The toxicology of dieldrin (HEOD). I. Long-term oral toxicity studies in mice. Food Cosmet. Toxicol. 11:415–432.

Weast, R.C. and M. Astle, eds. 1982. CRC handbook of chemistry and physics — a ready reference book of chemical and physical data, 63rd ed. Cleveland, OH: CRC Press.

WHO. 1982. World Health Organization. Guidelines for drinking water quality. Unedited final draft.

Windholz, M. 1983. The Merck index. 10th ed. Rahway, NJ: Merck and Co., Inc. pp. 450–451.

---

* Confidential Business Information submitted to the Office of Pesticide Programs.

Dimethrin

I.   General Information and Properties

A. CAS No. 70–38–2

B. Structural Formula

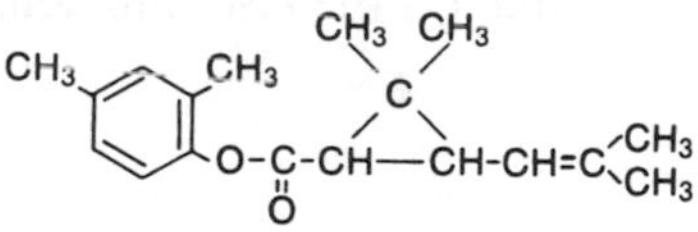

2,4-Dimethylbenzyl-2,2-dimethyl-3(2-methylpropenyl)-
cyclopropane carboxylate

C. Synonyms

- ENT 21,170; Chrysanthemumic acid; 2,4-Dimethylbenzylester.

D. Uses

- Dimethrin is an insecticide for use in ponds and swamps as a mosquito larvicide (Meister, 1986).

E. Properties (U.S. AEHA-MT, 1962)

Chemical Formula	$C_{18}H_{24}O_2$
Molecular Weight	286.39 (Ambrose, 1964)
Physical State (25°C)	Amber liquid
Boiling Point	175°C
Melting Point	--
Density	--
Vapor Pressure (25°C)	--
Specific Gravity	0.98
Water Solubility (25°C)	Insoluble (further details not provided)
Log Octanol/Water Partition Coefficient	--

Taste Threshold   --
Odor Threshold   --
Conversion Factor   --

F. Occurrence

- No information is available on the occurrence of dimethrin in water.

G. Environmental Fate

- No information is available on the environmental fate of dimethrin.

II.   Pharmacokinetics

A. Absorption

- In a preliminary metabolic study by Ambrose (1964), four rabbits were given 5 mL/kg (5 mg/kg) of undiluted dimethrin by intubation. Urine was collected every 24 hours over a 72-hour period. Identification of two possible metabolites in the urine indicated that dimethrin was absorbed. Sufficient data were not available to quantify the extent of absorption.

B. Distribution

- No information on the distribution of dimethrin was found in the available literature.

C. Metabolism/Excretion

- Information presented by Ambrose (1964) indicates that dimethrin (5 mg/kg), administered by intubation to rabbits, is metabolized (by reduction) and excreted in the urine as chrysanthemumic acid and the glucuronic ester of 2,4-dimethyl benzoic acid. Sufficient information was not presented to determine if these are the only metabolites of dimethrin or if any unchanged dimethrin is excreted.

III. Health Effects

A. Humans

- No information on the health effects of dimethrin in humans was found in the available literature.

B. Animals

1. *Short-term Exposure*

- The acute oral LD_{50} value of dimethrin for male and female Sherman rats was reported to be > 15,000 mg/kg (Gaines, 1969).

- Ambrose (1964) conducted an acute oral study in which male and female albino rabbits (2/sex/dose) and male albino Wistar-CWL rats (5/dose) were given a single dose of 10 or 15 mL/kg (9.8 or 14.7 mg/kg) of technical-grade dimethrin (98% pure) by gavage. Albino guinea pigs (4/sex) received a single dose of 10 mL/kg (9.8 mg/kg) by gavage. No effects were observed in rats or rabbits during a 2-week observation period. (Specific parameters observed were not identified). In guinea pigs, the only effect reported during a similar observation period was a refusal to eat or drink for 24 hours following dosing.

- Ambrose (1964) administered 10 mL/kg (9.8 mg/kg) of technical-grade dimethrin (98% pure) to 15 male albino Wistar-CWL rats by gavage, 5 days per week for 3 weeks. This corresponds to an average daily dose of 7 mg/kg. No adverse effects, as judged by general appearance, behavior and growth, were observed. At necropsy, no gross abnormalities were observed. No histopathological examinations were performed.

2. *Dermal/Ocular Effects*

- Ambrose (1964) conducted a dermal irritation study in which dimethrin (98% pure) was applied at a dose level of 10 mL/kg (9.8 mg/kg) to the intact or abraded skin of four albino rabbits (2/sex) for a 24-hour exposure period. No skin irritation was observed immediately after the removal of the dimethrin or during a 2-week observation period.

- Ambrose (1964) reported that single or multiple (3 consecutive days) instillations of 0.1 mL of undiluted dimethrin (98% pure) into the conjunctival sac of eight albino rabbits caused no visible irritation or chemosis and no injury to the cornea as detectable by means of fluorescein staining. When 0.2 mL of dimethrin was applied to the penile mucosa of five albino rabbits on two occasions 6 days apart, no irritation or sloughing of the mucosa was observed during a 1-week observation period.

- Masri et al. (1964) applied 3 mL of undiluted dimethrin to the shaved back and sides of three albino rabbits 10 times over a 2-week period (frequency of application not specified). The only reported reaction was the development of a slight scaliness which disappeared after cessation of application.

- Ambrose (1964) applied dimethrin (98% pure) to the skin of albino rabbits (5/dose) 5 days per week for 13 weeks (65 applications). Doses administered were 0.5 mL/kg undiluted dimethrin or 0.5 mL/kg of a 50% solution of dimethrin in cottonseed oil (equivalent to 0.25 mL/kg of dimethrin); controls received 0.5 mL/kg of cottonseed oil only. No evidence of any cutaneous reaction was observed. Occasionally, a slight, nonpersistent erythema was observed in all groups of rabbits. At necropsy, all organs from treated animals were indistinguishable from the controls. No histopathological differences between control and treated animals were observed.

3. *Long-term Exposure*

- Masri et al. (1964) administered dimethrin to male (5/dose) and female (6/dose) weanling albino rats for 16 weeks at dietary levels of 0, 0.2, 0.6, 1.5 or 3.0%. Based on food consumption and body weight data presented in the study, these dietary levels of dimethrin were calculated to correspond to about 0, 120, 320, 1,000 or 2,300 mg/kg/day for males, and 0, 130, 400, 1,100 or 2,500 mg/kg/day for females. Results indicated a significant reduction in body weight in males receiving 0.6 or 3.0% and females receiving 1.5 or 3.0%. Absolute liver weight and liver-to-body weight ratios were significantly higher in both the male and female 1.5% and 3.0%-dose groups. Kidney-to-body weight ratios were also significantly higher for these groups. Scattered gross pathologic changes did not appear to bear a relationship to dose. Histopathological examination revealed dose-related morphological changes in the liver that consisted of a round eosinophilic ring in the cytoplasm, approximately the size of the nucleus. Amorphous material within the ring stained less densely than the rest of the cytoplasm. Also, many hepatic cells of rats receiving 1.5 or 3.0% dimethrin appeared larger than those of controls and had less distinct basophilic cytoplasmic particles. Hepatic changes were less pronounced in the 0.6% group. No cell inclusions were seen in rats receiving 0.2% dimethrin. The effects of increased liver and kidney-to-body weight ratios as well as histopathological changes in the liver were shown to be reversible after withdrawal of dimethrin. The No-Observed-Adverse-Effect Level (NOAEL) identified in this study was 0.2% dimethrin (120 mg/kg/day for males; 130 mg/kg/day for females).

• Ambrose (1964) administered dimethrin to male and female albino Wistar-CWL rats (10/sex/dose) for 52 weeks at dietary levels of 0, 0.05, 0.1, 0.5, 1.0 or 2.0%. These dietary levels correspond to 0, 30, 60, 300, 600 or 1,200 mg/kg/day. The only statistically significant effect reported in this study was an increase in the liver-to-body weight ratios in both male and female animals receiving 1.0 or 2.0% dimethrin. Withdrawal of dimethrin from the diet for 6 weeks resulted in return of liver weights to levels indistinguishable from the controls. No differences in hemoglobin parameters were noted between the treated and control animals at any time during the 52-week period. Histologically, no significant changes or lesions that could be attributed to dimethrin in the diet were observed in any of the test groups of animals. A NOAEL of 300 mg/kg was identified from this study.

• As described in a review by Cohen and Grasso (1981), dimethrin has been implicated as a hypolipidemic agent and causes an increase in hepatic peroxisome proliferation. Dietary administration of certain hypolipidemic agents to rodents has resulted in the induction of liver carcinomas.

4. *Reproductive Effects*

• No information on the reproductive effects of dimethrin was found in the available literature.

5. *Developmental Effects*

• No information on the developmental effects of dimethrin was found in the available literature.

6. *Mutagenicity*

• No information on the mutagenicity of dimethrin was found in the available literature.

7. *Carcinogenicity*

• No information on the carcinogenicity of dimethrin was found in the available literature. However, the report by Cohen and Grasso (1981) implicating dimethrin as a hypolipidemic agent may indicate that dimethrin has carcinogenic potential in rodents. (It should be noted that the relationship between hypolipidemic agents and liver carcinomas in rodents has not been observed in humans.)

IV.   Quantification of Toxicological Effects

A. One-day Health Advisory

No information was found in the available literature that was suitable for determination of the One-day HA values for dimethrin. It is therefore recommended that the Longer-term HA for a 10-kg child (10 mg/L, calculated below) be used at this time as a conservative estimate of the One-day HA value.

B. Ten-day Health Advisory

No information was found in the available literature that was suitable for determination of the Ten-day HA values for dimethrin. It is therefore recommended that the Longer-term HA for a 10-kg child (10 mg/L, calculated below) be used at this time as a conservative estimate of the Ten-day HA value.

C. Longer-term Health Advisory

The 16-week rat study by Masri et al. (1964) has been selected to serve as the basis for determination of the Longer-term HA. In this study, male and female rats were administered dimethrin at dietary levels of 0, 0.2, 0.6, 1.5 or 3.0% for 16 weeks. Results of this study indicated a statistically significant reduction in body weights of males receiving 0.6 or 3.0%, and in females receiving 1.5 or 3.0%. Absolute liver weight and liver-to-body weight ratios were significantly higher in the 1.5% and 3.0%-dose groups. Kidney-to-body weight ratios were also significantly higher in those groups. Histopathological examinations revealed dose-related morphological changes in the liver occurring at dose levels as low as 0.6%. A NOAEL of 0.2% dimethrin (120 mg/kg/day for males; 130 mg/kg/day for females) was identified in this study.

Using a NOAEL of 120 mg/kg/day, the Longer-term HA for a 10-kg child is calculated as follows:

$$\text{Longer-term HA} = \frac{(120 \text{ mg/kg/day}) (10 \text{ kg})}{(100) (1 \text{ L/day})} = 12 \text{ mg/L} (10,000 \text{ } \mu\text{g/L})$$

Using a NOAEL of 120 mg/kg/day, the Longer-term HA for a 70-kg adult is calculated as follows:

$$\text{Longer-term HA} = \frac{(120 \text{ mg/kg/day}) (70 \text{ kg})}{(100) (2 \text{ L/day})} = 42 \text{ mg/L} (40,000 \text{ } \mu\text{g/L})$$

D. Lifetime Health Advisory

The 52-week study in rats by Ambrose (1964) has been selected to serve as the basis for determination of the Lifetime HA for dimethrin. In this study, dimethrin was administered to albino Wistar-CWL rats for 52 weeks at dietary levels of 0, 0.05, 0.1, 0.5, 1.0 or 2.0%. A statistically significant increase in the liver-to-body weight ratio was observed in both male and female rats receiving 1.0 or 2.0% dimethrin (600 and 1,200 mg/kg/day). Histologically, no changes that could be attributed to dimethrin were observed in any of the test groups. No adverse effects were reported in rats receiving dimethrin at 0.5% (300 mg/kg/day for males) or lower.

Using a NOAEL of 300 mg/kg/day, the Lifetime HA is derived as follows:

Step 1: Determination of the Reference Dose (RfD)

$$RfD = \frac{(300 \text{ mg/kg/day})}{(1,000)} = 0.3 \text{ mg/kg/day}$$

Step 2: Determination of the Drinking Water Equivalent Level (DWEL)

$$DWEL = \frac{(0.3 \text{ mg/kg/day}) (70 \text{ kg})}{(2 \text{ L/day})} = 10.5 \text{ mg/L } (10,000 \text{ }\mu g/L)$$

Step 3: Determination of the Lifetime Health Advisory

$$\text{Lifetime HA} = (10.5 \text{ mg/L}) (20\%) = 2.1 \text{ mg/L } (2,000 \text{ }\mu g/L)$$

It should be noted that the Lifetime HA of 2.1 mg/L apparently exceeds the water solubility of dimethrin (insoluble).

E. Evaluation of Carcinogenic Potential

- No information on the carcinogenicity of dimethrin was found in the available literature. However, the report by Cohen and Grasso (1981) implicating dimethrin as a hypolipidemic agent may indicate that dimethrin has carcinogenic potential in rodents. (It should be noted that the relationship between hypolipidemic agents and liver carcinomas in rodents has not been observed in humans.)

- The International Agency for Research on Cancer has not evaluated the carcinogenicity of dimethrin.

- Applying the criteria described in EPA's guidelines for assessment of carcinogenic risk (U.S. EPA, 1986), dimethrin may be classified in Group D: not classifiable. This category is for substances with inadequate animal or human evidence of carcinogenicity.

V.   Other Criteria, Guidance and Standards

- No information on existing criteria, guidance or standards pertaining to dimethrin was found in the available literature. However, tolerances for pyrethroids, of which dimethrin is a member, range from 0.05 ppm in potatoes (post-harvest) to 3 ppm in wheat, barley, rice and oats (CFR, 1985).

VI.   Analytical Methods

- Dimethrin is a cyclopropane carboxylate pesticide which, as such, can be analyzed by EPA Method 616 (U.S. EPA, 1984). This method covers CHO pesticides such as cycloprate and resmethrin, which are chemically identical to dimethrin. The method is similar to other 600 series methods in that 1 liter of sample is extracted with methylene chloride and reduced to 1 mL or less. Analysis is by flame-ionization gas chromatography (FID/GC). A cleanup procedure is provided in case interferences are noted.

- While Method 616 can be used for monitoring dimethrin, the analyst should demonstrate precision and accuracy data for this compound before proceeding. An estimated detection limit (EDL) should also be determined as specified by regulation (CFR, 1984). The detection limit should fall into the range of 20 to 50 μg/L.

VII.   Treatment Technologies

- The manufacture of this compound was discontinued (Meister, 1986). No information was found in the available literature on treatment technologies capable of effectively removing dimethrin from contaminated water.

VIII. References

Ambrose, A.M. 1964. Toxicologic studies on pyrethrin-type esters of chrysanthemumic acid. II. Chrysanthemumic acid, 2,4-dimethylbenzyl ester. Toxicol. Appl. Pharmacol. 6:112–120.

CFR. 1984. Code of Federal Regulations. 40 CFR Part 136, Appendix B. Fed. Reg. Vol. 49, No. 209. October 26.

CFR. 1985. Code of Federal Regulations. 40 CFR 180.128.

Cohen, A.J. and P. Grasso. 1981. Review of hepatic response to hypolipidemic drugs in rodents and assessment of its toxicological significance to man. Food Cosmet. Toxicol. 4:585–605.

Gaines, T.B. 1969. Acute toxicity of pesticides. Toxicol. Appl. Pharmacol. 14:515–534.

Lehman, A.J. 1959. Appraisal of the safety of chemicals in foods, drugs, and cosmetics. Assoc. Food Drug Off. U.S., Q. Bull.

Masri, M.S., A.P. Henderson, A.J. Cox and F. De, eds. 1964. Subacute toxicity of two chrysanthemumic acid esters: barthrin and dimethrin. Toxicol. Appl. Pharmacol. 6:716–725.

Meister, R., ed. 1986. Farm chemicals handbook. Willoughby, OH: Meister Publishing Company. p. C81.

U.S. AEHA-MT. 1962. U.S. Army Environmental Hygiene Agency-MT. Project no. 2727T6–57162, February 1. Toxicological Studies on 2,4-dimethyl benzyl chrysanthemumate.

U.S. EPA. 1984. U.S. Environmental Protection Agency. EPA Method 616— Determination of CHO compounds. Available from EPA's Environmental Monitoring and Support Laboratory, Cincinnati, OH.

U.S. EPA. 1986. U.S. Environmental Protection Agency. Guidelines for carcinogen risk assessment. Fed. Reg. 51(185): 33992–34003. September 24.

U.S. EPA. No date. Registration materials, Office of Pesticides Programs.

---

* Confidential Business Information submitted to the Office of Pesticide Programs.

Dinoseb

I. General Information and Properties

A. CAS No. 88–85–7

B. Structural Formula

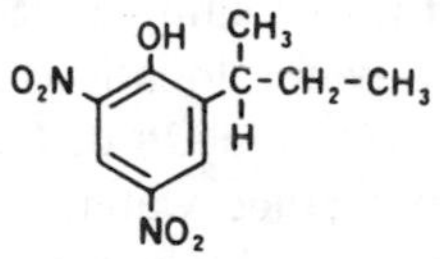

$$O_2N - \text{(ring)} - OH, \ CH_3, \ C-CH_2-CH_3, \ H, \ NO_2$$

2-sec-butyl-4,6-dinitrophenol

C. Synonyms

- DNBP, dinitro, dinoseb (BSI, ISO, WSSA); dinosebe (France); Basanite (BASF Wyandotte); Caldon, Chemox General, Chemox PE, Chemsect DNBP, DN-289 (product discontinued), Dinitro, Dinitro-3, Dinitro General, Dynamite (Drexel Chemical); Elgetol 318, Gebutox, Hel-Fire (Helena); Kiloseb, Nitropone C, Premerge 3 (Agway); Sinox General (FMC Corp.); Subitex, Unicrop DNBP, Vertac Dinitro Weed Killer 5, Vertac General Weed Killer, Vertac Selective Weed Killer (Meister, 1984).

D. Uses

- Dinoseb is used as a herbicide, desiccant and dormant fruit spray (Meister, 1984).

E. Properties (WSSA, 1983)

Chemical Formula	$C_{10}H_{12}N_2O_5$
Molecular Weight	240
Physical State (room temp.)	Dark amber crystals
Boiling Point	--
Melting Point	32°C
Density (°C)	1.2647 (45°C)

323

Vapor Pressure (151°C)	1 mm Hg
Specific Gravity	--
Water Solubility	52 mg/L (25°C)
Log Octanol/Water Partition Coefficient	--
Taste Threshold	--
Odor Threshold	--
Conversion Factor	--

F. Occurrence

- Dinoseb has been found in 1 of 89 surface water samples analyzed and in none of 1,270 ground-water samples (STORET, 1988). Samples were collected at 89 surface water locations and 1,184 ground-water locations, and dinoseb was found in Ohio. The 85th percentile of all nonzero samples was 1 μg/L in surface water. This information is provided to give a general impression of the occurrence of this chemical in ground and surface waters as reported in the STORET database. The individual data points retrieved were used as they came from STORET and have not been confirmed as to their validity. STORET data are often not valid when individual numbers are used out of the context of the entire sampling regime, as they are here. Therefore, this information can only be used to form an impression of the intensity and location of sampling for a particular chemical.

- Dinoseb has been found in New York ground water; typical positives were 1 to 5 ppb (Cohen et al., 1986).

G. Environmental Fate

- Dinoseb was stable to hydrolysis at pH 5, 7 and 9 at 25°C over a period of 30 days (Dzialo, 1984).

- With natural sunlight on a California sandy loam soil, dinoseb had a half-life of 14 hours; with artificial light, it had a half-life of 30 hours, indicating that dinoseb is subject to photolytic degradation (Dinoseb Task Force, 1985a).

- In water with natural sunlight, dinoseb had a half-life of 14 to 18 days; with artificial light, it had a half-life of 42 to 58 days (Dinoseb Task Force, 1985b).

- With soil thin-layer chromatography plates, dinoseb was intermediate to very mobile in silt loam, sand, sandy loam and silty clay loam (Dinoseb Task Force, 1985c).

- Soil adsorption studies gave a K_d of less than 5 for four soils: a silt loam, sand, sandy loam and silty clay loam, with organic matter content of 0.8 to 3% (Dinoseb Task Force, 1985d).

II. Pharmacokinetics

A. Absorption

- Following oral administration of dinoseb to rats (Bandal and Casida, 1972) and mice (Gibson and Rao, 1973) (specific means of administration not specified), approximately 25% of the administered dose appeared in the feces. However, following intraperitoneal (i.p.) administration in the mouse, approximately 40% appeared in the feces, thus suggesting to Gibson and Rao (1973) that dinoseb is initially completely absorbed following oral administration with subsequent secretion into the gut.

B. Distribution

- Following oral administration of dinoseb in the mouse (specific means of administration not specified), no appreciable amounts accumulated in the blood, liver or kidney (Gibson and Rao, 1973).

C. Metabolism

- While the metabolism of dinoseb has not been completely characterized, a number of metabolites have been identified, including 2-(2-butyric acid)-4,6-diaminophenol, 2-(2-butyric acid)-4,6-dinitrophenol, 2-sec-butyl-4-nitro-6-aminophenol, 2-sec-butyl-4-acetamido-6-nitrophenol and 2-(3-butyric acid)-4,6-dinitrophenol (Ernst and Bar, 1964; Froslie and Karlog, 1970; Bandal and Casida, 1972).

D. Excretion

- In mice, dinoseb is excreted in both urine (20%) and feces (30%) following oral administration (specific means of administration not specified) (Gibson and Rao, 1973).

III. Health Effects

A. Humans

1. *Short-term Exposure*

• While minimal data are available concerning human toxicity, at least one death has been attributed to an accidental exposure of a farm worker to sprayed dinoseb and dinitro-ortho-cresol (Heyndrickx et al., 1964).

2. *Long-term Exposure*

• No information on the long-term health effects of dinoseb in humans was found in the available literature.

B. Animals

1. *Short-term Exposure*

• In rats and mice, the acute oral LD_{50} of dinoseb ranges from 20 to 40 mg/kg (Bough et al., 1965).

2. *Dermal/Ocular Effects*

• In rats, the acute dermal toxicity of dinoseb ranges from 67 to 134 mg/kg (Noakes and Sanderson, 1969).

• No information on the ocular effects of dinoseb in animals was found in the available literature.

3. *Long-term Exposure*

• Hall et al. (1978) reported the results (abstract only) of a feeding study in male and female rats. Eight groups of rats, each group composed of 14 males and 14 females, were exposed to levels of 0, 50, 100, 150, 200, 300, 400 or 500 ppm of dinoseb (80% pure) in the diet for 153 days, respectively. Assuming that 1 ppm in the diet of rat is equivalent to 0.05 mg/kg/day (Lehman, 1959), these levels correspond to 0, 2.5, 5.0, 7.5, 10.0, 15.0, 20.0 and 25.0 mg/kg/day. Mortality was observed at 300 ppm (15 mg/kg/day) and above, and growth was depressed at all dose levels. The LOAEL for this study was identified as 50 ppm (2.5 mg/kg/day), the lowest dose tested.

• In a 6-month dietary study by Spencer et al. (1948), groups of male rats were exposed to dinoseb (99% pure) at levels of 0 (30 animals), 1.35, 2.7, 5.4 (20 animals) and 13.5 mg/kg/day (10 animals). Based on increased mortality at the highest dose and increased liver weight at intermediate doses, the No-Observed-Adverse-Effect Level (NOAEL) for dinoseb was identified as 2.7 mg/kg/day.

- In a study submitted to EPA in support of the registration of dinoseb (Hazelton, 1977), four groups of rats (60/sex/dose) were exposed to dinoseb (purity not specified) in their diets for periods up to 2 years at dose levels of 0, 1, 3 and 10 mg/kg/day, respectively. Although no evidence of dose-related changes in histopathology, hematology, blood chemistry or certain other parameters were observed, a dose-related decrease in mean thyroid weight was observed in all treated males. The LOAEL in this study was identified as 1 mg/kg/day.

4. Reproductive Effects

- In a reproduction study by Linder et al. (1982), four groups of 10 male rats each were exposed to dinoseb (97% pure) in the diet at levels of 0, 3.8, 9.1 or 15.6 mg/kg/day over an 11-week period, respectively. In addition, a group of five animals was exposed to 22.2 mg/kg/day. The fertility index was reduced to 0 at 22.2 mg/kg and to 10% at 15.6 mg/kg/day; in neither case did the fertility index improve in 104 to 112 days following treatment. A variety of other effects were seen at levels of 9.1 mg/kg/day and higher, including decreased weight of the seminal vesicles, decreased sperm count and an increased incidence of abnormal sperm. The NOAEL for dinoseb in this study was 3.8 mg/kg/day based on a decrease in sperm count and other effects at higher levels.

- In a two-generation rat reproduction study (Irvine, 1981), four groups of rats (25/sex/dose) were exposed to 0, 1, 3 and 10 mg/kg/day of dinoseb in the diet for 29 weeks. Although no reproductive effects were observed in this study per se, a decrease in pup body weight was observed at day 21 postparturition for all dose levels. Thus, based on a compound-related depression in pup body weight at all dose levels, the LOAEL in this study was 1 mg/kg/day.

5. Developmental Effects

- Although dinoseb has been reported to be teratogenic (e.g., oligodactyly, imperforate anus, hydrocephalus, etc.) when administered to mice intraperitoneally (Gibson, 1973), it was not teratogenic when administered orally to mice (Gibson, 1973; Gibson and Rao, 1973) or rats (Spencer and Sing, 1982).

- Dinoseb (95% pure), administered to pregnant rats in the diet on days 6 through 15 of gestation, produced a marked reduction in fetal survival at doses of 9.2 mg/kg/day and above but not at doses of 6.9 mg/kg/day (NOAEL) and below (Spencer and Sing, 1982).

- Dinoseb (purity not specified) was without effect in a study in which pregnant mice were orally exposed to a single dose of 15 mg/kg/day (Chernoff and Kavlock, 1983).

- In a developmental toxicity study by Research and Consulting Company (1986), four groups of 16 Chinchilla rabbits were exposed to dinoseb (98% pure) by oral gavage at levels of 0, 1, 3 or 10 mg/kg/day from day 6 to 18 of gestation. At the highest dose level dinoseb produced a statistically significant increase in malformations and/or anomalies when compared to the controls, with external, internal (body cavities and cephalic viscera) and skeletal defects being observed in 11/16 litters examined. Neural tube defects, the major developmental toxic effect, included dyscrania associated with hydrocephaly, scoliosis, kyphosis, malformed or fused caudal and sacral vertebrae and encephalocele. The NOAEL for dinoseb in this study was identified as 3.0 mg/kg/day, based on the occurrence of neural tube defects at the highest dose level.

- In a study by the Dinoseb Task Force (1986), developmental toxicity was observed in Wistar/Han rats. Groups of 25 rats received dinoseb (purity 96.1%) by gavage at levels of 0, 1, 3 or 10 mg/kg/day from day 6 to 15 of gestation. Developmental toxicity was observed at the high dose as evidenced by a slight depression in fetal body weight, increased incidence of absence of skeletal ossification for a number of sites and an increase in the number of supernumerary ribs. Slight to moderate decreases in body weight gain and food consumption were observed in dams at the intermediate- and high-dose levels. Based on the occurrence of developmental effects at the highest dose level, a NOAEL of 3.0 mg/kg/day was identified.

6. *Mutagenicity*

- With the exception of an increase in DNA damage in bacteria (Waters et al., 1982), dinoseb was not mutagenic in a number of organisms including *Salmonella typhimurium*, *Escherichia coli*, *Saccharomyces cerevisiae*, *Drosophila melanogaster* or *Bacillus subtilis* (Simmon et al., 1977; Waters et al., 1982; Moriya et al., 1983).

7. *Carcinogenicity*

- No evidence of a carcinogenic response was observed in a 2-year chronic feeding study in which dinoseb was administered to rats at levels as high as 10 mg/kg/day (Hazelton, 1977).

IV.   Quantification of Toxicological Effects

A. One-day Health Advisory

No information was found in the available literature that was suitable for determination of the One-day HA value. It is therefore recommended that the Ten-day HA value for a 10-kg child (0.3 mg/L, calculated below) be used as a conservative estimate of the One-day HA value.

B. Ten-day Health Advisory

The rabbit developmental toxicity study (Research and Consulting Company, 1986) in which dinoseb produced neural tube defects at doses greater than 3 mg/kg/day (NOAEL) was selected as the basis for determination of the Ten-day HA. While it is reasonable to base a Ten-day HA for the adult on a positive developmental toxicity study, there is some question as to whether it is appropriate to base the Ten-day HA for a 10-kg child on such a study. However, since this study is of appropriate duration and since the fetus may be more sensitive than a 10-kg child, it was judged that, while it may be overly conservative, it is reasonable to base the Ten-day HA for a 10-kg child on such a study.

Using a NOAEL of 3.0 mg/kg/day, the Ten-day HA for a 10-kg child is calculated as follows:

$$\text{Ten-day HA} = \frac{(3.0 \text{ mg/kg/day}) (10 \text{ kg})}{(100) (1 \text{ L/day})} = 0.3 \text{ mg/L } (300 \text{ } \mu\text{g/L})$$

C. Longer-term Health Advisory

The Hall et al. (1978) 153-day dietary dinoseb study in rats was originally selected to serve as the basis for determination of the Longer-term HA (decreased growth was observed at all exposure levels with a LOAEL of 2.5 mg/kg/day). Subsequently, however, a two-generation reproduction study in rats (Irvine, 1981) was identified with a LOAEL of 1 mg/kg/day (based on a decrease in pup body weight at all dose levels). Since a reproduction study is of appropriate duration, the Irvine (1981) study has been selected to serve as the basis for determination of the Longer-term HA.

Using a LOAEL of 1 mg/kg/day, the Longer-term HA for a 10-kg child is calculated as follows:

$$\text{Longer-term HA} = \frac{(1.0 \text{ mg/kg/day}) \ (10 \text{ kg})}{(1{,}000) \ (1 \text{ L/day})} = 0.010 \text{ mg/L } (10 \text{ } \mu\text{g/L})$$

The Longer-term HA for a 70-kg adult is calculated as follows:

$$\text{Longer-term HA} = \frac{(1.0 \text{ mg/kg/day}) \ (70 \text{ kg})}{(1{,}000) \ (2 \text{ L/day})} = 0.035 \text{ mg/L } (40 \text{ } \mu\text{g/L})$$

D. Lifetime Health Advisory

The 2-year dietary rat study by Hazelton (1977) was selected to serve as the basis for determination of the Lifetime HA. In this study, a compound-related decrease in mean thyroid weights was observed in all males (LOAEL = 1 mg/kg/day) treated with dinoseb (purity not specified).

Using a LOAEL of 1 mg/kg/day, the Lifetime HA for a 70-kg adult is calculated as follows:

Step 1: Determination of the Reference Dose (RfD)

$$\text{RfD} = \frac{(1 \text{ mg/kg/day})}{(1{,}000)} = 0.001 \text{ mg/kg/day}$$

Step 2: Determination of the Drinking Water Equivalent Level (DWEL)

$$\text{DWEL} = \frac{(0.001 \text{ mg/kg/day}) \ (70 \text{ kg})}{(2 \text{ L/day})} = 0.035 \text{ mg/L } (40 \text{ } \mu\text{g/L})$$

Step 3: Determination of the Lifetime Health Advisory

$$\text{Lifetime HA} = (0.035 \text{ mg/L}) \ (20\%) = 0.007 \text{ mg/L } (10 \text{ } \mu\text{g/L})$$

E. Evaluation of Carcinogenic Potential

- No evidence of carcinogenicity was found in a 2-year dietary study in which dinoseb was administered to rats at levels as high as 10 mg/kg/day (Hazelton, 1977).

- The International Agency for Research on Cancer has not evaluated the carcinogenic potential of dinoseb.

- Applying the criteria described in EPA's guidelines for assessment of carcinogenic risk (U.S. EPA, 1986), dinoseb is classified in Group D: not classifiable. This group is for agents with inadequate human and animal evidence of carcinogenicity.

V.   Other Criteria, Guidance and Standards

- Tolerances have been established for dinoseb (CFR, 1985) at 0.1 ppm on a wide variety of agricultural commodities.

- The EPA RfD Workgroup approved a 0.001 mg/kg/day RfD for dinoseb. The EPA RfD Workgroup is an EPA-wide group whose function is to ensure that consistent RfD values are used throughout the EPA.

VI.   Analytical Methods

- Analysis of dinoseb is by a gas chromatographic (GC) method applicable to the determination of certain chlorinated acid pesticides in water samples (U.S. EPA, 1985). In this method, approximately 1 liter of sample is acidified. The compounds are extracted with ethyl ether using a separatory funnel. The derivatives are hydrolyzed with potassium hydroxide, and extraneous organic material is removed by a solvent wash. After acidification, the acids are extracted and converted to their methyl esters using diazomethane as the derivatizing agent. Excess reagent is removed, and the esters are determined by electron capture GC. The method detection limit has been estimated at 0.07 μg/L for dinoseb.

VII.   Treatment Technologies

- The treatment technologies which will remove dinoseb from water include activated carbon and ion exchange. No data were found for the removal of dinoseb from drinking water by conventional treatment or by aeration. However, limited data suggest that aeration would not be effective in the removal of dinoseb from drinking water (ESE, 1984).

- Becker and Wilson (1978) reported on the treatment of a contaminated lake water with three activated carbon columns operated in series. The columns processed about 2 million gallons of lake water and achieved a 99.98 percent removal of dinoseb. Weber and Gould (1966) performed successful isotherm tests using Columbia LC carbon, which is coconut based, and reported the following Langmuirian equilibrium constants:

$$Q = 444 \text{ mg dinoseb per g of carbon}$$

$$1/b = 1.39 \text{ mg/L}$$

Though the Langmuir equation provides a good fit over a broad concentration range, greater adsorption would probably be achieved at lower concentrations (less than 100 μg/L) than predicted by using these constants.

- Weber (1972) has classified dinoseb as an acidic pesticide; and such compounds have been readily adsorbed in large amounts by ion exchange resins. Harris and Warren (1964) studied the adsorption of dinoseb from aqueous solution by an anion exchanger (Amberlite® IRA-400) and a cation exchanger (Amberlite® IR-200). The anion exchanger adsorbed dinoseb to less than detectable limits in solution.

VIII.   References

Bandal, S.K. and J.E. Casida. 1972. Metabolism and photoalteration of 2-sec-butyl-4,6-dinitrophenol (DNBP herbicide) and its isopropyl carbonate derivative (dinobuton acaricide). J. Agr. Food Chem. 20:1235–1245.

Becker, D.L. and Wilson, S.C. 1978. The use of activated carbon for the treatment of pesticides and pesticidal wastes. In: Carbon adsorption handbook. Cheremisinoff, D.H., and F. Ellerbusch, eds. Ann Arbor, MI: Ann Arbor Science Publishers.

Bough, R.G., E.E. Cliffe and B. Lessel. 1965. Comparative toxicity and blood level studies on binapacryl and DNBP. Toxicol. Appl. Pharmacol. 7:353–360.

CFR. 1985. Code of Federal Regulations. 40 CFR 180.281. July 1.

Chernoff, N. and R.J. Kavlock. 1983. A teratology test system which utilizes postnatal growth and viability in the mouse. Environ. Sci. Res. 27:417–427.

Cohen, S.Z., C. Eiden and M.N. Lorber. 1986. Monitoring ground water for pesticides in the USA. In: Evaluation of pesticides in ground water, American Chemical Society Symposium Series. In press.

Dinoseb Task Force. 1985a. Photodegradation of dinoseb on soil. Report No. 6015–191 (Tab 3). Prepared by Hazelton Laboratories America, Inc. July 19.

Dinoseb Task Force. 1985b. Photodegradation of dinoseb in water. Report No. 6015–190 (Tab 4). Prepared by Hazelton Laboratories America, Inc. July 19.

Dinoseb Task Force. 1985c. Determination of the mobility of dinoseb in selected soils by soil TLC. Report No. 6015–192 (Tab 1). Prepared by Hazelton Laboratories America, Inc. July 19.

Dinoseb Task Force. 1985d. The adsorption/desorption of dinoseb on representative agricultural soils. Report No. 6015-193 (Tab 2). Prepared by Hazelton Laboratories America, Inc. July 19.

Dinoseb Task Force. 1986. Probe embryotoxicity study with dinoseb technical grade in Wistar rats. Project No. 045281. Prepared by Research and Consulting Company. April 22.

Dzialo, D. 1984. Hydrolysis of dinoseb. Project No. 84239. Unpublished study. Prepared by Uniroyal Inc.

ESE. Environmental Science and Engineering. 1984. Review of treatability data for removal of twenty-five synthetic organic chemicals from drinking water. Washington, DC: U.S. Environmental Protection Agency, Office of Drinking Water.

Ernst, W. and F. Bar. 1964. Die umwandlung des 2,4-dinitro-6-secbutylphenols and seiner ester im tierischen organismus. Arzenimittel Forschung. 14:81-84. (Ger.)

Froslie, A. and O. Karlog. 1970. Ruminal metabolism of DNOC and DNBP. Acta. Vet. Scand. 11:31-43.

Gibson, J.E. 1973. Teratology studies in mice with 2-sec-butyl-4,6-dinitrophenol (dinoseb). Fd. Cosmet. Toxicol. 11:31-43.

Gibson, J.E. and K.S. Rao. 1973. Disposition of 2-sec-butyl-4,6-dinitrophenol(dinoseb) in pregnant mice. Fd. Cosmet. Toxicol. 11:45-52.

Hall, L., R. Linder, T. Scotti, R. Bruce, R. Moseman, T. Heidersheit, D. Hinkle, T. Edgerton, S. Chaney, J. Goldstein, M. Gage, J. Farmer, L. Bennett, J. Stevens, W. Durham and A. Curley. 1978. Subchronic and reproductive toxicity of dinoseb. Toxicol. Appl. Pharmacol. 45:235-236. Abstract.

Harris, C.I. and G.F. Warren. 1964. Adsorption and desorption of herbicides by soil. Weeds. 12:120.

Hazelton.* 1977. Hazelton Laboratories of America, Inc. 104-Week dietary study in rats: dinoseb DNBP. Final Report. Unpublished study. MRID 00211.

Heyndrickx, A., R. Maes and F. Tyberghein. 1964. Fatal intoxication by man due to dinitro-ortho-cresol (DNOC) and dinitro butylphenol (DNBP). Mededel Lanbovwhoge School Opzoekingstaa Staa Gent. 29:1189-1197.

Irvine, L.F.H.* 1981. 3-Generation reproduction study. Hazelton Laboratories Europe, Ltd.

Lehman, A. J. 1959. Appraisal of the safety of chemicals in foods, drugs and cosmetics. Assoc. Food Drug Off. U.S., Q. Bull.

Linder, R.E., T.M. Scotti, D.J. Svendsgaard, W.K. McElroy and A. Curley. 1982. Testicular effects of dinoseb in rats. Arch. Environ. Toxicol. 11:475-485.

Meister, R., ed. 1984. Farm chemicals handbook. Willoughby, OH: Meister Publishing Co.

Moriya, M., T. Ohta, T. Watanabe, K. Kato and Y. Shirasu. 1983. Further mutagenicity studies on pesticides in bacterial reversion assay systems. Mutat. Res. 116:185–216.

Noakes, D.N. and D.M. Sanderson. 1969. A method for determining the dermal toxicity of pesticides. Brit. J. Ind. Med. 26:59–64.

Research and Consulting Company. 1986. Embryotoxicity study with dinoseb technical grade in the rabbit (oral administration). Unpublished study.

Simmon, V.F., A.D. Mitchell and T.A. Jorgenson. 1977. Evaluation of selected pesticides as chemical mutagens *in vitro* and *in vivo* studies. EPA 600/1-77-028. Research Triangle Park, NC: U.S. Environmental Protection Agency.

Spencer, F. and L.T. Sing. 1982. Reproductive toxicity in pseudopregnant and pregnant rats following postimplantational exposure: effects of the herbicide dinoseb. Pestic. Biochem. Physiol. 18:150–157.

Spencer, H.C., V.K. Rowe, E.M. Adams and D.D. Irish. 1948. Toxicological studies on laboratory animals of certain alkyldinitrophenols used in agriculture. J. Ind. Hyg. Toxicol. 30:10–25.

STORET. 1988. STORET Water Quality File. Office of Water. U.S. Environmental Protection Agency (data file search conducted in May, 1988).

U.S. EPA. 1985. U.S. EPA Method 615 – Chlorinated phenoxy acids. Fed. Reg. 50:50701. October 4.

U.S. EPA. 1986. U.S. Environmental Protection Agency. Guidelines for carcinogen risk assessment. Fed. Reg. 51(185):33992–34003. September 24.

Waters, M.D., S. Shahbeg, S. Sandhu et al. 1982. Study of pesticide genotoxicity. Basic Life Sci. 21:275–326.

Weber, J.B. 1972. Interaction of organic pesticides with particulate matter in aquatic and soil systems. In: Gould, R.F., ed. Advances in Chemistry Series 111. Washington, DC: American Chemical Society.

Weber, W.J., Jr. and J.P. Gould. 1966. Sorption of organic pesticides from aqueous solution. In: Gould, R.F., ed. Advances in Chemistry Series 60. Washington, DC: American Chemical Society.

WSSA. 1983. Weed Science Society of America. Herbicide handbook, 5th ed. Champaign, IL: WSSA.

---

* Confidential Business Information submitted to the Office of Pesticide Programs.

Diphenamid

I. General Information and Properties

A. CAS No. 957–51–7

B. Structural Formula

N,N-Dimethyl-alpha-phenyl-benzeneacetamide

C. Synonyms

- Dymid; Enide (Meister, 1986).

D. Uses

- Diphenamid is used as a preemergent and selective herbicide for tomatoes, peanuts, alfalfa, soybean, cotton and other crops (Meister, 1986).

E. Properties (Windholz et al., 1983)

Chemical Formula	$C_{16}H_{17}ON$
Molecular Weight	239.30
Physical State (at 25°C)	White crystalline solid
Boiling Point	--
Melting Point	135°C
Density	--
Vapor Pressure (25°C)	--
Specific Gravity	--
Water Solubility (27°C)	260 mg/L
Log Octanol/Water Partition Coefficient	--
Taste Threshold	--

Odor Threshold   --
Conversion Factor   --

F. Occurrence

- Diphenamid has not been found in the 3 surface water samples collected at 2 locations or in 678 ground-water samples taken at 676 locations (STORET, 1988).

G. Environmental Fate

- Diphenamid is stable to hydrolysis at pH 5, 7 and 9 for 7, 12 and 10 days, respectively, at elevated temperatures (49°C or 120°F) (NOR-AM, 1986).

- Diphenamid is intermediately mobile (class 3) on silt loam and silty clay loam soil thin-layer chromatography (TLC) plates; on sandy loam, it is in class 5, indicating that it would leach readily in this soil (Helling and Turner, 1968).

II.   Pharmacokinetics

A. Absorption

- McMahon and Sullivan, (1965) administered 50 mg/kg diphenamid-carbonyl ^{14}C to rats by 3 different routes: oral, intraperitoneal (ip) or subcutaneous (sc) injection. Diphenamid was well absorbed by each of the three routes with peak blood levels being greatest for ip, followed by sc and oral. The half-lives of diphenamid in blood for each of the three routes were comparable.

B. Distribution

- No information was found in the available literature on the distribution of diphenamid.

C. Metabolism

- In the McMahon and Sullivan (1965) study described above, the main route of metabolism of diphenamid was reported as N-dealkylation to nondiphenamid. The nondiphenamid is excreted as the N-glucuranide. p-Hydroxylation was reported as a minor route of metabolism.

D. Excretion

• No specific information on the excretion of diphenamid was found in the available literature. However, in the above study, the authors evaluated the urinary metabolites of diphenamid and reported that diphenamid was readily absorbed and metabolized into excretable metabolites.

III.   Health Effects

A. Humans

• No information was found in the available literature on the health effects of diphenamid in humans.

B. Animals

1. *Short-term Exposure*

• RTECS (1987) reported the acute oral LD_{50} values in the rat, mouse, dog, monkey and rabbit to be 685, 600, 1,000, 1,000 and 1,500 mg/kg, respectively.

2. *Dermal/Ocular Effects*

• Weddon and Brown (1976) applied Enide 90W (a 93% wettable powder formulation of diphenamid) to the intact or abraded skin of New Zealand White rabbits (2/sex/dose) for 24 hours at 0, 200, 1,000 or 2,000 mg/kg. Neither skin irritation nor systemic effects were observed in any of the exposed animals.

3. *Long-term Exposure*

• Woodard et al. (1966b) administered technical diphenamid (purity not specified) in the feed to beagle dogs (3/sex/dose) at dose levels of 0, 3, 10 or 30 mg/kg/day for 103 weeks. No pathological effects were reported at 3 mg/kg/day for clinical chemistry, hematology, urinalysis, gross pathology and histopathology. Liver weights were slightly increased in the 10- and 30-mg/kg/day dosage groups of both sexes, and there were slight increases in numbers of portal macrophages and/or fibroblasts when compared to untreated controls. Liver enzyme levels were normal in all treated groups, except for elevation of serum glutamic-oxaloacetic transaminase (SGOT) after 8 weeks in one female dosed with 3 mg/kg/day. A No-Observed-Adverse-Effect Level

(NOAEL) of 3 mg/kg/day and a Lowest-Observed-Adverse-Effect Level (LOAEL) of 10 mg/kg/day were identified by this study.

- Hollingsworth et al. (1966) fed technical diphenamid (> 98% pure) to Charles River rats (30/sex/dose) at dose levels of 0, 3, 10 or 30 mg/kg/day for 101 weeks. A slight increase in the mean absolute liver weights of males and the relative liver and thyroid weights of females in the high-dose groups was observed. No other adverse effects were reported at 10 mg/kg/day or less in general behavior, food consumption, body and organ weights, hematology, gross pathology and histopathology. A NOAEL of 10 mg/kg/day and a LOAEL of 30 mg/kg/day are identified by this study.

4. *Reproductive Effects*

- In a three-generation reproduction study, Woodard et al. (1966a) administered technical diphenamid to Charles River albino rats (10 males and 20 females/dose) at dose levels of 0, 10 or 30 mg/kg/day. No reproductive or pathological effects were observed for the parental generations (F_0, F_{1b}, F_{2b}) at any dose tested. Weanlings of the F_{3b} generation dosed with 30 mg/kg/day showed reversible liver changes, including slight congestion, glycogen depletion and irregular size of the hepatocytes. Based on reproductive end points, this study identifies a NOAEL of 30 mg/kg/day. Based on fetal toxicity, a NOAEL of 10 mg/kg/day and a LOAEL of 30 mg/kg/day are identified.

5. *Developmental Effects*

- Woodard et al. (1966a) reported no developmental effects in rat pups at any dose level. Reversible liver changes were observed in weanling pups of the F_{3b} generation dosed with 30 mg/kg/day. A NOAEL based on fetotoxicity of 10 mg/kg/day can be identified.

6. *Mutagenicity*

- Moriya et al. (1983) reported that diphenamid (up to 5,000 μg/plate) did not increase reversion frequency in *S. typhimurium* or *E. coli* test systems, either with or without metabolic activation.

- Shirasu et al. (1976) reported that diphenamid (1%) was not mutagenic in a recombination assay utilizing *B. subtilis* or in reversion assays with *E. coli* or *S. typhimurium*.

7. *Carcinogenicity*

- In a 2-year feeding study in rats by Hollingsworth et al. (1966), diphenamid was administered to Charles River albino rats (30/sex/dose) at dose levels of 0, 3, 10 or 30 mg/kg/day for 101 weeks. Based on histopathological examination of a variety of tissues and organs, the authors reported that the type and incidence of neoplasms were comparable in treated and control rats.

- In a 2-year feeding study in dogs by Woodard et al. (1966b), diphenamid was administered in the feed to beagle dogs (3/sex/dose) at dosage levels of 0, 3, 10 or 30 mg/kg/day for 103 weeks. Histopathological examinations were performed on a variety of tissues and organs, and no evidence of increased tumor frequency was reported.

IV.   Quantification of Toxicological Effects

A. One-day Health Advisory

No information was found in the available literature that was suitable for determination of the One-day HA value for diphenamid. It is therefore recommended that the Drinking Water Equivalent Level (DWEL), adjusted for a 10-kg child (0.3 mg/L, calculated below), be used at this time as a conservative estimate of the One-day HA value.

For a 10-kg child, the adjusted DWEL is calculated as follows:

$$\text{DWEL} = \frac{(0.03 \text{ mg/kg/day}) (10 \text{ kg})}{(1 \text{ L/day})} = 0.3 \text{ mg/L}$$

where: 0.03 mg/kg/day = RfD (see Lifetime Health Advisory section).

B. Ten-day Health Advisory

No information was found in the available literature that was suitable for determination of the Ten-day HA value for diphenamid. It is therefore recommended that the DWEL, adjusted for a 10-kg child (0.3 mg/L) be used at this time as a conservative estimate of the Ten-day HA value.

C. Longer-term Health Advisory

No information was found in the available literature that was suitable for determination of the Longer-term HA value for diphenamid. It is therefore recommended that the DWEL value, adjusted for a 10-kg child (0.3 mg/L) be used at this time as a conservative estimate of the Longer-term HA value.

D. Lifetime Health Advisory

The feeding study in dogs by Woodard et al. (1966b) has been selected to serve as the basis for determination of the Lifetime HA value for diphenamid. In this study, dogs were administered technical diphenamid (0, 3, 10 or 30 mg/kg/day) in the diet for 103 weeks. Based on clinical chemistry, hematology, urinalysis, gross pathology and histopathology, this study identified a NOAEL of 3 mg/kg/day and a LOAEL of 10 mg/kg/day. The study by Hollingsworth et al. (1966), which identified a NOAEL of 10 mg/kg/day in a 101-week experiment in rats, was not selected since the rat appears to be somewhat less sensitive than the dog (the NOAEL in the rat is the same as the LOAEL in the dog).

Using a NOAEL of 3 mg/kg/day, the Lifetime HA is calculated as follows:

Step 1: Determination of the Reference Dose (RfD)

$$RfD = (3 \text{ mg/kg/day}) = 0.03 \text{ mg/kg/day}$$

Step 2: Determination of the Drinking Water Equivalent Level (DWEL)

$$DWEL = \frac{(0.03 \text{ mg/kg/day}) (70 \text{ kg})}{(2 \text{ L/day})} = 1.0 \text{ mg/L } (1{,}000 \text{ } \mu g/L)$$

Step 3: Determination of the Lifetime Health Advisory

$$\text{Lifetime HA} = (1.0 \text{ mg/L}) (20\%) = 0.2 \text{ mg/L } (200 \text{ } \mu g/L)$$

E. Evaluation of Carcinogenic Potential

- No evidence of carcinogenic potential was detected in rats (30/sex/dose) fed diphenamid in the diet for 2 years at a dose level of 30 mg/kg/day (Hollingsworth et al., 1966), or in dogs (3/sex/dose) fed diphenamid in the diet for 2 years, also at a dose of 30 mg/kg/day (Woodward et al., 1966b). These studies are limited by the low doses and the small number of animals employed.

- The International Agency for Research on Cancer has not evaluated the carcinogenic potential of diphenamid.

- Applying the criteria described in EPA's guidelines for assessment of carcinogenic risk (U.S. EPA, 1986), diphenamid is classified in Group D: not classifiable. This category is for substances with inadequate animal evidence of carcinogenicity.

V.   Other Criteria, Guidance and Standards

- Tolerances in or on raw agricultural commodities of 0.01 ppm for milk to 2 ppm for peanut hay and forage have been set for diphenamid (U.S. EPA, 1985).

VI.   Analytical Methods

- Analysis of diphenamid is by a gas chromatographic (GC) method applicable to the determination of certain nitrogen-phosphorus containing pesticides in water samples (U.S. EPA, 1988). In this method, approximately 1 liter of sample is extracted with methylene chloride. The extract is concentrated and the compounds are separated using capillary column GC. Measurement is made using a nitrogen phosphorus detector. The method detection limit has not been determined for diphenamid but it is estimated that the detection limits for analytes included in this method are in the range of 0.1 to 2 μg/L. This method has been validated in a single laboratory, and the estimated detection limit for the analytes, including diphenamid, is 0.6 μg/L.

VII.   Treatment Technologies

- Available data indicate that granular activated carbon (GAC) adsorption will remove diphenamid from water.

- Whittaker (1980) experimentally determined adsorption isotherms for diphenamid on GAC.

- Whittaker (1980) reported the results of GAC columns operating under bench-scale conditions. At a flow rate of 0.8 gpm/ft^2 and an empty bed contact time of 6 minutes, diphenamid breakthrough (when effluent concentration equals 10% of influent concentration) occurred after 500 bed volumes (BV). When two bi-solute diphenamid solutions were passed over the same column, diphenamid breakthrough occurred after 235 BV for diphenamid-propham solution and after 290 BV for diphenamidfluometuron solution.

- GAC adsorption appears to be the most effective treatment technique for the removal of diphenamid from contaminated water. However, selection of individual or combinations of technologies to attempt diphenamid removal from water must be based on a case-by-case technical evaluation, and an assessment of the economics involved.

VIII.   References

Helling, C.S. and B.C. Turner. 1968. Pesticide mobility: determination by soil TLC. Science. 16:562–563.

Hollingsworth, R.L., M.W. Woodard and G. Woodard.* 1966. Diphenamid safety evaluation by dietary feeding to rats for 101 weeks. Final Report. Unpublished study. MRID 00076381.

McMahon, R.E. and H.R. Sullivan. 1965. The metabolism of the herbicide diphenamid in rats. Biochem. Pharmacol. 14:1085–1092.

Meister, R.T., ed. 1986. Farm chemicals handbook. Willoughby, OH: Meister Publishing Co.

Moriya, M., T. Ohta, K. Watanabe, T. Miyazawa, K. Kato and Y. Shirasu. 1983. Further mutagenicity studies on pesticides in bacterial reversion assay systems. Mutat. Res. 116:185–216.

NOR-AM. 1986. NOR-AM Chemical Company. Diphenamid: hydrolysis study (groundwater data call-in). Unpublished study submitted to the Office of Pesticide Programs. Wilmington, DE.

RTECS. 1987. Registry of Toxic Effects of Chemical Substances. National Institute for Occupational Safety and Health. Washington, DC: National Library of Medicine On-Line File.

Shirasu, Y., M. Moriya, K. Kato, A. Furuhashi and T. Kada. 1976. Mutagenicity screening of pesticides in the microbial system. Mutat. Res. 40:19–30.

STORET. 1988. STORET Water Quality File. Office of Water. U.S. Environmental Protection Agency (data file search conducted in May, 1988).

TDB. 1985. Toxicology Data Bank. MEDLARS II. National Library of Medicine's National Interactive Retrieval Service.

U.S. EPA. 1985. U.S. Environmental Protection Agency. Code of Federal Regulations. 40 CFR 180.230.

U.S. EPA. 1986. U.S. Environmental Protection Agency. Guidelines for carcinogen risk assessment. Fed. Reg. 51(185):33992–34003. September 24.

U.S. EPA. 1988. U.S. Environmental Protection Agency. U.S. EPA Method 507 — Determination of nitrogen and phosphorus containing pesticides in water by GC/NPD. Draft. April 15. Available from U.S. EPA's Environmental Monitoring and Support Laboratory, Cincinnati, OH.

Weddon T.E. and P.K. Brown.* 1976. Enide 90 W — Dermal LD_{50} and skin irritation evaluation in New Zealand rabbits. Technical Report No. 124–9610-MWG-76-6. Unpublished study. MRID 00054611.

Whittaker, K.F. 1980. Adsorption of selected pesticides by activated carbon using isotherm and continuous flow column systems. Ph.D. Thesis. Purdue University.

Windholz, M., S. Budavari, R.F. Blumetti and E.S. Otterbein, eds. 1983. The Merck index, 10th ed. Rahway, NJ: Merck and Co., Inc.

Woodard M.W., G. Woodard and M.T. Cronin.* 1966a. Diphenamid: three-generation reproduction study in rats. Unpublished study. MRID 00076383.

Woodard M.W., G. Woodard and M.T. Cronin.* 1966b. Diphenamid safety evaluation by dietary feeding to dogs for 103 weeks. Final Report. Unpublished study. MRID 00076382.

---

* Confidential Business Information submitted to the Office of Pesticide Programs.

Disulfoton

I. General Information and Properties

A. CAS No. 298-04-4

B. Structural Formula

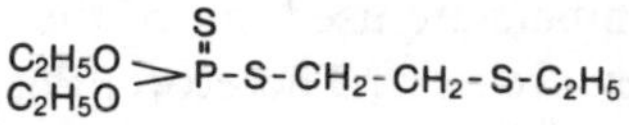

$$\begin{array}{c} C_2H_5O \\ C_2H_5O \end{array} \!\!> \overset{\overset{S}{\|}}{P} - S - CH_2 - CH_2 - S - C_2H_5$$

O,O-Diethyl-S-[2-(ethylthio)-ethyl], phosphorodithioate

C. Synonyms

- Disulfoton; Disyston; Disystox: Dithiodemeton; Bayer 19639; Disyston; Ethyl thiometon; Frumin AL; M-74 (Meister, 1983).

D. Uses

- Disulfoton is used as a systemic insecticide-acaricide (Meister, 1983).

E. Properties (Meister, 1983; Windholz et al., 1983)

Chemical Formula	$C_8H_{19}O_2PS_3$
Molecular Weight	274.38
Physical State (at 25°C)	Pale Yellow liquid
Boiling Point	108°C (0.01 mm Hg); 132 to 133°C (1.5 mm Hg)
Melting Point	--
Density (20°C)	1.144
Vapor Pressure (at 20°C)	1.8×10^{-4} mm Hg
Specific Gravity	--
Water Solubility (at 23°C)	25 mg/L
Log Octanol/Water Partition Coefficient	--
Taste Threshold	--
Odor Threshold	--
Conversion Factor	--

F. Occurrence

- Disulfoton has been found in only 1 of 368 surface water samples and none of 1,182 ground-water samples analyzed (STORET, 1988). The concentration of the surface water sample was 0.34 μg/L. Samples were collected at 93 surface water locations and 1,080 ground-water locations. This was also the maximum concentration detected. Disulfoton was detected only in California. This information is provided to give a general impression of the occurrence of this chemical in ground and surface waters as reported in the STORET database. The individual data points retrieved were used as they came from STORET and have not been confirmed as to their validity. STORET data are often not valid when individual numbers are used out of the context of the entire sampling regime, as they are here. Therefore, this information can only be used to form an impression of the intensity and location of sampling for a particular chemical.

G. Environmental Fate

- Disulfoton has a low mobility in Hugo sandy loam soil; 28% of the pesticide applied to a 6-inch-high soil column was eluted with a total of 110 feet of dilute buffer (McCarty and King, 1966). In another study, disulfoton sulfoxide and disulfoton sulfone were more mobile in sandy loam, clay loam and silty clay loam soils than the parent compound. Aging ^{32}P-disulfoton prior to elution increased the adsorption to 10 to 20 times that of unaged ^{32}P-disulfoton, meaning that the degradates are probably less mobile than the parent compound. Mobility of disulfoton in soil appears to decrease as organic matter content and cation exchange capacity (EC) increase (Kawamori et al., 1971a, 1971b).

- Insecticidal disulfoton residues are mobile in a silt loam soil as determined by a mosquito bioassay (Lichtenstein et al., 1966). Based on soil thin-layer chromatography (TLC) plates, disulfoton has low mobility in sand (R_f 0.18), sandy loam (R_f 0.16) and silt loam (R_f 0.11 and 0.33), and intermediate mobility in sandy clay loam soil (R_f 0.39) (Thornton et al., 1976).

- When applied to subirrigated soil columns at 20 lb active ingredient (a.i.) per acre, disulfoton exhibited slight upward mobility in a Hagerstown silty clay loam and a Lakeland sandy loam soil (Mobay Chemical Corporation, 1972). Disulfoton (6 lb/gal EC), applied at 4 lb a.i./A to sloping field plots (1 inch/ft) of sandy loam, silt loam and highly organic silt loam soils, was slightly mobile in runoff water. Disulfoton concentrations measured about 1.6% of applied amounts over a 28-day period in which 1.5 to 2.5 inches of irrigation were applied (Flint et al., 1970).

• Disulfoton (granular, G) dissipates rapidly in field plots of sandy loam soil treated at 2 kg/ha (incorporated to a depth of 10 cm), with a half-life of 1 week, and 90% loss after 5 weeks. Disulfoton sulfoxide has a half-life of 8 to 10 weeks, while disulfoton sulfone remains fairly stable over a 42-week period. Disulfoton sulfone was detected at a depth of 20 cm (Suett, 1975). Disulfoton residues dissipated with half-lives of 1 to 6 months in muck-sand, silt loam, and clay soils treated with disulfoton 10% G or 6 lb/gal EC at 10 ppm. Dissipation from the upper 6 inches was enhanced by increasing amounts of rainfall (Chemagro Corporation, 1969; Loeffler, 1969; Mobay Chemical Corporation, 1964).

• Disulfoton is likely to be found in runoff water and sediment from treated and cultivated fields. In monitoring studies conducted in 1973 and 1974, disulfoton was found at average concentrations of 13.8 parts per billion (ppb) in sediment samples taken from tailwater pits receiving irrigation and rainfall runoff water from cultivated silt loam corn fields. The maximum concentration in sediment samples was 32.7 ppb. The compound was also detected in soil samples from a tailwater pit draining silt loam corn and sorghum fields at an average concentration of 11 ppb. Sediment samples in tailwater pits draining sorghum fields contained disulfoton at a concentration of 117.2 ppb (Kadoum and Mock, 1978).

II.   Pharmacokinetics

A. Absorption

• Puhl and Fredrickson (1975) administered by gavage single oral doses of disulfoton-o-ethyl-1-[14]C (99% purity) to Sprague-Dawley rats (12/sex/dose). Males received 1.2 mg/kg and females received 0.2 mg/kg. In the 10 days following dosing, averages of 81.6, 7.0 and 9.2% of the dose were recovered in the urine, feces and expired air, respectively. Males excreted 50% of the administered dose in the urine in the first 4 to 6 hours; females required 30 to 32 hours. These data indicate that disulfoton is absorbed readily from the gastrointestinal tract.

B. Distribution

- In the study by Puhl and Fredrickson (1975), described above, 4.1 and 16.1% of the administered dose were detected in the livers of males and females, respectively, and 0.4 and 1.2% of the dose were detected in the kidneys of males and females, respectively, 48 hours postdosing.

C. Metabolism

- March et al. (1957) studied the metabolism of disulfoton *in vivo* and *in vitro* in mice (strain not specified). In the *in vivo* portion of the study, mice received radiolabeled disulfoton intraperitoneally (dose not specified). Results indicated that unspecified urinary metabolites consisted mainly of hydrolysis products. *In vitro* metabolism data indicated the presence of dithio-systox sulfoxide and sulfone, and the thiol analog sulfoxide and sulfone. The dithiosystox sulfoxide was present in the greatest quantity followed by thiol analog sulfoxide, dithio-systox sulfone and thiol analog sulfone. Based on a review of these data (U.S. EPA, 1984), it was concluded that the metabolism of disulfoton in mice involves at least two reactions: (1) the sequential oxidation of the thioether sulfur and/or oxidative desulfuration; and (2) hydrolytic cleavage of the ester, producing phosphoric acid, thiophosphoric acid and dithiophosphoric acid.

- In the above study by Puhl and Fredrickson (1975), the major urinary metabolites detected in both sexes were diethylphosphate (DFP) and diethylphosphorothioate (DEPT). These products were formed from hydrolysis of disulfoton and/or its oxidation products. Minor urinary metabolites included the oxygen analog sulfoxide, oxygen analog sulfone and disulfoton sulfoxide.

D. Excretion

- In the above study by Puhl and Fredrickson (1975), 97.8% of the administered dose was recovered (81.6% in urine, 7.0% in feces and 9.2% as expired carbon dioxide) during a 10-day postdosing period. Excretory pathways were similar for males and females, but the rate of excretion was slower for females.

III. Health Effects

A. Humans

1. *Short-term Exposure*

- No significant anticholinesterase effects were observed in human subjects (five test subjects, two controls) who received disulfoton in doses of 0.75 mg/day (orally) for 30 days (Rider et al., 1972).

- Quinby (1977) reported that three carpenters were sprayed accidentally with disulfoton while the compound was being applied by airplane to a wheat field adjacent to their work site. The individuals were reexposed as they handled contaminated building materials in the days following spraying. Exposure levels were not identified. The older two carpenters experienced coronary attacks and one had at least two severe cerebral vascular effects subsequent to exposure. The author postulated that the effects may have been due to disturbances of clotting mechanisms.

2. *Long-term Exposure*

- No long-term human studies were identified for disulfoton.

B. Animals

1. *Short-term Exposure*

- Reported acute oral LD_{50} values for adult rats administered disulfoton (approximately 94 to 96% purity when identified) ranged from 1.9 to 2.6 mg/kg for females and 6.2 to 12.5 mg/kg for males (Crawford and Anderson, 1974; Bombinski and DuBois, 1957); a value of 5.4 mg/kg was reported for weanling male rats (Brodeur and DuBois, 1963).

- In guinea pigs, acute oral LD_{50} values ranged from 8.9 to 12.7 mg/kg (Bombinski and DuBois, 1957: Crawford and Anderson, 1973).

- Mihail (1978) reported acute oral LD_{50} values of 7.0 mg/kg and 8.2 mg/kg in male and female NMRI mice, respectively.

- Hixson (1982) reported that the acute oral LD_{50} of disulfoton (97.8% pure) in white Leghorn hens was 27.5 mg/kg. Hixson (1983) reported the results of an acute delayed neurotoxicity study in which 20 white Leghorn hens were administered technical disulfoton (97.8% pure) by gavage at a dose level of 30 mg/kg on two occasions, 22 days apart. The study also employed groups of 5, 10 and 5 hens for positive controls, antidote controls and negative controls, respectively. Disulfoton did not produce acute delayed neurotoxicity under the conditions of this study.

Based on this information, a No-Observed-Adverse-Effect Level (NOAEL) of 30 mg/kg (the only dose tested) was identified in this study.

- Taylor (1965) reported the results of a demyelination study in which white Leghorn hens (6/dose) were administered disulfoton in the diet for 30 days at concentrations of 0, 2, 10 or 25 ppm; these dietary levels are equivalent to 0, 0.1, 0.6 and 1.5 mg/kg/day. The author indicated that no evidence of demyelination was observed in any of the tissues examined. Based on this information, a NOAEL of 1.5 mg/kg/day (the highest dose tested) was identified.

2. *Dermal/Ocular Effects*

- DuBois (1957) reported that the acute dermal LD_{50} of technical disulfoton in male Sprague-Dawley rats was 20 mg/kg. Mihail (1976) reported acute dermal LD_{50} values of 15.9 mg/kg and 3.6 mg/kg in male and female Wistar rats, respectively.

- No information was found in the available literature on the effects of ocular exposure to disulfoton.

3. *Long-term Exposure*

- Hayes (1983) presented the results of a 23-month feeding study in which CD-1 mice (50/sex/dose) were administered disulfoton (98.2% pure) at dietary concentrations of 0, 1, 4 or 16 ppm. Assuming that 1 ppm in the diet of mice is equivalent to 0.15 mg/kg/day (Lehman, 1959), these dietary levels correspond to doses of about 0, 0.15, 0.6 and 2.4 mg/kg/day. No treatment-related effects were observed in terms of body weight, food consumption or hematology. A statistically significant increase in mean kidney weight and kidney-to-body weight ratio was noted in high-dose females; this increase may have been associated with a nonsignificant increase in the incidence of malignant lymphomas of kidneys in this group. Plasma, red blood cell and brain cholinesterase (ChE) activity was decreased significantly in both sexes at the highest dose tested (16 ppm). However, since ChE activity was measured only in the control and high-dose groups, a NOAEL for this effect could not be determined.

- In a study by Hoffman et al. (1975), beagle dogs (4/sex/dose) were administered disulfoton (95.7% pure) at dietary concentrations of 0, 0.5 or 1.0 ppm for 2 years. Assuming that 1 ppm in the diet of dogs is equivalent to 0.025 mg/kg/day (Lehman, 1959), these dietary levels correspond to doses of about 0, 0.0125 and 0.025 mg/kg/day. A fourth group of animals received disulfoton in the diet at 2 ppm for 69 weeks, then 5 ppm for weeks 70 to 72, and finally 8 ppm from week 73 until termination (104 weeks); these doses correspond to 0.05, 0.125 and 0.2 mg/kg/day, respectively. No treatment-related effects were observed in terms of general appearance, behavior, ophthalmoscopic examinations, food consumption, body weight, organ weight, hematology, clinical chemistry or histopathology. Additionally, no effects on ChE activity were observed in animals that received 0.5 or 1.0 ppm (0.0125 or 0.025 mg/kg/day). However, exposure at 2.0 ppm (0.05 mg/kg/day) for 69 weeks caused ChE inhibition in plasma and red blood cells in both sexes. Maximum inhibition occurred at week 40, when males exhibited 50% and 33% inhibition of ChE in red blood cells and plasma, respectively; females exhibited 22 and 36% inhibition of ChE in red blood cells and plasma, respectively. At a dose level of 8 ppm (0.2 mg/kg/day), males exhibited 56 to 66% and 63 to 70% inhibition of red blood cell and plasma ChE, respectively; females exhibited 46 to 53% and 54 to 64% inhibition of red blood cell and plasma ChE, respectively. Based on these data, a NOAEL of 1.0 ppm (0.025 mg/kg/day) was identified.

• Carpy et al. (1975) presented the results of a 2-year feeding study in which Sprague-Dawley rats (60/sex/dose) were administered disulfoton (95.7% pure) at dietary concentrations of 0, 0.5, 1.0 or 2.0 ppm. Based on data presented by the authors, these dietary levels correspond to doses of about 0, 0.02, 0.05 and 0.1 mg/kg/day for males and 0, 0.03, 0.04 and 0.1 mg/kg/day for females. At week 81 of the study, the 0.5-ppm dose was increased to 5.0 ppm (0.2 and 0.3 mg/kg/day for males and females, respectively) since no adverse effects were observed in the 1.0-ppm dose group. No treatment-related effects were observed in terms of food consumption, body weight, hematology, clinical chemistry, urinalysis and histopathology. A trend was observed at all dose levels toward increased absolute and relative spleen, liver, kidney and pituitary weights in males and toward decreased weights of these organs in females. In males receiving 5 ppm, the increases were statistically significant (p < 0.05) for absolute spleen and liver weights. In females receiving 5 ppm, the decrease in absolute and relative kidney weights was statistically significant (p < 0.05). At all levels tested, the brain showed a trend toward decreased absolute and relative weights in males and increased weights in females. Additionally, statistically significant inhibition of plasma, red blood cell and brain ChE was observed in both sexes at 2.0 and 5.0 ppm. At 1.0 ppm brain ChE in females was inhibited 11% (p < 0.01). Based on this information, a Lowest-Observed-Adverse-Effect Level (LOAEL) of 1.0 ppm (0.04 mg/kg/day for females) was identified for ChE inhibition. It was concluded (U.S. EPA, 1984) that a NOAEL for systemic toxicity could not be identified due to the inadequacy of histopathology and necropsy data.

- Hayes (1985) presented the results of a 2-year feeding study in which Fischer 344 rats (60/sex/dose) were administered disulfoton (97.91% pure) at dietary concentrations of 0, 0.8, 3.3 or 13 ppm. Assuming that 1 ppm in the diet of rats is equivalent to 0.05 mg/kg/day (Lehman, 1959), these dietary levels correspond to doses of about 0, 0.04, 0.17 and 0.65 mg/kg/day. Mortality was generally low for all groups with the exception of increased mortality in high-dose females during the last week of the study. No compound-related effects were observed in terms of clinical chemistry, hematology or urinalysis. A dose-related trend in ChE inhibition was observed in both sexes at all dose levels. Statistically significant inhibition of plasma, red blood cell and brain ChE occurred in all dose groups throughout the study. Histopathologically, a statistically significant increase ($p < 0.05$) in corneal neovascularization was observed in both sexes at 13 ppm (0.65 mg/kg/day). A dose-related increase in the incidence of optic nerve degeneration was also observed at all doses tested. This effect was statistically significant ($p < 0.05$) in mid-dose males and in mid- and high-dose females. Additionally, a significantly ($p < 0.05$) higher incidence of cystic degeneration of the Harderian gland was observed in females at all doses and in mid-dose males. A significantly ($p < 0.05$) increased incidence of atrophy of the pancreas also was observed in high-dose males. This study identified a LOAEL of 0.8 ppm (0.04 mg/kg/day) (lowest dose tested) for both optic nerve degeneration and ChE inhibition.

4. Reproductive Effects

- Taylor (1966) conducted a three-generation reproduction study in which albino Holtzman rats (20 females and 10 males) were administered disulfoton (98.5% pure) at dietary concentrations of 0, 2, 5 or 10 ppm. Assuming that 1 ppm in the diet of rats is equivalent to 0.05 mg/kg/day (Lehman, 1959), these dietary levels correspond to doses of about 0, 0.1, 0.25 and 0.5 mg/kg/day. At 10 ppm (0.5 mg/kg/day), litter size was reduced by 21% in the F_a and 33% in the F_b in both the first and third generations. Also in these generations, a 10 to 25% lower pregnancy rate was noted for F_a matings. Histopathologically, F_{3b} litters at 10 ppm (0.5 mg/kg/day) exhibited cloudy swelling and fatty infiltration of the liver (both sexes), mild nephropathy in kidneys (both sexes) and juvenile hypoplasia of the testes. No histopathological examinations were conducted on the 2- and 5-ppm dose groups. Cholinesterase determinations revealed a 60 to 70% inhibition of red blood cell ChE in F_{3b} litters and their parents at 5 and 10 ppm (0.25 and 0.5 mg/kg/day). At 2 ppm (0.1 mg/kg/day), the inhibition was insignificant in males and moderate (32 to 42%) in females. Based on these data, a LOAEL of 2 ppm (0.1 mg/kg/day) was identified for ChE inhibition. It was concluded (U.S. EPA, 1984a) that a reproductive NOAEL could not be determined due to deficiencies in data reporting (e.g., insufficient data on reproductive parameters, no statistical analyses, incomplete necropsy report and insufficient histopathology data).

5. Developmental Effects

• Lamb and Hixson (1983) conducted a study in which CD rats (25/dose) were administered disulfoton (98.2% pure) by gavage at levels of 0, 0.1, 0.3 or 1 mg/kg/day on days 6 through 15 of gestation. Mean plasma and red blood cell ChE activities were decreased significantly in dams receiving 0.3 and 1 mg/kg/day. Examination of the fetuses after Cesarean section reflected no increases in the incidence of soft tissue, external or skeletal abnormalities. However, at the 1.0 mg/kg/day dose level, increased incidences of incompletely ossified parietal bones and sternebrae were observed. This is considered a fetotoxic effect, since it is indicative of retarded development. Based on the information presented in this study, a developmental NOAEL of 0.3 mg/kg/day was identified based on fetotoxic effects. A NOAEL of 0.1 mg/kg/day was identified for ChE inhibition in treated dams.

• Tesh et al. (1982) conducted a teratogenicity study in which New Zealand White rabbits were administered disulfoton (97.3% pure) at initial doses of 0, 0.3, 1.0 or 3.0 mg/kg on days 6 through 18 of gestation. Due to mortality and signs of toxicity, the high dose was reduced to 2.0 mg/kg/day and finally to 1.5 mg/kg/day. The control group consisted of 15 animals, the low- and mid-dose groups consisted of 14 does each and the high-dose group contained 22 animals. No signs of maternal toxicity were observed in the low- or mid-dose groups. In the high-dose group, signs of maternal toxicity included muscular tremors, unsteadiness and incoordination, increased respiratory rate and increased mortality. No compound-related effects on maternal body weight or fetal survival, growth and development were observed. Based on this information, a NOAEL of 1.0 mg/kg/day was identified for maternal toxicity. The NOAEL for teratogenic and fetotoxic effects was 1.5 mg/kg/day (the highest dose tested).

6. *Mutagenicity*

- Hanna and Dyer (1975) reported that disulfoton (99.3% pure) was mutagenic in *Salmonella typhimurium* strains C 117, G 46, TA 1530 and TA 1535, and in *Escherichia coli* strains WP 2, Wp 2uvrA, CM 571, CM 611, WP 67 and WP 12. These tests were performed without metabolic activation; however, demeton-S-methyl sulphoxide, the major metabolite of disulfoton, was also mutagenic in these microbial tests (U.S. EPA, 1984a).

- Simmon (1979) presented the results of an unscheduled DNA synthesis assay using human fibroblasts (W I38). Disulfoton (purity not specified) was positive in this assay only in the absence of metabolic activation.

7. *Carcinogenicity*

- Carpy et al. (1975) presented the results of a 2-year feeding study in which Sprague-Dawley rats (60/sex/dose) were administered disulfoton (95.7% pure) at dietary concentrations of 0, 0.5, 1.0 or 2.0 ppm. Based on data presented by the authors, these dietary levels correspond to doses of about 0, 0.02, 0.05 and 0.1 mg/kg/day for males and 0, 0.03, 0.04 and 0.1 mg/kg/day for females. At week 81 of the study, the 0.5-ppm dose was increased to 5.0 ppm (reported to be equivalent to 0.2 and 0.3 mg/kg/day for males and females, respectively) since no adverse effects were observed in the 1.0-ppm dose group. The number of tumor-bearing animals at all dose levels was comparable to that of controls suggesting that, under the conditions of this study, disulfoton is not oncogenic. However, a review of this study (U.S. EPA, 1984) concluded that due to numerous deficiencies (e.g., invalid high dose, insufficient necropsy data, insufficient histology data), the data presented were inadequate for an oncogenic evaluation.

- Hayes (1983) presented the results of a 23-month feeding study in which CD-1 mice (50/sex/dose) were administered disulfoton (98.2% pure) at dietary concentrations of 0, 1, 4 or 16 ppm. Assuming that 1 ppm in the diet of mice is equivalent to 0.15 mg/kg/day (Lehman, 1959), these dietary levels correspond to doses of about 0, 0.15, 0.6 and 2.4 mg/kg/day. The incidence of specific neoplasms was similar among treated and control animals. There was an increased incidence of malignant lymphoma (the most frequently observed neoplastic lesion) in both males and females at 16 ppm (2.4 mg/kg/day) when compared with controls, but this change was not statistically significant. Therefore, under the conditions of this study, disulfoton was not oncogenic in mice at dietary concentrations up to 16 ppm (2.4 mg/kg/day).

• Hayes (1985) presented the results of a 2-year feeding study in which Fischer 344 rats (60/sex/dose) were administered disulfoton (97.91% pure) at dietary concentrations of 0, 0.8, 3.3 or 13 ppm, corresponding to doses of about 0, 0.04, 0.17 and 0.65 mg/kg/day (Lehman, 1959). The most commonly occurring neoplastic lesions included leukemia, adenoma of the adrenal cortex, pheochromocytoma, fibroadenoma of the mammary glands, adenoma and carcinoma of the pituitary glands, interstitial cell adenoma of the testes, and uterine stromal polyps. The incidences of these lesions showed no dose-related trend and were not significantly different in treated versus control animals. Therefore, under the conditions of this assay, disulfoton was not oncogenic in male or female Fischer 344 rats at dietary concentrations up to 13 ppm (0.65 mg/kg/day).

IV.   Quantification of Toxicological Effects

A. One-day Health Advisory

No suitable information was found in the available literature for the determination of a One-day HA value for disulfoton. It is therefore recommended that the Ten-day HA value for a 10-kg child of 0.01 mg/L (10 μg/L), calculated below, be used at this time as a conservative estimate of the One-day HA value.

B. Ten-day Health Advisory

The developmental toxicity study by Lamb and Hixson (1983) has been selected to serve as the basis for the Ten-day HA value for disulfoton. In this study, CD rats were administered disulfoton (98.2% pure) by gavage at doses of 0, 0.1, 0.3 or 1 mg/kg/day on days 6 through 15 of gestation. Mean plasma and red blood cell ChE activities were decreased significantly in dams receiving 0.3 and 1 mg/kg/day. Based on this information, a NOAEL of 0.1 mg/kg/day was identified. The only other study of comparable duration was a rabbit teratology study (Tesh et al., 1982). This study identified NOAELs of 1.0 mg/kg/day for maternal toxicity and 1.5 mg/kg/day (the highest dose tested) for developmental toxicity. The rabbit appeared to be less sensitive to disulfoton than the rat, therefore the rat study was selected for this calculation.

Using a NOAEL of 0.1 mg/kg/day, the Ten-day HA for a 10-kg child is calculated as follows:

$$\text{Ten-day HA} = \frac{(0.1 \text{ mg/kg/day}) (10 \text{ kg})}{(100) (1 \text{ L/day})} = 0.01 \text{ mg/L} (10 \text{ } \mu\text{g/L})$$

C. Longer-term Health Advisory

The 2-year dog feeding study by Hoffman et al. (1975) has been selected to serve as the basis for the Longer-term HA values for disulfoton. In this study, beagle dogs were administered disulfoton (95.7% pure) at dietary concentrations of 0, 0.5 or 1.0 ppm (0, 0.0125 and 0.025 mg/kg/day). A fourth group of dogs received disulfoton at 2.0 ppm (0.05 mg/kg/day) for 69 weeks, then 5.0 ppm (0.125 mg/kg/day) for weeks 70 to 72, and finally 8.0 ppm (0.2 mg/kg/day) from weeks 73 to 104. Exposure to 2.0 ppm (0.05 mg/kg/day) for 69 weeks caused plasma and red blood cell ChE inhibition in both sexes. Brain ChE inhibition was also noted at termination in this group. Based on this information, a NOAEL of 1.0 ppm (0.025 mg/kg/day) was identified. No other suitable studies were available for consideration for the Longer-term HA. Since the effects in the study by Hoffman et al. (1975) were observed following 69 weeks of exposure, the study is considered to be of appropriate duration for derivation of a Longer-term HA.

Using a NOAEL of 0.025 mg/kg/day, the Longer-term HA for a 10-kg child is calculated as follows:

$$\text{Longer-term HA} = \frac{(0.025 \text{ mg/kg/day}) (10 \text{ kg})}{(100) (1 \text{ L/day})} = 0.0025 \text{ mg/L} (3 \text{ } \mu\text{g/L})$$

Using a NOAEL of 0.025 mg/kg/day, the Longer-term HA for a 70-kg adult is calculated as follows:

$$\text{Longer-term HA} = \frac{(0.025 \text{ mg/kg/day}) (70 \text{ kg})}{(100) (2 \text{ L/day})} = 0.0088 \text{ mg/L} (9 \text{ } \mu\text{g/L})$$

D. Lifetime Health Advisory

The studies in rats by Hayes (1985) and Carpy et al. (1975) have been selected to serve as the bases for the Lifetime HA values for disulfoton. Each of these studies identifies a LOAEL of 0.04 mg/kg/day. In the Hayes (1985) study, Fischer 344 rats were administered disulfoton (98.1% pure) at dietary concentrations of 0, 0.8, 3.3 or 13 ppm (0, 0.04, 0.17 and 0.65 mg/kg/day) for 2 years. Dose-related, statistically significant inhibition of ChE in plasma, red blood cell and brain was observed in both sexes at all doses; also, a dose-related optic nerve degeneration was observed in females. Based on this information, a LOAEL of 0.04 mg/kg/day was identified. In the Carpy et al. (1975)

2-year study, Sprague-Dawley rats were administered disulfoton (95.7% pure) at dietary concentrations of 0, 0.5, 1.0 or 2.0 ppm (0, 0.02, 0.05 and 0.1 mg/kg/day for males and 0, 0.03, 0.04 and 0.1 mg/kg/day for females). At week 81 of the study, the 0.5 ppm dose was increased to 5.0 ppm (equivalent to 0.2 and 0.3 mg/kg/day for males and females, respectively). Statistically significant inhibition of plasma and red blood cell ChE was observed in both sexes at 2.0 and 5.0 ppm. Additionally, at 1 ppm (0.04 mg/kg/day), brain ChE was inhibited significantly ($p < 0.01$) in females. Since the initial low dose used in the study (0.5 ppm) was raised to 5.0 ppm, the 1.0-ppm dose is the lowest dose tested and represents the study LOAEL.

The Carpy et al. (1975) study supports the LOAEL of 0.04 mg/kg/day for ChE inhibition and for optic nerve degeneration in the Hayes (1985) study. Using this LOAEL, the Lifetime HA is calculated as follows:

Step 1: Determination of the Reference Dose (RfD)

$$RfD = \frac{(0.04 \text{ mg/kg/day})}{(1,000)} = 0.00004 \text{ mg/kg/day}$$

Step 2: Determination of the Drinking Water Equivalent Level (DWEL)

$$DWEL = \frac{(0.00004 \text{ mg/kg/day}) (70 \text{ kg})}{(2 \text{ L/day})} = 0.0014 \text{ mg/L } (1 \text{ } \mu g/L)$$

Step 3: Determination of the Lifetime Health Advisory (HA)

$$\text{Lifetime HA} = (0.0014 \text{ mg/L})(20\%) = 0.0003 \text{ mg/L } (0.3 \text{ } \mu g/L)$$

E. Evaluation of Carcinogenic Potential

- Three studies were available on the carcinogenicity of disulfoton. The chronic study in rats by Carpy et al. (1975) was inadequate for an oncogenic evaluation. The remaining two studies presented results indicating that disulfoton was not carcinogenic in mice (Hayes, 1983) or in rats (Hayes, 1985).

- The International Agency for Research on Cancer has not evaluated the carcinogenicity of disulfoton.

- Applying the criteria described in EPA's guidelines for assessment of carcinogenic risk (U.S. EPA, 1986), disulfoton may be classified in Group E: no evidence of carcinogenicity in humans. This category is used for substances that show no evidence of carcinogenicity in at least two adequate animal tests or in both epidemiologic and animal studies. However, disulfoton and its metabolites are mutagenic compounds (see section on Mutagenicity).

V. Other Criteria, Guidance and Standards

- The National Academy of Sciences (NAS, 1977) has calculated an ADI of 0.0001 mg/kg/day, based on a NOAEL of 0.01 mg/kg/day from a subchronic dog feeding study on phorate (a closely related organophosphorus insecticide) and an uncertainty factor of 100, with a Suggested-No-Adverse-Response-Level (SNARL) of 0.0007 mg/L.

- The World Health Organization (WHO, 1976) has identified an ADI of 0.002 mg/kg/day based on chronic data from a 2-year chronic feeding study in dogs (Hoffman et al., 1975) with a NOAEL of 0.025 mg/kg/day.

- U.S. EPA Office of Pesticide Programs (OPP) has established residue tolerances for disulfoton at 0.1 to 0.75 ppm in or on a variety of raw agricultural commodities (U.S. EPA, 1985). At the present time, these tolerances are based on a Provisional ADI (PADI) of 0.00004 mg/kg/day. As for the RfD calculation, this PADI is calculated based on a LOAEL of 0.8 ppm (0.04 mg/kg/day) for both ChE inhibition and optic nerve degeneration that were identified in the 2-year rat feeding study by Hayes (1985) and using a safety factor of 1,000.

VI. Analytical Methods

- Analysis of disulfoton is by a gas chromatographic (GC) method applicable to the determination of certain nitrogen-phosphorus-containing pesticides in water samples (U.S. EPA, 1988). In this method, approximately 1 L of sample is extracted with methylene chloride. The extract is concentrated and the compounds are separated using capillary column GC. Measurement is made using a nitrogen-phosphorus detector. The method has been validated in a single laboratory and the estimated detection limit for the analytes in this method, including disulfoton, is 0.3 μg/L.

VII. Treatment Technologies

- No information was found in the available literature regarding treatment technologies used to remove disulfoton from contaminated water.

VIII.   References

Bombinski, T.J. and K.P. DuBois.* 1957. The acute mammalian toxicity and pharmacological actions of Di-Syston. Report No. 1732. Unpublished study received Nov. 20, 1957 under 3125-58. Prepared by University of Chicago, Dept. of Pharmacology. Submitted by Mobay Chemical Corp., Kansas City, MO. CDL:100153-B. MRID 00068347.

Brodeur, J. and K.P. Dubois.* 1963. Comparison of acute toxicity of anticholinesterase insecticides to weanling and adult male rats. In Proceedings of the Society for Experimental Biology and Medicine. Vol. 114. MRID 05004291. New York, NY: Academic Press. pp. 509-511.

Carpy, S., C. Klotzsche and A. Cerioli.* 1975. Disulfoton: 2-year feeding study in rats: AGRO DOK CBK 1854/74. Report No. 47069. Unpublished study received December 15, 1976 under 3125-58. Prepared by Sandoz, Ltd., Switzerland. Submitted by Mobay Chemical Corp., Kansas City, MO. CDL:095641-C. MRID 00069966.

Chemagro Corporation. 1969. Di-Syston soil persistence studies. Unpublished study.

Crawford, C.R. and R.H. Anderson.* 1973. The acute oral toxicity of Di-Syston technical to guinea pigs. Report No. 39113. Unpublished study received December 15, 1976 under 3125-58. Submitted by Mobay Chemical Corp., Kansas City, MO. CDL:095640-F. MRID 00071872.

Crawford, C.R. and R.H. Anderson.* 1974. The acute oral toxicity of several Di-Syston metabolites to female and male rats. Report No. 39687. Unpublished study received December 15, 1976 under 3125-58. Submitted by Mobay Chemical Corp., Kansas City, MO. CDL:095640-G. MRID 00071873.

Doull, J.* 1957. The acute inhalation toxicity of Di-Syston to rats and mice. Report No. 1802. Unpublished study received November 20, 1957 under 3125-58. Prepared by University of Chicago, Dept. of Pharmacology. Submitted by Mobay Chemical Corp., Kansas City, MO. CDL:100153-D. Fiche Master ID 00069349.

DuBois, K.P.* 1957. The dermal toxicity of Di-Syston to rats. Report No. 2063. Unpublished study received January 23, 1958 under unknown admin.no. Prepared by University of Chicago, Dept. of Pharmacology. Submitted by Mobay Chemical Corp., Kansas City, MO. CDL:109216-8. MRID 00043213.

Flint, D.R., D.D. Church, H.R. Shaw and J. Armour II. 1970. Soil runoff, leaching and adsorption, and water stability studies with Di-Syston: Report No. 2899. Unpublished study submitted by Mobay Chemical Corp., Kansas City, MO.

Hanna, P.J. and K.F. Dyer. 1975. Mutagenicity of organophosphorus compounds in bacteria and *Drosophila*. Mutat. Res. 28:405–420.

Hayes, R.H.* 1983. Oncogenicity study of disulfoton technical on mice. Unpublished Toxicology Report No. 413 of Study No. 80–271–04. Prepared by the Corporate Toxicology Department, Mobay Chemical Corp., Stilwell, KS. Dated Aug. 10, 1983.

Hayes, R.H.* 1985. Chronic feeding/oncogenicity study of technical disulfoton (Di-Syston) with rats. Unpublished study No. 82–271–01. Prepared by Mobay Chemical Corp. Accession No. 258557.

Hixson, E.J.* 1982. Acute oral toxicity of Di-Syston technical in hens. Unpublished report. No. 341. Study No. 82–018–01. Prepared by the Environmental Health Research Institute of Mobay Chemical Corp., Stilwell, KS. Dated Oct. 25, 1982. MRID 00139596.

Hixson, E.J.* 1983. Acute delayed neurotoxicity study on disulfoton. Toxicology Report No. 365. Study No. 82–418–01. Prepared by Agricultural Chemicals Division, Mobay Chemical Corp., Kansas City, MO. Dated Mar. 7, 1983. MRID 00129384.

Hoffman, K., C.H. Weischer, G. Luckhaus et al.* 1975. S 276 (Disulfoton) chronic toxicity study on dogs (two-year feeding experiment). Report No. 5618. Report No. 45287. Unpublished study received Dec. 15, 1976 under 3125–58. Prepared by A.G. Bayer, W. Germany. Submitted by Mobay Chemical Corp., Kansas City, MO. CDL:095640-N. MRID 00073348.

Kadoum, A.M. and D.F. Mock. 1978. Herbicide and insecticide residues in tailwater pits: water and pit bottom soil from irrigated corn and sorghum fields. J. Agric. Food Chem. 26:45–50.

Kawamori, I., T. Saito and K. Iyatomi, 1971a. Fate of organophosphorus insecticides in soils. Part I. Botyu-Kagaku. 36:7–12.

Kawamori, I., T. Saito and K. Iyatomi. 1971b. Fate of organophosphorus insecticides in soils. Part II. Botyu-Kaoaku. 36:12–17.

Lamb, D.W., and E.J. Hixson.* 1983. Embryotoxic and teratogenic effects of disulfoton. Unpublished Study No. 81–611–02. Prepared by Mobay Chemical Co. MRID 00129458.

Lehman, A.J. 1959. Appraisal of the safety of chemicals in foods, drugs and cosmetics. Assoc. Food Drug Off. U.S.

Lichtenstein, F., K. Schulz, R. Skrentny and Y. Tsukano. 1966. Toxicity and fate of insecticide residues in water: insecticide residues in water after direct application or by leaching of agricultural soil. Arch. Environ. Health. 12:199–212.

Loeffler, W.W. 1969. A summary of Dasanit and Di-Syston soil persistence data: Report No. 25122. Unpublished study submitted by Mobay Chemical Corp., Kansas City, MO.

March, R.B., T.R. Fukuto and R.L. Metcalf.* 1957. Metabolism of P-32-

dithio-systox in the white mouse and American cockroach: Submitter 1830. Unpublished study. MRID 00083215.

McCarty, P.L. and P.H. King. 1966. The movement of pesticides in soils. In Proceedings of the 21st Industrial Waste Conference: May 3–5, 1966, Lafayette, IN. Purdue University Engineering Extension Series No. 121. pp. 156–176.

Meister, R., ed. 1983. Farm chemicals handbook. Willoughby, OH: Meister Publishing Company.

Mihail, F. 1976. S 276 (Disyston active ingredient) acute toxicity studies. Report No. 7062b. Prepared by A.G. Bayer, Institut Fur Toxikologie, for Mobay Chemical Corp. June 12, 1978.

Mihail, F.* 1978. S 276 (Disyston active ingredient) acute toxicity studies. Report No. 7602a. Prepared by A.G. Bayer, Institut Fur Toxikologie, for Mobay Chemical Corp. June 12, 1978.

Mobay Chemical Corporation. 1964. Synopsis of analytical and residue information on Di-Syston (clover). Includes method dated Mar. 5, 1964.

Mobay Chemical Corporation. 1972. Dasanit – Di-Syston: analytical and residue information on tobacco. Includes methods dated Mar. 5, 1964; Mar. 28, 1966; Oct. 27, 1967; and others. Unpublished study, including published data.

NAS. 1977. National Academy of Sciences. Volume I. Drinking water and health. Washington, DC: National Academy Press.

Puhl, R.J. and D.R. Fredrickson.* 1975. The metabolism and excretion of Di-Syston by rats. Unpublished report submitted by Mobay Chemical Corp. Report No. 44261. Prepared by Chemagro Agricultural Division, Mobay Chemical Corp. Dated May 6, 1975.

Quinby, G.E. 1977. Poisoning of construction workers with disulfoton. Clin. Toxicol. 10:479.

Rider, J.A., J. I. Swader and E.J. Pulette. 1972. Anticholinesterase toxicity studies with Guthion, Phosdrin, Di-syston and Trithion in human subjects. Proc. Fed. Am. Soc. Exp. Biol. 31:520.

Simmon, V.F.* 1979. In vitro microbiological mutagenicity and unscheduled DNA synthesis studies of eighteen pesticides. Report No. EPA-600/1-79-042. Unpublished study including submitter summary. Received April 3, 1980 under 279-2712. Prepared by SRI International. Submitted by FMC Corp., Philadelphia, PA. CDL:099350-A. MRID 00028625.

STORET. 1988. STORET Water Quality File. Office of Water. U.S. Environmental Protection Agency (data file search conducted in May, 1988).

Suett, D.L. 1975. Persistence and degradation of chlorfenvinphos, chlormephos, disulfoton, phorate and primiphos-ethyl following spring and late-summer soil application. Unpublished study submitted by ICI Americas, Inc., Wilmington, DE.

Taylor, R.F.* 1965. Letter sent to Chemagro Corporation dated Jan. 5, 1965: report on demyelination studies on hens. Report No. 15107. Unpublished study received March 24, 1965 under 6F0478. Prepared by Harris Laboratories, Inc. Submitted by Mobay Chemical Corp., Kansas City, MO. CDL:090534-C. MRID 00057265.

Taylor, R.E.* 1966. Letter sent to D. MacDougall dated May 5, 1966: Di-Syston, three generation rat breeding studies: submitter 18154. Unpublished study received March 7, 1977 under 3125–252. Prepared by Harris Laboratories, Inc. Submitted by Mobay Chemical Corp., Kansas City, MO. CDL:096021-L. MRID 00091104.

Tesh, J.M. et al.* 1982. S 276: Effects of oral administration upon pregnancy in the rabbit. An unpublished report (Bayer NO. R 2351) prepared by Life Science Research, Essex, England and submitted to A.G. Bayer, Wuppertal, Germany. Dated Dec. 22, 1982.

Thornton, J.S., J.B. Hurley and J.J. Obrist. 1976. Soil thin-layer mobility of twenty-four pesticide chemicals: Report No. 51016. Unpublished study submitted by Mobay Chemical Corp., Kansas City, MO.

U.S. EPA.* 1984. U.S. Environmental Protection Agency. Disulfoton (Di-Syston) Registration Standard. Washington, DC: Office of Pesticide Programs.

U.S. EPA. 1985. U.S. Environmental Protection Agency. Code of Federal Regulations. 40 CFR 180.183. July 1.

U.S. EPA. 1986. U.S. Environmental Protection Agency. Guidelines for carcinogen risk assessment. Fed. Reg. 51(185):33992–34003. September 24.

U.S. EPA. 1988. U.S. Environmental Protection Agency. U.S. EPA Method 507 — Determination of nitrogen and phosphorus containing pesticides in water by GC/NPD. Draft. Available from U.S. EPA's Environmental Monitoring and Support Laboratory, Cincinnati, OH. April 15

WHO. 1976. World Health Organization. Pesticide Residues Series No. 5, City, Country or State: World Health Organization. p. 204.

Windholz, M., S. Budavari, R.F. Blumetti, and F.S. Otterbein, eds. 1983. The Merck index — an encyclopedia of chemicals and drugs, 10th ed. Rahway, NJ: Merck and Company, Inc.

---

* Confidential Business Information submitted to the Office of Pesticide Programs.

Diuron

I. General Information and Properties

A. CAS No. 330–54–1

B. Structural Formula

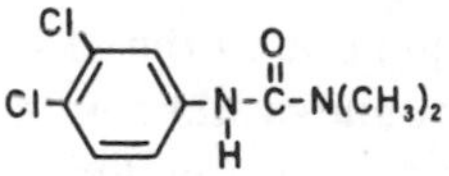

N'-(3,4-Dichlorophenyl)-N,N-dimethylurea

C. Synonyms

- Cekiuron; Crisuron; Dailon; Diater; Di-on; Direx 4L; Ditox-800; Diurex; Diurol; Drexel Diuron 4L; Dynex; Farmco Diuron; Karmex; Unidron; Vonduron. Discontinued names: Sup'r Flo; Urox 'D' (Meister, 1988).

D. Uses

- Diuron is used at low rates as a selective herbicide to control germinating broadleaf and grass weeds in numerous crops such as sugarcane, pineapple, alfalfa, grapes, cotton and peppermint. It is used as a general weed killer at higher rates. As a soil sterilant, it is more persistent and preferred over monuron on lighter soil and/or in areas of heavy rainfall. Drexel Diuron 4L and Farmco Diuron Flowable control weeds in wheat, barley, citrus, vineyards, bananas, pineapple, cotton and noncrop areas (Meister, 1988).

E. Properties (Meister, 1988; Windholz et al., 1983)

Chemical Formula	$C_9H_{10}N_2OCl_2$
Molecular Weight	233.10
Physical State (at 25°C)	White crystalline solid
Boiling Point	--
Melting Point	158–159°C
Density	--

Vapor Pressure (50°C)	3.1×10^{-6} mm Hg
Specific Gravity	--
Water Solubility (25°C)	42 mg/L
Log Octanol/Water Partition Coefficient	--
Taste Threshold	--
Odor Threshold	--
Conversion Factor	--

F. Occurrence

- Diuron has been found in none of the 25 surface water samples analyzed and in none of 1,337 ground-water samples (STORET, 1988). Samples were collected at 22 surface water locations and 1,292 ground-water locations.

G. Environmental Fate

- Radiolabeled diuron and its degradation products, 3-(3,4-dichlorophenyl)-1-methylurea (DCPMU) and 3-(3,4-dichlorophenyl)urea (DCPU), had half-lives of 4 to 8, 5 and 1 month, respectively, in aerobic soils maintained at 18 to 29°C and moisture levels at approximately field capacity (Walker and Roberts, 1978; Elder, 1978). 3,4-Dichloroaniline (DCA) was identified as a minor degradation product of diuron (Belasco, 1967; Belasco and Pease, 1969; Elder, 1978). Increasing soil organic matter content appears to increase the rate of decline of diuron phytotoxic residues (McCormick, 1965; Corbin and Upchurch, 1967; McCormick and Hiltbold, 1966; Liu et al., 1970).

- Degradation of diuron phytotoxic residues is much (28 to 50%) slower in flooded soil than in aerobic soil (Imamliev and Bersonova, 1969; Wang et al., 1977).

- Diuron has a low-to-intermediate mobility in fine to coarse-textured soils and freshwater sediments (Hance 1965a,b; Harris and Sheets, 1965; Harris, 1967; Helling and Turner, 1968; Grover and Hance, 1969; Gerber et al., 1971; Green and Corey, 1971; Helling, 1971; Guth, 1972; Grover, 1975; Helling, 1975). Mobility is correlated with organic matter content and cation exchange capacity (CEC). Soil texture apparently is not, by itself, a major factor governing the mobility of diuron in soil.

- In a study using radiolabeled material, the diuron degradation products (96% pure) had K_d values of 66 and 115 in silty clay loam soils, indicating that they are relatively immobile or less mobile than diuron (Elder, 1978).

- In the field, diuron residues (nonspecific method used) generally persisted for up to 12 months in soils that ranged in texture from sand to silt loam treated with diuron at 0.8 to 4 lb/A (Cowart, 1954; Hill et al., 1955; Weed et al., 1953; Weed et al., 1954; Miller et al., 1978). These residues may leach in soil to a depth of 120 cm (4 feet). Diuron was detectable (3 to 74 ppb) in runoff-water sediment and soil samples for up to 3 years after the last application to pineapple-sugarcane fields in Hawaii (Mukhtar, 1976; Green et al., 1977). Diuron may have the potential to leach into ground water.

- Phytotoxic residues persisted for up to 12 months in soils ranging in texture from sand to silty clay loam to boggy meadow soil following the last of one to six annual applications of diuron at 1 to 18 lb/A (Weldon and Timmons, 1961; Dalton et al., 1965; Bowmer, 1972; Dawson et al., 1968; Arle et al., 1965; Wang and Tsay, 1974; Spiridonov et al., 1972; Addison and Bardsley, 1968; Cowart, 1954; Hill et al., 1955; Weed et al, 1953; Weed et al., 1954). Diuron persistence in soil appears to be a function of application rate and amount of rainfall and/or irrigation water. Three degradation products (DCPMU, DCPU, and DCA) were identified in soil (planted with cotton) that had received multiple applications of diuron (80% wettable powder totaling 5 to 5.7 lb/A (Dalton et al., 1965).

- Diuron persists in irrigation-canal soils for 6 or more months following application at 33 to 46 kg/ha (Evans and Duseja, 1973a; Evans and Duseja, 1973b; Bowmer and Adeney, 1978a; Bowmer and Adeney, 1978b). The relative percentages of diuron and its degradates DCPMU and DCPU were 60–90:10–25:1–30 in clay and sandy clay soils, 4.5 to 17 weeks after treatment. Diuron levels in water samples were highest (0.5 to 8 ppm) in the initial flush of irrigation water. These levels declined rapidly, probably as a function of dilution and not degradation.

II. Pharmacokinetics

A. Absorption

- Diuron is absorbed through the gastrointestinal tract of rats and dogs. Hodge et al. (1967) fed diuron to rats and dogs at dietary levels from 25 to 2,500 ppm and from 25 to 1,250 ppm active ingredient (a.i.), respectively, for periods up to 2 years. These doses are equivalent to 1.25 to 125 mg/kg/day for the rat and 0.625 to 31.25 mg/kg/day for the dog. Urinary and fecal excretion products after 1 week to 2 years accounted for about 10% of the daily dose ingested. The excretion data provided evidence that gastrointestinal absorption occurred in rats and dogs.

B. Distribution

- Hodge et al. (1967) fed diuron (80% wettable powder) for 2 years to rats at dietary levels of 25 to 2,500 ppm a.i. and to dogs at dietary levels of 25 to 1,250 ppm a.i. Assuming that 1 ppm in the diet is equivalent to 0.05 mg/kg/day in rats and 0.025 mg/kg/day in dogs, this corresponds to doses of 1.25 to 125 mg/kg/day in rats and 0.625 to 31.25 mg/kg/day in dogs (Lehman, 1959). Analysis of tissue samples for diuron residues revealed levels ranging from 0.2 to 56 ppm, depending on dose. This constituted only a minute fraction of the total dose ingested. The authors concluded that there was little diuron storage in tissues.

C. Metabolism

- Geldmacher von Mallinckrodt and Schlussier (1971) analyzed the urine of a woman who had ingested a dose of 38 mg/kg of diuron along with 20 mg/kg of aminotriazole. The urine was found to contain 1-(3,4-dichlorophenyl)-3-methylurea and 1-(3,4-dichlorophenyl)-urea, and may also have contained some 3,4-dichloroaniline. No unaltered diuron was detected.

- Hodge et al. (1967) fed diuron (80% wettable powder) to male beagle dogs at a dietary level of 125 ppm active ingredient for 2 years. Assuming that 1 ppm in the diet is equivalent to 0.025 mg/kg/day (Lehman, 1959), this corresponds to a dose of 3.1 mg/kg/day. Analysis of urine at weeks 1 to 4 or after 2 years revealed the major metabolite was N-(3,4-dichlorophenyl)-urea. Small amounts of N-(3,4-dichlorophenyl)-N'-methylurea, 3,4-dichloroanaline, 3,4-dichlorophenol and unmetabolized diuron also were detected.

D. Excretion

- Hodge et al. (1967) fed diuron (80% wettable powder) for 2 years to rats at dietary levels of 25 to 2,500 ppm and to dogs at dietary levels of 25 to 1,250 ppm. Assuming that 1 ppm in the diet is equivalent to 0.05 mg/kg/day in rats and 0.025 mg/kg/day in dogs, this corresponds to doses of 1.25 to 125 mg/kg/day in rats and 0.625 to 31.25 mg/kg/day in dogs (Lehman, 1959). In rats, urinary excretion (6.3 to 492 ppm, depending on dose) was consistently greater than fecal excretion (1.0 to 204 ppm). In dogs, urinary excretion (6.3 to 307 ppm) was similar to fecal excretion (7.9 to 308 ppm). For both rats and dogs, combined urinary and fecal excretion accounted for only about 10% of the daily diuron ingestion.

III. Health Effects

A. Humans

- No information was found in the available literature on the health effects of diuron in humans.

B. Animals

1. *Short-term Exposure*

- Acute oral LD_{50} values of 1,017 mg/kg and 3,400 mg/kg have been reported in albino rats by Boyd and Krupa (1970), NIOSH (1987) and Taylor (1976a), respectively. Signs of central nervous system depression were noted after treatment.

- Hodge et al. (1967) administered single oral doses of recrystallized diuron in peanut oil to male CR rats. The approximate lethal dose was 5,000 mg/kg, and the LD_{50} was 3,400 mg/kg. Survivors sacrificed after 14 days showed large and dark-colored spleens with numerous foci of blood.

- Hodge et al. (1967) administered oral doses of 1,000 mg/kg of recrystallized diuron five times a week for 2 weeks (10 doses) to six male CR rats. At necropsy, 3 or 11 days after the final dose, the spleens were large, dark and congested, and foci of blood were noted in both the spleen and bone marrow.

- Hodge et al. (1967) fed Wistar rats (5/sex/dose) diuron (purity not specified) in the diet for 42 days at dose levels of 0, 200, 400, 2,000, 4,000 or 8,000 ppm a.i. Assuming that 1 ppm in the diet is equivalent to

0.05 mg/kg/day (Lehman, 1959), this corresponds to doses of 0, 10, 20, 100, 200 or 400 mg/kg/day. Following treatment body weight, clinical chemistry, food consumption, hematology, urinalysis and histology were evaluated. No effects were observed at 400 ppm (20 mg/kg/day) or less. At 2,000 ppm (100 mg/kg/day) or greater, red blood cell counts and hemoglobin values were decreased. A marked inhibition of growth occurred in the 4,000 ppm (200 mg/kg/day) or greater dosage groups, and there was increased mortality at 8,000 ppm. Based on these data, a No-Observed-Adverse-Effect Level (NOAEL) of 400 ppm (20 mg/kg/day) and a Lowest-Observed-Adverse-Effect Level (LOAEL) of 2,000 ppm (100 mg/kg/day) were identified.

2. *Dermal/Ocular Effects*

- Taylor (1976b) applied diuron to the intact or abraded skin of eight albino rabbits at dose levels of 1,000 to 2,500 mg/kg for 24 hours. After treatment, a slight erythema was observed, but no other symptoms of toxicity were noted during a 14-day observation period. The dermal LD_{50} was reported as $> 2,500$ mg/kg.

- Larson (1976) applied technical diuron at doses of 1, 2.5, 5 or 10 mg/kg to intact abraded skin of New Zealand White rabbits for 24 hours. Adverse effects were not detected in exposed animals.

- In studies conducted by DuPont (no date), diuron (50% water paste) was not irritating to intact skin and was moderately irritating to abraded skin of guinea pigs. No data were available on skin sensitization. See also DuPont (1961).

- In studies conducted by Larson and Schaefer (1976), 10 mg of a fine dry powder of diuron was instilled into the conjunctival sac of one eye of each of six New Zealand White rabbits. Ocular irritation was not detected within 72 hours.

3. *Long-term Exposure*

- Hodge et al. (1967) fed albino Charles River rats (5/sex/dose) diuron (98% pure) for 90 days at dietary levels of 0, 50, 500 or 5,000 ppm. Assuming that 1 ppm in the diet is equivalent to 0.05 mg/kg/day (Lehman, 1959), this corresponds to doses of 0, 2.5, 25 or 250 mg/kg/day. Following treatment, body weight, food consumption, clinical chemistry and histopathology were evaluated. No adverse effects were observed in any parameter at 50 ppm. At 500 ppm there were no effects on males, but females gained less weight than controls and appeared cyanotic. At the 5,000-ppm dose level, body weights were reduced in both sexes, spleens were enlarged and exhibited hemosiderosis, and there was clinical and pathological evidence of chronic methemoglobinemia. Based on these data, a NOAEL of 50 ppm (2.5 mg/kg/day) and a LOAEL of 500 ppm (25 mg/kg/day) were identified.

* Hodge et al. (1967) fed diuron (80% wettable powder) to groups of Charles River rats (20/sex/dose) for 90 days at dietary levels of 0, 250 or 2,500 ppm a.i. Assuming that 1 ppm in the diet is equivalent to 0.05 mg/kg/day (Lehman, 1959), this corresponds to doses of 0, 12.5 or 125 mg/kg/day. At 2,500 ppm, both males and females ate less and gained less weight than did controls. There was a slight decrease in red blood cell count, more pronounced in females than in males. No effect on food consumption or weight gain was noted at 250 ppm, but hematological changes were evident in females. This study identified a LOAEL of 250 ppm (12.5 mg/kg/day), the lowest dose tested.

* In a 2-year feeding study conducted by Hodge et al. (1964a, 1967), beagle dogs (two males/dose and three females/dose) were administered technical diuron (80% a.i.) in the diet at dose levels of 0, 25, 125, 250 or 1,250 ppm a.i. Assuming that 1 ppm in the diet of dogs is equivalent to 0.025 mg/kg/day (Lehman, 1959), this corresponds to doses of diuron of 0, 0.625, 3.12, 6.25 or 31.25 mg/kg/day. Following treatment, body weight, clinical chemistry, hematology, organ weight, gross pathology and histopathology were evaluated. No adverse effects were reported at 25 ppm in any parameter measured. Abnormal blood pigment was observed at 125 ppm or greater. Hematological alterations [depressed red blood cells (RBC), hematocrit and hemoglobin] were observed at 250 ppm or greater. In the 1,250 ppm group, a slight weight loss occurred as well as increased erythrogenic activity in bone marrow and hemosiderosis of the spleen. Based on these data, a NOAEL of 25 ppm (0.625 mg/kg/day) and a LOAEL of 125 ppm (3.12 mg/kg/day) were identified.

* Hodge et al. (1964b, 1967) administered technical diuron (80% a.i.) in the diet of rats (35/sex/dose) for 2 years at dose levels of 0, 25, 125, 250 or 2,500 ppm a.i. Assuming that 1 ppm in the diet of rats is equivalent to 0.05 mg/kg/day (Lehman, 1959), this corresponds to doses of diuron of 0, 1.25, 6.25, 12.5 or 125 mg/kg/day. Following treatment, body weight, clinical chemistry, hematology, food consumption, urinalysis, organ weights and histopathology were evaluated. No adverse effects were reported at 25 ppm (1.25 mg/kg/day) for any parameters measured. Abnormal blood pigments (sulfhemoglobin) were observed at 125 ppm (6.25 mg/kg/day) or greater. Hematological changes (decreased RBC, reduced hemoglobin), growth depression, hemosiderosis of the

spleen and increased mortality were observed at 250 ppm (12.5 mg/kg/day) or greater. Based on these data, a NOAEL of 25 ppm (1.25 mg/kg/day) and a LOAEL of 125 ppm (6.25 mg/kg/day) were identified.

4. *Reproductive Effects*

- Hodge et al. (1964b, 1967) studied the effects of diuron (80% wettable powder) in a three-generation reproduction study in rats. Animals were supplied food containing 0 and 125 ppm a.i. Assuming that 1 ppm in the diet of rats is equivalent to 0.05 mg/kg/day (Lehman, 1959), this corresponds to a dose of 6.25 mg/kg/day. Fertility rate, body weight, hematology and histopathology were monitored. No effect was seen on any parameter except body weight, which significantly decreased in the F_{2b} and F_{3a} litters. A LOAEL of 125 ppm (6.25 mg/kg/day) was identified.

5. *Developmental Effects*

- Khera et al. (1979) administered by gavage a formulation containing 80% diuron at dose levels of 125, 250 or 500 mg of formulation per kg body weight to pregnant Wistar rats (14 to 18/dose) on days 6 through 15 of gestation. Vehicle (corn oil) controls (19 dams) were run concurrently. No maternal or teratogenic effects were observed at 125 mg/kg/day. Developmental effects appeared to increase in all treatment groups, i.e., wavy ribs, extra ribs and delayed ossification. The incidence of wavy ribs was statistically significant at 250 mg/kg ($p < 0.05$) and greater. Maternal and fetal body weights decreased significantly at 500 mg/kg ($p < 0.05$). A NOAEL was not determined from this study for fetotoxic effects; hence, a LOAEL of 125 mg/kg of formulation per day was identified.

6. *Mutagenicity*

- Andersen et al. (1972) reported that diuron did not exhibit mutagenic activity in T_4 bacteriophage test systems (100 μg/plate) or in tests with eight histidine-requiring mutants of *Salmonella typhimurium* (small crystals applied directly to surface of plate).

- Fahrig (1974) reported that diuron (purity not specified) was not mutagenic in a liquid holding test for mitotic gene conversion in *Saccharomyces cerevisiae*, in a liquid holding test for forward mutation to streptomycin resistance in *Escherichia coli*, in a spot test for back mutation in *S. marcescens* or in a spot test for forward mutation in *E. coli*.

- Recent studies by DuPont (1985) did not detect evidence of mutagenic activity for diuron in reversion tests in several strains of *S. typhimurium* (with or without metabolic activation), in a Chinese hamster ovary/

hypoxanthine guanine phosphoribosyl-transferase (CHO/HGPRT) forward gene mutation test or in unscheduled DNA synthesis tests in primary rat hepatocytes. However, in an *in vivo* cytogenetic test in rats, diuron was observed to cause clastogenic effects.

7. *Carcinogenicity*

- Hodge et al. (1964b, 1967) fed Wistar rats (35/sex/dose) diuron (80% wettable powder) in the diet at levels of 0, 25, 125, 250 or 2,500 ppm a.i. for 2 years. Assuming that 1 ppm in the diet of rats corresponds to 0.05 mg/kg/day (Lehman, 1959), this corresponds to doses of 0, 1.25, 6.25, 12.5 or 125 mg/kg/day. There was some early mortality in males at 250 and 2,500 ppm, but the authors ascribed this to viral infection. Histological examination of tissues showed no evidence of changes related to diuron; however, only 10 animals or fewer were examined per group. Tumors and neoplastic changes observed were similar in exposed and control groups, and the authors concluded there was no evidence that diuron was carcinogenic in rats.

IV.   Quantification of Toxicological Effects

A. One-day Health Advisory

No suitable information was found in the available literature for use in the determination of the One-day HA value for diuron. It is, therefore, recommended that the Ten-day HA value for a 10-kg child, calculated below as 1.0 mg/L (1,000 μg/L), be used at this time as a conservative estimate of the One-day HA value.

B. Ten-day Health Advisory

The study by Khera et al. (1979) has been selected to serve as the basis for the Ten-day HA for diuron. In this study, pregnant rats were administered diuron (80%) on days 6 through 15 of gestation at dose levels of 0, 125, 250 or 500 mg/kg/day. Developmental effects appeared to increase in the diuron-treated groups as compared to the control group, i.e. wavy ribs, extra ribs and delayed ossification. The incidence of wavy ribs was statistically significant at 250 mg/kg/day ($p < 0.05$). Fetal and maternal body weights were decreased at 500 mg/kg ($p < 0.05$). A NOAEL was not determined based on developmental toxicity; hence, the LOAEL for this study was 125 mg/kg/day, the lowest dose tested.

Using a LOAEL of 125 mg/kg/day, the Ten-day HA for a 10-kg child is calculated as follows:

$$\text{Ten-Day HA} = \frac{(125 \text{ mg/kg/day})(10 \text{ kg})(0.80)}{(1,000) (1 \text{ L/day})} = 1.0 \text{ mg/L } (1,000 \text{ } \mu g/L)$$

C. Longer-term Health Advisory

The 90-day feeding study in rats by Hodge et al. (1967) has been chosen to serve as the basis for determination of the Longer-term HA values for diuron. In this study, five animals per sex were fed diuron (98% pure) at dose levels of 0, 2.5, 25 or 250 mg/kg/day. Based on decreased weight gain and methemoglobinemia, this study identified a NOAEL of 2.5 mg/kg/day and a LOAEL of 25 mg/kg/day. These values are supported by the 42-day feeding study of Hodge et al. (1964b), in which a NOAEL of 20 mg/kg/day and a LOAEL of 100 mg/kg/day were identified. This study was not selected, however, since the duration of exposure was only 42 days.

Using a NOAEL of 2.5 mg/kg/day, the Longer-term HA for a 10-kg child is calculated as follows:

$$\text{Longer-term HA} = \frac{(2.5 \text{ mg/kg/day}) (10 \text{ kg})}{(100) (1 \text{ L/day})} = 0.25 \text{ mg/L } (300 \text{ } \mu g/L)$$

The Longer-term HA for a 70-kg adult is calculated as follows:

$$\text{Longer-term HA} = \frac{(2.5 \text{ mg/kg/day}) (70 \text{ kg})}{(100) (2 \text{ L/day})} = 0.875 \text{ mg/L } (900 \text{ } \mu g/L)$$

D. Lifetime Health Advisory

The 2-year feeding study in dogs by Hodge et al. (1964a, 1967) has been selected to serve as the basis for estimating the Lifetime HA for diuron. In this study, dogs (2 to 3/sex/dose) were fed diuron at doses of 0.625, 3.12, 6.25 or 31.15 mg/kg/day of active ingredient. Hematological alterations were observed at 3.12 mg/kg/day or greater, and this was identified as the LOAEL. No effects were reported at 0.625 mg/kg/day in any parameter measured, and this was identified as the NOAEL. This value is supported by a lifetime study in rats by the same authors (Hodge et al., 1964b). In this study, rats were fed diuron at dose levels of 0, 1.25, 6.25, 12.5 or 125 mg/kg/day for 2 years. Hematological changes were observed at 6.25 mg/kg/day or greater, and a NOAEL of 1.25 mg/kg/day was identified.

Using a NOAEL of 0.625 mg/kg/day, the Lifetime HA is calculated as follows:

Step 1: Determination of the Reference Dose (RfD)

$$RfD = \frac{(0.625 \text{ mg/kg/day})}{(100)\,(3)} = 0.002 \text{ mg/kg/day}$$

where: 3 = additional uncertainty factor. This factor is used to account for a lack of adequate chronic toxicity studies in the database preventing establishment of the most sensitive toxicological end point.

Step 2: Determination of the Drinking Water Equivalent Level (DWEL)

$$DWEL = \frac{(0.002 \text{ mg/kg/day})\,(70 \text{ kg})}{(2 \text{ L/day})} = 0.07 \text{ mg/L } (70 \text{ } \mu g/L)$$

Step 3: Determination of the Lifetime Health Advisory

Lifetime HA = $(0.07 \text{ mg/L})\,(20\%) = 0.014 \text{ mg/L } (10 \text{ } \mu g/L)$

E. Evaluation of Carcinogenic Potential

- Hodge et al. (1964b, 1967) fed rats (35/sex/dose) diuron in the diet at ingested doses of up to 125 mg/kg/day for 2 years. Histological examinations did not reveal increased frequency of tumors; however, fewer than half of the survivors were examined.

- The International Agency for Research on Cancer has not evaluated the carcinogenic potential of diuron.

- Applying the criteria described in EPA's guidelines for assessment of carcinogenic risk (U.S. EPA, 1986a), diuron may be classified in Group D: not classifiable. This category is for substances with inadequate animal evidence of carcinogenicity.

- A structurally related analogue of diuron (linuron) appears to reflect some oncogenic activity. In addition, a Russian study by Rubenchik et al. (1973) reported gastric carcinomas and pancreatic adenomas in rats (strain not designated) given diuron at 450 mg/kg/day for 22 months. However, the actual data for the study are unavailable for Agency review.

V.   Other Criteria, Guidance and Standards

- An Acceptable Daily Intake (ADI) of 0.002 mg/kg/day, based on a NOAEL of 0.625 mg/kg from a dog study (Hodge et al., 1964a, 1967) and an uncertainty factor of 300 has been calculated (U.S. EPA, 1986b).

- Residue tolerances that range from 0.1 to 7 ppm have been established for diuron in or on agricultural commodities (U.S. EPA, 1985).

VI.   Analytical Methods

- Analysis of diuron is by a high-performance liquid chromatographic (HPLC) method applicable to the determination of certain carbamate and urea pesticides in water samples (U.S. EPA, 1986c). This method requires a solvent extraction of approximately 1 L of sample with methylene chloride using a separatory funnel. The methylene chloride extract is dried and concentrated to a volume of 10 mL or less. HPLC is used to permit the separation of compounds, and the measurement is conducted with an ultraviolet (UV) detector. The method has been validated in a single laboratory, and the estimated detection limit for diuron is 0.07 μg/L.

VII.   Treatment Technologies

- Available data indicate that granular-activated carbon (GAC) and powdered activated carbon (PAC) adsorption and chlorination effectively remove diuron from water.

- El-Dib and Aly (1977b) determined experimentally the Freundlich constants for diuron on GAC. Although the values do not suggest a strong adsorption affinity for activated carbon, diuron is better adsorbed than other phenylurea pesticides.

- El-Dib and Aly (1977b) calculated, based on laboratory tests, that 66 mg/L of PAC would be required to reduce diuron concentration by 98%, and 12 mg/L of PAC would be required to reduce diuron concentration by 90%.

- Conventional water treatment techniques of coagulation with ferric sulfate, sedimentation and filtration proved to be only 20% effective in removing diuron from contaminated water (El-Dib and Aly, 1977a). Aluminum sulfate was reportedly less effective than ferric sulfate.

- Oxidation with chlorine for 30 minutes removed 70% of diuron at pH 7. Under the same conditions, 80% of diuron was oxidized by chlorine dioxide (El-Dib and Aly, 1977a). Chlorination, however, will produce several degradation products whose environmental toxic impact should be evaluated prior to selection of oxidative chlorination for treatment of diuron-contaminated water.

- The treatment technologies cited above for the removal of diuron from water are available and have been reported to be effective. However, selection of an individual technology or combinations of technologies to attempt diuron removal from water must be based on a case-by-case technical evaluation and an assessment of the economics involved.

VIII. References

Addison, D.A. and C.E. Bardsley. 1968. *Chlorella vulgaris* assay of the activity of soil herbicides. Weed Sci. 16:427–429.

Andersen, K.J., E.G. Leighty and M.T. Takahasi. 1972. Evaluation of herbicides for possible mutagenic properties. J. Agr. Food Chem. 20:649–656.

Arle, H.F., J.H. Miller and T.J. Sheets. 1965. Disappearance of herbicides from irrigated soils. Weeds. 13(1):56–60.

Belasco, I.J. 1967. Absence of tetrachloroazobenzene in soils treated with diuron and linuron. Unpublished study. Submitted by E.I. duPont de Nemours & Company, Inc., Wilmington, DE.

Belasco, I.J. and H.L. Pease. 1969. Investigation of diuron- and linuron-treated soils for 3,3',4,4'-tetrachloroazobenzene. J. Agric. Food Chem. 17:1414–1417.

Bowmer, K.H. 1972. Measurement of residues of diuron and simazine in an orchard soil. Aust. J. Exp. Agric. Anim. Husb. 12(58):535–539.

Bowmer, K.H. and J.A. Adeney. 1978a. Residues of diuron and phytotoxic degradation products in aquatic situations. I. Analytical methods for soil and water. Pestic. Sci. 9(4):342–353.

Bowmer, K.H. and J.A. Adeney. 1978b. Residues of diuron and phytotoxic degradation products in aquatic situations. II. Diuron in irrigation water. Pestic. Sci. 9(4):354–364.

Boyd, E.M. and V. Krupa. 1970. Protein deficient diet and diuron toxicity. J. Agric. Food Chem. 18:1104–1107.

Corbin, F.T. and R.P. Upchurch. 1967. Influence of pH on detoxication of herbicides in soil. Weeds. 15(4):370–377.

Cowart, L.E. 1954. Soil-herbicidal relationships of 3-(p-chlorophenyl)-1,1-dimethylurea and 3-(3,4-dichlorophenyl)-1,1-dimethylurea. In: Proceedings of the Western Weed Control Conference, Vol. 14. Salt Lake City, UT: Western Weed Control Conference. pp. 37–45.

Dalton, R.L., A.W. Evans and R.C. Rhodes. 1965. Disappearance of diuron in cotton field soils. In: Proceedings of the Southern Weed Conference, Vol. 18. Athens, GA: Southern Weed Science Society. pp. 72–78.

Dawson, J.H., V.G. Bruns and W.J. Clore. 1968. Residual monuron, diuron, and simazine in a vineyard soil. Weed Sci. 16(1):63–65.

DuPont.* 1961. E. I. duPont de Nemours & Co., Inc. Condensed technical information (diuron).

DuPont.* No date. E. I. duPont de Nemours & Co., Inc. Toxicity of 3-(3,4-dichlorophenyl)-1,1-dimethylurea. Unpublished study. Medical Research Project Nos. MR-48 and MR-263. MRID 00022036.

DuPont.* 1985. E. I. duPont de Nemours & Co., Inc. Mutagenicity studies with diuron: *Salmonella* test, No. HLR 471–84 (7185); CHO/HGPRT forward gene mutation assay, HLR No. 282–85 (06/28/85); unscheduled DNA synthesis test in primary rat hepatocytes, HLR No. 349–85 (07/10/85); and *in vivo* cytogenetic test, No. 36685 (06/20/85).

Elder, V.A. 1978. Degradation of specifically labeled diuron in soil and availability of its residues to oats. Doctoral dissertation. Honolulu, HI: University of Hawaii. Available from University Microfilms, Ann Arbor, MI. Report No. 79–13776.

El-Dib, M.A. and O.A. Aly. 1977a. Removal of phenylamide pesticides from drinking waters. I. Effect of chemical coagulation and oxidants. Water Res. 11:611–616.

El-Dib, M.A. and O.A. Aly. 1977b. Removal of phenylamide pesticides from drinking waters. II. Adsorption on powdered carbon. Water Res. 11:617–620.

Evans, J.O. and D.R. Duseja. 1973a. Herbicide contamination of surface runoff waters. EPA-R2-73-266. Washington, DC: U.S. Environmental Protection Agency, Office of Research and Monitoring. Available from National Technical Information Service, Springfield, VA. PB-222283.

Evans, J.O. and D.R. Duseja. 1973b. Results and discussion: field experiments. In: Herbicide contamination of surface runoff waters. EPA-R2-73-266. Project No. 13030 FDJ. Utah State University. pp. 33–35, 38–43. Available from Superintendent of Documents, U.S. Government Printing Office, Washington, DC.

Fahrig, R. 1974. Comparative mutagenicity studies with pesticides. Sci. Pub. 10. Lyon, France: International Agency for Research on Cancer (IARC). pp. 161–181.

Geldmacher von Mallinckrodt, M. and F. Schlussier.* 1971. Metabolism and toxicity of 1-(3,4-dichlorophenyl)-3,3-dimethylurea (diuron) in man. Arch. Toxicol. 27(3):311–314. Cited in Weed Abst. 21:331. MRID 00028010.

Gerber, H.R., P. Ziegler and P. Dubah. 1971. Leaching as a tool in the evaluation of herbicides. In: Proceedings of the 10th British Weed Control Conference (1970), Vol. 1. Droitwich, England: British Weed Control Conference. pp. 118–125.

Green, R.E. and J.C. Corey. 1971. Pesticide adsorption measurement by flow equilibration and subsequent displacement. Proc. Soil Sci. Soc. Am. 35:561–565.

Green, R.E., K.P. Goswami, M. Mukhtar and H.Y. Young. 1977. Herbicides from cropped watersheds in stream and estuarine sediments in Hawaii. J. Environ. Qual. 6(2):145–154.

Grover, R. 1975. Adsorption and desorption of urea herbicides on soils. Can. J. Soil Sci. 55:127–135.

Grover, R. and R.J. Hance. 1969. Adsorption of some herbicides by soil and roots. Can. J. Plant Sci. 40:378–380.

Guth, J.A. 1972. Adsorption and leaching characteristics of pesticides in soil. Unpublished study including German text. Prepared by Ciba-Geigy, AG. Submitted by Shell Chemical Company, Washington, DC.

Hance, R.J. 1965a. Observations on the relationship between the adsorption of diuron and the nature of the adsorbent. Weed Res. 5:108–114.

Hance, R.J. 1965b. The adsorption of urea and some of its derivatives by a variety of soils. Weed Res. 5:98–107.

Harris, C.I. 1967. Movement of herbicides in soil. Weeds. 15(3):214–216.

Harris, C.I. and T.J. Sheets. 1965. Influence of soil properties on adsorption and phytotoxicity of CIPC, diuron, and simazine. Weeds. 13(3):215–219.

Helling, C.S. 1971. Pesticide mobility in soils. II. Applications of soil thin-layer chromatography. Proc. Soil Sci. Soc. Am. 35:737–748.

Helling, C.S. 1975. Soil mobility of three Thompson-Hayward pesticides. Interim Report. Unpublished study. U.S. Agricultural Research Service, Pesticide Degradation Laboratory.

Helling, C.S. and B.C. Turner. 1968. Pesticide mobility: determination by soil thin-layer chromatography. Method dated Nov. 1, 1968. Science. 162:562–563.

Hill, G.D., J.W. McGahen, H.M. Baker, D.W. Finnerty and C.W. Bingeman. 1955. The fate of substituted urea herbicides in agricultural soils. Agron. J. 47(2):93–104.

Hodge, H.C., W.L. Downs, E.A. Maynard et al.* 1964a. Chronic feeding studies of diuron in dogs. Unpublished study. MRID 00017763.

Hodge, H.C., W.L. Downs, E.A. Maynard et al.* 1964b. Chronic feeding studies of diuron in rats. Unpublished study. MRID 00017764.

Hodge, H.C., W.L. Downs, B.S. Panner, D.W. Smith and E.A. Maynard. 1967. Oral toxicity and metabolism of diuron (N-(3,4-dichlorophenyl)-N',N'dimethylurea) in rats and dogs. Food Cosmet. Toxicol. 5:513–531.

Imamliev, A.I. and K.A. Bersonova. 1969. Movement of detoxication of dalapon and diuron in soil. In: Imamliev, A.I and E.A. Popova, eds. Problems of physiology and biochemistry of the cotton plant. Tashkent, USSR: Akademii Nauk Uzbekskoi, Institut Eksperimental'noi Biologii Rastenii. pp. 266–274.

Khera, K.S., C. Whalen, G. Trivett and G. Angers. 1979. Teratogenicity studies on pesticidal formulations of dimethoate, diuron and lindane in rats. Bull. Environ. Contam. Toxicol. 22:522–529.

Larson, K.A.* 1976. Acute dermal toxicity—diuron. Unpublished study. MRID 00017795.

Larson, K.A. and J.H. Schaefer.* 1976. Eye irritation study using the pesticide diuron. Prepared for Colorado International Corporation. Unpublished study. MRID 00017797.

Lehman, A.J. 1959. Appraisal of the safety of chemicals in foods, drugs and cosmetics. Assoc. Food Drug Off. U.S., Q. Bull.

Liu, L.C., H.R. Cibes-Viade and J. Gonzalez-Ibanez. 1970. The persistence of atrazine, ametryne, prometryne, and diuron in soils under greenhouse conditions. J. Agric. Univ. Puerto Rico. 54(4):631–639.

McCormick, L.L. 1965. Microbiological decomposition of atrazine and diuron in soil. Doctoral dissertation. Auburn, AL: Auburn University. Available from University Microfilms, Ann Arbor, MI. Report No. 65-6892.

McCormick, L.L. and A.E. Hiltbold. 1966. Microbiological decomposition of atrazine and diuron in soil. Weeds. 14(1):77–82.

Meister, R., ed. 1988. Farm chemicals handbook. Willoughby, OH: Meister Publishing Company.

Miller, J.H., P.E. Keeley, R.J. Thullen and C.H. Carter. 1978. Persistence and movement of ten herbicides in soil. Weed Sci. 26(1):20–27.

Mukhtar, M. 1976. Desorption of adsorbed ametryn and diuron from soils and soil components in relation to rates, mechanisms, and energy of adsorption reactions. Doctoral dissertation. Honolulu, HI: University of Hawaii. Available from University Microfilms, Ann Arbor, MI. Report No. 77-14,601.

NIOSH. 1987. National Institute for Occupational Safety and Health. Registry of toxic effects of chemical substances (RTECS). Microfiche edition. July.

Rubenchik, B.L., N.E. Botsman, G.P. Gorman and L.I. Loevskaya. 1973. Relation between the chemical structure and carcinogenic activity of urea derivatives. Oukalogiya (Kiev) 4:10–16.

Spiridonov, Y.Y., V.S. Skhiladze and G.S. Spiridonova. 1972. The effects of diuron and monuron in a meadow-bog soil of the moist subtropics of Adzhariia. Subtrop. Crops. (1):150–155.

STORET. 1988. STORET Water Quality File. Office of Water. U.S. Environmental Protection Agency (data file search conducted in May, 1988).

Taylor, R.E.* 1976a. Acute oral toxicity (LD_{50}). Project T100l. Unpublished study. MRID 00028006.

Taylor, R.E.* 1976b. Primary skin irritation study. Project T1002. Unpublished study. MRID 00028007.

U.S. EPA. 1985. U.S. Environmental Protection Agency. Code of Federal Regulations. 40 CFR 180.106, p. 252. July 1.

U.S. EPA. 1986a. U.S. Environmental Protection Agency. Guidelines for carcinogen risk assessment. Fed. Reg. 51(185):33992–34003. September 24.

U.S. EPA. 1986b. U.S. Environmental Protection Agency. Acceptable Daily Intake data; tolerances printout. Office of Pesticide Programs. Office of Pesticides and Toxic Substances. February 21.

U.S. EPA. 1986c. U.S. Environmental Protection Agency. U.S. EPA Method 4 — Determination of pesticides in ground water by HPLC/UV. Draft. January. Available from U.S. EPA's Environmental Monitoring and Support Laboratory, Cincinnati, OH.

U.S. EPA. 1987. U.S. Environmental Protection Agency. Interim guidance for establishing RfD. Addendum to TOX SOP #1002. Office of Pesticide Programs. May 1.

Walker, A. and M.G. Roberts. 1978. The degradation of methazole in soil. II. Studies with methazole, methazole degradation products, and diuron. Pestic. Sci. 9(4):333–341.

Wang, C.C. and J.S. Tsay. 1974. Accumulative residual effect and toxicity of some persistent herbicides in multiple cropping areas. Med. Coll. Med. Natl. Taiwan Univ. 14(1):1–13.

Wang, Y.S., T.C. Wang and Y.L. Chen. 1977. A study on the degradation of herbicide diuron in soils and under the light. J. Chinese Agric. Chem. Soc. 15(1/2):23–31.

Weed, M.B., R. Sutton, G.D. Hill and L.E. Cowart. 1953. Substituted ureas for pre-emergence weed control in cotton. Unpublished study. Submitted by E.I. duPont de Nemours & Co. Inc., Wilmington, DE.

Weed, M.B., A.W. Welch, R. Sutton and G.D. Hill. 1954. Substituted ureas for pre-emergence weed control in cotton. In: Proceedings of the Southern Weed Conference, Vol. 7. Athens, GA: Southern Weed Science Society. pp. 68–87.

Weldon, L.W. and F.L. Timmons. 1961. Penetration and persistence of diuron in soil. Weeds. 9(2):195–203.

Windholz, M., S. Budavari, R.F. Blumetti and E.S. Otterbein, eds. 1983. The Merck index — an encyclopedia of chemicals and drugs, 10th ed. Rahway, NJ: Merck and Company, Inc.

---

* Confidential Business Information submitted to the Office of Pesticide Programs.

Endothall

I. General Information and Properties

A. CAS No. 145-73-3

B. Structural Formula

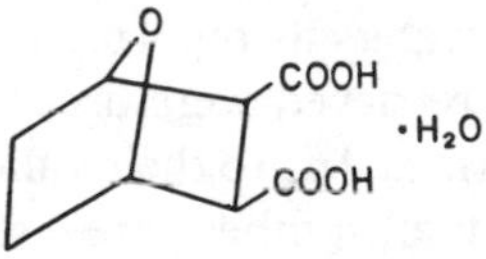

7-Oxabicyclo-(2,2,1)-heptane-2,3-dicarboxylic acid

C. Synonyms

- 1,2-Dicarboxy-3,6-endoxocyclohexane; Aquathol® K; Hydrothol 191; Hydrothol 191 Granular; Endothall Truf Herbicide; Herbicide 273; Des-i-cate; Accelerate (Meister, 1988).

D. Uses

- Endothall is used as a preemergence and postemergence herbicide, defoliant, dessicant, aquatic algicide, and a growth regulator (Meister, 1988).

E. Properties (Reinert and Rodgers, 1984; Neely and Mackay, 1982; Chiou et al., 1977; Carlson et al., 1978; Simsiman et al., 1976)

Chemical Formula	$C_8H_{10}O_5$
Molecular Weight	186.06
Physical State (25°C)	White crystalline solid
Boiling Point	--
Melting Point	144°C to the anhydride
Density	--
Vapor Pressure (25°C)	Negligible
Specific Gravity	--
Water Solubility (25°C)	100 g/L (acid monohydrate) 1.228 g/L (dipotassium salt)

381

Log Octanol/Water   1.91 (acid)
  Partition Coefficient   1.36 (dipotassium salt)
Taste Threshold   --
Odor Threshold   --
Conversion Factor   --

F. Occurrence

- Endothall was not found in any of the 3 surface water or 604 ground-water samples analyzed that were taken from 2 surface water locations and 600 ground-water locations (STORET, 1988). This information is provided to give a general impression of the occurrence of this chemical in ground and surface waters as reported in the STORET database. The individual data points retrieved were used as they came from STORET and have not been confirmed as to their validity. STORET data are often not valid when individual numbers are used out of the context of the entire sampling regime, as they are here. Therefore, this information can only be used to form an impression of the intensity and location of sampling for a particular chemical.

G. Environmental Fate

- The low K_{ow} values indicate that endothall would not significantly partition to sediments. K_{oc} values for equilibrium sorption studies using the dipotassium salt were 110 and 138 for sediment-water systems from a small eutrophic pond and oligomesotrophic reservoir, respectively (Reinert and Rodgers, 1984). An overall K_p value of 0.958 was calculated from this study, which compares favorably with K_p values for the acid ranging from 0.41 to 0.9 calculated from a flask system containing Lake Tomahawk water and sediment (Simsiman and Chesters, 1975). A K_p value of approximately 0.4 was calculated from the data presented by Sikka and Rice (1973) in which a Syracuse, NY, farm pond was treated with dipotassium endothall. Therefore, sorption would not be considered a significant environmental fate process for endothall in the environments studied.

- Endothall has been shown to not significantly bioconcentrate. In laboratory and field studies, consistently low endothall levels have been observed. A BCF for endothall in mosquito fish (*Gambusia affinis*) of 10 was observed in a modified Metcalf model ecosystem (Isensee, 1976). In a field study by Serns (1977), a 5 mg/L dipotassium endothall concentration resulted in BCF values in bluegills ranging from 0.003 to 0.008. After 72 hours, fish flesh residues were not detectable. Endothall residues in caged bluegills were consistently below the minimum detectable level of 0.1 mg/kg in a reservoir study by Reinert and Rodgers (1986). Similar results were seen after an application of the diamine salt (Walker, 1963). Comparable fish BCF values calculated from regression equations were 0.65 (Neely et al., 1974) and 1.05 (Chiou ct al., 1982).

• Some organisms will exhibit temporary endothall residues that exceed the water concentration by more than an order of magnitude. Isensee (1976) observed BCF values of 150 for the water flea, 63 for green alga (*Oedogonium*) and 36 for a snail (Physa); however, the endothall concentrations in the organisms were transient and were not passed along trophic levels. A BCF of 0.73 was calculated for the dipotassium salt of endothall in duckweed (*Lemna minor*) using the endothall K_{ow} and the regression equation found in Lockhart et al. (1983).

• Volatilization, hydrolysis and oxidation are not significant fate processes affecting the persistence of endothall in aquatic environments (Reinert and Rodgers, 1984). Endothall is also not subject to photochemical degradation. In a laboratory study using the disodium salt of endothall, no degradation was observed when a lamp of 254 nm wavelength was employed (Mitchell, 1961).

• Biotransformation and biodegradation are the dominant fate processes affecting the persistence of endothall in aquatic environments (Simsiman and Chesters, 1975; Holmberg and Lee, 1976; Simsiman et al., 1976). A biotransformation half-life of 8.35 days was observed in a shake-flask study using three [^{14}C]endothall concentrations and water from an oligomesotrophic reservoir (Reinert and Rodgers, 1986). Overall aqueous decay rates are considered a good estimate of endothall biotransformation in that other fate processes are insignificant. After a review of the literature, Keckemet (1980) determined an aqueous endothall half-life of about 6.7 days. The persistence of both the dipotassium and amine salts was less than 7 days in Gatan Lake, Panama Canal (Gangstad, 1983). A half-life for the dipotassium salt of 7.3 days was calculated from field studies using farm ponds (Yeo, 1970). Reinert et al. (1985) observed 4.1-day endothall half-life in 133 L plastic greenhouse pools containing water, sediment and Eurasian water milfoil. Dipotassium endothall half-lives in a marginally treated north Texas reservoir ranged from 1.1 to 1.2 days (Rodgers et al., 1984; Reinert and Rodgers, 1986). The results presented in Holmberg and Lee (1976) compare favorably with the above endothall half-lives. A 4.1-day half-life was calculated from a Wisconsin pond treated with dipotassium endothall.

- Endothall persistence in sediments ranged from 0 to 7 days in work reported by Keckemet (1980) and less than 4 days after a nominal 2 mg/L dosage in a Texas reservoir (Rodgers et al., 1984; Reinert and Rodgers, 1986). Gangstad (1983) observed endothall persistence in Gatan Lake sediment < 3 days for the dipotassium salt, but > 21 days for the amine salt when treated with 2 mg/L. In a pond study, Langeland and Warner (1986) observed an overall endothall (Aquathol® K) persistence of 26 days after an initial concentration of 2.1 mg/L.

II. Pharmacokinetics

A. Absorption

- Few data exist regarding endothall pharmacokinetics in mammals. Soo et al. (1967) performed pharmacokinetic experiments with male and female Wistar rats. Taking ^{14}C levels in urine and exhaled air as a crude estimate of absorption, approximately 10% of a 5 mg/kg oral dose of ^{14}C-labeled (at carbons 1 or 2 of the ring) endothall (dissolved in 20% ethanol to a concentration of 1 mg/mL) was absorbed by the rats within 72 hours. The rats had received 5 mg/kg of unlabeled endothall in the diet for 2 weeks prior to treatment with ^{14}C-endothall. Endothall measurements in this study were as ^{14}C label.

- Deaths in rabbits directly exposed to endothall in the eye or on the skin (Pharmacology Research, Inc., 1975a, 1975b) indicate the potential for absorption by these routes.

B. Distribution

- In the Soo et al. (1967) study, the absorbed endothall was distributed in low levels through most body tissues. Peak levels in all tissues were observed 1 hour after dosing, with most of the dose (about 95%) found in the stomach and intestine. Otherwise, the tissues with the highest concentrations after 1 hour were the liver and kidney (1.1 and 0.9% respectively), with lower concentrations (0.02 to 0.1%) in heart, lung, spleen and brain. Very low concentrations were observed in muscle, and endothall was not detected in fat. No marked preferential accumulation was apparent.

C. Metabolism

- The metabolism of endothall is not known to be characterized.

D. Excretion

- Soo et al. (1967) described excretion as follows:

  Clearance of ^{14}C-endothall was biphasic in the stomach ($t_{1/2}$ = 2.2 and 14.2 hours) and kidney ($t_{1/2}$ = 1.6 and 34.6 hours) and monophasic in the intestine and liver ($t_{1/2}$ = 14.4 and 21.6 hours, respectively). Total excretion of the ^{14}C label was over 95% complete by 48 hours and over 99% complete by 72 hours, suggesting that no significant bioaccumulation occurred.

- Approximately 90% of the administered dose was excreted in the feces. Urinary excretion accounted for approximately 7% of the dose, and approximately 3% of the radioactive label was recovered in expired carbon dioxide.

- Recycling of endothall by biliary circulation was ruled out as a major excretory route because ^{14}C activity in liver was small relative to that in the original dose.

- Approximately 20% of the dose excreted in the feces was unchanged endothall with the remaining radioactivity presumed to be an endothall conjugate.

- Soo et al. (1967) also found no radioactivity in pups from lactating dams given oral doses of ^{14}C-endothall.

III.   Health Effects

A. Humans

1. *Short-term Exposure*

- No information was found in the available literature on the health effects of endothall in humans except for one case history of a young male suicide victim who ingested an estimated 7 to 8 g of disodium endothall in solution (approximately 100 mg endothall ion/kg). Repeated vomiting was evident. Autopsy revealed focal hemorrhages and edema in the lungs and gross hemorrhage of the gastrointestinal (GI) tract (Allender, 1983).

2. *Long-term Exposure*

- No information was found in the available literature on long-term human exposure to endothall.

A. Animals

1. *Short-term Exposure*

- Early acute studies report cardiac arrest (Goldstein, 1952) or respiratory failure (Srensek and Woodard, 1951) as causes of death in dogs and rabbits. Endothall was injected intravenously in both studies with these effects observed at doses of 5 mg/kg (lowest) and higher.

- The available acute oral dose studies are essentially restricted to mortality data without biochemical or histopathological observations. The acute toxicity of endothall acid appeared to be greater than that of the salt forms normally used in herbicide formulations. In rats, the oral LD_{50} of endothall was reported as 35 to 51 mg/kg for the acid form and 182 to 197 mg/kg for the sodium salt (Simsiman et al., 1976; Tweedy and Houseworth, 1976).

- Rats were given 1,000 or 10,000 ppm disodium endothall in the diet (Brieger, 1953a) and doses were calculated by assuming a body weight of 0.4 kg and daily food consumption of 20 g. Slight liver degeneration and focal hemorrhagic areas in the kidney were reported for male and female rats dosed orally with approximately 40 mg endothall ion/kg/day for 4 weeks; most of the rats receiving approximately 400 mg endothall ion/kg/day died within 1 week.

- Nine male dogs (one dog/dose) were dosed orally with capsules containing 1 to 50 mg disodium endothall/kg/day (0.8 to 40 mg endothall ion/kg/day) for 6 weeks (Brieger, 1953b). All dogs that were administered 20 to 50 mg disodium endothall/kg/day died within 11 days. Vomiting and diarrhea were observed in the group given 20 mg disodium endothall/kg/day. Pathological changes in the GI tract, described as congested and edematous stomach walls and edematous upper intestines, were indicated as common in all dogs. Erosion and hemorrhages in the stomach were observed with doses of 20 mg/kg/day or more.

2. *Dermal/Ocular Effects*

- Goldstein (1952) reported that a 1% solution of endothall applied to the unbroken skin of rabbits produced no effects. The same solution applied to scarified skin resulted in mild skin lesions. Ten to twenty percent solutions or applications of the pure, powdered material to intact or scarified skin resulted in more severe damage, including necrosis, and the deaths of some treated animals.

- Topical exposure of six rabbits to 200 mg endothall technical/kg resulted in the death of all rabbits within 24 hours (Pharmacology Research, Inc., 1975a).

- Technical endothall (0.1 g equivalent to 80 mg endothall ion) produced severe eye irritation in three rabbits when directly applied to the conjunctiva. Effects included corneal opacity, conjunctival irritation and iridic congestion. Furthermore, technical endothall apparently produced systemic effects by this route of absorption, since several animals died within 24 hours as a result of this exposure. Eyes were rinsed with water 20 to 30 seconds after treatment in three rabbits; conjunctival irritation and iridic congestion reversed in 4 days in two rabbits but persisted along with corneal opacity in one rabbit for 7 days (Pharmacology Research, Inc., 1975b).

3. *Long-term Exposure*

- Beagle dogs (4/sex/group) fed diets containing 0, 100, 300 or 800 ppm disodium endothall (equivalent to 0, 2, 6 or 16 mg endothall ion/kg/day for 24 months showed no gross signs of toxicity (Keller, 1965). Values for hematology, urinalysis, weight gain and food consumption were within normal limits and comparable to those for control animals. Increased stomach and small intestine weights were observed in the intermediate and high-dose groups. However, microscopic examination of essentially all tissues in the high-dose group revealed no pathological changes that could be attributed to endothall ingestion. A No-Observed-Adverse-Effect Level (NOAEL) of 2 mg endothall ion/kg/day is identified from this study.

- Brieger (1953b) reported no toxic effects in female rats given dietary levels as high as 2,500 ppm disodium endothall (about 100 mg endothall ion/kg/day, assuming food intake of 20 g/day and mean body weight of 0.4 kg) for 2 years.

4. *Reproductive Effects*

- A three-generation study in rats was reported by Scientific Associates (1965). Groups of male and female rats were fed diets containing 0, 100, 300 or 2,500 ppm disodium endothall (equivalent to 0, 4, 12 or 100 mg endothall ion/kg/day) until they were 100 days old and were then mated. Three successive generations of offspring were maintained on the test diet for 100 days and then bred to produce the next test generation. Pups in the 4 mg/kg/day dose group were normal, pups in the 12-mg/kg/day group had decreased body weights at 21 days of age and pups in the 100 mg/kg/day group did not survive more than 1 week. A NOAEL for reproductive effects of 4 mg endothall ion/kg/day was identified from this study.

5. *Developmental Effects*

- A short-term teratology study in rats by Science Applications, Inc. (1982) indicated no observable signs of developmental toxicity at dose levels that were fatal to the dams. This study suggests that the dams are more susceptible to endothall than are the embryos or fetuses. Groups of 25 or 26 female rats were mated and then orally dosed with 0, 10, 20 or 30 mg/kg/day of aqueous endothall technical (0, 8, 16 or 24 mg endothall ion/kg/day) on days 6 to 19 of gestation. Two dams died from the 20 mg/kg/day dose, and 10 dams died from the 30 mg/kg/day dose. No clinical signs were noted prior to death, and no lesions were observed at necropsy. The researchers concluded that endothall technical was not embryotoxic or teratogenic at maternal doses of 30 mg/kg/day or below. A NOAEL of 10 mg endothall technical/kg/day based on maternal effects was identified.

6. *Mutagenicity*

- Mutagenicity results from short-term *in vitro* tests are mixed, with various forms of endothall reported as test agents. Mutagenicity studies utilizing *Salmonella* with and without metabolic activation resulted in negative findings for endothall technical (Andersen et al., 1972; Microbiological Associates, 1980a). Mutagenic activity was not found in BALB/3T3 Clone A31 mouse cells exposed to endothall technical (Microbiological Associates, 1980b).

- For the following studies, Wilson et al. (1956) used "commercial Endothall" with no further description, whereas the remaining investigators used Aquathol® K, a commercial formulation containing dipotassium endothall at a level of 28.6% acid equivalent. In *Drosophila melanogaster*, mutagenic results were mixed, with Wilson et al. (1956) and Sandler and Hamilton-Byrd (1981) reporting positive and negative results, respectively. Sandler and Hamilton-Byrd (1981) reported negative results in a mutagenicity assay with the mold *Neurospora crassa*. A sister chromatid exchange study in human lymphocytes was negative (Vigfusson, 1981). Transformation was induced in a BALB/c 3T3 test for malignant transformation (Litton Bionetics, Inc., 1981).

7. *Carcinogenicity*

- No statistically significant numbers or types of tumors were observed in rats fed as much as 100 mg endothall ion/kg/day for 2 years (Brieger, 1953b).

IV.   Quantification of Toxicological Effects

Because the test materials in the various toxicity studies were salt or acid forms of endothall, the HAs described herein are expressed in terms of endothall ion.

A. One-Day Health Advisory

No studies were located in available literature that were suitable for calculation of the One-day HA. The single-dose studies measured mortality as the toxicological end point and are not suitable for use in calculating an HA. The value of 0.8 mg/L calculated as the Ten-day HA can be used as a conservative estimate of the One-day HA.

B. Ten-day Health Advisory

The teratology study by Science Applications, Inc. (1982) has been selected as the basis for the Ten-day HA. It is the only study that defined a short-term NOAEL (8 mg endothall ion/kg/day, based on maternal toxicity).

The Ten-day HA for a 10-kg child is calculated as follows:

$$\text{Ten-day HA} = \frac{(8 \text{ mg/kg/day}) (10 \text{ kg})}{(100) (1 \text{ L/day})} = 0.8 \text{ mg/L } (800 \text{ } \mu g/L)$$

C. Longer-term Health Advisory

There is concluded to be insufficient data for calculation of a Longer-term HA. Therefore, the DWEL adjusted for a 10-kg child (0.2 mg/L) is proposed as a conservative estimate for a Longer-term HA.

D. Lifetime Health Advisory

The 2-year feeding study in dogs by Keller (1965), which identified a NOAEL of 2 mg endothall ion/kg/day, has been selected to serve as the basis for the Lifetime HA for endothall. The study by Scientific Associates (1965) was of shorter duration (100 days/generation) and did not as completely define a NOAEL (except for 4 mg endothall ion/kg/day for reproductive effects); however, the NOAEL in this study approximates that in the Keller (1965) study. The 2-year study in rats by Brieger (1953b) showed no adverse effects from doses up to 100 mg endothall ion/kg/day, but no information was provided on the parameters tested and the levels at which effects did occur.

Using the NOAEL of 2 mg/kg/day, the Lifetime HA for endothall is calculated as follows:

Step 1: Determination of the Reference Dose (RfD)

$$RfD = \frac{(2 \text{ mg/kg/day})}{(100)} = 0.02 \text{ mg/kg/day}$$

Step 2: Determination of the Drinking Water Equivalent Level (DWEL)

$$DWEL = \frac{(0.02 \text{ mg/kg/day}) (70 \text{ kg})}{(2 \text{ L/day})} = 0.7 \text{ mg/L } (700 \text{ } \mu g/L)$$

Step 3: Determination of the Lifetime Health Advisory

Lifetime HA = (0.7 mg/L) (20%) = 0.14 mg/L (100 μg/L)

E. Evaluation of Carcinogenic Potential

- Available toxicity data do not show endothall as carcinogenic. Endothall can be placed in Group D (inadequate evidence in humans and animals) by the EPA's guidelines for carcinogenic risk assessment (U.S. EPA, 1986).

- The International Agency for Research on Cancer has not evaluated the carcinogenic potential of endothall (WHO, 1982).

V.   Other Criteria, Guidance and Standards

- An interim tolerance of 200 μg/L has been published for residues of endothall, used to control aquatic plants, in potable water (CFR, 1979).

- Residue tolerances for endothall published by the U.S. EPA (CFR, 1977) include 0.1 ppm in or on cottonseed, 0.1 ppm in or on potatoes, 0.05 ppm in or on rice grain and 0.05 ppm in or on rice straw.

- A tolerance is a derived value based on residue levels, toxicity data, food consumption levels, hazard evaluation and scientific judgment; it is the legal maximum concentration of a pesticide in or on a raw agricultural commodity or other human or animal food (Paynter et al., undated).

- The ADI set by the U.S. EPA Office of Pesticide Programs is 0.02 mg/kg/day based on the 2 mg/kg/day NOAEL in the 2-year dog study by Keller (1965) and a 100-fold uncertainty factor.

VI.   Analytical Methods

- Endothall is a dicarboxylic herbicide used to control aquatic vegetation. It presents a difficult analytical challenge since its highly polar structure does not allow simple solvent extraction from water. It lacks any elements which would allow use of the selective detectors commonly used in pesticide analysis. To obtain any sensitivity, chemical derivatization is required both for detection and concentration of the extract.

A draft method has been developed by EPA's Environmental Monitoring and Support Laboratory, Cincinnati (EMSL-CI) (U.S. EPA, 1988) which allows a small volume of sample to be derivatized with pentafluorophenylhydrazine (PFPH), and cleaned up by solid phase absorbant-extraction. Analysis is by electron-capture gas chromatography. The single laboratory estimated detection limit is 9.1 μg/L. The single operator recovery and precision for this method in tap water is 69% and 4%, respectively.

VII.   Treatment Technologies

- No information was found in the available literature on treatment technologies capable of effectively removing endothall from contaminated water.

VIII.   References

Allender, W.J. 1983. Suicidal poisoning by endothall. J. Anal. Toxicol. 7:79–82.

Andersen, K.J., E.G. Leighty and M.T. Takahashi. 1972. Evaluation of herbicides for possible mutagenic properties. J. Agr. Food Chem. 20:649–654.

Brieger, H.* 1953a. Preliminary studies on the toxicity of endothall (disodium). EPA Pesticide Petition No. 6G0503, redesignated No. 7F0570, 1966. EPA Accession No. 246012.

Brieger, H.* 1953b. Endothall, long term oral toxicity test—rats. EPA Pesticide Petition No. 6G0503, redesignated No. 7F0570, 1966. EPA Accession No. 246012.

Carlson, R., R. Whitaker and A. Landskov. 1978. Endothall. Chapter 31. In: Zweig, G. and J. Sherma, eds. Analytical methods for pesticides and plant growth. New York, NY: Academic Press, pp. 327–340.

Chiou, C.T., V.H. Freed, D.W. Schmedding and M. Manes. 1982. Partitioning of organic compounds in octanol-water systems. Environ. Sci. Technol. 16:4–9.

CFR. 1977. Code of Federal Regulations. 40 CFR 180.293.

CFR. 1979. Code of Federal Regulations. 21 CFR 193.180. April 1, 1979.

Gangstad, E.O. 1983. Herbicidal, environmental, and health effects of endothall. OCE-NRM-23. Washington, DC: Office, Chief of Engineers, U.S. Army Corps of Engineers. 25 pp.

Goldstein, F. 1952. Cutaneous and intravenous toxicity of endothall (disodium-3-endohexahydrophthalic acid). Pharmacol. Exp. Ther. 11:349.

Holmberg, D.J. and G.F. Lee. 1976. Effects and persistence of endothall in the aquatic environment. J. Water Poll. Control Fed. 48:2738–2746.

Isensee, A.R. 1976. Variability of aquatic model ecosystem-derived data. Int. J. Environ. Stud. 10:35–41.

Keckemet, O. 1980. Endothall-potassium and environment. Proc. 1980 Brit. Crop Protection Conf. Weeds, 8 pp.

Keller, J.* 1965. Two-year chronic feeding study of disodium endothall to beagle dogs. Scientific Associates report. EPA Pesticide Petition 6G0503, redesignated No. 7F0570, June 1966. EPA Accession No. 24601.

Langeland, K.A. and J.P. Warner. 1986. Persistence of diquat, endothall and fluridone in ponds. J. Aquat. Plant Mgmt. 24:43–46.

Litton Bionetics, Inc. 1981. Evaluation of Aquathol® K in the *in vitro* transformation of BALB/3T3 cells with and without metabolic activation assay. Project No. 20992. Report to Municipality of Metropolitan Seattle, Seattle, WA, by Litton Bionetics, Inc., Rockville, MD. EPA Accession No. 245680.

Lockhart, W.L., B.N. Billeck, G.G.E. de March and D.C.G. Muir. 1983. Uptake and toxicity of organic compounds: studies with an aquatic macrophyte (*Lemna minor*). In: Bishop, W.E., R.E. Cardwell and B.B. Heidolph, eds. Aquatic Toxicology and Hazard Assessment: Sixth Symposium. ASTM STP 802. Philadelphia, PA: ASTM. pp. 460–468.

Meister, R., ed. 1988. Farm chemicals handbook. Willoughby, OH. Meister Publishing Company.

Microbiological Associates.* 1980a. Activity of T1604 in the *Salmonella/* microsomal assay for bacterial mutagenicity. Unpublished final report for Pennwalt Corp. by Microbiological Associates, Bethesda, MD. EPA Accession No. 244126.

Microbiological Associates.* 1980b. Activity of T1604 in the *in vitro* mammalian cell point mutation assay in the absence of exogenous metabolic activation. Unpublished final report for Pennwalt Corp. by Microbiological Associates, Bethesda, MD.

Mitchell, L.C. 1961. The effects of ultraviolet light (2537A) on 141 pesticide chemicals by paper chromatography. J. Assoc. Off. Analyt. Chem. 44:643–712.

Neely, W.B., and D. Mackay. 1982. An evaluative model for estimating environmental fate. In: Dickson, K.L., A.W. Maki and J. Cairns, Jr., eds. Modeling the fate of chemicals in the aquatic environment. Ann Arbor, MI: Ann Arbor Science Publishers, Inc. pp. 127–143.

Neely, W.B., D.R. Branson and G.E. Blau. 1974. Partition coefficient to measure bioconcentration potential of organic chemicals in fish. Environ. Sci. Tech. 8:1113–1115.

Paynter, O.E., J.G. Cummings and M.H. Rogoff. Undated. United States Pesticide Tolerance System. Unpublished draft report. Washington, DC: U.S. EPA, Office of Pesticide Programs.

Pharmacology Research, Inc.* 1975a. U.S. EPA Pesticide Resubmission File 4581-EIE. Summary data on acute oral toxicity and dermal irritation in rabbits (Endothall). EPA Accession No. 244125.

Pharmacology Research, Inc.* 1975b. U.S. EPA Pesticide Resubmission File. Summary data, primary eye irritation in the rabbit and inhalation toxicity in several species (endothall). EPA Accession No. 246012.

Reinert, K.H. and J.H. Rodgers, Jr. 1984. Influence of sediment types on the sorption of endothall. Bull. Environ. Contam. Toxicol. 32:557–564.

Reinert, K.H. and J.H. Rodgers, Jr. 1986. Validation trial of predictive fate models using an aquatic herbicide (endothall). Environ. Toxicol. Chem. 5:449–461.

Reinert, K.H., J.H. Rodgers, Jr., M.L. Hinman and T.J. Leslie. 1985. Compartmentalization and persistence of endothall in experimental pools. Ecotox Environ. Safety. 10:86–96.

Rodgers, J.H., Jr., K.H. Reinert and M.L. Hinman. 1984. Water quality monitoring in conjunction with the Pat Mayse Lake aquatic plant management program. In: Proceedings, 18th Annual Meeting, Aquatic Plant Control, Research Program. November 14–17, 1983. Misc. Paper A-84-1. Raleigh, NC: U.S. Army Corps of Engineers. pp. 17–24.

Sandler, L. and E.L. Hamilton-Byrd. 1981. The induction of sex-linked recessive ethal mutations in *Drosophila melanogaster* by Aquathol® K, as measured by the Muller-5 test. Report to Municipality of Metropolitan Seattle, Seattle, WA.

Scientific Associates.* 1965. Three-generation rat reproductive study, disodium endothall. EPA Pesticide Petition No. 6G0503, redesignated 7F0570, 1966. EPA Accession No. 246012.

Science Applications, Inc.* 1982. A dose range-finding teratology study of endothall technical and disodium endothall in albino rats. Resubmission of Pesticide Application for Aquathol® K Aquatic Herbicide (EPA Registration No. 4581-204) and Hydrothal 191 Aquatic Algicide and Herbicide (EPA Registration No. 4581-174). EPA Accession No. 071249.

Serns, S.L. 1977. Effects of dipotassium endothall on rooted aquatics and adult and first generation bluegills. Water Res. Bull. 13:71–80.

Sikka, H.C. and C.P. Rice. 1973. Persistence of endothall in the aquatic environment as determined by gas-liquid chromatography. J. Agric. Food Chem. 21:842–846

Simsiman, G.V. and G. Chesters. 1975. Persistence of endothall in the aquatic environment. Water, Air, Soil Poll. 4:399–413

Simsiman, G.V., T.C. Daniel and G. Chesters. 1976. Diquat and endothall: their fates in the environment. Res. Rev. 62:131–174.

Soo, A., I. Tinsley and S.C. Fang. 1967. Metabolism of ^{14}C-endothall in rats. J. Agric. Food Chem. 15:1018–1021.

Srensek, S.E. and G. Woodard. 1951. Pharmacological actions of "endothall" (disodium-3,6-endoxo-hexahydrophthalic acid). Fed. Proc. 10:337. Abstract.

STORET. 1988. STORET Water Quality File. Office of Water. U.S. Environmental Protection Agency (data file search conducted in May, 1988).

Tweedy, B.G. and L.D. Houseworth. 1976. Miscellaneous herbicides. In: Herbicides-chemistry, degradation and mode of action. Kearney, P.C., and D.D. Kaufman, eds. Chapter 17. New York, NY: Marcel Dekker, Inc. pp. 815–833.

U.S. EPA. 1986. U.S. Environmental Protection Agency. Guidelines for carcinogen risk assessment. Fed. Reg. 51(185):33992–34003. September 24.

U.S. EPA. 1988. U.S. Environmental Protection Agency. Draft test method XXX for endothall. Environmental Monitoring and Support Laboratory, Cincinnati, Ohio. June.

Vigfusson, N.V. 1981. Evaluation of the mutagenic potential of Aquathol® K by induction of sister chromatid exchanges in human lymphocytes *in vitro*. Report to Municipality of Metropolitan Seattle, Seattle, WA. EPA Accession No. 245680.

Walker, C.R. 1963. Endothall derivatives as aquatic herbicides in fishery habitats. Weeds. 11:226–232.

WHO. 1982. World Health Organization. IARC Monographs on the evaluation of the carcinogenic risk of chemicals to humans: chemicals, industry processes and industries associated with cancer to humans. International Agency for Research on Cancer Monographs Vol. 1 to 29. Supplement 4. Geneva, Switzerland: World Health Organization.

Wilson, S.M., A. Daniel and G.B. Wilson. 1956. Cytological and genetical effects of the defoliant endothall. J. Hered. 47:151–154.

Yeo, R.R. 1970. Dissipation of endothall and effects on aquatic weeds and fish. Weed Sci. 18:282–284.

---

* Confidential Business Information submitted to the Office of Pesticide Programs.

Ethylene Thiourea

I. General Information and Properties

Ethylene thiourea (ETU) is no longer used in commerce but is a common degradation product of the ethylene bisdithiocarbamate (EBDC) pesticides.

Although the toxicity of ETU may be similar to the toxic effects observed with the EBDCs, the One-day, Ten-day, Longer-term and Lifetime HAs for ETU should not necessarily be considered protective of exposure to individual EBDCs at this time. The mechanisms of toxicity for these substances are still under evaluation.

A. CAS No. 96–45–7

B. Structural Formula

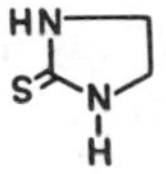

2-Imidazolidinethione

C. Synonyms

- ETU.

D. Uses

- Ethylene thiourea is a degradation product of several EBDC pesticides.

E. Properties (Windholz et al., 1983; U.S. EPA, 1982; U.S. EPA, 1986a.)

Chemical Formula	$C_3H_6N_2S$
Molecular Weight	102.2
Physical State (25°C)	White crystals
Boiling Point	--
Melting Point	203°
Density	--
Vapor Pressure	--
Specific Gravity	--

Water Solubility (30°C)   20 g/L
Log Octanol/Water
   Partition Coefficient   --
Taste Threshold   --
Odor Threshold   --

F. Occurrence

- ETU was not found in sampling performed at 264 ground-water stations, according to the STORET database (STORET, 1988).

G. Environmental Fate

- Ethylene thiourea can be degraded by photolysis (U.S. EPA, 1982).

- ^{14}C-Ethylene thiourea was intermediately mobile (R_f 0.61) to very mobile (Rf 1.00) in muck and sandy loam soils, respectively, as determined by soil thin-layer chromatography (TLC) (U.S. EPA, 1986a). Adsorption was correlated to organic matter. Following 6 days of incubation in dry silty clay loam soil, ETU residues were immobile; however, ETU residues subjected to a wet-dry cycle were slightly mobile (R_f 0.2).

- Levels of ETU (purity unspecified) declined at an unspecified rate in sand, with a half-life of 1 to 6 days (U.S. EPA, 1986a). Concentrations of ETU declined from 220 ppm at day 0 to 116 ppm by day 1 and 86 ppm by day 6.

- Mancozeb has been shown to have a half-life of less than 1 day in sterile water before degrading to ETU (U.S. EPA, 1982). The ethylene bis-dithiocarbamates (EBDCs) are generally unstable in the presence of moisture and oxygen, and the EBDCs decompose rapidly in water as well as in biological systems (U.S. EPA, 1982).

- Photolysis is a major degrading pathway for ETU (U.S. EPA, 1982).

II.   Pharmacokinetics

A. Absorption

- Allen et al. (1978) reported a very high rate of absorption of ^{14}C-ETU gastrically administered at 40 mg/kg to female Rhesus monkeys and female Sprague-Dawley rats. In both species, feces accounted for less than 1.5% of the excreted radioactivity at 48 hours after administration.

- Absorption was also high in male Sprague-Dawley rats orally administered ^{14}C-ETU at 4 mg/kg, with 82.7% of the total administered dose detected in the urine at 24 hours (Iverson et al., 1980).

B. Distribution

- Allen et al. (1978) reported that in Rhesus monkeys administered ^{14}C-ETU at 40 mg/kg by gastric intubation, total tissue distribution at 48 hours was approximately 25% of the administered dose; approximately half of that was concentrated in muscle, with measurable amounts noted in blood, skin and liver. In Sprague-Dawley rats, however, total tissue distribution was less than 1% of the administered dose.

- Except in the thyroid, ETU was not found to accumulate in rats given an oral dose (amount not specified) (U.S. EPA, 1982). Up to 80% of the absorbed dose was eliminated in the urine 24 hours after administration.

C. Metabolism

- Iverson et al. (1980) identified the 24-hour urinary metabolites of ^{14}C-ETU orally administered to male Sprague-Dawley rats at 4 mg/kg. Imidazoline was present at 1.9% of the total recovered dose, imidazolone at 4.9%, ethylene urea at 18.3% and unchanged ETU at 62.6%. In female cats, intravenous (iv) administration of this dose resulted in unchanged ETU present in the urine at only 28% of the total recovered dose, with S-methyl ETU at 64.3% and ethylene urea at 3.5%.

- One hundred percent of the ETU (dose not specified) fed to mice was recovered rapidly (time not specified) with 50% recovered in the urine (U.S. EPA, 1982). Of the urinary products, 52% was unchanged ETU, 12% was ethylene urea and 37% were polar products.

- All animals that have been tested appear to metabolize EBDCs rapidly. ETU and ethylene bidiisothiocyanato sulfide (EBIS) are the major metabolites formed (U.S. EPA, 1982). Approximately 7% of an EBDC dose is converted to ETU *in vivo* in the rat and 2% in the mouse (Nelson, 1987; Jordan and Neal, 1979).

D. Excretion

- Allen et al. (1978) reported that 48 hours after gastric administration of ^{14}C-ETU at 40 mg/kg to Rhesus monkeys, approximately 55% of the administered dose was detected in the urine and 0.5% in the feces. In

Sprague-Dawley rats dosed identically, 82% was recovered in the urine and 1.3% in the feces.

- Iverson et al. (1980) reported that 82.7 and 80.6% of the total radioactivity of a single 4 mg/kg dose of ^{14}C-ETU was eliminated in the 24-hour urine of orally treated male Sprague-Dawley rats and iv-treated female cats, respectively.

III.   Health Effects

A. Humans

- No suitable information was found in the available literature on the health effects of ETU in humans.

B. Animals

1. *Short-term Exposure*

- The acute oral LD_{50} for ETU is 1,832 mg/kg in rats (U.S. EPA, 1982).

- Graham and Hansen (1972) measured ^{131}I uptake in male Osborne-Mendel rats administered ETU (purity not stated) in the diet at 50, 100, 500 or 750 ppm for various time periods (30, 60, 90 or 120 days). Assuming that 1 ppm in the diet of younger rats is equivalent to approximately 0.1 mg/kg/day (Lehman, 1959), these levels correspond to doses of about 5, 10, 50 or 75 mg/kg/day. Four hours after the injection of ^{131}I, uptake was decreased significantly in rats that had ingested ETU at 500 or 750 ppm for all time periods. At 24 hours after ^{131}I injection, uptake was decreased significantly in rats that had ingested 100, 500 or 750 ppm for all time periods. Histologically, the thyroid glands of rats ingesting ETU at approximately 5.0 mg/kg, the No-Observed-Adverse-Effect Level (NOAEL) for this study, were not different from those of control rats. There was slight hyperplasia of the thyroid in rats given 100 ppm (10 mg/kg/day). At doses of 500 or 750 ppm (50 or 75 mg/kg/day), the thyroid had moderate to marked hyperplasia.

- In an 8-day maximum tolerated dose (MTD) study by Plasterer et al. (1985), dose levels of 0, 75, 150, 300, 600 and 1,200 mg/kg ETU were given by gavage to mice (10/group, sex not specified). Body weight and mortality were evaluated. No significant effects were noted on body weight at the end of the eighth day. Based on mortality, ETU was considered moderately toxic by the authors. An MTD of 600 mg/kg was determined.

- In a study by Freudenthal et al. (1977), ETU (> 95% pure) was fed to rats (20/sex/group) in the diet at levels of 0, 1, 5, 25, 125 or 625 ppm for 30 days. Assuming that 1 ppm in the diet of a young rat is equivalent to 0.1 mg/kg (Lehman, 1959), these levels correspond to doses of about 0, 0.1, 0.5, 2.5, 12.5 or 62.5 mg/kg. Thyroid function, food consumption, body weight gain and histopathology were assessed in the animals. Rats in the 625-ppm groups showed signs of toxicity after 8 days of exposure. Hair loss, dry skin, increased salivation and decreased food consumption and body weight gain were observed. Other effects noted in the 625-ppm dose group were decreased iodine uptake and percent triiodothyronine (T_3) bound to thyroglobulin. Thyroid-stimulating hormone (TSH) was increased, and T_3 and thyroxine (T_4) decreased in the 625-ppm dose group. Thyroid hyperplasia was also noted in this group. Animals exposed to 125 ppm exhibited increased TSH, decreased T_4 and thyroid hyperplasia. Other thyroid parameters were not affected. Based on the absence of adverse effects in rats exposed to 25 ppm or less after 30 days, a NOAEL of 25 ppm (2.5 mg/kg) was identified.

- Arnold et al. (1983) showed that the thyroid effects of ETU (purity not stated) administered in the diet for 7 weeks to male and female Sprague-Dawley rats were reversible when ETU was removed from the diet. Dose-related significant decreases in body weight and increases in thyroid weight were observed in all treated animals, starting at dose levels of 75 ppm (approximately 7.5 mg/kg/day based on Lehman, 1959). This dose was identified as the Lowest-Observed-Adverse-Effect Level (LOAEL) for this study.

- In a 60-day study which was a continuation of the above study, by Freudenthal et al. (1977), 14/40 rats in the 625-ppm group died. Thyroid hyperplasia and altered thyroid function were observed in the two high-dose groups. Thyroid hyperplasia was also observed in the 25-ppm group. This effect, however, was not observed in this dose group when exposure was continued to 90 days. Thus, the NOAEL for this study is presumed to be 25 ppm, or 2.5 mg/kg.

2. *Dermal/Ocular Effects*

- No information was found in the above literature on the dermal/ocular effects of ETU.

3. *Long-term Exposure*

- Freudenthal et al. (1977) described alterations in thyroid function and changes in thyroid morphology when Sprague-Dawley rats were administered ETU (96.8% pure) in the diet at levels of 1 to 625 ppm (approximately 0.1 to 62.5 mg/kg/day based on Lehman, 1959) for up to 90 days. The NOAEL was reported to be 19.5 mg/kg/day at week 1 and 12.5 mg/kg/day at week 12.

- Graham and Hansen (1972) measured ^{131}I uptake in male Osborne-Mendel rats administered ETU (purity not specified) in the diet at 50, 100, 500 or 750 ppm for up to 120 days. Assuming that 1 ppm in the diet of older rats is equivalent to approximately 0.05 mg/kg/day (Lehman, 1959), these dosages are equivalent to approximately 2.5, 5, 25 and 37.5 mg/kg/day. Four hours after the injection of radioactive iodine, uptake was decreased significantly in rats ingesting ETU at 500 or 750 ppm (25 or 37.5 mg/kg/day) for all feeding periods. At 24 hours after ^{131}I injection, uptake was significantly decreased in rats ingesting the 100-, 500- and 750-ppm doses for all feeding periods. Histologically, the thyroid glands of rats ingesting ETU at approximately 2.5 mg/kg, the NOAEL for this study, were not different from those of control rats. There was slight hyperplasia of the thyroid in rats given 100 ppm (5 mg/kg/day). At doses of 500 or 750 ppm (25 or 37.5 mg/kg/day), the thyroid had moderate to marked hyperplasia.

- The thyroid appears to be the primary target organ for ETU toxicity in longer-term exposure studies. Graham et al. (1973) measured ^{131}I uptake in male and female Charles River rats fed ETU (purity not specified) in the diet at 0, 5, 25, 125, 250 or 500 ppm for up to 12 months. Assuming that 1 ppm in the diet of older rats is equivalent to approximately 0.05 mg/kg/day (Lehman, 1959), these levels correspond to doses of about 0.25, 1.25, 6.25, 12.5 or 25 mg/kg/day. Adverse effects were noted at 2, 6 and 12 months. At 12 months, significant decreases in body weight and increases in thyroid weight were seen at the 125-, 250-and 500-ppm levels. Uptake of ^{131}I was significantly decreased in male rats after 12 months at 500 ppm, but was increased in females. Microscopic examination of the thyroid revealed the development of nodular hyperplasia at dose levels of 125 ppm and higher. The NOAEL for thyroid effects in this study was 25 ppm (approximately 1.25 mg/kg/day).

- Ulland et al. (1972) reported a dose-related increased incidence of hyperplastic goiter in male and female rats fed ETU at 175 and 350 ppm in their diet for 18 months (approximately 8.75 and 17.5 mg/kg/day, based on Lehman, 1959). An increased incidence (significance not specified) of simple goiter was also reported in all treatment groups.

- In a 2-year study by Graham et al. (1975), Charles River rats were fed ETU (purity not specified) in the diet at 0, 5, 25, 125, 250 or 500 ppm (approximately 0.25, 1.25, 6.25, 12.5 or 25 mg/kg/day, based on

Lehman, 1959). Statistically significant (p < 0.01) decreases in body weight were observed in both sexes fed at 500 ppm. Increases in thyroid-to-body weight ratios were apparent at 250 and 500 ppm (p < 0.01). There was an increased iodine (^{131}I) uptake at 5 ppm and a decreased uptake at 500 ppm, as well as slight thyroid hyperplasia at the 5- and 25-ppm dose levels (significance not stated). Based on these results, a LOAEL for lifetime exposure of 5 ppm (0.25 mg/kg/day) was identified.

4. Reproductive Effects

* Plasterer et al. (1985) administered ETU (purity not specified) by gavage as a water slurry to CD-1 mice at 600 mg/kg/day on days 7 to 14 of gestation. At this dose level, maternal toxicity was not observed but the reproductive index was significantly decreased (p < 0.05), indicating severe prenatal lethality.

* New Zealand White rabbits were dosed with ETU (100% pure) at 10, 20, 40 or 80 mg/kg/day on days 7 to 20 of pregnancy (Khera, 1973). Observed effects included an increase (p < 0.05) in resorption sites at 80 mg/kg. No adverse effects on fetal weight or on the number of viable fetuses per pregnancy were noted at any dose level, and no signs of maternal toxicity were observed. Based on the results of this study, a NOAEL of 80 mg/kg/day for maternal toxicity and a NOAEL of 40 mg/kg/day for fetotoxicity were identified.

5. Developmental Effects

* The ability of ETU to induce various adverse effects, including teratogenicity and maternal toxicity, has been demonstrated by several investigators using various animal models. Available data indicate that rats are probably the most sensitive species.

* Khera (1973) orally administered ETU (100% pure) to Wistar rats at daily doses of 5, 10, 20, 40 or 80 mg/kg from 21 or 42 days before conception to pregnancy day 15 and on days 6 to 15 or 7 to 20 of pregnancy. Dose-dependent lesions of the fetal central nervous and skeletal systems were produced, irrespective of the time at which ETU was administered. Teratogenic effects seen at the two highest dose levels included meningoencephalocele; meningorrhagia; meningorrhea; hydrocephalus; obliterated neural canal; abnormal pelvic limb posture with equinovarus; micrognathia; oligodactyly; and absent, short or kinky tail. Less serious defects were seen at 20 mg/kg, and at 10 mg/kg, there was only a retardation of parietal ossification and of cerebellar Purkinje-cell migration. Retarded parietal ossification was the only

abnormality seen at 5 mg/kg (significance not stated), its incidence being limited to small areas and to a few large litters. No signs of maternal toxicity were observed in rats administered ETU at 40 mg/kg/day for 57 days (42 days preconception to day 15 of gestation). Based on the results of this phase of the study, the NOAEL for maternal toxicity was 40 mg/kg/day, and the NOAEL for developmental effects was 5 mg/kg/day.

- In the same study (Khera, 1973), New Zealand White rabbits were dosed with ETU at 10, 20, 40 or 80 mg/kg/day on days 7 to 20 of pregnancy. Observed effects included a reduction in fetal brain-to-body weight ratio at 10 and 80 mg/kg (p < 0.01). Renal lesions, characterized by degeneration of the proximal convoluted tubules, were noted microscopically (dose level not specified), but there were no skeletal abnormalities that were attributed by the authors to ETU. The results of this study are not useful for determining a NOAEL or LOAEL.

- Dose-related central nervous system (CNS) lesions in Wistar rat fetuses were reported by Khera and Tryphonas (1985). Ethylene thiourea (> 98% pure) was administered by gastric intubation at 0, 15 or 30 mg/kg to dams on day 13 of pregnancy. Observed lesions at 30 mg/kg included histopathological changes of the CNS such as karyorrhexis in the germinal layer of basal lamina extending from the thoracic spinal cord to the telencephalon, and obliteration and duplication of the central canal and disorganization of the germinal and mantle layers. In the brain, the ventricular lining was fully denuded, neuroepithelial cells were arranged in the form of rosettes and nerve cell proliferation was disorganized. In the 15 mg/kg/day group, cellular necrosis was less severe and consisted of small groups of cells dispersed in the germinal layers of the neuraxis. None of the dams treated with ETU at any level in this study showed any overt signs of toxicity. Based on the results of this study, the NOAEL for maternal toxicity was 30 mg/kg and the LOAEL for developmental toxicity was 15 mg/kg.

- Sato et al. (1985) investigated the teratogenic effects of ETU (purity not specified) on Long-Evans rats exposed by gastric intubation to a single dose of 80, 120 or 160 mg/kg on 1 day between days 11 and 19 of gestation. Fetal malformations were related to both the day of administration and the dosage level. A short or absent tail was noted, for example, in 100% of fetuses exposed to ETU on gestational day 11 to 14. On day 11, a dose-dependent incidence of spina bifida and myeloschisis with hind-brain crowding were observed. A high incidence (48 to 87.5%, not dose-related) of macrocephaly with occipital bossing was noted with administration of ETU on day 12, and an almost total incidence (96 to

100%) with administration on day 13. Other abnormalities seen in this study were exencephaly, microcephaly and hypognathia and extremely high incidences (100% in many groups) of hydroencephaly and hydrocephalus, especially associated with administration days 14 through 19. Maternal toxicity was not addressed by the authors. The results of this study are not useful in determining LOAELs or NOAELs for teratogenicity or maternal toxicity, but serve instead as evidence of the kinds of developmental and teratogenic effects that a single dose of ETU at 80 mg/kg can induce in rats.

- Khera and Iverson (1978) reported that there was no clear evidence of teratogenicity in kittens whose mothers had been administered ETU (purity not specified) at 5, 10, 30, 60 or 120 mg/kg by gelatin capsule for days 16 to 35 of gestation. However, fetuses from cats in a moribund state subsequent to ETU toxicosis (30 to 120 mg/kg dosage groups) did show a high incidence (11/35) of malformations including coloboma, umbilical hernia, spina bifida and cleft palate. Maternal toxicity and death were observed at dose levels of 10 mg/kg and above, manifesting signs of toxicity that were delayed in onset and characterized by progressive loss of body weight, ataxia, tremors and hind-limb paralysis. In this study, the NOAEL for maternal toxicity was identified as 5 mg/kg/day and the NOAEL for developmental effects was 10 mg/kg/day.

- Chernoff et al. (1979) demonstrated the teratogenic effects of ETU in Sprague-Dawley rats, CD-1 mice and golden hamsters. The rats were administered ETU (purity not specified) by gastric intubation at 80 mg/kg/day on days 7 to 21 of gestation. Gross defects of the skeletal system (micrognathia, micromelia, oligodactyly, kyphosis) and the CNS (hydrocephalus, encephalocele), as well as cleft palate were noted in a majority of fetuses at this dose level. No clear evidence of teratogenicity was seen in groups of rats administered dose levels of 5 to 40 mg/kg/day. No similar pattern of defects was observed in CD-1 mice dosed at 100 or 200 mg/kg/day on days 7 to 16 of gestation or in golden hamsters dosed at 75, 150 or 300 mg/kg/day on days 5 to 10 of gestation. Observations of maternal toxicity included a marked decrease in the average weight gain of pregnant rats dosed at 80 mg/kg/day (p < 0.001). No significant effects were observed in mice or hamsters. Based on the results of this study, the NOAELs for maternal and developmental toxicity were 40 mg/kg/day in the rat, 200 mg/kg/day in the mouse and 300 mg/kg/day in the hamster.

- Adverse developmental effects of orally administered ETU, including teratogenicity and/or maternal toxicity, have been reported at 60, 100 and 240 mg/kg in rats (Khera, 1982; Teramoto et al., 1980; Ruddick and Khera, 1975) and at 400 and 1,600 to 2,400 mg/kg in mice (Teramoto et al., 1980; Khera, 1984).

6. *Mutagenicity*

- Seiler (1973) described ETU as exhibiting weak but significant mutagenic activity in *Salmonella typhimurium* HIS G-46. A 2.5-fold increase in mutation frequencies ($p < 0.001$) was seen at intermediate concentrations (100 or 1,000 ppm/plate), but at higher concentrations (10,000 and 25,000 ppm), ETU was somewhat lethal to the test colonies resulting in lower relative mutagenic indices (1.60 and 1.16, respectively).

- Schupbach and Hummler (1977) reported that ETU induced mutations of the base-pair substitution type in *S. typhimurium* TA1530 *in vitro* as well as in a host-mediated assay. In the host-mediated assay, a dose of 6,000 mg/kg ($LD_{50} = 5,400$ mg/kg) resulted in a slight but significant increase of the reversion frequency by a factor of 2.37. Results of a micronucleus test were negative after twofold oral administrations of 700, 1,850 or 6,000 mg/kg to Swiss albino mice; it was concluded that ETU does not induce any chromosomal anomalies in the bone marrow. No dominant-lethal effect was observed after single oral doses of 500, 1,000 or 3,500 mg/kg were given to male mice.

7. *Carcinogenicity*

- Graham et al. (1975) reported that ETU was a follicular thyroid carcinogen in male and female Charles River rats that were fed the compound (purity not specified) for 2 years at dietary levels of 250 and 500 ppm (approximately 12.5 and 25 mg/kg/day based on Lehman, 1959).

- In a survey of several compounds for tumorigenicity, Innes et al. (1969) reported that ETU (purity not stated) administered by diet to two strains of specific pathogen-free hybrid mice at a daily dosage of 215 mg/kg/day for 18 months resulted in statistically significant ($p < 0.01$) increases in hepatomas (14/16 or 18/18 for males and 18/18 or 9/16 for females) and in total tumor incidence. Pulmonary tumors and lymphomas were also investigated, but were not found to occur in the ETU group. The thyroid was not evaluated in this study. No other dose level was tested.

- Dose-related incidences of follicular and papillary thyroid cancers with pulmonary metastases and related lesions such as thyroid solid-cell adenomas were reported in Charles River CD rats by Ulland et al. (1972). Ethylene thiourea (97% pure) was administered by diet for 18 months at 175 or 350 ppm followed by administration of a control diet for 6 months. Assuming that 1 ppm in the diet of older rats is equivalent to approximately 0.05 mg/kg/day (Lehman, 1959), these levels correspond to doses of about 8.75 and 17.5 mg/kg/day. The first tumor was found after 68 weeks, and most were detected after 18 to 24 months when the study was terminated. The statistical significance of the reported findings was not addressed.

IV.   Quantification of Toxicological Effects

A. One-day Health Advisory

No data located in the available literature were suitable for determination of the One-day HA value. It is therefore recommended that the Ten-day HA value for the 10-kg child (0.3 mg/L, calculated below) be used at this time as a conservative estimate of the One-day HA value.

B. Ten-Day Health Advisory

The study by Freudenthal (1977) has been selected to serve as the basis for determination of the Ten-day HA for a 10-kg child. ETU was fed to a group of rats (20/sex/group) for up to 90 days at levels of 0, 1, 5, 25, 125 or 625 ppm (0, 0.1, 0.5, 2.5, 12.5 or 62.5 mg/kg/day, assuming that 1 ppm in the diet of a young rat equals 0.1 mg/kg/day, based on Lehman, 1959). Toxic effects on thyroid function and morphology were observed after 30 days' exposure to 125 ppm or greater. No adverse effects were noted in the 25-ppm group (2.5 mg/kg). Developmental effects reported in other studies (delayed parietal ossification) have been reported in rats exposed *in utero* at 5 mg/kg (Khera, 1973). The adversity of this effect is unclear. Khera and Iverson (1978) have reported maternal toxicity and death in cats exposed to 10 mg/kg. Therefore, 2.5 mg/kg was selected as a conservative NOAEL for deriving the Ten-day HA.

Using the NOAEL of 2.5 mg/kg/day, the Ten-day HA for a 10-kg child is calculated as follows:

$$\text{Ten-day HA} = \frac{(2.5 \text{ mg/kg/day}) (10 \text{ kg})}{(100) (1 \text{ L/day})} = 0.25 \text{ mg/L} (300 \text{ } \mu\text{g/L})$$

C. Longer-term Health Advisory

The study by Graham et al. (1973) has been selected to serve as the basis for determination of the Longer-term HA. In a 12-month study, ^{131}I uptake was measured in male and female Charles River rats fed ETU (purity not specified) in the diet at 0, 5, 25, 125, 250 or 500 ppm for 2, 6 or 12 months. Assuming that 1 ppm in the diet of older rats is equivalent to approximately 0.05 mg/kg/day (Lehman, 1959), these levels correspond to doses of about 0, 0.25, 1.25, 6.25, 12.5 or 25 mg/kg/day.

Adverse effects were noted at all three test intervals. At 12 months, significant decreases in body weight and increases in thyroid weight were seen at the 125-, 250- and 500-ppm levels. Uptake of ^{131}I was significantly decreased in male rats after 12 months at 500 ppm but was increased in females. Microscopic examination of the thyroids revealed the development of nodular hyperplasia at dose levels of 125 ppm and higher. The NOAEL for thyroid effects in this study was 25 ppm (approximately 1.25 mg/kg/day).

The Longer-term HA for a 10-kg child is calculated as follows:

$$\text{Longer-term HA} = \frac{(1.25 \text{ mg/kg/day})(10 \text{ kg})}{(100) (1 \text{ L/day})} = 0.125 \text{ mg/L} (100 \text{ } \mu g/L)$$

The Longer-term HA for a 70-kg adult is calculated as follows:

$$\text{Longer-term HA} = \frac{1.25 \text{ mg/kg/day})(70 \text{ kg})}{(100) (2 \text{ L/day})} = 0.44 \text{ mg/L} (400 \text{ } \mu g/L)$$

D. Lifetime Health Advisory

The study by Graham et al. (1975) was selected as the most appropriate basis for the calculation of a DWEL. In this 2-year study [presumably a continuation of the Graham et al. (1973) study], Charles River rats were fed ETU (purity not stated) in the diet at 0, 5, 25, 125, 250 or 500 ppm (approximately 0, 0.25, 1.25, 6.25, 12.5 or 25 mg/kg/day, based on Lehman, 1959).

Statistically significant (p < 0.01) decreases in body weight were observed in both sexes fed at 500 ppm. Increases (p < 0.01) in thyroid-to-body weight ratios were apparent at 250 and 500 ppm. There was an increased iodine (^{131}I) uptake at 5 and 125 ppm and a decreased uptake at 500 ppm, as well as slight thyroid hyperplasia at the 5- and 25-ppm dose levels (statistical significance not stated). This effect is considered to be biologically significant. Tumors were evident in animals in the 125-ppm group. Based on these results, the LOAEL for lifetime exposure was identified as 5 ppm (approximately 0.25 mg/kg/day).

Using the LOAEL of 0.25 mg/kg/day, the DWEL is calculated as follows:

Step 1: Determination of the Reference Dose (RfD)

$$RfD = \frac{(0.25 \text{ mg/kg/day})}{(1,000)\,(10)} = 0.000025 \text{ mg/kg/day } (0.03 \text{ } \mu g/kg/day)$$

Step 2: Determination of the Drinking Water Equivalent Level (DWEL)

$$DWEL = \frac{(0.00003 \text{ mg/kg/day})\,(70 \text{ kg})}{(2 \text{ L/day})} = 0.00105 \text{ mg/L } (1 \text{ } \mu g/L)$$

Step 3: Determination of the Lifetime Health Advisory

According to EPA's guidelines for assessment of carcinogenic risk (U.S. EPA, 1986b), ETU is classified in Group B: probable human carcinogen. Therefore, a Lifetime Health Advisory is not recommended for ETU. The estimated cancer risk level associated with lifetime exposure to ETU at 1 $\mu g/L$ is approximately 4.3×10^{-6}.

E. Evaluation of Carcinogenic Potential

- Three studies that evaluated the carcinogenic potential of ETU were identified. The results of these studies indicate that ETU is a thyroid carcinogen in rats (Graham et al., 1975; Ulland et al., 1972) and increases the incidence of hepatomas as well as total tumor incidence in mice (Innes et al., 1969).

- Graham et al. (1975) reported ETU to be a thyroid carcinogen in male and female Charles River rats that were fed the compound (purity not specified) for 2 years at dietary levels of 250 and 500 ppm (approximately 12.5 and 25 mg/kg/day in the diet of older rats, based on Lehman, 1959). At 125 ppm (approximately 6.3 mg/kg/day), ETU was a thyroid oncogen.

- Dose-related incidences of follicular and papillary thyroid cancers with pulmonary metastases and related lesions such as thyroid solid-cell adenomas were reported in Charles River CD rats by Ulland et al. (1972). Ethylene thiourea (97% pure) was administered in the diet for 18 months at 175 and 350 ppm followed by administration of a control diet for 6 months. Assuming that 1 ppm in the diet of older rats is equivalent to approximately 0.05 mg/kg/day (Lehman, 1959), these levels correspond to doses of about 8.75 and 17.5 mg/kg/day. The first tumor was found after 68 weeks, and most were detected after 18 to 24 months when the study was terminated. The statistical significance of the reported findings was not addressed.

- Innes et al. (1969) reported that ETU (purity not stated) administered by diet to specific pathogen-free hybrid mice at a daily dosage of 215 mg/kg/day for 18 months resulted in statistically significant ($p < 0.01$) increases in hepatomas and in total tumor incidence. No other dose level was tested. (Pulmonary tumors and lymphomas were also investigated in this study.)

- Applying the criteria described in EPA's final guidelines for assessment of carcinogenic risk (U.S. EPA, 1986b), ETU may be classified in Group B2: probable human carcinogen based on sufficient evidence from animal studies.

- The EPA Carcinogen Assessment Group estimated a one-hit slope of 0.1428/mg/kg/day based on the Innes et al. (1969) study identifying male mouse liver tumors as the sensitive sex/species end point (U.S. EPA, 1979). An assumed consumption of 2 liters of water per day by a 70-kg adult over a lifetime results in drinking water concentrations of 24, 2.4 and 0.24 μg/L for 10^{-4}, 10^{-5} and 10^{-6} cancer risk levels, respectively.

- Data are not available to estimate excess cancer risks using other mathematical models.

V.   Other Criteria, Guidance and Standards

- No other data have been located for ETU.

VI.   Analytical Methods

- Ethylene thiourea is analyzed by a nitrogen-phosphorus detector/gas chromatographic method as described in Method 6 (U.S. EPA, 1988). In this procedure, the ETU sample is mixed with ammonium chloride and potassium fluoride and passed through an exchange column (Extrelut). The ETU is then eluted with methylene chloride, concentrated for exchange with ethyl acetate to a volume of 5 mL. The method describes conditions which permit the separation and measurement of ETU by GC with a nitrogen-phosphorus detector. This method has been validated by a single laboratory. The estimated detection limit for ETU by Method 6 is 5 μg/L.

VII.   Treatment Technologies

- No data were found on the removal of ethylene thiourea from drinking water by conventional treatment.

- No data were found on the removal of ethylene thiourea from drinking water by activated carbon adsorption. However, since ethylene thiourea has a high solubility and is hydrophilic, treatment with activated carbon probably would not be effective.

- No data were found on the removal of ethylene thiourea from drinking water by ion exchange. However, the structure of ethylene thiourea indicates it is not ionic and thus probably would not be amenable to ion exchange.

- No data were found on the removal of ethylene thiourea from drinking water by aeration. Since vapor pressure data are unavailable, Henry's Coefficient, and thus the effectiveness of aeration, cannot be estimated. However, the high melting point and the high solubility indicate that Henry's Coefficient would be low and that aeration or air stripping probably would not be an effective form of treatment.

VIII.   References

Allen, J.R., J.P. Von Miller and J.L. Seymour. 1978. Absorption, tissue distribution and excretion of ^{14}C ethylenethiourea by the Rhesus monkey and rat. Res. Comm. Chem. Path. Pharmacol. 20:109–115.

Arnold, D.L., D.R. Krewski, D.B. Junkins, P.F. McGuire, C.A. Moodie and I.C. Munro. 1983. Reversibility of ethylenethiourea-induced thyroid lesions. Toxicol. Appl. Pharmacol. 67:264–273.

CHEMLAB. 1985. The Chemical Information System, CIS, Inc. Baltimore, MD.

Chernoff, N., R.J. Kavlock, E.H. Rogers, B.D. Carver and S. Murray. 1979. Perinatal toxicity of Maneb, ethylene thiourea, and ethylenebisthiocyanate sulfide in rodents. J. Toxicol. Environ. Health. 5:821–834.

Freudenthal, R.I., G. Kerchner, R. Persing and R. Baron. 1977. Dietary sub-acute toxicity of ethylene thiourea in the laboratory rat. J. Env. Path. Toxicol. 1:147–161.

Graham, S.L. and W.H. Hansen. 1972. Effects of short-term administration of ethylene thiourea upon thyroid function of the rat. Bull. Environ. Contam. Toxicol. 7(1):19–25.

Graham, S.L., W.H. Hansen, K.J. Davis and C.H. Perry. 1973. Effects of one-year administration of ethylenethiourea upon the thyroid of the rat. J. Agr. Food Chem. 21:324–329.

Graham, S.L., K.J. Davis, W.H. Hansen and C.H. Graham. 1975. Effects of prolonged ethylene thiourea ingestion on the thyroid of the rat. Food Cosmet. Toxicol. 13:493–499.

Innes, J.R., B.M. Ulland, M.G. Valerio, L. Petrucelli, L. Fishbein, E.R. Hart and A.J. Pallotta. 1969. Bioassay of pesticides and industrial chemicals for tumorigenicity in mice: a preliminary note. J. Natl. Cancer Inst. 42:1101–1114.

Iverson, F., K.S. Khera and S.L. Hierlihy. 1980. *In vivo* and *in vitro* metabolism of ethylene thiourea in the rat and the cat. Toxicol. Appl. Pharmacol. 52:16–21.

Jordan, L.W. and R.A. Neal. 1979. Examination of the *in vivo* metabolism of maneb and zineb to ethylenethiorea (ETU) in mice. Bull. Environ. Contam. Toxicol. 22:271–277.

Khera, K.S. 1973. Ethylene thiourea: teratogenicity study in rats and rabbits. Teratology. 7:243–252.

Khera, K.S. 1982. Reduction of teratogenic effects of ethylenethiourea in rats by interaction with sodium nitrite *in vivo*. Food Cosmet. Toxicol. 20:273–278.

Khera, K.S. 1984. Ethylenethiourea-induced hindpaw deformities in mice and effects of metabolic modifiers on their occurrence. J. Toxicol. Environ. Health. 13:747–756.

Khera, K.S. and F. Iverson. 1978. Toxicity of ethylenethiourea in pregnant cats. Teratology. 18:311–314.

Khera, K.S. and L. Tryphonas. 1985. Nerve cell degeneration and progeny survival following ethylenethiourea treatment during pregnancy in rats. Neurol. Toxicol. 6:97–102.

Lehman, A.J. 1959. Appraisal of the safety of chemicals in foods, drugs and cosmetics. Assoc. Food and Drug Off. U.S., Q. Bull.

Meister, R., ed. 1983. Farm chemicals handbook. Willoughby, OH: Meister Publishing Co.

Nelson, S.S. 1987. Bioconversion of mancozeb to ETU in rats. Rohm and Haas Technical Report No. 31C-87-24. Submitted to EPA. MRID 40301101.

Plasterer, M.R., W.S. Bradshaw, G.M. Booth, M.W. Carter, R.L. Schuler and B.D. Hardin. 1985. Developmental toxicity of nine selected compounds following prenatal exposure in the mouse: naphthalene, p-nitrophenol, sodium selenite, dimethyl phthalate, ethylenethiourea, and four glycol ether derivatives. J. Toxicol. Environ. Health. 15:25–38.

Ruddick, J.A. and K.S. Khera. 1975. Pattern of anomalies following single oral doses of ethylenethiourea to pregnant rats. Teratology. 12:277–282.

Sato, K., N. Nakagata, C.F. Hung, M. Wada, T. Shimoji and S. Ishii. 1985. Transplacental induction of myeloschisis associated with hindbrain crowding and other malformations in the central nervous system in Long-Evans rats. Child. Nerv. Syst. 1:137–144.

Schupbach, M. and H. Hummler. 1977. A comparative study on the mutagenicity of ethylenethiourea in bacterial and mammalian test systems. Mut. Res. 56:111–120.

Seiler, J.P. 1973. Ethylenethiourea (ETU), a carcinogenic and mutagenic metabolite of ethylenebis-dithiocarbamate. Mut. Res. 26:189–191.

STORET. 1988. STORET Water Quality File. Office of Water. U.S. Environmental Protection Agency (data file search conducted in May, 1988).

Teramoto, S., R. Saito and Y. Shirasu. 1980. Teratogenic effects of combined administration of ethylenethiourea and nitrite in mice. Teratology. 21:71–78.

Ulland, B.M., J.H. Weisburger, E.K. Weisburger, J.M. Rice and R. Cypher. 1972. Brief communication: thyroid cancer in rats from ethylene thiourea intake. J. Natl. Cancer Inst. 49:583–584.

U.S. EPA. 1979. U.S. Environmental Protection Agency. The Carcinogen Assessment Group's risk assessment on ethylene bisdithiocarbamate.

U.S. EPA. 1982. U.S. Environmental Protection Agency. Ethylene bis-dithiocarbamate pesticides. Final resolution of rebuttable presumption against registration. Decision document. Office of Pesticide Programs.

U.S. EPA. 1986a. U.S. Environmental Protection Agency. Task 2: Environmental fate and exposure assessment. Final report. June 10.

U.S. EPA. 1986b. U.S. Environmental Protection Agency. Guidelines for carcinogen risk assessment. Fed. Reg. 51(185):33992–34003. September 24.

U.S. EPA. 1988. U.S. Environmental Protection Agency. Mcthod 6 - Determination of ethylene thiourea (ETU) in ground water by gas chromatography with a nitrogen-phosphorus detector. Available from U.S. EPA's Environmental Monitoring and Support Laboratory, Cincinnati, OH.

Windholz, M., S. Budavari, R.F. Blumetti and E.S. Otterbein, eds. 1983. The Merck index, 10th ed. Rahway, NJ: Merck and Co., Inc.

---

* Confidential Business Information submitted to the Office of Pesticide Programs.

Fenamiphos

I. General Information and Properties

A. CAS No. 22224-92-6

B. Structural Formula

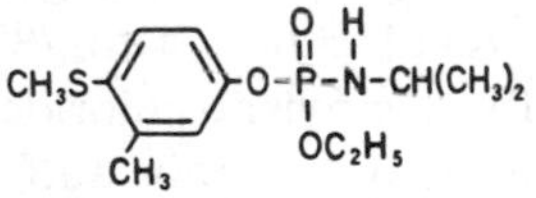

(l-Methylethyl)-ethyl-3-methyl-4-(methylthio)phenyl-phosphoramidate

C. Synonyms

- Nemacur; B 68138; Bay 68138; Bayer 68138; ENT 27572; Phenamiphos (Meister, 1983).

D. Uses

- Fenamiphos is used as a systemic nematicide (Meister, 1983).

E. Properties (Meister, 1983)

Chemical Formula	$C_{13}H_{22}O_3NSP$
Molecular Weight	303 (calculated)
Physical State (at 25°C)	Brown, waxy solid
Boiling Point	--
Melting Point	49.2°C
Density	--
Vapor Pressure (30°C)	7.5×10^{-7} mm Hg
Specific Gravity	--
Water Solubility (25°C)	400 mg/L
Log Octanol/Water Partition Coefficient	--
Taste Threshold	--
Odor Threshold	--
Conversion Factor	--

F. Occurrence

- Fenamiphos has not been detected in 664 ground-water samples analyzed from 659 locations (STORET, 1988). No surface water locations were tested.

G. Environmental Fate

- Ring-labeled ^{14}C-fenamiphos (radiochemical purity 94%), at 1 and 10 ppm, degraded with half-lives of 7 to 14 days in a buffered aqueous solution at pH 3 and > 30 days at pH 9, appeared to be stable at pH 7 when incubated in the dark at 30°C (McNamara and Wilson, 1981). In the pH 3 buffer solution, the primary degradation product was deaminated fenamiphos accounting for 74 to 78% of the applied material. Degradates identified in methylene chloride extracts from the pH 3, 7 and 9 solutions included fenamiphos sulfoxide, fenamiphos sulfone, fenamiphos phenol, fenamiphos sulfoxide phenol and fenamiphos sulfone phenol.

- Ring-labeled 14C-fenamiphos (radiochemical purity > 99%), at 12 ppm, degraded with a half-life of 2 to 4 hours in pH 7 buffered water irradiated with artificial light (approximately 5,200 μW/cm^2, 300 to 600 nm) (Dime et al., 1983). After 24 hours of irradiation, fenamiphos accounted for approximately 4% of the applied radioactivity, fenamiphos sulfonic acid phenol for approximately 19%, fenamiphos sulfoxide for approximately 17%, fenamiphos sulfonic acid for approximately 6% (tentative identification), and fenamiphos sulfoxide phenol for approximately 1%. In the dark control, fenamiphos accounted for approximately 94% of the applied compound at 24 hours posttreatment.

- Ring-labeled ^{14}C-fenamiphos (radiochemical purity > 99%), at approximately 20 ppm, degraded with a half-life of < 1 hour on sandy loam soil irradiated with artificial light (approximately 6,200 μW/cm^2, 300 to 600 nm) (Dime et al., 1983). After 48 hours of irradiation, fenamiphos and the degradates fenamiphos sulfoxide and fenamiphos sulfone accounted for approximately 12, 55 and 6% of the extractable radioactivity, respectively. In the dark control, fenamiphos accounted for approximately 93% of the extractable compound at 48 hours posttreatment.

- ^{14}C-Fenamiphos (purity 86%), at 3 ppm, degraded with a half-life of < 4 days in silty clay loam soil previously treated with fenamiphos (Green et al., 1982). Fenamiphos sulfoxide comprised up to approximately 74% of the applied radioactivity (maximum at 11 days posttreatment); fenamiphos sulfone comprised approximately 10% and volatile ^{14}C-residues comprised 17% of the applied material at 55 days posttreatment. At 55 days posttreatment, 1.13% of the applied fenamiphos remained undegraded in the soil previously treated with fenamiphos, 5.41% remained undegraded in soil with no prior history of fenamiphos treatment and 40.58% remained undegraded in sterile soil. Fenamiphos sulfoxide was the major degradate in all three treatments.

- ^{14}C-Fenamiphos (test substance uncharacterized), at 0.29 to 2.30 μg/mL of water, was adsorbed to sandy loam and clay loam soils with 26.3 to 30.0% and 42.2 to 52.3% of the applied radioactivity, respectively, adsorbed after 16 hours (Church, 1970).

- Fenamiphos (3 lb/gallon SC and 15% G), at approximately 20 lb active ingredient (a.i.)/A, was mobile in columns (16-cm length) of sandy soil eluted with 10 inches of water. Fenamiphos was detected throughout the columns, and 0.9 to 2.2% of the applied material was recovered in the leachate (Gronberg and Atwell, 1980).

- Aged (30 days) ^{14}C-fenamiphos residues, at approximately 4 lb a.i./A, were slightly mobile in a column (12-inch length) of sandy loam soil leached with 22.5 inches of water; approximately 2.3% of the applied radioactivity leached from the column and approximately 91% of the applied radioactivity remained in the top 5 inches of the soil column (Tweedy and Houseworth, 1980).

II.   Pharmacokinetics

A. Absorption

- Gronberg (1969) administered ^{14}C-labeled fenamiphos (99% purity) by oral intubation to rats. Only 5 to 7% was recovered in feces, indicating that 93 to 95% was absorbed from the gastrointestinal tract.

B. Distribution

- Gronberg (1969) administered single oral doses of 2 mg/kg of ethyl-^{14}C-fenamiphos (99% purity) by oral intubation to rats. Forty-eight hours after treatment, residues measured in tissues were: brain < 0.1 ppm; heart 0.1 ppm; liver 0.8 to 1.7 ppm; kidney 0.4 to 0.5 ppm; fat 0.2 to 0.4 ppm; muscle < 0.1 ppm; and gastrointestinal tract 0.2 ppm.

C. Metabolism

- In studies conducted by Gronberg (1969), rats were administered 2 mg/kg oral doses of fenamiphos (99% purity) using ethyl-^{14}C, methylthio-3H or isopropyl-^{14}C label. The authors proposed a pathway of fenamiphos metabolism involving oxidation to the sulfoxide and sulfone analogs. Subsequent hydrolysis, conjugation and excretion via urine gave high molecular-weight compounds (600 to 800). No other details were provided.

D. Excretion

- Gronberg (1969) administered ethyl-^{14}C, methylthio-3H or isopropyl-^{14}C-labeled fenamiphos (2 mg/kg, 99% purity) to rats by gavage. Thirty-nine to 42% or 50% of the administered radioactivity was expired as CO_2, respectively. Thirty-eight to 40% of the ethyl-^{14}C labels were in urine and 5% in feces, respectively. Eighty percent of the methylthio-^{3}H label was found in urine. The majority of the administered dose was excreted 12 to 15 hours after treatment.

III.   Health Effects

A. Humans

- No information on the health effects of fenamiphos in humans was found in the available literature, including any data on accidental poisoning.

B. Animals

1. *Short-term Exposure*

- NIOSH (1987) reported the acute oral LD_{50} of fenamiphos in the rat, mouse, dog, cat, rabbit and guinea pig as 8, 22.7, 10, 10, 10 and 75 mg/kg, respectively.

- Kimmerle and Lorke (1970) fed chickens (eight/dose) diets containing technical fenamiphos at levels of 0, 1, 3, 10 or 30 ppm a.i. for 30 days. This corresponded to doses of 0, 2, 5, 16 or 26 mg/kg/day. Following treatment, feed consumption, neurotoxicity and cholinesterase (ChE) activity were determined. Histopathological sections of the brain, spinal cord and peripheral nerves were also evaluated. No neuropathy was observed at any dose level tested. No ChE symptoms were reported, but ChE activity in whole blood was inhibited in a dose-dependent manner from 21% at 3 ppm to 65% at 30 ppm. Based on ChE inhibition, a No-Observed-Adverse-Effect Level (NOAEL) of 1 ppm (2 mg/kg/day) was identified.

2. Dermal/Ocular Effects

- DuBois et al. (1967) reported acute dermal LD_{50} values of 78 mg/kg for rats.

- Crawford and Anderson (1973) applied 120 mg of a spray concentrate of fenamiphos (37.47% a.i.) to shaved intact and abraded skin of six New Zealand White rabbits and reported slight erythema 24 and 72 hours posttreatment.

- In ocular studies conducted by Crawford and Anderson (1973), the instillation of 0.1 mL of a spray concentrate of fenamiphos (37.47% a.i.) into the eyes of New Zealand White rabbits resulted in corneal and conjunctival damage at 24 and 72 hours posttreatment. These effects had not subsided by 21 days posttreatment.

3. Long-term Exposure

- In feeding studies conducted by Mobay Chemical Corporation (1983), Fischer 344 rats (50/sex/dose) were administered technical fenamiphos (89% purity) at dose levels of 0, 0.36, 0.60 or 1.0 ppm a.i. for 90 days. Assuming that 1 ppm in the diet of rats is equivalent to 0.05 mg/kg/day (Lehman, 1959), this corresponds to dose levels of 0, 0.018, 0.03 or 0.05 mg/kg/day. Following treatment, brain, plasma and erythrocyte ChE levels were measured. Cholinesterase levels were not significantly reduced at any dose tested. Other parameters were not evaluated. The author reported a NOAEL of 1 ppm (0.05 mg/kg/day, the highest dose tested).

- Loser and Kimmerle (1968) fed Wistar rats (15/sex/dose) fenamiphos (82% a.i.) for 90 days in the diet at dose levels of 0, 4, 8, 16 or 32 ppm active ingredient. Assuming that 1 ppm in the diet is equivalent to 0.05 mg/kg/day (Lehman, 1959), this corresponds to doses of 0, 0.2, 0.4, 0.8 or 1.6 mg/kg/day. Following treatment, body weight, food consumption, hematology, ChE activity, urinalysis and gross pathology were evaluated. No histologic examination was performed. No effects on any end point were reported except for ChE inhibition. No effect was seen at 4 ppm (0.2 mg/kg/day). At 8 ppm (0.4 mg/kg/day), ChE in whole blood and plasma was decreased by 11% and 19%, respectively. Higher doses produced larger decreases in ChE. Based on these data, a NOAEL of 4 ppm (0.2 mg/kg/day) was identified.

- Loser (1970) administered technical fenamiphos (99.4% purity) in the feed of beagle dogs (two/sex/dose) for 3 months at dietary levels of 0, 1, 2 or 5 ppm. Assuming that 1 ppm in the diet of dogs is equivalent to 0.025 mg/kg/day (Lehman, 1959), this corresponds to doses of 0, 0.025, 0.05 or 0.125 mg/kg/day. Untreated controls (three/sex) were run concurrently. Following treatment, body weight, feed consumption, clinical chemistry, urinalysis, ChE activity and gross pathology were evaluated. At 5 ppm, there was a slight decrease in weight gain, although the author did not consider this to be important. No compound-related effects were reported in any other parameters measured except ChE activity. At 1 ppm, plasma ChE was inhibited 13% and 18%, and red blood cell ChE was inhibited 6% and 19% in males and females, respectively. At 2 ppm, plasma and red blood cell ChE was comparable to control levels in males, and was inhibited 13% in plasma and 16% in red blood cells in females. At 5 ppm, ChE in plasma was inhibited 44% and 41%, and red blood cell ChE was inhibited 26% and 22% (females and males, respectively). No brain ChE measurements were reported. Based on the absence of significant ($> 20\%$) ChE inhibition at 1 or 2 ppm, a NOAEL of 2 ppm (0.05 mg/kg/day) is identified.

- Hayes et al. (1982) administered fenamiphos (90% purity) in the diet to CD albino mice (50/sex/dose) at dose levels of 0, 2, 10 or 50 ppm for 20 months. Assuming that 1 ppm in the diet of mice is equivalent to 0.15 mg/kg/day (Lehman, 1959), this corresponds to doses of 0, 0.3, 1.5 or 7.5 mg/kg/day. Following treatment, body weight, food consumption, hematology and mortality were evaluated. Absolute brain weights were decreased at 2 ppm (0.3 mg/kg/day) or greater. At 50 ppm (7.5 mg/kg/day), there was a decrease in body weight. Based on these data, a Lowest-Observed-Adverse-Effect Level (LOAEL) of 2 ppm (0.3 mg/kg/day), the lowest dose tested, was identified, but not a NOAEL.

- Loser (1972a) administered technical fenamiphos (78.8% purity) in the diet of Wistar rats (40/sex/dose) for 2 years at dose levels of 0, 3, 10 or 30 ppm a.i. Assuming that 1 ppm in the diet of rats is equivalent to 0.05 mg/kg/day (Lehman, 1959), this corresponds to doses of 0, 0.15, 0.5 or 1.5 mg/kg/day. Untreated controls (50 males, 60 females) were run concurrently. Following treatment, body weight, food consumption, hematology, urinalysis, plasma and erythrocyte ChE activity, gross pathology and histopathology were evaluated. At the highest dose (30 ppm), a slight increase in female mortality (38% versus 29% in controls) was noted, but the author did not consider this significant. There were statistically significant ($p < 0.05$) increases in thyroid gland and lung weights in females and in heart weight in males. No compound-related effects were observed in any of the other parameters measured except an inactivation of plasma and erythrocyte ChE. At 10 ppm, ChE was decreased by 18 to 41%, and at 30 ppm, ChE was decreased by 28 to 60%. No brain ChE measurements were reported. Based on ChE inhibition, the author identified a NOAEL of 3 ppm (0.15 mg/kg/day). Based on organ weight changes, the NOAEL was 10 ppm (0.5 mg/kg/day).

• In chronic feeding studies by Loser (1972b), beagle dogs (four/sex/dose) were administered technical fenamiphos (78.8% purity) in the feed for 2 years at 0, 0.5, 1, 2, 5 or 10 ppm active ingredient. Assuming that 1 ppm in the diet of dogs is equivalent to 0.025 mg/kg/day (Lehman, 1959), this corresponds to doses of 0, 0.013, 0.025, 0.050, 0.125 or 0.250 mg/ kg/day. Following treatment, no compound-related effects were observed on appearance, general behavior, food consumption, clinical chemistry, hematology, gross pathology or histopathology at any dose tested. Plasma and erythrocyte ChE levels were inhibited about 26% at 2 ppm, about 21 to 57% at 5 ppm and about 32 to 51% at 10 ppm. Cholinesterase was not inhibited at 1 ppm (0.025 mg/kg/day) or below. Based on ChE inhibition, this study identified a NOAEL of 1 ppm (0.025 mg/kg/day) and a LOAEL of 2 ppm (0.05 mg/kg/day).

4. Reproductive Effects

• In a three-generation study conducted by Loser (1972c), FB30 rats (10 males or 20 females/dose) were fed technical fenamiphos (78.8%) in the diet at dose levels of 0, 3, 10 or 30 ppm active ingredient. Assuming that 1 ppm in the diet of rats is equivalent to 0.05 mg/kg/day (Lehman, 1959), this corresponds to doses of 0, 0.15, 0.5 or 1.5 mg/kg/day. Fertility, lactation performance, pup development and parental and litter body weights were evaluated. No compound-related effects were observed on any parameter in animals exposed to 10 ppm (0.5 mg/kg/ day) or less. At 30 ppm (1.5 mg/kg/day), one male of the F_{2b} generation showed a lower body weight gain than the untreated controls, but there were no differences in body weight gain in any other generation. Based on these data, a reproductive NOAEL of 30 ppm (1.5 mg/kg/day) was identified.

5. Developmental Effects

• MacKenzie et al. (1982) administered fenamiphos (88% a.i.) by gavage to pregnant New Zealand White rabbits (20/dose) at dose levels of 0, 0.1, 0.3 or 1.0 mg/kg/day on days 6 to 18 of gestation. Following treatment, there was a decrease in maternal body weight at 0.3 mg/kg/ day or above. At the 1.0-mg/kg/day level, eight dead pups and seven late resorptions were reported, and fetal weight was depressed. A significant (p < 0.05) increase in the incidence of chain-fused sternebrae was also observed at 1.0 mg/kg. Based on maternal body weight, a NOAEL of 0.1 mg/kg was identified. Based on fetotoxicity and teratogenicity, a NOAEL of 0.3 mg/kg/day was identified.

6. *Mutagenicity*

- Herbold (1979) reported that fenamiphos was not mutagenic in *Salmonella typhimurium* (TA1535, TA1537, TA98 or TA100) up to 2,500 μg/ plate, either with or without activation.

- In a dominant lethal test with male NMRI mice (Herbold and Lorke, 1980), acute oral doses of 5 mg/kg did not produce mutagenic effects.

7. *Carcinogenicity*

- Hayes et al. (1982) administered fenamiphos (90% purity) for 20 months in the diet to CD albino mice (50/sex/dose) at dose levels of 0, 2, 10 or 50 ppm (0, 0.3, 1.5 or 7.5 mg/kg/day). Based on gross and histopathologic examination, neoplasms in various tissues and organs were similar in type, organization, time of occurrence and incidence in control and treated animals.

- Loser (1972a) administered technical fenamiphos (78.8% purity) in the diet of Wistar rats (40/sex/dose) for 2 years at dose levels of 3, 10 or 30 ppm active ingredient. Assuming that 1 ppm in the diet of rats is equivalent to 0.05 mg/kg/day (Lehman, 1959), this corresponds to doses of 0.15, 0.5 or 1.5 mg/kg/day. Untreated controls (50 males, 60 females) were run concurrently. No evidence of carcinogenicity was detected either by gross or histological examination.

IV.   Quantification of Toxicological Effects

A. One-day Health Advisory

No information was found in the available literature that was suitable for determination of the One-day HA value for fenamiphos. It is therefore recommended that the Ten-day HA value for the 10-kg child of 0.009 mg/L (9 μg/L, calculated below) be used at this time as a conservative estimate of the One-day HA value.

B. Ten-day Health Advisory

The study by MacKenzie et al. (1982) has been selected to serve as the basis for determination of the Ten-day HA value for fenamiphos. In this study, pregnant rabbits (20/dose) were administered technical fenamiphos (88% purity) by gavage at dose levels of 0, 0.1, 0.3 or 1.0 mg/kg on days 6 through 18 of gestation. A decrease in maternal body weight was observed in animals dosed with 0.3 mg/kg/day or above. No maternal toxicity was reported at 0.1 mg/kg/day. No fetotoxicity or teratogenic effects were observed at 1.0 mg/kg or less or 0.3 mg/kg or less, respectively. Chain fusion of sternebrae were observed in the 1.0 mg/kg group. Based on maternal effects, a NOAEL of 0.1 mg/kg/day was identified.

Using a NOAEL of 0.1 mg/kg/day, the Ten-day HA for a 10-kg child is calculated as follows:

$$\text{Ten-day HA} = \frac{(0.1 \text{ mg/kg/day})(10 \text{ kg})(0.88)}{(100) (1 \text{ L/day})} = 0.009 \text{ mg/L } (9 \text{ } \mu\text{g/L})$$

where: 0.88 = correction factor to account for 88% active ingredient in administered doses.

C. Longer-term Health Advisory

The study by Loser (1970) has been selected to serve as the basis for determination of the Longer-term HA value for fenamiphos. In this study, beagle dogs (2/sex/dose) were fed technical fenamiphos (99.4% purity) in the diet at dose levels of 0, 1, 2 or 5 ppm (0, 0.025, 0.05 or 0.125 mg/kg/day) for 3 months. No effects were detected on body weight, food consumption, clinical chemistry, urinalysis and gross pathology. The only effect observed was inhibition of plasma and erythrocyte ChE activity at the 5 ppm dose level (0.125 mg/kg/day). No significant effect was seen at 2 ppm or less (0.05 mg/kg/day), which was identified as the NOAEL. The 90-day study in Fischer 344 rats by Mobay Chemical Corporation (1983) identified a NOAEL of 1 ppm (0.05 mg/kg/day), but this was not considered, since it was the highest dose tested and a LOAEL was not identified. The study by Loser and Kimmerle (1968) identified a NOAEL of 0.2 mg/kg/day in rats, but this was not chosen, since available data (Loser, 1972a,b) suggest that the rat is less sensitive than the beagle dog.

Using a NOAEL of 0.05 mg/kg/day, the Longer-term HA for a 10-kg child is calculated as follows:

$$\text{Longer-term HA} = \frac{(0.05 \text{ mg/kg/day})(10 \text{ kg})}{(100) (1 \text{ L/day})} = 0.005 \text{ mg/L } (5 \text{ } \mu\text{g/L})$$

The Longer-term HA for a 70-kg adult is calculated as follows:

$$\text{Longer-term HA} = \frac{(0.05 \text{ mg/kg/day}) (70 \text{ kg})}{(100) (2 \text{ L/day})} = 0.018 \text{ mg/L} (20 \text{ } \mu\text{g/L})$$

D. Lifetime Health Advisory

The study by Loser (1972b) in dogs has been selected to serve as the basis for determination of the Lifetime HA value for fenamiphos. In this study, dogs (4/sex/dose) were fed technical fenamiphos (78.8% purity) in the diet for 2 years at dose levels of 0, 0.5, 1, 2, 5 or 10 ppm active ingredient (0, 0.013, 0.025, 0.05, 0.125 or 0.25 mg/kg/day). The only effect detected was inhibition of plasma and erythrocyte cholinesterase at dose levels of 2, 5 or 10 ppm (0.05, 0.125 or 0.25 mg/kg/day). The NOAEL identified in this study was 1 ppm (0.025 mg/kg/day). The chronic studies in rats by Loser (1972a) and by Hayes et al. (1982) were not chosen, since the data indicate the rat is less sensitive than the dog.

Using a NOAEL of 0.025 mg/kg/day, the Lifetime HA is calculated as follows:

Step 1: Determination of the Reference Dose (RfD)

$$\text{RfD} = \frac{(0.025 \text{ mg/kg/day})}{(100)} = 0.00025 \text{ mg/kg/day}$$

Step 2: Determination of the Drinking Water Equivalent Level (DWEL)

$$\text{DWEL} = \frac{(0.00025 \text{ mg/kg/day}) (70 \text{ kg})}{(2 \text{ L/day})} = 0.009 \text{ mg/day} (9 \text{ } \mu\text{g/L})$$

Step 3: Determination of the Lifetime Health Advisory

$$\text{Lifetime HA} = (0.009 \text{ mg/L}) (20\%) = 0.0018 \text{ mg/L} (2 \text{ } \mu\text{g/L})$$

E. Evaluation of Carcinogenic Potential

- No evidence of carcinogenic potential was detected in chronic feeding studies in rats (Loser, 1972a) or mice (Hayes et al., 1982).

- The International Agency for Research on Cancer has not evaluated the carcinogenic potential of fenamiphos.

- Applying the criteria described in EPA's guidelines for assessment of carcinogenic risk (U.S. EPA, 1986), fenamiphos may be classified in Group D: not classifiable. This category is for substances with inadequate animal evidence of carcinogenicity.

V. Other Criteria, Guidance and Standards

• Residue tolerances have been established for fenamiphos and its cholinesterase-inhibiting metabolites in or on various agricultural commodities at 0.02 to 0.60 ppm based on an ADI for fenamiphos of 0.0025 mg/kg/day (U.S. EPA, 1985).

• The World Health Organization (WHO) calculated a PADI of 0.0003 mg/kg/day for fenamiphos (Vettorazzi and Van den Hurk, 1985).

VI. Analytical Methods

• Analysis of fenamiphos is by a gas chromatographic (GC) method applicable to the determination of certain nitrogen-phosphorus–containing pesticides in water samples (Method #507; U.S. EPA, 1988). In this method, approximately 1 liter of sample is extracted with methylene chloride. The extract is concentrated and the compounds are separated using capillary column GC. Measurement is made using a nitrogen-phosphorus detector. The method has been validated in a single laboratory. The estimated detection limits for analytes with this method, including fenamiphos, is 1.0 μg/L.

VII. Treatment Technologies

• No information was found in the available literature on treatment technologies used to remove fenamiphos from contaminated water.

VIII. References

Church, D.D.* 1970. Bay 68138 — leaching, runoff, and water stability. Report No. 26849. Unpublished study received May 27, 1970 under OF0982. Submitted by Chemagro Corp., Kansas City, MO. CDL:091690-H. MRID 00067117.

Crawford, C. and R. Anderson.* 1973. The eye and skin irritancy of Nemacur 3 lbs/gal spray concentrate to rabbits. Report No. 37549. Unpublished study. MRID 00119227.

Dime, R.A., C.A. Leslie and R.J. Puhl.* 1983. Photodecomposition of Nemacur in aqueous solution and on soil. Report No. 86171. Mobay Chemical Corp. 1983. Nemacur: the effects on the environment-environmental chemistry (Supplement No. 4 to brochure, dated Feb. 1, 1973). Document

No. AS83–2611. Compilation; unpublished study received Dec. 9, 1983 under 3125–236. CDL:251891-A. MRID 00133402.

DuBois, K.P., M. Flynn and F. Kinoshita.* 1967. The acute toxicity and anticholinesterase action of Bayer 68138. Unpublished study. MRID 00082807.

FDA. 1979. Food and Drug Administration. Pesticide analytical manual. Revised June 1979.

Green, R., C. Lee and W. Apt.* 1982. Processes affecting pesticides and other organics in soil and water systems: assessment of soil and pesticide properties important to the effective application of nematicides via irritation. Hawaii contributing project to Western Regional Research Project W-82. Unpublished study. MRID 00154533.

Gronberg, R.R.* 1969. The metabolic fate of (Bay 68138), (Bay 68138 sulfoxide), and (Bay 68138 sulfone) by white rats. Report No. 26759. Unpublished study. MRID 00052527.

Gronberg, R.R. and S.H. Atwell.* 1980. Leaching of Nemacur residues in Florida sand. Report No. 66409. Rev. Unpublished study received Aug. 28, 1980 under 3125–236. Submitted by Mobay Chemical Corp., Kansas City, MO. CDL:243126-Y. MRID 00045607.

Hayes, R.H., D.W. Lamb and D.R. Mallicoat.* 1982. Technical fenamiphos oncogenicity study in mice. Report No. 3037. Unpublished study. MRID 00098614.

Herbold, B.* 1979. Nemacur: *Salmonella*/microsome test for detection of point-mutagenic effects. Report No. 8730; 82210. Unpublished study. MRID 00121287.

Herbold, B. and D. Lorke.* 1980. SRA 3386: dominant lethal study on male mouse to test for mutagenic effects. Report No. 8838: 69377. Unpublished study. MRID 00086981.

Kimmerle, G. and D. Lorke.* 1970. Bay 68138: subchronic neurotoxicity studies on chickens. Report No. 1831; 27489. Unpublished study. MRID 00082105.

Lehman, A.J. 1959. Appraisal of the safety of chemicals in foods, drugs and cosmetics. Assoc. Food Drug Off.

Loser, E.* 1970. Bay 68138: subchronic toxicological studies on dogs (three months feeding test). Report No. 1655; Report No. 26906. Unpublished study. MRID 00064616.

Loser, E.* 1972a. Bay 68138: chronic toxicological studies on rats (two-year feeding experiment). Report No. 3539; Report No. 34344. Unpublished study. MRID 00038490.

Loser, E.* 1972b. Bay 68138: chronic toxicological studies on dogs (two-year feeding experiment). Report No. 3561; Report No. 34345. Unpublished study. MRID 00037965.

Loser, E.* 1972c. Bay 68138: generation studies on rats. Report No. 3424; Report No. 34029. Unpublished study. MRID 00037979.

Loser, E. and G. Kimmerle.* 1968. Bay 68138: subchronic toxicological study on rats. Report No. 74523307. Unpublished study. MRID 00082810.

MacKenzie, K., S. Dickie, B. Mitchell et al.* 1982. Teratology study with Nemacur in rabbits. Unpublished study. MRID 00121286.

McNamara, F.T. and C.M. Wilson.* 1981. Behavior of Nemacur in buffered aqueous solutions. Report No. 68582. Unpublished study received July 23, 1981 under 3125-236. Submitted by Mobay Chemical Corp., Kansas City MO. CDL:245613-A. (00079270).

Meister, R., ed. 1983. Farm chemicals handbook. Willoughby, OH: Meister Publishing Company.

Mobay Chemical Corporation.* 1983. Combined chronic toxicity/ oncogenicity study of technical fenamiphos with rats. Unpublished study. MRID 00130774.

NIOSH. 1987. National Institute for Occupational Safety and Health. Registry of toxic effects of chemical substances (RTFCS). Microfiche edition. July.

STORET. 1988. STORET Water Quality File. Office of Water. U.S. Environmental Protection Agency (data file search conducted in May, 1988).

Tweedy, B.G. and L.D. Houseworth.* 1980. Leaching of aged Nemacur residues in sandy loam soil. Report No. 40506. Unpublished study received Aug. 28, 1980 under 3125-236. Submitted by Mobay Chemical Corp., Kansas City, MO. CDL:243126-N. MRID 00045598.

U.S. EPA. 1985. U.S. Environmental Protection Agency. Code of Federal Regulations. 40 CFR 180.349, p. 324. July 1.

U.S. EPA. 1986. U.S. Environmental Protection Agency. Guidelines for carcinogen risk assessment. Fed. Reg. 51(185):33992-34003. September 24.

U.S. EPA. 1988. U.S. Environmental Protection Agency. Method 507 - Determination of nitrogen and phosphorus containing pesticides in water by GC/NPD. April 15 draft. Available from U.S. EPA's Environmental Monitoring and Support Laboratory, Cincinnati, OH.

Vettorazzi, G. and G.W. Van den Hurk. 1985. Pesticides reference index, JMPR 1961-1984. p. 10.

---

* Confidential Business Information submitted to the Office of Pesticide Programs.

Fluometuron

I. General Information and Properties

A. CAS No. 2164-17-2

B. Structural Formula

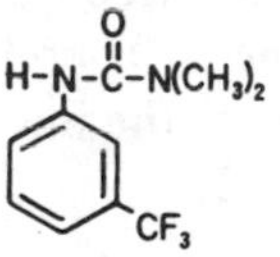

$$H-N-\overset{\overset{\displaystyle O}{\|}}{C}-N(CH_3)_2$$

CF₃ ... (structure)

N,N-Dimethyl-N-[3-(trifluoromethyl)phenyl]-urea

C. Synonyms

- C 2059; Cotoron; Cottonex (Meister, 1988).

D. Uses

- Fluometuron is used in preemergence and postemergence control of annual grasses and broadleaves in cotton and sugarcane (Meister, 1988).

E. Properties (Windholz et al., 1983; CHEMLAB, 1985; TDB, 1985)

Chemical Formula	$C_{10}H_{11}ON_2F_3$
Molecular Weight	232.21
Physical State (25°C)	White crystals
Boiling Point	--
Melting Point	163–164.5°C
Density	--
Vapor Pressure (20°C)	5×10^{-7} mm Hg
Specific Gravity	--
Water Solubility (25°C)	80 mg/L
Log Octanol/Water Partition Coefficient	1.88 (calculated)
Taste Threshold	--

Odor Threshold   --
Conversion Factor   --

F. Occurrence

- Fluometuron was not found in any of 156 ground-water samples analyzed from 125 locations or in 14 surface water samples analyzed from 14 locations (STORET, 1988). This information is provided to give a general impression of the occurrence of this chemical in ground and surface waters as reported in the STORET database. The individual data points retrieved were used as they came from STORET and have not been confirmed as to their validity. STORET data are often not valid when individual numbers are used out of the context of the entire sampling regime, as they are here. Therefore, this information can only be used to form an impression of the intensity and location of sampling for a particular chemical.

G. Environmental Fate

- ^{14}C-Fluometuron (test substance not characterized) was intermediately mobile ($R_f = 0.50$) in a silty clay loam soil (2.5% organic matter) based on thin-layer chromatography (TLC) tests of soil (Helling, 1971; Helling et al., 1971).

- ^{14}C-Fluometuron (test substance not characterized), at various concentrations, was very mobile in a Norge loam soil (1.7% organic matter) with a Freundlich K of 0.31 (Davidson and McDougal, 1973). Freundlich K values, determined in soil:water slurries (5 to 10 g/100 mL) treated with ^{14}C-fluometuron (test substance not characterized) at 0.05 to 10.0 ppm, were 0.37 for Uvrier sand (1% organic matter), 1.07 for Collombey sand (2.2% organic matter), 1.66 for Les Evouettes loam (3.6% organic matter), 3.16 for Vetroz sandy clay loam (5.6% organic matter), and 1.36 for Illarsatz high organic soil (22.9% organic matter) (Guth, 1972).

- Freundlich K values were positively correlated with the organic matter content of the soil. Fluometuron (test substance not characterized), at 10 to 80 μM/kg, was adsorbed at 10 to 51% of the applied amount to a loamy sand soil (1.15% organic matter) and 16 to 67% of the applied amount to a sandy loam soil (1.9% organic matter) in water slurries during test periods of 1 minute to 7 days, with adsorption increasing with time (LaFleur, 1979). Approximately 22% of the applied fluometuron desorbed in water from the loamy sand soil and 15% desorbed from the sandy loam soil during a 7-day test period.

- Fluometuron (50% wettable powder) dissipated from the 0- to 5-cm depth of a sandy clay loam soil (3.2% organic matter) in central Europe with a half-life of less than 30 days (Guth et al., 1969). Fluometuron residues (not characterized) dissipated with a half-life of 30 to 90 days.

II.   Pharmacokinetics

A. Absorption

- Boyd and Fogleman (1967) reported that fluometuron is slowly absorbed from the gastrointestinal (GI) tract of female CFE rats (200 to 250 g). Based on the radioactivity recovered in the urine and feces of four rats given 50 mg ^{14}C-labeled fluometuron after a 2-week pretreatment with 1,000 ppm unlabeled fluometuron [estimated as 100 mg/kg/day, assuming 1 ppm equals 0.1 mg/kg/day in the young rat (Lehman, 1959)], the test compound appears not to have been fully absorbed within 72 hours. Of an orally administered dose (50 mg/kg), up to 15% was excreted in the urine and 49% in the feces.

B. Distribution

- Boyd and Fogleman (1967) detected radioactivity in the liver, kidneys, adrenals, pituitary, red blood cells, blood plasma and spleen 72 hours after oral administration of ^{14}C-labeled fluometuron at dose levels of 50 or 500 mg/kg in rats. The highest concentration was detected in red blood cells.

C. Metabolism

- Boyd and Fogleman (1967) concluded by thin-layer chromatographic analysis that the urine of rats in their study contained m-trifluoromethylaniline, desmethyl-fluometuron, demethylated fluometuron, hydroxylated desmethyl-fluometuron, hydroxylated demethylated fluometuron and hydroxylated aniline.

- Lin et al. (1976) reported that after incubation of ^{14}CF$_3$-labeled fluometuron with cultured human embryonic lung cells for up to 72 hours, 95% of the compound remained unchanged. Human embryonic lung cell homogenate metabolized small amounts of fluometuron through oxidative pathways to N-(3-trifluoromethylphenyl)-N-formyl-N-methylurea, N-(3-trifluoromethylphenyl)-N-methylurea and N-(3-trifluoromethylphenyl) urea.

D. Excretion

- Boyd and Fogleman (1967) reported that urinary excretion of radioactive label peaked at 24 hours after administration of ^{14}C-fluometuron (50 mg/kg) and decreased during the remaining 48 hours. Seventy-two hours after oral administration of the radioactive label, up to 15% of the administered dose was eliminated in the urine.

- In the study by Boyd and Fogleman (1967), fecal excretion of fluometuron peaked by 48 hours postdosing and decreased over the remaining 24 hours. Forty-nine percent of the administered dose (50 mg/kg) was eliminated in the feces.

III. Health Effects

A. Humans

- No information was found in the available literature on the health effects of fluometuron in humans.

B. Animals

1. *Short-term Exposure*

- NIOSH (1985) reported the acute oral LD_{50} values of fluometuron as 6,416, 2,500, 900 and 810 mg/kg in the rat, rabbit, mouse and guinea pig, respectively.

- Sachsse and Bathe (1975) reported an acute oral LD_{50} value of 4,636 mg/kg for both male and female Tif RAl rats.

- Fogleman (1964a) reported the acute oral LD_{50} values for CFW albino mice as 2,300 mg/kg in females and 900 mg/kg in males.

2. *Dermal/Ocular Effects*

- Siglin et al. (1981) conducted a primary dermal irritation study in which undiluted fluometuron powder (80%) was applied to intact and abraded skin of six young adult New Zealand White rabbits for 24 hours. The test substance was severely irritating, with eschar formation observed at 24 and 72 hours.

- Fogleman (1964b) exposed the skin of eight albino rabbits (4/sex) to a 10% aqueous suspension of fluometuron (applied under rubber dental damming) for 6 hours/day for 10 days. No contact sensitization developed during the exposure period. Weight depression at day 130 was evident in the treated group.

- Galloway (1984) reported no sensitizing reactions in Hartley albino guinea pigs exposed to undiluted fluometuron on alternate days for 22 days and on day 36.

- Technical fluometuron was not found to be an eye irritant in rabbits (Fogleman, 1964c).

3. *Long-term Exposure*

- Fogleman (1965a) conducted a 90-day feeding study in which CFE rats (15/sex/dose) were administered technical fluometuron (purity not specified) in the diet at dose levels of 100, 1,000 or 10,000 ppm (reported as 7.5, 75 or 750 mg/kg/day). Following exposure, various parameters including hematology, clinical chemistry and histopathology were evaluated. Enlarged, darkened spleens were observed grossly in male rats given 75 mg/kg/day. At the highest dose level, a depression in body weight and congestion in the parenchyma of the spleen, adrenals, liver and kidneys were evident. A mild deposition of hemosiderin in the spleen was also evident. Spleens were large and dark, livers were brownish and muddy colored and kidneys were small with discolored pelvises in high-dose males. Histopathological findings were confined to mild congestion in various organs and mild hemosiderin deposits in the spleens of high-dose rats. No effects were evident in rats given the 7.5-mg/kg/day dose level for any parameter measured. This dose level was identified as the No-Observed-Adverse-Effect Level (NOAEL) for this study.

- Fogleman (1965b) administered technical fluometuron (purity not stated) in feed to three groups of beagle pups (3/sex/dose) at dose levels of 0, 40, 400 or 4,000 ppm (reported as 0, 1.5, 15 or 150 mg/kg/day) for 90 days. At 150 mg/kg/day, mild inflammatory-type reactions and congestion in the liver and kidneys and mild congestion and hemosiderin deposits in the spleen were observed. Also at this high dose, the spleen to body weight ratio was slightly increased. No adverse systemic effects were observed in dogs administered 1.5 or 15 mg/kg/day (NOAEL).

- In an NCI (1980) study, B6C3F$_1$ mice and Fischer 344 rats (10 of each sex) were given fluometuron (> 99% pure) in the diet for 90 days at 250, 500, 1,000, 2,000, 4,000, 8,000 and 16,000 ppm. Decreased body weight gain (> 10%) was apparent with doses above 2,000 ppm. Treatment-related splenomegaly was found in rats with doses above 1,000 ppm. Microscopic examination was done on spleens only from rats given more than 2,000 ppm, and this assessment indicated dose-related changes including hyperemia of red pulp with atrophy of Malpighian corpuscles and depletion of lymphocytic elements. Body weight gain was reduced (> 10%) in male and female mice given more than 2,000 ppm. Assuming that 1 ppm in the diet equals 0.05 mg/kg/day in the rat and 0.15 mg/kg/day in the mouse (Lehman, 1959), 1,000 ppm (NOAEL) corresponds to 50 mg/kg/day in rats and 2,000 ppm (NOAEL) corresponds to 300 mg/kg/day in mice.

- Hofmann (1966) administered 0, 3, 10, 30 or 100 mg/kg technical fluometuron (Cotoron = C-2059, purity not specified) as a suspension in 1% Mulgafarin six times per week for 1 year by pharnyx probe to four groups of Wistar rats (25/sex/dose). Following treatment, general behavior, mortality, growth, food consumption, clinical chemistry, blood, urine and histopathology were evaluated. Males dosed with 30 or 100 mg/kg/day and females dosed with 100 mg/kg/day showed significant (p < 0.05) reductions in body weight at the end of the study compared to controls. No toxicological effects were observed in rats administered 3 or 10 mg/kg/day (NOAEL).

- In the NCI (1980) study, Fischer 344 rats (10 of each sex) were given fluometuron (> 99% pure) at dietary levels of 250, 500, 1,000, 2,000 and 4,000 ppm in a repeat of the 90-day study to examine splenic effects more closely. Splenomegaly in all treated groups was noted. A dose-related increase in spleen weights and a dose-related decrease in circulating red blood cells were observed in females fed 250 ppm and higher. Increased spleen weights were evident in males given doses above 500 ppm. However, statistical analysis of the data was not done. Stated in the report without presentation of data is the observation of a dose-related increase in red blood cells with polychromasia and anisocytosis in male and female rats and congestion of red pulp with a corresponding decrease of white pulp in spleen. Assuming that 1 ppm equals 0.05 mg/kg/day in the rat (Lehman, 1959), a Lowest-Observed-Adverse-Effect Level (LOAEL) of 250 ppm (12.5 mg/kg/day) is suggested in this study.

- No noncarcinogenic effects (survival, body weight and pathological changes) in B6C3F$_1$ mice and Fischer 344 rats were found in the NCI (1980) bioassay discussed under Carcinogenicity.

4. *Reproductive Effects*

- No information was found in the available literature on the effects of fluometuron on reproduction.

- A reproduction study with technical fluometuron in rats is in progress to satisfy U.S. EPA Office of Pesticide Programs (OPP) data requirements.

5. *Developmental Effects*

- Fritz (1971) reported a teratology study in rats in which dams were given C-2059 suspension in carboxymethylcellulose during days 6 through 15 of gestation. Offspring were removed on day 20 of gestation for examination. The NOAEL was indicated as 100 mg/kg/day, and higher doses reduced fetal body weight. However, this study was invalidated by the U.S. EPA OPP because of inadequate reporting.

- A teratology study in which pregnant Spf New Zealand rabbits were given technical fluometuron (purity not specified) by gavage at dose levels of 50, 500 and 1,000 mg/kg/day during gestation days 6 through 19 was reported by Arhur and Triana (1984). Does were examined for body weight, food consumption and pathological and developmental effects, and laparohysterectomy was done on gestation day 29 for pathological evaluation of fetuses. Increased liver weights and increased mean number of resorptions were found with all doses ($p < 0.05$ at the low and mid doses; insufficient number of fetuses for statistical analysis at the high dose). A LOAEL of 50 mg/kg/day was identified. Reductions in body weights and food consumption occurred in does given 500 and 1,000 mg/kg/day. Deaths, abortions and perforated stomachs were observed in does given 1,000 mg/kg/day.

6. *Mutagenicity*

- In bacterial assays (Dunkel and Simmon, 1980), fluometuron (6.6 mg/ plate) was not mutagenic in *Salmonella* strains TA1535, TA1537, TA1538, TA98 and TA100, either with or without metabolic activation.

- Seiler (1978) reported that fluometuron (2,000 mg/kg bw) given as a single oral dose of an aqueous suspension by gavage resulted in a strong inhibition of mouse testicular DNA synthesis in mice killed 3.5 hours after treatment. Results were inconclusive in a subsequent micronucleus test.

- In yeast assays (Seibert and Lemperle, 1974), a commercial formulation of fluometuron was ineffective in inducing mitotic gene conversion in *Saccharomyces cerevisiae* strain D4 without exogenous metabolic activation.

7. *Carcinogenicity*

- In a long-term bioassay (NCI, 1980), fluometuron was administered in feed to Fischer 344 rats and B6C3F$_1$ mice. Groups of rats (50/sex/dose) were fed diets containing 125 or 250 ppm fluometuron for 103 weeks. Mice (50/sex/dose) were fed 500 or 1,000 ppm for an equivalent period of time. Assuming that 1 ppm equals 0.05 mg/kg/day in the older rat and 0.15 mg/kg/day in the mouse (Lehman, 1959), 125 and 250 ppm equaled 6.25 and 12.5 mg/kg/day in rats and 500 and 1,000 ppm equaled 75 and 150 mg/kg/day in mice. Results based on survival, body weights, and nonneoplastic pathology (including spleen) were negative in rats. Following treatment, there were no significant increases in tumor incidences in male or female Fischer 344 rats or in female B6C3F$_1$ mice compared to controls. In male B6C3F$_1$ mice, an increased incidence of hepatocellular carcinomas and adenomas was noted. The incidences were dose-related and were marginally higher than those in the corresponding matched controls or pooled controls from concurrent studies [matched control, 4/21 or 19%; low dose, 13/47 or 28%; high dose, 21/49 or 43% (p = 0.049); pooled controls, 44/167 or 26%]. NCI (1980) concluded that additional testing was needed because of equivocal findings for male mice and because both rats and mice may have been able to tolerate higher doses. The NOAELs identified for rats and mice are 12.5 and 75 mg/kg/day, respectively.

- Chronic feeding studies with technical fluometuron in rats and mice are ongoing to satisfy OPP data requirements.

IV. Quantification of Toxicological Effects

A. One-day Health Advisory

No information was found in the available literature that was suitable for determination of the One-Day HA value for fluometuron. The teratology study by Arhur and Triana (1984) was not selected because a NOAEL was not identified. It is therefore recommended that the Longer-term HA value for a 10-kg child (2 mg/L, calculated below) be used at this time as a conservative estimate of the One-day HA value.

B. Ten-day Health Advisory

No information was found in the available literature that was suitable for determination of the Ten-day HA value for fluometuron. The teratology study by Arhur and Triana (1984) was not selected because a NOAEL was not identified. It is therefore recommended that the Longer-term HA value for a 10-kg child (2 mg/L, calculated below) be used at this time as a conservative estimate of the Ten-day HA value.

C. Longer-term Health Advisory

The 90-day feeding study in dogs by Fogleman (1965b) has been selected to serve as the basis for the Longer-term HA value for fluometuron. In this study, dogs given technical fluometuron at dose levels of 0, 1.5, 15 or 150 mg/kg/day in the diet for 90 days showed pathological effects in spleen, liver and kidney at the highest dose and no observable effects at the lower doses. The 90-day feeding studies with rats by Fogleman (1965a) and NCI (1980) were not selected because the 15 mg/kg/day NOAEL in the Fogleman (1965b) study was below the lowest dose of 75 mg/kg/day in the Fogleman (1965a) study where effects were noted, and pathological changes in spleen found with the dietary level of 12.5 mg/kg/day in the repeat NCI (1980) 90-day study in rats were not found with this level in the initial 90-day study and in the 2-year bioassay in rats by the NCI (1980). Because 7.5 mg/kg/day in the Fogleman (1965a) study and 12.5 mg/kg/day (estimated) in the NCI (1980) carcinogenicity bioassay were NOAELs, it is concluded that 15 mg/kg/day would be consistent with a NOAEL in these 90-day studies in rats. The study by Hofmann (1966) in which rats were given technical fluometuron as a suspension by gavage at dose levels of 0, 3, 10, 30 and 100 mg/kg, six times per week for 1 year, was not selected because feeding the substance in the diet is preferred over giving it as a suspension by gavage for estimating exposure from drinking water, although the 10 mg/kg NOAEL in this study approximates the 15 mg/kg/day NOAEL in the Fogleman (1965b) study. The 90-day feeding study in mice by NCI (1980) was not selected because the NOAEL of 300 mg/kg/day (estimated) is above the effect levels in the other studies considered. The 15 mg/kg/day dose level in dogs was, therefore, identified as the NOAEL.

Using a NOAEL of 15 mg/kg/day, the Longer-term HA for a 10-kg child is calculated as follows:

$$\text{Longer-term HA} = \frac{(15 \text{ mg/kg/day}) (10 \text{ kg})}{(100) (1 \text{ L/day})} = 1.5 \text{ mg/L} (2,000 \text{ } \mu g/L)$$

The Longer-term HA for a 70-kg adult is calculated as follows:

$$\text{Longer-term HA} = \frac{(15 \text{ mg/kg/day}) (70 \text{ kg})}{(100) (2 \text{ L/day})} = 5.3 \text{ mg/L} (5,000 \text{ } \mu g/L)$$

D. Lifetime Health Advisory

The NCI (1980) carcinogenicity bioassay in Fischer 344 rats has been selected to serve as the basis for determination of the Lifetime HA value for fluometuron. Rats were exposed to dose levels of 0, 125 and 250 ppm fluometuron in the diet (estimated as 6.25 and 12.5 mg/kg/day) for 103 weeks. No observable effects were evident in this study. Although pathological changes in spleens of rats given 250 ppm fluometuron in the diet (estimated as 12.5 mg/kg/day) were noted in the repeat 90-day study in rats by NCI (1980), it appears that splenic lesions were either not evident or were able to reverse in the rats given the 250-ppm dietary level for 2 years (only one rat died by 1 year into the bioassay). Furthermore, pathological changes in the spleen were not evident with doses below 2,000 ppm in the initial 90-day study in Fischer 344 rats by NCI (1980). The 90-day and 1-year studies discussed under Longer-term Health Advisory have not been selected for calculation of a Lifetime HA because of their short duration compared to the 103-week NCI (1980) bioassay and because, although not as many end points were assessed in the NCI (1980) bioassay as in these studies, major effects observed in these studies (pathology, body weight) were evaluated in the NCI (1980) bioassay. The NCI (1980) bioassay in B6C3F$_1$ mice was not considered because higher dose levels (500 and 1,000 ppm, estimated as 75 and 150 mg/kg/day) were used.

Using the NCI (1980) bioassay in rats with a NOAEL of 12.5 mg/kg/day, the Lifetime HA is calculated as follows:

Step 1: Determination of the Reference Dose (RfD)

$$RfD = \frac{(12.5 \text{ mg/kg/day})}{(100)\,(10)} = \frac{0.0125 \text{ mg/kg/day}}{(\text{rounded to } 0.013 \text{ mg/kg/day})}$$

where: 10 = additional uncertainty factor used by U.S. EPA OPP to account for data gaps (chronic feeding studies in rats and dogs, reproduction study in rats, teratology studies in rats and rabbits).

Step 2: Determination of the Drinking Water Equivalent Level (DWEL)

$$DWEL = \frac{(0.013 \text{ mg/kg/day})\,(70 \text{ kg})}{(2 \text{ L/day})} = 0.455 \text{ mg/L } (500 \text{ μg/L})$$

Step 3: Determination of the Lifetime Health Advisory

$$\text{Lifetime HA} = (0.46 \text{ mg/L})\,(20\%) = 0.09 \text{ mg/L } (90 \text{ μg/L})$$

E. Evaluation of Carcinogenic Potential

- NCI (1980) determined that fluometuron was not carcinogenic in male and female Fischer 344 rats and female B6C3F$_1$ mice. The marginal increase in the incidence of hepatocellular carcinomas and adenomas in male B6C3F$_1$ mice was concluded to be equivocal evidence in the NCI (1980) report on its bioassay.

- IARC (1983) has classified fluometuron in Group 3: this chemical cannot be classified as to its carcinogenicity for humans.

- Applying the criteria described in EPA's guidelines for assessment of carcinogenic risk (U.S. EPA, 1986a), fluometuron may be classified in Group D: not classifiable. This category is used for substances with inadequate animal evidence of carcinogenicity.

V.   Other Criteria, Guidance and Standards

- The U.S. EPA/OPP previously calculated an ADI of 0.008 mg/kg/day based on a NOAEL of 7.5 mg/kg/day in a 90-day feeding study in rats (Fogleman, 1965a) and an uncertainty factor of 1,000 (used because of data gaps). This has been updated to 0.013 mg/kg/day, based on a 2-year feeding study in rats using a NOAEL of 12.5 mg/kg/day and an uncertainty factor of 1,000.

- Tolerances have been established for negligible residues of fluometuron in or on cottonseed and sugar cane at 0.1 ppm (U.S. EPA, 1985). A tolerance is a derived value based on residue levels, toxicity data, food consumption levels, hazard evaluation and scientific judgment, and it is the legal maximum concentration of a pesticide in or on a raw agricultural commodity or other human or animal food (Paynter et al., no date).

VI.   Analytical Methods

- Analysis of fluometuron is by a high-performance liquid chromatographic (HPLC) method applicable to the determination of certain carbamate and urea pesticides in water samples (U.S. EPA, 1986b). This method requires a solvent extraction of approximately 1 liter of sample with methylene chloride using a separatory funnel. The methylene chloride extract is dried and concentrated to a volume of 10 mL or less. HPLC is used to permit the separation of compounds and measurement is conducted with a UV detector. This method has been validated in a single laboratory and estimated detection limits have been determined for the analytes in the method, including fluometuron. The estimated detection limit is 0.10 μg/L.

VII. Treatment Methodologies

• Available data indicate that granular activated carbon (GAC) adsorption will remove fluometuron from water.

• Whittaker (1980) experimentally determined adsorption isotherms for fluometuron on GAC.

• Whittaker (1980) reported the results of GAC columns operating under bench-scale conditions. At a flow rate of 0.8 gpm/ft^2 and an empty bed contact time of 6 minutes, fluometuron breakthrough (when effluent concentration equals 10% of influent concentration) occurred after 1,640 bed volumes (BV). When a bisolute solution of fluometuron diphenamide was passed over the same column, fluometuron breakthrough occurred after 320 BV.

• GAC adsorption appears to be the most promising treatment technique for the removal of fluometuron from contaminated water. However, selection of individual or combinations of technologies to attempt fluometuron removal from water must be based on a case-by-case technical evaluation, and an assessment of the economics involved.

VIII. References

Arhur, A. and V. Triana.* 1984. Teratology study (with fluometuron) in rabbits. Ciba-Geigy Corporation. Report No. 217–84. Unpublished study. MRID 842096.

Boyd, V.F. and R.W. Fogleman.* 1967. Metabolism of fluometuron (1,1-dimethyl-3-(alpha, alpha, alpha-trifluoro-m-tolyl) urea) in the rat. Ciba Agrochemical Company. Research Report CF-1575. Unpublished study. MRID 00022938.

CHEMLAB. 1985. The Chemical Information System. CIS, Inc., Baltimore, MD.

Davidson, J. and J. McDougal. 1973. Experimental and predicted movement of three herbicides in a water-saturated soil. J. Environ. Qual. 2(4):428–433.

Dunkel, V.C. and V.F. Simmon. 1980. Mutagenic activity of chemicals previously tested for carcinogenicity in the National Cancer Institute bioassay. PROGRAM. IARC. Sci. Publ. 27:283–302.

Fogleman, R.W.* 1964a. Compound C-2059 technical – acute oral toxicity – male and female mice. AME Associates for CIBA Corporation. Project No. 20–042. Research Report CF-735. Unpublished study. MRID 00019093.

Fogleman, R.W.* 1964b. Compound C-2059 80 WP-repeated rabbit dermal toxicity. AME Associates for CIBA Corporation. Project No. 20–0242. Research Project CF-740. Unpublished study. MRID 00018593.

Fogleman, R.W.* 1964c. Compound C-2059 technical – acute eye irritation – rabbits. AME Associates for CIBA Corporation. Project No. 20–042. Unpublished study. MRID 0019032.

Fogleman, R.W.* 1965a. Cotoran – 90-day feeding rats. AME Associates for CIBA Corporation. Project No. 20–042. Unpublished study. MRID 00019034.

Fogleman, R.W.* 1965b. Subacute toxicity – 90-day administration – dogs. AME Associates for CIBA Corporation. Project No. 20–042. Unpublished study. MRID 00019035.

Fritz, H.* 1971. Reproduction study: Segment II. Preparation C-2059. Experiment No. 22710100. Ciba-Geigy, Ltd. Unpublished study. MRID 000019211.

Galloway, D.* 1984. Guinea pig skin sensitization. Project No. 3397–84. Unpublished study. Stillmeadow, Inc. for Ciba-Geigy Corporation. MRID 00143601.

Guth, J.A. 1972. Adsorption and elution behavior of plant protective agents in soils. Translation of: Adsorptions – und Einwasch Verhalten von Pflanzenschutzmitteln in Boeden. Schriftenreihe des Vereins fuer Wasser, Boeden, and Lufthygiene, Berlin-Dahlem. (37):143–154.

Guth, J.A., H. Geissbuehler and L. Ebner. 1969. Dissipation of urea herbicides in soil. Meded. Rijksfac. Landbouwwet. XXXIV(3):1027–1037.

Helling, C.S. 1971. Pesticide mobility in soils: II. Applications of soil thin-layer chromatography. Soil Sci. Soc. Am. Proc. 35:737–738.

Helling, C.S., D.D. Kaufman and C.T. Dieter. 1971. Algae bioassay detection of pesticides mobility in soils. Weed Sci. 19(6):685–690.

Hofmann, A.* 1966. Examinations on rats of the chronic toxicity of preparation Bo-27 690 (Cotoran = C-2059). Hofmann-Battelle-Geneva. (Translation; unpublished study). MRID 00019088.

IARC. 1983. International Agency for Research on Cancer. Vol. 30. IARC monographs on the evaluation of carcinogenic risk of chemicals to man. Lyon, France: IARC.

LaFleur, K. 1979. Sorption of pesticides by model soils and agronomic soils: rates and equilibria. Soil Sci. 127(2):94–101.

Lehman, A.J. 1959. Appraisal of the safety of chemicals in foods, drugs, and cosmetics. Assoc. Food Drug Off. U.S., Q. Bull.

Lin, T.H., R.E. Menzer and H.H. North. 1976. Metabolism in human embryonic lung cell cultures of three phenylurea herbicides: chlorotoluron, fluometuron and metobromuron. J. Agric. Food Chem. 24:759–763.

Meister, R., ed. 1988. Farm chemicals handbook. Willoughby, OH: Meister Publishing Co.

NCI. 1980. National Cancer Institute. Bioassay of fluometuron for possible carcinogenicity. NCI-CG-TR-195. Bethesda, MD.

NIOSH. 1985. National Institute for Occupational Safety and Health. Registry of Toxic Effects of Chemical Substances. National Library of Medicine Online File.

Paynter, O.E., J.G. Cummings and M.H. Rogoff. No date. United States Pesticide Tolerance System. Unpublished draft report. Washington, DC: U.S. EPA Office of Pesticide Programs.

Sachsse, K. and R. Bathe.* 1975. Acute oral LD_{50} of technical fluometuron (C-2059) in the rat. Project No. Siss. 4574. Unpublished study. MRID 00019213.

Seiler, J.P. 1978. Herbicidal phenylalkylurea as possible mutagens. I. Mutagenicity tests with some urea herbicides. Mutat. Res. 58:353–359.

Siebert, D., and E. Lemperle. 1974. Genetic effects of herbicides: induction of mitotic gene conversion in *Saccharomyces cerevisiae*. Mutat. Res. 22:111–120.

Siglin, J.C., P.J. Becci and R.A. Parent.* 1981. Primary skin irritation in rabbits (EPA—FIFRA). FDRL: Study No. 6817A. Food and Drug Research Laboratories for Ciba-Geigy Corporation. Unpublished study. MRID 00068040.

STORET. 1988. STORET Water Quality File. Office of Water. U.S. Environmental Protection Agency (data file search conducted in May, 1988).

TDB. 1985. Toxicology Data Bank. Medlars II. National Library of Medicine's National Interactive Retrieval Service.

U.S. EPA. 1985. U.S. Environmental Protection Agency. Code of Federal Regulations. 40 CFR 180.229. July 1. p. 293.

U.S. EPA. 1986. U.S. Environmental Protection Agency. Guidelines for carcinogen risk assessment. Fed. Reg. 51(185):33992–34003. September 24.

U.S. EPA. 1986b. U.S. Environmental Protection Agency. U.S. EPA Method 4—Determination of pesticides in ground water by HPLC/UV. Draft. January. Available from U.S. EPA's Environmental Monitoring and Support Laboratory, Cincinnati, OH.

Whittaker, K.F. 1980. Adsorption of selected pesticides by activated carbon using isotherm and continuous flow column systems. Ph.D. Thesis. Purdue University.

Windholz, M., S. Budavari, R.F. Blumetti and E.S. Otterbein, eds. 1983. The Merck index — an encyclopedia of chemicals and drugs, 10th ed. Rahway, NJ: Merck and Company, Inc.

---

* Confidential Business Information submitted to the Office of Pesticide Programs.

Fonofos

I. General Information and Properties

A. CAS No. 944–22–9

B. Structural Formula

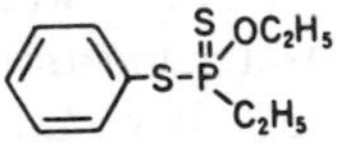

O-Ethyl-S-phenylethylphosphonodithioate

C. Synonyms

- Difonate; Difonatal; Dyfonate®; FNT 25, 796; Fonophos: Stauffer N2790 (Meister, 1983).

D. Uses

- Fonofos is used as a soil insecticide (Meister, 1983).

E. Properties (Windholz et al., 1983; TDB, 1985)

Chemical Formula	$C_{10}H_{15}OS_2P$
Molecular Weight	246.32
Physical State (25°C)	Light yellow liquid
Boiling Point	130°C
Melting Point	--
Density	--
Vapor Pressure (25°C)	2.1×10^{-4} mm Hg
Specific Gravity (20°C)	1.154
Water Solubility (25°C)	Practically insoluble
Log Octanol/Water Partition Coefficient	--
Taste Threshold	--
Odor Threshold	--
Conversion Factor	--

F. Occurrence

- Fonofos has been found in tailwater pit sediment and water samples. Monitoring studies conducted in 1973 and 1974 in Haskell County, Kansas, showed that the highest concentrations found were 770 ppb for sediment and 5.9 ppb for water during 1974. Mean peak concentrations were highest in June and July (Kadoum and Mock, 1978).

- Fonofos (Dyfonate) has been found in Iowa ground water; a typical positive sample found was 0.1 ppb (Cohen et al., 1986).

- Fonofos has been found in 194 of 3,399 surface water samples analyzed and in 2 of 570 ground-water samples (STORET, 1988). Samples were collected from 183 surface water locations and 456 ground-water locations. The 85th percentile of all nonzero samples was 0.32 μg/L in surface water and 8,900 μg/L in ground-water sources. The maximum concentration found in surface water was 33 μg/L and in ground water was 8,900 μg/L. Fonofos was found in surface water in Iowa, Illinois, Michigan and Ohio, and in ground water in Alabama. This information is provided to give a general impression of the occurrence of this chemical in ground and surface waters as reported in the STORET database. The individual data points retrieved were used as they came from STORET and have not been confirmed as to their validity. STORET data are often not valid when individual numbers are used out of the context of the entire sampling regime, as they are here. Therefore, this information can only be used to form an impression of the intensity and location of sampling for a particular chemical.

G. Environmental Fate

- Under aerobic conditions, fonofos at 10 ppm was degraded at a moderate rate with a half-life ranging from 3 to more than 16 weeks in soils varying in texture from loamy sand to clay loam to peat (McBain and Menn, 1966; Hoffman et al., 1973; Hoffman and Ross, 1971; Miles et al., 1979). The major degradate identified was O-ethylethane phosphonothioic acid; other degradates identified included fonofos oxon, O-ethylethane phosphonic acid, O-ethyl O-methylethyl phosphonate, diphenyl disulfide, methylphenyl sulfoxide, and methylphenyl sulfone (Hoffman et al., 1973; Hoffman and Ross, 1971). The soil fungus, R*hizopus japonicus*, rapidly degraded ^{14}C-fonofos to yield dyfoxon, thiophenol, ethylethoxy phosphonic acid and methylphenyl sulfoxide (Lichtenstein et al., 1977).

- Fonofos is relatively immobile in a silt loam and sandy loam soil but relatively mobile in quartz sand. After 7 to 12 inches of water were added to 7-inch soil columns, 2 to 9% of the applied ^{14}C-fonofos leached from the treated soil layer in Plano silt loam and Fox fine sandy loam columns. When a quartz sand was leached with 7 inches of water, 50% of the applied radioactivity was detected in the leachate. Dyfoxon, a fonofos degradate, and two unidentified compounds were found in the leachate of the silt loam soil (Lichtenstein et al., 1972).

- Fonofos is relatively mobile in runoff water from loam sand. After 30 days, only 0.54 to 1.2% of the applied ^{14}C-fonofos was recovered in runoff water from drenching a Sorrento loam soil on an inclined plane at a 15-degree slope. Fonofos accounted for most of the recovered radioactivity, which was primarily adsorbed to the silt fraction (Hoffman et al., 1973).

- Fonofos is not volatile from soil but is fairly volatile from water. Within 24 hours after application, 15 to 16% of the ^{14}C-fonofos applied volatilized from soil water (a suspension of fine sand in tap water or tap water alone); 1% volatilized from a silt loam soil alone. ^{14}C-Fonofos volatilized from soil water with a half-life of 5 days; 80% of the applied radioactivity was volatilized at the end of 10 days (Lichtenstein and Schulz, 1970).

- In the field, fonofos dissipated with a half-life of 28 to 40 days when either a 10% G or a 4-lb/gal EC formulation was applied at 4.8 to 10 lb active ingredient (a.i.)/A to a sandy loam and two silt loam soils (Kiigemagi and Terriere, 1971; Schulz and Lichtenstein, 1971; Talekar et al., 1977). Using a root maggot bioassay, toxic fonofos residues in a sandy loam field soil were detected up to 17 weeks after the 10% G formulation was applied at 2 to 5 lb a.i./A. Residues were detected up to 28 weeks after treatment when the same soil was maintained in a greenhouse (Ahmed and Morrison, 1972).

II.   Pharmacokinetics

A. Absorption

- McBain et al. (1971) administered ^{14}C-phenyl-labeled fonofos (99% purity, dissolved in corn oil) orally to albino rats (two/dose) at doses of 2, 4 or 8 mg/kg. Only 7% of the label was recovered in feces, indicating that absorption was nearly complete (about 93%). Hoffman et al. (1971) reported essentially identical results in rats dosed with 0.8 mg/kg fonofos. Measurements of urinary, fecal and biliary excretion indicated that about 80 to 90% of the dose was absorbed from the gastrointestinal tract.

* Hoffman et al. (1971) administered single oral doses of ^{35}S-labeled fonofos (2.0 mg/kg; 99% purity) to rats. About 32% of the label was excreted in feces. Measurement of biliary excretion indicated that 15% of the label in the feces came from the bile. The authors concluded that about 17% had not been absorbed.

B. Distribution

* Hoffman et al. (1971) administered ^{35}S-labeled fonofos (2.0 mg/kg, 13.4 mCi/mM; 99% purity) to rats by gavage (in safflower oil); the levels of label in blood and tissues were measured for 16 days. Higher levels of radioactivity were found in the kidneys, blood, liver and intestines, and lower levels were found in bone, brain, fat, gonads and muscle. Concentration values at 2 days ranged from about 400 ppb in the kidneys to about 70 ppb in other tissues. All values were 10 ppb or lower by day 8. Tissue levels declined in first-order fashion, with near total (99.3%) elimination during 2 to 16 days after dosing.

C. Metabolism

* McBain et al. (1971) administered single oral doses of 2, 4 or 8 mg/kg of ethyl or phenyl-^{14}C-labeled fonofos (97.5% or 99% purity) to male albino rats (two/dose). Only 2.6 to 7.1% was recovered as unchanged fonofos in the urine. The remainder was converted to a variety of terminal metabolites, including O-ethylethane phosphonothioic acid, O-ethylethane phosphonic acid, and O-conjugates of 3- and 4-(hydroxphenyl)methyl sulfone. McBain et al. (1971) reported that fonofos was converted by rat liver microsomes *in vitro* to the more toxic fonofos oxon, but only traces of this compound were excreted by the intact animal.

D. Excretion

* McBain et al. (1971) administered single oral doses of 2, 4 or 8 mg/kg of ^{14}C-labeled fonofos (97.5% or 99% purity) orally to male albino rats (two rats/dose). When the label was on the phenyl ring, recovery of label was 90.7% in urine and 7.4% in feces. When the label was on the ethyl group, recovery of label was 62.8% in urine and 31.8% in feces. Of this fecal label, 15.3% was found to be excreted in the bile.

- Hoffman et al. (1971) dosed rats orally with ^{14}C-ethyl-labeled fonofos (0.8 mg/kg; 98% purity). After 15 days, average recovery of label was 91% in urine, 7.4% in feces and 0.35% in expired air. Essentially all of the excretion occurred within 4 days. In rats dosed with ^{35}S-labeled fonofos (2 mg/kg: 99% purity), average recovery of label after 4 days was 62.5% in the urine, 31.8% in feces and 0.1% in expired air. Bile duct cannulation studies indicated that about 15% of the label in feces arose from biliary excretion.

III.   Health Effects

A. Humans

1. *Short-term Exposure*

- The Pesticide Incident Monitoring System (PIMS) database reported 21 cases between 1966 and 1979 of human toxicity resulting from exposure to fonofos. Fourteen of the cases involved fonofos only, and seven involved mixtures. Two fatalities occurred, and four individuals required medical treatment. No quantitative exposure data and no description of adverse effects were provided.

- One reported case of accidental ingestion involved a 19-year-old woman who ate pancakes prepared with ingredients containing fonofos. No quantitative estimate of the dose level was provided. The individual developed nausea, vomiting, salivation, sweating and suffered cardio-respiratory arrest. She was treated at a hospital and was found to have muscle fasciculation, blood pressure of 64/0 mm Hg, a pulse rate of 46, pinpoint pupils, and profuse salivary and bronchial secretions. The patient also developed a pancreatic pseudocyst. The woman was discharged after 2 months of treatment. A second individual who also ate the contaminated pancakes died (Hayes, 1982).

2. *Long-term Exposure*

- No information on the long-term exposure effects of fonofos on humans was found in the available literature.

B. Animals

1. *Short-term Exposure*

- Reported values for the oral LD_{50} of fonofos for female rats range from 3.2 to 7.9 mg/kg, and values for male rats range from 6.8 to 18.5 mg/kg (Horton, 1966a,b; Dean, 1977).

- Horton (1966a) administered single oral doses of fonofos (purity not specified) to rats (strain not specified). Doses of 1.0 or 2.15 mg/kg did not produce visible symptoms. Doses of 4.6 to 46 mg/kg elicited rapid appearance of fasciculations and tremors, salivation, exophthalmia and labored respiration, with females being somewhat more sensitive than males. Gross autopsy of animals that died revealed congested liver, kidneys and adrenals and lung erythema. Autopsy of survivors showed no effects. Based on gross changes, a No-Observed-Adverse-Effect Level (NOAEL) of 2.15 mg/kg was identified by this study.

- Cockrell et al. (1966) fed fonofos in the diet to dogs at levels of 0 or 8 ppm for 5 weeks. These levels were stated by the authors to be equivalent to 0 and 0.2 mg/kg/day. Plasma and red blood cell cholinesterase (ChE) were measured at 2 and 4 weeks; organ weights, brain ChE and changes in gross pathology were measured at termination (5 weeks). Following treatment, no systemic toxicity was observed; brain and plasma or red blood cell cholinesterase levels were unaffected. No other details were provided. This study identified a NOAEL of 8 ppm (0.2 mg/kg/day).

- In a demyelination study, groups of 10 adult hens each received technical fonofos (99.8% pure) in the diet for 46 days (Woodard and Woodard, 1966). Levels fed were equivalent to 0, 2, 6.32 or 20 mg/kg/day. Only hens at 20 mg/kg showed impairment of locomotion and equilibrium, and one showed histological evidence of possible demyelination of the peripheral nerves. A NOAEL for demyelination of 6.32 mg/kg/day was indicated by the study.

2. *Dermal/Ocular Effects*

- Reported dermal LD_{50} values of fonofos for the rabbit (both sexes) ranged from 121 to 147 mg/kg (Horton, 1966a,b). However, Dean (1977) determined a different LD_{50} in rabbits: 25 mg/kg for females and 100 mg/kg for males.

- Instillation of 0.1 mL undiluted fonofos (about 23 mg/kg/day) in one eye of each of three rabbits caused negligible local irritation, but was lethal to all within 24 hours (Horton, 1966a,b; Dean, 1977).

- Dean (1977) applied 0.5 mL undiluted fonofos to closely clipped intact skin of rabbits; no dermal irritation was reported but all animals died within 24 hours.

- Horn et al. (1966) applied fonofos (10% granular) to the intact or abraded skin of New Zealand rabbits (five/sex/dose; the five animals included both normal and abraded skin animals) 5 days per week for 3 weeks at doses of 0, 35 or 70 mg/kg. Following treatment, dermal effects, general appearance and behavior, hematology, organ weights, cholinesterase levels, gross pathology and histopathology were evaluated. No difference was observed in any of the responses between the intact or abraded skin animals. One normal and one abraded skin male and one normal skin female died in the 70-mg/kg group; and one intact skin male died in the 35-mg/kg group. No irritation of the skin was observed at any dose tested for either intact or abraded skin. In males, adrenal weights were increased by about 50% at 35 mg/kg, and by 70% at 70 mg/kg (p value not given). Similar but smaller (15 to 20%) increases in adrenal weights were seen in females. No hematological effects were observed at any dose tested. No histopathological changes occurred except slight to moderate liver glycogen depletion at 70 mg/kg. Reductions were observed in red blood cell, plasma and brain cholinesterase activity for both sexes of the treated groups. Only values for abraded skin were reported. At 35 mg/kg, ChE in red blood cells was inhibited 70% (for both sexes), while plasma ChE levels were inhibited 74% (males) and 91% (females), and brain ChE was inhibited 66% (males) and 89% (females). At 70 mg/kg, ChE in red blood cells was inhibited 36% (males) and 45% (females). ChE in plasma was inhibited 67% for both sexes. ChE in brain was inhibited 59% (males) and 57% (females).

3. Long-term Exposure

- Daily oral doses of fonofos in corn oil (at 0, 2, 4 or 8 mg/kg/day) for 90 days failed to elicit delayed neurotoxicity in adult hens (Miller et al., 1979, abstract only). A minimum NOAEL of 8 mg/kg/day for delayed neurotoxicity was indicated by these reported results.

- In a similar experiment (Cockrell et al., 1966), rats were fed diets containing difonate at 0, 10, 31.6 or 100 ppm for 13 weeks. Based on the assumption that 1 ppm in the diet is equivalent to 0.05 mg/kg/day, these doses correspond to 0, 0.5, 1.58 or 5 mg/kg/day (Lehman, 1959). Cholinesterase was measured in serum and red blood cells before and after exposure, and brain ChE was measured at termination. At 100 ppm, there was significant inhibition of ChE in serum (70%, females only), red blood cells (85%, females only) and brain (51% to 60%, both sexes). Decreases of over 50% in red blood cell ChE in both males and females were reported at the 31.6-ppm level. At 10 ppm, the largest difference detected was a 23% decrease in red blood cell ChE in females; the authors did not consider this to be significant. All other ChE measurements at this dose were comparable between exposed and control animals. Other observations were negative for compound effect, and there were no histopathological findings. Based on ChE inhibition, the NOAEL in rats was identified as 10 ppm (0.5 mg/kg/day).

- Cockrell et al. (1966) fed fonofos in the diet to dogs at levels of 0, 16, 60 or 240 ppm for 14 weeks. These levels were stated by the authors to be equivalent to 0, 0.4, 1.5 or 6 mg/kg/day. Dogs showed increased lacrimation and salivation plus convulsions (at 16 ppm), bloody diarrhea (at 60 ppm), or tremors and anxiety and increased liver weight (at 240 ppm). At 16 ppm, there was about 60% ChE inhibition in erythrocytes and slight ChE inhibition in the brain (female only). At 60 ppm, ChE in red blood cells was inhibited 60% or more, and plasma ChE was decreased about 20% (in males only) at week 13. At the high dose (240 ppm), ChE was nearly totally inhibited in red blood cells, about 50% inhibited in plasma and moderately inhibited in brain. Based on cholinesterase inhibition and systemic toxicity, a Lowest-Observed-Adverse-Effect Level (LOAEL) of 16 ppm (0.4 mg/kg/day), the lowest dose tested, was identified.

- Pure-bred beagle dogs were fed fonofos in the diet for 2 years (Woodard et al., 1969). Groups of four males and four females each received 0, 16, 60 or 240 ppm fonofos. Based on the assumption that 1 ppm in the diet is equivalent to 0.025 mg/kg/day, these doses correspond to 0, 0.4, 1.5 or 6 mg/kg/day (Lehman, 1959). After 14 weeks, the low dose (16 ppm) was reduced to 8 ppm (0.2 mg/kg/day), and this dose level was maintained for the duration of the study. Cholinesterase levels in plasma were inhibited about 50% at 240 ppm, about 25% to 50% at 60 ppm, and were not different from controls at the low dose (16 or 8 ppm). In red blood cells, ChE levels were inhibited almost completely at the 240-ppm level and about 65% at 60 ppm. In animals receiving 16 ppm for 14 weeks, ChE in red blood cells was inhibited about 30%. After reduction of the dose to 8 ppm, ChE levels returned to values comparable to controls. At sacrifice, no inhibition of ChE in brain was detected at any dose level. At 240 ppm, nervous, apprehensive behavior and tremors were seen, and three dogs died, each with marked acute congestion of tissues and hemorrhage of the small intestinal mucosa. At this dose level, also, serum alkaline phosphatase was increased, as were liver weights. Histopathological examination of animals receiving 240 ppm revealed a marked increase in basophilic granulation of the myofibril of the inner layer of the muscularis of the small intestine, and there were slight changes in the liver. At 60 ppm, increased liver weight was observed. At the low dose (16/8 ppm), the only effect was a single brief episode of fasciculation in one male dog at 5 months. The author judged that this could not be ascribed with certainty to fonofos exposure. For this study, the NOAEL for ChE inhibition and for systemic toxicity was 8 ppm (0.2 mg/kg/day).

- Albino rats received fonofos in the diet for 2 years at 0, 10, 31.6 or 100 ppm (0, 0.5, 1.58 or 5 mg/kg/day; Lehman, 1959) (Banerjee et al., 1968). Fonofos was judged not to have affected survival, food intake, body weight gain, organ weights or gross and histopathological findings. At 100 ppm, females showed tremors and nervous behavior, and males had reduced hemoglobin and packed-cell volume. At 100 ppm, ChE was markedly decreased in plasma (50 to 75%), red blood cells (close to 100%) and brain (about 40%, in females only). At 31.6 ppm, there was moderate (about 50%) inhibition of ChE in red blood cells and plasma (at weeks 26 and 52 only). At 10 ppm, no decrease in ChE was seen in brain or red blood cells, and no effect was seen in plasma, except for a moderate decrease (40 to 56%) in males at weeks 19 and 26 only. Based on cholinesterase inhibition, a NOAEL of 10 ppm (0.5 mg/kg/day) is identified.

4. Reproductive Effects

- Woodard et al. (1968) exposed three generations of rats to dietary fonofos at 0, 10 or 31.6 ppm. Based on the assumption that 1 ppm in the diet of a rat is equivalent to 0.05 mg/kg/day (Lehman, 1959), this corresponds to doses of 0, 0.5 or 1.58 mg/kg/day. No differences were detected in exposed dams with respect to mortality, body weight or uterine implantation sites. No effects were seen in offspring on conception ratio, litter size, number of live-born and stillborn, litter weight and weanling survival. Skeletal and visceral examination of offspring revealed no evidence of developmental defects. A NOAEL of 31.6 ppm (1.58 mg/kg/day, the highest dose tested) was identified.

5. Developmental Effects

- Groups of pregnant mice each received 10 daily doses of fonofos by gavage (0, 2, 4, 6 or 8 mg/kg/day) on gestational days 6 through 15 (Minor et al., 1982). At 8 mg/kg/day, maternal food intake and body weight gain were decreased. At 6 mg/kg/day, two dams experienced tremors and died. Increased incidences of variant ossifications of the sternebrae (8 mg/kg/day) and a slight dilatation of the fourth ventricle of the brain (4 and 8 mg/kg/day) were observed, but the authors did not interpret this as evidence of teratogenicity. Based on these findings, the NOAEL for fetotoxicity identified in this study was 2 mg/kg/day. The teratogenic NOAEL identified for this study was 8 mg/kg/day.

6. *Mutagenicity*

- Fonofos, with or without metabolic activation, was not mutagenic in each of five microbial assay systems [the Ames (*Salmonella typhimurium*) test; reverse mutation in an *Escherichia coli* strain; mitotic recombination in the yeast, *Saccharomyces cerevisiae* D3; and differential toxicity assays in strains of *E. coli* and *Bacillus subtilis*] and in a test for unscheduled DNA synthesis in human fibroblast (WI-38) cells (Simmon, 1979).

7. *Carcinogenicity*

- Groups of 30 male and 30 female CD albino rats (Charles River) each received 0, 10, 31.6 or 100 ppm fonofos in the diet (0, 0.5, 1.58 or 5 mg/kg/day) for 2 years (Banerjee et al., 1968). Based on gross and histological examination, the authors detected no carcinogenic effects.

IV. Quantification of Toxicological Effects

A. One-day Health Advisory

No information was found in the available literature that was suitable for determination of the One-day HA value for fonofos. It is therefore recommended that the adjusted DWEL for a 10-kg child of 0.02 mg/L (20 μg/L), calculated below, be used at this time as a conservative estimate of the One-day HA value.

For a 10-kg child, the adjusted DWEL is calculated as follows:

$$DWEL = \frac{(0.002 \text{ mg/kg/day}) \, (10 \text{ kg})}{1 \text{ L/day}} = 0.02 \text{ mg/L}$$

where: 0.002 mg/kg/day = RfD (see Lifetime Health Advisory section)

B. Ten-day Health Advisory

No information was found in the available literature that was suitable for determination of the Ten-day HA value for fonofos. It is therefore recommended that the adjusted DWEL for a 10-kg child of 0.02 mg/L (20 μg/L) be used at this time as a conservative estimate of the Ten-day HA value.

C. Longer-term Health Advisory

No information was found in the available literature that was suitable for determination of the Longer-term HA value for fonofos. It is therefore recommended that the adjusted DWEL for a 10-kg child of 0.02 mg/L (20 μg/L) be used at this time as a conservative estimate of the Longer-term HA and that the DWEL of 0.07 mg/kg (70 μg/L) be used for a 70-kg adult.

D. Lifetime Health Advisory

The 2-year feeding study in dogs by Woodard et al. (1969) has been selected to serve as the basis for the Lifetime HA for fonofos. Dogs received dietary fonofos at 0, 16, 60 or 240 ppm (0, 0.4, 1.5 or 6 mg/kg/day) for 14 weeks. Marginal (about 30%) inhibition of ChE was noted in red blood cells at the 16-ppm level; this dose was reduced to 8 ppm (0.2 mg/kg/day). Following dose reduction, ChE levels returned to control. At 60 ppm, dogs showed increased liver weights and significant inhibition (25 to 65%) of ChE activity in plasma and erythrocytes. At 240 ppm, there was increased ChE inhibition and increased mortality. There were no toxic effects in dogs at 8 ppm (0.2 mg/kg/day), with the possible exception of one brief episode of fasciculation in one dog at 5 months. This was not judged to be significant, and a NOAEL of 8 ppm (0.2 mg/kg/day) was identified. The 2-year feeding study in rats by Banerjee et al. (1968) has not been selected, since rats appear to be less sensitive than dogs when doses are calculated on a body weight (mg/kg) basis.

Step 1: Determination of the Reference Dose (RfD)

$$\text{RfD} = \frac{(0.2 \text{ mg/kg/day})}{(100)} = 0.002 \text{ mg/kg/day}$$

Step 2: Determination of the Drinking Water Equivalent Level (DWEL)

$$\text{DWEL} = \frac{(0.002 \text{ mg/kg/day}) (70 \text{ kg})}{(2 \text{ L/day})} = 0.07 \text{ mg/L} (70 \text{ } \mu\text{g/L})$$

Step 3: Determination of the Lifetime Health Advisory

$$\text{Lifetime HA} = (0.07 \text{ mg/L}) (20\%) = 0.014 \text{ mg/L} (10 \text{ } \mu\text{g/L})$$

E. Evaluation of Carcinogenic Potential

- Groups of 30 male and 30 female albino rats (Charles River, Cesarean-derived) each received 0, 10, 31.6 or 100 ppm fonofos in the diet (0, 0.5, 1.58 or 5 mg/kg/day) for 2 years (Banerjee et al., 1968). Based on gross and histological examination, the authors detected no carcinogenic effect.

- IARC (1982) has not evaluated the carcinogenic potential of fonofos.

- Applying the criteria described in EPA's guidelines for assessment of carcinogenic risk (U.S. EPA, 1986), fonofos may be classified in Group D: not classifiable. This category is for substances with inadequate animal evidence of carcinogenicity.

V.   Other Criteria, Guidance and Standards

- No existing criteria, guidelines or standards for oral exposure to fonofos were located.

- The U.S. EPA Office of Pesticide Programs (OPP) has calculated an ADI of 0.002 mg/kg/day for fonofos. This was based on a NOAEL of 0.2 mg/kg/day (8 ppm) for both ChE inhibition and systemic effects, in a 2-year feeding study in dogs (Woodard et al., 1969), and an uncertainty factor of 100.

- The Threshold Limit Value (TLV) for fonofos is 100 μg/m3 (ACGIH, 1984).

- The U.S. EPA (1985) has established tolerances for fonofos in or on raw agricultural commodities that range from 0.1 to 0.5 ppm.

VI.   Analytical Methods

- Analysis of fonofos is by a gas chromatographic (GC) method applicable to the determination of certain nitrogen-phosphorus containing pesticides in water samples (Method 507, U.S. EPA, 1988). In this method, approximately 1 liter of sample is extracted with methylene chloride. The extract is concentrated and the compounds are separated using capillary column GC. Measurement is made using a nitrogen-phosphorus detector. The method detection limit has not been determined for fonofos, but it is estimated that the detection limits for analytes included in this method are in the range of 0.1 to 2 μg/L.

VII.   Treatment Technologies

• No information on treatment technologies used to remove fonofos from contaminated water was found in the available literature.

VIII.   References

ACGIH. 1984. American Conference of Governmental Industrial Hygienists. Documentation of the threshold limit values for substances in workroom air, 3rd ed. Cincinnati, OH: ACGIH.

Ahmed, J. and F.O. Morrison. 1972. Longevity of residues of four organophosphate insecticides in soil. Phytoprotection. 53(2–3):71–74.

Banerjee, B.M., D. Howard and M.W. Woodard.* 1968. Dyfonate (N-2790) safety evaluation by dietary administration to rats for 105 weeks. Woodard Research Corporation. Unpublished study. MRID 00082232.

Cockrell, K.O., M.W. Woodard and G. Woodard.* 1966. N-2790 Safety evaluation by repeated oral administration to dogs for 14 weeks and to rats for 13 weeks. Woodard Research Corporation. Unpublished study. MRID 0090818.

Cohen, S.Z., C. Eiden and M.N. Lorber. 1986. Monitoring ground water for pesticides in the U.S.A. Evaluation of Pesticides in Ground Water, American Chemical Society Symposium Series #315.

Dean, W.P.* 1977. Acute oral and dermal toxicity (LD_{50}) in male and female albino rats. Study No. 153–047. International Research and Development Corporation. Unpublished study. MRID 00059860, 00059856 and 00059857.

Hayes, W.J. 1982. Pesticides studied in man. Baltimore, MD: Williams and Wilkins. p. 413.

Hoffman, L.J., J.M. Ford and J.J. Menn. 1971. Dyfonate metabolism studies. I. Absorption, distribution, and excretion of O-ethyl S-phenyl ethylphosphonodithioate in rats. Pesticide Biochem. Physiol. 1:349–355.

Hoffman, L.J., J.B. McBain and J.J. Menn. 1973. Environmental behavior of O-ethyl S-phenyl ethylphosphonodithioate (Dyfonate): ARC-B-35. Unpublished study submitted by Stauffer Chemical Company, Richmond, CA.

Hoffman, L.J. and J.H. Ross. 1971. Dyfonate soil metabolism. Project 038022. Unpublished study submitted by Stauffer Chemical Company, Richmond, CA.

Horn, H.J., G. Woodard and M.T. Cronin.* 1966. N-2790 10% granular: subacute dermal toxicity: 21-day experiment in rabbits. Unpublished study. MRID 00092438.

Horton, R.J.* 1966a. N-2790: Acute oral LD_{50}—rats; acute dermal toxicity—rabbits; acute eye irritation—rabbits. Technical Report T-986. Stauffer Chemical Company. Unpublished study. MRID 00090806.

Horton, R.J.* 1966b. N-2790: Acute oral LD_{50}—rats; acute dermal toxicity—rabbits; acute eye irritation—rabbits. Technical Report T-985. Stauffer Chemical Company. Unpublished study. MRID 00090807.

IARC. 1982. International Agency for Research on Cancer. IARC monographs on the evaluation of the carcinogenic risk of chemicals to humans. Chemicals, industrial processes and industries associated with cancer in humans. International Agency for Research on Cancer Monographs. Vols. 1 to 29, Supplement 4. Geneva, Switzerland: World Health Organization.

Kadoum, A.M. and D.F. Mock. 1978. Herbicide and insecticide residues in tailwater pits: water and pit bottom soil from irrigated corn and sorghum fields. J. Agric. Food Chem. 26(1):45–50.

Kiigemagi, U. and L.C. Terriere. 1971. The persistence of Zinophos and Dyfonate in soil. Bull. Environ. Contam. Toxicol. 6(4):355–361.

Lehman, A.J. 1959. Appraisal of the safety of chemicals in foods, drugs and cosmetics. Assoc. Food Drug Off. U.S.

Lichtenstein, E.P., H. Parlar, F. Korte and A. Suss. 1977. Identification of fonofos metabolites isolated from insecticide-treated culture media of the soil fungus *Rhizopus japonicus*. J. Agric. Food Chem. 25(4):845–848.

Lichtenstein, E.P. and K.R. Schulz. 1970. Volatilization of insecticides from various substrates. J. Agric. Food Chem. 18(5):814–818.

Lichtenstein, E.P., K.R. Schulz and T.W. Fuhremann. 1972. Movement and fate of Dyfonate in soils under leaching and nonleaching conditions. J. Agric. Food Chem. 20(4):831–838.

McBain, J.B. and J.J. Menn. 1966. Persistence of O-ethyl-S-phenyl ethylphosphonodithioate (Dyfonate) in soils: ARC-B-10. Unpublished study submitted by Stauffer Chemical Company, Richmond, CA.

McBain, J.B., L.J. Hoffman and J.J. Menn. 1971. Dyfonate metabolism studies. II. Metabolic pathway of O-ethyl S-phenyl ethylphosphonodithioate in rats. Pesticide Biochem. Physiol. 1:356–365.

Meister, R., ed. 1983. Farm chemicals handbook. Willoughby, OH: Meister Publishing Company.

Miles, J.R.W., C.M. Tu and C.R. Harris. 1979. Persistence of eight organophosphorus insecticides in sterile and non-sterile mineral and organic soils. Bull. Environ. Contam. Toxicol. 22:312–318.

Miller, J.L., L. Sandvik, G.L. Sprague, A.A. Bickford and T.R. Castles. 1979. Evaluation of delayed neurotoxic potential of chronically administered Dyfonate in adult hens. Toxic. Appl. Pharmacol. 48 (Part 2):A197.

Minor, J., J. Downs, G. Zwicker et al.* 1982. A teratology study in CD-1 mice with Dyfonate technical T-10192. Final report. Stauffer Chemical Company. Unpublished study. MRID 00118423.

Schulz, K.R. and E.P. Lichtenstein. 1971. Field studies on the persistence and movement of Dyfonate in soil. J. Econ. Entomology. 64(1):283–287.

Simmon, V.F. 1979. *In vitro* microbiological mutagenicity and unscheduled DNA synthesis studies of eighteen pesticides. National Technical Information Service, Springfield, Virginia, publication. EPA-600/1-79-041, Research Triangle Park, North Carolina. p. 164.

STORET. 1988. STORET Water Quality File. Office of Water. U.S. Environmental Protection Agency (data file search conducted in May, 1988).

TDB. 1985. Toxicology Data Bank. MEDLARS II. National Library of Medicine's National Interactive Retrieval Service.

Talekar, N.S., L.T. Sun, F.M. Lee and J.S. Chen. 1977. Persistence of some insecticides in subtropical soil. J. Agric. Food Chem. 25(2):348–352.

U.S. EPA. 1985. United States Environmental Protection Agency. Code of Federal Regulations. 40 CFR 180.221, p. 290.

U.S. EPA. 1986. U.S. Environmental Protection Agency. Guidelines for carcinogen risk assessment. Fed. Reg. 51(185):33992–34003. September 24.

U.S. EPA. 1988. U.S. Environmental Protection Agency. U.S. EPA Method 507 — Determination of nitrogen and phosphorus containing pesticides in ground water by GC/NPD. Draft. April 15. Available from U.S. EPA's Environmental Monitoring and Support Laboratory, Cincinnati, OH.

Windholz, M., S. Budavari, R.F. Blumetti and E.S. Otterbein, eds. 1983. The Merck index — an encyclopedia of chemicals and drugs, 10th ed. Rahway, NJ: Merck and Company, Inc.

Woodard, M.W., J. Donoso, J.P. Gray et al.* 1969. Dyfonate (N-2790) safety evaluation by dietary administration to dogs for 106 weeks. Woodard Research Corporation. Unpublished study. MRID 00082233.

Woodard, M.W., C.L. Leigh and G. Woodard.* 1968. Dyfonate (N-2790) three-generation reproduction study in rats. Woodard Research Corporation. Unpublished study. MRID 00082234.

Woodard, M.W. and G. Woodard.* 1966. N-2790 (Dyfonate): Demyelination study in chickens. Woodard Research Corporation. Unpublished study. MRID 00090819.

---

* Confidential Business Information submitted to the Office of Pesticide Programs.

Glyphosate

I. General Information and Properties

A. CAS No. 1071–83–6

B. Structural Formula

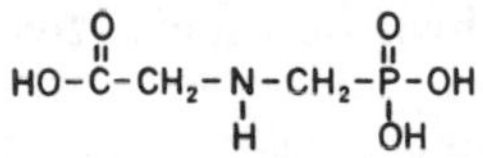

$$\text{HO}-\overset{\overset{\text{O}}{\|}}{\text{C}}-\text{CH}_2-\underset{\underset{\text{H}}{|}}{\text{N}}-\text{CH}_2-\overset{\overset{\text{O}}{\|}}{\underset{\underset{\text{OH}}{|}}{\text{P}}}-\text{OH}$$

Glycine, N-(Phosphonomethyl)

C. Synonyms

- Active ingredient (glyphosate) in Rodeo®, Roundup®, Landmaster®, Shakle®, Roundup L & G®, Polado®.

D. Uses

- Glyphosate is used as an herbicide for control of grasses, broadleaved weeds and woody brush (U.S. EPA, 1986b).

E. Properties (Meister, 1983)

Chemical Formula	$C_3H_8NO_5P$
Molecular Weight	169.07
Physical State (25°C)	White crystalline solid
Boiling Point	--
Melting Point	200°C (decomposes)*
Density	0.5 g/mL (bulk density of dried technical)*
Vapor Pressure	--
Water Solubility	11.6 g/L*
Log Octanol/Water Partition Coefficient	0.0006–0.0017*
Taste Threshold	--
Odor Threshold	--
Conversion Factor	--

* Monsanto, 12/13/87

F. Occurrence

- No glyphosate was detected in 6 surface water samples or 98 ground-water samples taken from 3 surface water stations and 97 ground-water stations, respectively (STORET, 1988).

G. Environmental Fate

- Biodegradation is considered the major fate process affecting glyphosate persistence in aquatic environments (Reinert and Rodgers, 1987). Glyphosate is biodegraded aerobically and anaerobically by microorganisms present in soil, water, hydrosoil and activated sludge.

- ^{14}C-Glyphosate and aminomethylphosphonic acid (94% glyphosate, 5.9% aminomethylphosphonic acid) were stable in sterile buffered water at pH 3, 6, and 9 during 32 days of incubation in the dark at 5 and 35°C (Brightwell and Malik, 1978).

- ^{14}C-Glyphosate (94% glyphosate, 5.9% aminomethylphosphonic acid) was adsorbed to Drummer silty clay loam, Ray silt, Spinks sandy loam, Lintonia sandy loam and Cattail Swamp sediment with Freundlich K values of 62, 90, 70, 22 and 175, respectively (Brightwell and Malik, 1978). For each soil preparation, the maximum percentages of applied glyphosate desorbed were 5.3, 3.7, 3.6, 11.5 and 0.9%, respectively. At concentrations ranging from 0.21 to 50.1 ppm, ^{14}C-Glyphosate was highly adsorbed to five soils, with organic matter contents ranging from 2.40 to 15.50% (Monsanto Company, 1975). Adsorption of glyphosate ranged from 71% (Soil E, 2.4% organic matter, pH 7.29) to 99% (Soil C, 15.5% organic matter, pH 5.35).

- ^{14}C-Glyphosate (94% glyphosate, 5.9% aminomethylphosphonic acid) was slightly mobile to relatively immobile, with less than 7% of the applied ^{14}C detected in the leachate from 30-cm silt, sand, clay, sandy clay loam, silty clay loam and sandy loam soil columns eluted with 20 inches of water (Brightwell and Malik, 1978). Aged (30 days) ^{14}C-glyphosate residues were relatively immobile in silt, clay and sandy clay loam soils with less than 2% of the radioactivity detected in the leachate following elution with 20 inches of water. Both glyphosate and aminomethylphosphonic acid were detected in the leachate of aged and unaged soil columns.

II. Pharmacokinetics

A. Absorption

- Feeding studies with chickens, cows and swine showed that ingestion of up to 75 ppm glyphosate resulted in nondetectable glyphosate residue levels (< 0.05 ppm) in muscle tissue and fat (Monsanto Company, 1983). The duration of exposure was for 30 days. Glyphosate residue levels were not detectable (< 0.025 ppm) in milk and eggs from cows and chickens on diets containing glyphosate.

B. Distribution/Metabolism

- The distribution, metabolism and retention of glyphosate by the tissues appear to be minimal since 60 to 90% of a single oral dose is rapidly eliminated unchanged in the feces (Duerson and Sipes, 1987).

C. Excretion

- After a single oral or intraperitoneal dose, less than 1% of the administered dose was retained after 120 hours of treatment (U.S. EPA, 1986b). In rats fed 1, 10 or 100 ppm of ^{14}C-glyphosate for 14 days, a steady-state equilibrium between intake and excretion of label was reached within about 8 days. The amount of radioactivity excreted in the urine decreased rapidly after withdrawal of treatment. Ten days after withdrawal, radioactivity was detectable in the urine and feces of rats fed 10 or 100 ppm of the test diet. Minimal residues of 0.1 ppm or less remained in the tissues of high-dose rats after 10 days of withdrawal. No single tissue showed a significant difference in the amount of label retained.

III. Health Effects

A. Humans

- Glyphosate, a widely utilized herbicide, was evaluated for acute irritation, cumulative irritation, photoirritation and allergic and photoallergic contact potential in 346 volunteers. The herbicide was less irritant than a standard liquid dishwashing detergent and a general all-purpose cleaner. There was no evidence for the induction of photoirritation, allergic or photoallergic contact dermatitis. Ten percent glyphosate in water is proposed as a diagnostic patch test concentration (Maibach, 1986).

B. Animals

1. *Short-term Exposure*

- An oral LD_{50} of 5,600 mg/kg in the rat is reported for glyphosate (Monsanto Company, 1985).

- Bababunmi et al. (1978) reported that daily intraperitoneal administration of 15, 30, 45 or 60 mg/kg to rats for 28 days resulted in reduced daily body weight gain, decreased blood hemoglobin, decreased red blood cell count and hematocrit values and elevated levels of serum glutamic-pyruvic transaminase and leucine-amino peptidase during the experimental period. The investigators did not specify the dose levels at which these effects were observed.

2. *Dermal/Ocular Effects*

- A dermal LD_{50} for glyphosate in the rabbit was reported to be > 5,000 mg/kg (Monsanto Company, 1985).

3. *Long-term Exposure*

- In subchronic studies reported by the Weed Science Society of America (1983), technical-grade glyphosate was fed to rats and dogs at dietary levels of 200, 600 or 2,000 ppm for 90 days. Mean body weights, food consumption, behavioral reactions, mortality, hematology, blood chemistry and urinalysis did not differ significantly from controls. There were no relevant gross or histopathological changes. No other details or data were provided.

- Bio/dynamics, Inc. (1981a) conducted a study in which glyphosate was administered in the diet to four groups of Sprague-Dawley rats (50/sex/dose) at dose levels of 0, 3.1, 10.3 or 31.5 mg/kg/day to males or 0, 3.4, 11.3 or 34.0 mg/kg/day to females. After 26 weeks, body weight, organ weight, organ-to-body weight ratios and hematological and clinical chemistry parameters were evaluated. No significant differences between control and exposed animals were observed at any dose level.

- Glyphosate was administered to mice at dietary levels of 5,000, 10,000 and 50,000 ppm for 3 months. Decreased body weight gains were observed in the high-dose group. No treatment-related effects in pathologic or histopathologic evaluations were observed. The no-effect level was considered to be 10,000 ppm (Monsanto Company, 1985).

- Beagle dogs were fed glyphosate at dietary concentrations of 30, 100 or 300 ppm for 2 years. No evidence of treatment-related chronic or carcinogenic effects were observed. The no-effect level was considered to be 300 ppm (Monsanto Company, 1985).

4. Reproductive Effects

- Bio/dynamics, Inc. (1981b) investigated the reproductive toxicity of glyphosate in rats. The glyphosate (98.7% purity) was administered in the diet at dose levels of 0, 3, 10 or 30 mg/kg/day to Charles River Sprague-Dawley rats for three successive generations. Twelve males and 24 females (the F_0 generation) were administered test diets for 60 days prior to breeding. Administration was continued through mating, gestation and lactation for two successive litters (F_{1a}, F_{1b}). Twelve males and 24 females from the F_{1b} generation were retained at weaning for each dose level to serve as parental animals for the succeeding generation. The following indices of reproductive function were measured: fetal, pup and adult survival; parental and pup body weight; food consumption; and mating, fertility or gestation. Necropsy and histopathologic evaluation were performed as well. No compound-related changes in these parameters were observed when compared to controls, although an addendum to the pathological report for this study reported an increase in unilateral focal tubular dilation of the kidney in the male F_{3b} pups when compared to concurrent controls. Based on data from this study, the authors concluded that the highest dose tested (30 mg/kg/day) did not affect reproduction in rats under the conditions of the study.

5. Developmental Effects

- Glyphosate was administered to pregnant rabbits by gavage at dose levels of 0, 75, 175 or 350 mg/kg/day on days 6 through 27 of gestation (Monsanto Company, 1985). No evidence of fetal toxicity or birth defects in the offspring was observed. However, at dose levels of 350 mg/kg/day, death, soft stools, diarrhea and nasal discharge were observed in the animals.

- No teratogenic effects were observed in the offspring of rats administered glyphosate by gavage at dosage levels of 300, 1,000 and 3,500 mg/kg/day on days 6 through 19 of gestation. Toxic effects were noted in the high-dose treated animals and their offspring (Monsanto Company, 1985).

6. *Mutagenicity*

- The Monsanto Company (1985) reported that glyphosate did not cause mutation in microbial test systems. A total of eight strains (seven bacterial and one yeast), including five *Salmonella typhimurium* strains and one strain each of *Bacillus subtilis*, *Escherichia coli* and *Saccharomyces cerevisiae*, were tested. No mutagenic effects were observed in any strain.

- Njagi and Gopalan (1980) found that glyphosate did not induce reversion mutations in *Salmonella typhimurium* histidine auxotrophs.

7. *Carcinogenicity*

- Bio/dynamics, Inc. (1981b) conducted a study to assess the oncogenicity of glyphosate (98.7% purity). The chemical was given in the diet to four groups of Sprague-Dawley rats at dose levels of 0, 3.1, 10.3 or 31.5 mg/kg/day to males or 0, 3.4, 11.3 or 34.0 mg/kg/day to females. After 26 months, animals were sacrificed and tissues were examined for histological lesions. A variety of benign and malignant tumors were observed in both the treated and control groups, the most common tumors occurring in the pituitary of both sexes and in the mammary glands of females. The total numbers of rats of both sexes that developed tumors (benign and malignant) were 72/100 (low dose), 79/100 (mid dose), 85/100 (high dose) and 87/100 (control). An increased rate of interstitial cell tumors of the testes was reported in the high-dose males when compared to concurrent controls (6/50 versus 0/50), but this was not considered to be related to compound administration. Based on the data from this study, the authors concluded that the highest dose level tested (31.5 and 34.0 mg/kg/day for males and females, respectively) was not carcinogenic in rats.

IV.   Quantification of Toxicological Effects

A. One-day Health Advisory

No information was found in the available literature that was suitable for determination of the One-day HA value for glyphosate. It is therefore recommended that the Ten-day HA value of 20 mg/L be used at this time as a conservative estimate of the One-day HA value.

B. Ten-day Health Advisory

The teratology study in pregnant rabbits (Monsanto Company, 1985) has been selected to serve as the basis for determination of the Ten-day HA for the 10-kg child. In this study, pregnant rabbits that received glyphosate at dose levels of 0, 75, 175 or 350 mg/kg/day on days 6 through 27 of gestation showed effects at 350 mg/kg/day; however, no treatment-related effects were reported at lower dose levels. The No-Observed-Adverse-Effect Level (NOAEL) identified in this study is, therefore, 175 mg/kg/day.

Using a NOAEL of 175 mg/kg/day, the Ten-day HA for a 10-kg child is calculated as follows:

$$\text{Ten-day HA} = \frac{(175 \text{ mg/kg/day}) (10 \text{ kg})}{(100) (1 \text{ L/day})} = 17.5 \text{ mg/L } (20,000 \text{ } \mu g/L)$$

C. Longer-term Health Advisory

No information was found in the available literature that was suitable for determination of the Longer-term HA value for glyphosate. It is therefore recommended that the adjusted DWEL for a 10-kg child (1 mg/L), calculated below, and 4 mg/L for an adult be used at this time as conservative estimates of the Longer-term HA values.

For a 10-kg child, the adjusted DWEL is calculated as follows:

$$\text{DWEL} = \frac{(0.1 \text{ mg/kg/day}) (10 \text{ kg})}{1 \text{ L/day}} = 1.0 \text{ mg/L}$$

where: 0.1 mg/kg/day = RfD (see Lifetime Health Advisory section).

D. Lifetime Health Advisory

The study by Bio/dynamics (1981b) has been selected to serve as the basis for determination of the Lifetime HA value for glyphosate. In this study, the reproductive toxicity of glyphosate in rats was investigated over three generations. Even though no compound-related changes in the reproductive indices were observed when compared to controls at a dose level of 30 mg/kg/day, there were pathological changes of renal focal tubular dilation in male F_{3b} weanling rats at this level. Therefore, the lower dose level of 10 mg/kg/day was identified as the NOAEL.

Using a NOAEL of 10 mg/kg/day, the Lifetime HA is calculated as follows:

Step 1: Determination of the Reference Dose (RfD)

$$RfD = \frac{(10 \text{ mg/kg/day})}{(100)} = 0.1 \text{ mg/kg/day}$$

Step 2: Determination of the Drinking Water Equivalent Level (DWEL)

$$DWEL = \frac{(0.1 \text{ mg/kg/day}) (70 \text{ kg})}{(2 \text{ L/day})} = 3.5 \text{ mg/L} (4,000 \text{ } \mu g/L)$$

Step 3: Determination of the Lifetime Health Advisory

$$Lifetime \text{ HA} = (3.5 \text{ mg/L}) (20\%) = 0.70 \text{ mg/L} (700 \text{ } \mu g/L)$$

E. Evaluation of Carcinogenic Potential

- Applying the criteria described in EPA's guidelines for assessment of carcinogenic risk (U.S. EPA, 1986a), glyphosate may be classified in Group D: not classifiable. This category is for substances with inadequate animal evidence of carcinogenicity.

- The evidence of carcinogenicity in animals is considered equivocal by the Science Advisory Panel (Pesticides), and has been classified in Category D [Office of Pesticide Programs has requested the manufacturer to conduct another study in animals (U.S. EPA, 1986b)].

V.   Other Criteria, Guidance and Standards

- No other criteria, guidelines or standards were found in the available literature pertaining to glyphosate.

- Tolerance of 0.1 ppm has been established for the combined residues of glyphosate and its metabolite in or on raw agricultural commodities resulting from the use of irrigation water following applications of glyphosate herbicide around aquatic sites (U.S. EPA, 1985).

VI.   Analytical Methods

- A method for glyphosate has been developed for ODW by EMSL-CI (U.S. EPA, 1988). Other methods are available for glyphosate. However, this modification has been directed toward water, since glyphosate is used as an aquatic herbicide and the media of concern is drinking water. In this procedure a water sample is filtered and a small aliquot (200 μL) is injected into a reverse phase high pressure liquid chromatograph column. Separation is by isocratic elution. Post-column derivatization involves oxidation of the glyphosate to glycine with reaction to o-phthalaldehyde to form a sensitive fluorophore. The method detection limit for this procedure for single laboratory validation is 6 $\mu g/L$ for tap water, and 9 $\mu g/L$ for ground water. Across four concentrations with 8 data points for each type of water (tap water, ground water), recoveries ranged from 95 to 110%, with a relative standard deviation ranging from 2 to 12%.

VII.   Treatment Methodologies

• No information was found in the available literature on treatment technologies capable of effectively removing glyphosate from contaminated water.

VIII.   References

Babaunmi, E.A., O.O. Olorunsogo and O. Bassir. 1978. Toxicology of glyphosate in rats and mice. Toxicol. Appl. Pharmacol. 45(1):319–320.

Bio/dynamics, Inc.* 1981a. A lifetime feeding study of glyphosate (Roundup Technical) in rats. Project No. 77–2062. Unpublished report. Prepared for Monsanto Co., St. Louis, MO. EPA Accession Nos. 246617 and 246621.

Bio/dynamics, Inc.* 1981b. A three-generation reproduction study in rats with glyphosate. Project 77–2063. Unpublished report. Prepared for Monsanto Co., St. Louis, MO. EPA Accession No. 245909. Cited in Monsanto Company, 1985.

Brightwell, B. and J. Malik. 1978. Solubility, volatility, adsorption and partition coefficients, leaching and aquatic metabolism of MON 0573 and MON 0101. Report No. MSL-0207.

Duerson, C.R. and I. Glenn Sipes. 1987. Absorption of glyphosate in the male Fischer rat. The Toxicologist. 7(1):Abstract 1986.

Maibach, H.I. 1986. Irritation, sensitization, photoirritation and photosensitization assays with a glyphosate herbicide. Contact Dermatitis. 15:152–156.

Meister, R.T., ed. 1983. Farm chemicals handbook. Willoughby, OH: Meister Publishing Company. p. C117.

Monsanto Company. 1975. Residue and metabolism studies in sugarcane and soils. St. Louis, MO: Monsanto Agricultural Products Company.

Monsanto Company. 1983. Rodeo herbicide: toxicological and environmental properties. Rodeo Herbicide Bulletin No. 1. St. Louis, MO: Monsanto Company.

Monsanto Company. 1985. Material safety data sheet, glyphosate technical. St. Louis, MO: Monsanto Company. MSDS No. 107–183–6.

Monsanto Company. 1987. Personal communication December 13, 1987.

NAS. 1977. National Academy of Sciences. Drinking water and health, Vol. I. Washington, DC: National Academy of Sciences.

NAS. 1980. National Academy of Sciences. Drinking water and health, Vol. 3. Washington, DC: National Academy Press, National Research Council. pp. 77–80.

Njagi, G.D.E. and H.N.B. Gopalan. 1980. Mutagenicity testing of some selected food preservatives, herbicides and insecticides. Bangladesh J. Bot. 9:141–146. Abstract.

Olorunsogo, O.O. 1981. Inhibition of energy-dependent transhydrogenase reaction by N-(phosphonomethyl)glycine in isolated rat liver mitochondria. Toxicol. Lett. 10:91–95.

Olorunsogo, O.O. and E.A. Bababunmi. 1980. Inhibition of succinate-linked reduction of pyridine nucleotide in rat liver mitochondria *"in vivo"* by N-(phosphonomethyl)glycine. Toxicol. Lett. 7:149–152.

Olorunsogo, O.O., E.A. Bababunmi and O. Bassir. 1977. Toxicity of glyphosate. In: G.L. Plaa and W.A.M. Duncan, eds. Proceedings of the lst International Congress on Toxicology. New York: Academic Press. p. 597. Abstract.

Olorunsogo, O.O., E.A. Bababunmi and O. Bassir. 1979a. Effect of glyphosate on rat liver mitochondria *in vivo*. Bull. Environ. Contam. Toxicol. 22:357–364.

Olorunsogo, O.O., E.A. Bababunmi and O. Bassir. 1979b. The inhibitory effect of N-(phosphonomethyl)glycine *in vivo* on energy-dependent, phosphate-induced swelling of isolated rat liver mitochondria. Toxicol. Lett. 4:303–306.

Reinert, K.H. and J.H. Rodgers. 1987. Fate and persistence of aquatic herbicides. Rev. Environ. Contam. Toxicol. 98:61–98.

Rueppel, M.L., B.B. Brightwell, J. Schaefer and J.T. Marvel. 1977. Metabolism and degradation of glyphosate in soil and water. J. Agric. Food Chem. 25:517–528.

Seiler, J.P. 1977. Nitrosation *in vitro* and *in vivo* by sodium nitrite, and mutagenicity of nitrogenous pesticides. Mutat. Res. 48:225–236.

Shoval, S. and S. Yariv. 1981. Infrared study of the fine structures of glyphosate and Roundup. Agrochimica. 25:377–386.

STORET. 1988. STORET Water Quality File. Office of Water. U.S. Environmental Protection Agency (data file search conducted in May, 1988).

U.S. EPA. 1985. U.S. Environmental Protection Agency. Code of Federal Regulations. 40 CFR 180.364. July 1.

U.S. EPA. 1986a. U.S. Environmental Protection Agency. Guidelines for carcinogen risk assessment. Fed. Reg. 51(185):33992–3002. September 24.

U.S. EPA. 1986b. U.S. Environmental Protection Agency. Guidance for the registration of pesticide products containing glyphosate as the active ingredient. Case No. 0178. June.

U.S. EPA. 1988. U.S. Environmental Protection Agency. U.S. EPA Draft Method — Analysis of glyphosate in drinking water by direct aqueous HPLC injection with post-column derivatization. EMSL-CI, Cincinnati, OH. June, 1988.

Weed Science Society of America. 1983. Herbicide handbook, 5th ed. Champaign, IL: Weed Science Society of America. pp. 258–263.

---

* Confidential Business Information submitted to the Office of Pesticide Programs.

Hexazinone

I. General Information and Properties

A. CAS No. 51235-04-2

B. Structural Formula:

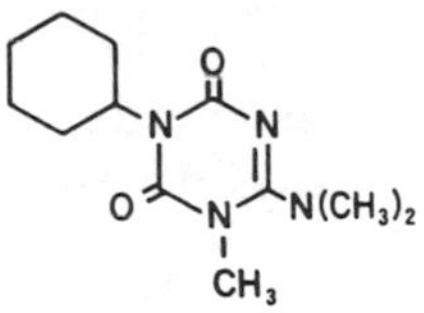

3-Cyclohexyl-6-(dimethylamino)-1 methyl-1,3,5-triazine-2,4(1H,3H)-dione

C. Synonyms

- Velpar; Hexazinone.

D. Use

- Hexazinone is used as a contact and residual herbicide (Meister, 1983).

- Usage areas include plantations of coniferous trees, railroad rights-of-way, utilities, pipelines, petroleum tanks, drainage ditches, and sugar and alfalfa (Kennedy, 1984).

E. Properties (Kennedy, 1984; CHEMLAB, 1985)

Chemical Formula	$C_{11}H_{20}SO_2N_3$
Molecular Weight	252
Physical State (25°C)	White crystalline solid
Boiling Point	--
Melting Point	115–117°C
Density	--
Vapor Pressure (86°C)	2×10^{-7} mm Hg
Specific Gravity	--
Water Solubility (25°C)	33,000 mg/L

Log Octanol/Water
   Partition Coefficient   −4.40 (calculated)
Taste Threshold   --
Odor Threshold   Odorless
Conversion Factor   --

F. Occurrence

- Hexazinone has not been found in any surface or ground-water samples analyzed from nine samples taken at two locations or six samples from six locations, respectively (STORET, 1988).

G. Environmental Fate

- Hexazinone did not hydrolyze in water within the pH range of 5.7 to 9 during a period of 8 weeks (Rhodes, 1975a).

- In a soil aerobic metabolism study, hexazinone was added to a Fallsington sandy loam and a Flanagan silt loam at 4 ppm. ^{14}C-Hexazinone residues had a half-life of about 25 weeks. Of the extractable ^{14}C-residues, half was present as parent compound and/or 3-cyclohexyl-1-methyl-6-methylamino-1,3,5-triazine-2,4-(1H,3H)-dione. Also present were 3-(4-hydroxycyclohexyl)-6-(dimethylamino)-1-methyl-1-(1H,3H)-dione and the triazine trione (Rhodes, 1975b).

- A soil column leaching study used ^{14}C-hexazinone, half of which was aged for 30 days and applied to Flanagan silt loam and Fallsington sandy loam. Leaching with a total of 20 inches of water showed that unaged hexazinone leached in the soils; however, leaching rates were slower for the aged samples, indicating that the degradation products may have less potential for contaminating ground water (Rhodes, 1975b).

- A field soil leaching study indicated that ^{14}C-hexazinone residues were leached into the lower sampling depths with increasing rainfall. A Keyport silt loam (2.75% organic matter; pH 6.5) and a Flanagan silt loam (4.02% organic matter; pH 5.0) were used. For the Keyport silt loam, ^{14}C-residues were found at all depths measured, including the 8- to 12-inch depth, when total rainfall equaled 8.43 inches, 1 month after application of hexazinone. For the Flanagan silt loam, ^{14}C-residues were found at all depths sampled, including the 12- to 15-inch depth, 1 month after application, when a total of 7.04 inches of rain had fallen (Rhodes, 1975c).

- A soil TLC test for Fallsington sandy loam and Flanagan silt loam gave R_f values for hexazinone of 0.85 and 0.68, respectively. This places hexazinone in Class 4, indicating it is very mobile in these soils (Rhodes, 1975c).

- In a terrestrial field dissipation study using a Keyport silt loam in Delaware, hexazinone had a half-life of less than 1 month. In a field study in Illinois (Flanagan silt loam), hexazinone had a half-life of more than 1 month (62% of the parent compound remained at 1 month) (Rhodes, 1975b). In a separate study with Keyport silt loam, some leaching of the parent compound to a depth of 12 to 18 inches was observed (Holt, 1979).

II.   Pharmacokinetics

A. Absorption

- Rapisarda (1982) reported that a dose of 14 mg/kg ^{14}C-labeled hexazinone (>99% pure) was about 80% absorbed in 3 to 6 days (77% recovery in urine, 20% in feces) when administered by gastric intubation to male and female Charles River CD rats with or without 3 weeks of dietary preconditioning with unlabeled hexazinone.

- Rhodes et al. (1978) administered 2,500 ppm (125 mg/kg) hexazinone in the diet to male rats for 17 days. This was followed by a single dose of 18.3 mg/300 g (61 mg/kg) ^{14}C-labeled hexazinone. The hexazinone was rapidly absorbed within 72 hours, with 61% detected in the urine and 32% in the feces. Trace amounts were found in the gastrointestinal (GI) tract (0.6%, tissues not specified) and expired air (0.08%).

B. Distribution

- Orally administered hexazinone has not been demonstrated to accumulate preferentially in any tissue (Rhodes et al., 1978; Holt et al., 1979; Rapisarda, 1982).

- Studies in rats by Rapisarda (1982) and Rhodes et al. (1978) showed that no detectable levels of ^{14}C-hexazinone were found in any body tissues when the animals were administered >14 mg/kg hexazinone by gastric intubation with or without dietary preconditioning.

- In a study with dairy cows by Holt et al. (1979) hexazinone was given in the diet at 0, 1, 5 or 25 ppm for 30 days. Assuming that 1 ppm in the diet of a cow equals 0.015 mg/kg (Lehman, 1959), these levels correspond to

0, 0.015, 0.075 or 0.37 mg/kg/day. The investigators reported no detectable residues in milk, fat, liver, kidney or lean muscle.

C. Metabolism

- Major urinary metabolites of hexazinone in rats identified by Rhodes et al. (1978) were 3-(4-hydroxycyclohexyl)-6-(dimethylamino)1-methyl-1,3,5-triazine-2,4-(1H,3H)-dione (metabolite A); 3-cyclohexyl-6-(methylamino)-1-methyl-1,3,5-triazine-2,4-(1H, 3H)-dione (metabolite B); and 3-(4-hydroxycyclohexyl)-6-(methylamino)-1-methyl-1,3,5-triazine-2,4-(1H,3H)-dione (metabolite C). The percentages of these metabolites detected in the urine were 46.8, 11.5 and 39.3%, respectively. The major fecal metabolites detected by Rhodes et al. (1978) were A (26.3%) and C (55.2%). Less than 1% unchanged hexazinone was detected in the urine or the feces. Similar results were reported by Rapisarda (1982).

D. Excretion

- Rapisarda (1982) and Rhodes et al. (1978) reported that excretion of ^{14}C-hexazinone and/or its metabolites occurs mostly in the urine (61 to 77%) and in the feces (20 to 32%).

III.   Health Effects

A. Humans

- The Pesticide Incident Monitoring System database (U.S. EPA, 1981) indicated that 3 of 43,729 incident reports involved hexazinone. Only one report cited exposure to hexazinone alone, without other compounds involved. A 26-year-old woman inhaled hexazinone dust (concentration not specified). Vomiting occurred within 24 hours. No other effects were reported and no treatment was administered. The other two reports did not involve human exposure.

B. Animals

1. *Short-term Exposure*

- Reported oral LD_{50} values for technical-grade hexazinone in rats range from 1,690 to >7,500 mg/kg (Matarese, 1977; Dashiell and Hinckle, 1980; Kennedy, 1984).

- Henry (1975) and Kennedy (1984) reported the oral LD_{50} value of technical-grade hexazinone in beagle dogs to be $> 3,400$ mg/kg.

- Reported oral LD_{50} values for hexazinone in guinea pigs range from 800 to 860 mg/kg (Dale, 1973; Kennedy, 1984).

- Kennedy (1984) studied the response of male rats to repeated oral doses of hexazinone (89 or 98% active ingredient). Groups of six rats were intubated with hexazinone, 0 or 300 mg/kg, as a 5% suspension in corn oil. Animals were dosed 5 days/week for 2 weeks (10 total doses). Clinical signs and body weights were monitored daily. At 4 hours to 14 days after exposure to the last dose, microscopic evaluation of lung, trachea, liver, kidney, heart, testes, thymus, spleen, thyroid, GI tract, brain and bone marrow was conducted. No gross or histological changes were noted in animals exposed to either active ingredient percentage of hexazinone.

- In an 8-week range-finding study (Kennedy and Kaplan, 1984), Charles River CD-1 mice (10/sex/dose) received hexazinone ($> 98%$ pure) in the diet for 8 consecutive weeks at concentrations of 0, 250, 500, 1,250, 2,500 or 10,000 ppm. Assuming 1 ppm in the diet of mice equals 0.15 mg/kg (Lehman, 1959), these dietary concentrations correspond to doses of about 0, 37.5, 75.0, 187.5, 375.0 or 1,500 mg/kg/day. No differences were observed in general behavior and appearance, mortality, body weights, food consumption or calculated food efficiency between control and exposed groups. No gross pathologic lesions were detected at necropsy. The only dose-related effects observed were increases in both absolute and relative liver weights in mice fed 10,000 ppm. A No-Observed-Adverse-Effect Level (NOAEL) of 2,500 ppm (375.0 mg/kg/day) was identified by the authors.

2. *Dermal/Ocular Effects*

- In an acute dermal toxicity test performed by McAlack (1976), up to 7,500 mg/kg of a 24% aqueous solution of hexazinone (reported to be 1,875 mg/kg of active ingredient) was applied occlusively for 24 hours to the shaved backs and trunks of male albino rabbits. No deaths were observed throughout a 14-day observation period. No other symptoms were reported.

- Morrow (1973) reported an acute dermal toxicity test in which 60 mL of a 24% aqueous solution of hexazinone (reported as 5,278 mg/kg) was applied occlusively to the shaved trunks of male albino rabbits for 24 hours. No mortalities were observed through an unspecified observation period. One animal exhibited a mild, transient skin irritation.

- In a 10-day study conducted by Kennedy (1984), semiocclusive dermal application of hexazinone for 6 hours/day for 10 days to male rabbits at 70 or 680 mg/kg/day resulted in no signs of skin irritation or toxicity. A trend toward elevated serum alkaline phosphatase (SAP) and serum glutamic pyruvic-transaminase (SGPT) activities was observed, but no hepatic damage was seen by microscopic evaluation. In a second 10-day study using 35, 150 or 770 mg/kg/day, the highest dose again resulted in elevated SAP and SGPT activities, but they returned to normal after 53 days of recovery. Histopathological evaluations were not performed in the second study.

- Edwards (1977) applied 6,000 mg/kg hexazinone as a 63% solution occlusively to the shaved backs and trunks of male albino rabbits. All rabbits showed moderate skin irritation which cleared 7 days after cessation of treatment. No treatment-related mortalities were reported after a 14-day observation period.

- Morrow (1972) reported the results of dermal irritation tests in which a single dose of 25 or 50% hexazinone was applied to the shaved, intact shoulder skin of each of 10 male guinea pigs. To test for sensitization, four sacral intradermal injections of 0.1 mL of a 15% solution were first given over a 3-week period. After a 2-week rest period, the guinea pigs were challenged with 25 or 50% hexazinone applied to the shaved, intact shoulder skin. The test material was found to be nonirritating and non-sensitizing at 48 hours postapplication.

- Using a 10% solution, Goodman (1976) repeated the Morrow (1972) study with guinea pigs and observed no irritation or sensitization.

- Dashiell and Henry (1980) reported that in albino rabbits, a single dose of hexazinone applied as 27% (vehicle not specified) solution to one eye per animal and unwashed was a severe ocular irritant. When applied at 27% (vehicle not specified) and washed or at 4% (aqueous solution) unwashed, mild to moderate corneal cloudiness, iritis and/or conjunctivitis resulted. By 21 days posttreatment with the higher dose, two of the three rabbit eyes had returned to normal; a small area of mild corneal cloudiness persisted through the 35-day observation period in one of the three eyes. Eyes treated with lower doses were normal within 3 days.

3. *Long-term Exposure*

- In a 90-day feeding study, Sherman et al. (1973) fed beagle dogs (4/sex/dose) hexazinone (97.5% active ingredient) in the diet at levels of 0, 200, 1,000 or 5,000 ppm. Assuming 1 ppm in the diet of a dog equals 0.025 mg/kg/day (Lehman, 1959), these levels correspond to about 0, 5, 25 or

125 mg/kg/day. At the highest dose level tested, decreased food consumption, weight loss, elevated alkaline phosphatase activity, lowered albumin/globulin ratios and slightly elevated liver weights were noted. No gross or microscopic lesions were observed at necropsy. Based on the results of this study, a NOAEL of 1,000 ppm (25 mg/kg/day) and a Lowest-Observed-Adverse-Effect Level (LOAEL) of 5,000 ppm (125 mg/kg/day) were identified.

- In a 90-day feeding study (Kennedy and Kaplan, 1984), Crl-CD rats (10/sex/dose) received hexazinone (>98% pure) at dietary levels of 0, 200, 1,000 or 5,000 ppm. Assuming 1 ppm in the diet of rats equals 0.05 mg/kg/day (Lehman, 1959), these levels correspond to about 0, 10, 50 or 250 mg/kg/day. Hematological and biochemical tests and urinalyses were conducted on subgroups of animals after 1, 2 or 3 months of feeding. Following 94 to 96 days of feeding, the rats were sacrificed and necropsied. The only statistically significant effect reported was a decrease in body weight in both males and females receiving 5,000 ppm. No differences in food consumption were reported. Results of histopathological examinations from the control and high-dose groups were unremarkable. The authors identified a NOAEL of 1,000 ppm (50 mg/kg/day).

- In a 1-year feeding study (Kaplan et al., 1975), weanling Charles River CD rats (36/sex/dose) received hexazinone (94 to 96% pure) at dietary levels of 0, 200, 1,000 or 2,500 ppm (which, according to the authors, corresponded to 0, 11, 60 or 160 mg/kg/day for males and 0, 14, 74 or 191 mg/kg/day for females). Results of this study indicated a decrease in weight gain by both sexes at 2,500 ppm and by females at 1,000 ppm. The authors indicated that various unspecified clinical, hematological and biochemical parameters revealed no evidence of adverse effects. No significant gross or histopathological changes attributable to hexazinone were noted. From the information presented in the study, a NOAEL of 200 ppm (11 mg/kg/day for males and 14 mg/kg/day for females) can be identified.

- In a 2-year study, Goldenthal and Trumball (1981) fed hexazinone (95 to 98% pure) to Charles River CD-1 mice (80/sex/dose) at dietary levels of 0, 200, 2,500 or 10,000 ppm. Assuming that 1 ppm in the diet of a mouse equals 0.15 mg/kg/day (Lehman, 1959), these levels correspond to 0, 30, 375 or 1,500 mg/kg/day. Corneal opacity, sloughing and discoloration of the distal tip of the tail were noted as early as the fourth week of the study in mice receiving 2,500 or 10,000 ppm. A statistically significant decrease in body weight was observed in male mice receiving 10,000 ppm and in female mice receiving 2,500 or 10,000 ppm. Statistically significant increases in liver weight were noted in male mice receiving 10,000 ppm; male and female mice in the 10,000-ppm dose group also displayed statistically significant increases in relative liver weight. Sporadic occurrence of statistically significant changes in hematological effects were considered by the authors to be unrelated to hexazinone treatment. Histologically, a number of liver changes were observed among mice fed 2,500 or 10,000 ppm. The most characteristic finding was hypertrophy of centrilobular parenchymal cells. Other histological changes included an increased incidence of hyperplastic liver nodules and an increased incidence and severity of liver cell necrosis. Mice fed 200 ppm showed no compound-related histopathological changes. A NOAEL of 200 ppm (30 mg/kg/day) was identified by the authors.

- Kennedy and Kaplan (1984) presented the results of a 2-year feeding study in which Crl-CD rats (36/sex/dose) received hexazinone (94 to 96% pure) at dietary levels of 0 (two groups), 200, 1,000 or 2,500 ppm (approximately 0, 10, 50 or 125 mg/kg/day, assuming that 1 ppm in the diet of a rat equals 0.05 mg/kg/day; Lehman, 1959). After 2 years of continuous feeding, all rats in all groups were sacrificed and examined. Males fed 2,500 ppm and females fed either 1,000 or 2,500 ppm had significantly lower body weights than controls (p < 0.05). Male rats fed 2,500 ppm had slightly elevated leukocyte counts with a greater proportion of eosinophils. Male rats fed either 1,000 or 2,500 ppm displayed decreased alkaline phosphatase activity. Statistically significant effects on organ weights included elevated relative lung weights in males fed 1,000 ppm; lower kidney and lower relative liver and heart weights in males fed 2,500 ppm; increased liver and spleen weights in females fed 200 ppm; and elevated stomach and relative brain weights in females fed 2,500 ppm. At necropsy, gross pathologic findings were similar among all groups. Changes attributed to hexazinone were not apparent in any of the tissues evaluated microscopically. The authors identified 200 ppm (10 mg/kg/day) as the NOAEL. However, the increased liver and spleen weights observed in females would indicate that 200 ppm might be more appropriately identified as a LOAEL.

4. *Reproductive Effects*

- In a one-generation reproduction study (Kennedy and Kaplan, 1984), Crl-CD rats (10/sex/dose) received hexazinone (>98% pure) for approximately 90 days at dietary levels of 0, 200, 1,000 or 5,000 ppm. Assuming that 1 ppm in the diet of rats equals 0.05 mg/kg/day (Lehman, 1959), this corresponds to approximately 0, 10, 50 and 250 mg/kg/day. Following the 90-day feeding period, six rats/sex/dose were selected to serve as the parental generation. The authors concluded that the rats had normal fertility. The young were delivered in normal numbers, and survival during the lactation period was unaffected. In the 5,000 ppm group, weights of pups at weaning (21 days) were significantly ($p < 0.01$) lower than controls or other test groups. The results of this study identify a NOAEL of 1,000 ppm (50 mg/kg/day) (no decrease in weanling weight).

- In a three-generation reproduction study (DuPont, 1979), Crl-CD rats (36/sex/dose) received hexazinone (98% pure) at dietary levels of 0, 200, 1,000 or 2,500 ppm for 90 days (approximately 0, 10, 50 or 125 mg/kg/day, assuming the above dietary assumptions for a rat). Following 90 days of feeding, 20 rats/sex/dose were selected to serve as the parental (F_0) generation. Reproductive parameters tested included the number of matings, number of pregnancies and number of pups per litter. Pups were weighed at weaning, and one male and female were selected from each litter to serve as parental rats for the second generation. Similar procedures were used to produce a third generation; the same reproductive parameters were collected for the second and third generations. The authors stated that there were no significant differences between the control and treated groups with respect to the various calculated indices (fertility, gestation, viability and lactation). However, body weights at weaning of pups in the 2,500 ppm dose group were significantly ($p < 0.05$) lower than those of controls for the F2 and F3 litters. The results of this study identify a NOAEL of 1,000 ppm (50 mg/kg/day).

5. Developmental Effects

- Kennedy and Kaplan (1984) presented the results of a study in which Charles River Crl-CD rats (25 to 27/dose) received hexazinone (97.5% pure) at dietary concentrations of 0, 200, 1,000 or 5,000 ppm (approximately 0, 10, 50 or 250 mg/kg/day, following the previously stated dietary assumptions for the rat) on days 6 through 15 of gestation. Rats were observed daily for clinical signs and were weighed on gestation days 6, 16 and 21. On day 21, all rats were sacrificed and ovaries and uterine horns were weighed and examined. The number and location of live fetuses, dead fetuses and resorption sites were noted. Fetuses from the 0- and 5,000-ppm dose groups were evaluated for developmental effects (gross, soft tissue or skeletal abnormalities). At sacrifice, no adverse effects were observed for the dams. No malformations were noted in the fetuses. However, pup weights in the high-dose group were significantly lower than in the controls. This study identified a NOAEL of 1,000 ppm (50 mg/kg/day).

- Kennedy and Kaplan (1984) presented the results of a study in which New Zealand White rabbits (14 to 17/dose) received hexazinone suspended in a 0.5% aqueous methyl cellulose vehicle by oral intubation on days 6 through 19 of gestation at levels of 0, 20, 50 or 125 mg/kg/day. Rabbits were observed daily and body weights were recorded throughout gestation. On day 29 of gestation, all rabbits were sacrificed, uteri were excised and weighed, and the number of live, dead and resorbed fetuses was recorded. Each fetus was examined externally and internally for gross, soft tissue and skeletal abnormalities. No clinical signs of maternal or fetal toxicity were observed. Pregnancy rates among all groups compared favorably. The numbers of corpora lutea and implantations per group were not significantly different. Resorptions and fetal viability, weight and length were also similar among all groups. Based on the information presented in this study, a minimum NOAEL of 125 mg/kg/day for maternal toxicity, fetal toxicity and teratogenicity can be identified.

6. *Mutagenicity*

- The ability of hexazinone to induce unscheduled DNA synthesis was assayed by Ford (1983) in freshly isolated hepatocytes from the livers of 8-week-old male Charles River/Sprague-Dawley rats. Hexazinone was tested at half-log concentrations from 1×10^{-5} to 10.0 mM and at 30.0 mM. No unscheduled DNA synthesis was observed.

- Valachos et al. (1982) conducted an *in vitro* assay for chromosomal aberrations in Chinese hamster ovary cells. Hexazinone was found to be clastogenic without S-9 activation at concentrations of 15.85 mM (4.0 mg/mL) or 19.82 mM (5.0 mg/mL); no significant increases in clastogenic activity were seen at 1.58, 3.94 and 7.93 mM (0.4, 1.0 and 2.0 mg/mL). With S-9 activation, significant increases in aberrations were noted only at a concentration of 15.85 mM (4.0 mg/mL). Concentrations above these yielded no analyzable metaphase cells due to cytotoxicity.

- In a study designed to evaluate the clastogenic potential of hexazinone in rat bone marrow cells (Farrow et al., 1982), Sprague-Dawley CD rats (12/sex/dose) were given a single dose of 0, 100, 300 or 1,000 mg/kg of hexazinone by gavage (vehicle not reported). No statistically significant increases in the frequency of chromosomal aberrations were observed at any of the dose levels tested. The authors concluded that, under the conditions of this study, hexazinone was not clastogenic.

- Hexazinone was tested for mutagenicity in *Salmonella typhimurium* strains TA1535, TA1537, TA1538, TA98 and TA100 at concentrations up to 7,000 μg/plate. The compound was not found to be mutagenic, with or without S-9 activation (DuPont, 1979).

7. *Carcinogenicity*

- Goldenthal and Trumball (1981) fed hexazinone (98% pure) for 2 years to mice (80/sex/dose) in the diet at 0, 200, 2,500 or 10,000 ppm (0, 30, 375 or 1,500 mg/kg/day, based on Lehman [1959]). A number of liver changes were observed histologically at the 2,500- and 10,000-ppm level. These included hypertrophy of the centrilobular parenchymal cells, increased incidence of hyperplastic liver nodules and liver cell necrosis. The authors concluded that hexazinone was not carcinogenic to mice.

- No carcinogenic effects were observed in Crl-CD rats (36/sex/dose) given hexazinone (94 to 96% pure) in the diet at 0, 200, 1,000, or 2,500 ppm (0, 10, 50, or 125 mg/kg/day) for 2 years (Kennedy and Kaplan, 1984). The authors concluded that none of the tumors were attributable to hexazinone.

IV. Quantification of Toxicological Effects

A. One-day Health Advisory

No information was found in the available literature that was suitable for determination of the One-day HA for hexazinone. It is, therefore, recommended that the Longer-term HA value of 3 mg/L (calculated below) for a 10-kg child be used at this time as a conservative estimate of the One-day HA value.

B. Ten-day Health Advisory

The study reported by Kennedy and Kaplan (1984) in which pregnant rabbits (14 to 17/dose) received hexazinone by oral intubation at levels of 0, 20, 50 or 125 mg/kg/day on days 6 through 19 of gestation was considered to serve as the basis for deriving the Ten-day HA for a 10-kg child. Since no signs of maternal or fetal toxicity were observed in this study, a NOAEL of 125 mg/kg/day (the highest dose tested) was identified. The NOAEL from this study is greater than that identified in a 90-day rat feeding study (50 mg/kg; Kennedy and Kaplan, 1984). The LOAEL from the one-generation rat reproduction study was 250 mg/kg based on decreased weanling weight. Effects at doses between 50 and 250 mg/kg have not been reported for the rat. However, in a

90-day dog study, a LOAEL of 125 mg/kg was identified (Sherman et al., 1973). Therefore, the rabbit study was not selected to derive a Ten-day value. It is, therefore, recommended that the Longer-term HA value of 3 mg/L for the 10-kg child be used at this time as a conservative estimate of the Ten-day HA value.

C. Longer-term Health Advisory

The 90-day feeding study in dogs (Sherman et al., 1973) has been selected to serve as the basis for determination of the Longer-term HA for hexazinone. In this study, dogs (4/sex/dose) received hexazinone in the diet at levels of 0, 200, 1,000 or 5,000 ppm (0, 5, 25, or 125 mg/kg/day) for 90 days. Decreased food consumption and body weight gain, elevated alkaline phosphatase activity, lowered albumin/globulin ratios and slightly elevated liver weights were observed at the highest dose. A NOAEL of 1,000 ppm (25 mg/kg/day) and a LOAEL of 5,000 ppm (125 mg/kg/day) were identified. This NOAEL is generally supported by a 90-day rat feeding study that reported a NOAEL of 50 mg/kg/day (Kennedy and Kaplan, 1984). Effects in dogs exposed to hexazinone at 50 mg/kg/day have not been reported.

Using a NOAEL of 25 mg/kg/day, the Longer-term HA for a 10-kg child is calculated as follows:

$$\text{Longer-term HA} = \frac{(25 \text{ mg/kg/day}) (10 \text{ kg})}{(100) (1 \text{ L/day})} = 2.5 \text{ mg/L} (3{,}000 \text{ } \mu\text{g/L})$$

The Longer-term HA for a 70-kg adult is calculated as follows:

$$\text{Longer-term HA} = \frac{(25 \text{ mg/kg/day}) (70 \text{ kg})}{(100) (2 \text{ L/day})} = 8.75 \text{ mg/L} (9{,}000 \text{ } \mu\text{g/L})$$

D. Lifetime Health Advisory

A 2-year rat feeding/oncogenicity study (Kennedy and Kaplan, 1984) was selected as the basis for determination of the Lifetime HA for hexazinone. Crl-CD rats (36/sex/dose) received 0, 200, 1,000 or 2,500 ppm hexazinone (0, 10, 50 or 125 mg/kg/day) for 2 years. Body weight gain in males and females in the 2,500-ppm group, and females in the 1,000-ppm group, was significantly lower than that in controls. No clinical, hematological or urinary evidence of toxicity was reported. Based on decreased body weight gain, a NOAEL of 200 ppm (10 mg/kg/day) and a LOAEL of 1,000 ppm (50 mg/kg/day) were identified.

Using a NOAEL of 10 mg/kg/day, the Lifetime HA is calculated as follows:

Step 1: Determination of the Reference Dose (RfD)

$$RfD = \frac{(10 \text{ mg/kg/day})}{(100)\,(3)} = 0.03 \text{ mg/kg/day}$$

where: 3 = modifying factor; to account for data gaps (chronic dog feeding study) in the total database for hexazinone.

Step 2: Determination of the Drinking Water Equivalent Level (DWEL)

$$DWEL = \frac{(0.03 \text{ mg/kg/day})\,(70 \text{ kg})}{(2 \text{ L/day})} = 1.05 \text{ mg/L} \ (1,000 \ \mu g/L)$$

Step 3: Determination of Lifetime Health Advisory

Lifetime HA = $(1.05 \text{ mg/L})\,(20\%) = 0.21 \text{ mg/L} \ (200 \ \mu g/L)$

E. Evaluation of Carcinogenic Potential

- No evidence of carcinogenicity in rats or mice has been observed.

- The International Agency for Research on Cancer has not evaluated the carcinogenic potential of hexazinone.

- The criteria described in EPA's guidelines for assessment of carcinogenic risk (U.S. EPA, 1986) place hexazinone in Group D: not classifiable. This category is for substances with inadequate or no animal evidence of carcinogenicity.

V.   Other Criteria, Guidance and Standards

- Residue tolerances range from 0.5 to 5.0 ppm for the combined residues of hexazinone and its metabolites in or on the raw agricultural commodities pineapple, pineapple fodder and forage (U.S. EPA, 1985a).

VI.   Analytical Methods

- Analysis of hexazinone is by a gas chromatographic method applicable to the determination of certain organonitrogen pesticides in water samples (U.S. EPA, 1985b). This method requires a solvent extraction of approximately 1 liter of sample with methylene chloride using a separatory funnel. The methylene chloride extract is dried and exchanged to acetone during concentration to a volume of 10 mL or less. The compounds in the extract are separated by gas chromatography, and measurement is made with a thermionic bead detector. The method detection limit for hexazinone is 0.72 μg/L.

VII.   Treatment Technologies

• No information was found in the available literature on treatment technologies used to remove hexazinone from contaminated water.

VIII.   References

CHEMLAB. 1985. The Chemical Information System, CIS, Inc. Baltimore, MD.

Dale, N.* 1973. Oral LD$_{50}$ test (guinea pigs). Haskell Laboratory Report No. 400–73. Unpublished study. MRID 00104973.

Dashiell, O.L. and J.E. Henry.* 1980. Eye irritation tests in rabbits — United Kingdom Procedure. Haskell Laboratory Report No. 839–80. Unpublished study. MRID 00076958.

Dashiell, O.L. and L. Hinckle.* 1980. Oral LD$_{50}$ test in rats — EPA proposed guidelines. Haskell Laboratory Report No. 943–80. Unpublished study. MRID 00062980.

DuPont.* 1979. E.I. duPont de Nemours & Co. Reproduction study in rats with sym-triazine-2,4(1H, 3H)-dione, 3-cyclohexyl-1-methyl-6-dimethyl-amino (INA 3674, hexazinone). Supplement to Haskell Laboratory Report. No. 352–77. Accession No. 97323.

Edwards, D.F.* 1977. Acute skin absorption test on rabbits LD$_{50}$. Haskell Laboratory Report No. 841–77. Unpublished study. MRID 00091140.

Farrow, M., T. Cartina, M. Zito et al.* 1982. *In vivo* bone marrow cytogenetic assay in rats. HLA Project No. 201–573. Final Report. Unpublished study. Received July 11, 1983 under 352–378. Submitted by E.I. duPont de Nemours & Co., Inc., Wilmington, DE. MRID 0013155.

Ford, L.* 1983. Unscheduled DNA synthesis/rat hepatocytes in vitro. (INA-3674-112). Haskell Laboratory Report No. 766–82. Unpublished study. MRID 00130708.

Goldenthal, E.I. and R.R. Trumball.* 1981. Two-year feeding study in mice. IRDC No. 125–026. Unpublished study. Submitted to the Office of Pesticide Programs. E.I. duPont de Nemours & Co., Inc. MRID No. 0079203.

Goodman, N.* 1976. Primary skin irritation and sensitization tests on guinea pigs. Report No. 434–76. Unpublished study. Submitted to the Office of Pesticide Programs. MRID 00104433.

Henry, J.E.* 1975. Acute oral test (dogs). Haskell Laboratory Report No. 617–75. Unpublished study. MRID 00076957.

Holt, R.F., F.J. Baude and D.W. Moore.* 1979. Hexazinone livestock feeding studies: milk and meat. Unpublished study. Submitted to the Office of Pesticide Programs. MRID 00028657.

Holt, R.F. 1979. Residues resulting from application of DPX-3674 to soil. Wilmington, DE: E. I. duPont de Nemours & Co., Inc.

Kaplan, A.M., Z.A. Zapp, Jr., C.F. Reinhardt et al.* 1975. Long-term feeding study in rats with sym-triazine-2,4(1H,3H)dione, 3-cyclohexyl-1-methyl-6-dimethylamino (INA-3674). One-year Interim Report. Haskell Laboratory Report No. 585–75. MRID 00078045.

Kennedy, G.L. 1984. Acute and environmental toxicity studies with hexazinone. Fund. Appl. Toxicol. 4:603–611.

Kennedy, G.L. and A.M. Kaplan. 1984. Chronic toxicity, reproductive, and teratogenic studies of hexazinone. Fund. Appl. Toxicol. 4:960–971.

Lehman, A.J. 1959. Appraisal of the safety of chemicals in foods, drugs, and cosmetics. Assoc. Food Drug Off. U.S., Q. Bull.

Matarese, C.* 1977. Oral LD_{50} test. Haskell Laboratory Report No. 1037–77. Unpublished study. MRID 0011477.

McAlack, J.W.* 1976. Skin absorption LD50. Haskell Laboratory Report No.353–76. Unpublished study. MRID 00063971.

Meister, R., ed. 1983. Farm chemicals handbook. Willoughby, OH: Meister Publishing Co.

Morrow, R.* 1972. Primary skin irritation and sensitization tests on guinea pigs. Haskell Laboratory Report No. 489–72. Unpublished study. MRID 00104978.

Morrow, R.* 1973. Skin absorption toxicity ALD and skin irritancy test. Haskell Laboratory Report No. 503–73. Unpublished study. MRID 00104974.

Rapisarda, C.* 1982. Metabolism of ^{14}C-labeled hexazinone in the rat. Document No. AMR-79–82. Unpublished study. Accession No. 247874. E.I.duPont de Nemours & Co., Inc.

Rhodes, Robert C. 1975a. Studies with "Velpar" weed killer in water. Wilmington, DE: Biochemicals Department Experimental Station, E. I. duPont de Nemours & Co., Inc.

Rhodes, Robert C. 1975b. Decomposition of "Velpar" weed killer in soil. Wilmington, DE: Biochemicals Department Experimental Station, E. I. duPont de Nemours & Co., Inc.

Rhodes, Robert C. 1975c. Mobility and adsorption studies with "Velpar" weed killer on soils. Wilmington, DE: Biochemicals Department Experimental Station, E. I. duPont de Nemours & Co., Inc.

Rhodes, R., R.A. Jewell and H. Sherman.* 1978. Metabolism of "Velpar®" weed-killer in the rat. Unpublished study. Wilmington, DE: E. I. duPont de Nemours & Co., Inc. MRID 00028864.

Sherman, H., N. Dale and L. Adams et al.* 1973. Three month feeding study in dogs with sym-triazine-2,4(1H,3H)-dione, 3-cyclohexyl-l-methyl-6-dimethylamino-(INA-3674). Haskell Laboratory Report No. 408–73. MRID 00104976; MRID 00114484.

STORET. 1988. STORET Water Quality File. Office of Water. U.S. Environmental Protection Agency (data file search conducted in May, 1988).

U.S. EPA. 1981. U.S. Environmental Protection Agency. Pesticide incident monitoring system. Washington, DC: Office of Pesticide Programs. February.

U.S. EPA. 1982. U.S. Environmental Protection Agency. Registration standard for hexazinone. Toxicology chapter. Washington, DC: Office of Pesticide Programs.

U.S. EPA. 1985a. U.S. Environmental Protection Agency. Code of Federal Regulations. 40 CFR 180.396.

U.S. EPA. 1985b. U.S. Environmental Protection Agency. U.S. EPA Method 633 – Organonitrogen pesticides. Fed. Reg. 50:40701. October 4.

U.S. EPA. 1986. U.S. Environmental Protection Agency. Guidelines for carcinogen risk assessment. Fed. Reg. 51(185):33992–34003. September 24.

Valachos, D., J. Martenis and A. Horst.* 1982. *In vitro* assay for chromosome aberrations in Chinese hamster ovary (CHO) cells. Haskell Laboratory Report No. 768–82. Unpublished study. MRID 00130709.

---

* Confidential Business Information submitted to the Office of Pesticide Programs.

Maleic Hydrazide

I. General Information and Properties

A. CAS No. 123–33–1

B. Structural Formula

1,2-Dihydro-3,6-pyridazinedione

C. Synonyms

- Antergon; Chemform; De-Sprout; MH; Retard; Slo-Gro; Sucker-Stuff; (Meister, 1983).

D. Uses

- Maleic hydrazide is used as a plant growth retardant (Meister, 1983).

E. Properties (Meister, 1983; CHEMLAB, 1985; TDB, 1985)

Chemical Formula	$C_4H_4O_2N_2$
Molecular Weight	112.09
Physical State (25°C)	Crystalline solid
Boiling Point	--
Melting Point	292°C
Density	1.60
Vapor Pressure (50°C)	0 mm Hg
Specific Gravity	--
Water Solubility (25°C)	6,000 mg/L
Log Octanol/Water Partition Coefficient	–3.67 (calculated)
Taste Threshold	--
Odor Threshold	--
Conversion Factor	--

F. Occurrence

- No information was found in the available literature on the occurrence of maleic hydrazide.

G. Environmental Fate

- Salts of maleic hydrazide will dissociate in solutions above pH 4.5 and exist only as maleic hydrazide. Maleic hydrazide is stable to hydrolysis at pHs of 3, 6 and 9. Photolysis potential has not been addressed (Registration Standard Science Chapter for Maleic Hydrazide; WSSA, 1983).

- In field dissipation studies using various soils from the eastern, southern and midwestern U.S., the half-lives for maleic hydrazide were reported to be between 14 and 100 days. There is no pattern, but the half-life may be related to organic matter content. Degradation by soil microorganisms appears to be rapid (Registration Standard Science Chapter for Maleic Hydrazide; WSSA, 1983).

- There is some indication that maleic hydrazide is highly mobile in unaged soils. Aerobic aging of maleic hydrazide results in a lowering of leaching potential (Registration Standard Science Chapter for Maleic Hydrazide; WSSA, 1983).

II.   Pharmacokinetics

A. Absorption

- Mays et al. (1968) administered single oral doses of ^{14}C-labeled maleic hydrazide to rats. After 6 days, only 12% had been excreted in the feces, suggesting that 88% had been absorbed.

B. Distribution

- Kennedy and Keplinger (1971) administered ^{14}C-labeled maleic hydrazide to pregnant rats in daily doses of either 0.5 or 5.0 mg/kg. Fetuses from dams sacrificed on day 20 were found to contain label equivalent to 20 to 35 ppb of the parent compound at the 0.5-mg/kg dose level, and 156 to 308 ppb at the 5.0-mg/kg dose level. Pups from females that were allowed to litter were sacrificed at 8 and at 24 hours, and stomach coagulum was analyzed to determine transfer through the milk. At the 0.5-mg/kg dose, the coagulum contained 4 to 7 ppb at 8 hours and 2 ppb at 24 hours; at the 5.0-mg/kg dose, the figures for 8 and 24 hours were 79 to 89 ppb and 7 to 8 ppb, respectively. These results suggest that maleic hydrazide crossed the placenta and was also transmitted to the pups via the milk.

C. Metabolism

- Barnes et al. (1957) reported that rabbits administered a single oral dose of 100 mg/kg of maleic hydrazide excreted 43 to 62% of the dose, unchanged, within 48 hours. The route of excretion (urinary or fecal) was not stated. The results were similar following a dose of 2,000 mg/kg, and no glucuronide or ethereal sulfate conjugates were found.

- Oral administration of maleic hydrazide labeled with ^{14}C to rats resulted in excretion of 0.2% labeled carbon dioxide in the expired air over a 6-day observation period (Mays et al., 1968). Urinary products (77% of the total dose) were largely unchanged maleic hydrazide (92 to 94% of the urinary total) and conjugates of maleic hydrazide (6 to 8%).

D. Excretion

- Mays et al. (1968) administered single oral doses of ^{14}C-maleic hydrazide to rats. Over a 6-day observation period, the animals excreted 0.2% of the label as carbon dioxide in the expired air, 12% in the feces and 77% in the urine. Only trace amounts were detected in tissues and blood after 3 days.

III.   Health Effects

A. Humans

- No information on human exposure to maleic hydrazide was found in the available literature.

B. Animals

1. *Short-term Exposure*

- The acute oral toxicity of maleic hydrazide (purity not specified) in rats was determined with administration of four dose levels to groups of five animals, with a 15-day observation period (Reagan and Becci, 1982). At dose levels of 5,000, 6,300, 7,940 or 10,000 mg/kg, deaths occurring in the male animals were 0/5, 0/5, 1/5 and 5/5, respectively, while those for female animals were 1/5, 1/5, 4/5 and 5/5, respectively. The LD_{50} values were calculated to be 6,300 mg/kg for males, 6,680 mg/kg for females and 7,500 mg/kg for both sexes combined. Adverse effects noted included ataxia, diarrhea, salivation, decreased motor activity and blood in the intestines and stomach.

- Sprague-Dawley rats (five males and five females) were fasted for 16 hours and then given a single oral dose of technical maleic hydrazide (purity not specified) at a level of 5,000 mg/kg and observed for 14 days (Shapiro, 1977a). No deaths occurred during this period. Necropsies were not performed, and no details were given with respect to adverse effects that may have been observed.

- The acute oral toxicity of the diethanolamine salt of maleic hydrazide (MH-DEA) (purity not specified) was determined in rats and rabbits (Uniroyal Chemical, 1971). In both species, MH-DEA was lethal at a level of 1,000 mg/kg, while doses between 300 and 500 mg/kg showed no toxicity in either species. The LD_{50} value for both species was calculated to be 700 mg/kg.

- Rats were used for a comparison of the acute oral toxicity of the sodium and diethanolamine salts (purities not specified) of maleic hydrazide (Tate, 1951). The diethanolamine salt showed an LD_{50} value of 2,350 mg/kg, while the LD_{50} for the sodium salt (H-Na) was 6,950 mg/kg. No details of the study were given.

- The acute oral LD_{50} value of technical-grade maleic hydrazide (purity not specified) for rabbits was greater than 4,000 mg/kg (Lehman, 1951). No details of the study were available.

2. Dermal/Ocular Effects

- Technical-grade maleic hydrazide was tested on male and female New Zealand White rabbits for both skin and eye irritation (Shapiro, 1977b,c). Applied at 0.5 mL, the maleic hydrazide was scored as a mild primary skin irritant. In the eye test, 100 mg of the material was used, and maleic hydrazide was judged not to be an eye irritant.

- The acute dermal toxicity of maleic hydrazide (purity and form not specified) was determined in five male and five female New Zealand White rabbits (Shapiro, 1977d). The skin of two males and three females was abraded. A single dose of 20,000 mg/kg was applied, and the animals were observed for 14 days. On the first day, two males (one with abraded skin) and one female died. The animals that died exhibited ataxia, shallow respiration and were comatose.

- In an evaluation of the acute dermal toxicity of Royal MH-30 (30^{97-} MH-DEA) and maleic hydrazide-technical, both formulations wer stated to be mild primary skin irritants and slight eye irritants (Uniroyal Chemical, 1977). Individual details of the study were not given.

3. *Long-term Exposure*

- Rats were fed MH-Na or MH-DEA (purity not specified) in the diet for 11 weeks (Tate, 1951). The MH-Na was given at dose levels of 0.5% or 5.0% (5,000 or 50,000 ppm). Assuming that 1 ppm in the diet of rats is equivalent to 0.05 mg/kg/day (Lehman, 1959), these doses correspond to 250 or 2,500 mg/kg/day. No significant mortality or other adverse effects were noted (no details given). The No-Observed-Adverse-Effect Level (NOAEL) for MH-Na in this study is 2,500 mg/kg (the highest dose tested). The MH-DEA was fed at a level of 0.1% (1,000 ppm) for 11 weeks. This is equivalent to a dose of 50 mg/kg/day (Lehman, 1959). At the end of 11 weeks, 21/24 animals had died. The author stated that after further investigation (details not given), it was concluded that the observed mortality was due to the DEA component of the formulation.

- The toxicity of maleic hydrazide in the diet for 1 year (320 to 360 days) was investigated in rats and dogs (Mukhorina, 1962). Rats received oral doses of maleic hydrazide at 0.7, 1.5 or 3 mg/kg/day, and a fourth group received 7 mg/kg MH-DEA. Dogs were administered an oral dose of 0.7 mg/kg/day maleic hydrazide. Other details in this translation on study design and conduct were not clear. Rats exposed at the high dose had hyperemia and hemorrhage of the lungs, myocardium, liver and brain; abnormal glucose-tolerance curves; lowered liver glycogen; dystrophic changes in the liver; nephritis; interstitial pneumonia; loss of hair; and significant reduction in weight gain compared with the controls (at 4 months, controls had gained 30%; those fed maleic hydrazide at 3 mg/kg/day had gained only 21%). Dogs fed 0.7 mg/kg/day maleic hydrazide showed no significant adverse changes. It appears that for both the rat and the dog the level of 0.7 mg/kg/day MH-DEA was a NOAEL.

- Mukhorina (1962) also reported on a study done in mongrel mice given 0.7 mg/kg/day maleic hydrazide (purity not specified) in the diet for 320 to 360 days. No pathological changes were found. Based on these data, the NOAEL for MH-DEA in the mouse is 0.7 mg/kg/day.

- In a study by Food Research Labs (1954), MH-Na was fed in the diet to rats (number not specified) from weaning for 2 years. Levels of MH-Na (expressed as the free acid) were 0.0, 0.5, 1.0, 2.0 or 5.0% (0, 5,000,

10,000, 20,000 or 50,000 ppm). Assuming that 1 ppm in the diet of rats corresponds to 0.05 mg/kg/day (Lehman, 1959), this is equivalent to doses of 0, 250, 500, 1,000 or 2,500 mg/kg/day. There were no changes in blood or urine and no dose- or time-dependent effects on longevity. Other study details were not presented. Based on these observations, the NOAEL identified from this study is 2,500 mg/kg/day (highest dose tested) for the rat.

- In a similar study in dogs (Food Research Labs, 1954), animals were fed doses of 0.0, 0.6, 1.2 or 2.4% maleic hydrazide (as MH-Na) in the diet for 1 year. Assuming 1% (10,000 ppm) in the diet of dogs corresponds to 250 mg/kg/day (Lehman, 1959), this is equivalent to a dose of 500 mg/kg/day. No effects attributable to exposure were detected.

- Van Der Heijden et al. (1981) fed technical maleic hydrazide, 99% active ingredient (a.i.) and containing less than 1.5 mg hydrazine/kg as an impurity, to rats at dietary levels of 1.0 or 2.0% (10,000 or 20,000 ppm) for 28 months. Assuming that 1 ppm in the diet of rats is equivalent to 0.05 mg/kg/day (Lehman, 1959), this corresponds to doses of 500 or 1,000 mg/kg/day. These two levels of maleic hydrazide in the diet caused proteinuria and increased the protein/creatinine ratio in the urine of both sexes, although there were no detectable histopathological changes in the kidney or the urinary tract. Based on the effects on kidney function, the no-effect level was considered by the authors to be lower than 1.0% maleic hydrazide in the diet of rats. On this basis, a Lowest-Observed-Adverse-Effect Level (LOAEL) of 500 mg/kg is identified.

4. *Reproductive Effects*

- In a two-generation reproduction study by Kehoe and MacKenzie (1983), Charles River CD(SD)BR rats (15 males and 30 females/dose) were administered the potassium salt of maleic hydrazide (K-MH) (purity not specified) at dietary concentrations of 0, 1,000, 10,000 or 30,000 ppm. Assuming that 1 ppm in the diet of rats is equivalent to 0.05 mg/kg/day (Lehman, 1959), these doses correspond to 0, 50, 500 and 1,500 mg/kg/day. No adverse effects on reproductive indices were observed at any dietary level. However, increased mortality was observed in F_1 parents that received 30,000 ppm. Also at this dose level, body weights were reduced in F_0 parents during growth and reproduction and in F_1 and F_2 pups during lactation. Based on the postnatal 50 mg decrease in the body weight of pups at 30,000 ppm (1,500 mg/kg/day), a reproductive NOAEL of 10,000 ppm (500 mg/kg/day) is identified.

- In a four-generation reproduction study in rats (Food Research Labs, 1954), animals were fed MH-Na (purity not specified) in the diet at dose levels of 0.5, 1.0, 2.0 or 5.0% (5,000, 10,000, 20,000 or 50,000 ppm) (expressed in terms of free acid). Assuming 1 ppm in the diet of rats corresponds to 0.05 mg/kg/day (Lehman, 1959), this is equivalent to 250, 500, 1,000 or 2,500 mg/kg/day. The authors reported that there were no effects on fertility, lactation or other reproductive parameters, but no data from the study were presented for an adequate assessment of these findings.

5. Developmental Effects

- Khera et al. (1979) administered maleic hydrazide (97% purity) to pregnant rats by gavage on days 6 to 15 of gestation at doses of 0, 400, 800, 1,200 or 1,600 mg/kg/day. Animals were sacrificed on day 22. No signs of toxicity or adverse effects on maternal weight gain were observed at any dose level tested. Values for corpora lutea, total implants, resorptions, dead fetuses, male/female ratio and fetal weight were within the control range. The number of live fetuses was decreased at the 1,200-mg/kg dose, but this was not statistically significant and did not occur at the highest dose tested. Fetuses examined for external, soft-tissue and skeletal abnormalities showed no increase in frequency of abnormalities at any dose level tested. Based on the results of this study, a NOAEL of 1,600 mg/kg/day (the highest dose tested) is identified for maternal effects, fetotoxicity and teratogenic effects.

- Hansen et al. (1984) studied the teratogenic effects of MH-Na and the monoethanolamine salt (MH-MEA) on fetuses from female rats exposed by gavage to doses of 500, 1,500 or 3,000 mg/kg/day in the diet at various stages of gestation. Replicate tests were run. No increased frequency of gross, skeletal or visceral abnormalities was observed in animals dosed by gavage on days 6 to 15 of gestation with 500 mg/kg/day of either MH-Na or MH-MEA. An increased frequency of minor skeletal variants (asymmetrical and bipartite sternebrae, wavy ribs, fused ribs, rudiment of cervical rib, single bipartite or other variations in thoracic vertebrae) was observed in animals receiving 1,500 ($p < 0.01$) or 3,000 ($p < 0.1$) mg/kg/day of MH-MEA on days 6 to 15, but this was observed neither in animals exposed to 3,000 mg/kg/day for days 1 to 21 of gestation nor in a replicate experiment. Similarly, MH-Na produced marginal increases in minor skeletal variants in one experiment at doses of 1,500 mg/kg/day for days 6 to 15 ($p < 0.1$) or 3,000 mg/kg/day for days 1 to 21 ($p < 0.1$), but this was not observed in a replicate experiment. Rats dosed with 3,000 mg/kg/day MH-MEA in the diet exhibited a significant decrease in maternal body weight and in weight gain compared to the controls. This effect was not observed when 3,000 mg/kg was given on days 1 to 21 by gavage, and there was no significant effect on food intake. Exposure to 3,000 mg/kg in the diet caused a significant increase in resorptions ($p < 0.001$) and a decrease in mean fetal weight ($p < 0.001$). Similar but less pronounced effects were observed when this dose was given by gavage. In addition, postimplantation loss was increased significantly ($p < 0.01$) in both experiments. The authors theorized that the more severe effects observed when the MH-MEA was fed in the diet (versus gavage) could be due to an alteration in the palatability of the diet, resulting in decreased food consumption. In contrast to the results with MH-MEA, MH-Na had no adverse effects on the dams except for a reduction in food consumption for days 1 to 6 in the group exposed from days 1 to 21 at 3,000 mg/kg. There were significant differences in body weight of the pups (up to age 35 days) of dams administered MH-MEA by gavage at 3,000 mg/kg/day from day 6 of gestation through day 21 of lactation; a significant delay in the pups' startle response to an auditory stimulus, significantly higher brain weight in both male and female pups and a delay in unfolding of the pinna were noted also. The authors attributed the increase in relative brain weight to the lower body weight. The delay in the startle response in MH-MEA–dosed offspring was considered the most significant effect, since it was observed in both sexes, but the authors noted that it cannot be explained. Based on these data, maternal, fetotoxic and teratogenic NOAELs of 1,500, 1,500 and 500 mg/kg/day, respectively, were identified for both MH-MEA and MH-Na.

- Aldridge (1983, cited in U.S. EPA, 1985a) administered K-MH by gavage at doses of 0, 100, 300 or 1,000 mg/kg/day to Dutch Belted rabbits (16/dose) on days 7 through 27 of gestation. No signs of maternal toxicity were reported, and the NOAEL for this effect is identified as 1,000 mg/kg/day (the highest dose tested). Malformed scapulae were observed in fetuses from the 300-and 1,000-mg/kg/day dose groups. An evaluation of this study by the Office of Pesticide Programs (U.S. EPA, 1985a) concluded that scapular malformations are rare and considered to be a major skeletal defect. Historical data for Dutch Belted rabbits from the testing laboratory (IRDC) indicated that scapular anomalies were observed in only 1 of 1,586 fetuses examined from 264 litters. Based on this information, a NOAEL of 100 mg/kg/day is identified for developmental effects.

6. *Mutagenicity*

- The mutagenic activity of maleic hydrazide and its formulations has been investigated in a number of laboratories. These studies are complicated by the fact that hydrazine (a powerful mutagen) is a common contaminant of these preparations, and N-nitrosoethanolamine (also a mutagen) may be present in MH-DEA. Present data are inadequate to determine with certainty whether any mutagenic activity of maleic hydrazide is due to impurities and not the maleic hydrazide itself.

- Tosk et al. (1979) reported that maleic hydrazide (purity not specified), at levels of 5, 10 and 20 mg, was not mutagenic in *Salmonella typhimurium* (TA1530). However, two formulations (MH-30 and Royal MH), at 50, 100 and 200 µL (undiluted), were highly mutagenic in this system.

- Moriya et al. (1983) reported that maleic hydrazide was not mutagenic in five strains of *S. typhimurium*.

- Ercegovich and Rashid (1977) observed a weak mutagenic response with maleic hydrazide (purity not specified) in five strains of *S. typhimurium*.

- Shiau et al. (1980) reported that maleic hydrazide was mutagenic, with and without activation, in several *Bacillus subtilis* strains.

- Epstein et al. (1972) reported that maleic hydrazide (500 mg/kg) was not mutagenic in a dominant-lethal assay in the mouse.

- Nasrat (1965) reported a slight increase in the frequency of sex-linked recessive lethals in the progeny of *Drosophila melanogaster* males reared on food containing 0.4% maleic hydrazide.

- Manna (1971) indicated that exposure to a 5% aqueous solution of maleic hydrazide produced chromosomal aberrations in the bone marrow of mice in a manner similar to that produced by x-rays and other known mutagens.

- Chaubey et al. (1978) reported that intraperitoneal injection of 100 or 200 mg/kg maleic hydrazide (purity not specified) did not affect the incidence of bone marrow erythrocyte micronuclei or the ratio of poly- to normochromatic erythrocytes in male Swiss mice.

- Sabharwal and Lockhard (1980) reported that at concentrations above 100 ppm, maleic hydrazide induced dose-related increases in sister chromatid exchange (SCE) in human lymphocytes and V79 Chinese hamster cells. Commercial formulations of maleic hydrazide (Royal MH and MH-30) at the 250- and 500-mg/kg doses did not cause an increase in micronucleated polychromatic erythrocytes in a mouse micronucleus test.

- Stetka and Wolff (1976) reported that maleic hydrazide (11 and 112 mg/L; purity not specified) caused no significant effect in an SCE assay.

- Nishi et al. (1979) reported that maleic hydrazide (1,000 μg/L; purity not specified), MH-DEA (20,000 μg/mL) and MH-K (20,000 μg/mL) produced cytogenetic effects in Chinese hamster V79 cells *in vitro*.

- Paschin (1981) reported that in the concentration range of 1,800 to 2,500 mg/L, maleic hydrazide (purity not specified) was mutagenic for the thymidine kinase locus of mouse lymphoma cells.

7. *Carcinogenicity*

- The carcinogenicity of maleic hydrazide (purity not specified) was evaluated in two hybrid strains of mice (C57BL/6 × AKR and C57BL/6 × C3H/Anf) (Kotin et al., 1968; Innes et al., 1969). Beginning at 7 days of age, mice were given maleic hydrazide at 1,000 mg/kg/day (suspended in 0.5% gelatin) by stomach tube. After 28 days of age, they were given maleic hydrazide in the diet at 3,000 ppm for 18 months. Assuming that 1 ppm in the diet of mice corresponds to 0.15 mg/kg/day (Lehman, 1959), this is equivalent to a dose of 450 mg/kg/day. These were the maximum tolerated doses. No evidence of increased tumor frequency was detected in gross or histologic examination.

- Barnes et al. (1957) fed maleic hydrazide at a level of 1% (10,000 ppm) in the diet of rats and mice (10 to 15/sex/dose) for a total of 100 weeks. Assuming that 1 ppm in the diet corresponds to 0.05 mg/kg/day in rats and 0.15 mg/kg/day in mice (Lehman, 1959), this is equivalent to a dose of 500 mg/kg/day in rats and 1,500 mg/kg/day in mice. A concurrent study was conducted in which the maleic hydrazide (500 mg/kg/week, corresponding to 71 mg/kg/day) was injected subcutaneously (sc) for the same length of time. No increase in the incidence of tumors was observed in animals exposed by either route when compared with controls (data were pooled).

- Cabral and Ponomarkov (1982) administered maleic hydrazide by gavage in weekly doses of 510 mg/kg in 0.2 mL olive oil to male and female C57BL/B6 mice for 120 weeks. Controls received 0.2 mL olive oil alone, and a third group served as untreated controls. A simultaneous study was conducted using sc injection as the route of administration. There was no evidence of carcinogenicity in the study.

- Van Der Heijden et al. (1981) fed maleic hydrazide (99% pure) containing less than 1.5 mg hydrazine/kg as impurity to rats at dietary levels of 1.0 or 2.0% (10,000 or 20,000 ppm) for 28 months. Assuming that 1 ppm in the diet of rats is equivalent to 0.05 mg/kg/day (Lehman, 1959), this corresponds to doses of 500 or 1,000 mg/kg/day. Histological examination revealed no increase in the tumor incidence in exposed animals compared with the control group.

- In a study by Uniroyal Chemical (1971), mice were administered maleic hydrazide (0.5% in water) by gavage twice weekly beginning at 2 months of age (weight 15 to 18 g) for a total of 2 years. A parallel study was conducted using sc administration. No carcinogenic effect was reported, but specific details of the study were not presented.

- Uniroyal Chemical (1971) reported a 2-year study in Wistar-derived rats in which MH-Na was included in the diet at levels of 0, 0.5, 1.0, 2.0 or 5.0% (0, 5,000, 10,000, 20,000 or 50,000 ppm). Assuming that 1 ppm in the diet of rats corresponds to 0.05 mg/kg/day (Lehman, 1959), this is equivalent to doses of 0, 250, 500, 1,000 or 2,500 mg/kg/day. Although no experimental details were presented, it was concluded that the MH-Na resulted in no blood dyscrasias or tissue pathology, and no indication of carcinogenic potential was detected.

- Epstein and Mantel (1968) used random-bred infant Swiss mice (ICR/Ha) to assess the carcinogenic effects of maleic hydrazide when administered during the neonatal period. The free acid form of maleic hydrazide (containing 0.4% hydrazine impurity) was prepared as an aqueous solution of 5 mg/mL, or as a suspension in redistilled tricaprylin at a concentration of 50 mg/mL. The mice were given injections in the nape of the neck on days 1, 7, 14 and 21 following birth. Six litters received the maleic hydrazide aqueous solution (total dose: 3 mg), and 16 litters received the maleic hydrazide suspension (total dose: 55 mg). One litter received one injection of the suspension at a higher dose (100 mg/mL, total dose: 10 mg), but this was lethal to all mice. A total of 16 litters served as controls (treated with solvents alone). The experiment was terminated between 49 and 51 weeks. The mice that received a total dose of 55 mg in the 3-week period had a high incidence of hepatomas: 65% of 26 male mice alive at 49 weeks, in contrast to solvent controls in which hepatomas occurred in 8% of 48 male mice. The males that received 3 mg total had an 18% incidence of hepatomas. In addition to these lesions, hepatic "atypia" was observed in five males (at 55 mg) and eight females, which the authors judged might be preneoplastic. At the 3-mg level, one atypia was seen in each sex. It was concluded that maleic hydrazide was highly carcinogenic in the male mice. The authors also noted that since there was a complete absence of multiple pulmonary adenomas and pulmonary carcinomas, it was unlikely that the carcinogenicity of maleic hydrazide was due to hydrazine (either present as trace contamination or formed by metabolism), since hydrazine is a potent lung carcinogen in several species of rats and mice (including CBA mice).

IV.   Quantification of Toxicological Effects

Several studies (Tate, 1951; Mukhorina, 1962; Hansen et al., 1984) indicate that the DEA ion is toxic and may contribute to the toxicity of the MH-DEA salt. For this reason, studies involving MH-DEA have not been considered as candidates in calculating HA values for maleic hydrazide.

A. One-day Health Advisory

No information was found in the available literature that was suitable for deriving a One-day HA value for maleic hydrazide. It is therefore recommended that the Ten-day HA value for a 10-kg child (10 mg/L, calculated below) be used at this time as a conservative estimate of the One-day HA value.

B. Ten-day Health Advisory

The developmental toxicity study by Aldridge (1983, cited in U.S. EPA, 1985a) has been selected to serve as the basis for the Ten-day HA. In this study, the potassium salt of maleic hydrazide (K-MH) was administered by gavage at doses of 0, 100, 300 or 1,000 mg/kg/day to Dutch Belted rabbits (16/dose) on days 7 through 27 of gestation. Malformed scapulae were observed in fetuses from the 300- and 1,000-mg/kg/day dose groups. Although the incidence of these malformations was not statistically significant and did not occur in a dose-related fashion, malformed scapulae are a rare, major skeletal defect. Additionally, historical data for this breed of rabbits indicate that scapular anomalies were observed in only 1 of 1,586 fetuses examined from 264 litters. For these reasons U.S. EPA (1985a) concluded that the possibility of teratogenic activity at these dose levels cannot be ruled out. The NOAEL for teratogenic effects is identified as 100 mg/kg/day.

Although a teratogenic response is clearly a reasonable basis upon which to base an HA for an adult, there is some question about whether the Ten-day HA for a 10-kg child can be based upon such a study. However, a teratogenic study is of appropriate duration and does supply some information concerning fetotoxicity. Since the fetus may be more sensitive to the chemical than a 10-kg child and since a teratogenic study is of appropriate duration, it is judged that, though possibly overly conservative, it is reasonable in this case to base the Ten-day HA for a 10-kg child on a developmental toxicity study.

Using a NOAEL of 100 mg/kg/day, the Ten-day HA for a 10-kg child is calculated as follows:

$$\text{Ten-day HA} = \frac{(100 \text{ mg/kg/day}) (10 \text{ kg})}{(100) (1 \text{ L/day})} = 10 \text{ mg/L} (10,000 \text{ } \mu\text{g/L})$$

C. Longer-term Health Advisory

No studies were found that were adequate for calculation of Longer-term HA values for maleic hydrazide. An 11-week feeding study in rats by Tate (1951) identified a NOAEL of 2,500 mg/kg/day, and 2-year feeding studies in rats and dogs by Food Research Laboratories (1954) identified NOAEL values of 2,500 and 500 mg/kg/day, respectively. These studies have not been selected because they provided too little experimental detail to be suitable for calculation of an HA value. It is, therefore, recommended that the Drinking Water Equivalent Level (DWEL) of 18 mg/L, calculated below, be used as a conservative estimate of the Longer-term HA for a 70-kg adult and that the modified DWEL of 5 mg/L, calculated below, be used as a conservative estimate of the Longer-term HA for a 10-kg child.

For a 10-kg child, the adjusted DWEL is calculated as follows:

$$\text{DWEL} = \frac{(0.5 \text{ mg/kg/day} (10 \text{ kg})}{1 \text{ L/day}} = 5 \text{ mg/L}$$

where: 0.5 mg/kg/day = RfD (see Lifetime Health Advisory section).

D. Lifetime Health Advisory

The 28-month feeding study in rats by Van Der Heijden et al. (1981) has been selected to serve as the basis for the Lifetime HA value for maleic hydrazide. Based on proteinuria (in the absence of visible histological effects in kidney), a LOAEL of 500 mg/kg/day was identified. This is a conservative selection, since 2-year feeding studies in dogs and rats by Food Research Laboratories (1954) identified NOAEL values of 500 and 2,500 mg/kg/day, respectively; those studies were not selected, however, because few data or details were provided.

Using the LOAEL identified by Van Der Heijden et al. (1981), the Lifetime HA is calculated as follows:

Step 1: Determination of the Reference Dose (RfD)

$$RfD = \frac{(500 \text{ mg/kg/day})}{(1,000)} = 0.5 \text{ mg/kg/day}$$

Step 2: Determination of the Drinking Water Equivalent Level (DWEL)

$$DWEL = \frac{(0.5 \text{ mg/kg/day}) (70 \text{ kg})}{(2 \text{ L/day})} = 17.5 \text{ mg/L } (18,000 \text{ } \mu g/L)$$

Step 3: Determination of the Lifetime Health Advisory

Lifetime HA = (17.5 mg/L) (20%) = 3.5 mg/L (4,000 μg/L)

E. Evaluation of Carcinogenic Potential

- No evidence of carcinogenic activity was detected in five studies in which maleic hydrazide was administered orally to mice or rats for periods from 18 months to more than 2 years (Kotin et al., 1968; Innes et al., 1969; Barnes et al., 1957; Cabral and Ponomarkov, 1982; Van Der Heijden et al., 1981; Uniroyal Chemical, 1971). Increased incidence of hepatomas has been reported in mice exposed by subcutaneous injection during the first 3 weeks of life (Epstein and Mantel, 1968).

- The International Agency for Research on Cancer has not evaluated the carcinogenic potential of maleic hydrazide.

- Applying the criteria described in EPA's guidelines for assessment of carcinogenic risk (U.S. EPA, 1986), maleic hydrazide may be classified in Group D: not classifiable. This group is used for substances with inadequate human or animal evidence of carcinogenicity.

V. Other Criteria, Guidance and Standards

- The U.S. EPA (1985b) has established residue tolerances for maleic hydrazide in or on raw agricultural commodities that range from 15.0 to 50.0 ppm.

VI. Analytical Methods

- There is no standardized method for the determination of maleic hydrazide in water samples. A procedure has been reported for the estimation of maleic hydrazide residues on various foods (U.S. FDA, 1975). In this procedure, the sample is boiled in alkaline solution to drive off volatile basic interferences. Distillation with zinc and nitrogen sweep expels hydrazine liberated from maleic hydrazide. Hydrazine is reacted in acid solution with p-dimethylaminobenzaldehyde to form a yellow compound which is measured spectrophotometrically.

VII. Treatment Technologies

• Currently available treatment technologies have not been tested for their effectiveness in removing maleic hydrazide from drinking water.

VIII. References

Aldridge, D.* 1983. Teratology study in rabbits with potassium salt of maleic hydrazide. Unpublished report. Prepared by International Research and Development Corporation for Uniroyal Chemical Company. Accession No. 250523. Cited in: U.S. EPA. 1985a.

Barnes, J.M., P.N. Magee, E. Boyland, A. Haddow, R.D. Passey, W.S. Bullough, C.N.D. Cruickshank, M.H. Salaman and R.T. Williams. 1957. The nontoxicity of maleic hydrazide for mammalian tissues. Nature. 180:62–64.

Cabral, J.R.P. and V. Ponomarkov. 1982. Carcinogenicity study of the pesticide maleic hydrazide in mice. Toxicology. 24:169–173.

Chaubey, R.C., B.R. Kavi, P.S. Chauhan and K. Sundaram. 1978. The effect of hycanthone and maleic hydrazide on the frequency of micronuclei in the bone-marrow erythrocytes of mice. Mutat. Res. 57:187–191.

CHEMLAB. 1985. The Chemical Information System, CIS, Inc., Baltimore, MD.

Epstein, S.S., E. Arnold, J. Andrea, W. Bass and Y. Bishop. 1972. Detection of chemical mutagens by the dominant lethal assay in the mouse. Toxicol. Appl. Pharmacol. 23:288–325.

Epstein, S.S. and N. Mantel. 1968. Hepatocarcinogenicity of the herbicide maleic hydrazide following parenteral administration to infant Swiss mice. Intl. J. Cancer. 3:325–335.

Ercegovich, C.D. and K.A. Rashid. 1977. Mutagenesis induced in mutant strains of *Salmonella typhimurium* by pesticides. Abstract. Am. Chem. Soc. 174:Pest 43.

Food Research Labs, Inc.* 1954. Chronic toxicity studies with sodium maleic hydrazide. Unpublished report. MRID 00112753.

Hansen, E., O. Meyer and E. Kristiansen. 1984. Assessment of teratological effect and developmental effect of maleic hydrazide salts in rats. Bull. Environ. Contam. Toxicol. 33:184–192.

Innes, J.R.M, B.M. Ulland, M.G. Valerio, L. Petrucelli, L. Fishbein, E.R.

Hart, A.J. Pallotta, R.R. Bates, H.L. Falk, J.J. Gart, M. Klein, I. Mitchall and J. Peters. 1969. Bioassay of pesticides and industrial chemicals for tumorigenicity in mice: a preliminary note. J. Natl. Cancer Inst. 42:1101–1114.

Kehoe, D.F. and K.M. MacKenzie.* 1983. Two-generation reproduction study with KMH in rats. Study No. 81065. Prepared by Hazelton Raltech, Inc. for Uniroyal Chemical Company. Accession No. 250522. Cited in: U.S. EPA. 1985a.

Kennedy, G. and M.L. Keplinger.* 1971. Placental and milk transfer of maleic hydrazide in albino rats. Unpublished report. MRID 00112778.

Khera, K.S., C. Whalen, C. Trivett and G. Angers. 1979. Teratologic assessment of maleic hydrazide and daminozide, and formulations of ethoxyquin, thiabendazole and naled in rats. J. Environ. Sci. Health (B). 14:563–577.

Kotin, P., H. Falk and A.J. Pallotta.* 1968. Evaluation of carcinogenic, teratogenic, and mutagenic activities of selected pesticides and industrial chemicals. Unpublished report. MRID 0017801.

Lehman, A.J. 1951. Chemicals in food: a report to the Association of Food and Drug Officials. Assoc. Food Drug Off. U.S., Q. Bull. 15:122. Cited in: Ponnampalam, R., N.I. Mondy and J.G. Babish. 1983. A review of environmental and health risks of maleic hydrazide. Regul. Toxicol. Pharmacol. 3:38–47.

Lehman, A.J. 1959. Appraisal of the safety of chemicals in foods, drugs and cosmetics. Assoc. Food Drug Off. U.S., Q. Bull.

Manna, G.K. 1971. Bone marrow chromosome aberrations in mice induced by physical, chemical and living mutagens. J. Cytol. Genet. (India) Congr. Suppl. 144–150.

Mays, D.L., G.S. Born, J.E. Christian and B.J. Liska. 1968. Fate of C^{14}-maleic hydrazide in rats. J. Agric. Food Chem. 16:356–357. Cited in: Swietlinska, Z. and J. Zuk. 1978. Cytotoxic effects of maleic hydrazide. Mutat. Res. 55:ic-30.

Meister, R., ed. 1983. Farm chemicals handbook. Willoughby, OH: Meister Publishing Company.

Moriya, M., T. Ohta, K. Watanabe, T. Miyazawa, K. Kato and Y. Shirasu. 1983. Further mutagenicity studies on pesticides in bacterial reversion assay systems. Mutat. Res. 116:185–216.

Mukhorina, K.V.* 1962. Action of maleic hydrazide on animal organisms. Unpublished report (translated from Russian). MRID 00106969.

Nasrat, G.E. 1965. Maleic hydrazide, a chemical mutagen in Drosophila melanogaster. Nature. 207:439.

Nishi, Y., M. Mori and N. Inui. 1979. Chromosomal aberrations induced by maleic hydrazide and related compounds in Chinese hamster cells in vitro. Mutat. Res. 67:249–257.

Paschin, Y.V. 1981. Mutagenicity of maleic acid hydrazide for the TK locus of mouse lymphoma cells. Mutat. Res. 91:359–362.

Reagan, E. and P. Becci.* 1982. Acute oral LD_{50} in rats of Royal-DRI-60-DG. Unpublished report. Food and Drug Research Labs. MRID 00110459.

Sabharwal, P.S. and J.M. Lockard. 1980. Evaluation of the genetic toxicity of maleic hydrazide and its commercial formulations by sister chromatid exchange and micronucleus bioassays. In Vitro. 16(3):205.

Shapiro, R.* 1977a. Acute oral toxicity. Report No. T-235. Unpublished report. MRID 00079657.

Shapiro, R.* 1977b. Primary skin irritation. Report No. T-212. Unpublished report. MRID 00079660.

Shapiro, R.* 1977c. Eye irritation. Report No. T-220. Unpublished report. MRID 00079661.

Shapiro, R.* 1977d. Acute dermal toxicity. Report No. T-242. Unpublished report. MRID 00079658.

Shiau, S.Y., R.A. Huff, B.C. Wells and I.C. Felkner. 1980. Mutagenicity and DNA-damaging activity for several pesticides with Bacillus subtilis mutants. Mutat. Res. 71:169–179.

Stetka, D.G. and S. Wolff. 1976. Sister chromatid exchange as an assay for genetic damage induced by mutagen-carcinogens. II. In vitro test for compounds requiring metabolic activation. Mutat. Res. 41:343–350.

Tate, H.D.* 1951. Progress report on mammalian toxicity studies with maleic hydrazide. Unpublished report. MRID 00106972.

TDB. 1985. Toxicity Data Bank. MEDLARS II. National Library of Medicine's National Interactive Retrieval Service.

Tosk, J., I. Schmeltz and D. Hoffmann. 1979. Hydrazines as mutagens in a histidine-requiring auxotroph of Salmonella typhimurium. Mutat. Res. 66:247–252.

Uniroyal Chemical Co.* 1971. Summary of toxicity studies on maleic hydrazide: Acute oral toxicity in rats and rabbits. Unpublished report. MRID 00087385. Bethany, CT: Uniroyal.

Uniroyal Chemical Co.* 1977. Results from acute toxicology tests run with Royal MH-30(R) and MH Technical (R). Unpublished report. MRID 00079651. Bethany, CT: Uniroyal.

U.S. EPA. 1985a.* U.S. Environmental Protection Agency. Maleic hydrazide, K-salt. Memorandum from G. Ghali to R. Taylor concerning EPA Reg. Numbers 400–84, 400–94 and 400–165. March 7.

U.S. EPA. 1985b. U.S. Environmental Protection Agency. Code of Federal Regulations. 40 CFR 180.175. July 1, p. 277.

U.S. EPA. 1986. U.S. Environmental Protection Agency. Guidelines for carcinogen risk assessment. Fed. Reg. 51(185):33992–34003. September 24.

U.S. FDA. 1975. U.S. Food and Drug Administration. Pesticide analytical manual. Vol. II. Washington, DC.

Van Der Heijden, C.A., E.M. Den Tonkelaar, J.M. Garbis-Berkvens and G.J. Van Esch. 1981. Maleic hydrazide, carcinogenicity study in rats. Toxicology. 19:139–150.

WSSA. 1983. Weed Science Society of America. Herbicide handbook, 5th ed. Champaign, IL: WSSA.

---

* Confidential Business Information submitted to the Office of Pesticide Programs.

MCPA

I. General Information and Properties

A. CAS No. 94–74–6

B. Structural Formula

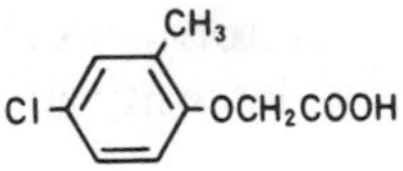

(4-Chloro-2-methylphenoxy)-acetic acid

C. Synonyms

- MCPA; MCP; Agroxone; Hormotuho; Metaxon.

D. Uses

- MCPA is a hormone-type herbicide used to control annual and perennial weeds in cereals, grassland and turf (Hayes, 1982).

E. Properties (CHEMLAB, 1985; Meister, 1983)

Chemical Formula	$C_9H_9O_3Cl$
Molecular Weight	200.63
Physical State (25°C)	Light brown solid
Boiling Point	--
Melting Point	118 to 119°C
Density (25°C)	1.56
Vapor Pressure (25°C)	--
Specific Gravity	--
Water Solubility (20° C)	825 mg/L
Log Octanol/Water Partition Coefficient	2.07 (calculated)
Taste Threshold	--
Odor Threshold	--
Conversion Factor	--

F. Occurrence

- MCPA has been found in 4 of 18 surface water samples analyzed and in none of 118 ground-water samples (STORET, 1988). Samples were collected at 13 surface water locations and 117 ground-water locations. MCPA was found only in California. The 85th percentile of all nonzero samples was 0.54 μg/L in surface water, and the range of concentrations was 0.04 to 0.54 μg/L. This information is provided to give a general impression of the occurrence of this chemical in ground and surface waters as reported in the STORET database. The individual data points retrieved were used as they came from STORET and have not been confirmed as to their validity. STORET data are often not valid when individual numbers are used out of the context of the entire sampling regime, as they are here. Therefore, this information can only be used to form an impression of the intensity and location of sampling for a particular chemical.

- MCPA has been detected in ground water in Montana. The highest level found was 5.5 ppb (5.5 μg/L).

- Frank et al. (1979) detected MCPA residues (1.1 to 1000 ppb) in 2 of 237 wells in Ontario, Canada, between 1969 and 1978.

G. Environmental Fate

- MCPA is not hydrolyzed at pH 7 and 34 to 35°C (Soderquist and Crosby, 1974, 1975). MCPA in aqueous solution (pH 8.3) has a photolytic half-life of 20 to 24 days in sunlight. With fluorescent light, MCPA in aqueous solution (pH 9.8) produced three minor (less than 10%) photolysis products in 71 hours: 4-chloro-2-methyl-phenol, 4-chloro-2-formylphenol and o-cresol (Soderquist and Crosby, 1974, 1975).

- MCPA is degraded more rapidly (1 day) in soils containing less than 10% organic matter than in soil containing higher levels (3 to 9 days) (Torstensson, 1975). This may be due to adsorption to the soil organic matter. MCPA, when applied a second time to soil, is degraded twice as fast (6 to 12 days) as it is after one application (15 to 28 days). Persistence does not depend greatly upon the soil type (Loos et al., 1979).

- Unlabeled MCPA in rice paddy water under dark conditions is totally degraded by aquatic microorganisms in 13 days (Soderquist and Crosby, 1974, 1975).

- MCPA would be expected to leach readily in most soils. Phytotoxic levels of MCPA leached 30 cm in a sandy soil column eluted with 50 cm of water (Herzel and Schmidt, 1979). Using soil thin-layer chromatographic techniques, MCPA was mobile (Rf 0.6 to 1.0) in calcium montmorillonite clay (Helling, 1971) and in sandy loam, silt loam and silty clay loam soils (Helling and Turner, 1968). Mobility increases as organic matter content decreases, possibly due to adsorption of MCPA to this soil component.

- MCPA does not volatilize from aqueous solution (pH 7.0) heated for 13 days at 34 to 35°C (Soderquist and Crosby, 1974, 1975).

- In the aquatic environment, MCPA dissipates rapidly (14 to 32 days) in water, but residue levels in the flooded soil remain unchanged (Soderquist and Crosby, 1974, 1975; Sokolov et al., 1974, 1975). A common metabolite, 5-chloro-o-cresol, is formed at low levels (1.3% or less) within 1 day of treatment.

- According to Eronen et al. (1979), in the forest ecosystem, MCPA remains in soil (0 to 3 cm) and leaf litter at 0.7 and 32 ppm, respectively, 10 months after application at 2.5 kg active ingredient per hectare (a.i./ ha). MCPA residues in moss decline to 7% of the initial levels within 40 days. Residues in soil (3 to 15 cm deep) are not detectable after 40 days.

II. Pharmacokinetics

A. Absorption

- No information on the absorption of MCPA was found in the available literature.

B. Distribution

- Elo and Ylitalo (1979) treated rats with 8 mg of ^{14}C-MCPA [98% active ingredient (a.i.)] intravenously and measured the distribution of radioactivity in nine tissues 1.5 hours after treatment. Highest levels were found in plasma, kidney, lung, liver and heart with lesser amounts found in brain/cerebrospinal fluid (CSF), testis and muscle. Prior treatment of rats with MCPA (intravenous injections of 25 to 500 mg/kg 3 hours before administration of radiolabeled compound or chronic exposure to 500 or 2,500 mg/L in drinking water) lead to decreased levels of ^{14}C-MCPA in the plasma and kidney and increased levels in brain/CSF.

- Elo and Ylitalo (1977) treated rats with 8 mg of ^{14}C-MCPA (purity not specified) intravenously and measured the distribution of radioactivity in brain, CSF, muscle, liver and kidney 1.5 to 120 hours after treatment. Prior treatment of rats with MCPA (subcutaneous injections of 250 or 500 mg/kg) caused a decrease in the amount of radioactivity found in the plasma. Increased levels were found in other tissues with the largest increases found in the CSF (39- to 67-fold) and brain (11- to 18-fold).

C. Metabolism

- MCPA is metabolized by the liver. Stimulation of microsomal oxidation by phenobarbital increases the rate of MCPA breakdown (Buslovich et al., 1979).

D. Excretion

- In studies by Fjeldstad and Wannag (1977), four healthy human volunteers each ingested a dose of 5 mg of MCPA (purity not specified). Approximately 50% (2.5 mg) of the dose was detected in the urine within several days. Urinary levels were not detectable on the fifth day following exposure.

- Rats treated orally with MCPA (purity not specified) excreted nearly all of the MCPA during the first 24 hours after intake (90% in urine and 7% in feces) (Elo, 1976).

III.   Health Effects

A. Humans

1. *Short-term Exposure*

- Case reports of attempted suicide by ingestion of MCPA have been published (Jones et al., 1967; Johnson and Koumides, 1965; Geldmacher et al., 1966). Symptoms included pinpoint pupils, diminished/absent reflexes, low blood pressure, spasms, unconsciousness and death. Dose estimates were not reported.

- Palva et al. (1975) reported one case of MCPA (purity not specified) exposure (dose and duration not specified) in a farmworker involved in spraying operations. The farmer exhibited reversible aplastic anemia, muscular weakness, hemorrhagic gastritis and signs of slight liver damage that were later followed by pancytopenia of all of the myeloid cell lines. In a followup study of the exposed farmer, Timonen and Palva (1980) reported the occurrence of acute myelomonocytic leukemia.

2. *Long-term Exposure*

- No information on the human health effects of chronic exposure to MCPA was found in the available literature.

B. Animals

1. *Short-term Exposure*

- Gurd et al. (1965) reported an acute oral LD_{50} value for MCPA (purity not specified) of 560 mg/kg in mice. Oral LD_{50} values for MCPA in mice of 550 mg/kg and 700 mg/kg/day in rats were reported in RTECS (1985).

- Elo et al. (1982) showed that MCPA (sodium salt; 99% a.i.) causes a selective damage of the blood-brain barrier. These authors observed that the penetration of intravenous tracer molecules such as ^{14}C-MCPA, ^{14}C-PABA, ^{14}C-sucrose, ^{14}C-antipyrine and iodinated human albumin (^{125}I-HA) in the brain and CSF of MCPA-intoxicated rats (200 to 500 mg/kg, sc) was increased compared to controls. The tissue-plasma ratios of ^{14}C-sucrose, ^{14}C-antipyrine and ^{125}I-HA treated rats were also increased in the brain and CSF of intoxicated animals, but the increases were less pronounced than those of ^{14}C-MCPA or ^{14}C-PABA.

- In oral studies by Vainio et al. (1983), Wistar rats administered the iso-octyl ester of MCPA (purity not specified) at doses of 0, 100, 150 or 200 mg/kg/day, 5 days per week for 2 weeks, showed hypolipidemia and peroxisome proliferation in the liver. A Lowest-Observed-Adverse-Effect Level (LOAEL) of 100 mg/kg was identified.

2. *Dermal/Ocular Effects*

- Raltech (1979) reported acute dermal LD_{50} values for MCPA (purity not specified) in rabbits of 4.8 g/kg for males and 3.4 g/kg for females.

- In acute dermal studies conducted by Verschuuren et al. (1975), an aqueous paste of MCPA (80.6% a.i.) (0.5 g) was applied to the abraded skin of five chinchilla rabbits. Slight erythema resulted; the skin became sclerotic after 5 to 6 days and healed by 12 days.

- In subacute dermal studies, Verschuuren et al. (1975) applied an aqueous paste of MCPA (80.6% active ingredient; 0, 0.5, 1.0 or 2.0 g) five times weekly for 3 weeks to the shaved skin of rabbits. Slight to moderate erythema occurred at all dose levels, and elasticity of the skin was decreased. The effects subsided at 2 weeks posttreatment. Weight loss was observed at all dose levels. High mortality and histopathological alterations were observed in the liver, kidneys, spleen and thymus at the 1.0- and 2.0-g dose levels.

3. *Long-term Exposure*

- Verschuuren et al. (1975) administered MCPA (80.6% a.i.) in the diet for 90 days to SPF weanling rats (10/sex/dose) at levels of 0, 50, 400 or 3,200 ppm. Assuming that 1 ppm in the diet of rats is equivalent to 0.05 mg/kg/day (Lehman, 1959), these levels correspond to doses of about 0, 2.5, 20 or 160 mg/kg/day. Following treatment, growth, food intake, mortality, hematology, blood and liver chemistry, organ weights and histopathology were measured. No compound-related effects were reported for any of these parameters except for growth retardation and elevated relative kidney weights at 400 ppm (20 mg/kg/day) or more. A No-Observed-Adverse-Effect Level (NOAEL) of 50 ppm (2.5 mg/kg/day) and a LOAEL of 400 ppm (20 mg/kg/day) were identified.

- Holsing and Kundzin (1970) administered MCPA technical (considered to be 100% a.i.) in the diet of Charles River CD rats (10/sex/dose) for 3 months. Doses were reported as 0, 4, 8 or 16 mg/kg/day; the concentration in the diet was not specified. Following treatment, no compound-related effects were observed in the physical appearance, behavior, growth, food consumption, survival, clinical chemistry, organ weights, organ-to-body weight ratios, gross pathology or histopathology at any dose tested, except for increases in kidney weight in males at 16 mg/kg/day. A NOAEL of 8 mg/kg/day and a LOAEL of 16 mg/kg/day were identified by this study.

- Holsing and Kundzin (1968) administered oral doses of MCPA technical to Charles River CD rats at dose levels of 0, 25, 50, and 100 mg/kg/day for 13 weeks. Cytopathological changes in the liver and kidneys were observed at all doses. Kidney effects included focal hyperplasia of the epithelial lining, interstitial nephritis, tubular dilation and/or hypertrophy. A LOAEL of 25 mg/kg/day (the lowest dose tested) is identified by this study.

- Reuzel and Hendriksen (1980) administered MCPA (94% a.i.) in feed to beagle dogs in two separate 13-week studies. Dosing regimens of 0, 3, 12 or 48 mg/kg/day, and 0, 0.3, 1 or 12 mg/kg/day, respectively, were employed. Decreased kidney and liver function, characterized by increases in blood urea, SGPT and creatinine were observed at doses as low as 3 mg/kg/day. Low prostatic weight and mucopurulent conjunctivitis were observed at higher doses. A NOAEL of 1 mg/kg/day and a LOAEL of 3 mg/kg/day were identified by these studies.

- Hellwig (1986) administered oral doses of MCPA (95% a.i.) to dogs at doses of 0, 6, 30, or 150 ppm for 1 year. Assuming that 1 ppm in the diet of dogs is equivalent to 0.025 mg/kg (Lehman, 1959), these levels correspond to doses of 0, 0.15, 0.75 or 3.75 mg/kg/day. Renal toxicity was observed at the two highest doses and was characterized by elevated serum levels of creatinine, urea and potassium, coloration of the kidneys and increased storage of pigment in the renal tubules. A NOAEL of 0.15 mg/kg/day and a LOAEL of 0.75 mg/kg/day were identified by this study.

- Holsing (1968) administered oral doses of MCPA (considered to be 100% a.i.) by capsule at 0, 25, 50 or 75 mg/kg/day to beagle dogs (3/sex/dose) for 13 weeks. Histopathological changes and alterations in various hematologic and biochemical parameters indicative of bone marrow, liver and kidney damage were observed at all dose levels. The hematological findings included decreased hematocrit, hemoglobin and erythrocyte counts. Several dogs had elevated blood urea nitrogen, serum glutamic-pyruvic transaminase, serum-oxaloacetic transaminase, alkaline phosphatase and serum bilirubin. Histopathological alterations were seen in the liver, kidney, lymph nodes, testes, prostate and bone marrow. All dogs of all three groups had various degrees of hepatic, renal and bone marrow injury. A LOAEL of 25 mg/kg/day (the lowest dose tested) was identified.

- Gurd et al. (1965) administered technical MCPA (purity not specified) in the feed to rats (5/sex/dose) for 7 months at dose levels of 0, 100, 400, 1,000 or 2,500 ppm. Assuming that 1 ppm in the diet of rats is equivalent to 0.05 mg/kg/day (Lehman, 1959), these levels correspond to doses of 0, 5, 20, 50 or 125 mg/kg/day. Following treatment, there was a marked decrease in body weight gain at 1,000 ppm (50 mg/kg/day) or 2,500 ppm (125 mg/kg/day), and some deaths occurred at 2,500 ppm (125 mg/kg/day). At 400 ppm (20 mg/kg/day) or greater, there was a reduction in numbers of red blood cells, hemoglobin content and hematocrit. Relative kidney weights were increased at 100 ppm (5 mg/kg/day), but no effects on body weight were evident. No histopathological changes were reported at any dose level tested. A LOAEL of 5 mg/kg/day (the lowest dose tested) was identified.

4. *Reproductive Effects*

- No effects on reproduction were found in rats exposed to doses of 0, 50, 150, or 450 ppm MCPA (95% a.i.) in the diet over a period of two generations (MacKenzie, 1986). Assuming that 1 ppm in the diet of rats corresponds to 0.05 mg/kg/day (Lehman, 1959), these levels correspond to doses of 0, 2.5, 7.5 or 22.5 mg/kg/day. Body weight depression was observed in the F_1 and F_2 generations at the two highest doses. A NOAEL of 22.5 mg/kg/day was identified for reproductive function, and a NOAEL of 2.5 mg/kg/day was identified for fetotoxicity (depressed weight gain).

5. *Developmental Effects*

- Irvine et al. (1980) administered MCPA (purity not specified) (0, 5, 12, 30 or 75 mg/kg/day) by gavage to rabbits (15 to 18/dose) on days 6 to 18 of gestation. No fetotoxicity or teratogenicity was observed at any dose level tested. Body weights of the does were markedly reduced in the 75 mg/kg/day dosage group. A fetal NOAEL of 75 mg/kg/day and a maternal NOAEL of 30 mg/kg/day were identified.

- Irvine (1980) administered MCPA (purity not specified) (0, 20, 50 or 125 mg/kg/day) by gavage to pregnant CD rats (16 to 38/dose) on days 6 to 15 of gestation. No maternal or fetal toxicity or teratogenic effects were observed. A NOAEL of 125 mg/kg/day (the highest dose tested) was identified.

- Palmer and Lovell (1971) administered oral doses of MCPA (75% a.i.; 0, 5, 25 or 100 mg/kg/day of the active ingredient) to mice (20/dose) on days 6 to 15 of gestation. Dams were monitored for pregnancy rate, body weight, and gross toxicity; no significant effects were observed. At 100 mg/kg/day, fetal weights were significantly reduced and there was delayed skeletal ossification. A NOAEL of 25 mg/kg/day and a LOAEL of 100 mg/kg/day based on fetal weights were identified.

6. *Mutagenicity*

- Moriya et al. (1983) reported that MCPA (purity not specified) (5,000 µg/plate) did not produce mutagenic activity in *Salmonella typhimurium* (TA100, TA98, TA1535, TA1537, TA1538) or in *Escherichia coli* (WP2 hcr) either with or without metabolic activation.

- In studies conducted by Magnusson et al. (1977), there were no effects on chromosome disjunction, loss or exchange in *Drosophila* fed MCPA at 250 or 500 ppm.

- In studies by Linnainmaa (1984), no increases were observed in the frequency of sister chromatid exchange (SCE) in blood lymphocytes from rats intragastrically administered MCPA (purity not specified) at 100 mg/kg/day for 2 weeks. A slight increase in SCE was observed in bone marrow cells from Chinese hamsters given daily oral doses of 100 mg/kg for 2 weeks. In Chinese hamster ovarian cell cultures, SCE was slightly increased following treatment with MCPA (10^{-5}, 10^{-4}, 10^{-3}M, 1 hour) with and without activation.

7. *Carcinogenicity*

- No information on the potential carcinogenicity of MCPA was found in the available literature. However, MCPA stimulates liver peroxisomal proliferation, which has been implicated in carcinogenicity (Vainio et al., 1983).

IV.   Quantification of Toxicological Effects

A. One-day Health Advisory

No information was found in the available literature that was suitable for determination of the One-day HA value for MCPA. It is therefore recommended that the Longer-term HA value for a 10-kg child (0.1 mg/L, calculated below) be used at this time as a conservative estimate of the One-day HA value.

B. Ten-day Health Advisory

Several reproductive/developmental toxicity studies have been performed in which rats or rabbits have been given oral doses of MCPA for acute duration (Irvine, 1980; Irvine et al., 1980; Palmer and Lovell, 1971; MacKenzie, 1986). The only sign of maternal toxicity observed in these studies was a reduction in body weight in rats exposed to 75 mg/kg (Irvine, 1980). Estimates of maternal NOAELs range from 30 to 125 mg/kg/day (Irvine, 1980; Irvine et al., 1980). In contrast, fetotoxicity has been observed at dose levels as low as 7.5 mg/kg/day (MacKenzie, 1986). These studies were judged to be inadequate for evaluating the toxicity of MCPA from acute oral exposure, especially with respect to kidney toxicity observed after longer durations of exposure. It is therefore recommended that the Longer-term HA value for a 10-kg child of 0.1 mg/L (calculated below) be used at this time as a conservative estimate of the Ten-day HA value.

C. Longer-term Health Advisory

Evidence of renal dysfunction has been observed in both 13-week (Reuzel and Hendriksen, 1980; Holsing, 1968) and 1-year (Hellwig, 1986) feeding studies in beagle dogs and serves as the basis for the Longer-term HA. In subchronic studies changes in blood urea and creatinine levels have been observed at doses of 25 mg/kg/day (Holsing, 1968) and 3 mg/kg/day (Reuzel and Hendriksen, 1980). Renal toxicity is not unique to dogs and has been observed in rats after 90-day exposure at dose levels of 20 mg/kg/day (Verschuuren et al., 1975) and 25 mg/kg/day (Holsing, 1968). The rat and dog may have similar sensitivities; a conservative estimate of the NOAEL was obtained from the studies described by Reuzel and Hendriksen (1980). In these studies, oral doses of 0, 3, 12 or 48 mg/kg/day, and 0, 0.3, 1 or 12 mg/kg/day, respectively, were administered to dogs for 13 weeks. Increases in blood urea, SGPT and creatinine levels were observed at dose levels as low as 3 mg/kg/day; low prostatic weight and mucopurulent conjunctivitis were observed at higher dose levels. A NOAEL of 1 mg/kg/day was identified by these studies.

Using a NOAEL of 1 mg/kg/day, the Longer-term HA for a 10-kg child is calculated as follows:

$$\text{Longer-term HA} = \frac{(1.0 \text{ mg/kg/day}) (10 \text{ kg})}{(100) (1 \text{ L/day})} = 0.1 \text{ mg/L } (100 \text{ } \mu\text{g/L})$$

The Longer-term HA for a 70-kg adult is calculated as follows:

$$\text{Longer-term HA} = \frac{(1.0 \text{ mg/kg/day}) (70 \text{ kg})}{(100) (2 \text{ L/day})} = 0.4 \text{ mg/L } (400 \text{ } \mu\text{g/L})$$

D. Lifetime Health Advisory

The chronic toxicity study in dogs (Hellwig, 1986) has been selected to serve as the basis for the determination of the Lifetime HA. Beagle dogs were exposed to 0, 6, 30 and 150 ppm (0, 0.15, 0.75 or 3.75 mg/kg/day) for 1 year. Renal toxicity was observed at the two highest doses and was characterized by elevated serum levels of creatinine, urea and potassium, coloration of the kidneys and increased storage of pigment in the renal tubules. A NOAEL of 0.15 mg/kg/day was identified, which is supported by the findings from subchronic feeding studies. From 90-day feeding studies, NOAELs of 1 mg/kg/day and 2.5 mg/kg/day have been identified for dogs (Reuzel and Hendriksen, 1980) and rats (Verschuuren et al., 1975) based on the absence of effects on the kidney seen at higher doses. In a 7-month feeding study, Gurd et al. (1965) observed increased kidney weight in rats exposed to doses as low as 5.0 mg/kg/day, the lowest dose tested.

Using a NOAEL of 0.15 mg/kg/day, the Lifetime HA is calculated as follows:

Step 1: Determination of the Reference Dose (RfD)

$$RfD = \frac{(0.15 \text{ mg/kg/day})}{(100)} = 0.0015 \text{ mg/kg/day}$$

Step 2: Determination of the Drinking Water Equivalent Level (DWEL)

$$DWEL = \frac{(0.0015 \text{ mg/kg/day}) (70 \text{ kg})}{(2 \text{ L/day})} = 0.05 \text{ mg/L } (50 \text{ } \mu g/L)$$

Step 3: Determination of the Lifetime Health Advisory

$$\text{Lifetime HA} = (0.05 \text{ mg/L}) (20\%) = 0.01 \text{ mg/L } (10 \text{ } \mu g/L)$$

E. Evaluation of Carcinogenic Potential

- Lifetime studies on the carcinogenic potential of MCPA in rats and mice are currently being evaluated.

- The International Agency for Research on Cancer (IARC, 1983) classified the potential carcinogenicity of MCPA in both humans and laboratory animals as indeterminate.

- Applying the criteria described in EPA's guidelines for assessment of carcinogenic risk (U.S. EPA, 1986), MCPA may be classified in Group D: not classifiable. This category is used for substances with inadequate animal evidence of carcinogenicity.

V.   Other Criteria, Guidance and Standards

- The National Academy of Sciences has recommended an ADI of 0.00125 mg/kg/day and a Suggested-No-Adverse-Response-Level (SNARL) of 0.009 mg/L, based on a LOAEL of 1.25 mg/kg/day in a 90-day study in rats (NAS, 1977).

- Residue tolerances have been established for MCPA at 0.1 ppm in milk and meat. Feed and forage residue tolerances range from 0.1 to 300 ppm (U.S. EPA, 1985a).

VI.   Analytical Methods

• Analysis of MCPA is by a gas chromatographic (GC) method applicable to the determination of certain chlorinated acid pesticides in water samples (U.S. EPA, 1985b). In this method, approximately 1 liter of sample is acidified. The compounds are extracted with ethyl ether using a separatory funnel. The derivatives are hydrolized with potassium hydroxide, and extraneous organic material is removed by a solvent wash. After acidification, the acids are extracted and converted to their methyl esters using diazomethane as the derivatizing agent. Excess reagent is removed, and the esters are determined by electron-capture GC. The method detection limit has been estimated at 249 μg/L for MCPA.

VII.   Treatment Technologies

• Oxidation by ozone of 500 mg/L MCPA, after 50 to 80% disappearance of initial compound, produced no identifiable degradation products (Legube et al., 1981). This indicates that oxidation by ozone may be a possible MCPA removal technique.

VIII.   References

Buslovich, S.Y., Z.A. Aleksashina and V.M. Kolosovskaya. 1979. Effect of phenobarbital on the embryotoxic action of 2-methyl-4-chlorophenoxyacetic acid (a herbicide). Russ. Pharmacol. Toxicol. 24(2):57–61.

CHEMLAB. 1985. The Chemical Information System, CIS, Inc., Baltimore, MD.

Elo, H.A. 1976. Distribution and elimination of 2-methyl-4-chlorophenoxyacetic acid (MCPA) in male rats. Scand. J. Work Environ. Health. 3:100–103.

Elo, H.A. and P. Ylitalo. 1977. Substantial increase in the levels of chlorophenoxyacetic acids in the CNS of rats as a result of severe intoxication. Acta Pharmacol. Toxicol. 41:280.

Elo, H.A. and P. Ylitalo. 1979. Distribution of 2-methyl-4-chlorophenoxyacetic acid and 2,4-dichlorophenoxyacetic acid in male rats: evidence for the involvement of the central nervous system in their toxicity. Toxicol. Appl. Pharm. 51:439–446.

Elo, H.A., P. Ylitalo, J. Kyottila and H. Hervonen. 1982. Increase in the penetration of tracer compounds into the rat brain during 2-methyl-4-

chlorophenoxyacetic acid (MCPA) intoxication. Acta Pharmacol. Toxicol. 50:104–107.

Eronen, L., R. Julkunen and A. Saarelainen. 1979. MCPA residues in developing forest ecosystem after aerial spraying. Bull. Environ. Contam. Toxicol. 21:791–798.

Fjeldstad, P. and A. Wannag. 1977. Human urinary excretion of the herbicide 2-methyl-4-chlorophenoxyacetic acid. Scand. J. Work Environ. Health. 3:100–103.

Frank, R., G.J. Siron and B.D. Ripley. 1979. Herbicide contamination of wellwaters in Ontario, Canada, 1969–78. Pestic. Monitor. J. 13:120–127.

Geldmacher, M., V. Mallinckrodt and L. Lautenbach. 1966. Zwei tödliche Vergiftungen (Suicid) mit chlorierten Phenoxyessigauren (2,4-D und MCPA). Archiv für Toxikologie. 21:261–278.

Gurd, M.R., G.L.M. Harmer and B. Lessel. 1965. Summary of toxicological data: acute toxicity and 7-month feeding studies with mecoprop and MCPA. Food Cosmet. Toxicol. 3:883–885.

Hayes, W.J. 1982. Pesticides studied in man. Baltimore, MD: Williams and Wilkins.

Helling, C.S. 1971. Pesticide mobility in soils. II. Application of soil thin-layer chromatography. Proc. Soil Sci. Soc. Am. 35(5):737–743.

Helling, C.S. and B.C. Turner. 1968. Pesticide mobility: determination by soil thin-layer chromatography. Science. 162(3853):562–563.

Hellwig, J. 1986. Report on the study of the toxicity of MCPA in beagle dogs after 12-month administration in the diet. Project No. 33D0046/8341. Unpublished study. MRID 164352.

Herzel, F. and G. Schmidt. 1979. Testing the leaching behavior of herbicides on lysimeters and small columns. WaBoLu-Berichte. (3):1–16.

Holsing, G.C.* 1968. Thirteen-week dietary/oral administration—dogs. Final Report. Project No. 517-101. Unpublished study. MRID 00004756.

Holsing, G.C. and M. Kundzin.* 1968. Three-month dietary administration rats. Project No. 517-100. Unpublished study. MRID 00004775.

Holsing, G.C. and M. Kundzin.* 1970. Final Report: three-month dietary administration—rats. Final Report. Project No. 517-106. Unpublished study. MRID 00004776.

IARC. 1983. International Agency for Research on Cancer. IARC monograph on the evaluation of carcinogenic risk to chemicals to man. Lyon, France: IARC.

Irvine, L.F.H., D. Whittaker, J. Hunter et al.* 1980. MCPA oral teratogenicity study in the Dutch belted rabbit. Report No. 1737R-277/5. Unpublished study. MRID 00041637.

Irvine, L.F.H.* 1980. MCPA oral teratogenicity study in the rat. Report No.1996-277/7b. Unpublished study. MRID 00066317.

Johnson, H.R.M. and O. Koumides. 1965. A further case of M.C.P.A. poisoning. Brit. Med. J. 2:629–630.

Jones, D.I.R., A.G. Knight and A.J. Smith. 1967. Attempted suicide with herbicide containing MCPA. Arch. Environ. Health. 14:363–366.

Legube, B., B. Langlaia, B. Sohm and M. Dore. 1981. Identification of ozonation products of aromatic hydrocarbon micropollutants: effect on chlorination and biological filtration. Ozone: Sci. Eng. 3(1):33–48.

Lehman, A.J. 1959. Appraisal of the safety of chemicals in foods, drugs and cosmetics. Assoc. Food Drug Off. U.S., Q. Bull.

Linnainmaa, K. 1984. Induction of sister chromatid exchanges by the peroxisome proliferators 2,4-D, MCPA, and clofibrate in vivo and in vitro. Carcinogenesis. 5(6):703–707.

Loos, M.A., I.F. Schlosser and W.R. Mapham. 1979. Phenoxy herbicide degradation in soils: quantitative studies of 2,4-D- and MCPA-degrading microbial populations. Soil Biol. and Biochem. 11(4):377–385.

MacKenzie, K.M. 1986. Two-generation reproductive study with MCPA in rats. Final report. Study No. 6148-100. Unpublished study.

Magnusson, J., C. Ramel and A. Friksson. 1977. Mutagenic effects of chlorinated phenoxyacetic acids in Drosophila melanogaster. Hereditas. 87:121–123.

Meister, R., ed. 1983. Farm chemicals handbook. Willoughby, OH: Meister Publishing Company.

Moriya, M., T. Ohta, K. Watanabe, T. Miyazawa, K. Kato and Y. Shirasu. 1983. Further mutagenicity studies on pesticides in bacterial reversion assay systems. Mutat. Res. 116:185–216.

NAS.1977. National Academy of Sciences. Drinking water and health, Vol. 1. Washington, DC: National Academy Press.

Palmer, A.K. and M.R. Lovell.* 1971. Effect of MCPA on pregnancy of the mouse. Unpublished study. MRID 00004447.

Palva, H.L.A., O. Koivisto and I.P. Palva. 1975. Aplastic anemia after exposure to a weed killer, 2-methyl-4-chlorophenoxyacetic acid. Acta. Haemat. 53:105–108.

Raltech.* 1979. Raltech Scientific Services, Inc. Defined dermal LD_{50}. Unpublished study. MRID 00021973.

Reuzel, P.G.J. and C.F.M. Hendriksen.* 1980. Subchronic (13-week) oral toxicity study of MCPA in beagle dogs. Final report. Project No. B77/1867. Report Nos. R6478 and R6337. Unpublished study prepared by Central Institute for Nutrition and Food Research.

RTECS. 1985. Registry of toxic effects of chemical substances. NIOSH, National Library of Medicine On-Line File.

Soderquist, C.J. and D.G. Crosby. 1974. The dissipation of 4-chloro-2-methylphenoxyacetic acid (MCPA) in a rice field. Unpublished study prepared by Univ. of California, Davis, Department of Environmental Toxicology. Submitted by Dow Chemical Company, Midland, MI.

Soderquist, C.J. and D.G. Crosby. 1975. Dissipation of 4-chloro-2-methylphenoxyacetic acid (MCPA) in a rice field. Pestic. Sci. 6(1):17–33.

Sokolov, M.S., L.L. Knyr, B.P. Strekozov, V.D. Agarkov, A.P. Chubenko and B.A. Kryzhko. 1974. The behavior of some herbicides under the conditions of a rice irrigation system. Khimiya v Sel'skom Khozyaistve (Chemistry in Agriculture). 13:224–234.

Sokolov, M.S., L.L. Knyr, B.P. Strekozov and V.D. Agarkov. 1975. Behavior of proanide, yalan, MCPA and 2,4-D in rice irrigation systems of the Kuban River. Agrokhimiya (Agricultural Chemistry). 3:95–106.

STORET. 1988. STORET Water Quality File. Office of Water. U.S. Environmental Protection Agency (data file search conducted in May, 1988).

Timonen, T.T. and I.P. Palva. 1980. Acute leukemia after exposure to a weed killer, 2-methyl-4-chlorophenoxyacetic acid. Acta. Haemat. 63:170–171.

Torstensson, N.T.L. 1975. Degradation of 2,4-D and MCPA in soils of low pH. In: Coulston, F. and F. Korte, eds. Pesticides: IUPAC Third International Congress, July 3–9, 1974, Helsinki, Finland. Stuttgart, West Germany: George Thieme (Environmental Quality and Safety, Supplement, Vol. 3). pp. 262–265.

Torstensson, N.T.L., J. Stark and B. Goransson. 1975. The effect of repeated applications of 2,4-D and MCPA on their breakdown in soil. Weed Res. 15(3):159–164.

U.S. EPA. 1985a. U.S. Environmental Protection Agency. Code of Federal Regulations. 40 CFR 180.339.

U.S. EPA. 1985b. U.S. Environmental Protection Agency. U.S. EPA Method 615 – Chlorinated phenoxy acids. Fed. Reg. 50:40701. October 4.

U.S. EPA. 1986. U.S. Environmental Protection Agency. Guidelines for carcinogen risk assessment. Fed. Reg. 51(185):33992–34002. September 24.

Vainio, H., K. Linnainmaa, M. Kahonen, J. Nickels, E. Hietanen, J. Marniemi and P. Peltonen. 1983. Hypolipidemia and peroxisome proliferation induced by phenoxyacetic acid herbicide in rats. Biochem. Pharmacol. 32(18):2775–2779.

Verschuuren, H.G., R. Kroes and E.M. den Tonkelaar. 1975. Short-term oral and dermal toxicity of MCPA and MCPP. Toxicology. 3:349–359.

---

* Confidential Business Information submitted to the Office of Pesticide Programs.

Methomyl

I. General Information and Properties

A. CAS No. 16752–77–5

B. Structural Formula

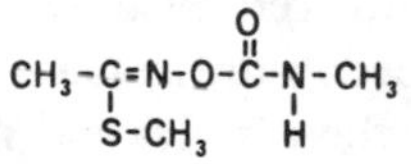

$$CH_3-\underset{\underset{S-CH_3}{|}}{C}=N-O-\underset{\underset{}{||}}{\overset{O}{C}}-\underset{\underset{H}{|}}{N}-CH_3$$

S-Methyl-N[(methylcarbamoyl)oxy]-thioacetimidate

C. Synonyms

- Dupont Insecticide 1179; Dupont 1179; Insecticide 1,179; Insecticide 1179; IN 1179, Lannate; Mesomile; Nudrin; SD 14999; WL 18236 (Meister, 1983).

D. Uses

- Methomyl is a carbamate insecticide used to control a broad spectrum of insects in agricultural and ornamental crops (Meister, 1983).

E. Properties (Meister, 1983; Windholz et al., 1983; Cohen, 1984; CHEMLAB, 1985; and TDB, 1985)

Chemical Formula	$C_5H_{10}O_2N_2S$
Molecular Weight	162.20
Physical State (25°C)	White crystalline solid
Boiling Point	--
Melting Point	78 to 79°C
Density (24°C)	1.29
Vapor Pressure (25°C)	5×10^{-5} mm Hg
Specific Gravity	1.29
Water Solubility (25°C)	10,000 mg/L
Log Octanol/Water Partition Coefficient	-3.56
Taste Threshold	--

Odor Threshold   --
Conversion Factor   --

F. Occurrence

- Methomyl has been found in 1 of 423 surface water samples analyzed at a concentration of 2 μg/L, and was not found in 1,374 ground-water samples analyzed (STORET, 1988). Samples were collected at 145 surface water locations and 1,326 ground-water locations. This information is provided to give a general impression of the occurrence of this chemical in ground and surface waters as reported in the STORET database. The individual data points retrieved were used as they came from STORET and have not been confirmed as to their validity. STORET data are often not valid when individual numbers are used out of the context of the entire sampling regime, as they are here. Therefore, this information can only be used to form an impression of the intensity and location of sampling for a particular chemical.

G. Environmental Fate

- In laboratory and greenhouse studies, methomyl was more rapidly degraded in a sandy loam and a California soil than in silt loam soils, with 21, 31, and 44 to 48% of the applied methomyl remaining in the respective soils 42 to 45 days after treatment. The major degradation product was carbon dioxide, which accounted for 23 to 47% of the applied methomyl after 42 to 45 days. A minor degradation product, S-methyl-N-hydroxy-thioacetimidate (a possible hydrolysis product), was also found. Methomyl half-lives were less than 30 days in sandy loam soil, less than 42 days in California soil, and approximately 45 days in muck and silt loam soils. In a sterilized Flanagan silt loam soil, 89% of the methomyl remained 45 days after application, indicating that methomyl degradation in soil is primarily a microbial process (Harvey, 1977a,b).

- The nitrogen-fixing ability of some bacteria was severely reduced (by as much as 85%) when methomyl was applied at 20 to 160 ppm (Huang, 1978).

- In another study, methomyl (18 ppm) had no effect on fungal and bacterial population or on carbon dioxide production in either silt loam or fine sand soils (Peeples, 1977).

- No methomyl residues were detected in a muck soil 7 to 32 days after treatment (E.I. DuPont de Nemours and Co., 1971).

II.   Pharmacokinetics

A. Absorption

- Single oral doses of 1-^{14}C-methomyl (purity not specified) were administered via gavage to female CD rats as a suspension in 1% aqueous methylcellulose. Ninety-five percent of the dose could be accounted for in excretory products or tissue residues, indicating virtually complete absorption from the gastrointestinal tract (Andrawes et al., 1976).

- Baron (1971) reported that in rats given a single oral dose of 5 mg/kg of 1-^{14}C-labeled methomyl (purity not specified), approximately 2% of the original label was excreted in the feces after 3 days, indicating essentially complete gastrointestinal absorption.

B. Distribution

- Baron (1971) fed a single oral dose of 1-^{14}C-labeled methomyl (5 mg/kg, purity not specified) to rats and analyzed 13 major tissues for residues at 1 and 3 days after dosing. Only 10% of the label was present in tissues 24 hours after dosing, with no evidence of accumulation at any site. By this time, over 40% of the label had been excreted via the lung. At 3 days after dosing, tissue residues were essentially unchanged from day 1, suggesting incorporation of label into tissue components.

- Baron (1971) reported that feeding methomyl to a lactating cow at levels of 0.2 or 20 ppm in the diet (duration not specified) resulted in very low residues (less than 0.02 ppm) in the milk, meat, fat, liver and kidney.

C. Metabolism

- According to Baron (1971), in 72 hours approximately 15 to 23% of a 5-mg/kg oral dose of 1-^{14}C-labeled methomyl in rats could be accounted for as carbon dioxide, 33% as another metabolite in expired air and 25% as metabolites in the urine.

- Harvey (1974) reported that in the rat, 1-^{14}C-labeled methomyl (dose and purity not specified) was metabolized to carbon dioxide (25%) or acetonitrile (50%) within 72 hours.

- Andrawes et al. (1976) reported that single oral doses of 4 mg/kg were rapidly metabolized in the rat. In exhaled air, carbon dioxide and aceto-nitrile were the major metabolites. In 24-hour urine samples, polar metabolites (80%) and acetonitrile (18%), both free and conjugated, were found with free methomyl, the oxime and the sulfoxide oxime detected at low levels.

- Dorough (1977), in a series of studies with ^{14}C-labeled isomeric forms of methomyl, confirmed the report by Harvey (1974) of the excretion of labeled carbon dioxide and acetonitrile in the expired air of treated rats. In addition, nearly complete (79 to 84%) hydrolysis of the ester linkage was apparent within 6 hours, prior to the major formation of carbon dioxide and acetonitrile from methomyl. The author suggested the following pathway: partial isomerization of methomyl is followed by hydrolysis of the two isomeric forms to yield two isomeric oximes that then break down to carbon dioxide and acetonitrile at different rates. No additional metabolites were identified.

D. Excretion

- Baron (1971) stated that within 72 hours after receiving a single oral dose of 1-^{14}C-labeled methomyl, rats excreted 15 to 23% as carbon dioxide, 33% as other metabolites in the expired air and approximately 16 to 27% as methomyl and metabolites in the urine.

- Harvey (1974) reported that 75% of an oral dose of 1-^{14}C-labeled metho-myl (dose and purity not specified) was excreted by rats within 72 hours, 50% as acetonitrile and 25% as carbon dioxide in the expired air. In contrast to other carbamates, sulfur-containing metabolites were not found in the urine.

- Andrawes et al. (1976) reported that single oral doses (4 mg/kg) of 1-^{14}C-labeled methomyl were rapidly excreted, with 32% of the dose recovered in urine, 19% in feces and 40% in exhaled air after 4 days.

III.   Health Effects

A. Humans

1. *Short-term Exposure*

- Liddle et al. (1979) reported a case of methomyl poisoning in Jamaica, W.I., involving five men who had eaten a meal that included unleavened bread. Methomyl was discovered in an unlabeled plastic bag in a tin can, and had evidently been used as salt in preparation of the bread. Approximately 3 hours after the meal, the men were found critically ill, frothing at the mouth, twitching and trembling. Three were dead on arrival at the hospital. One of the two survivors showed generalized twitching and spasms, fasciculation and respiratory impairment thought to be due to severe bronchiospasms. The other patient walked unaided and appeared generally normal. Both patients were given atropine intravenously, and the symptomatic patient recovered within 2 hours after treatment. Methomyl was confirmed in the stomach contents of each of the men who died, and analysis of the bread indicated that it contained 1.1% methomyl. It was stated that two of the victims had eaten about 75 to 100 g of bread each, or 0.82 to 1.1 g of methomyl. From these data it may be calculated that a dose of 12 to 15 mg/kg body weight can be fatal in humans.

• Araki et al. (1982) reported a case of a 31-year-old woman who committed suicide, giving methomyl in drinks to herself and her two children. The 9-year-old elder son survived. In autopsies performed on the mother and the 6-year-old son, the mucous membranes of the stomach were blackish-brown, markedly edematous and congested. The lungs were heavy and congested. On the basis of measured stomach contents and tissue levels, it was estimated that the total doses taken were 2.75 g (55 mg/kg) by the mother and 0.26 g (13 mg/kg) by the child.

2. Long-term Exposure

• Morse and Baker (1979) reported on a survey of the health of workers in a plant that manufactured methomyl. The plant had also manufactured propanil, an herbicide manufactured from 3,4-dichloroaniline. The plant employed 111 workers in seven job categories. A complete work history, symptoms or history of poisoning, personal habits, and sources of other chemical exposure were obtained. Blood samples were collected from 100 of the 111 workers (96% males). Blood chemistries, blood counts and cholinesterase (ChE) determinations were carried out. A routine urinalysis was also performed. Average employment at the plant was 2 years. Packaging workers had the highest rate of adverse symptoms: small pupils (46%), nausea and vomiting (46%), blurred vision (46%) and increased salivation (27%). Biomedical examination did not demonstrate significant effects, and acetylcholinesterase findings were normal. Other effects, such as chloracne, were reported but were considered related to propanil exposure.

B. Animals

1. *Short-term Exposure*

- The acute oral LD_{50} reported for methomyl in the fasted male and female rat ranged from 17 to 25 mg/kg (Bedo and Cieleszky, 1980; Dashiell and Kennedy, 1984; Kaplan and Sherman, 1977). The oral LD_{50} in the nonfasted rat was 40 mg/kg (Dashiell and Kennedy, 1984). Clinical signs in rats included chewing motions, profuse salivation, lacrimation, bulging eyes, fasciculations and tremors characteristic of ChE inhibition.

- The acute oral LD_{50} for methomyl in the mouse ranged from 27 to 35 mg/kg (Boulton et al., 1971; El-Sebae et al., 1979).

- The oral LD_{50} in hens was 28 mg/kg; in Japanese quail, it was 34 mg/kg (Kaplan and Sherman, 1977).

- The 4-hour inhalation LC_{50} of methomyl in rats was 300 mg/m^3. Animals showed the typical signs of ChE inhibition, including salivation, lacrimation and tremors (ACGIH, 1984).

- Bedo and Cieleszky (1980) administered single oral doses of methomyl (purity not specified) by gavage to stock colony rats at dose levels of 0, 2, 3 or 10 mg/kg. The high dose (10 mg/kg) produced tremors in rats, and brain ChE levels were decreased. In the liver, mixed-function oxidase and glucose-6-phosphatase activity, and levels of glycogen and vitamin A were unaffected. Apparently, dose levels of 2 or 3 mg/kg did not produce these effects, but did result in increased activities of chymotrypsin, lipase and amylase in pancreatic juice.

- Woodside et al. (1978) fed methomyl (purity not specified) in the diet to male and female Wistar rats for 7 days at dose levels of 0, 5.0, 17 or 41 mg/kg/day in males and 0, 6.3, 15 or 39 mg/kg/day in females. Body weight gain was depressed at doses of 17 and 41 mg/kg/day in the males and at 15 and 39 mg/kg/day in the females. Liver and kidney weight were also depressed at 41 mg/kg/day in the male rat and at 15 and 39 mg/kg/day in the female rat. No effects were noted at the lowest doses. This study did not mention clinical signs of toxicity, and no measurements of plasma or brain ChE activity were reported. The No-Observed-Adverse-Effect Level (NOAEL) identified in this study is 5.0 mg/kg/day.

- Bedo and Cieleszky (1980) fed methomyl (purity not specified) in the diet at levels of 0, 100, 400 or 800 ppm to young adult male and female stock colony rats for 10 days. Assuming that 1 ppm in the diet of rats is equivalent to 0.05 mg/kg/day (Lehman, 1959), these levels correspond to 0, 5, 20 or 40 mg/kg/day. Brain ChE inhibition could not be detected at any dietary level. The only findings were increased mixed-function oxidase activity in the livers of female rats at 400 and 800 ppm. This study identified a NOAEL of 800 ppm (40 mg/kg/day).

- Kaplan and Sherman (1977) administered methomyl (90% pure) to six male Charles River CD rats at 0 or 5.1 mg/kg/day, five times a week for 2 weeks. Following treatment, survival, clinical signs, ChE activity and histopathology were evaluated. All rats survived the dosing period. Clinical signs in treated rats included chewing motions, profuse salivation, lacrimation, bulging eyes, fasciculations and tremors characteristic of ChE inhibition. The authors reported that the signs became less pronounced after the first week of dosing, indicating some degree of adaptation. Plasma ChE was comparable to control levels, and no compound-related histopathologic effects were reported. A Lowest-Observed-Adverse-Effect Level (LOAEL) of 5.1 mg/kg/day was identified from this study.

2. Dermal/Ocular Effects

- Kaplan and Sherman (1977) applied a 52.8% aqueous suspension of methomyl to the clipped, intact skin of six adult male albino rabbits and covered the area with an occlusive patch for a 24-hour period. The lethal dose was found to be greater than 5,000 mg/kg, the maximum tolerated dose.

- McAlack (1973) reported a 10-day subacute exposure of rabbit skin to methomyl. Male albino rabbits, six per dosage group, were treated with 0, 50 or 100 mg/kg/day for 10 days. The compound was diluted in water (29% solution), placed on the skin and covered with an occlusive covering for 6 hours per day. No signs of ChE inhibition were noted in any of the animals.

- Rabbits (5/sex) survived 15 daily doses of 200 mg/kg/day of methomyl applied to intact skin. When the same dose of methomyl was applied to abraded skin, rabbits showed labored respiration, nasal discharge, salivation, excessive mastication, tremors, poor coordination, hypersensitivity and abdominal hypertonia. These effects occurred within 1 hour after dosing in most animals. One animal died after the first dose, and another died after the eighth application. These deaths appeared to be compound-related (Kaplan and Sherman, 1977).

3. *Long-term Exposure*

- Kaplan and Sherman (1977) reported a 90-day feeding study in Charles River CD rats (10/sex/group) given food containing methomyl (90% purity) at dietary levels of 0, 10, 50, 125 or 250 ppm active ingredient (a.i.). Assuming that 1 ppm in the diet of rats is equivalent to 0.05 mg/kg/day (Lehman, 1959), these levels correspond to doses of about 0, 0.5, 2.5, 6.2 or 12.5 mg/kg/day. After 6 weeks, the 125-ppm dose was increased to 500 ppm (25 mg/kg/day) for the remainder of the study. Clinical signs, biochemical analyses (including plasma ChE) and urinalyses were not abnormal. In a few cases, lower hemoglobin values were observed at 1 month in females receiving 50 ppm (2.5 μg/kg/day) and at 2 months in males receiving 250 ppm. At 3 months, the red cell count of female rats at 250 ppm was somewhat lower than controls, but still within normal limits. These findings were consistent with moderate increases of erythroid components observed histologically in the bone marrow. Microscopic examination of all other tissues showed no consistent abnormalities. Based on these observations, this study identified a NOAEL of 50 ppm (2.5 mg/kg/day) and a LOAEL of 250 ppm (12.5 mg/kg/day).

- In a 90-day study using dogs, Kaplan and Sherman (1977) fed methomyl (90% pure) to four males and four females, 11 to 13 months of age, at dietary levels of 0, 50, 100 or 400 ppm a.i. Assuming that 1 ppm in the diet of dogs is equivalent to 0.025 mg/kg/day (Lehman, 1959), these levels correspond to doses of about 0, 1.25, 2.5 or 10 mg/kg/day. Hematological, biochemical and urine analyses were conducted at least three times on each dog prior to the study and then at 1, 2 and 3 months during the exposure period. Body weight was monitored weekly. At necropsy, organ weights were recorded, and over 30 tissues were prepared for histopathologic examination. No effects attributable to methomyl were found during or at the conclusion of the study. Based on these data, a NOAEL of 10 mg/kg/day was identified.

- Homan et al. (1978) reported a 13-week dietary study of methomyl (purity not specified) in Fischer 344 rats. Dose levels were reported to be 0, 1, 3, 10.2 or 30.2 mg/kg/day for male rats, and 0, 1, 3, 9.9 or 29.8 mg/kg/day for female rats. There were no deaths or clinical signs of toxicity. The body weight gain of females (but not males) was significantly depressed at all dose levels from day 28 until completion of the study. Kidney weight to body weight ratios were significantly increased in female rats at the two highest dose levels, but absolute kidney weights were not significantly increased. Red blood cell ChE activity was elevated at the high dose levels, but plasma and brain ChE levels were normal at all dose levels. Histopathological examination of 31 tissues from representative high-dose and control animals revealed no significant effects. Weights of brain, liver, kidney, spleen, heart, adrenals and testes were not altered. This study identified a NOAEL of 3 mg/kg/day and a LOAEL of 9.9 mg/kg/day.

- Bedo and Cieleszky (1980) reported a 90-day feeding study of methomyl in male and female rats receiving dietary levels of 100 or 200 ppm. Assuming that 1 ppm in the diet of rats is equivalent to 0.05 mg/kg/day (Lehman, 1959), these levels correspond to doses of 5 or 10 mg/kg/day. At 200 ppm, the female rats showed decreased brain ChE activity, decreased liver vitamin A content and elevated total serum lipids. This study identified a NOAEL of 100 ppm (5 mg/kg/day).

- Kaplan and Sherman (1977) reported a 22-month dietary feeding study in which Charles River CD male and female rats were fed methomyl (90 or 100% pure) at dietary levels of 0, 50, 100, 200 or 400 ppm a.i. Assuming that 1 ppm in the diet of rats is equivalent to 0.05 mg/kg/day (Lehman, 1959), these levels correspond to doses of about 0, 2.5, 5, 10 or 20 mg/kg/day. Mortality data were not reported. At autopsy, 9 of 13 males and 21 of 23 females at the 400-ppm level had kidney tubular hypertrophy and vacuolization of epithelial cells of the proximal convoluted tubules. Compound-related histological alterations were also seen in the spleens of female rats at the 200-ppm dose level. No effects were seen on ChE levels in plasma or red blood cells. This study identified a LOAEL of 200 ppm (10 mg/kg/day) and a NOAEL of 100 ppm (5 mg/kg/day).

- Kaplan and Sherman (1977) performed a 2-year feeding study in beagle dogs (4/sex/dose). Methomyl (90 or 100% pure) was supplied at dietary levels of 0, 50, 100, 400 or 1,000 ppm a.i. Assuming that 1 ppm in the diet of dogs is equivalent to 0.025 mg/kg/day (Lehman, 1959), these levels correspond to doses of about 0, 1.25, 2.5, 10 or 25 mg/kg/day. Hematological, biochemical (including plasma- and red-blood-cell ChE activity) and urine analyses were conducted once on each dog prior to the start of the study, at 3, 6, 12 and 18 months during the exposure period and at 24-month sacrifice. At 1 year, one male and one female per dose group were sacrificed for histopathological examination. One female dog at the 1,000-ppm dose level died after 8 weeks in the study, and a replacement dog died after 18 days. Death was preceded by convulsive seizures and coma. These deaths appear to be compound-related. Two male dogs in the 1,000-ppm dose group showed clinical signs during week 13, including tremors, salivation, incoordination and circling movements. Hematological studies revealed slight-to-moderate anemia in five dogs (1,000-ppm dose group) at 3 months, which persisted in one dog to sacrifice. No compound-related signs or effects were noted with respect to appetite, body weight changes, biochemical studies (including ChE) and urinanalyses. Dose-related histopathological changes were seen in kidney and spleen of animals receiving 400 and 1,000 ppm. Changes were also seen in livers and bone marrow of animals receiving 1,000 ppm. Pigment deposition was noted in the epithelial cells of the proximal convoluted tubules of the kidney in males at 400 and 1,000 ppm and in females at 1,000 ppm. A minimal-to-slight increase in bile duct proliferation and a slight increase in bone marrow activity was seen in animals receiving 1,000 ppm. The authors concluded that histological results indicated a NOAEL of 100 ppm (2.5 mg/kg/day). Minimal histopathological changes seen in the kidneys and spleen of animals receiving 400 ppm (10 mg/kg/day) identified this level as the LOAEL.

- Hazelton Laboratories (1981) reported a 2-year study of methomyl (99% purity) in mice. Male and female CD-1 mice (80/sex/dose) were fed methomyl in the diet at dose levels of 0, 50, 100, or 800 ppm for 104 weeks. Assuming 1 ppm in the diet to be equivalent to 0.15 mg/kg/day (Lehman, 1959), these levels correspond to doses of about 0, 7.5, 15 or 120 mg/kg/day. Survival was significantly reduced (no details provided) in both males and females at the 800-ppm dose level by week 26. The 800-ppm dose level was reduced to 400 ppm (1.0 mg/kg/day) at week 28 and then further reduced to 200 ppm (30 mg/kg/day) at week 39. At week 39, the 100 ppm was decreased to 75 ppm (11.2 mg/kg/day). Survival was depressed in all groups of treated males at 104 weeks. No compound-related histopathological changes were noted in tissues of animals necropsied at 104 weeks. A LOAEL of 50 ppm (7.5 mg/kg/day, the lowest dose tested) may be identified based on decreased survival.

- Haskell Laboratories (1981) reported a 2-year study of methomyl (99% purity) in rats. Charles River CD rats (100/sex/dose) were fed methomyl in the diet at dose levels of 0, 50, 100 or 400 ppm for up to 2 years. Assuming 1 ppm in the diet is equivalent to 0.05 mg/kg/day (Lehman, 1959), these levels correspond to doses of 0, 2.5, 5 or 20 mg/kg/day. Effects seen in the 400-ppm group include reduced weight gain in both sexes, and, in females, lower erythrocyte counts, hemoglobin values and hematocrits. Blood and brain cholinesterase levels were within the range, as were other clinical chemistry parameters. A NOAEL of 5 mg/kg/day and a LOAEL of 20 mg/kg/day were identified from this study.

4. *Reproductive Effects*

- Male and female weanling Charles River CD rats were fed methomyl (90% pure) at dietary levels of 0, 50 or 100 ppm a.i. for 3 months. Assuming that 1 ppm in the diet of weanling rats is equivalent to 0.05 mg/kg/day (Lehman, 1959), these doses correspond to about 0, 2.5 or 5 mg/kg/day. Ten males and 20 females from each group were bred and continued on the diet through three generations. No adverse effects were reported on reproduction or lactation, and no pathologic changes were found in the weanling pups of the F_{3b} generation (Kaplan and Sherman, 1977). A NOAEL of 5 mg/kg/day was identified from the highest dose tested.

5. *Developmental Effects*

- New Zealand White rabbits, five per group, were dosed with 0, 2, 6 or 16 mg/kg of methomyl (98.7% pure) on days 7 through 19 of gestation. One animal died at the 16 mg/kg dose level, exhibiting characteristic signs of ChE inhibition, including tremors, excitability, salivation and convulsions. No adverse effects were observed at any dose level on embryo viability or on the frequency of soft-tissue or skeletal malformations (Feussner et al., 1983). This study identified a maternal NOAEL of 6 mg/kg and a teratogenic NOAEL of 16 mg/kg/day, the highest dose tested.

- Kaplan and Sherman (1977) fed methomyl (90% pure) to pregnant New Zealand White rabbits on days 8 to 16 of gestation at dietary levels of 0, 50 or 100 ppm active ingredient. Assuming that 1 ppm in the diet of rabbits is equivalent to 0.03 mg/kg/day (Lehman, 1959), these levels correspond to doses of about 0, 1.5 or 3 mg/kg/day. One-third of the fetuses were stained with Alizarin Red S and examined for skeletal defects. Since no soft-tissue or skeletal abnormalities were observed at any dose level tested, a NOAEL of 3 mg/kg/day was identified.

6. *Mutagenicity*

- Methomyl has been reported to be negative in the Ames test utilizing *Salmonella typhimurium* strains TA98, TA100, TA1535, TA1537 and TA1538 without metabolic activation (Blevins et al., 1977; Moriya et al., 1983). Waters et al. (1980) reported methomyl as negative with and without metabolic activation in strains TA100, TA1535, TA1537 and TA1538.

7. *Carcinogenicity*

- Kaplan and Sherman (1977) fed ChR-CD rats (35/sex/dose) methomyl (90% pure) in the diet at levels of 0, 50, 100, 200 or 400 ppm active ingredient for 22 months. Assuming that 1 ppm in the diet of rats is equivalent to 0.05 mg/kg/day (Lehman, 1959), these doses correspond to about 0, 2.5, 5, 10 or 20 mg/kg/day. Gross and histological examination revealed no increased tumor incidence in either male or female rats.

- Haskell Laboratories (1981) report on a study in which Charles River CD rats (100/sex/group) were administered methomyl (99% pure) in the diet at dietary levels of 0, 50, 100 or 400 ppm for 2 years. Gross and histological examination revealed no increase in tumor incidence in either sex.

- Hazelton Laboratories (1981) reported the results of a 2-year study of methomyl (purity not specified) in CD-1 mice (80/sex/dose). Initial dose levels were 0, 50, 100, or 800 ppm. Assuming that 1 ppm in the diet of mice is equivalent to 0.15 mg/kg/day (Lehman, 1959), these doses correspond to 0, 7.5, 15 or 120 mg/kg/day. Because of early mortality, the 800-ppm dose was reduced to 400 ppm (60 mg/kg/day) at week 28, and then to 200 ppm (30 mg/kg/day) at week 39. At week 29, the 100-ppm dose was reduced to 75 ppm (11.2 mg/kg/day). Histological examination at necropsy did not reveal any treatment-related effects on tumor incidence.

IV. Quantification of Toxicological Effects

A. One-Day Health Advisory

No information found in the available literature was suitable for determination of the One-day HA value for methomyl. It is, therefore, recommended that the Longer-term HA (0.3 mg/L), calculated below, be used at this time as a conservative estimate of the One-day HA value.

B. Ten-day Health Advisory

The health effects associated with acute and subchronic exposure to methomyl are primarily associated with cholinesterase (ChE) inhibition. Symptoms of ChE inhibition have been shown in rats administered methomyl via gavage at doses as low as 5.1 mg/kg/day for 2 weeks (Kaplan and Sherman, 1977). Methomyl incorporated into the diet may have less dramatic effects; no ChE effects were observed in rats exposed subchronically to methomyl at dietary levels of 100 ppm, equivalent to a dose of 5 mg/kg/day (Kaplan and Sherman, 1977; Bedo and Cieleszky, 1980). A similar NOAEL, 2.5 mg/kg/day, was found for lifetime studies in dogs (Kaplan and Sherman, 1977). No controlled human studies have been performed, but human fatalities from methomyl ingestion after a single exposure to an estimated dose of 12 mg/kg in bread or 13 mg/kg in drinks have been reported (Liddle et al., 1979; Araki et al., 1982).

Because the timing and nature of administration can profoundly affect the expression of methomyl toxicity, and little margin of safety can be expected between doses that are fatal and those that cause little or no acute toxicity, the available acute studies were judged to be inadequate for the basis of the Ten-day HA value. Therefore, it is recommended that the Longer-term HA for a 10-kg child (0.3 mg/L) be used at this time as a conservative estimate of the Ten-day HA value.

C. Longer-term Health Advisory

The onset of subchronic or chronic methomyl toxicity appears to occur at doses similar to those that cause acute toxicity. Acute ChE inhibition has been reported to occur at doses as low as 5.1 mg/kg/day for rats exposed via gavage, and human fatalities from ingestion of approximate methomyl doses of 12 mg/kg in bread and 13 mg/kg in drinks have been reported (Liddle et al., 1979; Araki et al., 1982). Kidney toxicity (increased kidney weight and hypertrophy) in acute, subchronic and chronic conditions has been reported at doses of 15, 9.9 and 10 mg/kg/day, respectively (Woodside et al., 1978; Homan et al., 1978; Kaplan and Sherman, 1977). Because of the severity of the toxic effects, the most conservative estimate of the NOAEL (2.5 mg/kg/day from the chronic study in dogs; Kaplan and Sherman 1977) was selected as the basis for the Longer-term HA. In this study, beagle dogs were exposed to methomyl in the diet at approximate doses of 0, 1.25, 2.5, 10 or 25 mg/kg/day for 2 years. Dogs receiving 1.25 or 2.5 mg/kg/day showed no evidence of toxic effects while those receiving the two highest doses exhibited histopathological changes in the kidney and spleen. Based on a NOAEL of 2.5 mg/kg/day, the Longer-term HA for the child and adult are calculated as follows:

$$\text{Longer-term HA}_{\text{child}} = \frac{(2.5 \text{ mg/kg/day}) (10 \text{ kg})}{(100)(1 \text{ L/day})} = 0.3 \ \mu\text{g/L} \ (300 \ \mu\text{g/L})$$

$$\text{Longer-term HA}_{\text{adult}} = \frac{(2.5 \text{ mg/kg/day}) (70 \text{kg})}{(100) (2 \text{ L/day})} = 0.9 \ \text{mg/L} \ (900 \ \mu\text{g/L})$$

D. Lifetime Health Advisory

Chronic exposure to methomyl in the diet induces renal toxicity in rats and dogs. Rats exposed to 900 ppm (20 mg/kg/day) for 22 months exhibited kidney tubular hypertrophy and vacuolation of the epithelial cells, and effects on the kidney (increased weight) have also been observed in rats exposed to 9.9 mg/kg/day in the diet for 13 weeks (Homan et al., 1978). Dogs exposed to 400 ppm (10 mg/kg/day) for 2 years exhibited swelling and increased pigmentation of the epithelial cells of the proximal tubules (Kaplan and Sherman, 1977). The NOAEL of 2.5 mg/kg/day identified from the dog study is a conservative estimate of the NOAEL and serves as the basis for the Lifetime HA.

In this study, beagle dogs (4/sex/dose) were exposed to 50, 100, 400 or 1,000 ppm methomyl in the diet for 2 years (1.25, 2.5, 10 and 25 mg/kg/day). Dogs receiving 1.25 or 2.5 mg/kg/day showed no evidence of toxic effects. Those receiving 10 mg/kg/day exhibited histopathological changes in the kidney and spleen. In addition to these effects, animals receiving the highest dose also exhibited symptoms of central nervous system (CNS) toxicity, as well as liver and bone marrow effects.

Using a NOAEL of 2.5 mg/kg/day, the Lifetime HA is calculated as follows:

Step 1: Determination of the Reference Dose (RfD)

$$RfD = \frac{(2.5 \text{ mg/kg/day})}{(100)} = 0.025 \text{ mg/kg/day}$$

Step 2: Determination of the Drinking Water Equivalent Level (DWEL)

$$DWEL = \frac{(0.025 \text{ mg/kg/day}) (70 \text{ kg})}{(2 \text{ L/day})} = 0.9 \text{ mg/L } (900 \text{ } \mu g/L)$$

Step 3: Determination of the Lifetime Health Advisory

$$\text{Lifetime HA} = (0.9 \text{ mg/L}) (20\%) = 0.2 \text{ mg/L } (200 \text{ } \mu g/L)$$

E. Evaluation of Carcinogenic Potential

- Two-year carcinogenicity studies in rats and mice (Kaplan and Sherman, 1977; Haskell Laboratories, 1981; Hazelton Laboratories, 1981) have not revealed any evidence of carcinogenicity.

- The International Agency for Research on Cancer has not evaluated the carcinogenic potential of methomyl.

- Applying the criteria described in EPA's final guidelines for assessment of carcinogenic risk (U.S. EPA, 1986), methomyl may be classified in Group E: no evidence of carcinogenicity. This group is used for substances that show no evidence of carcinogenicity in at least two adequate animal tests in different species or in both epidemiologic and animal studies. Two-year studies in rats and mice have not revealed any evidence of carcinogenicity (Kaplan and Sherman, 1977; Haskell Laboratories, 1981; Hazelton Laboratories, 1981).

V.   Other Criteria, Guidance and Standards

- The National Academy of Sciences (NAS, 1983) has a Suggested-No-Adverse-Response-Level (SNARL) of 0.175 mg/L, which was calculated using an uncertainty factor of 100 and a NOAEL of 2.5 mg/kg/day identified in the 2-year dog study by Kaplan and Sherman (1977).

- Residue tolerances have been established for methomyl in or on raw agricultural commodities (U.S. EPA, 1985). These tolerances are based on an ADI value of 0.025 mg/kg/day, which is based on a NOAEL of 2.5 mg/kg/day in dogs and an uncertainty factor of 100. Residues range from 0.1 (negligible) to 40 ppm.

- The World Health Organization identified a Temporary ADI of 0.01 mg/kg/day for methomyl (Vettorazzi and Van den Hurk, 1985), based on the same chronic toxicity data, but using a larger uncertainty factor than that used to derive the RfD.

- ACGIH (1984) has adopted a threshold limit value (TLV) of 0.2 mg/m^3 as a time-weighted average exposure to methomyl for an 8-hour day.

VI.   Analytical Methods

- Analysis of methomyl is by a high-performance liquid chromatographic (HPLC) procedure used for the determination of N-methyl carbamoyl-oximes and N-methylcarbamates in drinking water (U.S. EPA, 1988). In this method, the water sample is filtered and a 400-μL aliquot is injected into a reverse-phase HPLC column. Compounds are separated by gradient elution chromatography. After elution from the HPLC column, the compounds are hydrolyzed with sodium hydroxide. The methyl amine formed during hydrolysis is reacted with o-phthalaldehyde to form a fluorescent derivative that is detected using a fluorescence detector. This method has been validated in a single laboratory, and the estimated detection limit for methomyl is 0.5 μg/L.

VII.   Treatment Technologies

- Available data indicate that granular-activated carbon (GAC) adsorption will remove methomyl from water. Whittaker (1980) experimentally determined adsorption isotherms for methomyl solutions on GAC.

- Whittaker (1980) reported the results of GAC columns operating under bench-scale conditions. At a flow rate of 0.8 gpm/ft^2 and empty bed contact time of 6 minutes, methomyl breakthrough (when effluent concentration equals 10% of influent concentration) occurred after 124 bed volumes (BV). When a bi-solute methomyl-metribuzin solution was passed over the same column, methomyl breakthrough occurred after 55 BV.

- Treatment technologies for the removal of methomyl from water are available and have been reported to be effective (Whittaker, 1980). However, the selection of an individual technology or a combination of technologies must be based on a case-by-case technical evaluation, and an assessment of the economics involved.

VIII.   References

ACGIH. 1984. American Conference of Governmental Industrial Hygienists. Documentation of the threshold limit values for substances in workroom air, 3rd ed. Cincinnati, OH: ACGIH.

Andrawes, W.R., R.H. Bailey and G.C. Holsing.* 1976. Metabolism of acetyl-1-^{14}C-methomyl in the rat. Report No. 26946. Unpublished study.

Araki, M., K. Yonemitsu, T. Kambe, D. Idaka, S. Tsunenari, M. Kanda and T. Kambara. 1982. Forensic toxicological investigation on fatal cases of carbamate pesticide methomyl (Lannate) poisoning. Nippon Hoigaku Zasshi. 36:584–588.

Baron, R.L. 1971. Toxicological considerations of metabolism of carbamate insecticides: methomyl and carbaryl. Invited Paper, Int. Symp. on Pesticide Terminal Residues. Washington, DC. pp. 185–197.

Bedo, M. and V. Cieleszky. 1980. Nutritional toxicology in the evaluation of pesticides. Bibl. Nutr. Dieta. 29:20–31.

Blevins, R.D., M. Lee and J.D. Regan. 1977. Mutagenicity screening of five methyl carbamate insecticides and their nitroso derivatives using mutants of *Salmonella typhimurium* LT2. Mutat. Res. 56:1–6.

Boulton, J.J., C.B. Boyce, P.J. Jewess and R.F. Jones. 1971. Comparative properties of N-acetyl derivatives of oxime N-methylcarbamates and aryl N-methylcarbamates as insecticides and acetylcholinesterase inhibitors. Pestic. Sci. 2:10–15.

CHEMLAB. 1985. The Chemical Information System, CIS, Inc., Baltimore, MD.

Cohen, S.Z. 1984. List of potential groundwater contaminants. Memorandum to I. Pomerantz. Washington, DC: U.S. Environmental Protection Agency. August 28.

Dashiell, O.L. and G.L. Kennedy. 1984. The effects of fasting on the acute oral toxicity of nine chemicals in the rat. J. Appl. Toxicol. 4(5):320–325.

Dorough, H.W. 1977. Metabolism of carbamate insecticides. Available from the National Technical Information Service, Springfield, VA. PB-266 233.

E.I. duPont de Nemours and Co.* 1971. Methomyl decomposition in muck soil—a field study. Unpublished study.

El-Sebae, A.H., S.A. Soliman, A. Khalil and S. El-Fiki. 1979. Comparative selective toxicity of some insecticides to insects and mammals. Proc. Br. Crop Prot. Conf. Pests and Diseases. pp. 731–736.

Feussner, E., M. Christian, G. Lightkep et al.* 1983. Embryo-fetal toxicity and teratogenicity study of methomyl in the rabbit. Study No. 104–005. Unpublished study. MRID 00131257.

Han, J.C.* No date. Evaluation of possible effects of methomyl on nitrifying bacteria in soil. Unpublished study. Wilmington, DE: E.I. duPont de Nemours and Co.

Harvey, J.* No date(a). Decomposition of ¹⁴C-methomyl in a high organic matter soil in the laboratory. Unpublished study. Wilmington, DE: E.I. duPont de Nemours and Co.

Harvey, J.* No date(b). Exposure of S-methyl N-[(methylcarbamoyl)oxy]thioacetimidate in sunlight, water and soil. Unpublished study. Wilmington, DE: E.I. duPont de Nemours and Co.

Harvey, J., Jr. and H.L. Pease.* No date. Decomposition of methomyl in soil. Unpublished study. Wilmington, DE: E.I. duPont de Nemours and Co.

Harvey, J., Jr.* 1974. Metabolism of aldicarb and methomyl. Environmental quality and safety supplement, Vol. III. Pesticides. International Union of Pure and Applied Chemistry Third International Congress. Helsinki, Finland.

Harvey, J., Jr.* 1977a. Decomposition of ¹⁴C-methomyl in a sandy loam soil in the greenhouse. Unpublished study. Prepared in cooperation with the University of Delaware, Soil Testing Laboratory. Submitted by E.I. duPont de Nemours and Co., Wilmington, DE.

Harvey, J., Jr.* 1977b. Degradation of ¹⁴C-methomyl in Flanagan silt loam in biometer flasks. Unpublished study. Prepared in cooperation with the University of Delaware, Soil Testing Laboratory. Submitted by E.I. duPont de Nemours and Co., Wilmington, DE.

Haskell Laboratories. 1981. Long-term feeding study in rats with methomyl. INX-1179. Final report. Haskell Laboratory Report No. 235-81.

Hazelton Laboratories.* 1981. Final report: 104-week chronic toxicity and carcinogenicity study in mice. Project No. 201-510. Unpublished study. EPA ACC# 070241.

Homan, E.R., R.R. Maronpot and J.B. Reid.* 1978. Methomyl: inclusion in the diet of rats for 13 weeks. Project Report 41-64. Unpublished study. MRID 00044881.

Huang, C.Y. 1978. Effects of nitrogen fixing activity of blue-green algae on the yield of rice plants. Botanical Bull. Academia Sinica. 19(1):41-52.

Kaplan, M.A. and H. Sherman. 1977. Toxicity studies with methyl N-[[(methyl-amino)carbonyl]oxy]-ethanimidothioate. Toxicol. Appl. Pharmacol. 40:1-17.

Lehman, A.J. 1959. Appraisal of the safety of chemicals in foods, drugs and cosmetics. Assoc. Food and Drug Off. U.S., Q. Bull.

Liddle, J.A., R.D. Kimbrough, L.L. Needham, R.E. Cline, A.L. Smrek, L.W. Yert and D.D. Bayse. 1979. A fatal episode of accidental methomyl poisoning. Clin. Toxicol. 15:159-167.

McAlack, J.W.* 1973. 10-day subacute exposure of rabbit skin to lannate® insecticide. Haskell Laboratory Report No. 24-73. Unpublished study. MRID 00007032.

Meister, R., ed. 1983. Farm chemicals handbook. Willoughby, OH: Meister Publishing Company.

Moriya, M., T. Ohta, K. Watanabe, T. Miyazawa, K. Kato and Y. Shirasu. 1983. Further mutagenicity studies on pesticides in bacterial reversion assay systems. Mutat. Res. 116:185-216.

Morse, D.L. and E.L. Baker. 1979. Propanil-chloracne and methomyl toxicity in workers of a pesticide manufacturing plant. Clin. Toxicol. 15:13-21.

NAS. 1983. National Academy of Sciences. Drinking water and health. Vol. 5. Washington, DC: National Academy Press.

Natoff, I.L. and B. Reiff. 1973. Effects of oximes on the acute toxicity of anticholinesterase carbamates. Toxicol. Appl. Pharmacol. 25:569-575.

Peeples, J.L.* 1977. Effect of methomyl on soil microorganisms. Unpublished study. Submitted by E.I. duPont de Nemours and Co., Wilmington, DE.

STORET. 1988. STORET Water Quality File. Office of Water. U.S. Environmental Protection Agency (data file search conducted in May, 1988).

TDB. 1985. Toxicology Data Bank. MEDLARS II. National Library of Medicine's National Interactive Retrieval Service.

U.S. EPA. 1985. U.S. Environmental Protection Agency. Code of Federal Regulations. 40 CFR 180.253. July 1. p. 278.

U.S. EPA. 1986. U.S. Environmental Protection Agency. Guidelines for carcinogen risk assessment. Fed. Reg. 51(185):33992–34003. September 24.

U.S. EPA. 1988. U.S. Environmental Protection Agency. Method 531 — Measurement of N-methyl carbamoyloximes and N-methylcarbaxetes in drinking water by direct aqueous injection HPLC with post column derivatization. Cincinnati, OH: Environmental Monitoring and Support Laboratory, ECAO.

Vettorazzi, G. and G.W. Van den Hurk. 1985. Pesticides reference index. Joint Meeting of Pesticide Residues, 1961–1984. p. 10.

Waters, M.D., V.F. Summon, A.D. Mitchell, T.A. Jorgenson and R. Valencia. 1980. An overview of short-term tests for the mutagenic and carcinogenic potential of pesticides. J. Environ. Sci. Health. B15(6):867–906.

Whittaker, K.F. 1980. Adsorption of selected pesticides by activated carbon using isotherm and continuous flow column systems. Ph.D. Thesis. Purdue University.

Windholz, M., S. Budavari, R.F. Blumetti, and E.S. Otterbein, eds. 1983. The Merck index — an encyclopedia of chemicals and drugs, 10th ed. Rahway, NJ: Merck and Company, Inc.

Woodside, M.D., L.R. DePass and J.B. Reid.* 1978. UC 45650: results of feeding in the diet of rats for 7 days. Project 41–102. Unpublished study. MRID 00044880.

---

* Confidential Business Information submitted to the Office of Pesticide Programs.

Methyl Parathion

I. General Information and Properties

A. CAS No. 298–00–0

B. Structural Formula

0,0-Dimethyl-0-(4-nitrophenyl) phosphorothioic acid

C. Synonyms

- Metaphos; Dimethyl parathion; Folidol M; Metocide; Penncap M; Sinafid M-48; Wofatox; Cekumethion; Devithion; Drexel Methyl Parathion 4E; E601; Fosferno M50; Gearfos; Parataf; Partron-M; Tekwaisa; Parathion-methyl (Meister, 1988).

D. Uses

- Methyl parathion is a restricted-use pesticide for control of various insects of economic importance; especially effective for boll weevil control (Meister, 1988).

E. Properties (Hawley, 1981; Meister, 1988; CHEMLAB, 1985; TDB, 1985)

Chemical Formula	$C_8H_{10}O_5NSP$
Molecular Weight	263.23
Physical State (25°C)	White crystalline solid
Boiling Point	--
Melting Point	35 to 36°C
Density	--
Vapor Pressure (20°C)	0.97×10^{-5} mm Hg
Specific Gravity	--
Water Solubility (25°C)	55 to 60 mg/L

Log Octanol/Water
   Partition Coefficient   3.11 (calculated)
Taste Threshold   --
Odor Threshold   --
Conversion Factor   --

F. Occurrence

- Methyl parathion has been found in 1,070 of 27,082 surface water samples analyzed and in 8 of 2,836 ground-water samples (STORET, 1988). Samples were collected at 3,558 surface water locations and 2,111 ground-water locations, and methyl parathion was found in 20 states. The 85th percentile of all nonzero samples was 1.18 μg/L in surface water and 0.05 μg/L in ground-water sources. The maximum concentration found was 13 μg/L in surface water and 0.05 μg/L in ground-water. This information is provided to give a general impression of the occurrence of this chemical in ground and surface waters as reported in the STORET database. The individual data points retrieved were used as they came from STORET and have not been confirmed as to their validity. STORET data are often not valid when individual numbers are used out of the context of the entire sampling regime, as they are here. Therefore, this information can only be used to form an impression of the intensity and location of sampling for a particular chemical.

G. Environmental Fate

- Methyl parathion (99% pure) at 10 ppm was added to sea water and exposed to sunlight; some samples were also kept in the dark (controls). After 6 days, 57% of the parent compound had degraded but the degradates were not identified. Since only 27% of the parent compound had degraded in the dark controls, this indicates that methyl parathion is subject to photodegradation in sea water (U.S. EPA, 1981).

- The degradation rate of two formulations (EC and MCAP) of methyl parathion, applied at 0.04 ppm, was compared in a sediment/water system. Degradates were not identified; however, the parent compound had a half-life of 1 to 3 days in water. In the hydrosoil plus sediment, methyl parathion applied as an emulsifiable concentrate formulation had a half-life of 1 to 3 days, whereas for the microencapsulated formulation, the half-life was 3 to 7 days (Agchem, 1983).

- Methyl parathion was relatively immobile in 30-cm soil columns of sandy loam, silty clay loam and silt loam soils leached with 15.7 inches of water, with no parent compound found below 10 cm or in the column leachate, which was the case for the column of sand (Pennwalt Corporation, 1977).

- Methyl parathion (MCAP or EC formulation) at 5 lb a.i./A (active ingredient/acre) was detected in runoff water from field plots irrigated 4 to 5 days posttreatment. Levels found in soil and turf plots ranged from 0.13 to 21 ppm and 0.17 to 0.20 ppm, respectively (Pennwalt Corporation, 1972).

- A field dissipation study with methyl parathion (4 lb/gal EC) at 3 lb a.i./A, applied alone or in combination with Curacron, dissipated to nondetectable levels (< 0.05 ppm) within 30 days in silt loam and loamy sand soils (Ciba-Geigy Corporation, 1978).

II.   Pharmacokinetics

A. Absorption

- Braeckman et al. (1983) administered a single oral dose of ^{35}S-methyl parathion (20 mg/kg) by stomach tube to four mongrel dogs. Peak concentrations in plasma ranged from 0.13 to 0.96 μg/mL, with peak levels occurring 2 to 9 hours after dosing. In two dogs given single oral doses of ^{35}S-methyl parathion (3 mg/kg) in this study, absorption was estimated to be 77 and 79%, based on urinary excretion of label. The authors concluded that methyl parathion was well absorbed from the gastrointestinal tract.

- Hollingworth et al. (1967) gave a single oral dose of ^{32}P-labeled methyl parathion by gavage (3 or 17 mg/kg, dissolved in olive oil) to male Swiss mice. Recovery of label in the urine reached a maximum of about 85%, most of this occurring within 18 hours of dosing. The amount of label in the feces was low, never exceeding 10% of the dose. This indicated that absorption was at least 90% complete.

B. Distribution

- Ackermann and Engst (1970) administered methyl parathion to pregnant albino rats and examined the dams and fetuses for the distribution of the pesticide. The pregnant rats (weighing about 270 g each) were given 3 mg (11.1 mg/kg) of methyl parathion orally on days 1 to 3 of gestation and sacrificed 30 minutes after the last dose. Methyl parathion was detected in the maternal liver (25 ng/g); placenta (80 ng/g); and in fetal brain (35 ng/g), liver (40 ng/g) and back musculature (60 ng/g).

C. Metabolism

- Hollingworth et al. (1967) gave ^{32}P-labeled methyl parathion by gavage (3 or 17 mg/kg, dissolved in olive oil) to male Swiss mice. About 85% of the label appeared in the urine within 72 hours. Urinary metabolites identified 24 hours after the low dose were: dimethyl phosphoric acid (53.1%), dimethyl phosphorothioic acid (14.9%), desmethyl phosphate (14.1%), desmethyl phosphorothioate (11.7%), phosphoric acid (2.0%), methyl phosphoric acid (1.7%) and phosphate (0.6%). The radioactivity in the urine was fully accounted for by hydrolysis products and P = 0 activation products. No evidence was found for reduction of the nitro group to an amine, oxidation of the ring methyl group or hydroxylation of the ring. A generally similar pattern was observed at the high dose, except for a lower percentage of dimethyl phosphoric acid (31.9%) and higher percentages of desmethyl phosphate (23.1%) and desmethylphosphorothionate (18.8%). Based on this, the authors proposed a metabolic scheme involving oxidative desulfuration, oxidative cleavage of the phospho group from the ring and hydrolysis of the phosphomethyl esters.

- Neal and DuBois (1965) investigated the *in vitro* detoxification of methyl parathion and other phosphorothioates using liver microsomes prepared from adult male Sprague-Dawley rats. Metabolism was found to involve oxidative desulfuration followed by hydrolysis to yield p-nitrophenol. Extracts from livers of adult male rats exhibited higher metabolic activity than those of adult females (3.2 versus 1.9 units, where one unit equals 1 μg p-nitrophenol/50 mg liver extract) (p < 0.01). The activity of weanling rat liver (2.7 units) was intermediate between these two. In the case of adult CF-l mice, the activity of female liver (3.2 units) was significantly greater (p < 0.05) than that of the males (2.3 units). The activity of young adult male guinea pig liver extracts was 5.6 units. The authors noted that these differences in metabolic detoxification rates correlated with the sex and species differences in susceptibility to the acute oral toxic effects of this family of compounds.

- Nakatsugawa et al. (1968) investigated the degradation of methyl parathion using liver microsomes from adult male rats and rabbits (strains not specified). Metabolism occurred by two oxidative pathways: activation of the phosphorus-sulfur bond to the phosphorus-oxygen analog, and cleavage at the aryl phosphothioate bond to yield p-nitrophenol. These reactions occurred only in the presence of oxygen and $NADPH_2$. The amounts of phenol and oxygen analog formed were

3.8 and 3.7 μM in the rabbit liver extract and 2.5 and 5.4 μM in the rat liver extract, respectively.

D. Excretion

- Braeckman et al. (1983) administered individual doses of 3 mg/kg of ^{35}S-methyl parathion to two mongrel dogs. In each dog, the agent was given once intravenously and, 1 week later, once orally via stomach tube. This dosing pattern was repeated once in one dog. Urine was collected every 24 hours for 6 days after each treatment. Urinary excretion 6 days after oral dosing was 63% in the animal without repeated dosing and 70% and 78% in the other. Urinary excretion 6 days after intravenous dosing was 80% in the animal without repeated dosing and 95 to 96% in the other. Most of the label appeared in urine within 2 days. Other excretory routes were not monitored.

- Hollingworth et al. (1967) gave ^{32}P-labeled methyl parathion (3 or 17 mg/kg, dissolved in olive oil) by gavage to male Swiss mice. Recovery of label in the urine reached a maximum of about 85%, most of this occurring within 18 hours of dosing. The amount of label in the feces was low, never exceeding 10% of the dose. This indicated that absorption was at least 90% complete.

III.   Health Effects

A. Humans

1. *Short-term Exposure*

- Nemec et al. (1968) monitored cholinesterase (ChE) levels in two workers (entomologists) who examined plants in a cotton field after it had been sprayed with an ultra-low-volume (nonaqueous) preparation of methyl parathion (1.5 to 2 lb/acre). The men entered a cotton field to examine the plants on 3 different days over a 2-week period; two of these occasions were within 2 hours after the ultra-low-volume spraying, and the third occasion was 24 hours after a spraying. After each field trip their arms were washed with acetone and the adhering methyl parathion determined. It was found that contact with the plants 2 hours after spraying resulted in 2 to 10 mg of methyl parathion residue on the arms; exposure 24 hours after spraying resulted in a residue on the arms of 0.16 to 0.35 mg. The amount of pesticide absorbed was not estimated. No toxic symptoms were experienced by either man, but measurement of red blood cell ChE activity immediately after the third of these exposures showed a decrease in activity to 60 to 65% of preexposure levels. These values did not increase significantly over the next 24 hours. It was concluded that workers should not enter such a field until more than 24 hours, and preferably 48 hours, have elapsed after spraying with ultra-low-volume insecticide sprays. Water emulsion sprays were not tested, but the authors cautioned that it cannot be assumed that they are less hazardous than the ultra-low-volume spray residues.

- Rider et al. (1969, 1970, 1971) studied the toxicity of technical methyl parathion (purity not specified) in human volunteers. Each phase of the study was done with different groups of seven male subjects, five of whom were test subjects and two were vehicle controls (Rider et al., 1969). Each study phase was divided into a 30-day pretest period for establishing cholinesterase baselines, a 30-day test period when a specific dose of methyl parathion was given and a posttest period.

- Thirty-two different dosages were evaluated by Rider et al. (1969), ranging from 1 to 19 mg/day. Early in the study, several of the groups were given more than one dose level during a single phase. The initial amount was 1.0 mg with an increase of 0.5 mg during each succeeding test period up to 15.0 mg/day. At this point, the dose was increased by 1.0 mg/day to a total dose of 19.0 mg/day. Pesticide in corn oil was given orally in capsules, once per day for each test period of 30 days. At no time during any of the studies were there any significant changes in blood counts, urinalyses or prothrombin times, or was there any evidence of toxic side effects. Cholinesterase activity of the plasma and red blood cells (RBCs) was measured twice weekly prior to, during and after the dosing period. The authors considered a mean depression of 20 to 25% or greater in ChE activity below control levels to be indicative of the toxic threshold. At 11.0 mg/day, a depression of 15% in plasma ChE occurred, but doses up to and including 19 mg/day did not produce any significant ChE depression.

- Rider et al. (1970) studied the effects of 22, 24 and 26 mg/day technical methyl parathion. There were no effects observed at 22 mg/day. At 24 mg/day, plasma and RBC ChE depression was produced in two subjects, the maximum decreases being 24 and 23% for plasma, and 27 and 55% for RBC. The mean maximal decreases (in all five subjects) were 17% for plasma and 22% for RBC. With 26 mg/day RBC ChE depression was again produced in only two of the subjects, with maximum decreases of 25 and 37%. The mean maximum decrease was 18%. Plasma cholinesterase was not significantly altered.

- Rider et al. (1971) assessed the effects of 28 and 30 mg/day technical methyl parathion. At 28 mg/day, a significant decrease in RBC ChE was produced in three subjects (data not given), with a maximum mean decrease of 19%. With a dose of 30 mg/day, a mean maximum depression of 37% occurred. Based on their criteria of 20 to 25% average depression of ChE activity, the authors concluded that this was the level of minimal incipient toxicity. Body weights of the test subjects were not reported, but assuming an average body weight of 70 kg, a dose of 22 mg/day corresponds to a No-Observed-Adverse-Effect Level (NOAEL) of 0.31 mg/kg/day, and the 30 mg/day dose corresponds to 0.43 mg/kg/day. The NOAEL is considered to be 22 mg/day herein because of the apparent sensitivity of some individual subjects at higher doses to have met the 20 to 25% criteria for ChE depression as an effect.

2. Long-term Exposure

- No information was found in the available literature on the health effects of long-term exposure to methyl parathion in humans.

B. Animals

1. Short-term Exposure

- Reported oral LD_{50} values for methyl parathion include 14 and 24 mg/kg in male and female Sherman rats, respectively (Gaines, 1969); 14.5 and 19.5 mg/kg in male and female CD-1 mice, respectively (Haley et al., 1975); 30 mg/kg in male ddY mice (Isshiki et al., 1983); 18.0 and 8.9 mg/kg in male and female Sprague-Dawley rats, respectively (Sabol, 1985); and 9.2 mg/kg in rats of unreported strain (Galal et al., 1977).

- Galal et al. (1977) determined the subchronic median lethal dose (C-LD_{50}) of methyl parathion (purity not specified) in adult albino rats. Groups of 10 animals received an initial daily oral dose (by gavage) of 0.37 mg/kg (4% of the acute oral LD_{50}). Every fourth day the dose was increased by a factor of 1.5 (dose based on the body weight of the animals as recorded at 4-day intervals). Treatment was continued until death or termination at 36 days. Hematological and blood chemistry analyses were performed initially and on the 21st and 36th days of the study. Histopathological studies of the liver, kidneys and heart were also carried out on the 21st and 36th days of treatment. The C-LD_{50} obtained was 13 mg/kg. The authors concluded that the most predominant hazards of subchronic exposure to methyl parathion were weight loss, hyperglycemia and macrocytic anemia, all probably secondary to hepatic toxicity. Since an increasing dose protocol was used, this study does not identify a NOAEL or a Lowest-Observed-Adverse-Effect Level (LOAEL).

- Daly et al. (1979) administered methyl parathion (technical, 93.65% active ingredient) to Charles River CD-1 mice for 4 weeks at levels of 0, 25 or 50 ppm in the diet. Assuming that 1 ppm in the diet of mice corresponds to 0.15 mg/kg/day (Lehman, 1959), this is equivalent to doses of about 0, 3.75 or 7.5 mg/kg/day. Five animals of each sex were used at each dose level. Mean body weights were lower ($p < 0.05$) than control for all treated animals throughout the test period. Mean food consumption was lower ($p < 0.05$) throughout for all test animals except females at the 25-ppm level. Mortality, physical observations, and gross postmortem examinations did not reveal any treatment-related effects. Cholinesterase measurements were not performed. Based on body weight gain, the LOAEL for this study was identified as 25 ppm (3.75 mg/kg/day).

- Tegeris and Underwood (1977) examined the effects of feeding methyl parathion (94.32% pure) to beagle dogs (4 to 6 months of age, weighing 5 to 10 kg) for 14 days. Two animals of each sex were given doses of 0, 2.5, 5 or 10 mg/kg/day. All animals survived the 14-day test period. Mean food consumption and weight gain were significantly ($p < 0.05$) depressed for both sexes at the 5 and 10 mg/kg/day dose levels. After the 3rd day, animals in the high-dose group began vomiting after all meals. Vomiting was observed sporadically at the lower dose levels, particularly during the second week. The authors attributed this to acetylcholinesterase inhibition, but no measurements were reported. No other symptomatology was described. Based on weight loss and vomiting, this study identified a LOAEL of 2.5 mg/kg/day in the dog.

- Fan et al. (1978) investigated the immunosuppressive effects of methyl parathion administered orally to Swiss (ICR) mice. The pesticide (purity not specified) was fed in the diet at dose levels corresponding to 0, 0.08, 0.7 or 3.0 mg/kg/day for 4 weeks. Active immunity was induced by weekly injection of vaccine (acetone-killed *Salmonella typhimurium*) during the period of diet treatment. Defense against microbial infection was tested by intraperitoneal injection of a single LD_{50} dose of active *S. typhimurium* cells. Protection by immunization was stated to be decreased in methyl parathion-treated animals, but no dose-response data were provided. The authors stated that pesticide treatment extending beyond 2 weeks was required to obtain significant increases in mortality. Increased mortality was associated with an increased number of viable bacteria in blood, decreased total gamma-globulins and specific immunoglobins in serum, and reduced splenic blast transformation in response to mitogens.

* Shtenberg and Dzhunusova (1968) studied the effect of oral exposure to methyl parathion (purity not specified) on immunity in albino rats vaccinated with NIISI polyvaccine. Three tests (six animals each) were conducted in which: (a) the vaccination was done after the animals had been on a diet supplying 1.25 mg/kg/day metaphos (methyl parathion) for 2 weeks; (b) the diet and vaccinations were initiated simultaneously; and (c) the diet was initiated 2 weeks after vaccination. The titer of agglutins in immunized control rats was 1:1,200. This titer was decreased in all exposed groups as follows: 1:46 in series (a), 1:75 in series (b) and 1:33.3 in series (c). The authors judged this to be clear evidence of inhibition of immunobiological reactivity in the exposed animals. Changes in blood protein fractions and in serum concentration of albumins were not statistically significant. Based on immune suppression, a LOAEL of 1.25 mg/kg/day was identified.

2. Dermal/Ocular Effects

* Gaines (1969) reported a dermal LD_{50} of 67 mg/kg for methyl parathion in male and female Sherman rats.

* Galloway (1984a,b) studied the skin and eye irritation properties of methyl parathion (technical; purity not specified) using albino New Zealand White rabbits. In the skin irritation test, 0.5 mL undiluted pesticide was applied and the treated area occluded for 4 hours. This treatment resulted in dermal edema that persisted for 24 hours, and in erythema that lasted for 6 days. After a total observation period of 9 days, a score of 2.0 was derived, and technical methyl parathion was rated as a weak irritant. In the eye irritation test, 0.1 mL of the undiluted pesticide was applied to nine eyes. Three were washed after exposure, and six were left unwashed. Conjunctival irritation was observed starting at 1 hour and lasting up to 48 hours postexposure. Maximum average irritation scores of 11 and 10.7 were assigned for nonwashed and washed eyes, respectively, and technical methyl parathion was considered a weak irritant.

* Galloway (1985) used guinea pigs to examine the sensitizing potential of methyl parathion (technical; purity not stated). Ten doses of 0.5 mL of a 10% solution (w/v in methanol) were applied to the clipped intact skin of 10 male guinea pigs (albino Hartley strain) over a 36-day period. This corresponds to an average dose of 13.9 mg/kg/day. Another group was treated with 2,4-dinitrochlorobenzene as a positive control. No skin

sensitization reaction was observed in methyl parathion-treated animals.

- Skinner and Kilgore (1982) studied the acute dermal toxicity of methyl parathion in male Swiss-Webster mice, and simultaneously determined ED_{50} values for cholinesterase and acetylcholinesterase inhibition. Methyl parathion (analytical grade, 99% pure) was administered in acetone solution to the hind feet of the mice; the animals were muzzled to prevent oral ingestion through grooming. The dermal LD_{50} was 1,200 mg/kg. The ED_{50} was 950 mg/kg for cholinesterase inhibition and 550 mg/kg for acetylcholinesterase inhibition.

3. *Long-term Exposure*

- Daly and Rinehart (1980) conducted a 90-day feeding study of methyl parathion (93.65% pure) using Charles River CD-1 mice. Groups of 15 mice of each sex were given diets containing the pesticide at levels of 0, 10, 30 or 60 ppm. Assuming that 1 ppm in the diet of mice corresponds to 0.15 mg/kg/day (Lehman, 1959), this is equivalent to doses of about 0, 1.5, 4.5 or 9.0 mg/kg/day. All mice survived the test. Mean body weights were significantly ($p < 0.05$) depressed for both sexes at 60 ppm throughout the study and for males during the first 5 weeks at 30 ppm. Animals of both sexes had a slight but not significant ($p > 0.05$) increase in the mean absolute and relative brain weights at 60 ppm. There were dose-related decreases ($p < 0.05$) in the mean absolute and relative testes weights of all treated males and in the ovary weights of the females at 30 and 60 ppm. Gross and microscopic examination revealed no dose-related effects. Histological examination revealed no findings in the brain, testes or ovary to account for the observed changes in the weights of these organs. Measurements on ChE were not performed. Based on decreased testes weight, the LOAEL for this study was 10 ppm (1.5 mg/kg/day).

- Tegeris and Underwood (1978) investigated the toxicity of methyl parathion (94.32% a.i.) in beagle dogs fed the pesticide for 90 days at dose levels of 0, 0.3, 1.0 or 3.0 mg/kg/day. Four dogs (4 months old, 4.5 to 8.0 kg) of both sexes were used at each dose level. Soft stools were observed in all treatment groups throughout, and there was also occasional spontaneous vomiting. There were no persistent significant ($p < 0.05$) effects on body weight gain, feed intake, fasting blood sugar, BUN, SGPT, SGOT, hematological or urological indices. Organ weights were within normal limits, with the exception of pituitary weights of females at 3.0 mg/kg, which were significantly ($p < 0.05$) higher than the control values. Gross and microscopic examination revealed no compound-related abnormalities. Plasma ChE was significantly ($p <$ 0.05) depressed in both sexes at 6 and 13 weeks at 3 mg/kg/day, and in the males only at 1.0 mg/kg/day at 13 weeks; erythrocyte ChE was also significantly ($p < 0.05$) depressed in all animals at 6 and 13 weeks at 3 mg/kg/day, and in both sexes at 13 weeks at 1.0 mg/kg/day; and brain ChE was significantly ($p < 0.05$) depressed in both sexes at 3.0 mg/kg/ day. Based on ChE depression, the NOAEL and LOAEL for this study were identified as 0.3 mg/kg/day and 1.0 mg/kg/day, respectively.

- Ahmed et al. (1981) conducted a 1-year feeding study in beagle dogs. Methyl parathion (93.6% pure) was administered in the diet at ingested dose levels of 0, 0.03, 0.1 or 0.3 mg/kg/day. Eight animals of each sex were included at each dose level, with no overt signs of toxicity noted at any dose. There were no treatment-related changes in food consumption or body weight. Cholinesterase determinations in plasma, red blood cells and brain revealed marginal variations, but the changes were not consistent and were judged by the authors to be unrelated to dosing. Organ weight determinations showed changes in both males and females at 0.1 and 0.3 mg/kg/day, but the changes were neither dose-related nor consistent. It was concluded that there was no demonstrable toxicity of methyl parathion fed to the dogs at these levels. The NOAEL for this study was 0.3 mg/kg/day.

- NCI (1978) conducted a 2-year feeding study of methyl parathion (purity not specified) in Fischer 344 rats (50/sex/dose) at dose levels of 0, 20 or 40 ppm in the diet. Assuming that 1 ppm in the diet of rats corresponds to 0.05 mg/kg/day (Lehman, 1959), this is equivalent to dose levels of about 0, 1 or 2 mg/kg/day. Cholinesterase levels were not measured, but no remarkable clinical signs were noted, and no significant ($p < 0.05$) changes were observed in mortality, body weight, gross pathology or histopathology. Based on this, a NOAEL of 40 ppm (2 mg/kg/day) was identified in rats.

- NCI (1978) conducted a chronic (105-week) feeding study in B6C3F$_1$ mice (50/sex/dose). Animals were initially fed methyl parathion (94.6% pure) at dose levels of 62.5 or 125 ppm. Assuming that 1 ppm in the diet of mice corresponds to 0.15 mg/kg/day (Lehman, 1959), this is equivalent to doses of about 9.4 or 18.8 mg/kg/day. Because of severely depressed body weight gain in males, their doses were reduced at 37 weeks to 20 or 50 ppm, and the time-weighted averages were calculated to be 35 or 77 ppm. This corresponds to doses of about 5.2 or 11.5 mg/ kg/day, respectively. Females were fed at the original levels throughout. Mortality was significantly ($p < 0.05$) increased only in female mice at

125 ppm. Body weights were lower (p < 0.05) for both sexes throughout the test period and decreases were dose-related. No gross or histopathologic changes were noted, and ChE activity was not measured. Based on body weight, this study identified a LOAEL of 35 ppm (5.2 mg/kg/day) in male mice.

- Daly et al. (1984) conducted a chronic feeding study of methyl parathion (93.65% active ingredient) in Sprague-Dawley (CD) rats (60/sex/dose) at dose levels of 0, 0.5, 5 or 50 ppm in the diet. Using food intake/body weight data given in the study report, these levels approximate doses of about 0, 0.025, 0.25 or 2.5 mg/kg/day. At 24 months, five animals of each sex were sacrificed for qualitative and quantitative tests for neurotoxicity. Ophthalmoscopic examinations were conducted on females at 3, 12 and 24 months and terminally. Hematology, urinalysis and clinical chemistry analyses were performed at 6, 12, 18 and 24 months. Mean body weights were reduced (p < 0.05) throughout the study for both sexes at 50 ppm. At this dose level, food consumption was elevated (p < 0.05) for males during weeks 2 to 13, but reduced for females for most of the study. Hemoglobin, hematocrit and RBC count were significantly (p < 0.05) reduced for females at 50 ppm at 6, 12, 18 and 24 months. For males at 5 and 50 ppm at 24 months, hematocrit and RBC count were significantly (p < 0.05) reduced and hemoglobin was reduced, but not significantly (p < 0.05). At 50 ppm, plasma and erythrocyte ChE were significantly (p < 0.05) depressed for both sexes during the test, and brain ChE was significantly (p < 0.05) decreased at termination. Slight decreases in ChE activity were also observed in animals at 5 ppm, but these changes were not statistically significant (p > 0.05). For males, the absolute weight and the ratio to brain weight of the testes, kidneys and the liver were reduced by 10 to 16% (not significant, p > 0.05) in both the 5- and 50-ppm groups, while for females absolute and organ/body weights for the brain and heart (also heart/brain weight) were found to be elevated significantly (p < 0.05) at the same dose levels. Overt signs of cholinergic toxicity (such as alopecia, abnormal gait and tremors) were observed in the 50-ppm animals and in one female at 5 ppm. At 24 months, 15 females were observed to have retinal degeneration. There was also a dose-related occurrence of retinal posterior subcapsular cataracts, possibly related or secondary to the retinal degeneration, since 5 of the 10 cataracts occurred in rats with retinal atrophy. The incidence of retinal atrophy was 20/55 at 50 ppm, 1/60 at 5 ppm, 3/60 at 0.5 ppm and 3/59 in the control group. Examination of the sciatic nerve and other nervous tissue from five rats per sex killed at week 106 gave evidence of peripheral neuropathy (abnormal fibers, myelin corrugation, myelin ovoids) in both sexes at 50 ppm (p < 0.05). Too few fibers were examined at the lower doses to perform statistical analyses, but the authors stated that nerves from both sexes in low- and mid-dose groups could not be distinguished qualitatively from controls. Slightly greater severity of nerve changes found in two males was not clearly related to treatment. No other lesions were observed that appeared to be related to ingestion of methyl parathion. Based on hematology, body weight, organ weights, clinical chemistry, retinal degeneration and cholinergic signs, a NOAEL of 0.5 ppm (0.025 mg/kg/day) was identified in this study.

4. *Reproductive Effects*

- Lobdell and Johnston (1964) conducted a three-generation study in Charles River rats. Each parental dose group included 10 males and 20 females. The investigators incorporated methyl parathion (99% pure) in the diet of males and females at dose levels of 0, 10 or 30 ppm, except for reduction of each dose by 50% during the initial 3 weeks of treatment, to produce dose equivalents of 0, 1.0 and 3.0 mg/kg/day, respectively. There was no pattern with respect to stillbirths, although the 30-ppm groups had a higher total number of stillborn. Survival was reduced in weanlings of the F_{1a}, F_{1b} and F_{2a} groups at 30 ppm, and in weanlings of the F_{3a} group at 10 ppm. At 30 ppm, there was also a reduction in fertility of the F_{2b} dams at the second mating; the first mating resulted in 100% of the animals having litters, while at the second mating, only 41% had litters. Animals exposed to 10 ppm methyl parathion did not demonstrate significant deviations from the controls. A NOAEL of 10 ppm (1.0 mg/kg/day) was identified in this study.

- Daly and Hogan (1982) conducted a two-generation study of methyl parathion (93.65% pure) toxicity in Sprague-Dawley rats. Each parental dose group consisted of 15 males and 30 females. The compound was added to the diet at levels of 0, 0.5, 5.0 or 25 ppm. Using compound intake data from the study report, equivalent dose levels are about 0, 0.05, 0.5 or 2.5 mg/kg/day. Feeding of the diet was initiated 14 weeks prior to the first mating and then continued for the remainder of the study. Reduced body weight (p < 0.05) was observed in F_0 and F_1 dams at the 25-ppm dose level. A slight decrease in body weight was noted in F_{1a} and F_{2a} pups in the 25-ppm group, but this was not significant (p > 0.05). Overall, the authors concluded that there was no significant (p > 0.05) effect attributable to methyl parathion in the diet. Based on maternal weight gain, the NOAEL for this study was 5.0 ppm (0.5 mg/kg/day).

5. *Developmental Effects*

- Gupta et al. (1985) dosed pregnant Wistar-Furth rats (10 to 12 weeks of age) with methyl parathion (purity not specified) on days 6 to 20 of gestation. Two doses were used: 1.0 mg/kg (fed in peanut butter) or 1.5 mg/kg (administered by gavage in peanut oil). The low dose produced no effects on maternal weight gain, caused no visible signs of cholinergic toxicity and did not result in increased fetal resorptions. The high dose caused a slight but significant ($p < 0.05$) reduction in maternal weight gain (11% in exposed versus 16% in controls, by day 15) and an increase in late resorptions (25% versus 0%). The high dose also resulted in cholinergic signs (muscle fasiculation and tremors) in some dams. Acetylcholesterase (AChE) activity, choline acetyltransferase (CAT) activity and quinuclidinyl benzilate (QNB) binding to muscarinic receptors were determined in several brain regions of fetuses at 1, 7, 14, 21 and 28 days postnatal age, and in maternal brain at day 19 of gestation. Exposure to 1.5 mg/kg reduced ($p < 0.05$) the AChE and increased CAT activity in all fetal brain regions at each developmental period and in the maternal brain. Exposure to 1.0 mg/kg caused a significant ($p < 0.05$) but smaller and less persistent reduction of AChE activity in offspring, but no change in brain CAT activity. Both doses reduced QNB binding in maternal frontal cortex ($p < 0.05$), but did not alter the postnatal pattern of binding in fetuses. In parallel studies, effects on behavior (cage emergence, accommodated locomotor activity, operant behavior) were observed to be impaired in rats exposed prenatally to 1.0 mg/kg, but not to the 1.5-mg/kg dose. No morphological changes were observed in hippocampus or cerebellum. It was concluded that subchronic prenatal exposure to methyl parathion altered postnatal development of cholinergic neurons and caused subtle alterations in selected behaviors of the offspring. The fetotoxic LOAEL for this study was 1.0 mg/kg.

- Gupta et al. (1984) administered oral doses of 1.0 or 1.5 mg/kg/day of methyl parathion (purity not specified) to female Wistar-Furth rats on days 6 through 15 or on days 6 through 19 of gestation. Protein synthesis in brain and other tissues was measured on day 15 or day 19 by subcutaneous injection of radioactive valine. The specific activity of this valine in the free amino acid pool and protein-bound pool (measured 0.5, 1.0 and 2.0 hours after injection) was significantly ($p < 0.05$) reduced in various regions of the maternal brain and in maternal viscera, placenta and whole embryos (day 15), and in fetal brain and viscera (day 19). The inhibitory effect of methyl parathion on protein synthesis was dose dependent, greater on day 19 than on day 15 of gestation and more pronounced in fetal than in maternal tissues. With respect to protein synthesis in both maternal and fetal tissues, the LOAEL of this study was 1.0 mg/kg.

- Fuchs et al. (1976) reported a study in which oral administration of methyl parathion to pregnant Wistar rats on either days 5 to 9, 11 to 15, or 11 to 19 of gestation resulted in growth retardation of offspring and increased resorptions at 3 mg/kg. The NOAEL was 1 mg/kg.

6. *Mutagenicity*

- Van Bao et al. (1974) examined the lymphocytes from 31 patients exposed to various organophosphate pesticides for indications of chromosome aberrations. Five of the examined patients had been exposed to methyl parathion. Blood samples were taken 3 to 6 days after exposure and again at 30 and 180 days. A temporary, but significant ($p < 0.05$) increase was found in the frequency of chromatid breaks and stable chromosome-type aberrations in acutely intoxicated persons. Two of the methyl parathion-exposed persons were in this category, having taken large doses orally in suicide attempts. The authors concluded that the results of this study strongly suggest that the organic phosphoric acid esters exert direct mutagenic effects on chromosomes.

- Shigaeva and Savitskaya (1981) reported that metophos (methyl parathion) induced visible morphological mutations and biochemical mutations in *Pseudomonas aeruginosa* at concentrations between 100 and 1,000 μg/mL, and significantly ($p < 0.05$) increased the reversion rate in *Salmonella typhimurium* at concentrations between 5 and 500 μg/ mL.

- Grover and Malhi (1985) examined the induction of micronuclei in bone marrow cells of Wistar male rats that had been injected with methyl parathion at doses between one-third and one-twelfth of the LD_{50}. The increase in micronuclei formation led the authors to conclude that methyl parathion has high mutagenic potential.

- Mohn (1973) concluded that methyl parathion was a probable mutagen, based on the ability to induce 5-methyltryptophan resistance in *Escherichia coli*. Similar results were obtained using the streptomycin-resistant system of *E. coli* and the trp-conversion system of *Saccharomyces cerevisiae*.

- Rashid and Mumma (1984) found methyl parathion to be mutagenic to *S. typhimurium* strain TA100 after activation with rat liver microsomal and cytosolic enzymes.

- Chen et al. (1981) investigated sister-chromatid exchanges (SCE) and cell-cycle delay in Chinese hamster cells (line V79) and two human cell lines (Burkitt lymphoma B35M and normal human lymphoid cell Jeff), and found methyl parathion to be the most active pesticide of eight tested with respect to its induction potential.

- Riccio et al. (1981) found methyl parathion to be negative in two yeast assay systems (diploid strains D3 and D7 of *Saccharomyces cerevisiae*), based on mitotic recombination (in D3), and mitotic crossing over, mitotic gene conversion and reverse mutation (in D7).

- In a study for dominant lethality in mice by Jorgenson et al. (1976), males (20 per dose group) were given methyl parathion in the diet for 7 weeks at 3 dose levels (not reported). Positive controls given triethylene melamine and untreated controls were also studied. Following treatment, each male was mated to 2 adult females weekly for 8 weeks. Methyl parathion was ineffective in this test.

7. *Carcinogenicity*

- NCI (1978) conducted chronic (105-week) feeding studies of methyl parathion in Fischer 344 rats and B6C3F$_1$ mice (50/sex/dose). Rats were fed methyl parathion (94.6% pure) at dose levels of 0, 20 or 40 ppm (equivalent to doses of 0, 1 or 2 mg/kg/day). Mice were initially fed dose levels of 62.5 or 125 ppm, but because of severely depressed body weight gain in males, their doses were reduced at 37 weeks to 20 or 50 ppm, respectively. Time-weighted averages for males were calculated to be 35 or 77 ppm (about 5.2 or 11.5 mg/kg/day). Females received the original dose level throughout. Based on gross and histological examinations, no tumors were observed to occur at an incidence significantly higher than that of the control value in either the mice or rats. The authors concluded that methyl parathion was not carcinogenic in either species under the conditions of the test.

- Daly et al. (1984) fed Sprague-Dawley rats (60/sex/dose) methyl parathion (93.65%) in the diet for 2 years. Doses tested were 0, 0.5, 5 or 50 ppm, estimated as equivalent to doses of 0, 0.025, 0.25 or 2.5 mg/kg/day. There were no significant ($p > 0.05$) increases in neoplastic lesions between treated and control groups.

IV. Quantification of Toxicological Effects

A. One-day Health Advisory

No data were located in the available literature that were suitable for deriving a One-day HA value. It is recommended that the Ten-day HA value for the 10-kg child (0.3 mg/L, calculated below) be used at this time as a conservative estimate of the One-day HA value.

B. Ten-day Health Advisory

The studies by Rider (1969, 1970, 1971) have been selected to serve as the basis for calculation of the Ten-day HA for methyl parathion. In these studies, human volunteers ingested methyl parathion for 30 days at doses ranging from 1 to 30 mg/day. The most sensitive indicator of effects was inhibition of plasma ChE. No effects in any subject were observed at a dose of 22 mg/day (about 0.31 mg/kg/day with assumed 70-kg body weight), and this was identified as the NOAEL. Doses of 24 mg/day inhibited ChE activity in plasma and red blood cells in two of five subjects, maximum decreases being 23 and 24% in plasma and 27 and 55% in red blood cells. Higher doses (26 to 30 mg/day) caused greater inhibition. On this basis, 24 mg/day (0.34 mg/kg/day) was identified as the LOAEL. Short-term toxicity or teratogenicity studies in animals identified LOAEL values of 1.0 to 2.5 mg/kg/day (Gupta et al., 1984, 1985; Shtenberg and Dzhunusova, 1968; Tegeris and Underwood, 1977), but did not identify a NOAEL value.

Using a NOAEL of 0.31 mg/kg/day, the Ten-day HA for a 10-kg child is calculated as follows:

$$\text{Ten-day HA} = \frac{(0.31 \text{ mg/kg/day}) (10 \text{ kg})}{(10) (1 \text{ L/day})} = 0.31 \text{ mg/L} (300 \text{ } \mu g/L)$$

C. Longer-term Health Advisory

The 90-day feeding study in dogs by Tegeris and Underwood (1978) has been selected to serve as the basis for calculation of the Longer-term HA for methyl parathion. In this study, a NOAEL of 0.3 mg/kg/day was identified, based on absence of effects on body weight, food consumption, clinical chemistry, hematology, urinalysis, organ weights, gross pathology, histopathology and ChE activity. The LOAEL, based on ChE inhibition, was 1.0 mg/kg/day. These values are supported by the results of Ahmed et al. (1981), who identified a NOAEL of 0.3 mg/kg/day in a 1-year feeding study in dogs, and by the study of Daly and Rinehart (1980), which identified a LOAEL of 1.5 mg/kg/day (based on decreased testes weight) in a 90-day feeding study in mice.

Using a NOAEL of 0.3 mg/kg/day, the Longer-term HA for a 10-kg child is calculated as follows:

$$\text{Longer-term HA} = \frac{(0.3 \text{ mg/kg/day}) (10 \text{ kg})}{(100) (1 \text{ L/day})} = 0.03 \text{ mg/L } (30 \text{ } \mu g/L)$$

Using a NOAEL of 0.3 mg/kg/day, the Longer-term HA for a 70-kg adult is calculated as follows:

$$\text{Longer-term HA} = \frac{(0.3 \text{ mg/kg/day}) (70 \text{ kg})}{(100) (2 \text{ L/day})} = 0.10 \text{ mg/L } (100 \text{ } \mu g/L)$$

D. Lifetime Health Advisory

The 2-year feeding study in rats by Daly et al. (1984) has been selected to serve as the basis for calculation of the Lifetime HA for methyl parathion. In this study, a NOAEL of 0.025 mg/kg/day was identified, based on the absence of effects on body weight, organ weights, hematology, clinical chemistry, retinal degeneration and cholinergic signs. A LOAEL of 0.25 mg/kg/day was identified, based on decreased hemoglobin, red blood cell counts and hematocrit (males); changes in organ-to-body weight ratios (males and females); and one case of visible cholinergic signs. There was increased retinal degeneration at 2.5 mg/kg/day, but this was not greater than control at 0.25 or 0.025 mg/kg/day. This LOAEL value (0.25 mg/kg/day) is lower than most other NOAEL or LOAEL values reported in other reports. For example, NOAEL values of 0.3 to 3.0 mg/kg/day have been reported in chronic studies by Ahmed et al. (1981), NCI (1978), Lobdell and Johnston (1964) and Daly and Hogan (1982).

Using a NOAEL of 0.025 mg/kg/day, the Lifetime HA for a 70-kg adult is calculated as follows:

Step 1: Determination of the Reference Dose (RfD)

$$\text{RfD} = \frac{(0.025 \text{ mg/kg/day})}{(100)} = 0.00025 \text{ mg/kg/day}$$

Step 2: Determination of the Drinking Water Equivalent Level (DWEL)

$$\text{DWEL} = \frac{(0.00025 \text{ mg/kg/day}) (70 \text{ kg})}{(2 \text{ L/day})} = 0.009 \text{ mg/L } (9 \text{ } \mu g/L)$$

Step 3: Determination of the Lifetime Health Advisory

$$\text{Lifetime HA} = (0.009 \text{ mg/L}) (20\%) = 0.002 \text{ mg/L } (2 \text{ } \mu g/L)$$

E. Evaluation of Carcinogenic Potential

- No evidence of carcinogenic activity was detected in either rats or mice in a 105-week feeding study (NCI, 1978).

- Statistically significant ($p < 0.05$) increases in neoplasm frequency were not found in a 2-year feeding study in rats (Daly et al., 1984).

- The International Agency for Research on Cancer (IARC) has not evaluated the carcinogenicity of methyl parathion.

- Applying the criteria described in EPA's guidelines for assessment of carcinogenic risk (U.S. EPA, 1986a), methyl parathion may be classified in Group D: not classifiable. This category is for substances with inadequate animal evidence of carcinogenicity.

V.   Other Criteria, Guidance and Standards

- NAS (1977) concluded that data were inadequate for calculation of an ADI for methyl parathion. However, using data on parathion, NAS calculated an ADI for both parathion and methyl parathion of 0.0043 mg/kg/day, using a NOAEL of 0.043 mg/kg/day in humans (Rider et al., 1969) and an uncertainty factor of 10 (NAS, 1977). From this ADI, NAS calculated a chronic Suggested-No-Adverse-Response Level (SNARL) of 0.03 mg/L, based on water consumption of 2 L/day by a 70-kg adult, and assuming a 20% relative source contribution (RSC).

- The U.S. EPA Office of Pesticide Programs (EPA/OPP) previously calculated a provisional ADI (PADI) of 0.0015 mg/kg/day, based on a NOAEL of 0.3 mg/kg/day. This is based on the 90-day dog study by Tegeris and Underwood (1978) and a 200-fold uncertainty factor. This PADI has been updated to use a value of 0.0025 mg/kg/day based on a NOAEL of 0.0250 mg/kg/day in a 2-year rat chronic feeding study and a 100-fold uncertainty factor (Daly et al., 1984).

- ACGIH (1984) has proposed a time-weighted average threshold limit value of 0.2 mg/m^3.

- The National Institute for Occupational Safety and Health has recommended a standard for methyl parathion in air of 0.2 mg/m^3 (TDB, 1985).

- U.S. EPA has established residue tolerances for parathion and methyl parathion in or on raw agricultural commodities that range from 0.1 to 0.5 ppm (CFR, 1985). A tolerance is a derived value based on residue levels, toxicity data, food consumption levels, hazard evaluation and scientific judgment; it is the legal maximum concentration of a pesticide in or on a raw agricultural commodity or other human or animal food (Paynter et al., no date).

- The World Health Organization established an ADI of 0.02 mg/kg/day (Vettorazzi and van den Hurk, 1985).

VI.   Analytical Methods

- Analysis of methyl parathion is by a gas chromatographic (GC) method applicable to the determination of certain nitrogen-phosphorus-containing pesticides in water samples (U.S. EPA, 1986b). In this method, approximately 1 liter of sample is extracted with methylene chloride. The extract is concentrated and the compounds are separated using capillary column GC. Measurement is made using a nitrogen phosphorus detector. The method detection limit has not been determined for methyl parathion, but it is estimated that the detection limits for analytes included in this method are in the range of 0.1 to 2 μg/L.

VII.   Treatment Technologies

- Available data indicate that granular-activated carbon (GAC) and reverse osmosis (RO) will effectively remove methyl parathion from water, as described in the studies below.

- Whittaker (1980) experimentally determined adsorption isotherms for methyl parathion and methyl parathion diazinon bisolute solutions. As expected, the bisolute solution showed a lesser overall carbon capacity than that achieved by the application of pure solute solution.

- Under laboratory conditions, GAC removed 99+% of methyl parathion (Whittaker et al., 1982).

- Reverse osmosis is a promising treatment method for methyl parathion-contaminated water. Chian et al. (1975) reported 99.5% removal efficiency for two types of membrane operating at 600 psig and a flux rate of 8 to 12 gal/ft^2/day. Membrane adsorption, however, is a major concern and must be considered, as breakthrough of methyl parathion would probably occur once the adsorption potential of the membrane was exhausted.

- Oxidation with ozone and chlorine may be possible in the treatment of methyl parathion, as described in the studies below.

- Oxidation with 4.5 and 9.5 mg/L ozone reduced the methyl parathion by 95 to 99%. The same removal efficiency was achieved with 1 and 2 mg/L chlorine (Gabovich and Kurennoy, 1974).

- Ozonation with 0.32 mg ozone/mg methyl parathion reduced methyl parathion in drinking water by 90 to 95% (Shevchenko et al., 1982).

- Oxidation degradation by either ozone or chlorine produces several degradation products, whose environmental toxic impact should be evaluated prior to selecting oxidative degradation for treatment of methyl parathion-contaminated water (Shevchenko et al., 1982).

- Aeration does not seem to be a practical technique for removing methyl parathion from potable water (Saunders and Sieber, 1983).

- Treatment technologies for the removal of methyl parathion from water are available and have been reported to be effective. However, selection of individual or combinations of technologies for methyl parathion removal from water must be based on a case-by-case technical evaluation, and an assessment of the economics involved.

VIII.   References

ACGIH. 1984. American Conference of Governmental Industrial Hygienists, Inc. Documentation of the threshold limit values for substances in workroom air, 3rd ed. Cincinnati, OH: ACGIH.

Ackermann, H. and R. Engst. 1970. The presence of organophosphorus insecticides in the fetus. Arch. Toxikol. 26(1):17–22. (In German).

Agchem.* 1983. Persistence and release rate of Penncap M insecticide in water and hydrosoil. Project No. WT-5-82. Unpublished study.

Ahmed, F.E., J.W. Sagartz and A.S. Tegeris.* 1981. One-year feeding study in dogs. Prepared by Pharmacopathics Research Laboratories Inc., Laurel, MD, for Monsanto Company. Unpublished study. MRID 00093895.

Braeckman, R.A., F. Audenaert, J. L. Willems, F. M. Belpaire and M.G. Bogaert. 1983. Toxicokinetics of methyl parathion and parathion in the dog after intravenous and oral administration. Arch. Toxicol. 54:71–82.

CFR. 1985. Code of Federal Regulations. 40 CFR 180.121. July 1, 1985. p. 484.

CHEMLAB. 1985. The Chemical Information System. CIS, Inc., Baltimore, MD.

Chen, H.H., J.L. Hsueh, S.R. Sirianni and C.C. Huang. 1981. Induction of sister-chromatid exchanges and cell cycle delay in cultured mammalian cells treated with eight organophosphorus pesticides. Mut. Res. 88:307–316.

Chian, E.S., W.N. Bruce and H.H.P. Fang. 1975. Removal of pesticides by reverse osmosis. Environ. Sci. and Tech. 9(1):52–59.

Ciba-Geigy Corporation.* 1978. Residue of CGA-15324 Curacron ® +4E and methyl parathion 4E on soil. Unpublished study. Compilation including AG-A Nos. 4635 I, II, III and 5023.

Daly, I. and G. Hogan.* 1982. A two-generation reproduction study of methyl parathion in rats. Prepared by Bio/Dynamics, Inc., for Monsanto Company. Unpublished study. MRID 00119087.

Daly, I., G. Hogan and J. Jackson.* 1984. A two-year chronic feeding study of methyl parathion in rats. Prepared by Bio/Dynamics, Inc., for Monsanto Company. Unpublished study. MRID 00139023.

Daly, I.W. and W.E. Rinehart.* 1980. A three month feeding study of methyl parathion in mice. Prepared by Bio/Dynamics, Inc., for Monsanto Company. Unpublished study. MRID 00072513.

Daly, I.W., W.E. Rinehart and M. Cicco.* 1979. A four week pilot study in mice with methyl parathion. Prepared by Bio/Dynamics, Inc., for Monsanto Company. Unpublished study. MRID 00072514.

Fan, A., J.C. Street and R.M. Nelson. 1978. Immunosuppression in mice administered methyl parathion and carbofuran by diet. Toxicol. Appl. Pharmacol. 45(1):235.

Fuchs, V., S. Golbs, M. Kuhnert and F. Oswald. 1976. Studies into the prenatal toxic action of parathion methyl on Wistar rats and comparison with prenatal toxicity of cyclophosphamide and trypan blue. Arch. Exp. Vet. Med. 30:343–350. (In German, English abstract.)

Gabovich, R.D. and I.L. Kurennoy. 1974. Ozonation of water containing humic compounds, phenols and pesticides. Army Medical Intelligence and Information Agency. USAMIIA-K-4564.

Gaines, T.B. 1969. Acute toxicity of pesticides. Toxicol. Appl. Pharmacol.14:515–534.

Galal, E.E., H.A. Samaan, S. Nour El Dien, S. Kamel, M. El Saied, M. Sadek, A. Madkour, K.H. El Saadany and A. El-Zawahry. 1977. Studies on the acute and subchronic toxicities of some commonly used anticholinesterase insecticides rats. J. Drug Res. Egypt. 9(1–2):1–17.

Galloway, C.* 1984a. Rabbit skin irritation: methyl parathion technical (Cheminova). Prepared by STILLMEADOW, Inc., Houston, TX, for Gowan Company. Unpublished study. MRID 00142804.

Galloway, C.* 1984b. Rabbit eye irritation: methyl parathion technical (Cheminova). Prepared by STILLMEADOW, Inc., Houston, TX, for Gowan Company. Unpublished study. MRID 00142808.

Galloway, C.* 1985. Guinea pig skin sensitization: methyl parathion technical

(Cheminova). STILLMEADOW. Inc., Houston, TX, for Gowan Company. Unpublished study. MRID 00142005.

Grover, I.S., and P.K. Malhi. 1985. Genotoxic effects of some organophosphorus pesticides. I. Induction of micronuclei in bone marrow cells in rat. Mutat. Res. 155:131–134.

Gupta, R.C., R.H. Rech, K.L. Lovell, F. Welsch and J.E. Thornburg. 1985. Brain cholinergic, behavior, and morphological development in rats exposed *in utero* to methyl parathion. Toxicol. Appl. Pharmacol. 77:405–413.

Gupta, R.C., J.E. Thornburg, D.B. Stedman and D.B. Welsch. 1984. Effect of subchronic administration of methyl parathion on *in vivo* protein synthesis in pregnant rats and their conceptuses. Toxicol. Appl. Pharmacol. 72:457–468.

Haley, T.J., J.H. Farmer, J.R. Harmon and K.L. Dooley. 1975. Estimation of the LD_1 and extrapolation of the $LD_{0.1}$ for five organothiophosphate pesticides. J. Eur. Toxicol. 8(4):229–235.

Hawley, G.G. 1981. The condensed chemical dictionary, 10th ed. New York, NY: Van Nostrand Reinhold Company.

Hollingworth, R.M., R.L. Metcalf and I.R. Fukuto. 1967. The selectivity of sumithion compared with methyl parathion. Metabolism in the white mouse. J. Agr. Food Chem. 15:242–249.

Isshiki, K., K. Miyata, S. Matsui, M. Tsutsumi and T. Watanabe. 1983. Effects of post-harvest fungicides and piperonyl butoxide on the acute toxicity of pesticides in mice. Safety evaluation for intake of food additives. III. Shokuhin Eiseigaku Zasshi. 24(3):268–274.

Jorgenson, T.A., C.J. Rushbrook and G.W. Newell. 1976. *In vivo* mutagenesis investigations of ten commercial pesticides. Toxicol. Appl. Pharmacol. 37:109.

Lehman, A.J. 1959. Appraisal of the safety of chemicals in food, drugs and cosmetics. Assoc. Food Drug Off. U.S., Q. Bull.

Lobdell, J.L. and C.D. Johnston.* 1964. Methyl parathion: three-generation reproduction study in the rat. Prepared by Woodard Research Corporation, Virginia, for Monsanto Company. Unpublished study. MRID 0081923.

Meister, R., ed. 1988. Farm chemicals handbook. Willoughby, OH: Meister Publishing Company.

Mohn, G. 1973. 5-Methyltryptophan resistance mutations in *Escherichia coli* K-12: mutagenic activity of monofunctional alkylating agents including organophosphorus insecticides. Mut. Res. 20:7–15.

Nakatsugawa, T., N.M. Tolman and P.A. Dahm. 1968. Degradation and activation of parathion analogs by microsomal enzymes. Biochem. Pharmacol. 17:1517–1528.

NAS. 1977. National Academy of Sciences. Drinking water and health. Vol. 1. Washington, DC: National Academy Press.

NCI. 1978. National Cancer Institute. Bioassay of methyl parathion for possible carcinogenicity. Bethesda, MD: NCI, National Institutes of Health. NCI-CG-TR-157.

Neal, R.A. and K.P. DuBois. 1965. Studies on the mechanism of detoxification of cholinergic phosphorothioates. J. Pharmacol. Exp. Ther. 148(2):185–192.

Nemec, S.J., P.L. Adkisson and H.W. Dorough. 1968. Methyl parathion adsorbed on the skin and blood cholinesterase levels of persons checking cotton treated with ultra-low-volume sprays. J. Econ. Entomol. 61(6):1740–1742.

Paynter, O.E., J.G. Cummings and M.H. Rogoff. No date. United States Pesticide Tolerance System. Unpublished draft report. Washington, DC: U.S. EPA, Office of Pesticide Programs.

Pennwalt Corporation.* 1972. Soil and water run off test for Penncap M versus methyl parathion E.C. Compilation. Unpublished study.

Pennwalt Corporation.* 1977. Penncap-M® + and Penncap-E® + insecticides — soil leaching. Unpublished study.

Rashid, K.A. and R.O. Mumma. 1984. Genotoxicity of methyl parathion in short-term bacterial test systems. J. Environ. Sci. Health. B19(6):565–577.

Riccio, E., G. Shepherd, A. Pomeroy, K. Mortelmans and M.D. Waters. 1981. Comparative studies between the S. cerevisiae D3 and D7 assays of eleven pesticides. Environ. Mutagen. 3(3):327.

Rider, J.A., H.C. Moeller, E.J. Puletti and J.I. Swader. 1969. Toxicity of parathion, systox, octamethyl pyrophosphoramide and methyl parathion in man. Toxicol. Appl. Pharmacol. 14:603–611.

Rider, J.A., J.I. Swader and E.J. Puletti. 1970. Methyl parathion and guthion anticholinesterase effects in human subjects. Federation Proc. 29(2):349. Abstracts.

Rider, J.A., J.I. Swader and E.J. Puletti. 1971. Anticholinesterase toxicity studies with methyl parathion, guthion and phosdrin in human subjects. Federation Proc. 30(2):443. Abstracts.

Sabol, E.* 1985. Rat: Acute oral toxicity of methyl parathion technical (Cheminova). Prepared by STILLMEADOW, Inc., Houston, TX, for Gowan Company. Unpublished study. MRID 00142806.

Saunders, P.F. and J.N. Seiber. 1983. A chamber for measuring volatilization of pesticides from model soil and water disposal systems. Chemosphere. 12(7/8):999–1012.

Shevchenko, M.A., P.N. Taran and P.V. Marchenko. 1982. Modern methods of purifying water from pesticides. Soviet J. Water Chem. Technol. 4(4):53–71.

Shigaeva, M.K. and I.S. Savitskaya. 1981. Comparative study of the mutagenic activity of some organophosphorus insecticides in bacteria. Tsitol. Genet. 15(3):68–72.

Shtenberg, A.I. and R.M. Dzhunusova. 1968. Depression of immunological reactivity in animals by some organophosphorus pesticides. Bull. Exp. Biol. Med. 65(3):317–318.

Skinner, C.S. and W.W. Kilgore. 1982. Acute dermal toxicities of various organophosphate insecticides in mice. J. Toxicol. Environ. Health. 9(3):491–497.

STORET. 1988. STORET Water Quality File. Office of Water. U.S. Environmental Protection Agency (data file search conducted in May, 1988).

TDB. 1985. Toxicology Data Bank. MEDLARS II. National Library of Medicine's National Interactive Retrieval Service.

Tegeris, A.S. and P.C. Underwood.* 1977. Fourteen-day feeding study in the dog. Prepared by Pharmacopathics Research Laboratories, Laurel, MD, for Monsanto Company. Unpublished study. MRID 00083109.

Tegeris, A.S. and P.C. Underwood.* 1978. Methyl parathion: ninety-day feeding to dogs. Unpublished study. Laurel, MD: Pharmacopathics Research Laboratories, Inc. MRID 00072512.

U.S. EPA. 1981. U.S. Environmental Protection Agency. Acephate, aldicarb, carbophenothion, DEF, EPN, ethoprop, methyl parathion, and phorate: their acute and chronic toxicity, bioconcentration potential, and persistence as related to marine environment. Unpublished study. Report No. EPA-600/4-81-023. Environmental Research Laboratory.

U.S. EPA. 1986a. U.S. Environmental Protection Agency. Guidelines for carcinogen risk assessment. Fed. Reg. 51(185):33992–34003. September 24.

U.S. EPA. 1986b. U.S. Environmental Protection Agency. U.S. EPA Method 1 – Determination of nitrogen- and phosphorus-containing pesticides in ground water by GC/NPD. Draft. January. Available from U.S. EPA's Environmental Monitoring and Support Laboratory, Cincinnati, OH.

Van Bao, T., I. Szabo, P. Ruzicska and A. Czeizel. 1974. Chromosome aberrations in patients suffering acute organic phosphate insecticide intoxication. Human Genetik. 24(1):33–57.

Vettorazzi, G. and G.W. van den Hurk, eds. 1985. Pesticides reference index. J.M.P.R. p. 41.

Whittaker, K.F. 1980. Adsorption of selected pesticides by activated carbon using isotherm and continuous flow column systems. Ph.D. Thesis. Purdue University.

Whittaker, K.F., J.C. Nye, R.F. Weekash, R.J. Squires, A.C. York and H.A.

---

* Confidential Business Information submitted to the Office of Pesticide Programs.

Razemier. 1982. Collection and treatment of wastewater generated by pesticide application. EPA-600/2-82-028. Cincinnati, OH.

Metolachlor

I. General Information and Properties

A. CAS No. 51218–45–2

B. Structural Formula

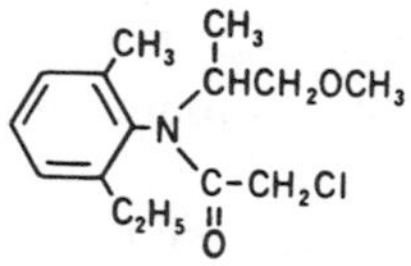

2-Chloro-N-(2-ethyl-6-methylphenyl)-N-(2-methoxy-l-methylethyl) acetamide

C. Synonyms

- o-Acetanilide; 2-chloro-6'-ethyl-N-(2-methoxy-l-methylphenyl); Dual®; Bicep®; Metetilachlor; Pimagram; Primextra; CGA-24705.

D. Uses (Meister, 1986)

- Metolachlor is used as a selective herbicide for preemergence and pre-plant incorporated weed control in corn, soybeans, peanuts, grain sorghum, pod crops, cotton, safflower, woody ornamentals, sunflowers and flax.

E. Properties (Meister, 1986; Ciba-Geigy, 1977; Windholz et al., 1983; Worthing, 1983)

Chemical Formula	$C_{15}H_{22}NO_2Cl$
Molecular Weight	283.46
Physical State	White to tan liquid
Boiling Point	100°C (at 0.001 mm Hg)
Melting Point	--
Density	--
Vapor Pressure (20°C)	1.3×10^{-5} mm Hg
Specific Gravity	--
Water Solubility (20°C)	530 mg/L

Log Octanol/Water
  Partition Coefficient   --
Taste Threshold   --
Odor Threshold   --
Conversion Factor   --

F. Occurrence

- Metolachlor has been found in 2,091 of 4,161 surface water samples analyzed and in 13 of 596 ground-water samples (STORET, 1988). Samples were collected at 332 surface water locations and 551 ground-water locations, and metolachlor was found in 7 states. The 85th percentile of all nonzero samples was 11.5 μg/L in surface water and 0.25 μg/L in ground-water sources. The maximum concentration found was 139 μg/L in surface water and 11.0 μg/L in ground-water. This information is provided to give a general impression of the occurrence of this chemical in ground and surface waters as reported in the STORET database. The individual data points retrieved were used as they came from STORET and have not been confirmed as to their validity. STORET data are often not valid when individual numbers are used out of the context of the entire sampling regime, as they are here. Therefore, this information can only be used to form an impression of the intensity and location of sampling for a particular chemical.

- Metolachlor residues resulting from agricultural use have also been detected in ground water in Iowa and Pennsylvania with concentrations ranging from 0.1 to 0.4 ppb.

G. Environmental Fate

- ^{14}C-Metolachlor (test substance uncharacterized), at approximately 4.6 lb active ingredient (a.i.)/acre (A), degraded with a half-life of 48 to 144 hours at 39 to 44°C when irradiated with artificial light (uncharacterized) and 6 to 8 days (approximately 3,000 Langley units) at 50 to 55°C when irradiated with natural sunlight (Aziz, 1974). The degradate, N-propen-1-ol-2-yl-N-chloroacetyl-2-methyl-6-ethylaniline (CGA-41638), was approximately 4 to 6% of the applied in both the artificial light- and natural sunlight-irradiated samples; three unknown degradates were also isolated. Nonextractable ^{14}C-residues accounted for approximately 40% of the applied, and volatiles accounted for approximately 7 to 10% of the applied after 168 hours of irradiation with artificial light or 8 days of irradiation with natural sunlight.

- ^{14}C-Metolachlor (purity unspecified) at approximately 8 ppm was essentially stable in loamy sand soil, over a period of 64 days (Kaiser, 1974). Sterilization of the soil had no appreciable effect on degradation. The soil was maintained at approximately 60% of field capacity (temperature unspecified).

- ^{14}C-Metolachlor (purity unspecified) at 1 to 10 ppm adsorbed to sandy clay loam, loam and two sand soils with Freundlich adsorption constants (K) ranging from 1.54 to 10 μg/g, indicating mobility in these soils (Burkhard, 1978). Except for in one sand soil, as soil organic matter content increased, adsorption increased. In loam soil ^{14}C-metolachlor desorbed with K values of 4.87 and 3.57 after 1 and 3 days, respectively. Adsorption and desorption occurred at about the same rate.

- Aged (30 days) ^{14}C-metolachlor (purity unspecified) residues were mobile in columns of loamy sand soil with approximately 26% of applied radioactivity leached from the columns, and approximately 87% of the applied radioactivity remaining in the soil (Dupre, 1974a). Radioactivity was concentrated (approximately 60% of applied) in the top 3 inches of soil.

- ^{14}C-Metolachlor (purity unspecified) residues were mobile in sandy loam, sand and silt loam soils (detected to the 12-inch depth in 12-inch columns) when leached with 20 inches of water (Houseworth, 1973). ^{14}C-Metolachlor residues were immobile in peat soil with approximately 98.8% of the applied radioactivity detected in the top 1 inch of soil. The leachate from the sandy loam, sand, loam, and silt loam soils contained 36.4, 20.9, 4.0 and 0.4% of the applied radioactivity, respectively.

- ^{14}C-Metolachlor (purity unspecified) residues at 1 lb a.i./A dissipated from a loamy sand soil (8° slope) in the leachate (0.25% of applied radioactivity), and the runoff water (3.15% of applied) and sediment (1.41% of applied) after the application of approximately 1.52 inches of simulated rainfall in 7 days (Dupre, 1974b). Of the applied rainfall, approximately 85% was collected as runoff. Greater than 85% and 5% of the applied radioactivity was detected in the treated and untreated soils, respectively. Total recovery of radioactivity was 95.8% of applied.

- Freundlich adsorption K values of 0.58, 1.46 and 7.83 were calculated for ^{14}C-metolachlor (test substance uncharacterized) in sand, silt loam and sandy loam soils, respectively (Harvey and Jordan, 1978). Adsorption was positively correlated to soil organic matter content.

- Lettuce planted 14 weeks posttreatment and harvested at 26 weeks contained 0.025 ppm of ^{14}C-metolachlor residues (Newby, 1979). Lettuce planted 41 weeks posttreatment and sampled at 13 and 15 weeks contained 0.144 and 0.065 ppm of ^{14}C-metolachlor residues, respectively. Soil at about 40 and 56 weeks posttreatment contained approximately 70 to 80 and 95% nonextractable ^{14}C-metolachlor residues, respectively.

- Residues of phenyl-labeled ^{14}C-metolachlor were taken up by greenhouse-grown, rotational winter wheat planted 168 days after a 2.0 lb a.i./A application to a silt loam soil (Sumner and Cassidy, 1974d). Total radioactivity was < 0.15 ppm in forage sampled 189 to 238 days posttreatment, and 0.60 and 0.03 ppm in straw and grain, respectively, harvested 273 days posttreatment.

- Residues of phenyl-labeled ^{14}C-metolachlor were taken up by greenhouse-grown, rotational oats planted 270 days after a 2.0 lb a.i./A application to a silt loam soil (Sumner and Cassidy, 1974c). Total radioactivity was < 0.17 ppm in forage sampled 300 to 330 days posttreatment, and 0.27 and 0.05 ppm in straw and grain, respectively, harvested 375 days posttreatment.

- ^{14}C-Metolachlor residues were detected in the whole plant (0.04 ppm), tops (< 0.09 ppm) and roots (< 0.06 ppm) of rotational carrots planted 9 months after an application of ^{14}C-metolachlor (purity unspecified) at 2 lb a.i./A to silt loam soil (Sumner and Cassidy, 1974a). Total residues of ^{14}C-metolachlor in the silt loam soil were 0.26 ppm in the 0 to 3 inch depth at carrot planting, and 0.04 to 0.35 ppm in the 0 to 9 inch profile in the subsequent samplings through harvest.

- ^{14}C-Metolachlor residues were detected in the whole plant (< 0.17 ppm), stalks (0.07 ppm), beans (0.04 ppm), meal (0.05 ppm) and oil (< 0.01 ppm) of rotational soybeans planted 9 months after an application of ^{14}C-metolachlor (purity unspecified) at 2 lb a.i./A to silt loam soil (Sumner and Cassidy, 1974b). Total residues of ^{14}C-metolachlor in the silt loam soil were 0.26 ppm in the 0 to 3 inch depth at soybean planting, and 0.06 to 0.35 ppm in the 0 to 9 inch profile in the subsequent samplings through harvest.

- *Photodegradation studies in water*: One study (Aziz and Kahrs, 1974; Aziz and Kahrs, 1975) cannot be validated because the experimental conditions were referenced, rather than described, and the reference was not available for review. In addition, this study would not fulfill data requirements because the test substance was uncharacterized, the incubation temperatures were not reported, it was not reported if the test solutions were buffered or if wavelengths < 290 nm were filtered out, the conditions under which dark controls were maintained were not reported, the natural sunlight conditions were not provided, the intensity and wavelength distribution of the artificial light source were not provided and degradates that comprised > 10% of the applied that were detected in the test solution exposed to artificial light were not characterized.

- *Leaching and adsorption/desorption studies*: Four studies were reviewed and considered to be scientifically valid. The first study (Burkhard, 1978) partially fulfills data requirements by providing information on the adsorption of parent metolachlor. However, the study was not conducted in a 0.01 N calcium ion solution, desorption data were reported for only one of the four soils tested and the majority of the study was conducted on foreign soils. The second study (Dupre, 1974a) partially fulfills data requirements by providing information on the mobility of aged (30 days) metolachlor residues. However, the purity of the test substance was unspecified, K_d values were not reported and ^{14}C-metolachlor residues were not characterized. The third study (Houseworth, 1973a) partially fulfills data requirements by providing information on the mobility of metolachlor residues. However, the purity of the test substance was unspecified and K_d values were not reported. The fourth study (Harvey and Jordan, 1978) partially fulfills data requirements by providing information on the adsorption of parent metolachlor. However, the test substance was uncharacterized and the study was not conducted in a 0.01 N calcium ion solution.

- *Ground and surface water monitoring studies*: Submitted data indicate that metolachlor is mobile in soil (Burkhard, 1978; Houseworth, 1973) and available in surface water runoff (Dupre, 1974b). Monitoring data (Ross and Balu, 1985) indicate metolachlor is present in surface waters.

II.   Pharmacokinetics

A. Absorption

- In studies conducted by Hambock (1974a,b), rats were administered a single oral dose (28.6 or 52.4 mg/kg) of metolachlor (purity not specified, but ^{14}C-labeled and unlabeled metolachlor were synthesized for these experiments). The chemical was readily absorbed, since 70 to 90% of the metolachlor was excreted as metabolites within 48 hours.

B. Distribution

- Data from rats given radioactive metolachlor (approximately 3.2 to 3.5 mg/kg) orally demonstrated that the chemical is rapidly metabolized. Residues in meat tissues and blood were very low and only blood contained residue levels in excess of 0.1 ppm (Hambock, 1974c).

C. Metabolism

- Studies conducted to identify urinary and fecal metabolites in the rat indicated that metolachlor is metabolized via dechlorination, O-methylation, N-dealkylation and side-chain oxidation (Hambock, 1974 a,b). Urinary metabolites included 2-ethyl-6-methylhydroxyacetanilide (MET-002) and N-(2-ethyl-6-methylphenyl)-N-(hydroxyacetyl)-DL-alanine (MET-004). Fecal metabolites included 2-chloro-N-(2-ethyl-6-methylphenyl)-N-(2-hydroxy-l-methylethyl) (MET-003) and MET-004.

D. Excretion

- When treated with ^{14}C-metolachlor (approximately 31 mg/kg orally), male rats excreted 21.5% and 51.4% of the administered dose in the urine and feces, respectively, in 48 hours (Hambock, 1974a,b). The excreta contained 1, 15 and 22% of the administered dose as MET-002, MET-003 and MET-004, respectively. No unchanged chemical was isolated, and no glucuronide or sulfate conjugates were identified.

III.   Health Effects

A. Humans

- Signs of human intoxication from metolachlor and/or its formulations (presumably following acute deliberate or accidental exposures) include abdominal cramps, anemia, ataxia, dark urine, methemoglobinemia, cyanosis, hypothermia, collapse, convulsions, diarrhea, gastrointestinal irritation, jaundice, weakness, nausea, shock, sweating, vomiting, CNS depression, dizziness, dyspnea, liver damage, nephritis, cardiovascular failure, skin irritation, dermatitis, sensitization dermatitis, eye and mucous membrane irritation, corneal opacity and adverse reproductive effects (HAZARDLINE, 1985).

A. Animals

1. *Short-term Exposure*

- The acute oral LD_{50} of technical metolachlor in the rat was reported to be 2,780 mg/kg (95% confidence range of 2,180 to 3,545 mg/kg; Bathe, 1973).

- Technical metolachlor in corn oil was shown to be emetic in beagle dogs, precluding the establishment of an LD_{50} (Roche, 1974). However, an "emetic dose" of 19 ± 9.7 mg/kg was established.

- Beagle dogs were fed technical metolachlor in the diet for 7 days in a range-finding study (Goldenthal et al., 1979). Each test group consisted of one male and one female. Doses were 1,000, 3,000 or 5,000 ppm with the controls receiving a basic diet plus the test material solvent (ethanol). The mean doses were 0, 13.7, 22.7 or 40.2 mg/kg. Decreased food consumption and body weight indicated that the two higher doses were unpalatable. No changes were observed at the lowest dose, although the animals exhibited soft stools and/or diarrhea over the study period. No other signs of overt toxicity, morbidity or mortality were observed in any animal. Accordingly, the lowest dose (13.7 mg/kg) is the No-Observed-Adverse-Effect Level (NOAEL) in this study.

2. *Dermal/Ocular Effects*

- The LD_{50} of technical metolachlor in the rabbit when tested by the unabraded dermal route is greater than 10,000 mg/kg (Bach, 1974).

- Sachsse (1973b) evaluated the dermal irritation potential of technical metolachlor (3/sex/dose) on the Russian strain rabbits (weighing 2 to 3 kg). The chemical was applied to abraded and unabraded skin for 24 hours and then observed for periods up to 72 hours. The results demonstrated that technical metolachlor is nonirritating to rabbit skin.

- Sachsse and Ullman (1977) studied skin sensitization in the albino guinea pig by the intradermal-injection method. Technical metolachlor (CG-24705) dissolved in the vehicle (propylene glycol) and the vehicle alone were intradermally injected into the skin of two groups of Pilbright guinea pigs. A positive reaction was observed in the animals injected with metolachlor in the vehicle, but not in animals treated with the vehicle alone. The authors concluded that technical metolachlor is a skin sensitizer.

- A study of eye irritation by technical metolachlor in the Russian strain rabbit (3/sex/dose) was conducted by Sachsse (1973a). The chemical was applied at a dose level of 0.1 mL/eye. Evaluation of both washed and unwashed eyes 24 hours and 7 days later revealed no evidence of irritation.

3. *Long-term Exposure*

- Beagle dogs (4/sex/dose) were administered technical metolachlor (> 90% a.i.) in their feed for up to 15 weeks (Coquet et al., 1974). Initial doses were 0, 50, 150 or 500 ppm (equivalent to 0, 4 to 5 or 14 to 19 mg/kg/day). However, after 8 weeks, the group receiving 50 ppm was switched to a diet that delivered 1,000 ppm (27 to 36 mg/kg/day) for the remaining 6 weeks. The dose was increased because no signs of toxicity were observed in the 500-ppm group after 8 weeks. No animals died during the study and no significant changes were observed in gross or histological pathology, blood or urine analyses. Except for a decrease in food consumption and associated slight weight loss at the 1,000-ppm dose, no compound-related effects were observed. The NOAEL for this study is 500 ppm (14 to 19 mg/kg/day).

- Tisdel et al. (1982) presented the results of a study in which metolachlor (95% a.i.) was administered to Charles River CD-1 mice (68/sex/dose) for 2 years at dietary concentrations of 0, 300, 1,000 or 3,000 ppm. Time-weighted average (TWA) concentrations, based upon diet analyses, were equal to 0, 287, 981 and 3,087 ppm. The dietary doses, from reported food intake and body weight data, were calculated to be equal to 0, 50, 170 or 526 mg/kg/day for the males and 0, 64, 224 or 704 mg/kg/day for the females. No treatment-related effects were observed in terms of physical appearance, food consumption, hematology, serum chemistry, urinalysis or gross histopathology. However, mortality was increased significantly in females fed 3,000 ppm (704 mg/kg/day). Statistically significant reductions in body weight gain were observed in both sexes at the highest dose. Also, statistically significant changes in absolute and organ-to-body weight ratios were noted occasionally (e.g., kidney- and liver-to-body weight ratios and decreased seminal vesicle-to-body weight ratio in high dose males). Based on this information, a NOAEL of 1,000 ppm (170 mg/kg/day for males and 224 mg/kg/day for females) is identified.

- Tisdel et al. (1983) presented the results of a study in which metolachlor (purity not specified) was administered to CD-Crl:CD (SD) BR rats for 2 years at dietary concentrations of 0, 30, 300 or 3,000 ppm. Assuming that 1 ppm in the diet of rats is equal to 0.05 mg/kg/day (Lehman,

1959), these dietary concentrations would be equal to 0, 1.5, 15 or 150 mg/kg/day. The control and 3,000-ppm groups consisted of 70 rats/sex. The 30- and 300-ppm groups consisted of 60 rats/sex. No treatment-related effects were noted in terms of mortality, organ weight and organ-to-body weight ratios. A variety of differences in clinical pathology measurements was found between control and treatment groups at various time intervals, but no consistent dose-related effects were apparent with the exception of a decrease in glutamic-oxaloacetic transaminase activity in high-dose males at 12 months, the significance of which is uncertain. Mean body weights of high-dose females were consistently less than controls from week 2 until termination of the study. This difference was statistically significant ($p < 0.01$) for 26 of the 59 intervals at which such measurements were made. Food consumption in high-dose females also was generally less than controls. Gross pathology findings were described by the investigators as being unremarkable. Based on this data, a NOAEL of 300 ppm (15 mg/kg/day) is identified.

4. Reproductive Effects

- A three-generation rat reproduction study was reported by Adler and Smith (1978). Targeted dietary exposures for metolachlor (96.5% a.i.) were 0, 30, 300 or 1,000 ppm. The actual exposures were analyzed to be 0, 30, 250 or 850 ppm. Assuming that 1 ppm equals 0.05 mg/kg/day (Lehman, 1959), the doses were calculated to be 0, 1.5, 22.5 or 42.5 mg/kg bw/day. No adverse effects were noted at any dose. A minimal NOAEL of 42.5 mg/kg is identified for reproductive effects.

- Smith et al. (1981) conducted a two-generation reproduction study in which Charles River CD rats (15 males and 30 females/dose) were administered technical-grade metolachlor (purity not specified) at dietary doses of 0, 30, 300 or 1,000 ppm. The TWA concentrations of metolachlor, based upon dietary analysis, were 0, 32, 294 or 959 ppm. Assuming that 1 ppm in the diet of rats is equivalent to 0.05 mg/kg/day (Lehman, 1959), these dietary concentrations are approximately equal to 0, 1.6, 14.7 or 48 mg/kg/day. Mating, gestation, lactation and female and male fertility indices were not affected in either generation. Additionally, pup survival was not affected. However, pup weights in the 959-ppm dose group, but not the 32- and 294-ppm dose groups, were significantly reduced in the F_{1a} and F_{2a} litters. Food consumption was reduced significantly for F_1 females receiving 32 ppm (1.6 mg/kg/day) and greater at various study intervals. Other effects that appeared to be treatment-related included increased liver-to-body weight ratios for both F_1 parental males and females at 1,000 ppm and increased thyroid-to-body weight and thyroid-to-brain weight in F_1 males at 1,000 ppm. Based on reduced pup weights, a reproductive NOAEL of 294 ppm (14.7 mg/kg/day) is identified.

- Tisdel et al. (1982) gave metolachlor (95% a.i.) to CD-l mice (68/sex/ dose) in the food for 2 years at concentrations of 0, 300, 1,000 or 3,000 ppm (the TWAs based on diet analyses were 0, 287, 981 or 3,087 ppm and corresponded to 0, 50, 170 or 520 mg/kg/day in males and to 0, 64, 224 or 704 mg/kg/day in the females). At the high dose, males were found to have a reduced seminal vesicle-to-body weight ratio.

- Tisdel et al. (1983) exposed CD-Crl:CD (SD) BR rats (70/sex/dose) to metolachlor (purity not specified) in the diet for 2 years at 0, 30, 300 or 3,000 ppm (the doses correspond to 0, 1.5, 15 or 150 mg/kg/day). They observed greater testicular atrophy and degeneration of the tubular epithelium in the 300- and 3,000-ppm groups than in the control group.

5. Developmental Effects

- Fritz (1976a) conducted a rat teratology study in which pregnant females (25/dose level) were administered doses of technical metolachlor (purity not specified) orally at 0, 60, 180 or 360 mg/kg/day during days 6 to 15 of gestation. No fetotoxic or developmental effects were noted.

- Lightkep et al. (1980) evaluated the teratogenic potential of metolachlor (95.4% pure) in New Zealand White rabbits (16/dose). The compound was administered via stomach tube as a suspension in aqueous 0.75% hydroxymethylcellulose at levels of 0, 36, 120 or 360 mg/kg/day. Single oral doses were given on days 6 to 18 of gestation. Abortions occurred in two rabbits: one in the 120 mg/kg/day group on day 25 (one early resorption) and one in the 360 mg/kg/day group on day 17 (one fetus) and day 20 (eight additional implantations). They did not consider these abortions to be treatment-related. Maternal toxicity (decreased food consumption and pupillary constriction) was observed in animals receiving the two highest doses. The highest dose group also exhibited blood in the cage pan and body weight loss over the treatment period. No significant developmental or fetotoxic effects were observed in the 319 fetuses, pups or late resorptions evaluated from all dose groups. Thus, a minimal NOAEL of 360 mg/kg/day for fetotoxicity and a NOAEL of 36 mg/kg/ day for maternal toxicity were identified.

6. *Mutagenicity*

- Technical metolachlor (purity not specified) was tested in the Ames *Salmonella* test system, using *S. typhimurium* strains TA1535, TA1537, TA98 and TA100 (Arni and Muller, 1976). No increase in mutagenic response was observed, with or without microsomal activation, at concentrations of 10, 100, 1,000 or 10,000 μg/plate. Toxicity was observed at 1,000 and 10,000 μg/plate without activation and at 10,000 μg/plate with activation.

- Fritz (1976b) reported the results of a dominant lethal study in male albino NMRI-derived mice using technical metolachlor (purity not specified). The compound was administered orally in single doses of 0, 100 or 300 mg/kg to males that then were mated to untreated females over a period of 6 weeks. No evidence of adverse effects were observed, as expressed by increased implantation loss or resorptions.

7. *Carcinogenicity*

- Marias et al. (1977) presented the results of a study in which technical-grade metolachlor was administered to Charles River CD-1 mice (50/sex/dose) at dietary concentrations of 0, 30, 300 or 3,000 ppm. Assuming that 1 ppm in the diet of the mouse is equal to 0.15 mg/kg/day (Lehman, 1959), these dietary levels are approximately 0, 4.5, 150 or 450 mg/kg/day. Males received the test material for 18 months; females received the test material for 20 months. Two samples of metolachlor (99.9% a.i. and 96.5% a.i.) were received. The 99.9% sample was used in dietary formulations for the first 33 weeks and the 96.5% pure sample was used for the rest of the study. Results of this study indicated no evidence of oncogenicity in either sex.

- Tisdel et al. (1982) presented the results of a study in which metolachlor (95% a.i.) was administered to Charles River CD-1 mice (68/sex/dose) for 2 years at dietary concentrations of 0, 300, 1,000 or 3,000 ppm. From food intake and body weight data, the doses were calculated to be equal to 0, 50, 170 or 526 mg/kg/day for the males and 0, 64, 224 or 704 mg/kg/day for the females. A statistically significant increase in the incidence of alveolar tumors in high-dose males was noted at the 18-month sacrifice; however, this effect was not confirmed by the final sacrifice at 24 months or by total tumor incidences for all animals.

- Ciba-Geigy (1979) reported the results of a study in which technical metolachlor was administered to Charles River albino rats in their diet for 2 years at doses equivalent to 0, 1.5, 15 or 50 mg/kg/day. A statistically significant increase in the incidence of primary liver tumors was observed in the high-dose females (15/60 compared with 5/60 at mid doses and 3/60 at the low dose and control). These tumors included hypertrophic-hyperplastic nodules, angiosarcoma, cystic cholangioma and hepatocellular carcinoma. The variety of tumor expression forms suggests that a variety of cell types and locations may be affected in the liver.

- Tisdel et al. (1983) presented the results of a study in which metolachlor (purity not specified) was administered to CD-Crl:CD (SD) BR rats for 2 years at dietary concentrations of 0, 30, 300 or 3,000 ppm. These doses were assumed to be equal to 0, 1.5, 15 or 150 mg/kg/day. An increased incidence of proliferative hepatic lesions (combined neoplastic nodules/ carcinomas) was found in the high-dose females at terminal sacrifice ($p < 0.018$ by the Fisher exact test). Six of the 60 had neoplastic nodules ($p < 0.05$) and 7 of the 60 had liver tumors (one additional tumor was diagnosed as a carcinoma; $p < 0.01$).

IV.   Quantification of Toxicological Effects

A. One-day Health Advisory

No suitable information was found in the available literature for determination of a One-day HA for metolachlor. Accordingly, it is recommended that the Longer-Term HA value for the 10-kg child (2.0 mg/L, calculated below) be used at this time as a conservative estimate of the One-day HA value.

B. Ten-day Health Advisory

There were no satisfactory studies found in the available literature to use for the calculation of the Ten-day HA. It is therefore recommended that the Drinking Water Equivalent Level (DWEL), adjusted for a 10-kg child, of 2 mg/L (2,000 mg/L), calculated below, be used for the Ten-day HA.

For a 10-kg child, the adjusted DWEL is calculated as follows:

$$\text{DWEL} = \frac{(0.15 \text{ mg/kg/day}) (10 \text{ kg})}{1 \text{ L/day}} = 1.5 \text{ mg/L (rounded to 2 mg/L)}$$

where: 0.15 mg/kg/day = RfD (see Lifetime Health Advisory section).

C. Longer-Term Health Advisory

No studies were found in the literature that were suitable for deriving the Longer-term HA value for metolachlor. In the lifetime study of Tisdel et al. (1983), a NOAEL of 15 mg/kg/day was found in rats. This NOAEL was similar to that NOAEL (13.7 mg/kg/day) found by Goldenthal et al. (1979). However, Goldenthal et al. (1979) used only one dog of each sex per dose group (0, 13.7, 22.7 or 40.2 mg/kg/day), while Tisdel et al. (1983) used 70/sex/dose at 0, 1.5, 15 or 150 mg/kg/day. Accordingly, more confidence can be placed in the Tisdel et al. study. It is therefore recommended that the DWEL of 5.0 mg/L (5,000 μg/L), calculated below, be used for the Longer-term HA value for an adult, and that the DWEL adjusted for a 10-kg child, 2.0 mg/L (2,000 μg/L), be used for the Longer-term HA for a child.

D. Lifetime Health Advisory

The study by Tisdel et al. (1983) has been selected to serve as the basis for the Lifetime HA. In this study, rats were given dietary doses of metolachlor equivalent to 0, 1.5, 15 or 150 mg/kg/day. No treatment-related effects were noted in terms of mortality, organ weight and organ-to-body weight ratios. The investigators noted a statistically significant decrease in glutamic-oxaloacetic transaminase activity in high-dose males at 12 months. Mean body weights of high-dose females were consistently less than controls from week 2 until termination of the study. This difference was significant (p < 0.01) for 26 of the 59 intervals at which such measurements were made. Food consumption in high-dose females also was generally less than controls. Gross pathology findings were described as unremarkable. Based on the decreased body weight gain at 3,000 ppm (150 mg/kg/day), a NOAEL of 300 ppm (15 mg/kg/day) was identified.

The Lifetime HA is calculated as follows:

Step 1: Determination of the Reference Dose (RfD):

$$RfD = \frac{15 \text{ mg/kg/day}}{100} = 0.15 \text{ mg/kg/day}$$

Step 2: Determination of the Drinking Water Equivalent Level (DWEL)

$$DWEL = \frac{(0.15 \text{ mg/kg/day})(70 \text{ kg})}{(2 \text{ L/day})} = 5.25 \text{ mg/L } (5,000 \text{ } \mu\text{g/L})$$

Step 3: Determination of the Lifetime Health Advisory

$$\text{Lifetime HA} = \frac{(5.25 \text{ mg/L}) (20\%)}{(10)} = 0.1 \text{ mg/L } (100 \text{ } \mu\text{g/L})$$

E. Evaluation of Carcinogenic Potential

- Four studies evaluating the carcinogenic potential of metolachlor have been identified. In two of these studies (Marias et al., 1977; Tisdel et al., 1982), no evidence of carcinogenicity in mice was observed. The other studies, both conducted using rats, showed an increased tumor incidence related to treatment. Ciba-Geigy (1979) reported a statistically significant increase in primary liver tumors in female Charles River rats exposed to 150 mg/kg/day in the diet for 2 years. Tisdel et al. (1983) also reported a statistically significant increase in the incidence of proliferative hepatic lesions (neoplastic nodules and carcinomas) in female rats at the same dietary dose over the same time period. Additionally, there was a nonstatistically significant increase in the frequency of adenocarcinoma of the nasal turbinates and fibrosarcoma of the nasal tissue in the high-dose males (150 mg/kg/day).

- The International Agency for Research on Cancer has not evaluated the carcinogenicity of metolachlor.

- Applying the criteria described in EPA's guidelines for the assessment of carcinogenic risk (U.S. EPA, 1986a), metolachlor is classified in Group C: possible human carcinogen. This category is for substances with limited evidence of carcinogenicity in animals and absence of human data.

V. Other Criteria, Guidance and Standards

- EPA/OPP has identified an Acceptable Daily Intake (ADI) for metolachlor of 0.015 mg/kg/day based on the NOAEL of 30 ppm (1.5 mg/kg/day) from the chronic rat feeding study (Tisdel et al., 1983) and an uncertainty factor of 100 (U.S. EPA, 1986b). Using this ADI and an assumed body weight of 60 kg, the maximum permissible intake has been calculated to be 0.9 mg/day. The total maximum residue concentration is 0.07209 mg/day or about 8% of the ADI.

- Residue tolerances ranging from 0.02 to 30 ppm have been established for a variety of agricultural products (CFR, 1985).

VI. Analytical Methods

- Analysis of metolachlor is by a gas chromatographic (GC) method applicable to the determination of certain nitrogen-phosphorus containing pesticides in water samples (U.S. EPA, 1988). In this method, approximately 1 liter of sample is extracted with methylene chloride. The extract is concentrated and the compounds are separated using a capillary column GC. Measurement is made using a nitrogen phosphorus detector. This method has been validated in a single laboratory, and estimated detection has been determined for the analytes using this method. The estimated detection limit for metolachlor is 0.75 μg/L.

VII. Treatment Technologies

- Whittaker (1980) experimentally determined adsorption isotherms for metolachlor on granular-activated carbon (GAC) Nuchar WV-G. Nuchar WV-G reportedly exhibited the following adsorption capacities at 20°C: 0.173, 0.148 and 0.105 mg metolachlor/mg carbon at concentrations of 79.84 mg/L, 10 mg/L and 1.74 mg/L, respectively.

- Holiday and Hardin (1981) reported the results of GAC treatment of wastewater contaminated with pesticides including metolachlor. The column, 3.5 feet in diameter, was packed with 10 feet of granular activated carbon, or 3,150 lb carbon/column. The column was operated at 1.04 gpm/ft^2 hydraulic load and 72 minutes contact time. Under these conditions, 99.5% of the metolachlor was removed from wastewater at an initial average concentration of 16.4 mg/L.

- GAC adsorption appears to be the most promising treatment technique for the removal of metolachlor from water. However, more actual data are required to determine the effectiveness of GAC in removing metolachlor from contaminated drinking water supplies.

VIII. References

Adler, G.L. and S.H. Smith.* 1978. Final report to Ciba-Geigy Corp: three generation reproduction study with CGA-24705 technical in albino rats. IBT No. 8533–07928. Received January 18, 1978, under 7F1913. Unpublished study. MRID 00015632.

Arni, P. and D. Muller.* 1976. Salmonella/mammalian-microsome mutagenicity test with CGA 24705. Test for mutagenic properties in bacteria. PH 2.632. Received January 19, 1977, under 7F1913. MRID 00015397.

Aziz, S.A.* 1974. Photolysis of CGA-24705 on soil slides under natural and artificial sunlight conditions. Report No. GAAC-74102. Unpublished study. Received on unknown date, under 5F1606. Submitted by Ciba-Geigy Corp., Greensboro, NC. CDL:094385-M. MRID 00016301.

Aziz, S.A. and R.A. Kahrs.* 1974. Photolysis of CGA-24705 in aqueous solution under natural and artificial sunlight conditions. Report No. GAAC-74041. Unpublished study. Received Sept. 26, 1974, under 5F1606. Submitted by Ciba-Geigy Corp., Greensboro, NC. CDL:094385-D. MRID 00016300.

Aziz, S.A. and R.A. Kahrs.* 1975. Photolysis of CGA-24705 in aqueous solution – additional information. Report No. GAAC-75021. Unpublished study. Received on unknown date under 5F1606. Submitted by Ciba-Geigy Corp., Greensboro, NC. CDL:094385-M. MRID 00016302.

Bach, K.J.* 1974. Acute dermal LD50 of CGA-24705-Technical in rabbits. Affiliated Medical Research, Inc. Contract No. 120–2255–34. Received September 26, 1974 under 5G1553. Unpublished study. MRID 00015526.

Bathe, R. 1973.* Acute oral LD_{50} of technical CGA-24705 in the rat. Project No. Siss 2979. Received September 26, 1974, under 5G1553. Unpublished study. MRID 00015523.

Burkhard, N.* 1978. Adsorption and desorption of metolachlor (Dual) in various soil types. Project Report 45/78. Unpublished study. Received July 23, 1981, under 100–587. Prepared by Ciba-Geigy, Ltd., Switzerland. Submitted by Ciba-Geigy Corp., Greensboro, NC. CDL:245627-D. MRID 00078291.

Ciba-Geigy Corporation.* 1977. Section A: general chemistry. Unpublished study. Received January 19, 1977, under 7F1913. MRID 00015392.

Ciba-Geigy Corporation.* 1979. Two-year chronic oral toxicity study with CGA-24705 technical in albino rats. Study No. 8532–07926. Conducted by Industrial Bio-Test Laboratories. Unpublished study. Received December 11, 1979, under 8F2098. MRID 00130776.

CFR. 1985. Code of Federal Regulations. 40 CFR 180.368. July 1.

Coquet, B., L. Galland, D. Guyot, X. Pouillet and J.L Rounaud.* 1974. Three-month oral toxicity study trial of CGA 24705 in the dog. IC-DREB-R 740119. Received September 26, 1974, under 5G1553. Unpublished study. MRID 00052477.

Dupre, G.D.* 1974a. Leaching characteristics of 14C-CGA-24705 and its degradation products following aging in sandy loam soil under greenhouse conditions. Report No. 73021–6. Unpublished study. Received Sept. 26, 1974, under 5G1553. Prepared by Bio/dynamics, Inc. Submitted by Ciba-Geigy Corp., Greensboro, NC. CDL:094222-C. MRID 00015657.

Dupre, G.D.* 1974b. Runoff characteristics of ^{14}C-CGA-24705 applied to sandy loam soil under greenhouse conditions. Report No. 73022–1. Unpublished study. Received Sept. 26, 1974, under 5G1553. Prepared by Bio/dynamics, Inc. Submitted by Ciba-Geigy Corp., Greensboro, NC. CDL:094222-D. MRID 00015658.

Fritz, H.* 1976a. Reproduction study on CGA-24705 technical: rat (Segment II test for teratogenic or embryotoxic effects) PH 2.632. Unpublished study, including addendum. Received January 19, 1977, under 7F1913. MRID 00015396.

Fritz, H.* 1976b. Dominant lethal study on CGA-24705 technical: mouse (test for cytotoxic or mutagenic effects on male germinal cells) PH 2.632.

Received January 18, 1978, under 7F1913. Unpublished study, including addendum. MRID 00015630.

Goldenthal, E.I., D.C. Jessup and J.S. Mehring.* 1979. Range-finding study with metolachlor technical in beagle dogs. IRDC No. 382–053. Unpublished study. Received December 11, 1979, under 100–597. MRID 00016631.

Hambock, H.* 1974a. Project 7/74: metabolism of CGA-24705 in the rat (status of results gathered up until June 10, 1974): AC 2.52. Unpublished study. MRID 00039193.

Hambock, H.* 1974b. Project 12/74: addendum to Project 7/74: metabolism of CGA-24705 in the rat: AC 2.52. Unpublished study. MRID 00015425.

Hambock, H.* 1974c. Project 1/74: distribution, degradation and excretion of CGA-24705 in the rat: AC 2.52. MRID 00039192.

Harvey, R.F. and G.L. Jordan.* 1978. Comparative study of the biological and physical properties of acetanilide herbicides in soil. Unpublished study. Received May 3, 1979, under 43142-1. Prepared by Univ. of Wisconsin. Submitted by Boots Hercules Agrochemicals Co., Wilmington, DE. CDL:098274-H. MRID 00031328.

HAZARDLINE. 1985. Bethesda, MD: National Library of Medicine. National Institutes of Health.

Holiday, A.D. and D.P. Hardin. 1981. Activated carbon removes pesticides from wastewater. Chem. Eng. 88:88–89.

Houseworth, L.D.* 1973. Report on parent leaching studies for CGA-24705. Report No. 1. Unpublished study. Received Sept. 26, 1974, under 5G1553. Prepared by Univ. of Missouri, Dept. of Plant Pathology. Submitted by Ciba-Geigy Corp., Greensboro, NC. CDL:094222-E. MRID 00015659.

Jessup, D.C., R.J. Arceo, F.L. Estes et al.* 1979. 6-Month chronic oral toxicity study in beagle dogs. IRDC No. 382–054. Unpublished study. MRID 00032174.

Kaiser, F.E.* 1974. Soil degradation study of Gba-Geigy (sic) ^{14}C-CGA-24705. Unpublished study. Received Mar. 27, 1975, under 5F1606. Prepared by Analytical Biochemistry Laboratories, Inc. Submitted by Ciba-Geigy Corp., Greensboro, NC. CDL:094376-N. MRID 00016296.

Lehman, A.J. 1959. Appraisal of the safety of chemicals in foods, drugs and cosmetics. Assoc. Food Drug Off. U.S., Q. Bull.

Lightkep, G.E., M.S. Christian, G.D. Christian et al.* 1980. Teratogenic potential of CGA-24705 in New Zealand White rabbits; Segment II evaluation—Project 203–001. Unpublished study. MRID 00041283.

Marias, A.J., J. Gesme, E. Albanese et al.* 1977. Carcinogenicity study with CGA-24705 technical in albino mice. Revised final report to Ciba-Geigy Corporation. IBT No. 622–07925 (8532-07925). Unpublished study. MRID 00084003.

Meister, R., ed. 1986. Farm chemicals handbook. Willoughby, OH: Meister Publishing Co.

Newby, L. 1979.* Dual rotational studies: additional information. Report No. ABR-79091. Unpublished study. Received Aug. 1, 1979, under 100583. Submitted by Ciba-Geigy Corp., Greensboro, NC. CDL:238899-A. MRID 00015538.

Roche, W.J.* 1974. Emetic dose 50 in beagle dogs. Affiliated Medical Research, Inc. Contract No. 120–2255–34. Received September 26, 1974. Greensboro, NC: Affiliated Medical Research, Inc. MRID 00015525.

Ross, R.H., and K. Balu. 1985. Summary of metolachlor water monitoring for 1979 – July 1985. Report #EIR-85024. Submitted by Safety Evaluation Dept., Agricultural Division, Ciba-Geigy Corp., Greensboro, NC. Accession No. 260602. (No MRID.)

Sachsse, K.* 1973a. Irritation of technical CGA-24705 in the rabbit eye. Project No. Siss 2979. MRID 00015528.

Sachsse, K.* 1973b. Skin irritation in the rabbit after single application of Technical CGA-24705. Project No. Siss 2979. Unpublished study. MRID 00015530.

Sachsse, K. and L. Ullmann.* 1977. Skin sensitizing (contact allergenic) effect in guinea pigs of Technical CGA-24705. Project No. Siss 5726. Unpublished study. MRID 00015631.

Smith, S.H., C.K. O'Loughlin, C.M. Salamon et al.* 1981. Two-generation reproduction study in albino rats with metolachlor technical. Study No. 450–0272. Final report. Unpublished study. MRID 00080897.

STORET. 1988. STORET Water Quality File. Office of Water. U.S. Environmental Protection Agency (data file search conducted in May, 1988).

Sumner, D.D. and J.E. Cassidy.* 1974a. The uptake of phenyl-^{14}C-CGA-24705 and its aged soil degradation products in rotation carrots. Report No. GAAC-74112. Unpublished study. Received June 30, 1978, under 100–583. Submitted by Ciba-Geigy Corp., Greensboro, NC. CDL:234216-H. MRID 00022882.

Sumner, D.D. and J.E. Cassidy.* 1974b. The uptake of phenyl-^{14}C-CGA-24705 and its aged soil degradation products in rotation soybeans. Report No. GAAC-74113. Unpublished study. Received June 30, 1978, under 100–583. Submitted by Ciba-Geigy Corp., Greensboro, NC. CDL:234216-G. MRID 00022881.

Sumner, D.D. and J.E. Cassidy.* 1974c. The uptake of phenyl-^{14}C-CGA-24705 and its aged soil degradation products in rotation oats. Report No. GAAC-74085. Unpublished study. Received July 23, 1981, under 100–587. Submitted by Ciba-Geigy Corp., Greensboro, NC. CDL:245627-K. MRID 00022883.

Sumner, D.D. and J.E. Cassidy.* 1974d. The uptake of phenyl-^{14}C-CGA-

24705 and its aged soil degradation products in rotation wheat. Report No. GAAC-74071. Unpublished study. Received Sept. 26, 1974, under 5F1606. Submitted by Ciba-Geigy Corp., Greensboro, NC. CDL:094385-H. MRID 00022878.

Tisdel, M., P. MacWilliams, R. Dahlgren et al. 1982. Carcinogenicity study with metolachlor in albino mice. Hazelton Raltech Study No. 79020. Unpublished study. MRID 0011759.

Tisdel, M., T. Jackson, P. MacWilliams et al.* 1983. Two-year chronic oral toxicity and oncogenicity study with metolachlor in albino rats. Raltech Study No. 80030. Final report. Unpublished study. MRID 00129377.

U.S. EPA. 1986a. U.S. Environmental Protection Agency. Guidelines for car-cinogen risk assessment. Fed. Reg. 51(185)33992–34003. September 24.

U.S. EPA. 1986b. U.S. Environmental Protection Agency. Draft guidance for the reregistration of products containing as the active ingredient: metolach-lor. Washington, DC: Office of Pesticide Programs.

U.S. EPA. 1988. U.S. Environmental Protection Agency. U.S. EPA Method 507 — Determination of nitrogen and phosphorus containing pesticides in water by GC/NPD. Draft. April 15. Available from U.S. EPA's Environ-mental Monitoring and Support Laboratory, Cincinnati, OH.

Whittaker, K.F. 1980. Absorption of selected pesticides by activated carbon using isotherm and continuous flow column systems. Ph.D. Thesis. Purdue University.

Windholz, M., S. Budavari, R.F. Blumetti and E.S. Otterbein, eds. 1983. The Merck index — an encyclopedia of chemicals and drugs. 10th ed. Rahway, NJ: Merck and Co., Inc.

Worthing, C.R., ed. 1983. The pesticide manual: a world compendium, 7th ed. London. England: BCPC Publishers.

---

* Confidential Business Information submitted to the Office of Pesticide Programs.

Metribuzin

I. General Information and Properties

A. CAS No. 21087–64–9

B. Structural Formula

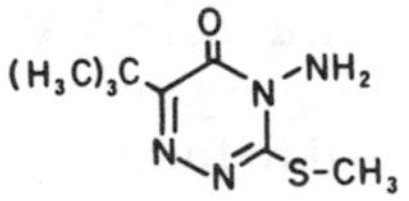

4-Amino-6-(1,1-dimethylethyl)-3-methylthio-1,2,4-triazin-5(4H)-one

C. Synonyms

- Bayer 6159; Bayer 6443H; Bayer 94337; Lexone; Sencor; Sencoral; Sencorer; Sencorex.

D. Uses

- Metribuzin is an herbicide used for the control of a large number of grass and broadleaf weeds infesting agricultural crops (Meister, 1983).

E. Properties (CHEMLAB, 1985)

Chemical Formula	$C_8H_{14}ON_4S$
Molecular Weight	214.28
Physical State (at 25°C)	White crystalline solid
Boiling Point	--
Melting Point	125–126°C
Density	--
Vapor Pressure (25°C)	10^{-5} mm Hg (20°C)
Specific Gravity	--
Water Solubility (25°C)	1,200 mg/L
Log Octanol/Water Partition Coefficient	–5.00 (calculated)
Taste Threshold	--
Odor Threshold	--
Conversion Factor	--

F. Occurrence

- Metribuzin has been found in 936 of 4,651 surface water samples analyzed and in none of 416 ground-water samples (STORET, 1988). These samples were collected at 376 surface water locations and 293 ground-water locations; metribuzin was found in 14 states. The 85th percentile of all nonzero samples was 4.79 μg/L in surface water. The maximum concentration found in surface water was 34.45 μg/L. This information is provided to give a general impression of the occurrence of this chemical in ground and surface waters as reported in the STORET database. The individual data points retrieved were used as they came from STORET and have not been confirmed as to their validity. STORET data are often not valid when individual numbers are used out of the context of the entire sampling regime, as they are here. Therefore, this information can only be used to form an impression of the intensity and location of sampling for a particular chemical.

- Metribuzin has been found in Iowa ground water resulting from agricultural uses; typical positives were 1 to 4.3 ppb (Cohen et al., 1986).

G. Environmental Fate

- The rate of hydrolysis of metribuzin is pH dependent. During a 28-day test, little or no degradation was observed at pH 6 or 9 at 25°C, or at pH 6 at 37°C or 52°C (Day et al., 1976).

- ^{14}C-Metribuzin on silty clay soil degraded, with a half-life of 15 days, when exposed to natural sunlight (Khasawinah, 1972). The half-life in control samples kept in the dark was 56 days. After 10 weeks, 20.6, 6.5 and 7.0% of the applied radioactivity was present in the irradiated soil as 6-t-butyl-1,2,4-triazin-3,5-(2H,4H)-dione (DADK), 6-t-butyl-3-(methylthio)-1,2,4-triazin-5(4H)-one (DA) and parent compound, respectively. A substantial portion of the applied radioactivity (56%) was bound to the soil. In the dark control, 4.6, 16.9, 44.0 and 34% of the applied radioactivity was present as DADK, DA, parent or bound compound, respectively.

- Under aerobic conditions, metribuzin at 10 ppm degraded with a half-life of 35 to 63 days in silt loam and sandy loam soils treated with a 50% wettable powder (WP) formulation, and 63 days in soils treated with a 4-lb/gal FlC formulation (Pither and Gronberg, 1976). Degradates found were 6-t-butyl-1,2,4-triazin-3,5-(2H,4H)-dione (DADK), 4-amino-6-butyl-1–2,4,-triazin-3,5-(2H,4H)-dione (DK), and 6-t-butyl-3-(methylthio)-1,2,4-triazin-5-(4H)-one (DA).

• ^{14}C-Metribuzin residues degraded slowly in silty clay soil under anaerobic conditions with a half-life of more than 70 days (Khasawinah, 1972). After 10 weeks of incubation, 10, 10.9, 57, and 19% of the applied radioactivity was present as DADK, DA, parent compound or bound to the soil, respectively.

• Metribuzin adsorption was significantly correlated to soil organic matter, clay and bar soil water contents (Savage, 1976). Calculated K_d values ranged from 0.27 for a sandy loam soil (0.75% organic matter, 11% clay and 12% of 0.33 bar soil water content) to 3.41 for a clay soil (42% organic matter, 71% clay and 42% of 0.33 bar soil water content).

• ^{14}C-Metribuzin residues were very mobile in Amarillo sandy loam and Louisiana Commerce silt loam soils; after leaching 12-inch soil columns with 20 inches of water, 96.6 and 91.6% of the applied radioactivity, respectively, was found in the leachate (Houseworth and Tweedy, 1973). ^{14}C-Metribuzin residues were relatively immobile in Indiana silt loam and New York muck soils; after leaching 12-inch soil columns, 90.6 and 89.4% of the applied radioactivity was detected in the top 3 cm of the Indiana silt loam and New York muck soil columns, respectively. No radioactivity was detected in column leachates.

• ^{14}C-Metribuzin residues (test substance not characterized) aged 30 days were moderately mobile in an Amarillo sandy loam soil column; after leaching a 12-inch column with 22.5 inches of water, 7.3% of the applied radioactivity was found in the leachate (Tweedy and Houseworth, 1974). In the soil column, 85.2% of the applied radioactivity remained within the top 2 inches.

• ^{14}C-Metribuzin residues (test substance not characterized) were intermediately mobile in sandy clay loam and silt loam soils (R_f 0.61 to 0.62) and mobile in sandy, sandy loam and two silty clay soils (R_f 0.68 to 0.77), based on soil thin-layer chromatography (TLC) tests (Thornton et al., 1976). ^{14}C-Metribuzin residues (test substance not characterized) were intermediately mobile in sand (R_f 0.61), sandy clay loam (R_f 0.64), two silty clay soils (R_f 0.62 and 0.71), silt loam (R_f 0.66) and sandy loam (R_f 0.82) soils, based on soil TLC tests (Obrist and Thorton, 1979). ^{14}C-Metribuzin (purity not specified) at 1.5 μg/spot had low mobility (R_f 0.13 to 0.26) in two muck soils and intermediate mobility (R_f 0.42 to 0.53) in six mineral soils ranging in texture from sand to clay, based on soil TLC plates developed in water (Sharon and Stephenson, 1976).

- In the field, metribuzin dissipates with half-lives of less than 1 month to 6 months. Three metribuzin degradates were detected: 6-t-butyl-1,2,4-triazin-3,5-(2H,4H)-dione (DADK); 4-amino-6-t-butyl-1,2,4-triazin-3,5-(2H,4H)-dione (DK), and 6-t-butyl-3-(methylthio)-1,2,4-triazin-5-(4H)-one (DA). Soil type and characteristics, chemical formulation or application rates did not discernibly affect the dissipation rate of metribuzin (Stanley and Schumann, 1969; Finlayson, 1972; Rockwell, 1972a,b,c; Rowehl, 1972a,b; Schultz, 1972; Mobay Chemical, 1973; Fisher, 1974; Murphy, 1974; United States Borax and Chemical Corp., 1974; Potts et al., 1975; Analytical Biochemistry Laboratories, 1976; Ballantine, 1976; and Ford, 1979).

II.   Pharmacokinetics

A. Absorption

- A study was conducted in four dogs using oral dosing of radiolabeled metribuzin (Khasawinah et al., 1972) to evaluate absorption, distribution and metabolites. Analysis of blood samples showed a peak level at 4 hours.

B. Distribution

- No information was found in the available literature on the distribution of metribuzin. Khasawinah et al. (1972) did not provide information on distribution of metribuzin in the dog study.

C. Metabolism

- No information was found in the available literature on the metabolism of metribuzin. The Khasawinah et al. (1972) study in dogs did not elaborate on metabolism in metribuzin.

D. Excretion

- Khasawinah et al. (1972) reported that 52 to 60% of the administered dose of metribuzin in dogs was excreted in the urine and 30% in the feces.

III.   Health Effects

A. Humans

- No information was found in the available literature on the health effects of metribuzin in humans.

B. Animals

1. *Short-term Exposure*

- Crawford and Anderson (1974) reported the acute oral LD_{50} values following the administration of technical metribuzin to guinea pigs and rats as 245 and 1,090 mg/kg, respectively, for male animals, and 274 and 1,206 mg/kg, respectively, for females.

- Mobay Chemical (1978) reported the acute oral LD_{50} values for a wettable granular formulation of metribuzin to be 2,379 and 2,794 mg/kg for male and female rats, respectively.

- Mobay Chemical (1978) reported the acute dermal LD_{50} for a wettable granular formulation of metribuzin to be > 5,000 mg/kg for both male and female rats.

- Mobay Chemical (1978) reported the acute (1-hour) inhalation LD_{50} in rats for a wettable granular formulation to be > 20 mg/L.

2. *Dermal/Ocular Effects*

- In studies conducted by Mobay Chemical (1978), metribuzin (wettable granular) was determined to be a very slight irritant to rabbit eyes and skin.

3. *Long-term Exposure*

- Loser et al. (1969) administered metribuzin to Wistar rats (15/sex/dose) for 3 months in their feed at levels of 0, 50, 150, 500 or 1,500 ppm (about 2.5, 7.5, 25 or 75 mg/kg/day, based on the dietary assumptions of Lehman et al., 1959). Following treatment, food consumption, growth, body weight, organ weight, clinical chemistry, hematology, urinalysis and histopathology were measured. No significant effects on these parameters were observed in either sex at 50 ppm (2.5 mg/kg/day). Among females, enlarged livers were found in the 150-, 500- or 1,500 ppm (7.5, 25 or 75 mg/kg/day) dosage groups ($p < 0.05$), and thyroid glands were also enlarged in the 500- or 1,500-ppm (25 or 75 mg/kg/day) groups ($p < 0.05$ and $p < 0.01$, respectively). In the males, enlarged thyroids were reported among the 500 (25 mg/kg/day) ($p < 0.05$) and

1,500 ppm (75 mg/kg/day) (p < 0.01) dosage groups, while an enlarged heart was reported at 1,500 ppm (75 mg/kg/day) (p < 0.05). At 1,500 ppm (75 mg/kg/day), lower body weights (p < 0.01) were reported in both sexes when compared to untreated controls.

- In studies conducted by Lindberg and Richter (1970), beagle dogs (4/sex/dose) administered oral doses of 50, 150 or 500 ppm (about 1.25, 3.75 or 12.5 mg/kg/day, based on the dietary assumptions of Lehman, 1959) technical metribuzin for 90 days showed no significant differences in body weights, food consumption, behavior, mortality, hematologic findings, urinalysis, gross pathology or histopathology.

- Loser and Mohr (1974) reported that dietary concentrations of 1.5, 2 or 20 mg/kg/day metribuzin did not significantly affect physical appearance, behavior, mortality, hematologic clinical chemistry, urinalysis or histopathology in rats (40/sex/dose) fed technical metribuzin in the diet for 24 months. The body weights of females at the 20 mg/kg/day dose level were usually lower (p < 0.05) than those of controls; at the end of the test period, however, no significant differences were noted.

- Hayes et al. (1981) administered technical metribuzin in the diet to albino CD mice (50/sex/dose) at 200, 800 or 3,200 ppm (about 30, 120 or 480 mg/kg/day, based on the dietary assumptions of Lehman, 1959) for 24 months. Following treatment, food consumption, general behavior, body and organ weights, mortality, hematology and histopathology were analyzed. No adverse effects were observed in these parameters in either sex at 800 ppm (120 mg/kg/day). However, a significant (p < 0.05) increase in absolute and relative liver and kidney weights was observed in female mice receiving 3,200 ppm (480 mg/kg/day).

- In studies conducted by Loser and Mirea (1974), four groups of beagle dogs (4/sex/dose) were administered metribuzin in the diet at dose levels of 0, 25, 100 or 1,500 ppm (about 0, 0.625, 2.5 or 37.5 mg/kg/day, based on the dietary assumptions of Lehman, 1959) for 24 months. Following treatment, food consumption, general behavior and appearance, clinical chemistry, hematology, urinalysis, body and organ weights and histopathology were evaluated. No toxicologic effects were reported in animals administered 100 ppm metribuzin (2.5 mg/kg/day) or less for any of the parameters measured. Necrosis of the renal tubular cells, slight iron deposition as well as slight hyperglycemia and temporary hypercholesterolemia were noted in animals administered 1,500 ppm (37.5 mg/kg/day).

4. *Reproductive Effects*

- In a three-generation reproduction study, Loser and Siegmund (1974) administered technical metribuzin in the feed at dose levels of 0, 35, 100 or 300 ppm (about 0, 1.75, 5 or 15 mg/kg/day, based on the dietary assumptions of Lehman, 1959) to FB30 (Filberfeld breed) rats during mating, gestation and lactation. Following treatment, fertility, lactation performance and pup development were evaluated. No treatment-related effects were reported at any dose tested.

5. *Developmental Effects*

- Unger and Shellenberger (1981) administered technical metribuzin by gastric intubation to pregnant female rabbits (16 to 17/dose) on days 6 through 18 of gestation at daily doses of 15, 45 or 135 mg/kg/day. Following treatment, there was a statistically significant (p < 0.05) decrease in body weight gain in the high-dose does (135 mg/kg). No maternal toxicity was reported in animals administered metribuzin at levels of 45 mg/kg/day or less. No treatment-related effects were reported at any dose level in fetuses based on gross, soft-tissue and skeletal examinations.

- Machemer (1972) reported no maternal toxicity, embryotoxicity or teratogenic effects following oral administration (via stomach tube) of technical metribuzin to FB30 rats (21 to 22/dose) on days 6 through 15 of gestation at dose levels of 5, 15, 50 or 100 mg/kg/day.

6. *Mutagenicity*

- Metribuzin showed no mutagenic activity in several bacterial assays (Inukai and Iyatomi, 1977; Shirasu et al., 1978) or in dominant lethal tests in mice (Machemer and Lorke, 1974, 1976). The results of microbial point mutation assays (Machemer and Lorke, 1974) did not indicate a mutagenic potential for metribuzin in the test systems utilized. The results of dominant lethal mutations in mice or chromosomal aberrations in hamster spermatogonia at dose levels of 300 mg/kg and 100 mg/kg, respectively, did not indicate any mutagenic effects of metribuzin.

7. *Carcinogenicity*

- Hayes et al. (1981) conducted studies in which technical metribuzin was administered in the diet to albino CD-1 mice (50/sex/dose) at 200, 800 or 3,200 ppm (approximately 30, 120 or 380 mg/kg/day) for 24 months. Minimal toxic effects were observed at the high-dose level in the form of increased liver weight and changes in the hematocrit and hemoglobin measurements. Although some increase in the number of tumor-bearing animals was observed in low- and mid-dose animals, significant increases in the incidence of specific tumor types were not observed at any dose level. It was concluded that, under the conditions of the test, there was no increase in the incidence of tumors in mice.

IV. Quantification of Toxicological Effects

A. One-day Health Advisory

No information was found in the available literature that was suitable for determination of a One-day HA for metribuzin. It is therefore recommended that the Ten-day HA value for a 10-kg child (5.0 mg/L, calculated below) be used at this time as a conservative estimate of the One-day HA value.

B. Ten-day Health Advisory

The study by Unger and Shellenberger (1981) has been selected to serve as the basis for determination of the Ten-day HA for metribuzin. In this study, pregnant rabbits (16 or 17/dose) that were administered technical metribuzin by gastric intubation at dosage levels of 0, 15, 45 or 135 mg/kg/day on days 6 through 18 of gestation showed a statistically significant (p < 0.05) decrease in body weight gain at the 135-mg/kg dose. No maternal toxicity was reported at or below the 45-mg/kg dose. No treatment-related effects were reported at any dose level in fetuses based on gross, soft-tissue and skeletal examinations. The NOAEL identified in this study was, therefore, 45 mg/kg/day.

Using a NOAEL of 45 mg/kg/day, the Ten-day HA for a 10-kg child is calculated as follows:

$$\text{Ten-day HA} = \frac{(45 \text{ mg/kg/day}) (10 \text{ kg})}{(100) (1 \text{ L/day})} = 4.5 \text{ mg/L} (5,000 \text{ mg/L})$$

C. Longer-term Health Advisory

The study by Loser et al. (1969) has been selected to serve as the basis for the Longer-term HA for metribuzin. In this study, rats (15/sex/dose) were fed diets containing metribuzin at doses of 50, 150, 500 or 1,500 ppm (about 2.5, 7.5, 25 or 75 mg/kg/day based on calculations in Lehman, 1959) for 90 days. Thyroid glands were enlarged in males in the 500- or 1,500-ppm (25 or 75 mg/kg/day) dosage groups, while the heart was enlarged at the 1,500-ppm (75 mg/kg/day) dose level. In females, enlarged livers were detected in the 150-, 500- or 1,500-ppm (7.5, 25 or 75 mg/kg/day) dosage groups, and the thyroid was enlarged in the 500- or 1,500-ppm (25 or 75 mg/kg/day) dosage groups. Body weights were reduced in both sexes at 1,500 ppm (75 mg/kg/day), compared to untreated controls. The NOAEL identified in this study was, therefore, 50 ppm (2.5 mg/kg/day). Lindberg and Richter (1970) determined a NOAEL of 12.5 mg/kg/day in dogs; however, this study was not chosen, since the NOAEL was higher than the LOAEL of 7.5 mg/kg/day identified by Loser et al. (1969) in the rat. Using a NOAEL of 2.5 mg/kg/day, the Longer-term HA for a 10-kg child is calculated as follows:

$$\text{Longer-term HA} = \frac{(2.5 \text{ mg/kg/day}) (10 \text{ kg})}{(100) (1 \text{ L/day})} = 0.25 \text{ mg/L} (300 \text{ } \mu\text{g/L})$$

Using a NOAEL of 2.5 mg/kg/day, the Longer-term HA for a 70-kg adult is calculated as follows:

$$\text{Longer-term HA} = \frac{(2.5 \text{ mg/kg/day}) (70 \text{ kg})}{(100) (2 \text{ L/day})} = 0.875 \text{ mg/L} (900 \text{ } \mu\text{g/L})$$

D. Lifetime Health Advisory

The study by Loser and Mirea (1974) has been selected to serve as the basis for the Lifetime HA for metribuzin. In this study, dogs (4/sex/dose) were administered metribuzin in the diet at dose levels of 0, 25, 100 or 1,500 ppm (0, 0.625, 2.5 or 37.5 mg/kg/day) for 24 months. Necrosis of the renal tubular cells was reported as well as slight and temporary changes in certain clinical chemistry parameters (e.g., blood glucose and cholesterol) at the high-dose level. No other toxicologic effects were reported. Based on this information, a NOAEL of 100 ppm (2.5 mg/kg/day) and a LOAEL of 1,500 ppm (37.5 mg/kg/day) were reported. Loser and Mohr (1974) reported a NOAEL of 20 mg/kg/day in rats. This study was not selected because no dose-related toxicologic responses were observed, and the rat may be less sensitive than the dog. Hayes et al. (1981) determined a NOAEL of 120 mg/kg/day in mice; however, this value exceeded the LOAEL (37.5 mg/kg/day) reported by Loser and Mirea (1974).

Using this study, the Lifetime HA is calculated as follows:

Step 1: Determination of the Reference Dose (RfD)

$$\text{RfD} = \frac{(2.5 \text{ mg/kg/day})}{(100)} = 0.025 \text{ mg/kg/day}$$

Step 2: Determination of the Drinking Water Equivalent Level (DWEL)

$$\text{DWEL} = \frac{(0.025 \text{ mg/kg/day}) (70 \text{ kg})}{(2 \text{ L/day})} = 0.875 \text{ mg/day} (900 \text{ } \mu\text{g/L})$$

Step 3: Determination of the Lifetime Health Advisory

Lifetime HA = (0.875 mg/L) (20%) = 0.175 mg/L (200 μg/L)

E. Evaluation of Carcinogenic Potential

- In a study by Hayes et al. (1981), metribuzin was administered in the feed of mice (50/sex/dose) at dose levels of 200, 800 or 3,200 ppm (30, 120 or 480 mg/kg/day) for 24 months. Following treatment, the incidence of tumor formation was analyzed in a variety of tissues. Neoplasms of various tissues and organs were similar in type, location, time of occurrence and incidence in control and treated animals. It was concluded that under the conditions of the test, there was no increase in the incidence of tumors in mice.

- The International Agency for Research on Cancer has not evaluated the carcinogenic potential of metribuzin.

- Applying the criteria described in EPA's guidelines for assessment of carcinogenic risk (U.S. EPA, 1986), metribuzin may be classified in Group D: not classifiable. This category is used for substances with inadequate human and animal evidence of carcinogenicity.

V.   Other Criteria, Guidance and Standards

- A Threshold Limit Value-Time-Weighted Average (TLV-TWA) of 5 mg/ m^3 was determined, based on animal studies substantiated by repeated inhalation tests, a safety factor of 5 and assuming a total pulmonary absorption (ACGIH, 1984).

VI.   Analytical Methods

- Analysis of metribuzin is by a gas chromatography (GC) method applicable to the determination of certain organonitrogen pesticides in water samples (U.S. EPA, 1985). This method requires a solvent extraction of approximately 1 L of sample with methylene chloride using a separatory funnel. The methylene chloride extract is dried and exchanged to acetone during concentration to a volume of 10 mL or less. The compounds in the extract are separated by GC and measurement is made with a thermionic bead detector. This method has been validated in a single laboratory, and estimated detection limits have been determined for the analytes, including metribuzin; the estimated detection limit is 0.15 μg/ L.

VII.   Treatment Technologies

- Available data indicate that granular-activated carbon (GAC) adsorption and a conventional treatment scheme will remove metribuzin from water.

- Whittaker (1980) experimentally determined adsorption isotherms for metribuzin on GAC.

- Whittaker (1980) reported the results of GAC columns operating under bench-scale conditions. At a flow rate of 0.8 gpm/ft^2 and an empty bed contact time of 6 minutes, metribuzin breakthrough (when effluent concentration equals 10% of influent concentration) occurred after 112 bed volumes (BV).

- In the same study, Whittaker (1980) reported the results for four metribuzin bisolute solutions when passed over the same GAC continuous flow column.

- Another study investigated the effectiveness of two different GAC columns in removing metribuzin from contaminated wastewater (Whittaker, et al., 1982). One type of GAC showed breakthrough for metribuzin (6 mg/L) from an initial concentration of 140 mg/L after 50 gallons of the wastewater had been treated. No pesticide was found in the effluent from the second type of GAC.

- Conventional water treatment, coagulation and sedimentation with alum and an anionic polymer removed more than 50% of the metribuzin present (Whittaker et al., 1980). The optimum alum dosage was 200 mg/L. Also equivalent dosages of ferric chloride were found to be equally effective.

- Treatment technologies for the removal of metribuzin from water are available and have been reported to be effective. However, selection of an individual technology or combination of technologies to attempt metribuzin removal from water must be by a case-by-case technical evaluation, and an assessment of the economics involved.

VIII.   References

ACGIH. 1984. American Conference of Governmental Industrial Hygienists. Documentation of the threshold limit values for substances in workroom air, 3rd ed. Cincinnati, OH: ACGIH.
Analytical Biochemistry Laboratories. 1976. Chemagro agricultural division—

Mobay Chemical Corporation soil persistence study: MW-HR-409–75. Report No. 50842. Unpublished study. Prepared in cooperation with Mobay Chemical Corp. Submitted by Ciba-Geigy Corp., Greensboro, NC.

Ballantine, L.G. 1976. Metolachlor plus metribuzin tank mix soil dissipation. Report No. ABR-76092. Summary of studies 095763-B through 095763-F. Unpublished study. Submitted by Ciba-Geigy Corp., Greensboro, NC.

CHEMLAB. 1985. The Chemical Information System, CIS, Inc. Baltimore, MD.

Cohen, S.Z., C. Eiden and M.N. Lorber. 1986. Monitoring ground water for pesticides in the U.S.A. In: Evaluation of pesticides in ground water. American Chemical Society Symposium Series. American Chemical Society. In press.

Crawford, C.R. and R.H. Anderson.* 1974. The acute oral toxicity of Sencor technical, several Sencor metabolites and impurities to rats and guinea pigs. Report No. 38927. Rev. unpublished study. MRID 00045270.

Day, E.W., W.L. Sullivan and O.D. Decker. 1976. A hydrolysis study of the herbicides oryzalin and metribuzin. Unpublished study. Submitted by Elanco Products Co., Div. of Eli Lilly Co., Indianapolis, IN.

Finlayson, D.G. 1972. Soil persistence study: Victoria, British Columbia, Canada. In: Sencor: the effects on the environment. Supplement No. 4. Document No. AS77–1968. Unpublished study. Submitted by Mobay Chemical Corp.

Fisher, R.A. 1974. Mobay Chemical Corporation residue experiment, Mentha, Michigan: Sencor residues in soil. Report No. 41395. Unpublished study, including Report Nos. 41625, 41626, 41627. Prepared in cooperation with Missouri Analytical Laboratories. Submitted by Mobay Chemical Corp., Kansas City, MO.

Ford, J.J. 1979. Herbicide combination—soil dissipation study involving Antor herbicide with three commercial herbicides: RI 47–003–06. Submitted by Hercules, Inc., Wilmington, DE.

Hayes, R.H., D.W. Lamb and D.R. Mallicoat. 1981. Metribuzin (R) (Sencor) oncogenicity study in mice: 80050. Unpublished study. MRID 00087795.

Houseworth, L.D. and B.G. Tweedy. 1973. Report on parent leaching studies for Sencor. Report No. 37180. Unpublished study. Prepared by Univ. of Missouri, Dept. of Plant Pathology. Submitted by Mobay Chemical Corp., Kansas City, MO.

Inukai, H. and A. Iyatomi.* 1977. Bay 94337: mutagenicity test on bacterial systems. Report No. 67; 54127. Unpublished study. MRID 00086770.

Khasawinah, A.M. 1972. The metabolism of Sencor (Bay 94337) in soil. Report No. 31043. Unpublished study. Submitted by Mobay Chemical Corp., Kansas City, MO.

Khasawinah, A. M., D. R. Flint, H. R. Shaw and D. D. Cox. 1972. The metabolic fate of carbonyl c^{ih}-SENCOR in dogs. Unpublished study. Submitted by Mobay Chemical Corp. MRID 00045264.

Lehman, A.J. 1959. Appraisal of the safety of chemicals in foods, drugs and cosmetics. Assoc. Food Drug Off. U.S., Q. Bull.

Lehman, W.J., W.F. Reehl and D.H. Rosenblatt. 1959. Handbook of chemical property estimation methods. New York, NY: McGraw Hill.

Lindberg, D. and W. Richter.* 1970. Report to Chemagro Corporation: 90-Day subacute oral toxicity of Bay 94337 in beagle dogs. IBT No. C776; 26488. Unpublished study. MRID 00106162.

Loser, E., D. Lorke and L. Mawdesley-Thomas.* 1969. Bay 94337: subchronic toxicological studies on rats (3-month feeding test). Report No. 1719; 26469. Unpublished study. MRID 00106161.

Loser, E. and D. Mirea.* 1974. Bay 94337: chronic toxicity studies on dogs (two-year feeding experiment). Report Nos. 4887; 41814. Unpublished study. MRID 00061260.

Loser, E. and U. Mohr.* 1974. Bay 94337: chronic toxicity studies on rats (two-year feeding experiment). Report Nos. 4888; 41816. Unpublished study. MRID 00061261.

Loser, E. and F. Siegmund.* 1974. Bay 94337: multigeneration study on rats. Report Nos. 4889; 41818. Unpublished study. MRID 00061262.

Machemer, L.* 1972. Sencor (Bay 94337): Studies for possible embryotoxic and teratogenic effects on rats after oral administration. Report Nos. 3678 and 35073. Unpublished study. MRID 00061257.

Machemer, L. and D. Lorke.* 1974. Evaluation of (R) Sencor for mutagenic effects on the mouse. Report Nos. 4942 and 43068. Unpublished study. MRID 00086766.

Machemer, L. and D. Lorke.* 1976. (R) Sencor: additional dominant lethal study on male mice to test for mutagenic effects by an improved method. Report Nos. 6110 and 49068. Unpublished study. MRID 00086768.

Meister, R., ed. 1983. Farm chemicals handbook. Willoughby, OH: Meister Publishing Company.

Mobay Chemical. 1973. Mobay Chemical Corporation. Sencor: metabolic, analytical, and residue information for sugarcane (Hawaii). Unpublished study. Prepared by Mobay Chemical Corp., Kansas City, MO.

Mobay Chemical.* 1978. Mobay Chemical Corporation. Supplement to synopsis of human safety of Sencor: Supplement No. 3. Summary of studies 235396-B through 235396-F. Unpublished study. MRID 00078084.

Murphy, H. 1974. Sencor residues in soils. Mobay Chemical Corporation residue experiment, Presque Island, Maine. Report No. 41395. Unpublished study, including Report Nos. 41625, 41626, 41627. Prepared in cooperation with Missouri Analytical Laboratories. Submitted by Mobay Chemical Corp., Kansas City, MO.

Obrist, J.J. and J.S. Thornton. 1979. Soil thin-layer mobility of Baycor (TM), Baytan, Drydene and Peropal (TM). Unpublished study. Prepared in cooperation with Agricultural Consultants, Inc. Submitted by Mobay Chemical Corp., Kansas City, MO.

Pither, K.M. and R.R. Gronberg. 1976. A comparison of the rate of metabolic degradation of Sencor in soil using the 50% wettable powder and 4 flowable formulations. Report No. 45990. Unpublished study. Submitted by Mobay Chemical Corp., Kansas City, MO.

Potts, C.R., M.M. Laporta, J. Devine et al. 1975. Prowl (CL 92, 553): determination of CL 92,553 N-(1-ethylpropyl)-3,4-dimethyl-2,6-dinitrobenzenamine and Sencor 4-amino-6-t-butyl-3-(methylthio)-1,2,4-triazin-5(4H)-one in soil. Report No. C-801. Unpublished study submitted by American Cyanamid Company, Princeton, NJ.

Rockwell, L.F. 1972a. Soil persistence study of BAY 94337: plot F-17, Research Farm, Stanley, Kansas. In: Sencor: the effects on the environment. Compilation. Unpublished study. Submitted by Mobay Chemical Corp., Kansas City, MO.

Rockwell, L.F. 1972b. Soil persistence study; plot F-2, Research Farm, Stanley, Kansas. In: Sencor: the effects on the environment. Supplement No. 4. Document No. AS77–1968. Unpublished study. Submitted by Mobay Chemical Corp., Kansas City, MO.

Rockwell, L.F. 1972c. Soil persistence study of DADK; plot F-17, Research Farm, Stanley, Kansas. In: Sencor: the effects on the environment. Supplement No. 4. Document No. AS77–1968. Compilation. Unpublished study. Submitted by Mobay Chemical Corp., Kansas City, MO.

Rowehl, E.R. 1972a. Soil persistence study of BAY 94337; Vero Beach, Florida. In: Sencor: the effects on the environment. Compilation. Unpublished study. Submitted by Mobay Chemical Corp., Kansas City, MO.

Rowehl, E.R. 1972b. Soil persistence study of DADK: Vero Beach, Florida. In: Sencor: the effects on the environment. Supplement No. 4. Document No. AS77–1968. Unpublished study. Submitted by Mobay Chemical Corp., Kansas City, MO.

Savage, K.E. 1976. Adsorption and mobility of metribuzin in soil. Weed Sci. 24(5):525–528.

Schultz, T.H. 1972. Soil persistence study. Report No. 33131. Unpublished study. Submitted by Chemagro. In Supplement No. 4 to brochure entitled: Sencor: the effects on the environment. Document No. AS77–1968. Compilation. Unpublished study.

Sharon, M. and G.R. Stephenson. 1976. Behavior and fate of metribuzin. Weed Sci. 24(2):153–160. Submitter Report No. 49127. In: unpublished study submitted by Mobay Chemical Corp., Kansas City, MO.

Shirasu, Y., M. Moriya and T. Ohta.* 1978. Metribuzin mutagenicity test on bacterial systems. Submitter Report No. 66748. Unpublished study. MRID 00109254.

Stanley, C.W. and S.A. Schumann. 1969. A gas chromatographic method for the determination of BAY 94337 residues in potatoes, soybeans, and corn. Report No. 25838. Unpublished study. Submitted by Mobay Chemical Corp., Kansas City, MO.

STORET. 1988. STORET Water Quality File. Office of Water. U.S. Environmental Protection Agency (data file search conducted in May, 1988).

Thornton, J.S., J.B. Hurley and J.J. Obrist. 1976. Soil thin-layer mobility of twenty-four pesticide chemicals. Report No. 51016. Unpublished study. Submitted by Mobay Chemical Corp., Pittsburgh, PA.

Tweedy, B.G. and L.D. Houseworth. 1974. Leaching of aged residues of Sencor-3-14C in sandy loam soil. Report No. 40567. Unpublished study. Prepared by Univ. of Missouri, Dept. of Plant Pathology. Submitted by Mobay Chemical Corp., Kansas City, MO.

Unger, T.M. and T.F. Shellenberger.* 1981. A teratological evaluation of Sencor (R) in mated female rabbits: 80051. Final report. Unpublished study. MRID 00087796.

United States Borax and Chemical Corp. 1974. Cobex plus Sencor (or Lexone): degradation in soil. Compilation. Unpublished study.

U.S. EPA. 1985. U.S. Environmental Protection Agency. U.S. EPA Method 507 – Organonitrogen pesticides. Fed. Reg. 50:40701. October 4, 1985.

U.S. EPA. 1986. U.S. Environmental Protection Agency. Guidelines for carcinogen risk assessment. Fed. Reg. 51(185):33992–34003. September 24.

Whittaker, K.F. 1980. Adsorption of selected pesticides by activated carbon using isotherm and continuous flow column systems. Ph.D. Thesis. Purdue University, Lafayette, IN.

Whittaker, K.F., J.C. Nye, R.F. Wukasch and H.A. Kazimier. 1980. Cleanup and collection of wastewater generated during cleanup of pesticide application equipment. Paper presented at National Hazardous Waste Symposium, Louisville, KY.

Whittaker, K.F., J.C. Nye, R.F. Wukasch, R.J. Squires, A.C. York and H.A.Kazimier. 1982. Collection and treatment of wastewater generated by pesticide application. EPA 600/2-82-028.

---

* Confidential Business Information submitted to the Office of Pesticide Programs.

Paraquat

I. General Information and Properties

Paraquat, with a chemical name 1,1'-dimethyl-4,4'-dipyridinium ion, is present mostly as the dichloride salt (CAS No. 1910–42–5) or as the dimethyl sulfate salt (CAS No. 2074–50–2, molecular weight 408.48) (Meister, 1988). Contents discussed below pertain to paraquat dichloride.

A. CAS No. 1910–42–5

B. Structural Formula

$$\left[CH_3-N^+ \langle\bigcirc\rangle\langle\bigcirc\rangle N^+-CH_3 \right] 2\,Cl^-$$

1,1'-Dimethyl-4,4'-bipyridinium-dichloride

C. Synonyms

* AT-5, Actor, Cekuquat, Crisquat, Herboxone, Dextrone, Dexuron, Esgram, Gramocil, Gramoxone, Gramuron, Goldquat 276, Herbaxon (Agro Quimicas), Herboxone, Mofisal, Osaquat Super, Paracol, Paracote, Pathclear, Preeglone, Priglone, Simpar, Sweep, Terraklene, Totacol, Total, Toxer, Weedol. Discontinued name: Ortho (Meister, 1988).

D. Uses

* Paraquat is a contact herbicide and desiccant used for desiccation of seed crops; for noncrop and industrial weed control in bearing and nonbearing fruit orchards, shade trees and ornamentals; for defoliation and desiccation of cotton; for harvest aid in soybeans, sugarcane, guar, and sunflowers; for pasture renovation; for use in "no-till" or before planting or crop emergence for dormant alfalfa and clover directed spray; and for killing potato vines. Paraquat is also effective for eradication of weeds on rubber plantations and coffee plantations and against paddy bund (Meister, 1988).

E. Properties (ACGIH, 1980; Meister, 1988; CHEMLAB, 1985; TDB, 1985)

Chemical Formula	$C_{12}H_{14}N_2 \cdot 2Cl$
Molecular Weight	257.18
Physical State	Colorless to yellow crystalline solid
Boiling Point	175 to 180°C
Melting Point	--
Density	1.24 g/mL (20°C)
Vapor Pressure	No measurable vapor pressure
Specific Gravity	1.24 at 20°C/20°C
Water Solubility	Very soluble
Log Octanol/Water Partition Coefficient	2.44 (calculated)
Taste Threshold	--
Odor Threshold	--
Conversion Factor	--

F. Occurrence

- Paraquat was undetectable in 843 ground-water samples analyzed (STORET, 1988). Samples were collected at 813 ground-water locations. No surface water samples were collected for analysis.

G. Environmental Fate

- ^{14}C-Paraquat dichloride (> 96.5% pure) at 91 mg/L was stable to hydrolysis at 25° and 40°C at pH 5, 7 and 9 for up to 30 days (Upton et al., 1985).

- Uniformly ring-labeled ^{14}C-paraquat (99.7% pure) at approximately 7.0 ppm in sand did not photodegrade when irradiated with natural sunlight for 24 months (Pack, 1982). No degradation products were detected at any sampling interval. After 24 months of irradiation, > 84% of the applied radioactivity was extractable and < 4% was unextractable.

- Paraquat was essentially stable to photolysis in soil (Day and Hemingway, 1981). Four degradation products, 1-methyl-4,4'-bipyridylium ion, 4-(1,2-dihydro-1-methyl-2-oxo-4-pyridyl)-1-methyl pyridylium ion, 4-carboxy-1-methyl pyridylium ion, and an unknown, individually constituted < 6.0% of the total radioactivity in either irradiated (undisturbed) or dark control soils.

- Paraquat (test substance uncharacterized) at 0.05 to 1.0 ppm in water plus soil declined with a half-life of > 2 weeks (Coats et al., 1964). In water only, paraquat declined with a half-life of approximately 23 weeks.

- ^{14}C-Paraquat (test substance uncharacterized) was immobile in silt loam and silty clay loam (R_f 0.00), and slightly mobile in sandy loam (R_f 0.13) soils, based on soil thin-layer chromatography (TLC) tests (Helling and Turner, 1968).

- Methyl-labeled ^{14}C-paraquat (test substance uncharacterized) at 1.0 ppm was stable to volatilization at room temperature over a 64-day period (Coats et al., 1964).

- In a pond treated with paraquat (test substance uncharacterized) at 1.14 ppm (Frank and Comes, 1967), paraquat residues (uncharacterized) declined from 0.55 ppm 1 day after treatment to nondetectable (< 0.001 ppm) 18 days after treatment. The dissipation of paraquat residues (uncharacterized) in water was accompanied by a concomitant increase of paraquat residues (uncharacterized) in the soil. Paraquat (test substance uncharacterized) at 0.04 ppm dissipated in pond water with a half-life of approximately 2 days (Coats et al., 1964). For more details, see Calderbank's chapter on paraquat in herbicides (Calderbank, 1976).

II. Pharmacokinetics

A. Absorption

- In Wistar rats given single oral doses of ^{14}C-paraquat dichloride or dimethyl sulfate by gavage (0.5 to 50 mg/kg, purity not stated), 69 to 96% was excreted unchanged, mostly in feces, and no radioactivity appeared in bile (Daniel and Gage, 1966). Some systemic absorption of the degradation products that were produced in the gut was noted. Approximately 30% of the administered dose appeared in feces in a degraded form.

- ^{14}C-Methyl-labeled paraquat (99.7% purity) was administered orally to a cow in a single dose of approximately 8 mg cation/kg (Leahey et al., 1972). A total of 95.6% of the dose was excreted in feces in the first 3 days. A small amount, 0.7% of the dose, was excreted in the urine, 0.56% during the first 2 days. Only 0.0032% of the dose appeared in the milk.

- A goat was administered ^{14}C-ring-labeled paraquat dichloride (> 99% purity) orally at 1.7 mg/kg for 7 consecutive days (Leahey et al., 1976a). At sacrifice, 2.4% and 50.3% of the radioactive dose had been excreted in the urine and feces, respectively, and 33.2% was recovered in the contents of the stomach and intestines. The radioactivity was associated with unchanged paraquat.

- In studies with pigs, ^{14}C-methyl-labeled (Leahey et al., 1976b) and ^{14}C-ring-labeled (Spinks et al., 1976) paraquat (> 99% purity) at dose levels of 1.1 and 100 mg ion/kg/day, respectively, was given for up to 7 days. At sacrifice, 69 to 72.5% and 2.8 to 3.4% of the total radioactive dose had been excreted in the feces and urine, respectively.

B. Distribution

- Pigs were given oral doses of ^{14}C-methyl-labeled (Leahey et al., 1976b) and ^{14}C-ring-labeled (Spinks et al., 1976) paraquat dichloride (> 99% purity) for up to 7 consecutive days at dose levels of 1.1 and 100 mg ion/kg/day, respectively. At sacrifice, radioactivity associated mostly with unchanged paraquat was identified in the lungs, heart, liver and kidneys, with trace amounts in the brain, muscle and fat.

- The distribution of radioactivity was studied in a goat fed ^{14}C-ring-labeled paraquat dichloride (1.7 mg/kg/day, 99.7% purity) in the diet for 7 consecutive days (Hendley et al., 1976). Most of the radioactivity was found in the lungs, kidneys and liver. The major residue was unchanged paraquat.

C. Metabolism

- Paraquat dichloride or paraquat dimethyl sulfate (radiochemical purity: 99.3 to 99.8%), labeled with ^{14}C in either methyl groups or in the ring, was poorly absorbed from the gastrointestinal tract of a cow (Leahey et al., 1972), goats (Hendley et al., 1976), pigs (Leahey et al., 1976b; Spinks et al., 1976) and rats (Daniel and Gage, 1966), and was excreted in the feces mostly as unchanged paraquat. However, after an oral dose, there was microbial degradation of paraquat in the gut. In one study with rats (Daniel and Gage, 1966), 30% of a dose of paraquat appeared in the feces in a degraded form. A portion of these microbial degradation products can be absorbed and excreted in the urine, whereas the remainder is excreted in the feces.

D. Excretion

- In studies with a cow (Leahey et al., 1972) and rats (Daniel and Gage, 1966), about 96% and 69 to 96%, respectively, of the administered radioactivity (single oral doses, ^{14}C-labeled) from paraquat was excreted in the feces within 2 to 3 days as unchanged paraquat.

- Goats (Hendley et al., 1976) and pigs (Leahey et al., 1976b; Spinks et al., 1976) that received single oral doses of ^{14}C-labeled paraquat (1.7 and 1.1 or 100 mg ion/kg/day, respectively) for up to 7 days excreted 50% and 69%, respectively, of the total administered dose in feces unchanged.

III.   Health Effects

A. Humans

1. *Short-term Exposure*

- The Pesticide Incident Monitoring System (U.S. EPA, 1979) indicated numerous cases of poisoning from deliberate or accidental ingestion of paraquat or by dermal and inhalation exposure from spraying, mixing and loading operations. Generally, the concentrations of the ingested doses or of amounts inhaled or spilled on the skin were not specified. Symptoms reported following these exposures included burning of the mouth, throat, eyes and skin. Other effects noted were nausea, pharyngitis, episcleritis and vomiting. No fatalities were reported following dermal or inhalation exposure. Deliberate and accidental ingestion of unspecified concentrations of paraquat resulted in respiratory distress and subsequent death. See also Cooke et al. (1973).

2. *Long-term Exposure*

- No information was found in the available literature on long-term human exposure to paraquat.

B. Animals

1. *Short-term Exposure*

- Acute oral LD_{50} values for paraquat (99.9% purity) were reported as 112, 30, 35 and 262 mg paraquat ion/kg in the rat, guinea pig, cat and hen, respectively. (Clark, 1965). Signs of toxicity included respiratory distress and cyanosis among rats and guinea pigs, bloodstained droppings among the hens, and muscular weakness, incoordination and frequent vomiting of frothy secretion among the cats.

- Acute (4-hour) inhalation LC_{50} values for paraquat ranged from 0.6 to 1.4 mg ion/m^3 paraquat (McLean Head et al., 1985).

2. *Dermal/Ocular Effects*

- Acute dermal LD_{50} values for rabbits (Standard Oil, 1977) were 59.9 mg/kg and 80 to 90 mg paraquat ion/kg for rats (FDA, 1970).

- Paraquat concentrate 3 (34.4% paraquat ion) was applied (0.5 mL or 172 mg paraquat ion) to intact and abraded skin of six male New Zealand White rabbits for 24 hours (Bullock, 1977b). Very slight, moderate or severe erythema and slight edema were noted during the 7-day observation period for both intact and abraded skin.

- Paraquat concentrate 3 (0.1 mL, 34.4% paraquat ion) was instilled into the conjunctival sac of one eye in each of six male New Zealand White rabbits (Bullock, 1977a). Untreated eyes served as controls. Unwashed eyes were examined for 14 days. Complete opacity of the cornea was reported in three of six rabbits. Roughened corneas, severe pannus, necrosis of the conjunctivae, purulent discharge, severe chemosis of the conjunctivae and mild iritis were also reported.

3. *Long-term Exposure*

- Beagle dogs (3/sex/dose) were fed technical o-paraquat (32.2% cation) in the diet for 90 days at dose levels of 0, 7, 20, 60 or 120 ppm (Sheppard, 1981). Assuming that 1 ppm is equivalent to 0.025 mg/kg/day, these levels correspond to doses of 0, 0.18, 0.5, 1.5 or 3 mg paraquat ion/kg/day (Lehman, 1959), respectively. Increased lung weight, alveolitis and alveolar collapse were observed at a 60 ppm Lowest-Observed-Adverse-Effect Level (LOAEL), and the No-Observed-Adverse-Effect Level (NOAEL) identified for this study was 20 ppm (0.5 mg paraquat ion/kg/day).

- Alderley Park beagle dogs (6/sex/dose) were fed diets containing technical paraquat (32.3% cation) daily for 52 weeks at dietary levels of 0, 15, 30 or 50 ppm (Kalinowski et al., 1983). Based on actual group mean body weights and food consumption, these values correspond to doses of 0, 0.45, 0.93 and 1.51 mg/kg/day paraquat ion for male dogs and 0, 0.48, 1.00 or 1.58 mg paraquat ion/kg/day for females. Clinical and behavioral abnormalities, food consumption, body weight, hematology, clinical chemistry, urinalysis, organ weights, gross pathology and histopathology were comparable for treated animals and controls at 15 ppm (the lowest dose tested). An increased severity and extent of chronic pneumonitis occurred at 30 ppm (LOAEL) in both sexes, but especially in the males. Based on the results of this study, the NOAEL identified was 15 ppm (0.45 mg paraquat cation/kg/day).

- Technical paraquat dichloride (32.7% cation) was fed to Alderley Park mice (60/sex/dose) for 97 to 99 weeks at levels of 0, 12.5, 37.5 and 100/125 ppm (100 ppm for the initial 35 weeks and then 125 ppm until termination of the study) (Litchfield et al., 1981). Based on the assumption that 1 ppm in the diet of mice is equivalent to 0.15 mg/kg/day (Lehman, 1959), these levels correspond to doses of 0, 1.87, 5.6 and 15/18.75 mg cation/kg. The animals were observed for toxic signs, body weights, and food consumption and utilization; urinalysis, gross pathology and histopathology were also evaluated. Renal tubular degeneration in the males, weight loss and decreased food intake in the females, were observed in the 37.5 ppm dose group. Based on these findings, a NOAEL of 12.5 ppm cation (1.87 mg/kg/day) was identified for both sexes.

- Fischer 344 rats (70/sex/dose) were fed diets containing 0, 25, 75 or 150 ppm of technical paraquat (32.7% cation) for 113 to 117 weeks (males) and 122 to 126 weeks (females) (Woolsgrove et al., 1983). Based on the assumption that 1 ppm in the diet is equivalent to 0.05 mg/kg/day (Lehman, 1959), these levels correspond to doses of 0, 1.25, 3.75 or 7.5 mg cation/kg/day. Clinical signs, food and water consumption, clinical chemistry, urinalysis, hematology, ophthalmoscopic effects, gross pathology and histopathology were evaluated. Increased incidences of slight hydrocephalus were noted in the female rats dying between week 53 and termination of the study; these incidences were 5/60, 8/30, 9/27 and 9/30 rats in the control, low-, mid- and high-dose groups, respectively. Also, increased incidences of spinal cord cysts and cystic spaces were noted in the male rats dying between week 53 and termination of the study. These incidences were 0/53, 6/36 and 4/35 rats at the control, low- and mid-level doses, respectively; no incidence was reported at the high-dose level. Eye opacities, cataracts and nonneoplastic lung lesions (alveolar macrophages and epithelialization, and slight peribronchiolar lymphoid hyperplasia) were observed at 75 ppm and above. Similar eye lesions occurred at 25 ppm (the lowest dose tested). These effects did not appear to be biologically significant, since they were either minimal or occurred after 104 weeks of treatment. Based on these results, an approximate NOAEL of 25 ppm (1.25 mg cation/kg/day) was identified.

4. *Reproductive Effects*

- Lindsay et al. (1982) fed Alderley Park rats technical paraquat dichloride (32.7% cation w/w) in unrestricted diet for three generations at dose levels of 0, 25, 75 or 150 ppm paraquat ion. Based on the assumption that 1 ppm in the diet of rats is equivalent to 0.05 mg/kg/day (Lehman, 1959), these levels correspond to doses of 0, 1.25, 3.75 or 7.5 mg/kg/day. No adverse reproductive effects were reported at 150 ppm (the highest dose tested) or less. An increased incidence of alveolar histiocytosis in the lungs of male and female parents (F_0, F_1 and F_2) was observed in the 75- and 150-ppm dose groups. Based on these results, a reproductive NOAEL of > 150 ppm (7.5 mg/kg/day, the highest dose tested) and a systemic NOAEL of 25 ppm (1.25 mg/kg/day, the lowest dose tested) were identified.

5. *Developmental Effects*

- Young adult Alderley Park mice (number not stated) were administered paraquat dichloride (100% purity) orally by gavage at dose levels of 0, 1, 5 or 10 mg paraquat ion/kg/day on days 6 through 15 of gestation (Hodge et al., 1977). No teratogenic responses were reported at 10 mg ion/kg/day (the highest dose tested) or lower. Partially ossified sternebrae in 26.3% of the fetuses in the high-dose group (10 mg ion/kg/day) and decreased maternal weight gain in the 5-mg ion/kg/day dose group were observed. Based on these results, the developmental NOAEL identified for this study was 5 mg/kg/day, while the maternal NOAEL was 1 mg/kg/day.

- Hodge et al. (1978) dosed Alderley Park rats (29 or 30/dose) by gavage with paraquat dichloride (100% purity) on days 6 through 15 of gestation at dose levels of 0, 1, 5 and 10 mg paraquat ion/kg/day. No teratogenic effects were reported at 10 mg ion/kg/day (the highest dose tested). Maternal body weight gain was significantly decreased ($p < 0.001$) at 5 mg ion/kg/day and above. Fetal body weight gain was significantly ($p = < 0.05$) decreased at the mid-dose level (5 mg/kg/day, LOAEL) and above. Based on these findings, the developmental and maternal NOAEL of 1 mg paraquat ion/kg/day was identified.

6. *Mutagenicity*

- Analytical-grade paraquat dichloride (99.6% purity) was weakly mutagenic in human lymphocytes, with and without metabolic activation, at cytotoxic concentrations (1,250 to 3,500 μg paraquat dichloride/mL blood) (Sheldon et al., 1985).

- Technical-grade, 45.7% active ingredient (a.i.) and analytical-grade (99.6% a.i.) paraquat dichloride were weakly positive in the L5178Y mouse lymphoma assay with and without metabolic activation in studies by Clay and Thomas (1985) and Cross (1985), respectively. Statistically significant increases in mutant colonies were observed only at doses below 29% cell survival (Cross, 1985).

- Analytical-grade paraquat dichloride (99.4% a.i.) increased sister-chromatid exchanges (SCE) at nontoxic doses ($\leq$ 124 μg/mL in nonactivated cultures and $\leq$ 245 μg/mL in S9-supplemented cultures. The induction of increased SCE was more marked in the absence of the S9 fraction (Howard et al., 1985).

- Mutagenic activity was detected in various assays with *Salmonella typhimurium* (Benigni et al., 1979), human embryo epithelial cells (Benigni et al., 1979) and *Saccharomyces cerevisiae* (Parry, 1977).

7. *Carcinogenicity*

- Technical paraquat dichloride (32.7% paraquat ion) fed to Alderley Park mice (60/sex/dose) for 99 weeks did not induce statistically significant dose-related oncogenic responses at dose levels of 0, 12.5, 37.5 or 100/125 ppm (100 ppm for the initial 35 weeks and then 125 ppm until termination of the study) (Litchfield et al., 1981). Based on the assumption that 1 ppm in food for mice is equivalent to 0.15 mg/kg/day (Lehman, 1959), these levels correspond to doses of 0, 1.87, 5.6 and 15/18.75 mg/kg. The study appeared to have been conducted properly. The results show that paraquat is not oncogenic at the dose levels tested.

- Woolsgrove et al. (1983) fed Fischer 344 rats (70/sex/dose) diets containing technical paraquat (32.69%) for 113 to 117 weeks (males) and 122 to 124 weeks (females) at dietary levels of 0, 25, 75 and 150 ppm. Based on the assumption that 1 ppm in the diet of rats is equivalent to 0.05 mg/kg/day (Lehman, 1959), these levels correspond to doses of 0, 1.25, 3.75 and 7.5 mg paraquat cation/kg/day. The predominant tumor types noted in this study were tumors of the lungs, endocrine glands (pituitary, thyroid and adrenal) and of the skin and subcutis. Both the lung and endocrine tumors occurred at a frequency similar to the incidence of these kinds of tumors in the historical control. Only when pooled, the skin tumors (squamous cell carcinomas in the head region including ear, nasal cavity, oral cavity and skin) appeared to provide some equivocal evidence of carcinogenicity in the high-dose males. However, the observed skin tumors were low in incidence and occurred at different sites in the head region. The different tumor sites were considered to be anatomically different and inappropriate for pooling by an independent pathology evaluation review (Experimental Pathology Laboratory, Inc.) as well as by the second EPA Review of Paraquat (1988a). Thus, the tumors could not be associated with oral exposure to the compound.

- Paraquat dichloride (98% pure) fed to JCR:ICR mice (60/sex/group) at 0, 2, 10, 30 or 100 ppm for 104 weeks did not induce statistically significant dose-related oncogenic responses (Toyashima et al., 1982a). At 100 ppm, increased mortality and statistically significant changes in hematology, clinical chemistry and organ weights were observed in both sexes, indicating a Maximum Tolerated Dose (MTD) was reached. The experiment identifies a NOAEL of 30 ppm.

- Paraquat dichloride (98% pure) fed to JCR:Wistar rats (50/sex/group) at 0, 6, 30, 100 or 300 ppm in the diet for 104 weeks did not cause compound-related neoplastic changes (Toyashima et al., 1982b). At 300 ppm, increased mortality, and statistically significant changes in hematology, clinical chemistry and organ weights were observed for males and females, indicating an MTD was reached. The experiment identifies a NOAEL of 100 ppm.

IV.   Quantification of Toxicological Effects

A. One-day Health Advisory

No suitable information was found in the available literature for the determination of the One-day HA value for paraquat. It is therefore recommended that the Ten-day HA value for the 10-kg child of 0.1 mg/L (100 μg/L), calculated below, be used at this time as a conservative estimate of the One-day HA value.

B. Ten-day Health Advisory

The rat developmental study (Hodge et al., 1978) has been selected to serve as the basis for the determination of the Ten-day HA value for paraquat. In this study, Alderley Park rats were administered paraquat (100% purity) during gestation days 6 through 15 at dose levels of 0, 1, 5 or 10 mg paraquat ion/kg/day. There was a statistically significant ($p \leq 0.001$; $p = 0.05$) decrease in maternal and fetal body weight gain at the 5-mg paraquat ion/kg/day dose; also at 5 mg/kg/day, there was a slight retardation in ossification. The fetotoxic and maternal NOAEL identified in this study was 1 mg paraquat ion/kg/day. Another adequate study of comparable duration reported a NOAEL of 5 mg/kg/day for developmental effects, and a NOAEL of 1 mg/kg/day for maternal effects supports this selection (Hodge et al., 1977).

Using a NOAEL of 1 mg/kg/day, the Ten-day HA for a 10-kg child is calculated as follows:

$$\text{Ten-day HA} = \frac{(1 \text{ mg/kg bw/day}) (10 \text{ kg})}{(100) (1 \text{ L/day})} = 0.1 \text{ mg/L } (100 \text{ } \mu g/L)$$

C. Longer-term Health Advisory

No subchronic studies were found in the available literature that were suitable for deriving the Longer-term HA value for paraquat. The 90-day oral study of dogs (Sheppard, 1981) reported a NOAEL (0.5 mg ion/kg/day) which is similar to the NOAEL (0.45 mg ion/kg/day) of the 52-week oral dog study (Kalinowski et al., 1983) used to derive the Lifetime HA. It is therefore recommended that the Drinking Water Equivalent Level (DWEL) of 0.2 mg/L (200 $\mu g/L$), calculated below, be used for the Longer-term HA value for an adult, and that the DWEL adjusted for a 10-kg child, 0.05 mg/L (50 $\mu g/L$), be used for the Longer-term HA value for a child.

D. Lifetime Health Advisory

The study by Kalinowski et al. (1983) has been selected to serve as the basis for the Lifetime HA value for paraquat. In this 52-week feeding study in beagle dogs, a NOAEL of 15 ppm (0.45 mg paraquat ion/kg/day) was identified based on the absence of hematological, biochemical, gross pathological and histological effects as well as the absence of any significant changes in food consumption, or in body and organ weights for treated and control groups. Adequate studies of comparable duration reported NOAELs higher than those of the critical study selected for derivation of the Lifetime HA. A lifetime oral study in rats (Woolsgrove et al., 1983) reported a NOAEL of 25 ppm (about 1.25 mg/kg/day); a NOAEL of 12.5 ppm (about 1.87 mg/kg/day) was identified for mice (Litchfield et al., 1981).

Step 1: Determination of the Reference Dose (RfD)

$$\text{RfD} = \frac{(0.45 \text{ mg ion/kg/day})}{(100)} = 0.0045 \text{ mg/kg/day}$$

Step 2: Determination of the Drinking Water Equivalent Level (DWEL)

$$\text{DWEL} = \frac{(0.0045 \text{ mg/kg/day}) (70 \text{ kg})}{(2 \text{ L/day})} = 0.16 \text{ mg/L } (200 \text{ } \mu g/L)$$

Step 3: Determination of the Lifetime Health Advisory

$$\text{Lifetime HA} = (0.16 \text{ mg/L}) (20\%) = 0.03 \text{ mg/L} (30 \text{ } \mu g/L)$$

E. Evaluation of Carcinogenic Potential

- Based on the recent EPA Peer Review of the data on paraquat (U.S. EPA, 1988a), the chemical is classified in Group E: evidence of noncarcinogenicity for humans (U.S. EPA, 1986). The chemical provided, at most, only equivocal evidence for skin tumors at high dose in male rats (but not in the females) in one study, and no evidence for carcinogenicity in three other studies in rats and mice. The observed skin tumors were low in incidence, occurred at different sites in the head region, and the tumor incidence was significantly different from controls only when pooled from all sites in the head region. Since pooling of the different tumor sites was considered by pathologists to be inappropriate, the tumors could not be associated with oral exposure to the compound.

- The International Agency for Research on Cancer has not evaluated the carcinogenic potential of paraquat.

V. Other Criteria, Guidance and Standards

- The Office of Pesticide Programs (OPP) has established tolerances on raw agricultural commodities for paraquat ion derived from either the bis(methyl sulfate) or dichloride salt ranging from 0.01 to 5 ppm (U.S. EPA, 1984). The tolerances are based on an ADI of 0.0045 mg/kg/day derived from a 1-year feeding study in dogs, with a NOAEL of 0.45 mg/kg/day and a safety factor of 100.

- The National Academy of Sciences (NAS, 1977) has a Suggested-No-Adverse-Response Level (SNARL) of 0.06 mg/L. This was calculated using an uncertainty factor of 1,000 and a NOAEL of 8.5 mg/kg/day identified in the 2-year rat study by Chevron Chemical Company (1975), with an assumed consumption of 2 L/day of water by a 70-kg adult, and with the assumption that 20% of total intake of paraquat was from water.

- The American Conference of Governmental and Industrial Hygienists has presented a threshold limit value of 0.1 mg/m^3 for paraquat of respirable particle sizes (ACGIH, 1980).

VI. Analytical Methods

• Paraquat is analyzed by a Draft EPA method (U.S. EPA, 1988b) for drinking water. In this procedure a measured quantity of sample is extracted with a solid phase adsorbent. After elution from the adsorbent tube, the paraquat residue is determined by high performance liquid chromatography with an ultraviolet detector (HPLC/UV). The method has a single laboratory validation maximum detection limit (MDL). For paraquat, the MDL is 0.80 μg/L.

VII. Treatment Technologies

• Weber et al. (1986) investigated the adsorption of paraquat and other compounds by charcoal and cation and anion exchange resins and their desorption with water. They developed Freundlich adsorption-desorption isotherms for paraquat on charcoal. When 250 mg of charcoal was added to paraquat solutions, it exhibited the following adsorptive capacities: 37.3 and 93.2 mg paraquat/g charcoal at concentrations of 0.373 mg/L and 37.3 mg/L, respectively. Paraquat was also adsorbed by IR-120 exchange resins (H^+ and Na^+ forms). The IR-120-H resin showed more affinity towards paraquat than the IR-120-Na resin. When 665 mg of paraquat in solution was added to 15 mg of resin, IR-120-H adsorbed 70% of paraquat while the IR-120-Na adsorbed 66% of paraquat.

• MacCarthy and Djebbar (1986) evaluated the use of chemically modified peat for removing paraquat from aqueous solutions under a variety of experimental conditions. Paraquat sorption isotherms on treated Irish peat were determined by equilibrating 100-mL volumes of 3.66 mg/L paraquat with 0.1 g of peat at ambient conditions. Tests indicated that equilibrium for paraquat was achieved after 6 days. Peat exhibited the following paraquat sorption capacities: 40, 55 and 60 mg paraquat/g peat at concentrations of 2, 4 and 6 mg/L, respectively. The effects of pH, ionic strength and flow rate on paraquat removal efficiency were also investigated. When 45 mL of 16-mg/L paraquat solution was gravity fed to a column with a diameter of 6 mm that had been packed with 700 mg treated peat, 95 to 99% paraquat removal efficiency was reported without a significant effect by variations in pH, ionic strength or flow rate.

- In summary, several techniques for the removal of paraquat from water have been examined. While data are not unequivocal, it appears that adsorption of paraquat by charcoal, ion exchange and modified peat are effective treatment techniques. However, selection of individual or combinations of technologies for paraquat removal from water must be based on a case-by-case technical evaluation and an assessment of the economics involved.

VIII.   References

ACGIH. 1980. American Conference of Governmental Industrial Hygienists. Documentation of the threshold limit values for substances in workroom air, 4th ed. Cincinnati, OH: ACGIH.

Benigni, R., M. Bignami, A. Carere, G. Conti, L. Conti, R. Crebelli, E. Dogliotti, G. Gualandi, A. Novelletto and V. Ortali. 1979. Mutational studies with diquat and paraquat in vitro. Mutat. Res. 68:183–193.

Bullock, C.H.* 1977a. S-1103: The eye irritation potential of ortho paraquat 3 lbs/gal concentrate. Standard Oil Company of California, Report No. SOCAL 1060/30:70. August 1. MRID 00054575.

Bullock, C.H.* 1977b. S-1104: The skin irritation potential of ortho paraquat 3 lbs/gal concentrate. Standard Oil Company of California, Report No. SOCAL 1061/30:71, August 1. MRID 00054576.

Calderbank, A. 1970. The fate of paraquat in water. Outlook Agric. 6(3):128–130.

Calderbank, A. 1976. In: G. Kearney, C. Phillips and D. Kaufman, eds. Herbicides: Chemistry, degradation and mode of action. 2nd ed. New York: Marcel Dekker.

CHEMLAB. 1985. The Chemical Information System. CIS, Inc., Baltimore, MD.

Chevron Chemical Company. 1975. Paraquat poisoning: a physician's guide for emergency treatment and medical management. San Francisco, CA: Chevron Environmental Health Center. (Cited in NAS, 1977.)

Clark, D.G.* 1965. The acute toxicity of paraquat. Imperial Chemical Industries Limited. Report No. IHR/170. January 1. MRID 00081825.

Clark, D.G., T.S. McElligott and E.W. Hurst. 1966. The toxicity of paraquat. Brit. J. Ind. Med. 23(2):126–132.

Clay, P. and M. Thomas.* 1985. Paraquat dichloride (technical liquor): assessment of mutagenic potential using L5178Y mouse lymphoma cells. Imperial Chemical Industries PLC, England. Report No. CTL/P/1398. September 24. MRID GS 0262-009.

Coats, G.E., H.H. Funderburk, Jr. and J.H. Lawrence et al.* 1964. Persistence of diquat and paraquat in pools and ponds. Proceedings, Southern Weed Control Conference. 17:308–320. Also in unpublished submission received Apr. 7, 1971 under unknown admin. no.; submitted by Chevron Chemical Co., Richmond, CA. CDL:180000-I. MRID 00055093.

Cooke, N.J., D.C. Flenley and H. Matthew. 1973. Paraquat poisoning. Serial studies of lung function. Q. J. Med. New Ser. 42:683–692.

Cross, M. 1985.* Paraquat dichloride: Assessment of mutagenic potential using L5178Y mouse lymphoma cells. Imperial Chemical Industries PLC, England. Report No. CTL/P/1374. September 17. MRID 00152691.

Daniel, J.W. and J.C. Gage.* 1966. Absorption and excretion of diquat and paraquat in rats. Imperial Chemical Industries Limited, England. Brit. J. Ind. Med. 23:133–136. MRID 00055107.

Day, S.R. and R.J. Hemingway.* 1981. ^{14}C-Paraquat: degradation on a sandy soil surface in sunlight. Report No. RJ 01688. Unpublished study submitted by Chevron Chemical Co. under Accession No. 257105.

FDA. 1970. Food and Drug Administration. Acute LD_{50}—rat. Project No. stated. Chambers, GA. MRID GS 0262-003. (Cited in U.S. EPA, 1985.)

Frank, P.A. and R.D. Comes.* 1967. Herbicidal residues in pond water and hydrosoil. Weeds. 15:210–213.

Helling, C. and B. Turner.* 1968. Pesticide mobility: determination by soil thin-layer chromatography. Science. 167:562–563.

Hendley, P., J.P. Leahey, C.A. Spinks, D. Neal and P.K. Carpenter.* 1976. Paraquat: metabolism and residues in goats. Huntingdon Research Centre, England. Project No. AR 2680A. July 16. MRID 00028597.

Hodge, M.C.E., S. Palmer, T.M. Weight and J. Wilson. 1977. Paraquat dichloride: teratogenicity study in the mouse. Imperial Chemical Industries Limited, England. Report No. CTL/P/364. June 12. MRID 00096338.

Hodge, M.C.E., S. Palmer, T.M. Weight and J. Wilson. 1978. Paraquat dichloride: teratogenicity study in the rat. Imperial Chemical Industries Limited, England. Report No. CTL/P/365. June 5. MRID 00113714.

Howard, C.A., J. Wildgoose, P. Clay and C.R. Richardson. 1985. Paraquat dichloride: an *in vitro* sister chromatid exchange study in Chinese hamster lung fibroblasts. Imperial Chemical Industries PLC, England. Report No. CTL/P/1392. September 24. MRID GS 0262-009.

Kalinowski, A.E., J.E. Doe, I.S. Chart, C.W. Gore, M.J. Godley, K. Hollis, M. Robinson and B.H. Woollen. 1983. Paraquat: one-year feeding study in dogs. Imperial Chemical Industries, England. Report No. CTL/P/734. April 20. MRID 00132474.

Leahey, J.P., R.J. Hemingway, J.A. Davis and R.E. Griggs. 1972. Paraquat metabolism in a cow. Imperial Chemical Industries Ltd, England. Report No. AR 2374A. November 17. MRID 00036297.

Leahey, J.P., C.A. Spinks, D. Neal and P.K. Carpenter. 1976a. Paraquat metabolism and residues in goats. Huntingdon Research Centre, England. Project No. AR 2680 A. July 16. MRID 00028597.

Leahey, J.P., P. Hendley and C.A. Spinks. 1976b. Paraquat metabolism and residues in pigs. Huntingdon Research Centre, England. Project No. AR 2694 A. October 4. MRID 00028598.

Lehman, A.J. 1959. Appraisal of the safety of chemicals in foods, drugs and cosmetics. Assoc. Food and Drug Off. U.S., Q. Bull.

Lindsay, S., P.B. Banham, M.J. Godley, S. Moreland, G.A. Wickramaratue and B.H. Woollen. 1982. Paraquat multigeneration reproduction study in rats: three generations. Imperial Chemical Industries PLC, England. Report No. CTL/P/719. December 22. Report No. CTL/P/719S. MRID 00126783. Chevron response to EPA comments on rat reproduction study. No date. Received by EPA on 9/10/85.

Litchfield, M.H., M.F. Sotheran, P.B. Banham, M.J. Godley, S. Lindsay, I. Pratt, K. Taylor and B.H. Woollen. 1981. Paraquat lifetime feeding study in the mouse. Imperial Chemical Industries Ltd., England. Report No. CTL/P/556. June 22. MRID 00087924.

MacCarthy, P. and K.E. Djebbar. 1986. Removal of paraquat, diquat and amitrole from aqueous solution by chemically modified peat. J. Environ. Qual. 15(2):103–107.

McLean Head, L., J.R. Marsh and S.W. Millward. 1985. Paraquat: 4-hour acute inhalation toxicity study in the rat. Imperial Chemical Industries Ltd. Report No. CTL/P/1325 and CTL/P/1325S. September 24. MRID GS 0262-004. (Cited in U.S. EPA, 1985.)

Meister, R., ed. 1988. Farm chemicals handbook. Willoughby, OH: Meister Publishing Company.

NAS. 1977. National Academy of Sciences. Drinking water and health. Washington, DC: National Academy Press.

Pack, D.E.* 1982. Long-term exposure of ^{14}C-paraquat on a sandy soil to California sunlight. Unpublished submission by Chevron Chemical Co. under Accession No. 257105.

Parry, J.M. 1977. The use of yeast cultures for the detection of environmental mutagens using a fluctuation test. Mutat. Res. 46:165–176.

Sheldon, T., C.A. Howard, J. Wildgoose and C.R. Richardson. 1985. Paraquat dichloride: a cytogenetics study in human lymphocytes *in vitro*. Imperial Chemical Industries PLC, England. Report No. CTL/P/1351. September 3. MRID GS 0262-009.

Sheppard, D.B. 1981. Paraquat thirteen week (dietary administration) toxicity study in beagles. Hazelton Laboratories Europe Ltd., England. Report No. CTL/C/1027. HLE Project No. 2481-72/111A. February 17. MRID 00072416.

Spinks, C.A., P. Hendley, J.P. Leahey and P.K. Carpenter. 1976. Metabolism and residues in pigs using ^{14}C-ring-labeled paraquat. Huntingdon Research Centre, England. Project No. AR 2692 A. October 1. MRID 00028599.

Standard Oil Company. 1977. Acute dermal LD_{50}—rabbit. Project No. SOCAL 1059/29:40. MRID 00054574.

STORET. 1988. STORET Water Quality File. Office of Water. U.S. Environmental Protection Agency (data file search conducted in May, 1988).

TDB. 1985. Toxicology Data Bank. MEDLARS II. National Library of Medicine's National Interactive Retrieval Service.

Toyashima, S., R. Sato, M. Kashima and M. Motoyama. 1982a. AT-5: Chronic toxicity study result—104-week dosing study in mouse. Ashi Chemical Industries Company, Ltd., Japan. March 10. MRID 402024-03.

Toyashima, S., R. Sato, M. Kashima, M. Motoyama and A. Ishikawa. 1982b. AT-5: Chronic toxicity study result—104-week dosing study in rat. Ashi Chemical Industries Company, Ltd., Japan. March 10. MRID 402024-03.

Upton, B.P., P. Hendley and M.W. Skidmore.* 1985. Paraquat: Hydrolytic stability in water pH 5, 7 and 9. ICI Plant Protection Division. Report Series RJ0436B. Submitted Sept. 3, 1985. Chevron Chemical Co., Richmond, CA.

U.S. EPA. 1979. U.S. Environmental Protection Agency. Summary of reported pesticide incidents involving paraquat. Pesticide Incident Monitoring System. Report No. 200. July.

U.S. EPA. 1984. U.S. Environmental Protection Agency. Code of Federal Regulations. 40 CFR 180.205. July 1.

U.S. EPA. 1986. U.S. Environmental Protection Agency. Guidelines for carcinogen risk assessment. Fed. Reg. 51(185):33992–34003. September 24.

U.S. EPA. 1988a. U.S. Environmental Protection Agency. Second peer review of paraquat. Memo from Reto Engler to Robert Taylor. June 28.

U.S. EPA. 1988b. U.S. Environmental Protection Agency. EPA Draft Method—Determination of diquat and paraquat in drinking water by high performance liquid chromatography with an ultraviolet detector. Available from EPA's Environmental Monitoring and Support Laboratory, Cincinnati, OH.

Weber, J.B., T.M. Ward and S.B. Weed. 1986. Adsorption and desorption of diquat, paraquat, prometone. Proc. Soil Sci. Soc. Amer. 32:197–200.

Windholz, M., S. Budvari, R.F. Blumetti and E.S. Otterbein, eds. 1983. The Merck Index, 10th edition. Rahway, NJ: Merck and Co., Inc.

Woolsgrove, B., R. Ashby, P. Hepworth, A.K. Whimmey, P.M. Brown, J.C. Whitney and J.P. Finn. 1983. Paraquat: combined toxicity and carcinogenicity study in rats. Life Sciences Research, England. Report No. 82/1LY217/328. October 27. MRID 00138637.

Worthing, C.R. 1983. The pesticide manual. Published by the British Crop
Council.

---

* Confidential Business Information submitted to the Office of Pesticide Programs.

Picloram

I. General Information and Properties

A. CAS No. 1918–02–01

B. Structural Formula

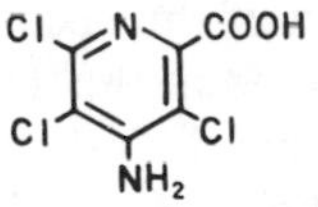

(4-Amino-3,5,6-trichloropicolinic acid)

C. Synonyms

- Amdon; ACTP; Borolin; K-PIN; Tordon (Meister, 1987).

D. Uses

- Picloram is used as a broad-spectrum herbicide for the control of broad-leaf and woody plants in rangelands, pastures and rights-of-way for powerlines and highways (Meister, 1987).

E. Properties (Meister, 1987)

Chemical Formula	$C_6H_3Cl_3N_2O_2$
Molecular Weight	241.6
Physical State (room temp.)	White powder
Boiling Point	Decomposes
Melting Point	215°C (decomposes)
Density	--
Vapor Pressure (35°C)	6.2×10^{-7} mm Hg
Specific Gravity	--
Water Solubility	0.043 g/100 mL (free acid)
	40 g/100 mL (salts)
Log Octanol/Water Partition Coefficient	--
Taste Threshold	--

Odor Threshold   Chlorine-like
Conversion Factor   --

F. Occurrence

- Picloram has been found in 420 of 744 surface water samples analyzed and in 3 of 64 ground-water samples (STORET, 1988). Samples were collected at 135 surface water locations and 30 ground-water locations, and picloram was found in seven states. The 85th percentile of all non-zero samples was 0.13 μg/L in surface water and 0.02 μg/L in ground-water sources. The maximum concentration found was 4.6 μg/L in surface water and 0.02 μg/L in ground water. This information is provided to give a general impression of the occurrence of this chemical in ground and surface waters as reported in the STORET database. The individual data points retrieved were used as they came from STORET and have not been confirmed as to their validity. STORET data are often not valid when individual numbers are used out of the context of the entire sampling regime, as they are here. Therefore, this information can only be used to form an impression of the intensity and location of sampling for a particular chemical.

G. Environmental Fate

- The main processes for dissipation of picloram in the environment are photodegradation and aerobic soil degradation. Field tests conducted in Texas with a liquid formulation of picloram have indicated that approximately 74% of the picloram originally contained in the test ecosystems, which included the soil, water and vegetation, was dissipated within 28 days after application (Scifres et al., 1977).

- Photodegradation of picloram occurs rapidly in water (Hamaker, 1964; Redemann, 1966; Youngson, 1968; Youngson and Goring, 1967), but is somewhat slower on a soil surface (Bovey et al., 1970; Merkle et al., 1967; Youngson and Goring, 1967). Hydrolysis of picloram is very slow (Hamaker, 1976).

- Laboratory studies have shown that under aerobic soil conditions, the half-life of picloram is dependent upon the applied concentration, and the temperature and moisture of the soil. The major degradation product is CO_2; other metabolites are present in insignificant amounts (McCall and Jefferies, 1978; Merkle et al., 1967; Meikle et al., 1970, 1974; Meikle, 1973; Hamaker, 1975). In the absence of light under anaerobic soil and aquatic conditions, picloram degradation is extremely slow (McCall and Jefferies, 1978).

- Following normal agricultural, forestry and industrial applications of picloram, long-term accumulation of picloram in the soil generally does not occur. In the field, the dissipation of picloram will occur at a faster rate in hot, wet areas compared to cool, dry locations (Hamaker et al., 1967). The half-life of picloram under most field conditions is a few months (Youngson, 1966). There is little potential for picloram to move off treated areas in runoff water (Fryer et al., 1979). Although picloram is considered to have moderate mobility (Helling, 1971a,b), leaching is generally limited to the upper portions of most soil profiles (Grover, 1977). Instances of picloram entering the ground water are largely limited to cases involving misapplications or unusual soil conditions (Frank et al., 1979).

II.   Pharmacokinetics

A. Absorption

- Picloram is readily absorbed from the gastrointestinal (GI) tract of rats (Nolan et al., 1980). Within 48 hours after dosing rats with 1,400 mg/kg body weight (bw), 80 to 84% of the dose was found in urine.

- A 500-kg Holstein cow was administered 5 ppm picloram in the feed for 4 days (approximately 0.23 mg/kg/day). Ninety-eight percent of the total dose was excreted in the urine, demonstrating nearly complete absorption (Fisher et al., 1965).

- Similar results were observed in three male Fischer CDF rats receiving ^{14}C-picloram (dose not specified), where 95% of the dose was absorbed (Dow, 1983).

B. Distribution

- Picloram appears to be distributed throughout the body, with the highest concentration in the kidneys (Redemann, 1964). In rats (strain, age and sex not specified) administered a single 20 mg/kg dose of ^{14}C-labeled picloram in food, radioactivity was found in abdominal fat, liver, muscle and kidneys with maximum levels occurring 2 to 3 hours after dosing.

- Hereford-Holstein steers fed picloram at daily doses of 3.2 to 23 mg/kg for 2 weeks had tissue concentrations of 0.05 to 0.32 mg/kg in muscle, 0.06 to 0.45 mg/kg in fat, 0.12 to 1.6 mg/kg in liver, 0.18 to 2.0 mg/kg in blood and 2 to 18 mg/kg in kidney (Kutschinski and Riley, 1969).

- In a similar study, two steers (strain not specified) fed 100 or 200 mg picloram (3 or 6 mg/kg bw/day) for 31 days had picloram concentrations of 4 or 10 mg/kg, respectively, in the kidneys, while concentrations in other tissues (muscle, omentum fat, heart, liver, brain) were less than 0.5 mg/kg (Leasure and Getzander, 1964).

C. Metabolism

- Picloram administered to rats or cattle was excreted in the urine in unaltered form (Fisher et al., 1965; Nolan et al., 1980; Dow, 1983), and no $^{14}CO_2$ was detected in expired air of rats given ^{14}C-carbon-labeled picloram (Redemann, 1964; Nolan et al., 1980; Dow, 1983). These studies indicate that picloram is not metabolized significantly

D. Excretion

- Picloram administered to rats is excreted primarily in the urine (Redemann, 1964; Nolan et al., 1980; Fisher et al., 1965).

- Male (Fischer 344) rats that were administered a single oral dose of picloram at 1,400 mg/kg bw, within 48 hours excreted 80 to 84% of the dose in the urine, 15% in the feces, less than 0.5% in the bile and virtually no measurable amount as expired CO_2 (Nolan et al., 1980).

- One Holstein cow administered 5 ppm picloram (approximately 0.23 mg/kg/day) in feed for 4 consecutive days excreted more than 98% of the dose in the urine (Fisher et al., 1965).

- In male Fischer 344 rats administered picloram at 10 mg/kg bw orally, clearance of picloram from the plasma was biphasic, showing half-lives of 29 and 228 minutes. When the same dose was administered intravenously, biphasic clearance occurred with half-lives of 6.3 and 128 minutes (Nolan et al., 1980).

- Cattle excrete picloram primarily in the urine (Fisher et al., 1965), although small amounts may appear in the milk (Kutschinski and Riley, 1969). In Holstein cows fed picloram for 6 to 14 days at doses of 2.7 mg/kg/day or less, no picloram could be found in the milk, while cows fed picloram at doses of 5.4 to 18 mg/kg/day had milk levels up to 0.28 mg/L. This corresponds to 0.02% of the ingested dose. When picloram feeding was discontinued, picloram levels in milk became undetectable within 48 hours.

- Nolan et al. (1983) investigated the excretion of picloram in humans. Six male volunteers (40 to 51 years old) ingested picloram at 0.5 or 5 mg/kg in approximately 100 mL of grape juice. Seventy-six percent of the dose was excreted unchanged in the urine within 6 hours (half-life of 2.9 hours). The remainder was eliminated with an average half-life of 27 hours. The authors did not report observations, if any, of adverse effects. Thus, excretion of picloram in humans was biphasic, as had been demonstrated in rats by Nolan et al. (1980).

III.   Health Effects

A. Humans

- No information on the health effects of picloram in humans was found in the available literature. In the excretion study by Nolan et al. (1983), described above, the authors did not address the presence of toxic effects in human volunteers ingesting picloram at 0.5 or 5 mg/kg.

B. Animals

1. *Short-term Exposure*

- The acute oral toxicity of picloram is low. Lethal doses have been estimated in a number of species, with LD_{50} values ranging from 2,000 to 4,000 mg/kg (NIOSH, 1980; Dow 1983).

- In a 7-day feeding study by Dow (1981a), picloram was fed to female dogs (1/dose) at dose levels of 400, 800 or 1,600 mg/kg bw/day. Picloram was acutely toxic (emesis, loss of body weight) to female dogs at the higher doses and not toxic at 400 mg/kg/day (the lowest dose tested), which was identified as the NOAEL.

- In a 7- to 14-day study by Dow (1981b), beagle dogs (one dog per dose) were administered picloram (79.4% Tordon) at dose levels of 0, 250, 500 or 1,000 mg/kg/day. Based on 79.4% active ingredient, actual doses administered were 200, 400 or 800 mg/kg/day. The No-Observed-Adverse-Effect Level (NOAEL) was determined to be 200 mg/kg/day, the lowest dose tested, based on the absence of reduced food intake.

- In a 32-day feeding study by Dow (1980), picloram was administered to mice at dose levels of 0, 90, 270, 580, 900 or 2,700 mg/kg/day. The NOAEL was 900 mg/kg/day, and the Lowest-Observed-Adverse-Effect Level (LOAEL) was 2,700 mg/kg/day, based on increased liver weight.

2. *Dermal/Ocular Effects*

- Most formulations of picloram have been evaluated for the potential to produce skin sensitization reactions in humans. Dow (1983) reported in summary data that Tordon 22K was not a sensitizer following an application as a 5% solution. A formulation of Tordon 101 containing 6% picloram acid and 2,4-D acid was not a sensitizer as a 5% aqueous solution in humans (Gabriel and Gross, 1964). When the triisopropanolamine salts of picloram and 2,4-D (Tordon 101) were applied as a 5% solution, sensitization occurred in several individuals; however, when applied alone, the individual components were nonreactive.

3. *Long-term Exposure*

- Subchronic studies with picloram have been conducted by Dow (1983), using three species (dogs, rats, mice) over periods of 3 to 6 months. A 6-month study was conducted with beagle dogs that received picloram at daily doses of 0, 7, 35 or 175 mg/kg/day (6/sex/dose group) (Dow, 1983). Increased liver weights were observed at the highest dose (175 mg/kg/day) for males and females, and at the intermediate dose (35 mg/kg/day) for males. Therefore, the 7 mg/kg/day dose level was considered to be a NOAEL.

- In a 13-week feeding study, CDF Fischer 344 rats (15/sex/dosage group) were fed picloram in their diet at dose levels of 0, 15, 50, 150, 300 or 500 mg/kg/day (Dow, 1983). Liver swelling was observed in both sexes at the 150 and 300 mg/kg/day dose levels. The NOAEL in this study was identified as 50 mg/kg/day.

- Ten male and female $B6C3F_1$ mice were administered picloram in their diet at dose levels of 0, 1,000, 1,400 or 2,000 mg/kg/day for 13 weeks (Dow, 1983). Liver weights were increased significantly (p values not reported) in females and males at all dose levels tested.

- Picloram (94%) was fed to Fischer 344 rats (50/sex/dose) for 2 years in the diet at dose levels of 0, 20, 60 or 200 mg/kg (Dow, 1986). There was a significant dose-related increase in size and altered tictorial properties of centrolobular hepatocytes and increased relative liver weights in males and females dosed at 60 and 200 mg/kg/day. The LOAEL based on histologic changes in the liver is 60 mg/kg/day and the NOAEL is 20 mg/kg/day.

- Osborne-Mendel rats receiving picloram at 370 or 740 mg/kg/day in the diet for 2 years had renal disease resembling that of the natural aging process (NCI, 1978). Increased indices of parathyroid hyperplasia, polyarteritis, testicular atrophy and thyroid hyperplasia and adenoma were observed. Polyarteritis may be indicative of an autoimmune effect.

4. *Reproductive Effects*

- As described above in the 2-year feeding study by NCI (1978), testicular atrophy was observed in male Osborne-Mendel rats receiving picloram at 370 or 740 mg/kg/day.

- Groups of 4 male and 12 female rats were maintained on diets containing 0, 7.5, 25 or 75 mg/kg/day of Tordon (95% picloram) through a three-generation (two litters per generation) fertility, reproduction, lactation and teratology study (McCollister et al., 1967). The rats were 11 weeks old at the start of the study and were maintained on the test diets for 1 month prior to breeding to produce the F_{1a} generation. Records were kept of numbers of pups born live, born dead or killed by the dam; litter size was culled to eight pups after 5 days. Lactation continued until the pups were 21 days old, when they were weaned and weighed. After a 7- to 10-day rest, the dam was returned for breeding the F_{1b} generation. The second generation (F_{2a} and F_{3b}) was derived from F_{2b} animals after 110 days of age. Two weanlings per sex per dose from litters of each generation were observed for gross pathology. Gross pathology was also assessed on all parent rats and all females not becoming pregnant. Five male and five female weanlings from each group of the F_{3b} litter were selected randomly for gross and microscopic examination (lung, heart, liver, kidney, adrenals, pancreas, spleen and gonads). Picloram reduced fertility in the F_{1b} generation which was fed a 75 mg/kg/day dose. No other effects were noted. Based on these results, a NOAEL of 25 mg/kg/day was identified.

5. *Developmental Effects*

- In the McCollister et al. (1967) study described above, the F_{1c}, F_{2c} and F_{3c} litters were used to study the teratogenic potential of picloram. The dams were sacrificed on day 19 or 20 of gestation, and offspring were inspected for gross abnormalities, including skeletal and internal structures, and placentas were examined for fetal death or resorptions. None were observed at any dose level.

- Thompson et al. (1972) administered picloram in corn oil to pregnant Sprague-Dawley rats on days 6 to 15 of gestation. Four groups of 35 rats (25 for the teratology portion and 10 for the postnatal portion of the study) received picloram at 0, 500, 750 or 1,000 mg/kg/day by gavage. Rats were observed daily for signs of toxicity. Prebreeding and gestation day 20 body weights were obtained on teratology rats, and prebreeding and postpartum day 21 body weights were obtained for signs of maternal toxicity. Rats given 750 or 1,000 mg/kg/day developed hyperesthesia and mild diarrhea after 1 to 4 days of treatment; 14 maternal deaths occurred between days 8 and 17 of gestation in these dose groups. Evidence of retarded fetal development, as reflected by an increase in unossified fifth sternebrae, was observed in all treatment groups but did not occur in a dose-related manner. The occurrence of bilateral accessory ribs was increased significantly in fetuses of dams given 1,000 mg/kg for 10 days during gestation. At this dose level, there was also maternal toxicity. Since adverse effects occurred at all doses, the LOAEL was 500 mg/kg, the lowest dose tested.

6. *Mutagenicity*

- The mutagenic activity of picloram has been studied in a number of microbial systems. Ames tests in several *Salmonella typhimurium* strains indicated that picloram was not mutagenic with or without activation by liver microsomal fractions (Andersen et al., 1972; Torracca et al., 1976; Carere et al., 1978).

- One study using the same system as above found picloram to be weakly mutagenic (Ercegovich and Rashid, 1977).

- Picloram was shown to be negative in the reversion of bacteriophage AP72 to T_4 phenotype (Andersen et al., 1972), but positive in the forward mutation spot test utilizing *Streptomyces coelicolor* (Carere et al., 1978).

- Irrespective of a weak mutagenic response in the *Salmonella typhimurium* test (Ercegovich and Rashid, 1977) and a positive forward mutation, the authors take the position that picloram is not mutagenic. In addition, in studies in male and female Sprague-Dawley rats fed picloram at dose levels of 20, 200 or 2,000 mg/kg/day, no cytological changes in bone marrow cells were observed (Mensik et al., 1976).

7. *Carcinogenicity*

- Picloram (at least 90% pure) was administered by diet to Osborne-Mendel rats and B6C3F$_1$ mice (NCI, 1978; also reviewed by Reuber, 1981). Pooled controls from carcinogenicity studies run in the same laboratory (at the Gulf South Research Institute) and overlapping this study by at least 1 year were used. Fifty male rats were dosed with picloram at 208 or 417 mg/kg/day and 50 female rats were dosed at 361 or 723 mg/kg/day. During the second year, rough hair coats, diarrhea, pale mucous membranes, alopecia and abdominal distention were observed in treated rats. In addition, a relatively high incidence of follicular hyperplasia, C-cell hyperplasia and C-cell adenoma of the thyroid occurred in both sexes. However, the statistical tests for adenoma did not show sufficient evidence for association of the tumor with picloram administration. An increased incidence of hepatic neoplastic nodules (considered to be benign tumors) was observed in treated animals. In male rats, the lesion appeared in only three animals of the low-dose treatment group and was not significant when compared to controls. However, the trend was significantly dose-related in females (p = 0.016). The incidence in the high-dose group was significant (p = 0.014) when compared with that of the pooled control group. The incidences of foci of cellular alteration of the liver were: female rats — matched controls 0/10, low-dose 8/50, high-dose 18/49; male rats — matched controls 0/10, low-dose 12/49, high-dose 5/49. Thus, there is evidence that picloram induced benign neoplastic nodules in the livers of rats of both sexes, but especially those of the females. Subsequent laboratory review by the National Toxicology Program (NTP) has questioned the findings of this study because animals with exposure to known carcinogens were placed in the same room with these animals and cross-contamination might have occurred.

- In the same study, NCI (1978), 50 male and 50 female mice received picloram at 208 or 417, and 361 or 723 mg/kg/day, respectively. Body weights of mice were unaffected, and no consistent clinical signs attributable to treatment were reported during the first 6 months of the study, except isolated incidences of tremors and hyperactivity. Later, particularly in the second year, rough hair coats, diarrhea, pale mucous membranes, alopecia and abdominal distention occurred. No tumors were found in male or female mice or male rats at incidences that could be significantly related to treatment. It was concluded that picloram was not a carcinogen for $B6C3F_1$ mice.

- Dow (1986) retested picloram (94% pure) in a 2-year chronic feeding/oncogenicity study in Fischer 344 rats. Rats (50/sex/dose) were fed 0, 20, 60 or 200 mg/kg/day. Oncogenic effects above those of controls were absent in this study.

IV.   Quantification of Toxicological Effects

A. One-day Health Advisory

No information was found in the available literature that was suitable for determination of the One-day HA value for picloram. It is therefore recommended that the Ten-day HA value for a 10-kg child (20 mg/L, calculated below) be used at this time as a conservative estimate of the One-day HA value.

B. Ten-day Health Advisory

The 7- to 14-day study in dogs by Dow (1981b) has been selected to serve as the basis for the Ten-day HA value for picloram because dogs appear to be the most sensitive species. Doses of 200, 400 or 800 mg/kg/day were used and the dose of 200 mg/kg/day was identified as the NOAEL for short-term exposures based on reduced food intake. Other short-term studies include a 7-day study in dogs by Dow (1981a) with a NOAEL of 400 mg/kg/day (the lowest dose tested) and a 32-day study in mice by Dow (1980) with a NOAEL of 900 mg/kg/day.

Using a NOAEL of 200 mg/kg/day, the Ten-day HA for a 10-kg child is calculated as follows:

$$\text{Ten-day HA} = \frac{(200 \text{ mg/kg/day}) \ (10 \text{ kg})}{(100) \ (1 \text{ L/day})} = 20 \text{ mg/L} \ (20{,}000 \ \mu g/L)$$

C. Longer-term Health Advisory

The study by Dow (1983) has been selected to serve as the basis for the Longer-term HA value for picloram because dogs have been shown to be the species most sensitive to picloram. In this study, picloram was fed for 6 months to beagle dogs (6/sex/group) in the diet at dose levels of 0, 7, 35 or 175 mg/kg/day. At 175 mg/kg/day, the following adverse effects were observed in both male and female dogs: decreased body weight gain, food consumption and alanine transaminase levels; and increased alkaline phosphatase levels, absolute liver weight and relative liver weight. At 35 mg/kg/day, increased absolute and relative liver weights were noted in males. No compound-related effects were detected in females at 35 mg/kg/day or in males or females at 7 mg/kg/day. Based on these data, 7 mg/kg/day was identified as the NOAEL for dogs for a 6-month exposure.

Using this study, the Longer-term HA for a 10-kg child is calculated as follows:

$$\text{Longer-term HA} = \frac{(7 \text{ mg/kg/day}) (10)}{(100) (1 \text{ L/day})} = 0.7 \text{ mg/L} (700 \text{ } \mu\text{g/L})$$

The Longer-term HA for a 70-kg adult is calculated as follows:

$$\text{Longer-term HA} = \frac{(7 \text{ mg/kg/day}) (70)}{(100) (2 \text{ L/day})} = 2.45 \text{ mg/L} (2,000 \text{ } \mu\text{g/L})$$

D. Lifetime Health Advisory

The study by Dow (1983), chosen for the Longer-term Health Advisory, has also been chosen to calculate the Lifetime HA value for picloram. In this study, picloram was fed for 6 months to beagle dogs (6/sex/group) in the diet at dose levels of 0, 7, 35 or 175 mg/kg/day. At 175 mg/kg/day, the following adverse effects were observed in both male and female dogs: decreased body alkaline phosphatase levels, absolute liver weight and relative liver weight. At 35 mg/kg/day, increased absolute and relative liver weights were noted in males. No compound-related effects were detected in females at 35 mg/kg/day or in males or females at 7 mg/kg/day. Based on these data, 7 mg/kg/day was identified as the NOAEL for dogs for a 6-month exposure. The results of this study were chosen, even though they reflect less than lifetime exposure, because the results indicate that the dog is more sensitive than the rat in a long-term exposure. Therefore, the Lifetime HA for picloram is determined as follows:

Step 1: Determination of the Reference Dose (RfD)

$$\text{RfD} = \frac{(7 \text{ mg/kg/day})}{(100)} = 0.07 \text{ mg/kg/day}$$

Step 2: Determination of the Drinking Water Equivalent Level (DWEL)

$$\text{DWEL} = \frac{(0.07 \text{ mg/kg/day}) (70)}{(2 \text{ L/day})} = 2.45 \text{ mg/L} (2,000 \text{ } \mu\text{g/L})$$

Step 3: Determination of the Lifetime Health Advisory

$$\text{Lifetime HA} = (2.45 \text{ mg/L}) (20\%) = 0.49 \text{ mg/L} (500 \text{ } \mu\text{g/L})$$

E. Evaluation of Carcinogenic Potential

- The National Cancer Institute conducted studies on the carcinogenic potential of picloram in rats and mice (NCI, 1978; this study was also reviewed by Reuber, 1981). In the study with mice, there was no indication of an oncogenic response from dietary exposure which included levels of more than 5,000 ppm picloram (723 mg/kg/day) for the greater part of their lifetime. The rat study, however, was negative for oncogenic effects in males, while female rats exhibited a statistically significant increase in neoplastic nodules in the liver. On a time-weighted average, exposures ranged up to 14,875 ppm (743 mg/kg/day) picloram in the diet. Results of the rat study are questionable since possible cross-contamination may have occurred.

- The International Agency for Research on Cancer has not evaluated the carcinogenic potential of picloram.

- Applying the criteria described in EPA's guidelines for assessment of carcinogenic risk (U.S. EPA, 1986a), picloram may be classified in Group D: not classifiable. This group is generally used for substances with inadequate human and animal evidence of carcinogenicity or for which no data are available.

V.   Other Criteria, Guidance and Standards

- The U.S. EPA Office of Pesticide Programs has set an RfD for picloram at 0.07 mg/kg/day (U.S. EPA, 1986b).

- Tolerances have been established for picloram in or on raw agricultural commodities (U.S. EPA, 1986c).

- The National Academy of Sciences (NAS, 1983) has calculated a chronic Suggested-No-Adverse-Response Level (SNARL) of 1.05 mg/L for picloram. An uncertainty factor of 1,000 was used because the issue of carcinogenicity had not yet been resolved and also because the Johnson (1971) study used by NAS does not provide enough information for a complete judgment of its adequacy.

VI.   Analytical Methods

- Analysis of picloram is by a gas chromatographic (GC) method applicable to the determination of certain chlorinated acid pesticides in water samples (U.S. EPA, 1988). In this method, approximately 1 liter of sample is acidified. The compounds are extracted with ethyl ether using a separatory funnel. The derivatives are hydrolyzed with potassium hydroxide and extraneous organic material is removed by a solvent wash. After acidification, the acids are extracted and converted to their methyl esters using diazomethane as the derivatizing agent. Excess reagent is removed, and the esters are determined by electron-capture

(EC) gas chromatography. This method has been validated in a single laboratory, and estimated detection limits have been determined for analytes in this method, including picloram. The estimated detection limit is 0.14 μg/L.

VII.   Treatment Technologies

- No information was found on treatment technologies capable of effectively removing picloram from contaminated water.

VIII.   References

Andersen, K.J., E.G. Leighty and M.T. Takahashi. 1972. Evaluation of herbicides for possible mutagenic properties. J. Agr. Food Chem. 20:649–658.

Bovey, R.W., M.I. Ketchersid and M.G. Merkle.* 1970. Comparison of salt and ester formulations of picloram. Weed Science. 18(4):447–451. MRID 00111466.

Carere, A., V.A. Ortali, G. Cardamone, A.M. Torracca and R. Raschetti. 1978. Microbiological mutagenicity studies of pesticides *in vitro*. Mutat. Res. 57:277–286.

Dow.* 1980. Picloram: Results of a 32-day toxicity tolerance in feed in $B_6C_3F_1$ mice. (No Dow number). Received May 16, 1980. EPA Accession No. 247156. Midland, MI: Dow Chemical Laboratories.

Dow.* 1981a. Results of range-finding study of picloram (4-amino-3,5,6-trichloropicolinic acid) administered orally in gelatin capsules to beagle dogs. TXT:K-38323(24). Freeport, TX: Dow Chemical.

Dow.* 1981b. Results of a short-term palatability study of picloram (4-amino-3,5,6-trichloropicolinic acid) fed in the diet to beagle dogs. TXT:K-38323(25). Freeport, TX: Dow Chemical.

Dow.* 1983. Dow Chemical Co. Toxicology profile of Tordon herbicides. Form No. 137-1640-83. Agr. Pro. Dept.

Dow.* 1986. Picloram: a two-year dietary chronic toxicity-oncogenicity study in Fischer 344 rats. EPA Accession Nos. 261129–261133.

Ercegovich, C.D. and K.A. Rashid. 1977. Mutagenesis induced in mutant strains of *Salmonella typhimurium* by pesticides. Am. Chem. Soc. Abstr. 174:43.

Fisher, D.E., L.E. St. John, Jr., W.H. Gutenmann, D.G. Wagner and D.J. Lisk. 1965. Fate of Bonvel T, Toxynil, Tordon and Trifluorilin in the dairy cow. J. Dairy Sci. 48:1711–1715.

Frank, R., G.J. Sirons and B.D. Ripley. 1979. Herbicide contamination and decontamination of well waters in Ontario, Canada, 1969–78. Pest. Mon. J. 13(3):120–127.

Fryer, J.D., P.D. Smith and J.W. Ludwig. 1979. Long-term persistence of picloram in a sandy loam soil. J. Env. Qual. 8(1):83–86.

Gabriel, K.L. and B.A. Gross. 1964. Repeated insult patch test study with Dow Chemical Company TORDON 101. Received November 16, 1964. MRID 0004117.

Grover, R.* 1977. Mobility of dicamba, picloram, and 2,4-D in soil columns. Weed Science. 25:159–162. MRID 00095247.

Hamaker, J.W.* 1964. Decomposition of aqueous TORDON® solutions by sunlight. Seal Beach, CA: The Dow Chemical Company. Bioproducts Research. MRID 00111477.

Hamaker, J.W., C.R. Youngson and C.A.I. Goring.* 1967. Prediction of the persistence and activity of Tordon herbicide in soils under field conditions. Down to Earth. 23(2):30–36. MRID Nos. 00109132–00111430.

Hamaker, J.W.* 1975. Distribution of picloram in a high organic sediment-water system: uptake phase. R&D Rep. Ag-Org. Res. Midland, MI: Dow Chemical Laboratories. MRID 00069075.

Hamaker, J.W. 1976. The hydrolysis of picloram in buffered, distilled water. GS-1460. Walnut Creek, CA: Dow Chemical Co. Agr. Prods. Dept.

Helling, C.S.* 1971a. Pesticide Mobility in Soils. I. Parameters of thin-layer chromatography. Soil Sci. Soc. Amer. Proc. 35:732–736. MRID 00111516.

Helling, C.S.* 1971b. Pesticide mobility in soils. II. Applications of soil thin-layer chromatography. Soil Sci. Soc. Amer. Proc. 35:737–743. MRID 00044017.

Johnson, J.E. 1971. The public health implication of widespread use of the phenoxy herbicides and picloram. Bioscience. 21:899–905.

Kutschinski, A.H. and V. Riley. 1969. Residues in various tissues of steers fed 4-amino-3,5,6-trichloropicolinic acid. J. Agric. Food Chem. 17:283–287.

Leasure, J.K. and M.E. Getzander. 1964. A residues study on tissues from beef cattle fed diets containing Tordon herbicide. Unpublished report. GS-P 141. Midland, MI: Dow Chemical Laboratories. Reviewed in NRCC.

McCall, P.J. and T.K. Jeffries.* 1978. Aerobic and anaerobic soil degradation of ^{14}C-picloram. Agricultural Products R&D Report GH-C 1073. Midland, MI: Dow Chemical Laboratories.

McCollister, D.D., J.R. Copeland and F. Oyen.* 1967. Results of fertility and reproduction studies in rats maintained on diets containing TORDON* herbicide. Received January 24, 1967 under OF0863. CDL:094525-H. MRID 00041098. EPA Accession No. 091152. Midland, MI: Dow Chemical Company, Toxicology Research Laboratory.

Meikle, R.W.* 1973. Comparison of the decomposition rates of picloram and

4-amino-2,3,5-trichloropyridine in soil. Unpublished report. MRID 00037883.

Meikle, R.W., C.R. Youngson, R.T. Hedlund, C.A.I. Goring and W.W. Addington.* 1974. Decomposition of picloram by soil microorganisms: a proposed reaction sequence. Weed Science. 22:263–268. MRID 00111505.

Meikle, R.W., C.R. Youngson and R.T. Hedlund.* 1970. Decomposition of picloram in soil: effect of a pre-moistened soil. Report of The Dow Chemical Company. GS-1097.

Meister, R., ed. 1987. Farm chemicals handbook. Willoughby, OH: Meister Publishing Company.

Merkle, M.G., R.W. Bovey and F.S. Davis.* 1967. Factors affecting the persistence of picloram in soil. Agronomy Journal: 39:413–415. MRID 00111441.

Mensik, D.C., R.V. Johnston, M.N. Pinkerton and E.B. Whorten.* 1976. The cytogenic effects of picloram on the bone marrow of rats. Unpublished report. Freeport, TX: Dow Chemical. 11 pp.

NAS. 1983. National Academy of Sciences. Drinking water and health. Vol. 5. Washington, DC: National Academy Press. pp. 60–63.

NCI. 1978. National Cancer Institute. Bioassay of picloram for possible carcinogenicity. Technical Report Series No. 23. Washington, DC: Department of Health, Education and Welfare.

NIOSH. 1980. National Institute for Occupational Safety and Health. RTECS, Registry of toxic effects of chemical substances. Vol. 2. DHHS Publ. (NIOSH) 81–116. U.S. Department of Health and Human Services. p. 354.

Nolan, R.J., F.A. Smith, C.J. Mueller and T.C. Curl. 1980. Kinetics of ^{14}C-labeled picloram in male Fischer 344 rats. Unpublished report. Midland, MI: Dow Chemical Laboratories. 34 pp.

Nolan, R.J., N.L. Freshour, P.E. Kastl and J.H. Saunders. 1983. Pharmacokinetics of picloram in human volunteers. Toxicologist. 4:10.

Redemann, C.T. 1964. The metabolism of 4-amino-3,5,6-trichloropicolinic acid by the rat. Unpublished report. GS-623. Seal Beach, CA: The Dow Chemical Co. Reviewed in NRCC. (National Research Council. 1974. Picloram: the effects of its use as a herbicide on environmental quality. Ottawa, Canada. NRCC No. 13684.)

Redemann, C.T.* 1966. Photodecomposition rate studies of 4-amino-3,5,6-trichloropicolinic acid. Walnut Creek, CA: The Dow Chemical Company, Bioproducts Research.

Reuber, M.D. 1981. Carcinogenicity of picloram. J. Tox. Environ. Health. 7:207–222.

Scifres, C.J., H.G. McCall, R. Maxey and H. Tai. 1977. Residual properties of 2,4,5-T and picloram in sandy rangeland soils. J. Env. Qual. 6(1):36–42.

STORET. 1988. STORET Water Quality File. Office of Water. U.S. Environmental Protection Agency (data file search conducted in May, 1988).

Thompson, D.J., J.L. Emerson, R.J. Strebing, C.C. Gerbig and V.B. Robinson. 1972. Teratology and postnatal studies on 4-amino-3,5,6-trichloropicolinic acid (picloram) in the rat. Food Cosmet. Toxicol. 10:797–803.

Torracca, A.M., G. Cordamone, V. Ortali, A. Carere, R. Raschette and G. Ricciardi. 1976. Mutagenicity of pesticides as pure compounds and after metabolic activation with rat liver microsomes. Atti. Assoc. Genet. Ital. 21:28–29. (In Italian; abstract in English)

U.S. EPA. 1986a. U.S. Environmental Protection Agency. Guidelines for carcinogen risk assessment. Fed. Reg. 51(185):33992–34003. September 24.

U.S. EPA. 1986b. U.S. Environmental Protection Agency. Registration standard for picloram. Washington, DC: Office of Pesticide Programs.

U.S. EPA. 1986c. U.S. Environmental Protection Agency. Code of Federal Regulations. 40 CFR 180.292.

U.S. EPA. 1988. U.S. Environmental Protection Agency. U.S. EPA Method 515.1 — Determination of chlorinated acids in water by GC/ECD. Draft. April 14. Available from U.S. EPA's Environmental Monitoring and Support Laboratory, Cincinnati, OH.

Youngson, C.R.* 1966. Residues of Tordon in soils from fields treated for selective weed control with tordon herbicide. Report. Walnut Creek, CA: The Dow Chemical Company, Bioproducts Research. MRID 00044023.

Youngson, C.R., and C.A.I. Goring.* 1967. Decomposition of Tordon herbicides in water and soil. Research Report. GS-850. Walnut Creek, CA: The Dow Chemical Company, Bioproducts Research. MRID 00111415.

Youngson, C.R.* 1968. Effect of source and depth of water and concentration of 4-amino-3,5,6-trichloropicolinic acid on rate of photodecomposition by sunlight. The Dow Chemical Company. Agricultural Products Research, Walnut Creek, CA. MRID 00059425.

---

* Confidential Business Information submitted to the Office of Pesticide Programs.

Prometon

I. General Information and Properties

A. CAS No. 1610–18–0

B. Structural Formula OCH$_3$

2,4-Bis(isopropylamino)-6-methoxy-s-triazine

C. Synonyms

- Prometon; Gesafram 50; Ontracic 800; Primatol 25E; Pramitol; Methoxypropazine (Meister, 1983).

D. Uses

- Prometon is a nonselective herbicide that controls most perennial broadleaf weeds and grasses (Meister, 1983).

E. Properties (Meister, 1983; TDB, 1985; CHEMLAB, 1985)

Chemical Formula	C$_{10}$H$_{19}$N$_5$O
Molecular Weight	225.34
Physical State (25°C)	White crystals
Boiling Point	--
Melting Point	91 to 92°C
Density	1.088 g/cm^3
Vapor Pressure (20°C)	2.3×10^{-6} mm Hg
Specific Gravity	--
Water Solubility (20°C)	750 mg/L
Log Octanol/Water Partition Coefficient	–1.06 (calculated)
Taste Threshold	--

Odor Threshold   --
Conversion Factor   --

F. Occurrence

- Prometon has been found in 386 of 1,419 surface water samples ana- lyzed and in 36 of 746 ground-water samples (STORET, 1988). Samples were collected at 250 surface water locations and 250 ground-water loca- tions and prometon was found in 12 states. The 85th percentile of all nonzero samples was 0.6 μg/L in surface water and 50 μg/L in ground- water sources. The maximum concentration found was 8.5 μg/L in sur- face water and 250 μg/L in ground-water. This information is provided to give a general impression of the occurrence of this chemical in ground and surface waters as reported in the STORET database. The individual data points retrieved were used as they came from STORET and have not been confirmed as to their validity. STORET data are often not valid when individual numbers are used out of the context of the entire sam- pling regime, as they are here. Therefore, this information can only be used to form an impression of the intensity and location of sampling for a particular chemical.

- Prometon residues resulting from agricultural practice have been detected in California ground waters at 0.21 to 80 ppb (Eiden, 1987).

G. Environmental Fate

- Prometon is stable to hydrolysis at pH 5, 7 and 9 at 25°C for 40 days (Ciba-Geigy, 1985a).

- Prometon in aqueous solution was stable to natural sunlight for 2 weeks (Ciba-Geigy, 1985b).

- Prometon has the potential to leach through soil, based on adsorption/ desorption tests and soil thin-layer chromatography (TLC). K_ds for five soils were: 2.61 (sandy loam), 2.90 (silt loam), 2.40 (silty clay loam), 1.20 (silt loam) and 0.398 (sand); organic matter content ranged from 0.8 to 5% (Ciba-Geigy, 1985c).

- R_f values for soil thin-layer chromatography (TLC) plates of five soils put prometon in Class 4 (Very Mobile), Class 3 (Intermediately Mobile) and Class 2 (Low Mobility). Prometon was very mobile in a Mississippi silt loam and Plainfield sand, intermediately mobile in a Hagerstown silty clay loam and Dubuque silt loam and had low mobility in a Califor- nia sandy loam (Ciba-Geigy, 1985d).

- In field dissipation studies, prometon was shown to have a half-life > 459 to 1,123 days at 3 different sites. Residues were found at all depths sampled, down to 18 inches. There was no deeper sampling. At 2 out of 3 sites, dealkylated prometon was found at the 0- to 18-inch depth (Ciba-Geigy, 1986)

II.   Pharmacokinetics

A. Absorption

- Prometon is rapidly absorbed from the gastrointestinal tract. Based on the radioactivity recovered in the urine and feces, prometon is completely absorbed within 72 hours in the rat (Bakke et al., 1967).

B. Distribution

- Seventy-two hours after intragastric intubation of ^{14}C-prometon in rats, no detectable levels of radioactivity were found in any of the tissues examined (Bakke et al., 1967).

C. Metabolism

- Eleven metabolites of prometon have been identified in the urine of rats treated with ^{14}C-prometon. 2-Methoxy-4,6-diamino-S triazine and ammeline represented 14% and 31%, respectively, of the radiolabel excreted in the urine (Ciba-Geigy, 1971).

- Based on the metabolites formed, triazine ring cleavage apparently does not occur during prometon metabolism (Ciba-Geigy, 1971).

D. Excretion

- Excretion of prometon and/or its metabolites in rats was most rapid during the first 24 hours after administration of ^{14}C-prometon and decreased to trace amounts at 72 hours. The radioactivity was quantitatively excreted in the urine (91%) and feces (9%) within 72 hours after dosing with ^{14}C-prometon (Bakke et al., 1967).

III.   Health Effects

A. Humans

- No information on the health effects of prometon in humans was found in the available literature.

B. Animals

1. *Short-term Exposure*

- The acute oral LD_{50} value for prometon ranges from 1,750 to 2,980 mg/kg in rats and is 2,160 mg/kg in mice (Meister, 1983; NIOSH, 1985).

- The acute inhalation LC_{50} value in rats is > 3.6 mg/L for 4 hours (Meister, 1983).

- Long-Evans rats of both sexes (5/sex/dose) were fed a diet containing 0, 10, 30, 100, 300, 600, 1,000, 3,000, 6,000 or 10,000 ppm prometon [technical, 97% active ingredient (a.i.)] for 4 weeks (Killeen et al., 1976a). This corresponds to doses of 0, 0.5, 1.5, 5, 15, 30, 50, 150, 300 or 500 mg/kg/day, assuming 1 ppm in the diet corresponds to 0.05 mg/kg/day (Lehman, 1959). Rats fed 3,000 or more ppm prometon showed a reduction in body weight during the treatment period; at 6,000 or 10,000 ppm (300 or 500 mg/kg/day) the reduction in body weight was statistically significant ($p < 0.05$ and 0.01, respectively). At 1,000 ppm or less, mean body weight of both males and females were comparable to controls. Gross pathology performed at the time of sacrifice did not show any compound-related effects. The No-Observed-Adverse-Effect Level (NOAEL) and Lowest-Observed-Adverse-Effect Level (LOAEL) identified in this study are 3,000 and 6,000 ppm (150 and 300 mg/kg/day), respectively.

- Beagle dogs (1/sex/dose) were administered 100, 300 or 3,000 ppm prometon (technical) in the diet (2.5, 7.5 or 75 mg/kg/day, assuming 1 ppm in the diet is equivalent to 0.025 mg/kg/day; Lehman, 1959) for 2 weeks after which the 100- and 300-ppm doses were changed to 1,000 and 2,000 ppm (25 and 50 mg/kg/day) for the next 2 weeks (Killeen et al., 1976b). Dogs that consumed 3,000 ppm showed a decrease in body weight and food consumption. The body weight of the females receiving 1,000 or 2,000 ppm (25 or 50 mg/kg/day) was also decreased slightly; food consumption was also slightly lower for the females receiving 2,000 ppm prometon (50 mg/kg/day). At 300 ppm and less, the body weight and food consumption for both males and females were comparable to those of the controls. The NOAEL and LOAEL identified in this study are 300 and 1,000 ppm (7.5 and 25 mg/kg/day), respectively.

2. Dermal/Ocular Effects

- Prometon is a minimal dermal irritant (Meister, 1983). Barely perceptible erythema was observed in rabbits exposed to 500 mg prometon (97%) applied to one abraded and one intact site for 24 hours. At 2,000 mg/kg, mild edema and slight desquamation were also observed (Ciba-Giegy, 1976).

3. Long-term Exposure

- Sprague-Dawley rats (30/sex/group) were fed a diet containing technical prometon (98% active ingredient) at levels of 0, 10, 50, 100 or 300 ppm for 90 days (Johnson and Becci, 1982). Based on the assumption that 1 ppm in the diet of rats is equivalent to 0.05 mg/kg/day (Lehman, 1959), these doses correspond to approximately 0, 0.5, 2.5, 5 or 15 mg/kg/day. Although female rats exposed to 300 ppm showed an increase in mean absolute weight of the kidneys, this was considered of no toxicological significance, since the relative kidney to body weight ratios were not changed. The NOAEL identified in this study is, therefore, 300 ppm (15.0 mg/kg/day, the highest dose tested).

4. Reproductive Effects

- Prometon (technical, 98% a.i.) in corn oil was administered to Charles River rats (25/dose) via gavage at levels of 0, 36, 120 or 360 mg/kg/day from days 6 through 15 of gestation (Florek et al., 1981). Rats treated with 120 or 360 mg/kg/day gained less body weight than the controls during treatment; body weight gain in the 36-mg/kg/day group was similar to that of the controls. Rats in all dosage groups exhibited excessive salivation. Increased respiratory rate and lacrimation were also seen in the 360 mg/kg/day group. No effects on implantation; litter size; fetal viability, resorption, average fetal body weight, or gross external, soft-tissue or skeletal variation in the fetuses were observed at any dose level. This study identified a maternal NOAEL of 36 mg/kg/day and a maternal LOAEL of 120 mg/kg/day.

- New Zealand White rabbits (16/dose) administered prometon at dose levels of 0, 0.5, 3.5 or 24.5 mg/kg/day (98% a.i.) from days 6 through 30 of gestation showed reduced pregnancy rates at all dosage levels (Lightkep et al., 1982). Pregnancy occurred in 16, 13, 13 and 11 rabbits given 0, 0.5, 3.5 and 24.5 mg/kg/day, respectively. Anorexia and excess lacrimation were observed more frequently in the high-dose group.

Maternal body weight was significantly retarded during treatment in the 24.5-mg/kg/day group. The maternal NOAEL identified in this study is 3.5 mg/kg/day and the maternal LOAEL is 24.5 mg/kg/day.

5. *Developmental Effects*

- Florek et al. (1981) reported no effects on fetal viability, resorption, average fetal body weight, or gross external, soft-tissue or skeletal variations in the fetuses of Charles River rats (25/dose) administered prometon via gavage at levels of 0, 36, 120 or 360 mg/kg/day (98% a.i.) in corn oil. Maternal weight gain was depressed in rats administered 120 and 360 mg/kg/day. A teratogenic NOAEL of 360 mg/kg/day (the highest dose tested) and a maternal-toxicity NOAEL of 36 mg/kg/day were identified.

- Lightkep et al. (1982) observed no gross, soft-tissue or skeletal variations in fetuses of New Zealand White rabbits (16/dose) administered prometon at dose levels of 0, 0.5, 3.5 or 24.5 mg/kg/day (98% a.i.) on days 6 through 30 of gestation. A teratogenic NOAEL of 24.5 mg/kg/day (the highest dose tested) and a maternal-toxicity NOAEL of 3.5 mg/kg/day were identified.

6. *Mutagenicity*

- No information on the mutagenicity of prometon was found in the available literature.

7. *Carcinogenicity*

- No information on the carcinogenicity of prometon was found in the available literature.

IV.   Quantification of Toxicological Effects

A. One-day Health Advisory

No information was found in the available literature that was suitable for determination of the One-day HA value for prometon. It is therefore recommended that the Drinking Water Equivalent Level (DWEL) value adjusted for the 10-kg child (0.2 mg/L, calculated below) be used at this time as a conservative estimate of the One-day HA value.

B. Ten-day Health Advisory

Prometon has been the subject of several acute toxicity assays, including 4-week feeding studies in rats and dogs and developmental toxicity studies with rats and rabbits in which pregnant animals were dosed for 10 days during gestation (Killeen et al., 1976a; Florek et al., 1981; Killeen et al., 1976b; Lightkep et al., 1982). The only toxic effect consistently observed in these studies was reduced weight gain in treated animals. Although of appropriate duration, these studies were judged unacceptable for deriving the Ten-day HA; fluctuations in weight gain may not be appropriately sensitive end points of toxicity for the basis of the HA. For this reason, it is recommended that the DWEL, adjusted for a 10-kg child (0.2 mg/L, calculated below) be used as a conservative estimate of the Ten-day HA value for prometon.

C. Longer-term Health Advisory

The only species to be tested in subchronic studies of prometon toxicity was the rat. In the study by Johnson and Becci (1982), rats were fed a diet containing 0, 10, 50, 100 or 300 ppm prometon (0, 0.5, 2.5, 5 or 15 mg/kg/day) for 90 days. A NOAEL of 15 mg/kg/day (the highest dose tested) was identified. Rats were also tested in a 4-week feeding study in which a NOAEL of 150 mg/kg and a LOAEL of 300 mg/kg/day were identified (Killeen et al. 1976a). Although the NOAEL from the subchronic study can be used as a basis for the Longer-term HA, lower NOAELs have been identified in acute studies of other species (3.5 mg/kg/day, rabbit, Lightkep et al., 1982; 7.5 mg/kg/day, dog, Killeen et al., 1976b). It is therefore recommended that the DWEL (0.5 mg/L, calculated below) and the DWEL, adjusted for the 10-kg child (0.2 mg/L, calculated below), be used as conservative estimates of the Longer-term HA for the adult and child, respectively.

D. Lifetime Health Advisory

No suitable chronic or lifetime studies were available for the calculation of a Lifetime HA for prometon. Effects on body weight gain have been observed in acute studies at doses as low as 120 mg/kg/day for rats (Florek et al., 1981), 25 mg/kg/day for dogs (Killeen et al., 1976b) and 24.5 mg/kg/day for rabbits (Lightkep et al., 1982). One subchronic study was available (Johnson and Becci, 1982). In this study, rats were fed diets containing 0, 10, 50, 100, or 300 ppm prometon for 90 days. No toxic effects were observed at any of the dose levels tested, and a NOAEL of 15 mg/kg/day was identified. This value may be a conservative estimate of the NOAEL for rats; a NOAEL of 150 mg/kg was identified from the 4-week feeding study by Killeen et al. (1976a). Taking into consideration both the acute and subchronic test results, the study of

Johnson and Becci (1982) has been selected to serve as the basis for determination of the RfD.

Step 1: Determination of the Reference Dose (RfD)

$$RfD = \frac{(15 \text{ mg/kg/day})}{(1,000)} = 0.015 \text{ mg/kg/day}$$

Step 2: Determination of the Drinking Water Equivalent Level (DWEL)

$$DWEL = \frac{(0.015 \text{ mg/kg/day}) (70 \text{ kg})}{(2 \text{ L/day})} = 0.5 \text{ mg/L } (500 \text{ } \mu\text{g/L})$$

The DWEL, modified for the 10-kg child and rounded to one significant figure, is derived as follows:

$$DWEL_{child} = \frac{(0.015 \text{ mg/kg/day}) (10 \text{ kg})}{(1 \text{ L/day})} = 0.2 \text{ mg/L } (200 \text{ } \mu\text{g/L})$$

Step 3: Determination of the Lifetime Health Advisory

Lifetime HA = (0.5 mg/L) (20%) = 0.1 mg/L (100 μg/L)

E. Evaluation of Carcinogenic Potential

- No carcinogenicity studies were found in the literature searched.

- The International Agency for Research on Cancer has not evaluated the carcinogenic potential of prometon.

- Applying the criteria described in EPA's final guidelines for assessment of carcinogenic risk (U.S. EPA, 1986a), prometon may be classified in Group D: not classifiable. This category is for substances with inadequate animal evidence of carcinogenicity.

V. Other Criteria, Guidance and Standards

- No information was found in the available literature on other existing criteria, guidelines and standards pertaining to prometon.

VI. Analytical Methods

- Analysis of prometon is by a gas chromatographic (GC) method applicable to the determination of certain nitrogen-phosphorus containing pesticides in water samples (U.S. EPA, 1986b). In this method, approximately 1 liter of sample is extracted with methylene chloride. The extract is concentrated and the compounds are separated using capillary column

GC. Measurement is made using a nitrogen-phosphorus detector. This method has been validated in a single laboratory, and the estimated detection limit for prometon is 0.3 μg/L.

VII.   Treatment Technologies

- Whittaker (1980) experimentally determined the adsorption isotherms for prometon on granular activated carbon (GAC).

- One study (Rees and Au, 1979) reported 95% removal efficiency when prometon-contaminated water was passed over a 1 by 20 cm column packed with resin.

- Available data indicate that GAC adsorption and resin adsorption will remove prometon from water (Whittaker, 1980; Rees and Au, 1979). However, selection of an individual technology or a combination of technologies to attempt prometon removal from water must be based on a case-by-case technical evaluation and an assessment of the economics involved.

VIII.   References

Bakke, J.E., J.D. Robbins and V.J. Fcil. 1967. Metabolism of 2-chloro-4,6-bis(isopropylamino)-s-triazine(propazine) and 2-methoxy-4,6-bis(iso-propylamino)-s-triazine (prometon) in the rat. Balance study and urinary metabolite separation. J. Agr. Food Chem. 15(4):628–631.

CHEMLAB. 1985. The Chemical Information System, CIS, Inc. Cited in U.S. EPA. 1985. Pesticide survey chemical profile. Final report. Contract No. 68–01–6750. Washington, DC: Office of Drinking Water.

Ciba-Geigy.* 1971. Metabolism of s-triazine herbicides. Unpublished study. EPA Accession No. 55672.

Ciba-Geigy.* 1976. Acute toxicity studies with prometon technical (97%) — primary skin irritation test—albino rabbits. Industrial Bio-Text Laboratories, Inc. IBT No. 8530–09308. Unpublished study. EPA Accession No. 231815.

Ciba-Geigy. 1985a. Hydrolysis of prometon (Hazelton Study 6015–165). In: Environmental fate data required by special ground water data call-in. Greensboro, NC. May 30.

Ciba-Geigy. 1985b. Photolysis of prometon in aqueous solution under natural sunlight and artificial sunlight conditions (1972). In: Environmental fate data required by special ground water data call-in. Ciba-Geigy Report No. 72127. Greensboro, NC. May 30.

Ciba-Geigy. 1985c. The adsorption/desorption of radiolabeled prometon on representative agricultural soils (Hazelton Study 6015–164). In: Environmental fate data required by special ground water data call-in. Greensboro, NC. May 30.

Ciba-Geigy. 1985d. Mobility determination of prometon in soils by TLC (Hazelton Study No. 6015–167). In: Environmental fate data required by special ground water data call-in. Greensboro, NC. May 30.

Ciba-Geigy.* 1986. Field disposition studies in California, Nebraska and New York. Prepared by Daniel Sumner. August 21.

Eiden, C. 1987. Assessing the leaching potential of pesticides: national perspectives. Draft report prepared by the U.S. Environmental Protection Agency, Office of Pesticide Programs, Washington, DC.

Florek, C., G.D. Christin, M.S. Christin and E. Marshall.* 1981. Teratogenicity study of prometon technical in pregnant rats. Argus Project 203–003. Unpublished study. EPA Accession No. 129983.

Haley, S.* 1972. Report to Geigy Agricultural Chemicals, Division of Ciba-Geigy Corporation. Teratogenic study with prometon technical in albino rats. IBT No. B904. Unpublished study.

Killeen, J.C., Jr., W.E. Rinehart, S. Munulkin et al.* 1976a. A four-week range-finding study with technical prometon in rats. Project No. 76–1445. Unpublished study. EPA Accession No. 54308.

Killeen, J.C., Jr., W.E. Rinehart, S. Munulkin et al.* 1976b. A four-week range-finding study with technical prometon in beagle dogs. Project No. 76–1446. Unpublished study. EPA Accession No. 54309.

Johnson, W. and P. Becci.* 1982. 90-Day subchronic feeding study with prometon technical in Sprague-Dawley rats. FDRL Study No. 6805. Unpublished study. EPA Accession No. 129985.

Lehman, A.J. 1959. Appraisal of the safety of chemicals in foods, drugs and cosmetics. Assoc. Food Drug Off. U.S., Q. Bull.

Lightkep, G., M. Christian, G. Christian et al.* 1982. Teratogenic potential of prometon technical in New Zealand White rabbits. Segment II – evaluation. Project No. 203–002. Final report. Unpublished study. EPA Accession No. 129984.

Meister, R., ed. 1983. Farm chemicals handbook. Willoughby. OH: Meister Publishing Company.

NIOSH. 1985. National Institute for Occupational Safety and Health. Registry of toxic effects of chemical substances. National Library of Medicine Online File.

Rees, G.A.V. and L. Au. 1979. Use of XAD-2 macroreticular resin for the recovery of ambient trace levels of pesticides and industrial organic pollutants from water. Bull. Environ. Contam. Toxicol. 22(4/5):561–566.

STORET. 1988. STORET Water Quality File. Office of Water. U.S. Environmental Protection Agency (data file search conducted in May, 1988).

TDB.1985. Toxicology Data Book. MEDLARS II. National Library of Medicine's National Interactive Retrieval Service.

U.S. EPA.* 1985. U.S. Environmental Protection Agency. Prometon, EPA I.D.No. 100–544. Caswell No. 96. EPA Accession No. 259108.

U.S. EPA. 1986a. U.S. Environmental Protection Agency. Guidelines for carcinogen risk assessment. Fed. Reg. 51(185):33992–34003. September 24.

U.S. EPA. 1986b. U.S. Environmental Protection Agency. U.S. EPA Method 507 — Determination of nitrogen and phosphorus containing pesticides in water by GC/NPD. Draft. Available from U.S. EPA's Environmental Monitoring and Support Laboratory, Cincinnati, OH. January.

Whittaker, K.F. 1980. Adsorption of selected pesticides by activated carbon using isotherm and continuous flow column systems. Ph.D. Thesis. Purdue University.

---

* Confidential Business Information submitted to the Office of Pesticide Programs.

Pronamide

I. General Information and Properties

A. CAS No. 23950–58–5

B. Structural Formula

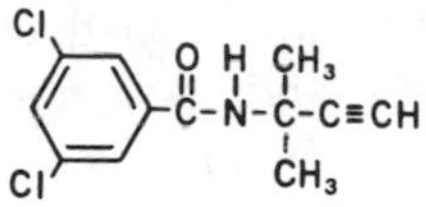

3,5-Dichloro(N-l,l-dimethyl-2-propynyl)benzamide

C. Synonyms

- Kerb®; Kerb®50W; Propyzamide; RH315 (Meister, 1983).

D. Uses

- Pronamide is used as an herbicide for pre- or postemergence weed and grass control in small, seeded legumes grown for forage or seed, southern turf, direct seeded or transplanted lettuce, endive, escarole, woody ornamentals, nursery stock and Christmas trees (Meister, 1983).

E. Properties (NIOSH, 1985; TDB, 1985)

Chemical Formula	$C_{12}H_{11}Cl_2ON$
Molecular Weight	256.14
Physical State (25°C)	White crystals
Boiling Point	--
Melting Point	154 to 156°C
Density	--
Vapor Pressure (25°C)	8.5×10^{-5} mm Hg
Specific Gravity	0.48 gm/cc
Water Solubility	15 ppm (mg/L)
Log Octanol/Water Partition Coefficient	3.05 to 3.27
Taste Threshold	--

Odor Threshold   --

Conversion Factor   --

F. Occurrence

- Pronamide has been found in 20 of 391 ground-water samples analyzed, all in California (STORET, 1988). No surface water samples were collected. Ground-water samples were collected at 391 locations. The concentration of all samples found to contain pronamide was 1.0 $\mu g/L$. STORET contains no information on surface water sampling for pronamide. This information is provided to give a general impression of the occurrence of this chemical in ground and surface waters as reported in the STORET database. The individual data points retrieved were used as they came from STORET and have not been confirmed as to their validity. STORET data are often not valid when individual numbers are used out of the context of the entire sampling regime, as they are here. Therefore, this information can only be used to form an impression of the intensity and location of sampling for a particular chemical.

G. Environmental Fate

- ^{14}C-Pronamide (100% radiopurity) at 1.5 ppm hydrolyzes very slowly (10% of applied material) in sterile, deionized water buffered to pH 5, 7 and 9 and incubated at 20°C for 28 days in the dark (Rohm and Haas Bristol Research Laboratories, 1973). The following minor hydrolysis products were identified: RH-24,644 (2-(3,5-dichlorophenyl)-4,4-dimethyl-5-methyleneoxazoline); RH-24,580 (3,5-dichloro-N-(1,1-dimethylacetonyl) benzamide); and RH-25,891 (2-(3,5-dichlorophenyl) 4,4-dimethyl-5-hydroxymethyl-oxazoline). Similar results were obtained in other hydrolysis studies (Rohm and Haas Bristol Research Laboratories, 1970).

- Pronamide has a half-life of 10 to 120 days in aerobic soils (Fisher, 1971; Walker, 1976; Walker and Thompson, 1977; Walker, 1978; Hance, 1979). Complete experimental conditions and purity were not specified, and/or a formulated product was applied. The degradation rate does not appear to depend upon soil texture. However, increasing soil temperature and, to a lesser extent, soil moisture and pH enhance pronamide degradation. The major degradates are RH-24,580 and RH-24,644. Soil sterilization greatly reduced the degradation rate of pronamide. Pronamide (at a recommended application rate of 0.5 to 2 lb/A) does not inhibit the growth or CO_2 evolution of bacteria and fungi (Lashen, 1970).

- Pronamide is moderately mobile in soils ranging in texture from loamy sand to clay based on preliminary soil column and adsorption/desorption tests (Walker and Thompson, 1977; Rohm and Haas Company, 1971; Fisher and Satterthwaite, 1971). The two major degradates of pronamide (RH-24,580 and RH-24,644) are mobile in sand and clay soils (Fisher, 1973). The mobility of pronamide and its two major degradates tends to decrease as the organic matter content, clay content and cation exchange capacity of the soil increases.

- The dissipation rate of pronamide from terrestrial field sites is quite variable, with half-lives ranging from 10 to 90 days (Benson, 1973; Walker, 1976; Hance et al., 1978a,b; Kostowska et al., 1978; Walker, 1978; Zandvoort et al., 1979). Data are insufficient to determine the effect, if any, of meteorological conditions or the role leaching may play in pronamide dissipation.

- The environmental fate of pronamide is the subject of several unpublished, undated reports (Cummings and Yih; Fisher and Cummings; Rohm and Haas Company; Satterthwaite and Fisher; Yih).

II. Pharmacokinetics

A. Absorption

- No information on the absorption of pronamide was found in the available literature. Data on the systemic toxicity of oral doses indicate that pronamide is absorbed following ingestion.

B. Distribution

- No information on the distribution of pronamide was found in the available literature.

C. Metabolism

- About 54 and 0.6% of the radioactivity were recovered as unmetabolized Kerb® in the feces and urine, respectively, of rats treated orally with (^{14}C-carbonyl)-pronamide (dose not specified) (Yih and Swithenbank, no date). The major metabolite in the feces was 2-(3,5-dichlorophenyl)-4,4-dimethyl-5-hydroxymethyloxazoline (15%), and the major metabolites in the urine were α-(3,5-dichlorobenzamido) isobutyric acid (22.4%), β-(3,5-dichlorobenzamido)-α-hydroxy-β-methyl-butyric acid (19.2%) and two unknown metabolites (24.1 and 16.7%).

- Unmetabolized Kerb® did not appear in the urine of cows treated orally with (^{14}C-carbonyl) Kerb®; the major metabolite was β-(3,5-dichlorobenzamido)-α-hydroxy-β-methyl-butyric acid (71.4%) (Yih and Swithenbank, no date).

D. Excretion

- After oral ingestion of radiolabeled Kerb® by rats, unmetabolized Kerb® accounted for 54 and 0.6% of the radioactivity recovered in feces and urine, respectively. In the cow, oral ingestion of Kerb® produced no unmetabolized Kerb® in the urine (Yih and Swithenbank, no date).

III.   Health Effects

A. Humans

- No information on the health effects of pronamide in humans was found in the available literature.

B. Animals

1. *Short-term Exposure*

- The acute oral LD$_{50}$ in rats for pronamide (technical) is in the range of 5,620 mg/kg bw (Meister, 1983) to 16,000 mg/kg bw (Powers, 1970).

2. *Dermal/Ocular Effects*

- Pronamide is not a primary dermal irritant to albino rabbits. In two separate studies, an aqueous paste of 500 mg pronamide [50% active ingredient (a.i.)] was applied to the skin of six rabbits for 24 hours (Powers, 1970; Regel, 1972). No signs of irritation were observed by Powers (1970). Twenty-four hours after exposure, Regel (1972) observed erythema, which subsided at 72 hours.

- Powers (1970) administered 100 mg of Kerb® (50% a.i.) in the conjunctival sac of 12 rabbits. Eye irritation and chemosis were noted at 24 hours but disappeared by day 7, as confirmed by fluorescein examination.

3. *Long-term Exposure*

- Charles River CD rats (10/sex/dose) were fed a diet containing 0, 50, 150, 450, 1,350 or 4,050 ppm pronamide (100% a.i.) for 3 months (Larson and Borzelleca, 1967a). These levels correspond to 0, 2.5, 7.5, 22.5, 67.5 or 202.5 mg/kg/day, respectively, assuming 1 ppm in the diet of rats is equivalent to 0.05 mg/kg/day (Lehman, 1959). Significant body weight depression was observed at the 4,050-ppm dose level. Initial significant body weight depression also occurred in the rats fed 1,350 ppm, but disappeared on continued feeding. At the 150-ppm dose, absolute and relative liver weights in females were significantly higher than in controls; no histological lesions were seen, and no dose-related trend was observed for this increase in relative liver weight. Individual data were not presented for organ weights and several other parameters, clinical observations were not presented and analytical determination of the test compound was not reported. The No-Observed-Adverse-Effect Level (NOAEL) identified in this study was 2.5 mg/kg/day.

- Beagle dogs (10 months old; 1/sex/dose) were fed a diet containing 0, 450, 1,350 or 4,050 ppm pronamide (100% a.i.) for 3 months (Larson and Borzelleca, 1967b). These levels correspond to approximate doses of 0, 11, 34 or 101 mg/kg/day, assuming 1 ppm in the diet of dogs is equivalent to 0.025 mg/kg/day (Lehman, 1959). A decrease in weight gain and food consumption and an increase in serum alkaline phosphatase, liver weight and liver-to-body weight ratios, as compared to controls, were seen in the animals dosed at 4,050 ppm. No histological changes were seen in the livers. The hematological and urinalysis findings were within normal ranges. The NOAEL identified in this study was 34 mg/kg/day.

- In a 2-year feeding study in beagle dogs (4/sex/dose) the addition of pronamide (97% a.i.) to the diet at dose levels of 0, 30, 100 or 300 ppm (0, 0.75, 2.5 or 7.5 mg/kg/day, assuming 1 ppm in the diet of dogs is equivalent to 0.025 mg/kg/day; Lehman, 1959) caused no adverse effects at any of the doses tested (Larson and Borzelleca, 1970b). A NOAEL of 7.5 mg/kg/day (the highest dose tested) was identified in this study.

- Smith (1974) administered Kerb® (97% a.i.) to 6-week-old (C57 BL16×C3H Anf)F$_1$ male and female mice (100/sex/dose) for 78 weeks at dietary pronamide concentrations of 0, 1,000 or 2,000 ppm (0, 150 or 300 mg/kg/day, assuming 1 ppm in the diet of mice is equivalent to 0.15 mg/kg/day; Lehman, 1959). Male and female mice that ingested 2,000 ppm gained significantly less weight (p < 0.05); males also exhibited adenomatous hyperplasia, degeneration, hyperplasia, intrahepatic cholestasis, necrosis and/or fatty changes of the liver. Liver weights were significantly increased over controls for males and females in both treatment groups. Based on this information, a Lowest-Observed-Adverse-Effect Level (LOAEL) of 1,000 ppm (150 mg/kg/day) was identified.

- Newberne et al. (1982) administered pronamide (94% a.i.) to male B6C3F$_1$ mice at dose levels of 0, 13, 70, 329 or 2,260 ppm (0, 2, 10, 49 or 340 mg/kg/day, assuming 1 ppm in the diet of mice is equivalent to 0.15 mg/kg/day; Lehman, 1959) for up to 24 months. Another group was fed 2,500 ppm pronamide for 6 months. The mean body weight of the mice fed 2,500 ppm was significantly depressed at 14 days and thereafter throughout the study. At the 24-month sacrifice, the mean body weight of this group was approximately 70% of the control group. Survival of the mice was unaffected. The highest dose level (2,500 ppm) resulted in liver lesions including bile duct hyperplasia, parenchymal cell hypertrophy, parenchymal cell necrosis, hyperplasia and cholestasis at all time periods examined. Based on this information, a NOAEL of 329 ppm (49 mg/kg/day) was identified.

4. *Reproductive Effects*

- Costlow and Kane (1985) administered pronamide (technical, 97% pure) to New Zealand White rabbits (18/dose) at doses of 0, 5, 20 or 80 mg/kg/day during gestation days 7 to 19. An increased incidence of gross and microscopic liver lesions, one maternal death, five abortions and a significant ($p < 0.05$) decrease in the maternal body weight gain were observed at the 80 mg/kg/day dose. At the 20 mg/kg/day dose, rabbits exhibited anorexia, vacuolation of hepatocytes and a slight decrease in body weight gain. There were no compound-related effects on the incidence of implantations, resorptions, fetal deaths or fetal body weight at any dose tested. A NOAEL of 20 mg/kg/day was identified based upon the absence of developmental/reproductive effects and a NOAEL of 5 mg/kg/day was identified based upon the absence of maternal toxicity observed at higher doses.

- In a three-generation reproduction study, 20 to 25 albino CD rats were fed a diet containing pronamide (RH-315; 97% a.i.) at dose levels of 0, 30, 100 or 300 ppm (Larson and Borzelleca, 1970c). Assuming 1 ppm in the diet is equivalent to 0.05 mg/kg/day, these levels correspond to doses of 0, 1.5, 5 or 15 mg/kg/day (Lehman, 1959). The authors reported no adverse reproductive effects in parents or pups, but individual animal data were not available to validate the above conclusions. Based on this information a NOAEL of 300 ppm (15 mg/kg/day, the highest dose tested) was identified.

5. Developmental Effects

- Costlow and Kane (1985) administered pronamide (technical, 97% pure) to New Zealand White rabbits (18/dose) at doses of 0, 5, 20 or 80 mg/kg/day (technical, 97% pure) during gestation days 7 to 19. There were no compound-related effects on the incidence of implantations, resorptions, fetal deaths or fetal body weight at any dose tested. The NOAEL in this study was 80 mg/kg/day based on the absence of developmental effects at any dose tested. The NOAELs for developmental effects and maternal toxicity, described above, are 20 and 5 mg/kg/day, respectively.

- In a study designed to evaluate fetal development, adult female rats (FDRL) were administered 0, 7.5 or 15 mg/kg/day pronamide by gavage in corn oil from days 6 through 16 of gestation (Vogin, 1971). No adverse effects were reported for the mean number of implantation sites, the number of live or dead fetuses or the mean fetal weight. The authors concluded that pronamide administered orally to rats at doses up to 15 mg/kg/day was not teratogenic. Individual animal data were not available to validate these conclusions, and, therefore, the study was not validated. Based on this information a NOAEL of 15 mg/kg/day (the highest dose tested) was identified.

6. Mutagenicity

- In a cytogenetic study, pronamide (Kerb®, analytical) administered by intragastric intubation at dose levels of 5, 50 or 500 mg/kg to rats did not produce any aberrations of the bone marrow chromosomes (Fabrizio, 1973).

7. Carcinogenicity

- In a study evaluating the carcinogenic potential of Kerb®, 6-week-old (C57 BL16 × C3H Anf)F_1 male and female mice (100/sex/dose) were fed pronamide (97% a.i.) in the diet at doses of 0, 1,000 or 2,000 ppm (0, 150 or 300 mg/kg/day, assuming 1 ppm in feed is equivalent to 0.15 mg/kg/day; Lehman, 1959) for 78 weeks (Smith, 1974). Male and female mice that ingested 2,000 ppm gained significantly less weight (p < 0.05); the animals also gained slightly less weight at the 1,000-ppm level, but the change was not significant. No increase in tumors was observed for female mice treated with pronamide over controls. For male mice, a total of 35 of the 99 animals in the high-dose group, 21 of the 100 animals in the low-dose group and 7 of the 100 animals in the control group developed hepatic neoplasms, of which 24, 18 and 7 were carcinomas in the high-dose, low-dose and control groups, respectively. A total of 28 of 99 male mice that ingested 2,000 ppm exhibited intrahepatic cholestasis, but did not have carcinomas of the liver.

- In a 2-year study in male B6C3F$_1$ mice (Newberne et al., 1982), technical pronamide (97%) was fed to the animals (63 animals/dose) at target doses of 0, 20, 100, 500 or 2,500 ppm. Actual measured levels were 0, 13, 70, 329, or 2,262 ppm (0, 2, 10, 49 or 340 mg/kg/day, assuming 1 ppm in the diet of mice is equivalent to 0.15 mg/kg/day; Lehman, 1959). Another group was fed a target dose of 2,500 ppm pronamide for 6 months. The mean body weight of mice fed 2,500 ppm was significantly depressed at 14 days and thereafter throughout the study. At the 24-month sacrifice, the mean body weight of this group was approximately 70% of the control group. Survival of the mice was unaffected. The highest dose (2,500 ppm) resulted in liver lesions, including bile duct hyperplasia, parenchymal cell hypertrophy, parenchymal cell necrosis, hyperplasia and cholestasis at all time periods examined. At 18 months, the 2,500-ppm target dose group had increased parenchymal cell neoplasms, but this was not statistically different from the controls. At 24 months, there was a statistically significant increased incidence of hepatic adenomas and carcinomas in the 500- and 2,500-ppm dose groups. The incidence of hepatic carcinomas was 5/63, 9/63, 12/63, 18/63 and 14/61 in the control, 20-ppm, 100-ppm, 500-ppm and 2,500-ppm groups, respectively. Thus, the liver appears to be the target organ for neoplasia. According to the authors, hypertrophy and hyperplasia are not uncommon in untreated older mice of this strain. However, pronamide tended to shift the onset of these lesions to an earlier age.

- Pronamide in the diet at dose levels of 0, 30, 100 or 300 ppm (0, 1.5, 5 or 15 mg/kg/day, assuming 1 ppm in the diet of rats is equivalent to 0.05 mg/kg/day; Lehman, 1959) fed to rats (30/sex/group) for 2 years did not produce any carcinogenic effects (Larson and Borzelleca, 1970a). However, doses used in this study were too low to assess the carcinogenic potential of pronamide.

IV.   Quantification of Toxicological Effects

A. One-day Health Advisory

No information was found in the available literature that was suitable for determination of the One-day HA value for pronamide. It is therefore recommended that the Longer-term HA value of 0.8 mg/L (800 μg/L) be used at this time as a conservative estimate of the One-day HA value for pronamide.

B. Ten-day Health Advisory

Toxicity from acute exposure to pronamide has been assessed in three reproductive/developmental toxicity studies. No effects were observed in rats exposed to pronamide via gavage (Vogin, 1971) or in feed (Larson and Borzelleca, 1967b) at doses as high as 15 mg/kg/day. No higher doses were tested in the rat, but higher doses have been tested in the rabbit (Costlow and Kane, 1985). In this study, New Zealand White rabbits were administered pronamide during gestation days 7 through 19 at dose levels of 0, 5, 20 or 80 mg/kg/day. Toxic effects observed at the highest dose include a statistically significant decrease in maternal body weight gain and an increased incidence of gross and microscopic liver lesions. Minor effects on body weight and liver toxicity were observed at the 20 mg/kg/day dose, and a NOAEL of 5 mg/kg/day was identified. This value is similar to the NOAEL used to derive the Longer-term HA (7.5 mg/kg/day; Larson and Borzelleca, 1970b), indicating little difference may exist between doses which cause acute and chronic toxicity. It is therefore recommended that the Longer-term HA value of 0.8 mg/L (800 μg/L), calculated below, be used at this time as a conservative estimate of the Ten-day HA value for pronamide.

C. Longer-term Health Advisory

Liver toxicity has been observed after both acute and chronic administration of pronamide to experimental animals. Adverse effects on the liver have been observed after acute exposure of rabbits to 80 mg/kg/day via gavage (Costlow and Kane, 1985), subchronic exposure of dogs to 90 mg/kg/day (Larson and Borzelleca, 1967b) and chronic feeding of 300 and 375 mg/kg/day to mice (Smith, 1974; Newberne et al, 1982). Data on rats are equivocal. Signs of hepatoxicity from subchronic exposure (increased liver weight after 90 days' exposure to 7.5 mg/kg/day; Larson and Borzelleca, 1970a) were not observed in a three-generation study with doses as high as 15 mg/kg/day (Larson and Borzelleca, 1970c).

These data indicate that there may be little difference between doses that cause acute, subchronic or chronic toxicity. Therefore, it is recommended that the basis for the Lifetime HA be used to derive a conservative estimate of the Longer-term HA. In this study (Larson and Borzelleca, 1970b), beagle dogs fed a diet containing pronamide at dose levels of 0, 30, 100 or 300 ppm (0, 0.75, 2.5 or 7.5 mg/kg/day) for 2 years showed no adverse effects at any of the doses tested. A NOAEL of 7.5 (the highest dose tested) was identified in this study.

The Longer-term HA for a 10-kg child is calculated as follows:

$$\text{Longer-term}_{child} = \frac{(7.5 \text{ mg/kg/day})(10 \text{ kg})}{(100)(1 \text{ L/day})} = 0.8 \text{ mg/L}$$

The Longer-term HA for a 70-kg adult is calculated as follows:

$$\text{Longer-term}_{adult} = \frac{(7.5 \text{ mg/kg/day})(70 \text{ kg})}{(100)(2 \text{ L/day})} = 3.0 \text{ mg/L}$$

D. Lifetime Health Advisory

Two-year chronic pronamide feeding studies have been performed in three species: the rat (Larson and Borzelleca, 1970a), mouse (Newberne et al., 1982) and dog (Larson and Borzelleca, 1970b). For the rat and dog studies, only low doses were used and no toxic effects were observed. The highest doses tested, 15 mg/kg/day (rat) and 7.5 mg/kg/day (dog), were identified as NOAELs for these studies. Because of deficiencies in the rat study (data on individual animals not provided, insufficient number of animals routinely monitored for clinical chemistry parameters), this study was not validated (U.S. EPA, 1985a), and is therefore not acceptable as the basis for the Lifetime HA value. The 2-year study performed on mice (Newberne et al., 1982) was rejected as the basis for the Lifetime HA because of the relative insensitivity of mice to pronamide compared to other species. The NOAEL of 75 mg/kg/day identified in this study was higher than doses causing liver toxicity in subchronic feeding studies in both the rat and dog (Larson and Borzelleca, 1967a,b). Taking all of these studies into consideration, the 2-year feeding study in dogs (Larson and Borzelleca, 1970b) was selected as the basis for determination of the Lifetime HA for pronamide. In this study, beagle dogs fed a diet containing pronamide at dose levels of 0, 30, 100 or 300 ppm (0, 0.75, 2.5 or 7.5 mg/ kg/day) for 2 years showed no adverse effects at any of the doses tested. A NOAEL of 7.5 mg/kg/day (the highest dose tested) was identified in this study.

Using a NOAEL of 7.5 mg/kg/day, the Lifetime HA is calculated as follows:

Step 1: Determination of the Reference Dose (RfD)

$$\text{RfD} = \frac{(7.5 \text{ mg/kg/day})}{(100)} = 0.075 \text{ mg/kg/day}$$

Step 2: Determination of the Drinking Water Equivalent Level (DWEL)

$$\text{DWEL} = \frac{(0.075 \text{ mg/kg/day})(70 \text{ kg})}{2 \text{ L/day}} = 2.6 \text{ mg/L}(2{,}600 \text{ } \mu g/L)$$

Step 3: Determination of the Lifetime Health Advisory

$$\text{Lifetime HA} = \frac{(2.6 \text{ mg/L}) (20\%)}{(10)} = 0.05 \text{ mg/L} (50 \text{ } \mu\text{g/L})$$

E. Evaluation of Carcinogenic Potential

- Applying the criteria described in EPA's final guidelines for assessment of carcinogenic risk (U.S. EPA, 1986), pronamide may be classified in Group C: possible human carcinogen. This category is for substances with limited evidence of carcinogenicity in animals in the absence of human data.

V.   Other Criteria, Guidance and Standards

- A Provisional Acceptable Daily Intake (PADI) of 0.0750 mg/kg/day and a calculated Theoretical Maximum Residue Concentration (TMRC) of 0.0409 mg/day that utilizes 0.91% of the PADI has been established (U.S. EPA, 1985a).

- Residue tolerances have been established for pronamide and its metabolites in or on raw agricultural commodities that range from 0.02 ppm to 10.0 ppm (U.S. EPA, 1985b).

VI.   Analytical Methods

- Analysis of pronamide is by a gas chromatographic (GC) method applicable to the determination of certain nitrogen-phosphorus containing pesticides in water samples (U.S. EPA, 1988). In this method, approximately 1 liter of sample is extracted with methylene chloride. The extract is concentrated and the compounds are separated using capillary column GC. Measurement is made using a nitrogen-phosphorus detector. This method has been validated in a single laboratory, and the estimated detection limit for pronamide is 0.76 μg/L.

VII.   Treatment Technologies

- Reverse osmosis (RO) is a promising treatment method for pesticide-contaminated water. As a general rule, organic compounds with molecular weights greater than 100 are candidates for removal by RO. Larson et al. (1982) report 99% removal efficiency of chlorinated pesticides by a thin-film composite polyamide membrane operating at a maximum pressure of 1,000 psi and at a maximum temperature of 113°F. More operational data are required, however, to specifically determine the effectiveness and feasibility of applying RO for the removal of pronamide from water. Also, membrane adsorption must be considered when evaluating RO performance in the treatment of pronamide-contaminated drinking water supplies.

VIII.   References

Benson, N.R. 1973. Efficacy, leaching and persistence of herbicides in apple orchards. Bulletin 863. Washington State University, College of Agriculture Research Center.

Costlow, R.D. and W.W. Kane.* 1985. Teratology study with Kerb technical (no clay) in rabbits. Unpublished study. No. 83R-026. Prepared and submitted by Rohm and Haas Company, Spring House, PA. Accession No. 256590.

Cummings, T.L. and R.Y. Yih. No date. Metabolism of Kerb (3,5-dichloro-N-(1,1-dimethyl-2-propynyl)benzamide) in different types of soil. Unpublished report. Prepared by Rohm and Haas Company, Philadelphia, PA. Memorandum Report No. 52.

Fabrizio, P.D.A.* 1973. Final report: cytogenetic study: Kerb analytical. Unpublished report. No. CDL:093756-D. Prepared by Litton Bionetics, Inc., Kensington, MD for Rohm and Haas Company, Philadelphia, PA. April 16. MRID 00038031.

Fisher, J.D. 1971. Dissipation and metabolism study of Kerb in soil and its effects on microbial activity. Unpublished report. Prepared by Rohm and Haas Company, Philadelphia, PA. Lab. 11 Research Report No. 11-229.

Fisher J.D. 1973. Soil leaching study with Kerb degradation products RH-24,580 and RH-24,644. Unpublished report. Prepared by Rohm and Haas Company, Philadelphia, PA. Tech. Report No. 3923-73-4.

Fisher, J.D. and T.L. Cummings. No date. Biodegradation study of carbonyl-^{14}C-Kerb and ring-^{14}C-3,5-dichlorobenzoate in a semicontinuous activated sludge test. Unpublished study. Prepared by Rohm and Haas Company, Philadelphia, PA. Report No. 16.

Fisher, J.D. and S.T. Satterthwaite. 1971. Leaching and metabolism studies of ^{14}C-Kerb in soils. Unpublished report. Prepared by Rohm and Haas Company, Philadelphia, PA. Lab. 11 Research Report No. 11-228.

Hance, R.J. 1979. Effect of pH on the degradation of atrazine, dichlorprop, linuron and propyzamide in soil. Pestic. Sci. 10(1):83-63.

Hance, R.J., P.D. Smith, T.H. Byast and E.G. Cotterill. 1978a. Effects of cultivation on the persistence and phytotoxicity of atrazine and propyzamide. Proc. Br. Crop Prot. Conf. Weeds. 14(2):541–547.

Hance, R.J., P.D. Smith, E.G. Cotterill and D.C. Reid. 1978b. Herbicide persistence: effects of plant cover, previous history of the soil and cultivation. Med. Fac. Landbouww. Rijksuniv. Gent. 43(2):1127–1134.

Kostowska, B., J. Rola and H. Slawinska. 1978. Decomposition dynamics of propyzamide in experiments with winter rape. Pamiet. Pulawski. 70:199–205.

Larson, P.S. and J.F. Borzelleca.* 1967a. Toxicologic study on the effect of adding RH-315 to the diet of rats for a period of three months. Unpublished study. No. CDL:091422-D. Prepared by the Medical College of Virginia, Dept. of Pharmacology, for Rohm and Haas Company, Philadelphia, PA. November 27. MRID 00085506.

Larson, P.S. and J.F. Borzelleca.* 1967b. Toxicologic study on the effect of adding RH-315 to the diet of beagle dogs for a period of three months. Unpublished study. No. CDL:091422-E. Prepared by the Medical College of Virginia, Dept. of Pharmacology, for Rohm and Haas Company, Philadelphia, PA. November 22. MRID 00085507.

Larson, P.S. and J.F. Borzelleca.* 1970a. Toxicologic study on the effect of adding RH-315 to the diet of rats for a period of two years. Unpublished study. No. CDL:004357-A. Prepared by the Medical College of Virginia, Dept. of Pharmacology, for Rohm and Haas Company, Philadelphia, PA. June 11. MRID 00133111.

Larson, P.S. and J.F. Borzelleca.* 1970b. Toxicologic study on the effect of adding RH-315 to the diet of beagle dogs for a period of two years. Unpublished study. No. CDL:090918-A. Prepared by the Medical College of Virginia, Dept. of Pharmacology, for Rohm and Haas Company, Philadelphia, PA. June 12. MRID 00107949.

Larson, P.S. and J.F. Borzelleca.* 1970c. Three-generation reproduction study on rats receiving RH-315 in their diets. Unpublished study. Prepared by the Medical College of Virginia, Dept. of Pharmacology, for Rohm and Haas Company, Philadelphia, PA. April 11. MRID 00107950.

Larson, R.E., P.S. Cartwright, P.K. Eriksson and R.J. Petersen. 1982. Applications of the FT-30 reverse osmosis membrane in metal finishing operations. Paper presented at Yokohama, Japan.

Lashen, E.S. 1970. Inhibitory effects of Kerb and Kerb transformation products on typical soil microorganisms. Unpublished report. Prepared by Rohm and Haas Company, Philadelphia, PA. Memorandum Report No. 22.

Lehman, A.J. 1959. Appraisal of the safety of chemicals in foods, drugs and cosmetics. Assoc. Food Drug Off. U.S., Q. Bull.

Meister, R., ed. 1983. Farm chemicals handbook. Willoughby, OH: Meister Publishing Co.

Newberne, P.M., R.G. McConnell and E.A. Essigmann.* 1982. Chronic study in the mouse. Final report. No. 81RC-157. Prepared by the MIT Animal Pathology Laboratory. Submitted by Rohm and Haas Company. August 10. Accession No. 248233.

NIOSH. 1985. National Institute for Occupational Safety and Health. Registry of Toxic Effects Chemical Substances.

Powers, M.B.* 1970. Final report: acute oral—rats; draize eye—rabbits; acute dermal—rabbits; acute inhalation exposure—rats. Unpublished study. Project No. 417-337. Prepared by TRW, Inc., Vienna, VA, for Rohm and Haas Company, Philadelphia, PA. October 6.

Regel, L.* 1972. Primary skin irritation study in albino rabbits. Unpublished study. No. 2060619. Prepared by WARF Institute, Inc., Madison, WI, for O.M. Scott & Sons, Marysville, OH. June 28. MRID 0001265.

Rohm and Haas Bristol Research Laboratories. 1970. Fate and persistence of Kerb (3,5-dichloro-N-(l,l-dimethyl-2-propynyl)benzamide) in aqueous systems. Unpublished report. Prepared by Rohm and Haas Company, Philadelphia, PA. RAR Report No. 597.

Rohm and Haas Bristol Research Laboratories. 1973. A study of the hydrolysis of the herbicide Kerb in water. Unpublished report. Prepared by Rohm and Haas Company, Philadelphia, PA. Lab. 23. Technical Report No. 23-73-8.

Rohm and Haas Company. No date. Field dissipation studies. Unpublished report. Philadelphia, PA: Rohm and Haas Company. Research Report No. XXXXVI.

Rohm and Haas Company. 1971. Soil adsorption studies with Kerb. Unpublished report. Philadelphia, PA: Rohm and Haas Company. Lab. 23 Technical Report No. 23-71-12.

Satterthwaite, S.T. and J.D. Fisher. No date. Photodecomposition of Kerb in water. Unpublished report. Prepared by Rohm and Haas Company, Philadelphia, PA. Lab. 11 Memorandum Report No. 7.

Satterthwaite, S.T.* 1977. ^{14}C-Kerb mouse feeding study. Unpublished study. No. 34H-77-3. Prepared and submitted by Rohm and Haas Company, Philadelphia, PA. February 19. MRID 0062604.

Smith, J.* 1974. Eighteen month study on the carcinogenic potential of Kerb (RH-315: pronamide) in mice. Unpublished study. Received September 16 under 3F1317. Prepared in cooperation with the Medical College of Virginia. Submitted by Rohm and Haas Company, Philadelphia, PA. CDL:094304-A. MRID 008201601.

STORET. 1988. STORET Water Quality File. Office of Water. U.S. Environmental Protection Agency (data file search conducted in May, 1988).

TDB. 1985. Toxicology Data Book. MEDLARS II. National Library of Medicine's National Interactive Retrieval Service.

U.S. EPA. 1985a. U.S. Environmental Protection Agency. Pronamide registration standard. Office of Pesticide Programs.

U.S. EPA. 1985b. U.S. Environmental Protection Agency. Code of Federal Regulations. 40 CFR 180.106. July 1. p. 252.

U.S. EPA. 1986. U.S. Environmental Protection Agency. Guidelines for carcinogen risk assessment. Fed. Reg. 51(185):33992–34003. September 24.

U.S. EPA. 1988. U.S. Environmental Protection Agency. U.S. EPA Method 507 — Determination of nitrogen- and phosphorus-containing pesticides in ground water by GC/NPD. April 15. Available from U.S. EPA's Environmental Monitoring and Support Laboratory, Cincinnati, OH.

Vogin, E.E.* 1971. Effects of RH-315 on the development of fetal rats. Unpublished study. No. 0512. Prepared by Food and Drug Research Laboratories, Inc., Maspeth, NY, for Rohm and Haas Company, Spring House, PA. October 22. MRID 00125789.

Walker, A. 1976. Simulation of herbicide persistence in soil. III. Propyzamide in different soil types. Pestic. Sci. 7:59–64.

Walker, A. 1978. Simulation of the persistence of eight soil-applied herbicides. Weed Res. 18:305–313.

Walker, A. and J.A. Thompson. 1977. The degradation of simazine, linuron and propyzamide in different soils. Weed Res. 17(6):399–405.

Yih, R.Y. and C. Swithenbank.* No date. Identification of metabolites of N-(1,1-dimethylpropynyl)-3,5-dichlorobenzamide in rat and cow urine and rat feces. Unpublished report. Prepared by Rohm and Haas Company, Spring House, PA. MRID 00107954.

Yih, R.Y. No date. Metabolism of N-(1,1-dimethylpropynyl)-3,5-dichlorobenzamide (Rh-315) in soil, plants and mammals. Unpublished report. Prepared by Rohm and Haas Company, Philadelphia, PA. Lab. 11 Research Report No. 11–210.

Zandvoort, R., D.C. van Dord, M. Leistra and J.G. Verlaat. 1979. The decline of propyzamide in soil under field conditions in the Netherlands. Weed Res. 19:157–164.

---

* Confidential Business Information submitted to the Office of Pesticide Programs.

Propachlor

I. General Information and Properties

A. CAS No. 1918–16–7

B. Structural Formula

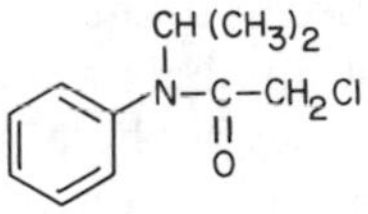

2-chloro-N-isopropylacetinilide

C. Synonyms

- Bexton; Prolex; Ramrod (Meister, 1983).

D. Uses

- Propachlor is used as a selective postemergence herbicide for control of many grasses and certain broadleaf weeds (Meister, 1983).

E. Properties (Rao and Davidson, 1982; HSDB, 1986)

Chemical Formula	$C_{11}H_{14}ClNO$
Molecular Weight	211.69
Physical State (room temp.)	White crystalline solid
Boiling Point	110°C at 0.03 mm HG
Melting Point	77 to 78°C
Density (25°C)	1.13 g/mL
Vapor Pressure (25°C)	7.9×10^{-5} mm Hg
Specific Gravity	1.242
Water Solubility (20°C)	580 mg/L
Log Octanol/Water Partition Coefficient	2.30
Taste Threshold	--
Odor Threshold	--
Conversion Factor	--

F. Occurrence

- Propachlor has been found in 34 of 1,690 surface water samples analyzed and in 2 of 99 ground-water samples (STORET, 1988). Samples were collected at 475 surface water locations and 94 ground-water locations, and propachlor was found in eight states. The 85th percentile of all nonzero samples was 2 μg/L in surface water and 0.12 μg/L in ground-water sources. The maximum concentration found was 10 μg/L in surface water and 0.12 μg/L in ground-water. This information is provided to give a general impression of the occurrence of this chemical in ground and surface waters as reported in the STORET database. The individual data points retrieved were used as they came from STORET and have not been confirmed as to their validity. STORET data are often not valid when individual numbers are used out of the context of the entire sampling regime, as they are here. Therefore, this information can only be used to form an impression of the intensity and location of sampling for a particular chemical.

G. Environmental Fate

- Propachlor is degraded in aerobic soils in the laboratory and in the field with half-lives of 2 to approximately 14 days, when the soils are treated with propachlor at recommended application rates. However, degradation was relatively slower in soil treated at 500 ppm, and 90% of the applied material remained after 21 days (Registrant CBI data).

- The major propachlor degradates produced under aerobic soil conditions are [(l-methylethyl)phenylamino]oxoacetic acid and [(2-methylethyl)phenylamino]-2-oxoethane sulfonic acid. These degradates are recalcitrant to further degradation in soil under anaerobic conditions. The half-life of propachlor in anaerobic soil is < 4 days (Registrant CBI data).

- Propachlor degrades very slowly (84.5% remaining after 30 days) in lake water (Registrant CBI data).

- Propachlor is moderately mobile to very mobile in soils ranging in texture from sand to clay. Mobility appears to be correlated with clay content and to a lesser degree with organic matter content and CEC. Aged ^{14}C-propachlor residues were mobile in a silt loam soil (Registrant CBI data).

- The rapid degradation of low levels of propachlor in soils is expected to result in a low potential for ground-water contamination by propachlor degradates. ^{14}C-Propachlor residues are taken up by rotated corn planted under confined conditions; < 3% of the radioactivity remains in soil at the time of planting (Registrant CBI data).

II. Pharmacokinetics

A. Absorption

- No direct data on rate of gastrointestinal absorption of propachlor were found in the available literature. Based on recovery studies, propachlor appears to be rapidly absorbed by the oral route of administration. An estimated 68% of a single dose of 10 mg of ring-labeled ^{14}C-propachlor administered to 12 rats was recovered in urine 56 hours after compound administration (Malik, 1986). These results are supported by other studies in which 54 to 64% (Lamoureux and Davison, 1975) and 68.8% (Bakke et al., 1980) of the administered dose was recovered in urine 24 hours and 48 hours after dose administration, respectively.

B. Distribution

- Fifty-six hours following oral administration of 10 mg of ring-labeled ^{14}C-propachlor (purity not specified) to rats, no detectable levels of radioactivity were identified in any tissue samples (Malik, 1986).

C. Metabolism

- Metabolism of propachlor occurs by initial glutathione conjugation followed by conversion via the mercapturic acid pathway; oxidative metabolism also occurs (Lamoureux and Davison, 1975; Malik, 1986).

- Eleven urinary metabolites have been identified as the result of propachlor metabolism in rats. The primary metabolic end products of propachlor are mercapturic acid and glucuronic acid conjugates (approximately 20 to 25%), methyl sulfones (30 to 35%), and phenols and alcohols (Lamoureux and Davison, 1975; Malik, 1986).

D. Excretion

- Propachlor (purity not specified) was excreted in the form of metabolites in the urine (68%) and feces (19%) of rats within 56 hours after dosing with ring-labeled ^{14}C-propachlor. Methyl sulfonyl metabolites accounted for 30 to 35% of the administered dose (Malik, 1986).

- In studies with germ-free rats, 98.6% of the administered dose (not specified) for propachlor (purity not specified) was identified in the urine (68.8%) and feces (32.1%) within 48 hours. The major metabolite was mercapturic acid conjugate, which accounted for 66.8% of the administered dose (Bakke et al., 1980).

III.   Health Effects

A. Humans

1. *Short-term Exposure*

- Schubert (1979) reported a case study in which occupational exposure to propachlor for 8 days resulted in erythemato-papulous (red pimply) contact eczema on the hands and forearms.

2. *Long-term Exposure*

- No information was found in the available literature on the health effects of long-term exposure to propachlor in humans.

B. Animals

1. *Short-term Exposure*

- The acute oral LD_{50} values for technical-grade (approximately 96.5%) and wettable powder (WP) (65%) propachlor range from 1,200 to 4,000 mg/kg in rats. Technical-grade and wettable powder propachlor both produced a low LD_{50} value of 1,200 mg/kg (Keeler et al., 1976; Heenehan et al., 1979; Auletta and Rinehart, 1979; Monsanto Company, (no date).

- Beagle dogs (2/sex/dose) were administered propachlor (65% WP) in the diet for 90 days at dose levels of 0, 1.3, 13.3 or 133.3 mg/kg/day (Wazeter et al., 1964). Body weight, survival rates, food consumption, behavior, general appearance, hematology, biochemical indices, urinalysis, histopathology and gross pathology were comparable in treated and control animals. The No-Observed-Adverse-Effect Level (NOAEL) identified for this study is 133.3 mg/kg/day (the highest dose tested).

- Naylor and Ruecker (1985) fed propachlor [96.1% active ingredient (a.i.)] to beagle dogs (6/sex/dose) in the diet for 90 days at dose levels of 0, 100, 500 or 1,500 ppm. Based on the assumption that 1 ppm in food is equivalent to 0.025 mg/kg/day (Lehman, 1959), these doses are equivalent to 0, 2.5, 12.5 or 37.5 mg/kg/day. Clinical signs, ophthalmoscopic, clinicopathologic, gross pathologic and histopathologic effects were comparable for treated and control groups. The reduction in food consumption and concomitant reductions in body weight gain at all test levels were considered by the author to be due to poor diet palatability. Based on these responses, a NOAEL of 1,500 ppm (37.5 mg/kg/day)— the highest dose tested—was identified.

2. Dermal/Ocular Effects

- The acute dermal LD_{50} value of technical propachlor and WP (65% propachlor) in the rabbit ranges from 380 mg/kg to 20 g/kg (Keeler et al., 1976; Monsanto Company, no date; Braun and Rinehart, 1978). Wettable powder produced the lowest LD_{50} in rabbits (380 mg/kg); the lowest LD_{50} produced in rabbits by technical propachlor was between 1,000 and 1,260 mg/kg.

- Propachlor (94.5% a.i.) (1 g/mL) applied to abraded and intact skin of New Zealand White rabbits (3/sex) for 24 hours produced erythema and slight edema at treated sites 72 hours posttreatment (Heenehan et al., 1979).

- Heenehan et al. (1979) instilled single applications (0.1 cc) of propachlor into one eye of tested New Zealand White rabbits for 30 seconds. Corneal opacity with stippling and ulceration, slight iris irritation, conjunctival redness, chemosis, discharge and necrosis were reported at 14 days. Similar responses were reported by Keeler et al. (1976) for a corresponding observation period and by Auletta (1984) during 3 to 21 days posttreatment.

3. Long-term Exposure

- Albino rats (25/sex/dose) administered 0, 1.3, 13.3 or 133.3 mg/kg/day propachlor (65% WP = 65% a.i.) in the diet for 90 days showed decreased weight gain (10 to 12% less than control levels) and increased liver weights in both sexes (10% greater than control levels) at 133.3 mg/kg/day (the highest dose tested) (Wazeter et al., 1964). The body and liver weights of rats of both sexes that received the low dose and mid dose were comparable to control levels. Survival, biochemical indices, hematology, urinalysis, gross pathology and histopathology did not differ significantly between treated and control groups. The NOAEL identified in this study is 13.3 mg/kg/day. The Lowest-Observed-Adverse Effect Level (LOAEL) is 133.3 mg/kg/day (the highest dose tested).

- Reyna et al. (1984a) administered propachlor (96.1% a.i.) to rats (30/sex/dose) in the diet for 90 days at mean dose levels of 0, 240, 1,100 or 6,200 ppm. Assuming that 1 ppm is equivalent to 0.05 mg/kg/day, these concentrations correspond to 0, 12, 55 or 310 mg/kg/day (Lehman, 1959). Body weights and food consumption were significantly decreased (no p value specified) at 55 mg/kg/day and 310 mg/kg/day in both sexes. Final body weights for females were 7 and 36% less than controls at the mid- and high-dose levels, respectively. In males, final body weights were 8 and 59% less than control levels for mid- and high-dose levels, respectively. However, histopathological examination showed no changes. Mid- and high-dose levels produced increased platelet counts, decreased white blood cell counts and mild liver cell dysfunction. Mild hypochromic, microcytic anemia was reported at the high dose. A NOAEL of 12 mg/kg/day can be identified for this study.

- Albino mice (30/sex/dose) were fed propachlor (96.1% a.i.) in the diet for 90 days at mean dose levels of 0, 385, 1,121 or 3,861 ppm (Reyna et al., 1984b). Based on the assumption that 1 ppm in food is equivalent to 0.15 mg/kg/day (Lehman, 1959), these doses correspond to 0, 58, 168 or 579 mg/kg/day. Reduced body weight gain, decreased white blood cell count, liver and kidney weight changes and increased incidences of centrolobular hepatocellular enlargement were reported at the mid (168 mg/kg/day) and high (579 mg/kg/day) doses when compared to controls. Based on these responses, a NOAEL of 385 ppm (58 mg/kg/day) can be identified.

4. Reproductive Effects

- No information on the reproductive effects of propachlor was found in the available literature.

5. Developmental Effects

- Miller (1983) reported no signs of maternal toxicity in New Zealand White female rabbits (16/dose) that were administered propachlor (96.5%) by gavage at doses of 0, 5, 15 or 50 mg/kg/day on days 7 to 19 of gestation. Statistically significant increases in mean implantation loss with corresponding decreases in the mean number of viable fetuses were reported at 15 and 50 mg/kg/day when compared to controls. Two low-dose and one mid-dose rabbit aborted on gestation days 22 to 25. These effects, however, do not appear to be treatment-related since no abortions occurred in the high-dose animals. No treatment-related effects were present in the 5 mg/kg/day group (the lowest dose tested). The authors reported that the maternal and embryonic NOAELs were 50 and 5 mg/kg/day, respectively.

- Schardein et al. (1982) administered technical propachlor by gavage to rats (25/dose) at dose levels of 0, 20, 60 or 200 mg/kg/day during days 6 to 19 of gestation. There were no adverse fetotoxic or maternal effects reported at any dose level. Based on this information, the NOAEL identified in this study was 200 mg/kg/day (the highest dose tested).

6. *Mutagenicity*

- Technical propachlor was not genotoxic in assays of *Salmonella typhimurium* with or without plant and animal activation; however, genotoxic activity was reported in yeast assays (*Saccharomyces cerevisiae*) at 1.3×10^{-3} M and 3 mg per plate after plant activation (Plewa et al., 1984).

- In a cytogenic study, propachlor administered by intraperitoneal injection at dose levels of 0.05, 0.2 or 1.0 mg/kg to Fischer 344 rats did not induce chromosomal aberrations in bone marrow cells (Ernst and Blazak, 1985).

- Gene mutation was not detected in assays employing Chinese hamster ovary (CHO) cells at levels up to 60 μg/mL. Propachlor was cytotoxic at levels of 40 μg/mL and higher (Flowers, 1985). Primary rat hepatocytes exposed to 0.1 to 5,000 μg/mL technical-grade propachlor showed no effect on unscheduled DNA synthesis when compared to controls (Steinmetz and Mirsalis, 1984).

7. *Carcinogenicity*

- No information was found in the available literature to evaluate the carcinogenic potential of propachlor. However, several chemicals analogous to this compound, i.e., alachlor and acetochlor, were found to be oncogenic in two animal species.

IV. Quantification of Toxicological Effects

A. One-day Health Advisory

No information was found in the available literature that was suitable for determination of the One-day HA value for propachlor. It is therefore recommended that the Ten-day HA value for the 10-kg child (0.5 mg/L, calculated below) be used at this time as a conservative estimate of the One-day HA value.

B. Ten-day Health Advisory

The developmental toxicity study in rabbits by Miller (1983) has been selected as the basis for determination of the Ten-day HA value for propachlor. Pregnant rabbits administered propachlor (96.5%) by gavage at a dose level of 5 mg/kg/day showed no clinical signs of toxicity in the adult animals and no reproductive or developmental effects in the fetuses. The study identified a NOAEL of 5 mg/kg/day. These results are contrasted by a developmental study reported by Schardein et al. (1982) in which rats were administered doses ranging from 20 to 200 mg/kg/day during gestation, with no adverse fetotoxic or maternal effects reported at any dose level. The NOAEL identified in that study was 200 mg/kg/day (the highest dose tested). Since the rabbit appears to be the more sensitive species, the NOAEL identified in the rabbit study will be used to derive the Ten-day HA.

Using a NOAEL of 5 mg/kg/day, the Ten-day HA for a 10-kg child is calculated as follows:

$$\text{Ten-day HA} = \frac{(5 \text{ mg/kg/day}) \ (10 \text{ kg})}{(100) \ (1 \text{ L/day})} = 0.5 \text{ mg/L} \ (500 \ \mu g/L)$$

C. Longer-term Health Advisory

While there are two 90-day studies suitable for developing a Longer-term HA, it was decided that it would be more appropriate to use the Reference Dose of 0.013 mg/kg/day, calculated under the Lifetime Health Advisory section, and adjust this number to protect a 10-kg child and a 70-kg adult, since acute data indicate toxicological concern at low levels (Miller, 1983). The resulting Longer-term HA thus becomes 0.1 mg/L and 0.5 mg/L for a 10-kg child and a 70-kg adult, respectively.

D. Lifetime Health Advisory

While there are no suitable studies to calculate a Lifetime HA, the 90-day study by Wazeter et al. (1964) has been selected to serve as the basis for determination of the Lifetime HA value for propachlor since it is the only data available of extended duration. Based on body and liver weight effects, a NOAEL of 13.3 mg/kg/day was identified. These results were further verified by the results of a similar study with rats conducted by Reyna et al. (1984a) in which a NOAEL of 12 mg/kg/day was identified.

Step 1: Determination of the Reference Dose (RfD)

$$\text{RfD} = \frac{(13.3 \text{ mg/kg/day})}{(1,000)} = 0.013 \text{ mg/kg/day} \ (10 \ \mu g/kg/day)$$

Step 2: Determination of the Drinking Water Level (DWEL)

$$\text{DWEL} = \frac{(0.013 \text{ mg/kg/day}) (70 \text{ kg})}{(2 \text{ L/day})} = 0.46 \text{ mg/L} (500 \text{ μg/L})$$

Step 3: Determination of the Lifetime Health Advisory

$$\text{Lifetime HA} = (0.46 \text{ mg/L}) (20\%) = 0.092 \text{ mg/L} (90 \text{ μg/L})$$

E. Evaluation of Carcinogenic Potential

• No studies on the carcinogenic potential of propachlor were found in the available literature. However, other structurally similar compounds such as alachlor and acetochlor have been found to be probable carcinogens.

• Applying the criteria described in EPA's final guidelines for assessment of carcinogenic risk (U.S. EPA, 1986), propachlor may be classified in Group D: not classified. This category is for substances with inadequate human and animal evidence of carcinogenicity.

V. Other Criteria, Guidance and Standards

• Residue tolerances ranging from 0.02 to 10.0 ppm have been established for propachlor in or on agricultural commodities (U.S. EPA, 1985).

• NAS (1977) has recommended an ADI of 0.1 mg/kg/day and a Suggested-No-Adverse-Effect Level (SNARL) of 0.7 mg/L, based on a NOAEL of 100 mg/kg/day in a rat study (duration of study not available).

VI. Analytical Methods

• Analysis of propachlor is by a gas chromatographic (GC) method applicable to the determination of certain chlorinated pesticides in water samples (U.S. EPA, 1988). In this method, approximately 1 liter of sample is extracted with methylene chloride. The extract is concentrated and the compounds are separated using capillary column GC. Measurement is made using an electron capture detector. This method has been validated in a single laboratory, and the estimated detection limit for analytes such as propachlor is 0.5 μg/L.

VII.   Treatment Technologies

- No data were found for the removal of propachlor from drinking water by conventional treatment or by activated carbon treatment.

- No data were found for the removal of propachlor from drinking water by aeration. However, the Henry's Coefficient can be estimated from available data on solubility (700 mg/L at 20°C) and vapor pressure (7.9×10^{-5} mm Hg at 25°C). Propachlor probably would not be amenable to aeration or air stripping because its Henry's Coefficient is approximately 3.8×10^{-9} m^3 atm Mol^{-1}. Baker and Johnson (1984) reported the results of water and pesticide volatilization from a waste disposal pit. Over a 2-year period, approximately 66.4 mg of propachlor evaporated for every liter of water which evaporated and only 8.3% of the propachlor was removed. These results support the assumption that aeration would not effectively remove propachlor from drinking water.

- Propachlor is similar in structure to alachlor and has similar physical properties. The effectiveness of various processes for removing propachlor would probably be similar to that for alachlor.

- Alachlor is amenable to the following processes:
    - GAC (Miltner and Fronk, 1985; DeFilippi et al., 1980).
    - PAC (Miltner and Fronk, 1985; Baker, 1983).
    - Ozonation (Miltner and Fronk. 1985).
    - Reverse osmosis (Miltner and Fronk, 1985).

- Chlorine and chlorine dioxide oxidation were partially effective in removing alachlor from drinking water (Miltner and Fronk, 1985).

- The following processes were not effective in removing alachlor from drinking water:
    - Diffused aeration (Miltner and Fronk, 1985).
    - Potassium permanganate oxidation (Miltner and Fronk, 1985).
    - Hydrogen peroxide oxidation (Miltner and Fronk, 1985).
    - Conventional treatment (Miltner and Fronk, 1985; Baker, 1983).

VIII.   References

Auletta, C. and W. Rinehart.* 1979. Acute oral toxicity in rats. Project No. 4891–77. BDN-77–431. Unpublished study. MRID 104342.

Auletta, C.* 1984. Eye irritation study in rabbits: propachlor. Project No. 5050–84. Unpublished study. Biodynamics, Inc. MRID 151787.

Baker, D. 1983. Herbicide contamination in municipal water supplies in northwestern Ohio. Final draft report. Prepared for Great Lakes National Program Office, U.S. Environmental Protection Agency, Tiffin, OH.

Baker, J.L. and L.A. Johnson. 1984. Water and pesticide volatilization from a waste disposal pit. Transactions of the American Society of Agricultural Engineers. 27:809–816. May/June.

Bakke, J., J. Gustafsson and B. Gustafsson. 1980. Metabolism of propachlor by the germ-free rat. Science. 210:433–435. October.

Braun, W. and W. Rinehart.* 1978. Acute dermal toxicity in rabbits [due to propachlor (technical)]. Project No. 4888–77. BDN-77-430. Unpublished study. Biodynamics, Inc. MRID 104351.

DeFilippi, R.P., V.J. Kyukonis, R.J. Robey and M. Modell. 1980. Supercritical fluid regeneration of activated carbon for adsorption of pesticides. EPA 600/2-80-054. Research Triangle Park, NC: U.S. Environmental Protection Agency.

Ernst, T. and W. Blazak.* 1985. An assessment of the mutagenic potential of propachlor utilizing the acute in vivo rat bone marrow cytogenetics assay (SR 84-180). Final report. SRI Project LSC-7405. Unpublished study. SRI International. MRID 00153940.

Flowers, L.* 1985. CHO/HGPRT gene mutation assay with propachlor. Final report. EWL 840083. Unpublished study. MRID 00153939.

Heenehan, P., W. Rinehart and W. Braun.* 1979. Acute oral toxicity study in rats. Project No. 4887–77. BDN-77-430. Biodynamics, Inc. MRID 104350.

HSDB. 1986. Hazardous Substances Database. Bethesda, MD: National Library of Medicine.

Keeler, P.A., D.J. Wroblewski and R.J. Kociba.* 1976. Acute toxicological properties and industrial handling: hazards of technical grade propachlor. Unpublished study. MRID 54786.

Lamoureaux, G. and K. Davison.* 1975. Mercapturic acid formation in the metabolism of propachlor, CDAA, Fluorodifen in the rats. Pesticide Biochem. Physiol. 5:497–506.

Lehman, A.J. 1959. Appraisal of the safety of chemicals in foods, drugs and cosmetics. Assoc. Food Drug Off. U.S., Q. Bull.

Malik, J.* 1986. Metabolism of propachlor in rats. Report No. MSL-5455. Job/Project No. 7815 (Summary). Unpublished study. MRID 157495.

Meister, R., ed. 1983. Farm chemicals handbook. Willoughby, OH: Meister Publishing Company.

Miller, L.* 1983. Teratology study in rabbits. (IR-82-224):401-190. International Research and Development Corporation. Unpublished study. MRID 00150936.

Miltner, R.J. and C.A. Fronk. 1985. Treatment of synthetic organic contaminants for Phase II regulations. Internal report. U.S. Environmental Protection Agency, Drinking Water Research Division. December.

Monsanto Company.* No date. Toxicology. Summary of studies 241666-C through 241666-E. Unpublished study. MRID 25527.

NAS. 1977. National Academy of Sciences. Drinking water and health. Washington, DC: National Academy Press.

Naylor, M. and F. Ruecker.* 1985. Subchronic study of propachlor administered in feed to dogs. DMEH Project No. ML-84-092. Unpublished study. MRID 00157852.

Plewa, M.J. et al. 1984. An evaluation of the genotoxic properties of herbicides following plant and animal activation. Mutat. Res. 136(3):233-246.

Rao, P.S.C. and J.M. Davidson. 1982. Retention and transformation of selected pesticides and phosphorus in soil-water systems: a critical review. EPA 600/53-82-060. Athens, GA: U.S. EPA.

Reyna, M., W. Ribelin, D. Thake et al.* 1984a. Three month feeding study of propachlor to albino rats. Project No. ML-83-083. Unpublished study. MRID 00152151.

Reyna, M., W. Ribelin, D. Thake et al.* 1984b. Three month feeding study of propachlor to albino rats. Project No. ML-81-72. Unpublished study. MRID 00152865.

Schardein, J., D. Wahlberg, S. Allen et al.* 1982. Teratology study in rats. (IR-81-264):401-171. Unpublished study. MRID 00115136.

Schubert, H. 1979. Allergic contact dermatitis due to propachlor. Dermatol. Monatsschr. 165(7):495-498. (Ger.) (PESTAB 80:115)

Steinmetz, K. and J. Mirsalis.* 1984. Evaluation of the potential of propachlor to induce unscheduled DNA synthesis in primary rat hepatocyte culture. Final report. Study No. LSC-7538. Unpublished study. MRID 00144512.

STORET. 1988. STORET Water Quality File. Office of Water. U.S. Environmental Protection Agency (data file search conducted in May, 1988).

U.S. EPA. 1985. U.S. Environmental Protection Agency. Code of Federal Regulations. 40 CFR 180.211. July 1.

U.S. EPA. 1986. U.S. Environmental Protection Agency. Guidelines for carcinogen risk assessment. Fed. Reg. 51(185):33992-34003. September 24.

U.S. EPA. 1988. U.S. Environmental Protection Agency. Method 508 - Determination of chlorinated pesticides in ground water by GC/ECD. Draft. April 15. Available from U.S. EPA's Environmental Monitoring and Support Laboratory, Cincinnati, OH.

Wazeter, F.X., R.H. Buller and R.G. Geil.* 1964. Ninety-day feeding study in the rat; ninety-day feeding study in the dog. 138-001 and 138-002. Unpublished study. MRID 00093270.

* Confidential Business Information submitted to the Office of Pesticide Programs.

Propazine

I. General Information and Properties

A. CAS No. 139–40–2

B. Structural Formula

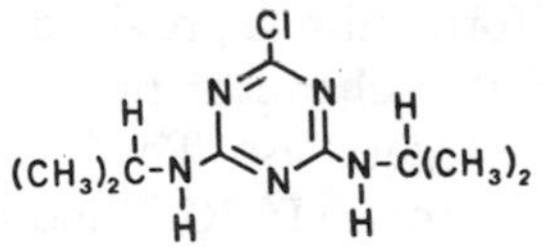

6-Chloro-N,N'-bis(l-methylethyl)-1-3,5-triazine-2,4-diamine

C. Synonyms

- Geigy 30,028; Gesomil; Milogard; Plantulin; Primatol P; Propasin; Prozinex (Meister, 1983).

D. Uses

- Propazine is a selective preemergence and preplant herbicide used for the control of most annual broadleaf weeds and annual grasses in milo and sweet sorghum (Meister, 1983).

E. Properties (Meister, 1983; IPC, 1984; CHEMLAB, 1985; TDB, 1985)

Chemical Formula	$C_9H_{16}N_5Cl$
Molecular Weight	230.09
Physical State (25°C)	Colorless crystals
Boiling Point	--
Melting Point	212 to 214°C
Density	--
Vapor Pressure (20°C)	2.9×10^{-8} mm Hg
Specific Gravity	--
Water Solubility (29°C)	8.6 mg/L
Log Octanol/Water Partition Coefficient	−1.21
Taste Threshold	--

Odor Threshold   --
Conversion Factor   --

F. Occurrence

- Propazine has been found in 33 of 1,097 surface water samples analyzed and in 15 of 906 ground-water samples (STORET, 1988). Samples were collected at 244 surface water locations and 607 ground-water locations. The 85th percentile of all nonzero samples was 2.3 $\mu g/L$ in surface water and 0.2 $\mu g/L$ in ground-water sources. The maximum concentration found was 13 $\mu g/L$ in surface water and 300 $\mu g/L$ in ground water. Propazine was found in five states in surface water and in four states in ground water. This information is provided to give a general impression of the occurrence of this chemical in ground and surface waters as reported in the STORET database. The individual data points retrieved were used as they came from STORET and have not been confirmed as to their validity. STORET data are often not valid when individual numbers are used out of the context of the entire sampling regime, as they are here. Therefore, this information can only be used to form an impression of the intensity and location of sampling for a particular chemical.

- Propazine was detected in ground water in California at trace levels (< 0.1 ppb) (U.S.G.S., 1985).

G. Environmental Fate

The following data were submitted by Ciba-Geigy and reviewed by the Agency (U.S. EPA, 1987):

- Hydrolysis studies show propazine to be resistant to hydrolysis. After 28 days, at pH 5, 60% remains; at pH 7, 92% remains; and at pH 9, 100% remains. Hydroxypropazine (2-hydroxy-4,6-bis-isopropylamino)-s-(triazine) is the hydrolysis product.

- Propazine at 2.5 ppm in aqueous solution was exposed to natural sunlight for 17 days. In that time, 5% degraded to hydroxypropazine.

- Under aerobic conditions, 10 ppm propazine was applied to a loamy sand (German) soil with 2.2% organic carbon. The soil was incubated at 25°C in the dark and kept at 70% of field capacity. Propazine degraded with a half-life of 15 weeks. Hydroxypropazine was the major degradate from aerobic soil metabolism; its concentration increased from 14% at 12 weeks to a maximum of 31% after 52 weeks of incubation. Trapped volatiles identified as CO_2 accounted for 1% of the applied propazine after 52 weeks. Bound residues increased up to 35% after 12 weeks of incubation.

- Under anaerobic conditions, further degradation of propazine was slight.

- Freundlich soil-water partition coefficient (K_d) values for propazine and hydroxypropazine were determined for four soils: two sandy loams (0.7% OM) and (1.4% OM), a loam soil (2.9% OM) and a clay loam (8.3% OM). The K_d values were: 0.34, 1.13, 2.69 and 3.19, respectively, for propazine. On the same four soils the K_d values for hydroxypropazine were: 1.13, 2.94, 31.8 and 10.6, respectively. All K_d values have units of mL/gm.

- Leaching studies for propazine performed on four soils under worst-case conditions (30-cm columns leached with 20 inches of water) for propazine indicate propazine's mobility in soil-water systems. In a loamy sand (0.7% OM), a sandy loam (1.4% OM), a loam (1.7% OM) and a silt loam (2.4% OM), 82.5%, 18%, 69.5% and 23.6% leached, respectively.

- In column studies using aged propazine, degradation products leached from a loamy sand soil with 2.2% OM. About 25% of the aged propazine added to the columns leached. In a loam soil with 3.6% OM, < 0.05% of the aged propazine added to the columns leached.

- In field dissipation studies, propazine was found at 18 inches, the deepest depth in the soil sampled. Hydroxypropazine was found at all depths and sites up to 3 years after application. Field half-lives for propazine were 5 to 33 weeks in the 0- to 6-inch depth, and 17 to 51 weeks at the 6- to 12-inch depth.

II. Pharmacokinetics

A. Absorption

- Bakke et al. (1967) administered single oral doses of ring-labeled ^{14}C-propazine to Sprague-Dawley rats. After 72 hours, about 23% of the label was recovered in the feces and about 66% was excreted in the urine. This indicates that gastrointestinal absorption was at least 77% complete.

B. Distribution

- Bakke et al. (1967) administered ring-labeled ^{14}C-propazine (41 to 56 mg/kg) to rats by gastric intubation. Radioactivity in a variety of tissues was observed to decrease from an average of 46.7 ppm 2 days posttreatment to 22.3 ppm after 8 days. Radioactivity was detected in the lung (30 ppm), spleen (25 ppm), heart (27 ppm), kidney (17 ppm) and brain (13 ppm) for up to 8 days. After 12 days, the only detectable quantities were in hide and hair (3.35% of administered dose), viscera (0.1%) and carcass (2.22%).

C. Metabolism

- Eighteen metabolites of propazine have been identified in the urine of rats given single oral doses of ^{14}C-propazine (Bakke et al., 1967). No other details were provided. Based on metabolites found in urine, Bakke et al. (1967) reported that dealkylation is one reaction in the metabolism of propazine. No other details were provided.

D. Excretion

- Bakke et al. (1967) administered single oral doses of ^{14}C-ring-labeled propazine to rats. Most of the radioactivity was excreted in the urine (65.8%) and feces (23%) within 72 hours. Excretion of propazine and/or metabolites was most rapid during the first 24 hours after administration, decreasing to smaller amounts at 72 hours.

III.   Health Effects

A. Humans

- Contact dermatitis was reported in workers involved in propazine manufacturing (Hayes, 1982). No other information on the health effects of propazine in humans was found in the available literature.

B. Animals

1. Short-term Exposure

- The reported acute oral LD_{50} values for propazine (purity not specified) were > 5,000 mg/kg in mice (Stenger and Kindler, 1963b) and 1,200 mg/kg in guinea pigs (NIOSH, 1987).

- Stenger and Kindler (1963a) reported that dietary administration of propazine (purity not specified) to rats (5/sex/dose) at doses of 1,250 or 2,500 mg/kg for 4 weeks resulted in a decrease in body weight, but there were no pathological alterations in organs or tissues. No other details were provided.

2. *Dermal/Ocular Effects*

- The acute dermal LD_{50} value in rabbits for propazine (90% water dispersible granules) was reported as $>$ 2,000 mg/kg (Cannelongo et al., 1979).

- Stenger and Huber (1961) reported that rats were unaffected when a 5% gum arabic suspension of propazine (0.4 mL/animal) was applied once a day for 5 consecutive days to shaved and intact skin of five rats then washed away 3 hours after application.

- Palazzolo (1964) reported that propazine (1 or 2 g/kg/day) applied to intact or abraded skin of albino rabbits (5/sex/dose) for 7 hours produced mild erythema, drying, desquamation and thickening of the skin. Body weights, mortality, behavior, hematology, clinical chemistry and pathology of the treated and untreated groups were similar.

3. *Long-term Exposure*

- In 90-day feeding studies by Wazeter et al. (1967a), beagle dogs (12/sex/dose) were fed propazine (80 WP) in the diet at 0, 50, 200 or 1,000 ppm active ingredient (a.i.). Based on the assumption that 1 ppm in the diet of dogs is equivalent to 0.025 mg/kg/day (Lehman, 1959), these doses correspond to 0, 1.25, 5.0 or 25 mg/kg/day. No compound-related changes were observed in general appearance, behavior, hematology, urinalysis, clinical chemistry, gross pathology or histopathology at any dose tested. In the 1,000 ppm dose group, four dogs lost 0.3 to 1.1 kg in body weight, which the author suggested may have been compound-related (no p value reported). Based on these results, a No-Observed-Adverse-Effect Level (NOAEL) of 200 ppm (5 mg/kg/day) and a LOAEL of 1,000 ppm (25 mg/kg/day) were identified.

- Wazeter et al. (1967b) supplied CD rats (80/sex/dose) with propazine (80 WP) in the diet for 90 days at dose levels of 0, 50, 200 or 1,000 ppm a.i. Based on the assumption that 1 ppm in the diet is equivalent to 0.05 mg/kg/day (Lehman, 1959), these doses correspond to 0, 2.5, 10 or 50 mg/kg/day. No compound-related changes were observed in appearance, general behavior, hematology, clinical chemistry, urinalysis, gross pathology and histopathology. There was a 12% reduction ($p < 0.01$)

in body weight of females at 1,000 ppm (50 mg/kg/day) at the end of the study. Based on body weight loss, a NOAEL of 200 ppm (10 mg/kg/day) and a Lowest-Observed-Adverse-Effect Level (LOAEL) of 1,000 ppm (50 mg/kg/day) were identified.

- Geigy (1960) dosed rats (12/sex/dose) of an unspecified strain with propazine (50% a.i.) by stomach tube for 90 days at 0, 250 or 2,500 mg/kg/day (a.i.) or for 180 days at 0 or 250 mg/kg/day (a.i.). In the 90-day study, a reduction in body weight and food consumption were reported at 2,500 mg/kg/day, but no effects were seen at 250 mg/kg/day. No histopathological evaluations were performed at the high-dose level. After 180 days, rats administered propazine at 250 mg/kg/day were similar to untreated controls in growth rates, daily food consumption, gross appearance and behavior, mortality, gross pathology and histopathology. This study identified a NOAEL of 250 mg/kg/day and a LOAEL of 2,500 mg/kg/day.

- Jessup et al. (1980a) fed CD mice (60/sex/dose) technical propazine (purity not specified) for 2 years at dose levels of 0, 3, 1,000 or 3,000 ppm. Based on the assumption that 1 ppm in the diet of mice is equivalent to 0.15 mg/kg/day (Lehman, 1959), these doses correspond to 0, 0.45, 150 or 450 mg/kg/day. The general appearance, behavior, survival rate, body weights, organ weights, food consumption and incidence of inflammatory, degenerative or proliferative alterations in various tissues and organs did not differ significantly from untreated controls. The author identified a NOAEL of 3,000 ppm (450 mg/kg/day, the highest dose tested).

- Jessup et al. (1980b) fed CD rats (60 to 70/sex/dose) technical propazine (purity not specified) in the diet for 2 years at dose levels of 0, 3, 100 or 1,000 ppm. Based on the assumption that 1 ppm in the diet of rats is equivalent to 0.05 mg/kg/day (Lehman, 1959), this corresponds to doses of 0, 0.15, 5 or 50 mg/kg/day. No compound-related effects were observed in behavior, appearance, survival, food consumption, hematology, urinalysis or nonneoplastic alterations in various tissues and organs. Mean body weight gains appeared to be lower in the treatment groups than the control groups. Body weights at 104 weeks were lower than controls at all dose levels. The percent decreases in males and females were as follows: –6.3 and –3.9% (3 ppm); –4.6 and –5.6% (100 ppm); –13.1 and –11.4% (1,000 ppm). These decreases were statistically significant in males at 3 and 1,000 ppm, and in females at 100 and 1,000 ppm. The decreases at 3 or 100 ppm appeared to be so small that they may not be considered biologically significant; a NOAEL was identified at 100 ppm (5 mg/kg/day).

4. *Reproductive Effects*

*   Jessup et al. (1979) conducted a three-generation study in which CD rats (20 females and 10 males/dose) were administered technical propazine in the diet at 0, 3, 100 or 1,000 ppm. Based on the assumption that 1 ppm in the diet is equivalent to 0.05 mg/kg/day (Lehman, 1959), this corresponds to doses of 0, 0.15, 5 or 50 mg/kg/day. No compound-related effects were observed in any dose group in general behavior, appearance or survival of parental rats or pups. The mean parental body weights were statistically lower at 1,000 ppm (50 mg/kg/day). No differences were reported in food consumption for treated and control animals. No treatment-related effects were observed in fertility, length of gestation or viability and survival of the pups through weaning. Mean pup weights at lactation were not adversely affected at 3 or 100 ppm (0.15 or 5 mg/kg/day). However, at 1,000 ppm (50 mg/kg/day), there was a statistically significant decrease in mean pup weights for all generations except F_{1a}. Based on these data, a NOAEL of 100 ppm (5 mg/kg/day) was identified.

5. *Developmental Effects*

*   Fritz (1976) administered technical propazine (0, 30, 100, 300 or 600 mg/kg/bw) orally by intubation to pregnant Sprague-Dawley rats (25/dose) on days 6 through 15 of gestation. No maternal toxicity, fetotoxicity or teratogenic effects were observed at 100 mg/kg/day or lower. Maternal body weight and food consumption were reduced at 300 mg/kg/day or higher. Fetal body weight was reduced, and there was delayed skeletal ossification (of calcanei) at 300 mg/kg/day or higher. Based on body weights, a maternal NOAEL of 100 mg/kg/day and a fetal NOAEL of 100 mg/kg/day were identified.

*   Salamon (1985) dosed pregnant CD rats (21 to 23 animals per dose group) with technical propazine (99.1% pure) by gavage at dose levels of 0, 10, 100 or 500 mg/kg/day on days 6 through 15 of gestation. Decreases in maternal body weight and food consumption were statistically significant ($p < 0.05$) at doses of 100 mg/kg/day or higher. Fetal body weight was reduced, and ossification of cranial structures was delayed at 500 mg/kg/day. Based on maternal toxicity, a NOAEL of 10 mg/kg/day was identified.

6. *Mutagenicity*

- Puri (1984a) reported that propazine (0, 0.4, 20, 100 or 500 μg/mL) did not produce DNA damage in human fibroblasts *in vitro*.

- Puri (1984b) reported that propazine (0, 0.50, 2.5, 12.5 or 62.5 μg/mL) did not cause DNA damage in rat hepatocytes *in vitro*.

- Strasser (1984) reported that propazine administered to Chinese hamsters by gavage (0, 1,250, 2,500 or 5,000 mg/kg) did not cause anomalies in nuclei of somatic interphase cells.

7. *Carcinogenicity*

- Innes et al. (1969) fed propazine in the diet to 72 mice [(C57BL16 × AKR)F$_1$ or (C57BL16 × C3H/ANf)F$_1$] for 18 months at dose levels of 0 or 46.4 mg/kg/day. Based on histopathological examination of tissues (no data reported), the authors stated that propazine, at the one dose tested, did not cause a statistically significant increase in the frequency of any tumor type in any sex-strain subgroup or combination of groups.

- Jessup et al. (1980b) fed CD rats (60 to 70/sex/dose) technical propazine (purity not specified) in the diet for 2 years at dose levels of 0, 3, 100 or 1,000 ppm. Based on the assumption that 1 ppm in the diet of rats is equivalent to 0.05 mg/kg/day (Lehman, 1959), this corresponds to doses of 0, 0.15, 5 or 50 mg/kg/day. Tumor incidence was evaluated for a variety of organs and tissues. The most commonly occurring tumors were mammary gland tumors in female rats. At the highest dose tested (1,000 ppm, 50 mg/kg/day), the authors reported an increase in adenomas (10/55, 18%), adenocarcinomas (9/55, 16%) and papillary carcinomas (8/55, 15%) compared to corresponding tumor levels in untreated controls of 3/55 (5%), 6/55 (11%) and 4/55, (7%), respectively. Also, it was reported that the percentage of tumor-bearing rats was 73% in the high-dose treated group compared to 50% in corresponding untreated controls. The authors did not consider these increases to be statistically significant. However, in 1981, Somers reported historical control values of 122/1,248 (10%) for rats with adenomas and of 769/1,528 (50%) for percentage of tumor-bearing animals. Further evaluations by Somers (1981) of the above data (control and treated) and historical control data indicated that the increase in mammary gland adenomas and the number of rats bearing one or more tumor was statistically significant (p < 0.02).

- Jessup et al. (1980a) fed CD mice (60/sex/dose) technical propazine (purity not stated) for 2 years at dose levels of 0, 3, 1,000 or 3,000 ppm. Assuming that 1 ppm in the diet of mice is equivalent to 0.15 mg/kg/day (Lehman, 1959), this corresponds to doses of 0, 0.45, 150 or 450 mg/kg/day. The incidence of proliferative and neoplastic alterations in the treated groups did not differ significantly from the control group at any dose level.

IV. Quantification of Toxicological Effects

A. One-day Health Advisory

No information was found in the available literature that was suitable for determination of the One-day HA value for propazine. It is therefore recommended that the Ten-day HA value for a 10-kg child, 1.0 mg/L (1,000 μg/L, calculated below), be used at this time as a conservative estimate of the One-day HA value.

B. Ten-day Health Advisory

The study by Salamon (1985) has been selected to serve as the basis for the determination of the Ten-day HA value for propazine. In this teratogenicity study in rats, body weight was decreased in dams dosed on days 6 to 15 of gestation with 100 mg/kg/day or greater. No adverse effects were observed in either dams or fetuses at 10 mg/kg/day. The rat study by Fritz (1976) reported maternal and fetal toxicity at 300 mg/kg/day, but not at 100 mg/kg/day. This NOAEL was not selected, since maternal weight loss was noted by Salamon (1985) at this dose.

Using a NOAEL of 10 mg/kg/day, the Ten-day HA for a 10-kg child is calculated as follows:

$$\text{Ten-day HA} = \frac{(10 \text{ mg/kg/day}) (10 \text{ kg})}{(100) (1 \text{ L/day})} = 1.0 \text{ mg/L} (1,000 \text{ } \mu\text{g/L})$$

C. Longer-term Health Advisory

The 90-day feeding study in dogs by Wazeter et al. (1967a) has been selected to serve as the basis for the Longer-term HA for propazine. In this study, body weight loss occurred at 1,000 ppm (25 mg/kg). A NOAEL of 200 ppm (5 mg/kg/day) was identified. This is supported by the 90-day rat feeding study by Wazeter et al. (1967b), which identified a NOAEL of 10 mg/kg/day and a LOAEL of 50 mg/kg/day. The 90-day study in rats by Geigy (1960) has not been selected, since the NOAEL (250 mg/kg/day) is higher than the LOAEL values reported above.

Using a NOAEL of 5 mg/kg/day, the Longer-term HA for the 10-kg child is calculated as follows:

$$\text{Longer-term HA} = \frac{(5 \text{ mg/kg/day}) (10 \text{ kg})}{(100) (1 \text{ L/day})} = 0.5 \text{ mg/L } (500 \text{ } \mu\text{g/L})$$

The Longer-term HA for a 70-kg adult is calculated as follows:

$$\text{Longer-term HA} = \frac{(5 \text{ mg/kg/day}) (70 \text{ kg})}{(100) (2 \text{ L/day})} = 1.75 \text{ mg/L } (2,000 \text{ } \mu\text{g/L})$$

D. Lifetime Health Advisory

The 2-year feeding study in rats by Jessup et al. (1980b) has been selected to serve as the basis for determination of the Lifetime HA for propazine. No effects were detected on behavior, appearance, mortality, food consumption, hematology, urinalysis or body weight gain at doses of 5 mg/kg/day. At 50 mg/kg/day, decreased weight gain was noted, and there was evidence of increased tumor frequency in the mammary gland. This NOAEL value (5 mg/kg/day) is supported by the NOAEL of 5 mg/kg/day in the three-generation reproduction study in rats by Jessup et al. (1979). The 2-year feeding study in mice by Jessup et al. (1980a) has not been selected, since the data suggest that the mouse is less sensitive than the rat.

The Lifetime HA is calculated as follows:

Step 1: Determination of the Reference Dose (RfD)

$$\text{RfD} = \frac{(5 \text{ mg/kg/day})}{(100) (3)} = 0.02 \text{ mg/kg/day}$$

where: 3 = additional uncertainty factor to account for data gaps (chronic feeding dog study) in the propazine database.

Step 2: Determination of the Drinking Water Equivalent Level (DWEL)

$$\text{DWEL} = \frac{(0.02 \text{ mg/kg/day}) (70 \text{ kg})}{(2 \text{ L/day})} = 0.70 \text{ mg/L } (700 \text{ } \mu\text{g/L})$$

Step 3: Determination of the Lifetime Health Advisory

$$\text{Lifetime HA} = \frac{(0.70 \text{ mg/L}) (20\%)}{(10)} = 0.014 \text{ mg/L } (10 \text{ } \mu\text{g/L})$$

E. Evaluation of Carcinogenic Potential

• No evidence of increased tumor frequency was detected in a 2-year feeding study in mice at doses up to 450 mg/kg/day (Jessup et al., 1980a) or in an 18-month feeding study in mice at a dose of 46.4 mg/kg/day (Innes et al., 1969).

• Jessup et al. (1980b) reported that the occurrence of mammary gland tumors in female rats administered technical propazine in the diet for 2 years at 1,000 ppm (50 mg/kg/day) was increased but did not differ significantly from concurrent controls. However, a reevaluation of the data by Somers (1981) that considered historical control data indicated that the increase in mammary gland adenomas and the number of rats bearing one or more tumors was statistically significant ($p < 0.02$).

• The International Agency for Research on Cancer has not evaluated the carcinogenic potential of propazine.

• Applying the criteria described in EPA's guidelines for assessment of carcinogenic risk (U.S. EPA, 1986a), propazine may be classified in Group C: possible human carcinogen. This category is for substances with limited evidence of carcinogenicity in animals in the absence of human data.

V.   Other Criteria, Guidance and Standards

• The U.S. EPA (1986b) has established residue tolerances of 0.25 ppm for propazine in or on various agricultural commodities (negligible) based on a Provisionary Acceptable Daily Intake (PADI) of 0.005 mg/kg/day.

• NAS (1977) determined an Acceptable Daily Intake (ADI) of 0.0464 mg/kg/day, based on a NOAEL of 46.4 mg/kg identified in an 80-week feeding study in mice with an uncertainty factor of 1,000.

• NAS (1977) calculated a chronic Suggested-No-Adverse-Effect Level (SNARL) of 0.32 mg/L, based on an ADI of 0.0464 mg/kg/day and a relative source contribution factor of 20%.

VI. Analytical Methods

- Analysis of propazine is by gas chromatographic (GC) Method 507, a method applicable to the determination of certain nitrogen-phosphorus containing pesticides in water samples (U.S. EPA, 1988). In this method, approximately 1 liter of sample is extracted with methylene chloride. The extract is concentrated and the compounds are separated using capillary column GC. Measurement is made using a nitrogen-phosphorus detector. This method has been validated in a single laboratory and the limit of detection for propazine was 0.13 μg/L.

VII. Treatment Technologies

- No information regarding treatment technologies applicable to the removal of propazine from contaminated water was found in the available literature.

VIII. References

Bakke, J.E., J.D. Robbins and V.J. Feil. 1967. Metabolism of 2-chloro-4,6-bis(isopropylamine), s-triazine (propazine) and 2-methoxy-4,6-bis (isopropylamino)-s-triazine (prometone) in the rat. Balance study and urinary metabolite separation. J. Agr. Food Chem. 15(4):628–631.

Cannelongo, B., E. Sabol, R. Sabol et al.* 1979. Rabbit acute dermal toxicity. Project No. 1132–79. Unpublished study. MRID 00111700.

CHEMLAB. 1985. The Chemical Information System, CIS, Inc., Baltimore, MD.

Fritz, H.* 1976. Reproduction study: G30028 technical rat study. Segment II. Test for teratogenic or embryotoxic effects. Experiment No. 227642. Unpublished study. MRID 00087879.

Geigy, S.A.* 1960. Chronic toxicity of propazine 50 WP. Unpublished study. MRID 00111671.

Hayes, W.J. 1982. Pesticides studied in man. Baltimore, MD: Williams and Wilkins. p. 564.

IPC.* 1984. Industria Prodotti Chimici. Atrazine product chemistry data. Unpublished compilation. MRID 00141156.

Innes, J., B. Ulland, M.G. Valerio, L. Petrucelli, L. Fishbein, E. Hart and A. Pallotta. 1969. Bioassay of pesticides and industrial chemicals for tumorigenicity in mice: a preliminary note. J. Natl. Can. Inst. 42(6):1101–1114.

Jessup, D.C., R.J. Arceo and J.E. Lowry.* 1980a. Two-year carcinogenicity study in mice. IRDC No. 382–004. Unpublished study. MRID 00044335.

Jessup, D.C., G. Gunderson and L.J. Ackerman.* 1980b. Two-year chronic oral toxicity study in rats. IRDC No. 382–007. Unpublished study. MRID 00041408.

Jessup, D.C., C. Schwartz, R.J. Arceo et al.* 1979. Three-generation study in rat. IRDC No. 382–010. Unpublished study. MRID 00041409.

Lehman, A.J. 1959. Appraisal of the safety of chemicals in foods, drugs and cosmetics. Assoc. Food Drug Off.

Meister, R., ed. 1983. Farm chemicals handbook. Willoughby, OH: Meister Publishing Company.

NAS. 1977. National Academy of Sciences. Drinking water and health. Washington, DC: National Academy Press.

NIOSH. 1987. National Institute for Occupational Safety and Health. Microfiche edition. Registry of Toxic Effects of Chemical Substances (RTECS). July.

Palazzolo, R.* 1964. Repeated dermal toxicity of propazine 80 W. Report to Geigy Research Laboratories. Unpublished study. MRID 00111670.

Puri, E.* 1984a. Autoradiographic DNA repair test on human fibroblasts with G30028 technical. Test No. 831373. Unpublished study. MRID 00150024.

Puri, E.* 1984b. Autoradiographic DNA repair test on rat hepatocytes with G30028 technical. Test Report No. 831371. Unpublished study. MRID 00150623.

Salamon, C.* 1985. Teratology study in albino rats with technical propazine. Report No. 450–1788. Unpublished study. American Biogenics Corporation. MRID 00150242.

Somers, J.A.* 1981. Propazine herbicide chemical No. 080808, 6(a)(2): submission of treated vs. control data involving mammary tumors in rats in IRDC Study No. 382–007; response to November 18, 1980. Letter sent to Robert J. Taylor. MRID 00076955. Dated April 14, 1981.

Stenger and Kindler.* 1963a. Subchronic oral toxicity in the rat. Translation of subchronische toxizitat – ratte p.o. Unpublished study. MRID 00111678.

Stenger and Kindler.* 1963b. Acute toxicity – mouse, oral. Translation of akute toxizitat – maus per OS. Unpublished study. MRID 00111675.

Stenger and Huber.* 1961. Subchronic toxicity – rat skin. Translation of subchronische toxizitat – ratte, haut. Unpublished study, including German text. MRID 00111677.

STORET. 1988. STORET Water Quality File. Office of Water. U.S. Environmental Protection Agency (data file search conducted in May, 1988).

Strasser, F.* 1984. Nucleus anomaly test in somatic interphase nuclei of Chinese hamster. Test Report No. 831372. Unpublished study. MRID 00150622.

TDB. 1985. Toxicology Data Bank. MEDLARS II. National Library of Medicine's National Interactive Retrieval Service.

U.S. EPA. 1986a. U.S. Environmental Protection Agency. Guidelines for carcinogen risk assessment. Fed. Reg. 51(185):33992–34003. September 24.

U.S. EPA. 1986b. U.S. Environmental Protection Agency. Code of Federal Regulations. 40 CFR 180.243. July 1, 1985. p. 296.

U.S. EPA. 1987. U.S. Environmental Protection Agency. Environmental fate of propazine. Memo from C. Eiden to D. Tarkas. June 9.

U.S. EPA. 1988. U.S. Environmental Protection Agency. U.S. EPA Method 507 — Determination of nitrogen and phosphorus containing pesticides in ground water by GC/NPD. Draft. April 15. Available from U.S. EPA's Environmental Monitoring and Support Laboratory, Cincinnati, OH.

U.S.G.S. 1985. U.S. Geological Survey. Regional assessment project. C. Eiden.

Wazeter, F., R. Buller, R. Geil et al.* 1967a. Ninety-day feeding study in the beagle dog. Propazine 80W. Report No. 248–002. Unpublished study. MRID 00111680.

Wazeter, F., R. Buller, R. Geil et al.* 1967b. Ninety-day feeding study in albino rats. Propazine 80W. Report No. 248–001. Unpublished study. MRID 00111681.

---

* Confidential Business Information submitted to the Office of Pesticide Programs.

Propham

I. General Information and Properties

A. CAS No. 122–42–9

B. Structural Formula

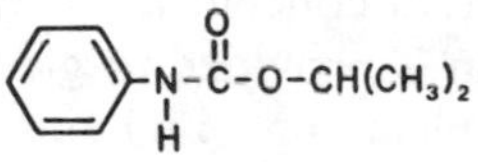

Phenyl l-methylethyl carbamate; isopropyl-N-phenylcarbamate;
Isopropyl carbanilate

C. Synonyms

- IPC, IFC, Ban-Hoe, Beet-Kleen, Chem-Hoe, Premalox, Triherbide-IPC, Tuberite (Meister, 1988).

D. Uses

- Propham is used as a pre- and postemergence herbicide for control of weeds in alfalfa, clover, flax, lettuce, safflower, spinach, sugarbeets, lentils and peas and on fallow land. It prevents cell division and acts on meristematic tissue (Meister, 1988).

E. Properties (Meister, 1988; Cohen, 1984; CHEMLAB, 1985; TDB, 1985)

Chemical Formula	$C_{10}H_{13}O_2N$
Molecular Weight	179.21
Physical State (25°C)	White crystals
Boiling Point (at 25 mm Hg)	--
Melting Point	87°C
Density	1.09 g/mL (20°C)
Vapor Pressure (25°C)	(sublimes slowly at 25°C)
Specific Gravity (20°C/20°C)	1.09
Water Solubility (25°C)	250 mg/L

Log Octanol/Water
   Partition Coefficient   1.22 (calculated)
Taste Threshold   --
Odor Threshold   --
Conversion Factor   --

F. Occurrence

- Propham has been found in 1 of 392 surface water samples analyzed and was undetectable in 583 ground-water samples (STORET, 1988). Samples were collected at 131 surface water locations and 572 ground-water locations, and propham was found in Texas. The 85th percentile of the sample and the maximum concentration found in surface water was 2 μg/L. This information is provided to give a general impression of the occurrence of this chemical in ground and surface waters as reported in the STORET database. The individual data points retrieved were used as they came from STORET and have not been confirmed as to their validity. STORET data are often not valid when individual numbers are used out of the context of the entire sampling regime, as they are here. Therefore, this information can only be used to form an impression of the intensity and location of sampling for a particular chemical.

G. Environmental Fate

- Ring-labeled ^{14}C-propham (purity unspecified) at 4 ppm in unbuffered distilled water declined to 2.4 ppm during 14 days of irradiation with a Pyrex-filtered light (uncharacterized) at 25°C (Gusik, 1976). Degradation products included isopropyl 4-hydroxycarbanilate (3.5% of applied propham), isopropyl 4-aminobenzoate (approximately 0.1%), 1-hydroxy-2-propylcarbanilate (approximately 0.1%), and polymeric materials (10 to 12%). No degradation occurred in the dark control during the same period.

- Under aerobic conditions, ring-labeled ^{14}C-propham (test substance uncharacterized) at 2 ppm degraded with a half-life of 2 to 7 days in silt loam soil, (Hardies, 1979; Hardies and Studer, 1979a), 4 to 7 days in loam soil (Hardies and Studer, 1979b), and 7 to 14 days in sandy loam soil (Hardies and Studer, 1979c) when incubated in the dark at approximately 25°C and 60% of water holding capacity.

- Under anaerobic conditions, ring-labeled ^{14}C-propham (test substance uncharacterized) declined from 8.5 to < 5% of the applied radioactivity during 60 days of incubation in silt loam soil in the dark at approximately 25°C and 60% of water holding capacity (Hardies, 1979; Hardies and Studer, 1979a). Under anaerobic conditions, ring-labeled ^{14}C-propham (test substance uncharacterized) declined from approximately 0.08 to approximately 0.04 ppm during 61 days of incubation in loam soil in the dark at approximately 25°C and 60% of water holding capacity (Hardies and Studer, 1979b); in sandy loam soil, the decline was from approximately 0.06 to 0.03 ppm during 63 days of incubation (Hardies and Studer, 1979c).

• ^{14}C-Propham (purity unspecified) at 0.2 to 20 ppm was adsorbed to two silt loams, a silty clay loam, a sandy clay loam, and two sandy loam soils with Freundlich K values of 0.74 and 2.72, 1.77, 0.65, and 0.27 and 1.58, respectively (Hardies and Studer, 1979d). Ring-labeled ^{14}C-propham (purity unspecified) was very mobile (> 98% of applied propham in leachate) in 30.5-cm columns of sandy clay loam and sandy loam soil leached with 20 inches of water (Hardies and Studer, 1979e). It was less mobile in columns of Babcock silt loam (42.3% in leachate), silty clay loam (approximately 62% at 11- to 27-cm depth), and Wooster silt loam (approximately 54% at 7.6- to 15-cm depth) soils. Aged (30-day) residues were relatively immobile in Wooster silt loam soil; < 1% of the applied radioactivity moved from the treated soil.

• Propham residues dissipated from the upper 6 inches of sandy loam, sandy clay loam, silty loam, and silty clay loam field plots with half-lives of 42 to 94, 57 to 160, 42 to 147, and approximately 21 to 42 days, respectively, following application of propham (ChemHoe 135, 3 lb/gal FlC) at 4 and 8 lb active ingredient (a.i.) per acre in September to November, 1977 (Pensyl and Wiedmann, 1979). Residues were nondetectable (< 0.02 ppm) within 164 to 283 days after treatment at all rates and sites. In general, propham residues in the 6- to 12-inch depth were < 0.04 ppm. Propham (3 lb/gal FlC) applied at 6 lb a.i./acre in mid-May dissipated with a half-life of 10 to 15 days in the 0- to 6-inch depth of silt loam soil (Wiedmann and Pensyl, 1981). Ring-labeled ^{14}C-propham (formulated as ChemHoe 135) applied at 4 lb a.i./acre dissipated with a half-life of < 7 days in the upper 3 inches of silt loam soil treated in November, 1981 (Wiedmann et al., 1982). The second half-life occurred approximately 133 days post-treatment.

II. Pharmacokinetics

A. Absorption

- After oral administration of 1,100 mg/kg ^{14}C-isopropyl-labeled pro-pham (99% a.i.) to rats (1,100 mg/kg), 88% of the label appeared in urine within 4 days. After oral doses of 1,100 mg/kg of ^{14}C-phenyl-labeled propham, 96% was excreted in urine and 2% was excreted in feces (Chen, 1979).

- Fang et al. (1972) reported that in rats given oral doses (ranging from less than 4 mg/kg to 200 mg/kg) of ^{14}C-propham (99% a.i.) 80 to 85% was excreted in urine and 5% was expired in air, indicating that pro-pham is well absorbed (85 to 98%) from the gastrointestinal tract.

B. Distribution

- Chen (1979) administered single oral doses of ^{14}C-phenyl- or ^{14}C-isopropyl-labeled propham (1,100 mg/kg; 99% a.i.) to rats. Trace amounts of both ^{14}C-phenyl- and ^{14}C-isopropyl-labeled (0.5 to 1.2%) propham were present in the liver, kidneys, muscle and carcass after 48 hours.

- Paulson and Jacobsen (1974) administered single oral doses of ^{14}C-propham (100 mg/kg; 99% a.i.) to goats. Six hours later, only low levels (0.2%) were detectable in milk.

C. Metabolism

- Chen (1979) administered single oral doses of ^{14}C-phenyl-labeled pro-pham (1,100 mg/kg; 99% a.i.) to rats by gavage. Most of the dose (96%) was excreted in urine as metabolites. The primary metabolites identified were the sulfate ester conjugate and the glucuronide conjugate of isopropyl 4-hydroxycarbanilate, which accounted for 78 and 1.3%, respectively, of the total primary metabolites recovered. Similar studies in rats (single oral dose of 100 mg/kg) by Paulson et al. (1972) support the rapid metabolism and excretion of propham. In these studies a third metabolite (the sulfate ester of 4-hydroxyacetanilide) and a fourth (uni-dentified) metabolite were found to account for 12.3% and 8.9%, respectively, of the total metabolites detected in urine. The data demon-strate that ring hydroxylation at the 4-position and subsequent conjuga-tion as well as hydrolysis and subsequent N-acetylation occurred prior to excretion.

D. Excretion

- ^{14}C-Propham is rapidly excreted primarily in the urine of rats. Peak urinary concentrations were reached 6 hours posttreatment. It was found that 96% and 2% of the administered dose of ^{14}C-propham (100 mg/kg; 99% a.i.) was excreted in the urine and feces, respectively (Chen, 1979; Paulson et al., 1972).

- Fang et al. (1972) reported that after oral administration of ring- or chain-^{14}C-labeled propham (99% a.i.) to rats, 80 to 85% of the administered dose was excreted in the urine over a 3-day period. In animals dosed with ^{14}C-isopropyl-labeled propham, 5% was detected as expired carbon dioxide.

III.   Health Effects

A. Humans

- No information was found in the available literature on the health effects of propham in humans.

B. Animals

1. *Short-term Exposure*

- Terrell and Parke (1977) administered single oral doses of propham (technical grade, purity not specified) to groups of 10 male and 10 female rats and monitored adverse effects for 14 days. Doses of 2,000 mg/kg produced loss of righting reflex, ptosis, piloerection, decreased locomotor activity, chronic pulmonary disease and rugae and irregular thickening of the stomach. The acute oral LD_{50} values in male and female rats were reported to be 3,000 ± 232 mg/kg and 2,360 ± 118 mg/kg, respectively. A No-Observed-Adverse-Effect Level (NOAEL) cannot be derived from the study because the doses used were too high, and adverse effects were found at all doses tested.

- Brown and Gross (1949) reported that when a single dose of 1,136 mg/kg propham (purity not specified) was administered orally to 8 Sprague-Dawley rats, no adverse effects were observed. A dose of 2,174 mg/kg resulted in periods of light anesthesia; a higher dose of 3,348 mg/kg resulted in light anesthesia and death (one death in six tested). Deep anesthesia was produced when 4,425 mg/kg of propham was administered to rats with subsequent death of 38% of the test animals (three deaths in eight animals tested).

- The acute inhalation LC_{50} value in albino rats was reported to be 10.71 mg/L (or 10,710 mg/m^3, PPG Industries, 1978).

2. *Dermal/Ocular Effects*

- The acute dermal LD_{50} value in albino rabbits was reported to be greater than 3,000 mg/kg (PPG Industries, 1978).

- Propham (3% aqueous solution) was slightly irritating when applied to the skin and eyes of albino rabbits (PPG Industries, 1978).

3. *Long-term Exposure*

- Tisdel et al. (1979) fed Sprague-Dawley rats (30/sex/dose) propham (technical grade, purity not specified) in the diet at 0, 250, 1,000 or 2,000 ppm for 91 days. Assuming that 1 ppm in the diet of rats is equivalent to 0.05 mg/kg/day (Lehman, 1959), these levels are equivalent to 0, 12.5, 50 or 100 mg/kg/day. Following treatment, body weight, organ weight, growth, clinical chemistry, gross pathology and histopathology were evaluated. No effects were reported at 1,000 ppm (50 mg/kg/day) or lower in any parameters measured. At the highest dose (2,000 ppm or 100 mg/kg/day), there was a significant increase in spleen weight (p < 0.05) and in spleen-to-body weight ratio (p < 0.01) in males, and a 70% inhibition of plasma cholinesterase (p < 0.01) in females at 45 days. Based on the above data, a NOAEL of 1,000 ppm (50 mg/kg/day) was identified.

4. *Reproductive Effects*

- In a report of a three-generation rat reproduction study, Ravert (1978) reported data from the P_2 to weaning of the F_{2b} generation. Sprague-Dawley rats (10 males or 20 females/dose) were administered technical grade propham (purity not specified) in the diet at dose levels of 0, 87.5, 250, 750 or 1,500 ppm for 9 weeks prior to breeding for each parental generation. Assuming that 1 ppm in the diet of rats is equivalent to 0.05 mg/kg/day (Lehman, 1959), these levels are equivalent to 0, 4.4, 12.5, 37.5 or 75 mg/kg/day. It was not clear whether the test animals were also fed propham-containing diets during pregnancies or through weaning of offspring. No effects were reported on fertility, mortality or pup development at any dose level tested.

5. *Developmental Effects*

- Ravert and Parke (1977) administered technical propham (purity not specified) by gavage to pregnant Sprague-Dawley rats (16 to 20/dose), at levels of 0, 37.6, 376 or 1,879 mg/kg/day on days 6 through 15 of gestation. End points monitored included maternal and fetal body weight and the number of corpora lutea, implants, live fetuses and dead fetuses. Fetuses were also examined for soft-tissue and skeletal anomalies. The only effects detected were reduced maternal and fetal body weights and higher resorption rates at the highest dose tested (1,879 mg/kg) and increased incidences of incomplete ossification of the parietal and frontal bones of the skull at 375.8 and 1,879 mg/kg. An apparent NOAEL appears to be 37.6 mg/kg/day. However, in this experiment, the high dose (1,879 mg/kg/day) is too high (i.e., one-half of the LD_{50}); nearly two-thirds of the pregnant rats at this dose died prior to scheduled sacrifice. Further, the dose intervals are also relatively large. Therefore, a reliable NOAEL cannot be determined accurately due to the large difference in dosages tested and the marginal effect noted at 376 mg/kg/day. (For more information on the developmental effects, see Worthing, 1979).

6. *Mutagenicity*

- Using the Ames *Salmonella* test, Margard (1978) reported that propham (purity not specified, 1,000 µg/plate) did not show any indications of mutagenic activity either with or without activation.

- When propham (100 µg/mL, purity not specified) was applied to cultures containing BALB/c 3T3 cell lines, no clonal transformation was evident (Margard, 1978).

- Friedrick and Nass (1983) reported that propham (1.1 to 2.2 mM) did not induce mutation in S49 mouse lymphoma cells.

7. *Carcinogenicity*

- Innes et al. (1969) administered propham to (C57BL/6 × C3H/ANf) or (C7BL/6 × AKR) mice (18/sex) in the diet at 0 or 560 ppm for 18 months. According to the author, this corresponds to a dose of about 0 to 215 mg/kg. The incidence of tumors was not significantly increased (p > 0.05) for any tumor type in any sex-strain subgroup or in the combined sexes of either strain. This duration of exposure and this dose level may not be sufficient for detecting late-occurring tumors.

- Hueper (1952) fed 15 Osborne-Mendel rats (sex not specified) dietary propham (20,000 ppm, purity not specified) for 18 months. The animals were alternately placed 1 to 2 months on the diet followed by 1 to 2 weeks on a normal diet. Assuming that 1 ppm in the diet of rats is equivalent to 0.05 mg/kg/day (Lehman, 1959), the dietary level was equivalent to 1,000 mg/kg/day. The Time-Weighted Average cannot be calculated due to a lack of detailed reporting of the study design. No tumors were observed in 6 of 8 surviving rats that were evaluated histologically. This study is limited by the low number of animals used, the poor survival rate, short duration, limited histopathological examination and method of treatment.

- Van Esch and Kroes (1972) fed groups of 23 to 26 golden hamsters 0 or 0.2% propham (2,000 ppm, purity not specified) in the diet for 33 months. Assuming that dietary assumptions appropriate for guinea pigs are also appropriate for hamsters and that 1 ppm in the diet of hamsters is equivalent to 0.04 mg/kg/day (Lehman, 1959), these levels are equivalent to 0 or 80 mg/kg/day. Based on histological examination, the authors reported no significant increase in tumor incidence.

IV.   Quantification of Toxicological Effects

A. One-day Health Advisory

No information was found in the available literature that was suitable for determination of the One-day HA value for propham. It is therefore recommended that the Longer-term HA value for a 10-kg child, 5 mg/L, calculated below, be used at this time as a conservative estimate of the One-day HA value.

B. Ten-day Health Advisory

The Longer-term HA of 5 mg/L for a 10-kg child, calculated below, is used for the Ten-day HA since the apparent NOAEL (37.6 mg/kg/day) in the teratology study by Ravert and Parke (1977) was not necessarily the highest NOAEL, due to the large difference between the doses selected (a 10-fold difference between 37.6 and 376 mg/kg/day).

C. Longer-term Health Advisory

The study by Tisdel et al. (1979) has been selected to serve as the basis for the Longer-term HA value for propham. In this study, rats were fed propham in the diet for 91 days. At 100 mg/kg/day, plasma cholinesterase was inhibited (70%) and spleen-to-body weight ratios were increased. No effects were observed at 50 mg/kg/day. This NOAEL is supported by the NOAEL of 75 mg/kg/day identified in the three-generation reproduction study in rats by Ravert (1978).

Using a NOAEL of 50 mg/kg/day, the Longer-term HA for a 10-kg child is calculated as follows:

$$\text{Longer-term HA} = \frac{(50 \text{ mg/kg/day}) (10 \text{ kg})}{(100) (1 \text{ L/day})} = 5.0 \text{ mg/L} (5,000 \text{ } \mu\text{g/L})$$

The Longer-term HA for a 70-kg adult is calculated as follows:

$$\text{Longer-term HA} = \frac{(50 \text{ mg/kg/day})(70 \text{ kg})}{(100) (2 \text{ L/day})} = 17.5 \text{ mg/L} (20,000 \text{ } \mu\text{g/L})$$

D. Lifetime Health Advisory

No chronic study was found in the available literature that was suitable for determination of the Lifetime HA value for propham. The chronic studies by Innes et al. (1969), Hueper (1952) and Van Esch and Kroes (1972) did not provide adequate data on noncarcinogenic end points. In the absence of appropriate chronic data, the 91-day study by Tisdel et al. (1979), which identified a NOAEL of 50 mg/kg/day and was selected to serve as the basis for the Longer-term HA, has also been selected for deriving the Lifetime HA.

Using this study, the Lifetime HA is calculated as follows:

Step 1: Determination of the Reference Dose (RfD)

$$\text{RfD} = \frac{(50 \text{ mg/kg/day})}{(1,000) (3)} = 0.02 \text{ mg/kg/day (rounded from } 0.017 \text{ mg/kg/day)}$$

where: 3 = additional uncertainty factor used to account for a lack of adequate chronic toxicity studies in the data base, preventing establishment of the most sensitive toxicological end point.

Step 2: Determination of the Drinking Water Equivalent Level (DWEL)

$$\text{DWEL} = \frac{(0.017 \text{ mg/kg/day}) (70 \text{ kg})}{(2 \text{ L/day})} = 0.595 \text{ mg/L} (600 \text{ } \mu\text{g/L})$$

Step 3: Determination of the Lifetime Health Advisory

$$\text{Lifetime HA} = (0.595 \text{ mg/L}) (20\%) = 0.12 \text{ mg/L} (100 \text{ } \mu\text{g/L})$$

E. Evaluation of Carcinogenic Potential

- The International Agency for Research on Cancer (IARC, 1976) evaluated propham and concluded that the carcinogenic potential currently cannot be determined.

- Applying the criteria described in EPA's guidelines for assessment of carcinogenic risk (U.S. EPA, 1986a), propham may be classified in Group D: not classifiable. This category is for substances with inadequate animal evidence of carcinogenicity.

V. Other Criteria, Guidance and Standards

- No information on other existing criteria, guidelines and standards was found in the available literature.

VI. Analytical Method

- Analysis of propham is done by Method #4 of the NPS survey methods, "Determination of Pesticides in Ground Water by High Performance Liquid Chromatography with an Ultraviolet Detector" (U.S. EPA, 1986b). In this method a 1 liter sample is extracted with methylene chloride, reduced to 5 mL with methanol substitution, and analyzed by HPLC with a UV detector. The method is being validated in a single laboratory and the estimated detection limit for propham is 0.75 μg/L.

VII. Treatment Technologies

- Whittaker (1980) experimentally determined adsorption isotherms for propham on GAC.

- Whittaker (1980) reported the results of studies with GAC columns operating under bench scale conditions. At a flow rate of 0.8 gal/min/ft^2 and an empty bed contact time of 6 minutes, propham breakthrough (when effluent concentration equals 10% of influent concentration) occurred after 720 bed volumes (BV).

- In the same study, Whittaker (1980) reported the results for seven propham bisolute solutions when passed over the same GAC continuous-flow column.

- The studies cited above indicate that GAC adsorption is the most promising treatment technique for the removal of propham from water. However, selection of an individual technology or combination of technologies for propham removal from water must be based on a case-by-case technical evaluation and an assessment of the economics involved.

VIII. References

Brown, J.H. and P. Gross.* 1949. Acute toxicity study of isopropyl n-phenylcarbamate. Unpublished study. MRID 00075264.

CHEMLAB. 1985. The Chemical Information System, CIS, Inc., Bethesda, MD.

Chen, Y.* 1979. Summary of animal metabolism of IPC. Unpublished study. MRID 00115438.

Cohen, S.Z. 1984. List of potential groundwater contaminants. Memorandum to I. Pomerantz. Washington, D.C.: U.S. Environmental Protection Agency. August 28.

Fang, S.C., E. Fallin, M.L. Montgomery et al.* 1972. Metabolic studies of ^{14}C-labeled propham and chloropropham in female rats. Unpublished study. MRID 00037854.

Fang, S.C. and E. Fallin. 1974. Metabolic studies of ^{14}C-labeled propham and chloropropham in the female rat. Pest. Biochem. Physiol. 4:1–11.

Friedrick, U. and G. Nass. 1983. Evaluation of a mutation test using S49 mouse lymphoma cells and monitoring simultaneously the induction of dexamethasone resistance, 6-thioguanine resistance and ouabain resistance. Mutat. Res. 110:147–162.

Gusik, F.F.* 1976. Photolysis of carbon 14 ring-labeled isopropyl carbanilate (IPC) in water. Unpublished study received Sept. 17, 1979 under 748–224. Submitted by PPG Industries, Inc., Barberton, OH. CDL:240988-C. MRID 00115466.

Hardies, D.E.* 1979. Metabolism of isopropyl carbanilate on a Wooster silt loam soil: BR 21422. Unpublished study received Sept. 17, 1979 under 748–224. Submitted by PPG Industries, Inc., Barberton, OH. CDL:240988-I. MRID 00115472.

Hardies, D.E. and D.Y. Studer.* 1979a. Metabolism of isopropyl carbanilate on a Woodburn silt loam soil: BR 21448. Unpublished study received Sept. 17, 1979 under 748–224. Submitted by PPG Industries, Inc., Barberton, OH. CDL:240998-F. MRID 00115469.

Hardies, D.E. and D.Y. Studer.* 1979b. Metabolism of isopropyl carbanilate on an Altvan loam soil: BR 21531. Unpublished study received Sept. 17, 1979 under 748–224. Submitted by PPG Industries, Inc., Barberton, OH. COL:240988-H. MRID 00115471.

Hardies, D.E. and D.Y. Studer.* 1979c. Metabolism of isopropyl carbanilate on a Hanford sandy loam: BR 21566. Unpublished study received Sept. 17, 1979 under 748–224. Submitted by PPG Industries, Inc., Barberton, OH. CDL:240988-G. MRID 00115470.

Hardies, D.E. and D.Y. Studer.* 1979d. Absorption of isopropyl carbanilate on five soil types: BR 21590. Unpublished study received Sept. 17, 1979 under 748–224. Submitted by PPG Industries, Inc., Barberton, OH. CDL:240987-C. MRID 00038945.

Hardies, D.E. and D.Y. Studer.* 1979e. A laboratory study of the leaching of isopropyl carbanilate in soils. Unpublished study prepared and submitted on Nov. 1, 1984, by PPG Industries, Inc., Chemical Division, Barberton, OH. Accession No. 255364.

Hueper, W.C.* 1952. Carcinogenic studies on isopropyl-n-phenyl-carbamate. Indus. Med. Surg. 21(2):71–74. Also unpublished submission. MRID 00091228.

IARC. 1976. International Agency for Research on Cancer. IARC monographs on the evaluation of carcinogenic risk of chemicals to man. Vol. 12. Lyon, France: IARC.

Innes, J., B. Ulland, M.G. Valerio, L. Petrucelli, L. Fishbein, E. Hart and A. Pallotta. 1969. Bioassay of pesticides and industrial chemicals for tumorigenicity in mice: a preliminary note. J. Natl. Can. Inst. 42:1101–1114.

Lehman, A.J. 1959. Appraisal of the safety of chemicals in foods, drugs and cosmetics. Assoc. Food Drug Off. U.S.

Margard, W.* 1978. Summary report on *in vitro* bioassay of selected compounds. Unpublished study. MRID 00115428.

Meister, R., ed. 1988. Farm chemicals handbook. Willoughby, OH: Meister Publishing Company.

Paulson, G. and A. Jacobsen.* 1974. Isolation and identification of propham metabolites from animal tissues and milk. Unpublished study. MRID 00115440.

Paulson, G., A. Jacobsen and R. Zaylskie.* 1972. Propham metabolism in the rat and goat: Isolation and identification of urinary metabolites. Unpublished study. MRID 00115397.

Pensyl, J. and J.L. Wiedmann.* 1979. Field dissipation of IPC and PPG-124 from soil treated with ChemHoe 135 FL3: BR 21574. Unpublished study received Sept. 17, 1979 under 748–224. Submitted by PPG Industries, Inc., Barberton, OH. CDL:240987-E. MRID 00038947.

PPG Industries, Inc.* 1970. Primary rabbit eye irritation study. International Bio-Test Laboratories. (#A-9252D). Unpublished study. EPA Accession No. 097066.

PPG Industries, Inc.* 1978. Study: IPC toxicity to test subjects. Unpublished study. MRID 00115420.

Ravert, J.* 1978. Three generation reproduction study of IPC in Sprague-Dawley rats. Unpublished study. MRID 00115425.

Ravert, J. and G. Parke.* 1977. Investigation of teratogenic and toxic potential of IPC-50%-rats. Unpublished study. MRID 00115434.

Ryan, A.J. 1971. The metabolism of carbamate pesticides. CRC Crit. Rev. Toxicol. 1:33–51.

STORET. 1988. STORET Water Quality File. Office of Water. U.S. Environmental Protection Agency (data file search conducted in May, 1988).

TDB. 1985. Toxicology Data Bank. Medlars II. National Library of Medicine's National Interactive Retrieval Service.

Terrell, Y. and G. Parke.* 1977. Acute oral toxicity in rats (IPC technical). Unpublished study. MRID 00115421.

Tisdel, M., G. Rao, G. Thompson et al.* 1979. IPC (propham) subchronic oral dosing study in rats. Unpublished study. MRID 00128777.

U.S. EPA. 1986a. U.S. Environmental Protection Agency. Guidelines for carcinogen risk assessment. Fed. Reg. 51(185)33992-34003. September 24.

U.S. EPA. 1986b. U.S. Environmental Protection Agency. U.S. EPA Method 4—Determination of pesticides in ground water by HPLC/UV, January 1986 draft. Available from U.S. EPA's Environmental Monitoring and Support Laboratory, Cincinnati, OH.

Van Esch, G.J. and R. Kroes. 1972. Long-term toxicity studies of chloropropham and propham in mice and hamsters. Food Cosmet. Toxicol. 10:373-381.

Whittaker, K.F. 1980. Adsorption of selected pesticides by activated carbon using isotherm and continuous flow column systems. Ph.D. Thesis. Lafayette, IN: Purdue University.

Wiedmann, J.L. and J. Pensyl.* 1981. Dissipation of IPC and PPG-124 in soil treated with ChemHoe 135—Spring 1980: BR 22412. Unpublished study received Dec. 20, 1982 under 748-224. Submitted by PPG Industries, Inc., Barberton, OH. CDL:249100-A. MRID 00121299.

Wiedmann, J.L., D. Mattle, D.R. Coffman and J. Pensyl.* 1982. Determination of IPC and PPG-124 residues in soil treated with carbon 14 labeled ChemHoe 135: BR 22882. Unpublished study received Dec. 20, 1982 under 748-224. Submitted by PPG Industries, Inc., Barberton, OH. CDL:249100-B. MRID 00121300.

Worthing, C.R. 1979. Pesticide manual, 6th ed. British Crop Protection Council, Worchestershire, England.

---

* Confidential Business Information submitted to the Office of Pesticide Programs.

Simazine

The information used in preparing this Health Advisory was collected primarily from the open literature and the Simazine Registration Standard (U.S. EPA, 1983).

I. General Information and Properties

A. CAS No. 122–34–9

B. Structural Formula

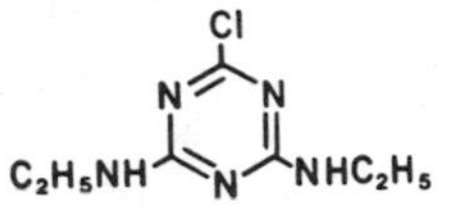

2-Chloro-4,6-bis(ethylamino)-1,3,5-triazine

C. Synonyms

- Aquazine, Cekusan, Framed (discontinued by Farmoplant), G-27692, Gesatop, Primatol, Princep, Simadex, Simanex, Tanzene (Meister, 1984).

D. Uses

- Simazine is used as a selective preemergence herbicide for control of most annual grasses and broadleaf weeds in corn, alfalfa, established Bermuda grass, cherries, peaches, citrus, different kinds of berries, grapes, apples, pears, certain nuts, asparagus, certain ornamental and tree nursery stock and in turf grass soil production (Meister, 1984). It is also used to inhibit the growth of most common forms of algae in aquariums, ornamental fish ponds and fountains. At higher rates, it is used for nonselective weed control in industrial areas.

E. Properties (Meister, 1984; Windholz et al., 1983; Reinert and Rogers, 1987)

Chemical Formula	$C_7H_{12}ClN_5$
Molecular Weight	201.69
Physical State (room temperature)	White, crystalline solid
Boiling Point	--
Melting Point	225 to 227°C
Density	1.302 g/cm^3
Vapor Pressure (20°C)	6.1×10^{-9} mm Hg
Water Solubility (20°C)	3.5 mg/L
Log Octanol/Water Partition Coefficient	2.51
Taste Threshold	--
Odor Threshold	--
Conversion Factor	--

F. Occurrence

- Simazine has been found in 922 of 5,873 surface water samples analyzed and in 202 of 2,654 ground-water samples (STORET, 1988). Samples were collected at 620 surface water locations and 2,128 ground-water locations. The 85th percentile of all nonzero samples was 2.18 μg/L in surface water and 1.60 μg/L in ground-water sources. The maximum concentration found in surface water was 1,300 μg/L, and in ground water it was 800 μg/L. Simazine was found in surface water in 16 states and in ground water in 8 states. This information is provided to give a general impression of the occurrence of this chemical in ground and surface waters as reported in the STORET database. The individual data points retrieved were used as they came from STORET and have not been confirmed as to their validity. STORET data are often not valid when individual numbers are used out of the context of the entire sampling regime, as they are here. Therefore, this information can only be used to form an impression of the intensity and location of sampling for a particular chemical.

- Simazine has been found in ground water in California, Pennsylvania and Maryland; typical positives were 0.2 to 3.0 ppb (Cohen et al., 1986).

G. Environmental Fate

- Simazine did not hydrolyze in sterile aqueous solutions buffered at pH 5, 7 or 9 (20°C) over a 28-day test period (Gold et al., 1973).

- Under aerobic soil conditions, the degradation of simazine depends largely on soil moisture and temperature (Walker, 1976). In a sandy loam soil, half-lives ranged from 36 days to 234 days. Simazine applied to loamy sand and silt loam soils and incubated (25 to 30°C) for 48 weeks, dissipated with half-lives of 16.3 and 25.5 weeks, respectively (Monsanto Company, no date). Simazine degradation products 2-chloro-4-ethylamino-6-amino-s-triazine (G-28279), 2-chloro-4,6 bis(amino)-s-triazine, and several unidentified polar compounds were detected 32 and 70 days after a sandy loam soil had been treated with ^{14}C-simazine (Beynon et al., 1972). The degradates 2-hydroxy-4,6-bis(ethylamino)-s-triazine and 2-hydroxy-4-ethylamino-6-amino-s-triazine were also detected in aerobic soil (Keller, 1978).

- Under anaerobic conditions, ^{14}C-simazine had a half-life of 8 to 12 weeks in a loamy sand soil (Keller, 1978). The treated soil (10 ppm) was initially maintained for 1 month under aerobic conditions, followed by 8 weeks under anaerobic conditions (flooded with water and nitrogen). Degradates found included G-28279, 2-chloro-4,6-bis(amino)-s-triazine, 2-hydroxy-4,6-bis(ethylamino)-s-triazine, and 2-hydroxy-4-ethylamino-6-amino-s-triazine.

- Simazine is expected to be slightly to very mobile in soils ranging in texture from clay to sandy loam, based on column leaching, soil thin-layer chromatography (TLC) and adsorption/desorption (batch equilibrium) studies. Using batch equilibrium tests, K_d values determined for 25 Missouri soils ranged from 1.0 for a sandy loam to 7.9 for a silty loam (Talbert and Fletchall, 1965). Simazine adsorption was correlated with soil organic matter content and, to a lesser extent, with cation exchange capacity (CEC) and clay content (Talbert and Fletchall, 1965; Helling and Turner, 1968; Helling, 1971). Simazine exhibited low mobility in peat and peat moss (K_d more than 21) and a higher mobility in clay fractions (K_d values ranged from 0.0 for kaolinite to 12.2 for montmorillonite (Talbert and Fletchall, 1965). Freundlich K and n values were determined to be 7.25 and 0.88, respectively, for a silty clay loam soil.

- Simazine, as determined by soil TLC, is mobile to very mobile in sandy loam soil (R_f 0.80 to 0.96), and of low to intermediate mobility in loam and silty clay loam (R_f 0.45), sandy clay loam (R_f 0.51), silt loam (R_f 0.16 to 0.51), clay loam (R_f 0.32 to 0.45) and silty clay (R_f 0.36) soils. R_f

values were positively correlated with soil organic matter and clay content (Helling and Turner, 1968; Helling, 1971).

- Based on results of soil column leaching studies, simazine phytotoxic residues were slightly mobile to mobile in soils ranging in texture from clay loam to sand (Harris, 1967; Ivey and Andrews, 1965; Rodgers, 1968). Upon application of 18 inches of water to 30-inch soil columns containing clay loam, loam, silt loam or fine sandy loam soils, simazine phytotoxic residues leached to depths of 4 to 6, 10 to 12, 22 to 24, and 26 to 28 inches, respectively (Ivey and Andrews, 1965).

- In field studies, simazine had a half-life of about 30 to 139 days in sandy loam and silt loam soils (Mattson et al., 1969; Martin et al., 1975; Walker, 1976). The degradate, 2-chloro-4-ethylamino-6-amino-s-triazine (G-28279), was detected at the 0- to 6-inch depth and at the 6- to 12-inch depth (Mattson et al., 1969; Martin et al., 1975).

- Simazine residues (uncharacterized) may persist up to 3 years in soil under aquatic field conditions. Dissipation of simazine in pond and lake water was variable, with half-lives ranging from 50 to 700 days. The degradation compound G-28279 was identified in lake water samples, but was no more persistent than the parent compound (Larsen et al., 1966; Flanagan et al., 1968; Kahrs, 1969; LeBaron, 1970; Smith et al., 1975; Kahrs, 1977).

- A recent review of simazine environmental fate is provided by Reinert and Rogers (1987). This review may provide additional information to the data summarized in the above sections.

II.   Pharmacokinetics

A. Absorption

- Orr and Simoneaux (1986) studied the metabolism of ^{14}C-simazine (uniformly labeled in the triazine ring) in groups of five male and five female Sprague-Dawley rats following oral administration in single doses at 0.5 or 200 mg/kg. At the low dose, 51 to 62% of the dose was eliminated as simazine equivalents in the urine and about 12% was found in the tissues, suggesting that about 63 to 74% (0.315 to 0.37 mg/kg) of the dose was absorbed. At the high dose, only about 22% (44 mg/kg) of the dose was found in urine and 2% (4 mg/kg) in the tissues of males or females. A third group of animals received daily single oral doses of unlabeled simazine at 0.5 mg/kg/day for 14 days followed by a single ^{14}C dose.

Elimination of ^{14}C in urine ranged from 59 to 66% of the dose, and 8% was found in the tissues. Bakke and Robbins (1968) reported that, in goats and sheep, from 67 to 77% of a dose (the dose was not reported in this abstract) of ^{14}C-simazine (given orally in gelatin capsules) was excreted in urine. This suggests absorption was around 70%.

B. Distribution

- Orr and Simoneaux (1986) reported that 7 days following oral administration of ^{14}C-simazine to male and female Sprague-Dawley rats at 0.5 or 200 mg/kg, the highest ^{14}C residue levels, expressed as simazine equivalents found in the red blood cells, were 16.3 to 19.9 ppm at the high dose and 0.18 to 0.23 ppm at the low dose. Residue levels in red blood cells were slightly higher in males than in females. Lower concentrations were found in the other tissues ranging from 0.0 to 0.16 ppm at the low dose and from 0.78 to 5.2 ppm at the high dose. Relatively higher residues were found in the liver and kidney of rats receiving the low dose (0.1 to 0.16 ppm), with lower levels found in fat, plasma and bone (0.0 to 0.03 ppm). A similar pattern was observed in rats receiving the high dose, except that the spleen contained higher ^{14}C residues (4.1 to 5.2 ppm) than the liver and kidney (2.9 to 4.0 ppm). Residue levels in tissues of animals receiving the low ^{14}C dose following 14 days of repeated dosing were generally lower than those in rats receiving a single low dose, except in red blood cells. The authors suggested that repeated dosing with unlabeled simazine significantly reduced the number of available binding sites in all tissues, except the red blood cells.

- In rats receiving the high dose, ^{14}C residue levels, expressed as ppm, were 29-fold (kidney) to 516-fold (spleen) higher than those in animals receiving the low dose (Orr and Simoneaux, 1986).

C. Metabolism

- Simoneaux and Sy (1971) conducted a preliminary investigation on the metabolites of ^{14}C ring-labeled simazine found in the urine of female white rats. The rats were dosed once orally at 1.5 mg/kg, and the urine eliminated within 24 hours was collected and analyzed by thin-layer chromatography and thin-layer electrophoresis. The metabolites identified by comparison to authentic compounds were conjugated mercapturates of the following 3 metabolites: hydroxysimazine, 2-hydroxy-4-amino-6-ethylamino-s-triazine and 2-hydroxy-4,6-diamino-s-triazine. These metabolites accounted for 6.8, 6.1 and 14% of the administered radioactivity, respectively. When the urine was hydrolyzed with performic acid (oxidizing agent) to cleave sulfur-carbon bonds of the conjugated mercapturates, prior to thin-layer chromatography, about 8.1, 7.7 and 31.3% of the radiolabel was detected as the three identified metabolites, respectively. Approximately half of the radioactivity in the urine was not identified.

- Bradway and Moseman (1982) administered simazine to male Charles River rats by gavage. Two doses of 0.017, 1.7, 17 or 167 mg/kg were given 24 hours apart. In 24-hour urine samples, the di-N-dealkylated metabolite (2-chloro-4,6-diamino-s-triazine) appeared to be the major product, ranging from 1.6% at the 1.7 mg/kg-dose to 18.2% at the 167-mg/kg dose, while the mono-N-dealkylated metabolite ranged from 0.35% at the 1.7-mg/kg dose to 2.8% at the 167-mg/kg dose.

- Guddewar and Dauterman (1979) purified glutathione S-transferase 61-fold from mouse liver. This enzyme conjugates chloro-s-triazine herbicides. A chloro group at the C2-position was found to be necessary for conjugation to occur. When this 2-chloro was replaced by a methylmercapto group, no conjugation occurred. The reaction rate decreased with triazine analogs lacking the alkyl side-chains. Atrazine was conjugated faster than either simazine or propazine.

- Similar results were obtained by Bohme and Bar (1967), who fed simazine (formulation and purity not stated) at levels of 200 or 800 mg/kg to albino rats and at 240 to 400 mg/kg to rabbits. Of the several metabolites identified, all retained the triazine ring intact. The principal species were the mono- and di-N-dealkylated metabolites.

- Bakke and Robbins (1968) administered ^{14}C-simazine orally by gelatin capsules to goats and sheep. The sheep were given simazine labeled on the triazine ring or on the ethylamino side-chain, while goats were given the ring-labeled compound only. Based on the metabolites identified in the urine of animals receiving the ring-labeled compound, there was no evidence to suggest that the triazine ring was metabolized. In sheep that received chain-labeled triazines, at least 40% of the ethylamino side-chains were removed. Using ion-exchange chromatography, 18 labeled metabolites were found in urine.

- Bohme and Bar (1967) and Larsen and Bakke (1975) observed that rat and rabbit urinary metabolites from the 2-chloro-s-triazines were all 2-chloro analogs of their respective parent molecules and none of the metabolites contained the 2-hydroxy moiety. Total N-dealkylation, partial N-dealkylation and N-dealkylation with N-alkyl oxidation were suggested as the major routes of the metabolism of 2-chloro-s-triazines in rats and rabbits.

D. Excretion

- Orr and Simoneaux (1986) studied the metabolism of ^{14}C-simazine (uniformly labeled in the triazine ring) in Sprague-Dawley rats following oral administration in single doses. Males and females receiving 0.5 mg/kg eliminated 50.5 and 62.1% of the dose, respectively, in the urine, and 19.1 and 13.3% of the dose in the feces, 7 days after dosing. In animals receiving a single dose of ^{14}C-simazine following repeated dosing at 0.5 mg/kg/day for 14 days with unlabeled simazine, males eliminated 58.5 and 24.5% of the dose in the urine and feces, respectively, whereas females eliminated 66.0 and 17.8% of the dose. A third group of males and females receiving a single dose of 200 mg/kg eliminated about 21% of the dose in the urine and about 2.0% in the feces. The elimination pattern with time after dosing was biphasic with rates being faster in females than males. Most of the radioactivity was eliminated in 72 hours with a half-life of 9 to 15 hours. Elimination of the remaining radioactivity occurred at a slower rate with half-life values ranging from 21 to 32 hours.

- In a preliminary study with two female white rats dosed orally at 1.5 mg/kg ring-labeled ^{14}C-simazine, Simoneaux and Sy (1971) found less than 0.05% of the dose being eliminated as ^{14}C-carbon dioxide 96 hours after dosing. Approximately 49.3% of the dose was eliminated in the urine and 40.8% in the feces.

- Bakke and Robbins (1968) studied the excretion of simazine in goats and sheep using triazines labeled with ^{14}C on the ring or on the ethylamino side-chains. No ^{14}CO$_2$ was detected from animals that received the ring-labeled compounds, which suggested that the triazine ring was not metabolized. Approximately 67 to 77% of the administered ring-labeled activity was found in the urine, and 13 to 25% was found in the feces. About 0.16 ppm of ^{14}C residues were present in the milk immediately after dosing, but the level decreased to 0.04 ppm within 48 hours.

- St. John et al. (1965) fed unlabeled simazine (5 ppm) to a lactating cow for 3 days. Urine and milk were collected and analyzed during the feeding period and for 3 days thereafter. No simazine was detected in the milk (sensitivity of method approximately 0.1 ppm), and only about 1% of the administered simazine was excreted in the urine as the parent compound. Milk, excreta and body tissues were not analyzed for simazine metabolites.

- Hapke (1968) reported that simazine residues were present in the urine of sheep for up to 12 days after administration of a single oral dose. The maximum concentration in the urine occurred from 2 to 6 days after administration.

III.   Health Effects

A. Humans

- There were 124 cases of contact dermatitis noted by Yelizarov (1977) in the Soviet Union among workers manufacturing simazine and propazine. Mild cases lasting 3 or 4 days involved pale pink erythema and slight edema. Serious cases lasting 7 to 10 days involved greater erythema and edema and a vesiculopapular reaction that sometimes progressed to the formation of bullae.

B. Animals

1. Short-term Exposure

- Oral LD_{50} values for simazine have been reported to be greater than 5,000 mg/kg in the rat (Martin and Worthing, 1977), the mouse and the rabbit (USDA, 1984).

- Mazaev (1964) administered a single oral dose of simazine (formulation and purity not stated) to rats at 4,200 mg/kg. Anorexia and weight loss were observed, with some of the animals dying in 4 to 10 days. When 500 mg/kg was administered daily, all the animals died in 11 to 20 days, with the time of death correlating with the loss of weight.

- Sheep and cattle seem to be much more susceptible than laboratory animals to simazine toxicity. Hapke (1968) reported that a single oral dose of simazine, 50% active ingredient (a.i.), as low as 500 mg/kg, was fatal to sheep within 6 to 25 days after administration. The animals that survived the exposure were sick for 2 to 4 weeks after treatment and showed loss of appetite, increased intake of water, incoordination, tremor and weakness. Some of the animals exhibited cyanosis and clonic convulsions.

- Palmer and Radeleff (1969) orally exposed cattle by drench to 10 doses of simazine 80W (purity not stated) at 10, 25 or 50 mg/kg/day and sheep by drench or capsule to 10 doses at 25, 50, 100 or 250 mg/kg. The number of test animals in each group was not stated, and the use of controls was not indicated. Anorexia, signs of depression, muscle spasms, dyspnea, weakness and uncoordinated gait were commonly observed in treated animals. Necropsy showed congestion of lungs and kidneys; swollen, friable livers; and small, hemorrhagic spots on the surface of the lining of the heart.

- Palmer and Radeleff (1964) found that repeated oral administration of simazine 80W (purity not stated) at either 31 daily doses of 50 mg/kg or 14 daily doses of 100 mg/kg was fatal to sheep. Simazine was also lethal when administered at 100 mg/day for 14 days by drench (Palmer and Radeleff, 1969).

- The acute inhalation LC_{50} value of simazine is reported to be more than 2.0 mg/L of air (4-hour exposure) (Weed Science Society of America, 1983).

2. Dermal/Ocular Effects

- The acute dermal toxicity in rabbits is greater than 8,000 mg/kg (NAS, 1977).

- In a 21-day subacute dermal toxicity study in rabbits, Ciba-Geigy (1980) reported that 15 dermal applications of technical simazine at doses up to 1 g/kg produced no systemic toxicity or any dose-related alterations of the skin.

- In primary eye irritation studies in rabbits, simazine caused transient inflammation of conjunctivae, but was not irritating to the iris or cornea (USDA, 1984).

3. Long-term Exposure

- Tai et al. (1985a) conducted a 13-week subacute oral toxicity study in Sprague-Dawley rats fed technical simazine at 0, 200, 2,000 or 4,000 ppm in their diets. Assuming that 1 ppm in the diet of rats is equivalent to 0.05 mg/kg/day (Lehman, 1959), these levels correspond to doses of about 0, 10, 100 or 200 mg/kg/day. Significant dose-related reductions in food intake, mean body weight and weight gain occurred in all treated groups. Significant weight loss occurred in mid- and high-dose animals during the first week of dosing. At 13 weeks, various dose-related effects were noted in hematological parameters (decreased mean erythrocyte and leukocyte counts and increased neutrophil and platelet counts), clinical chemistry [lowered mean blood glucose, sodium, calcium, blood urea nitrogen (BUN), lactic dehydrogenase (LDH), serum glutamic-oxaloacetic transaminase (SGOT) and creatinine and increased cholesterol and inorganic phosphate levels] and urinalysis determinations (elevated ketone levels and decreased protein levels). Relative and absolute adrenal, brain, heart, kidney, liver, testes and spleen weights increased, and ovary and heart weights decreased. Necropsies revealed no gross lesions attributable to simazine. A dose-related incidence of renal calculi and renal epithelial hyperplasia was detected microscopically in treated rats, primarily in the renal pelvic lumen and rarely in the renal tubules. Microscopic examinations revealed no other lesions that could be attributed to simazine. It appeared to the authors that reduced mean food intake in treated rats was most likely due to the unpalatability of simazine. Lower individual body weights and reduced body weight gains paralleled mean food intake in treated rats. The majority of the alterations in clinical chemistry values may have been related to reduced food consumption. Since these dietary levels of simazine seriously affected the nutritional status of treated rats, the results of this study are of limited value.

- Tai et al. (1985b) also conducted a 13-week dietary study with beagle dogs fed technical simazine at 0, 200, 2,000 or 4,000 ppm. Assuming that 1 ppm in the diet of dogs is equivalent to 0.025 mg/kg/day (Lehman, 1959), these levels correspond to doses of about 0, 5, 50 or 100 mg/kg/day. As in the previously described study in rats, reduced daily food consumption was attributed to the palatability of simazine in the diet and corresponded with weight loss, decreased weight gain and various effects on hematology, clinical chemistry, and urinalysis determinations. Changes in these parameters were generally similar to those noted in the rat study (Tai et al., 1985a). Due to the seriously affected nutritional status of the test animals, the results of this study are of limited value.

- Dshurov (1979) studied the histological changes in the organs of 21 sheep following exposures to simazine (50% a.i.) by gavage at 0, 1.4, 3.0, 6.0, 25, 50, 100 or 250 mg/kg/day for various time durations up to about 22 weeks. Fatty and granular liver degeneration, diffuse granular kidney degeneration, neuronophagia, diffuse glial proliferation and degeneration of ganglion cells in the cerebrum and medulla were found. In sheep that died, spongy degeneration, hyperemia and edema were observed in the cerebrum; the degree of severity was related to the dose of simazine and the duration of exposure. The thyroid showed hypofunction after daily doses of 1.4 to 6.0 mg/kg were administered for periods of 63 to 142 days. The most severe antithyroid effect followed one or two doses of 250 mg/kg, which in one sheep produced parenchymatous goiter and a papillary adenoma. This type of goiter was also seen in sheep administered simazine at 50 or 100 mg/kg once per week for approximately 22 weeks. Based on these data, a Lowest-Observed-

Adverse-Effect Level (LOAEL) of 1.4 mg/kg can be identified; however, it is not clear from the study whether the authors considered the 50% formulation when providing the dosage levels.

4. Reproductive Effects

- Woodard Research Corporation (1965) reported that no adverse effects on reproductive capacity were observed in a three-generation study in rats. In this study, two groups of 40 weanling rats (20/sex) were used; one served as the control and the other was fed simazine 80W at 100 ppm. This corresponds to a dose of about 5 mg/kg/day, based on the assumption that 1 ppm in the diet of rats corresponds to 0.05 mg/kg/day (Lehman, 1959). After 74 days of dosing, animals were paired and mated for 10 days, resulting in F_{1a} litters. After weaning first litters, parents were remated to produce F_{1b} litters. Weanlings of parents in the 100-ppm group were divided into two groups and fed simazine at 50 ppm (approximately 2.5 mg/kg/day) or at 100 ppm. After 81 days they were mated to produce the F_{2a} and F_{2b} litters. F_{2b} weanlings were fed the same dietary levels of simazine (0, 50 or 100 ppm). F_{2b} rats were mated to produce F_{3a} and F_{3b} litters. Reproductive performance of male rats could not be adequately evaluated in this study. Noted decreases in body weight for the animals (parents and progeny) did not appear to be significant at the end of the study period because the final mean body weights were within the acceptable changes for the maximum tolerated dose. Reproductive performance of female rats fed simazine was basically similar to that of controls, and no developmental changes were detected. The No-Observed-Adverse-Effect Level (NOAEL) for this study is approximately 5 mg/kg/day.

- Dshurov (1979) reported that repeated administration of simazine (50% a.i.) to sheep (6.0 mg/kg for 142 days or 25 mg/kg for 37 to 111 days) caused changes in the germinal epithelium of the testes and disturbances of spermatogenesis.

5. Developmental Effects

- Simazine (97% a.i.) was administered by gavage to a group of 19 New Zealand White rabbits (2.5 to 4.5 kg) at daily doses of 5, 75 or 200 mg/kg for days 7 through 19 of presumed gestation (Ciba-Geigy, 1984). The herbicide was delivered as a suspension in 3% cornstarch containing 0.5 percent Tween 80. A control group was given the vehicle only. One death was observed in each group; however, death of the low-dosed dam was assumed to be due to a dosing accident and not compound-related. Maternal toxicity at 75 and 200 mg/kg was indicated by abortions, tremors, decreased motor control and activity, ataxia (in one high-dosed animal), few or no stools, anorexia, weight loss, and decreased body weight gain. Embryotoxicity was not observed, but fetal toxicity, which was believed to be the consequence of maternal toxicity, was evident in the intermediate- and high-dose groups as indicated by decreased numbers of viable fetuses. Fetal toxicity at 200 mg/kg was also reflected by reduced fetal body weights, increased occurrence of floating and fully formed ribs and decreased ossification of the patellae. No malformations were associated with any dose level of simazine, and at 5 mg/kg the herbicide exerted neither toxic nor teratogenic effects. The authors concluded that at doses of 75 mg/kg/day and higher, simazine was very toxic to fetuses and dams but was neither embryotoxic nor teratogenic.

- Chen (1981) studied the teratogenic effects of simazine in rats. The herbicide was administered by gastric intubation to pregnant rats from the 6th to 15th day of gestation at dose levels of 78, 312, 1,250 and 2,500 mg/kg bw/day. Ossification of the sternum and cranium was delayed at all four dose levels. At doses of 312 mg/kg and above, body weight gain was inhibited in the dams, the number of live fetuses was reduced and the rate of fetal resorption increased. At the higher dose level (2,500 mg/kg), simazine caused delayed fetal development. The author concluded that simazine was toxic to rat embryos but was not teratogenic.

- No treatment-related developmental effects were observed by Newell and Dilley (1978) in the offspring of rats exposed to simazine at 0, 17, 77 and 317 mg/m³ via inhalation for 1 to 3 hours/day on days 7 through 14 of gestation.

6. *Mutagenicity*

- Simazine has shown negative results in a variety of microbial mutagenicity assay systems including tests with the following organisms: *Salmonella typhimurium* (Simmons et al., 1979; Eisenbeis et al., 1981; Anderson et al., 1972); *Escherichia coli* (Simmons et al., 1979; Fahrig, 1974); *Bacillus subtilis* (Simmons et al., 1979); *Serratia marcescens* (Fahrig, 1974); and *Saccharomyces cerevisiae* (Simmons et al., 1979).

- Simazine induced lethal mutations in the sex-linked recessive lethal test using the fruitfly *Drosophila melanogaster* (Valencia, 1981). In a study reported by Murnik and Nash (1977), simazine increased X-linked lethals when injected into male *D. melanogaster*, but failed to do so when fed to larvae.

- There are contradictory data concerning the ability of simazine to cause DNA damage. According to Simmons et al. (1979), simazine induced unscheduled DNA synthesis in a human lung fibroblast assay. However, in the same test conducted by Waters et al. (1982), simazine showed a negative response.

- Simazine does not produce chromosomal effects as indicated by the sister-chromatid exchange test and mouse micronucleus assay (Waters et al., 1982).

7. Carcinogenicity

- Simazine was not tumorigenic in an 18-month feeding study in mice at the highest tolerated dose of 215 mg/kg/day (Innes et al., 1969). In this bioassay of 120 compounds, male and female mice of two hybrid strains (C57BL/6 × C3H/Anf)F_1 and (C57BL/6 × AKR)F_1 were exposed to simazine (purity not stated) at the maximum tolerated dose of 215 mg/kg by gavage from ages 7 to 28 days. For the remainder of the study, the animals were maintained on a diet with simazine at 215 mg/kg/day. Based on information presented only in tabular form, gross necropsy and histological examination revealed no significant increase in tumors related to treatment with simazine. Other toxicological data were not provided. This study is not considered to provide adequate data to fully assess the carcinogenic potential of simazine.

- Hazelton Laboratories (1960) conducted a 2-year dietary study in Charles River rats administered simazine 50W (49.9% a.i.) in the feed at 0, 1, 10 and 100 ppm (expressed on the basis of 100% a.i.). Based on the dietary assumptions of Lehman (1959), these levels are equivalent to approximately 0, 0.05, 0.5 and 5 mg/kg/day. These authors reported an excess of thyroid and mammary tumors in high-dose females. However, complete histopathological details were not provided and statistical significance was not evaluated. Furthermore, the high incidence of respiratory and ear infections in all groups renders this study unsuitable for evaluating the carcinogenic potential of simazine.

- Simazine was found to produce sarcomas at the site of subcutaneous injection in both rats and mice (Pliss and Zabezhinsky, 1977; abstract only).

- An interim report by Ciba-Geigy Corporation (1987) on a new chronic feeding/oncogenicity study in Sprague-Dawley rats indicates that simazine may cause mammary gland tumors. However, when this study is complete, an evaluation of these data will be performed.

IV.   Quantification of Toxicological Effects

A. One-day Health Advisory

No suitable studies were found in the available literature for the determination of the One-day HA value for simazine. It is therefore recommended that the Ten-day HA of 0.5 mg/L (500 μg/L), calculated below, be used at this time as a conservative estimate of the One-day HA value.

B. Ten-day Health Advisory

No suitable studies were found in the available literature for the determination of the Ten-day HA value for simazine, with the exception of a rabbit teratology study by Ciba-Geigy (1984). In this study, simazine was administered by gavage to female rabbits during gestation days 7 to 19 (13 days). A NOAEL of 5 mg/kg/day for maternal toxicity was established and a LOAEL of 75 mg/kg/day was noted based on abortions and tremors, decreased motor control and activity, anorexia and weight loss. The NOAEL of 5 mg/kg/day in the rabbit is selected for use as the basis for calculation of the Ten-day HA.

The Ten-day HA for a 10-kg child is calculated as follows:

$$\text{Ten-day HA} = \frac{(5.0 \text{ mg/kg/day}) (10 \text{ kg})}{(1 \text{ L/day}) (100)} = 0.5 \text{ mg/L (500 μg/L)}$$

C. Longer-term Health Advisory

No suitable studies were found in the available literature for the determination of the Longer-term HA values for simazine. It is therefore recommended that the adjusted DWEL of 0.05 mg/L (50 μg/L), calculated below, be used at this time as a conservative estimate of the Longer-term HA value for a 10-kg child and that the DWEL of 0.2 mg/L (200 μg/L) be used for a 70-kg adult.

For a 10-kg child, the adjusted DWEL is calculated as follows:

$$\text{DWEL} = \frac{(0.005 \text{ mg/kg/day}) (10 \text{ kg})}{1 \text{ L/day}} = 0.05 \text{ mg/L}$$

where: 0.005 mg/kg/day = RfD (see Lifetime Health Advisory Section)

D. Lifetime Health Advisory

The three-generation reproduction study in rats by Woodard Research Corporation (1965) has been selected to serve as the basis for calculation of the DWEL and Lifetime HA for simazine. In this study, two groups of 40 weanling rats (20/sex) were used; one served as the control, and the other was fed simazine 80W at 100 ppm (approximately 5 mg/kg/day). After 74 days of dosing, animals were paired and mated for 10 days, resulting in F_{1a} litters. After weaning first litters, parents were remated to produce F_{1b} litters. Weanlings of parents in the 100-ppm group were divided into two test groups: one group was fed simazine at 50 ppm (about 2.5 mg/kg/day) and the other at 100 ppm. After 81 days of dosing, animals were mated to produce the F_{2a} and F_{2b} litters. The F_{2b} weanlings were then divided into 50- and 100-ppm dosage groups. F_{2b} rats were mated to produce F_{3a} and F_{3b} litters. Reproductive performance of female rats fed simazine was the same as that of controls, and no teratological changes were detected. The NOAEL for this study is 5 mg/kg/day. It is important to note that, in this study, rats in the F_0 generation were exposed to simazine at the high dose (100 ppm) only. However, considering that the F_1 and F_2 generations treated with 100 ppm did not reflect any adverse reproductive effects, this feature of the study design did not seem to affect the results. Therefore, the NOAEL of 5 mg/kg/day is used for calculation of the RfD.

Step 1: Determination of the Reference Dose (RfD)*

$$RfD = \frac{5 \text{ mg/kg/day}}{(1,000)} = 0.005 \text{ mg/kg/day}$$

Step 2: Determination of the Drinking Water Equivalent Level (DWEL)

$$DWEL = \frac{(0.005 \text{ mg/kg/day}) (70 \text{ kg})}{(2 \text{ L/day})} = 0.175 \text{ mg/L} (200 \text{ μg/L})$$

Step 3: Determination of the Lifetime Health Advisory

$$\text{Lifetime HA} = \frac{(0.175 \text{ mg/L}) (20\%)}{10} = 0.0035 \text{ mg/L} (4 \text{ μg/L})$$

E. Evaluation of Carcinogenic Potential

- An interim report by Ciba-Geigy Corporation (1987) on a new chronic feeding/oncogenicity study in Sprague-Dawley rats indicates that simazine reflects some oncogenic activity in the mammary glands as previously noted with atrazine and propazine. Apparently this positive study has been completed recently and submitted to the Office of Pesticide Programs. An evaluation of this study will be performed in the near future. Neither the study in mice by Innes et al. (1969) nor the study in

---

*This RfD is considered as an interim RfD until the Agency's identified data gaps for chronic effects in rats and dogs are filled, and until a new reproduction study is conducted, so that the flaws noted in the present study (i.e., number of doses tested, male fertility, etc.) are corrected.

rats by Hazelton Laboratories (1960) is considered adequate for assessment of the carcinogenicity of this substance. However, the Hazelton rat study (1960) still reflected a potential positive oncogenic effect in the same target organ, the mammary glands.

- Simazine is a chloro-s-triazine derivative, with a chemical structure analogous to atrazine and propazine. Both of these structurally related compounds were found to significantly ($p > 0.05$) increase the incidence of mammary tumors in rats. The structure-activity relationship of this group of chemicals indicates that simazine is likely to reflect a similar pattern of oncogenic response in rats as atrazine and propazine.

- Applying the criteria described in EPA's guidelines for the assessment of carcinogenic risk (U.S. EPA, 1986a), simazine may be preliminarily classified in Group C: possible human carcinogen. This category is used for substances with limited evidence of carcinogenicity in animals in the absence of human data. This classification is considered preliminary until the Office of Pesticide Programs completes a peer review of the weight of the evidence for simazine and its analogs.

V.   Other Criteria, Guidance and Standards

- A tolerance level of 10 μg/L has been established for simazine and its metabolites in potable water when present as a result of application to growing aquatic weeds (U.S. FDA, 1979).

- Residue tolerances have been established for simazine alone and the combined residues of simazine and its metabolites in or on various raw agricultural commodities (U.S. EPA, 1986b). These tolerances range from 0.02 ppm (negligible) in animal products to 15 ppm in various animal fodders.

VI.   Analytical Methods

- Analysis of simazine is by a gas chromatographic (GC) method applicable to the determination of certain nitrogen-phosphorus containing pesticides in water samples (U.S. EPA, 1988). In this method, Method #507, approximately 1 L of sample is extracted with methylene chloride. The extract is concentrated and the compounds are separated using capillary column GC. Measurement is made using a nitrogen-phosphorus detector. The method detection limit has not been determined for the analytes in this method, including simazine. The estimated detection limit is 0.075 μg/L.

VII.   Treatment Technologies

- Treatment technologies that will remove simazine from water include activated carbon adsorption; ion exchange; and chlorine, chlorine dioxide, ozone, hydrogen peroxide and potassium permanganate oxidation. Conventional treatment processes were relatively ineffective in removing simazine (Miltner and Fronk, 1985a). Limited data suggest that aeration would not be effective in simazine removal (ESE, 1984; Miltner and Fronk, 1985a).

- Baker (1983) reported that a 16.5-inch granular-activated carbon (GAC) filter cap using F-300, which was placed upon the rapid sand filters at the Fremont, Ohio, water treatment plant and had been in service for 30 months, reduced the simazine levels by 35 to 89% in the water from the Sandusky River. Miltner and Fronk (1985a) developed adsorption capacity data using spiked, distilled water treated with Filtrasorb 400. The following Freundlich isotherm values were reported for simazine: K = 490 mg/g; l/n = 0.56.

- At the Bowling Green, Ohio, water treatment plant, powder-activated carbon (PAC) in conjunction with conventional treatment achieved an average reduction of 47% of the simazine levels in the water from the Maumee River (Baker, 1983). Miltner and Fronk (1985b) monitored simazine levels at water treatment plants, which utilized PAC, in Bowling Green and Tiffin, Ohio. Applied at dosages ranging from 3.6 to 33 mg/L, the PAC achieved 43 to 100% removal of simazine with higher percent removals reflecting higher PAC dosages. Andersen (1968) reported that activated charcoal (wood charcoal, 300-mesh A.C. from Harrison Clark, Ltd.) was effective in "inactivating" simazine when mixed into simazine-treated soils, although no quantitative data on simazine concentrations were reported.

- Rees and Au (1979) reported that an adsorption column containing XAD-2 resin removed 81 to 95% of the simazine in spiked tap water.

- Turner and Adams (1968) reported that, in a study on the adsorption of simazine by ion exchange resins (Sheets, 1959), duolite C-3 cation exchange resin removed from solution up to 2,000 μg of simazine per gram of resin. Little adsorption was observed with Duolite A-2 anion exchange resin.

- Miltner and Fronk (1985b) reported the bench-scale testing results of the addition of various oxidants to spiked, distilled water. Chlorine oxidation achieved 62 to 74% removal of simazine. However, when spiked Ohio River water was treated with smaller chlorine dosages during shorter time intervals, less than 17% removal was achieved. Chlorine dioxide oxidation of spiked, distilled water achieved only a 22% removal and achieved 8 to 27% removal of simazine in spiked Ohio River water when applied at a smaller dosage over a shorter time interval. Ozonation of spiked, distilled water resulted in a 92% removal of simazine. Oxidation of spiked, distilled water with hydrogen peroxide obtained a 19 to 42% removal of simazine, and in spiked Ohio River water, a smaller dosage over a shorter time interval obtained a simazine removal of 1 to 25%. Potassium permanganate oxidized up to 26% of the simazine present in spiked, distilled water.

VIII. References

Andersen, A.H. 1968. The inactivation of simazine and linuron in soil by charcoal. Weed Res. 8:58–60.

Anderson, K.J., E.G. Leighty and M.T. Takahashi. 1972. Evaluation of herbicides for possible mutagenic properties. J. Agric. Food Chem. 20:649–656.

Baker, D. 1983. Herbicide contamination in municipal water supplies in northwestern Ohio. Final Draft Report 1983. Prepared for Great Lakes National Program Office, U.S. Environmental Protection Agency. Tiffin, OH.

Bakke, J.E. and J.D. Robbins. 1968. Metabolism of atrazine and simazine by the goat and sheep. Abstr. Pap. 155th National Meeting Am. Chem. Soc. (A43).

Beynon, K.I., G. Stoydin and A.N. Wright. 1972. A comparison of the breakdown of the triazine herbicides cyanazine, atrazine, and simazine in soils and in maize. Pestic. Biochem. Physiol. 2:153–161.

Bohme, C. and F. Bar. 1967. The transformation of triazine herbicides in the animal organism. Food Cosmet. Toxicol. 5:23–28.

Bradway, D.E. and R.F. Moseman. 1982. Determination of urinary residue levels of the N-dealkyl metabolites of triazine herbicides. J. Agric. Food Chem. 30:244–247.

Chen, B.Q. 1981. Experimental studies on toxicity and teratogenicity of simazine. Chung Hua Fang I Hsueh Tsa Chih. 15(2):83–85.

Ciba-Geigy Corporation. 1980. 21-Day subacute dermal toxicity study in rabbits. Bio-Research Laboratories. Project No. 12017. January 14. MRID 00057567.

Ciba-Geigy Corporation. 1984. A teratology study of simazine technical in

New Zealand White rabbits. MRID 161407. Accession Nos. 252938 and 252938 (a 1986 amendment).

Ciba-Geigy Corporation. 1987. Rat chronic feeding/oncogenicity study (in progress). MRID 401514-0. Accession No. 40154.

Cohen, S.Z., C. Eiden and M.N. Lorber. 1986. Monitoring ground water for pesticides in the U.S.A. In: Evaluation of pesticides in ground water, American Chemical Society Symposium Series. No. 315.

Dshurov, A. 1979. Histological changes in organs of sheep in chronic simazine poisoning. Zentralbl. Veterinaermed. Reihe A. 26:44–54. (In German with English abstract.)

Eisenbeis, S.J., D.L. Lynch and A.E. Hampel. 1981. The Ames mutagen assay tested against herbicides and herbicide combinations. Soil Sci. 131:44–47.

ESE. 1984. Environmental Science and Engineering. Review of treatability data for removal of 25 synthetic organic chemicals from drinking water. Washington, DC: U.S. Environmental Protection Agency, Office of Drinking Water.

Fahrig, R. 1974. Comparative mutagenicity studies with pesticides. IARC Sci. Publ. 10:161–181.

Flanagan, J.H., J.R. Foster, H. Larsen et al. 1968. Residue data for simazine in water and fish. Unpublished study. Prepared in cooperation with the University of Maryland and others. Submitted by Geigy Chemical Company, Ardsley, NY.

Gold, B., K. Balu and A. Hofberg. 1973. Hydrolysis of simazine in aqueous solution. Report No. GAAC-73044. Unpublished study. Submitted by Ciba-Geigy Corporation, Greensboro, NC.

Guddewar, M.B. and W.C. Dauterman. 1979. Studies on a glutathione S-transferase preparation from mouse liver which conjugates chloro-s-triazine herbicides. Pestic. Biochem. Physiol. 12:1–9.

Hapke, H. 1968. Research into the toxicology of weedkiller simazine. Berl. Tieraerztl. Wochenschr. 81:301–303. Abstract.

Harris, C.I. 1967. Fate of 2-chloro-s-triazine herbicides in soil. J. Agric. Food Chem. 15:157–162.

Hazelton Laboratories. 1960. A two-year dietary feeding study—albino rats. Unpublished study. Submitted by Ciba-Geigy Corporation. MRID 00037752, 00025441, 00025442, 00042793 and 00080626.

Helling, C.S. 1971. Pesticide mobility in soils. II. Applications of soil thin-layer chromatography. Proc. Soil Sci. Soc. 35:737–748.

Helling, C.S. and B.C. Turner. 1968. Pesticide mobility: determination by soil thin-layer chromatography. Method dated Nov. 1, 1968. Science. 162:562–563.

Innes, J.R.M., B.M. Ulland, M.G. Valerio et al. 1969. Bioassay of pesticides and industrial chemicals for tumorigenicity in mice: a preliminary note. J. Natl. Cancer Inst. 42:1101–1114.

Ivey, M.J. and H. Andrews. 1965. Leaching of simazine, atrazine, diuron, and DCPA in soil columns. Unpublished study. Prepared by the University of Tennessee. Submitted by American Carbonyl, Inc., Tenafly, NJ.

Kahrs, R.A. 1969. Determination of simazine residues in fish and water by microcoulometric gas chromatography. Method No. AG-111 dated Aug. 22, 1969. Unpublished study. Submitted by Geigy Chemical Company, Ardsley, NY.

Kahrs, R.A. 1977. Simazine lakes – 1975 EUP Program: status report – 1977. Report No. ABR-77082. Unpublished study. Submitted by Ciba-Geigy Corporation, Greensboro, NC.

Keller, A. 1978. Degradation of simazine (Gesatop) in soil under aerobic-anaerobic and sterile-aerobic conditions. Project Report 05/78. Unpublished study. Submitted by Ciba-Geigy Corporation, Greensboro, NC.

Larsen, G.L. and J.E. Bakke. 1975. Metabolism of 2-chloro-4-cyclo-propylamino-6-isopropylamino-s-triazine (cyprazine) in the rat. J. Agric. Food Chem. 23:388–392.

Larsen, H., D.L. Sutton, A.R. Eaton et al. 1966. Summary of residue studies – simazine 80W. Unpublished study. Prepared in cooperation with U.S. Fish and Wildlife, Fish Control Laboratory and others. Submitted by Ciba-Geigy Corporation, Greensboro, NC.

LeBaron, H.M. 1970. Fate of simazine in the aquatic environment. Report No. GAAC-70013. Unpublished study. Submitted by Geigy Chemical Company, Ardsley, NY.

Lehman, A.J. 1959. Appraisal of the safety of chemicals in foods, drugs and cosmetics. Assoc. Food Drug Off. U.S., Q. Bull.

Martin, H. and C.R. Worthing, eds. 1977. Pesticide manual. Worcester, England: British Crop Protection Council.

Martin, V., L. Motko, B. Gold et al. 1975. Simazine residue tests. AG-A No. 1022. Unpublished study. Prepared in cooperation with University of Missouri. Submitted by Ciba-Geigy Corporation, Greensboro, NC.

Mattson, A.M., R.A. Kahrs and R.T. Murphy. 1969. Quantitative determination of triazine herbicides in soils by chemical analysis: GAAC-69014. Method dated Mar. 18, 1969. Unpublished study. Submitted by Ciba-Geigy Corporation, Greensboro, NC.

Mazaev, V.T. 1964. Experimental determination of the maximum permissible concentrations of cyanuric acid, simazine, and a 2-hydroxy derivative of simazine in water reservoirs. Chem. Abstr. 62:15304.

Meister, R., ed. 1984. Farm chemicals handbook. Willoughby, OH: Meister Publishing Co.

Miltner, R.J. and C.A. Fronk. 1985a. Treatment of synthetic organic contaminants for Phase II regulations. Progress report. U.S. Environmental Protection Agency, Drinking Water Research Division. July.

Miltner, R.J. and C.A. Fronk. 1985b. Treatment of synthetic organic contaminants for Phase II regulations. Internal report. U.S. Environmental Protection Agency, Drinking Water Research Division. December.

Monsanto Company. No date. The soil dissipation of glyphosate, alachlor, atrazine and simazine herbicides. Unpublished study.

Murnik, M.R. and C.L. Nash. 1977. Mutagenicity of the triazine herbicides atrazine, cyanazine and simazine in *Drosophila melanogaster*. J. Toxicol. Environ. Health. 3:691–697.

NAS. 1977. National Academy of Sciences. Drinking water and health, Part 1, Chap. 1–5. Washington, DC: National Academy Press. pp. V-184-V-348.

Newell, G.W. and J.V. Dilley. 1978. Teratology and acute toxicology of selected chemical pesticides administered by inhalation. EPA 600/ 1-78-003. Washington, DC: U.S. Environmental Protection Agency.

Orr, G.R. and B.J. Simoneaux. 1986. Disposition of simazine in rats. MRID 262646.

Palmer, J.S. and R.D. Radeleff. 1964. The toxicologic effects of certain fungicides and herbicides on sheep and cattle. Ann. N.Y. Acad. Sci. 111:729–736.

Palmer, J.S. and R.D. Radeleff. 1969. The toxicity of some organic herbicides to cattle, sheep, and chickens. Production Research Report No. 106, U.S. Department of Agriculture, Agricultural Research Service. pp. 1–26.

Pliss, G.B. and M.A. Zabezhinsky. 1977. Carcinogenicity of symmetric triazine derivatives. Pest. Abstr. 5:72–1017.

Rees, G.A.V. and L. Au. 1979. Use of XAD-2 macroreticular resin for the recovery of ambient trace levels of pesticides and industrial organic pollutants from water. Bull. Environ. Contam. Toxicol. 22:561–566.

Reinert, K.H. and J.H. Rogers. 1987. Fate of aquatic herbicides. In: Reviews of environmental contamination and toxicology, Vol. 93. pp. 87–89.

Rodgers, E.G. 1968. Leaching of seven s-triazines. Weed Sci. 16:117–120.

Sheets, T.J. 1959. The uptake, distribution, and phytotoxicity of 2-chloro-4,6-bis(ethylamine)-s-triazine. Ph.D. Thesis. University of California. Cited by Turner, M.A. and R.S. Adams, Jr. 1968.

Simmons, V.F., D.C. Poole, E.S. Riccio, D.E. Robinson, A.D. Mitchell and M.D. Waters. 1979. *In vitro* mutagenicity and genotoxicity assays of 38 pesticides. Environ. Mutagen. 1:142–143.

Simoneaux, B.J. and A. Sy. 1971. Metabolism of simazine and its metabolites in female rats. MRID 262646.

Smith, A.E., R. Grover, G.S. Emmond and H.C. Korven. 1975. Persistence and movement of atrazine, bromacil, monuron, and simazine in intermittently filled irrigation ditches. Can. J. Plant Sci. 55:809–816.

St. John L.E., W.J. Ammering, D.G. Wagner, R.G. Warner and D.J. Lisk. 1965. Fate of 4,6-dinitro-2-isobutylphenol, 2-chloro-4,6-bis-(ethylamino)-s-triazine, and pentachloronitrobenzene in the dairy cow. J. Dairy Sci. 48:502–503.

STORET. 1988. Storet Water Quality File. Office of Water. U.S. Environmental Protection Agency (data file search conducted in March, 1988).

Tai, C.N., C. Breckenridge and J.D. Green.* 1985a. Simazine technical subacute oral 13-week toxicity study in rats. Ciba-Geigy Pharmaceuticals Division. Report No. 85018. Accession No. 257693.

Tai, C.N., C. Breckenridge and J.D. Green.* 1985b. Subacute oral 13-week toxicity study in dogs. Simazine technical. Ciba-Geigy Pharmaceuticals Division. Report No. 85022. Accession No. 257692.

Talbert, R.E. and O.H. Fletchall. 1965. The adsorption of some s-triazines in soils. Weeds. 13:46–52.

Turner, M.A. and R.S. Adams, Jr. 1968. The adsorption of atrazine and atratone by anion- and cation-exchange resins. Soil Sci. Amer. Proc. 32:62–63.

USDA. 1984. United States Department of Agriculture. Pesticide background statements. In: Herbicides. Agriculture handbook Number 633, Vol. 1. USDA Forest Service.

U.S. EPA. 1983. U.S. Environmental Protection Agency. Simazine registration standard. Washington, DC: Office of Pesticide Programs. November 7.

U.S. EPA. 1986a. U.S. Environmental Protection Agency. Guidelines for carcinogen risk assessment. Fed. Reg. 51(185)33992–34003. September 24.

U.S. EPA. 1986b. U.S. Environmental Protection Agency. Code of Federal Regulations. Protection of the environment: tolerances and exemptions from tolerances for pesticide chemicals in or on raw agricultural commodities. 40 CFR 180.213.

U.S. EPA. 1988. U.S. Environmental Protection Agency. Method #507— Determination of nitrogen- and phosphorus-containing pesticides in ground water by GC/NPD. Draft. April 15. Available from U.S. EPA's Environmental Monitoring and Support Laboratory, Cincinnati, OH.

U.S. FDA. 1979. U.S. Food and Drug Administration. Code of Federal Regulations. 21 CFR 193.400. April 1.

Valencia, R. 1981. Mutagenesis screening of pesticides using *Drosophila*. Project summary. EPA 600/S1-81-017. Research Triangle Park, NC: Health Effects Research Laboratory, U.S. Environmental Protection Agency.

Walker, A. 1976. Simulation of herbicide persistence in soil. Pestic. Sci. 7:41–49.

Waters, M.D., S.S. Saindhu, Z.S. Simmon et al. 1982. Study of pesticide genotoxicity. In: Fleck, R.A. and A. Nollaender, eds. Genetic toxicology:

an agricultural perspective. 21:275–326. New York, NY: Plenum Press. Basic Life Sciences Series.

Weed Science Society of America. 1983. Herbicide handbook, 5th ed. Champaign, IL: Weed Science Society of America. pp. 433–437.

Windholz, M., S. Budavari, R.F. Blumetti and E.S. Otterbein, eds. 1983. The Merck index—an encyclopedia of chemicals and drugs, 10th ed. Rahway, NJ: Merck and Company, Inc.

Woodard Research Corporation.* 1965. Three-generation reproduction study of simazine in the diet of rats. MRID 00023365, 00080631.

Yelizarov, G.P. 1977. Occupational skin diseases caused by simazine and propazine. Pest. Abstr. 6:73–0352.

---

* Confidential Business Information submitted to the Office of Pesticide Programs.

Tebuthiuron

I. General Information and Properties

A. CAS No. 34014–18–1

B. Structural Formula

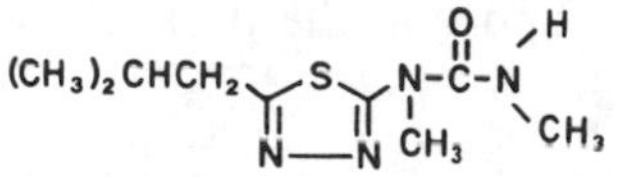

$$(CH_3)_2CHCH_2 \overset{S}{\underset{N \text{---} N}{\diagdown}} N \text{-} \overset{O}{\overset{\parallel}{C}} \text{-} N \overset{H}{\underset{CH_3}{\diagup}}$$

N-[5-(1,1-Dimethyl ethyl)-1,3,4-thiadiazol-2-yl]-N,N′-dimethylurea

C. Synonyms

- Combine; Herbic; Graslan; Perflan; Spike.

D. Uses

- Tebuthiuron is used as an herbicide for total vegetation woody plant control in noncrop areas and for brush and weed control in rangeland (Meister, 1983).

E. Properties (Meister, 1983)

Chemical Formula	$C_9H_{16}ON_4S$
Molecular Weight	228 (calculated)
Physical State (25°C)	White crystalline, odorless powder; colorless solid
Boiling Point	--
Melting Point	159 to 161°C
Density	--
Vapor Pressure (25°C)	2×10^{-6} mm Hg
Specific Gravity	--
Water Solubility (25°C, pH 7)	2,500 mg/L
Log Octanol/Water Partition Coefficient	1.79
Taste Threshold	--

Odor Threshold   --
Conversion Factor   --

F. Occurrence

- No occurrence data have been found in the STORET database (STORET, 1988).

G. Environmental Fate

- Tebuthiuron is resistant to hydrolysis. ^{14}C-Tebuthiuron, at 10 and 100 ppm, did not degrade during 64 days of incubation in sterile aqueous solutions at pH 3, 6 and 9 in the dark at 25°C (Mosier and Saunders, 1976).

- After 23 days of irradiation with artificial light (20-W black light), tebuthiuron accounted for 87 to 89% of the applied radioactivity in deionized (pH 7.1) and natural (pH 8.1) water treated with thiadiazole ring-labeled ^{14}C-tebuthiuron at 25 ppm (Elanco Products Company, 1972; Rainey and Magnussen, 1976b). After 15 days of irradiation with a black light or a sunlamp, tebuthiuron accounted for approximately 82 and 53%, respectively, of the applied compound in natural water treated with ^{14}C-tebuthiuron at 2.5 ppm.

- Thiadiazole ring-labeled ^{14}C-tebuthiuron in loam soil degraded from 8 ppm immediately posttreatment, to 5.7 ppm at 273 days posttreatment indicating a half-life greater than 273 days (Rainey and Magnussen, 1976a, 1978).

- ^{14}C-Tebuthiuron, at 1.0 ppm, degraded with a half-life of greater than 48 weeks in a loam soil maintained under anaerobic conditions in the dark at 23°C (Berard, 1977). N-[5-(1,1-Dimethylethyl)-1,3,4-thiadiazol-2-yl]-N-methylurea was the major degradate.

- Ring-labeled ^{14}C-tebuthiuron was very mobile (> 94% of that applied was found in the leachate) in a 12-inch column of Lakeland fine sand soil leached with 20 inches of water (Holzer et al., 1972). It was mobile in columns of loamy sand (approximately 73% at 6 to 10 inches), loam (approximately 84% at 1 to 8 inches) and muck (100% at 0 to 4 inches) soils leached with 4 to 8 inches of water.

- Based on column leaching studies, tetuthiuron is mobile to very mobile in loam, loamy sand and Lakeland sand soils and has low mobility in silty loam soil (Day, 1976a).

- ^{14}C-Tebuthiuron residues aged 30 days were mobile in a column of sandy loam soil; 39% of ^{14}C-residues were found in the soil and 40% of ^{14}C-residues were in the leachate (Day, 1976b).

- ^{14}C-Tebuthiuron degraded with half-lives of greater than 33 months in field plots in California (loam soil), 12 to 15 months in Louisiana (clay soil) and 12 to 15 months in Indiana (loam soil). The three sites were treated with thiadiazole ring-labeled ^{14}C-tebuthiuron at 8.96, 2.24 and 8.96 kg/ha, respectively (Rainey and Magnussen, 1976a, 1978). N-[5-(1,1-Dimethylethyl)-1,3,4-thiadiazol-2-yl]-N-methylurea was the major degradate at all three sites. Radioactivity was detected in the 15- to 30-cm depth of soil (10.2% of the applied compound at 18 months) at the California site, in the 30- to 45-cm depth of soil (1.3% of the applied compound at 33 months) at the Louisiana site and in the 30- to 45-cm depth of soil (4.7% of the applied compound at 15 months) at the Indiana site. ^{14}C-Tebuthiuron residues did not appear to accumulate in silt loam soil in Louisiana after three applications of ^{14}C-tebuthiuron (0.84 kg/ha at zero time; 1.4 kg/ha at 22 and 73 weeks).

II.   Pharmacokinetics

A. Absorption

- Morton and Hoffman (1976) reported that 94 to 96% of a single oral dose of tebuthiuron (10 mg/kg) was excreted in the urine of rats, rabbits and dogs. In mice, 66% was excreted in the urine, and 30% in the feces. These data indicate that tebuthiuron was well absorbed (about 70 to 96%) from the gastrointestinal tract.

B. Distribution

- No quantitative data were found in the available literature on the tissue distribution of tebuthiuron in exposed animals.

- Adams et al. (1982) administered tebuthiuron in the diet to 20 pregnant Wistar rats at levels of 100 or 200 ppm for 6 days prior to delivery. Forty-eight hours after delivery, radiolabeled tebuthiuron was reintroduced into the diet at the same levels as before. Radioactive label was detected in the milk at mean levels of 2.7 and 6.2 ppm for the 100- and 200-ppm groups, respectively.

C. Metabolism

- Morton and Hoffman (1976) reported that tebuthiuron was metabolized extensively by mice, rats, rabbits and dogs. Tebuthiuron was administered by gavage to male and female ICR mice, Harlan rats, Dutch Belted rabbits and beagle dogs at a dose of 10 mg/kg. Examination of urine extracts by thin-layer chromatography (TLC) showed the presence of eight radioactively labeled metabolites in rat, rabbit and dog urine and seven in mouse urine. Small amounts of unchanged tebuthiuron also were detected in each case (except for the mouse). The major metabolites were formed by N-demethylation of the substituted urea side-chain in each species examined. Oxidation of the dimethylethyl group also occurred in all species examined.

D. Excretion

- Morton and Hoffman (1976) reported that tebuthiuron was excreted rapidly in the urine of several species. Radiolabeled tebuthiuron was administered to male and female ICR mice, Harlan rats, Dutch Belted rabbits and beagle dogs at a dose of 10 mg/kg by gavage. Elimination of radioactivity was virtually complete within 72 hours and recovery values at 96 hours were 96.3, 94.5, 94.3 and 95.7% in the mouse, rat, rabbit and dog, respectively. In the rats, rabbits and dogs, the radioactivity was excreted almost exclusively in the urine. In the mice, 30% of the radioactivity was excreted in the feces.

III.   Health Effects

A. Humans

- No information on the health effects of tebuthiuron in humans was found in the available literature.

B. Animals

1. Short-term Exposure

- Todd et al. (1974) reported the acute oral LD_{50} values of tebuthiuron in rats, mice and rabbits to be 644, 579 and 286 mg/kg, respectively. In cats, oral doses of 200 mg/kg were not lethal, while 500 mg/kg given orally was not lethal to dogs, quail, ducks or chickens.

- Todd et al. (1972a) supplied 40 weanling Sprague-Dawley rats (105 to 146 g) with food containing tebuthiuron (purity not stated) at levels of 2,500 ppm for 15 days. At various time periods, five rats were necropsied and evaluated. Based on the dietary assumptions of Lehman (1959), 1 ppm in the diet of a rat is equivalent to 0.05 mg/kg/day; therefore, this level corresponds to 125 mg/kg/day. The animals were observed for an additional 15-day recovery period. All the animals exhibited reduced body weight gain during the treatment period. Light and electron microscopic evaluation revealed formation of vacuoles containing electron-dense bodies and myeloid figures in pancreatic acinar cells. This condition was rapidly reversed during the recovery period.

2. Dermal/Ocular Effects

- Todd et al. (1974) administered 200 mg/kg tebuthiuron to the shaved, abraded backs of male and female New Zealand White rabbits. During the study, one rabbit died following development of diarrhea and emaciation. All surviving rabbits gained weight over the 14-day observation period and were without signs of dermal irritation.

- Todd et al. (1974) tested tebuthiuron for sensitization in 2- to 3-month-old female albino guinea pigs. Each animal received topical applications of 0.1 mL of an ethanolic solution containing 2% tebuthiuron to the region of the flank three times per week for 3 weeks. Ten days after the last of the nine treatments, a challenge application was made, followed by a second challenge 15 days after the first. Tebuthiuron induced no dermal or systemic responses indicative of contact sensitization.

- Todd et al. (1974) instilled 0.1 mL (71 mg) of tebuthiuron into one eye and the conjunctival sac of each of six New Zealand White rabbits (2 to 3 months old). No irritation of the cornea or iris was observed, but there was slight transient hyperemia of the conjunctiva. All eyes were normal by the end of the 7-day test period.

3. Long-term Exposure

- Todd et al. (1972b) administered tebuthiuron (purity not stated) in the diet to groups of male and female Harlan rats (10/sex/group, 28 to 35 days old, 74 to 156 g) at levels of 0, 40, 100 or 250 mg/kg/day for 3 months. Body weights and food consumption were measured weekly. Blood obtained prior to necropsy was evaluated for blood sugar, blood urea nitrogen (BUN) and serum glutamic-pyruvic transaminase (SGPT). Sections of organs and tissues were prepared for gross and microscopic evaluation. There were no clinical signs of toxicity or mortality in any of the groups. A moderate reduction in body weight gain and a decrease in efficiency of food utilization in males and females in the highest dose group (250 mg/kg/day) was evident from week 1 of the study. Tebuthiuron had no clinically important effects on any of the hematological or clinical chemistry parameters measured. All rats receiving 250 mg/kg/day tebuthiuron showed diffuse vacuolation of the pancreatic acinar cells. The degree of this change ranged from slight to moderate, but the effect was not associated with necrosis or with the presence of an inflammatory response. One female rat receiving 100 mg/kg/day tebuthiuron showed very slight pancreatic changes. Based on these results, a No-Observed-Adverse-Effect Level (NOAEL) of 40 mg/kg/day and a Lowest-Observed-Adverse-Effect Level (LOAEL) of 100 mg/kg/day were identified.

- Todd et al. (1972c) administered tebuthiuron (purity not stated) in gelatin capsules to groups of four beagle dogs (2/sex/group, 13 to 23 months old, 7 to 23 kg) at dose levels of 0, 12.5, 25 or 50 mg/kg/day for 3 months. The physical condition of the animals was assessed daily, and body weights were recorded weekly. Gross and microscopic histopathology examinations were performed. Anorexia was noted, especially in the high-dose animals, leading to some weight loss. There was no mortality. Behavior and appearance were unremarkable at all test levels. No abnormalities were seen in the hematological or urinalysis studies. Clinical chemistry findings indicated increased BUN in the 50-mg/kg females. In addition, this group and the 50-mg/kg males exhibited increasing levels of alkaline phosphatase, up to fourfold over those of controls; however, these levels had returned to normal at the terminal sampling. There were no urinary abnormalities. The 25-mg/kg females and males demonstrated increased thyroid-to-body weight ratios, and the 50-mg/kg females also showed increased spleen-to-body weight ratios. Histopathological findings were unremarkable. The LOAEL was identified as 25 mg/kg, based on increased thyroid-to-body weight ratios and increased alkaline phosphatase values. A NOAEL of 12.5 mg/kg was identified.

- Todd et al. (1976a) administered tebuthiuron (purity > 97%) in the diet to groups of Harlan rats (40/sex/dose) for 2 years at dietary levels of 0, 400, 800 or 1,600 ppm. Based on the dietary assumptions of Lehman (1959), 1 ppm in the diet of a rat is equivalent to 0.05 mg/kg/day; therefore, these doses correspond to 20, 40 or 80 mg/kg/day. Physical appearance, behavior, food intake, body weight gain and mortality were recorded. Hematologic and blood chemistry values were obtained throughout the study; urinalysis was also performed. At necropsy, organ weights were determined and organs and tissues were examined grossly and histologically. Mortality in exposed animals was similar to or less than that observed in the control group. Variations in hematology, blood chemistry and urinalysis data from all groups were slight and unrelated to the test compound. Reduced body weight gain (10% or greater) was observed in the highest-dose group animals. There was also a slight increase in the kidney weights of the high-dose males. Microscopic examination revealed a low incidence of slight vacuolation of the pancreatic acinar cells in animals in the highest dose group. The NOAEL for this study, based on acinar vacuolation, was 40 mg/kg.

- Todd et al. (1976b) administered tebuthiuron (purity not stated) in the diet for 2 years to groups of Harlan ICR mice (40/sex/dose) at levels of 0, 400, 800 or 1,600 ppm. Based on the dietary assumptions of Lehman (1959), 1 ppm in the diet of a mouse is equivalent to 0.150 mg/kg/day; therefore, these dietary levels correspond to approximately 60, 120 or 240 mg/kg/day. Physical appearance, behavior, appetite, body weight gain and mortality were recorded. Hematologic, blood chemistry and organ weight values were obtained for animals surviving the test period. Gross and microscopic evaluations were conducted on organs and tissues obtained at necropsy. No important differences were observed between treated and control groups for any of the parameters evaluated. The vacuolation of pancreatic acinar cells noted in the Todd et al. (1976a) rat studies was not evident in this study in mice. Based on this, the NOAEL for this study was identified as 240 mg/kg/day.

4. Reproductive Effects

- Hoyt et al. (1981) studied the effects of tebuthiuron (98% active ingredient) in a two-generation reproduction study in rats. Weanling Wistar rats (25/sex/dose, F_0 generation) were maintained on diets containing tebuthiuron at 0, 100, 200 and 400 ppm based on the active ingredient (0, 7, 14 or 28 mg/kg/day, based on actual food consumption) for a period of 101 days preceding two breeding trials. First generation (F_1) offspring were maintained on the same diets for a period of 124 days preceding two breeding trials. Spermatogenesis and sperm morphology were examined in 10 F_0 males per treatment group. In addition, representative F_{1a} and F_{2a} weanlings and F_1 adults were necropsied and given histopathologic examinations after live-phase observations were completed. No changes in the efficiency of food utilization (EFU) were noted during the F_0 growth period, but during the F_1 growth period, a statistically significant ($p \leq 0.05$) depression in cumulative (124 days) EFU values occurred in both male and female rats receiving 28 mg/kg/day. EFU was not affected at the other dose levels. A dose-related depression in mean body weight occurred among female rats of the F_1

generation receiving 14 or 28 mg/kg/day; mean body weight was depressed significantly ($p \leq 0.05$) only in the high-dose females. In the 7-mg/kg/day group, body weights of either sex were not affected. The reproductive capacity of the animals was not affected at any level; no dose-related conditions or lesions were found in any offspring. In adult males from the F_0 generation, no dose-related histologic lesions were found, and sperm morphology and spermatogenesis were normal. A LOAEL of 14 mg/kg/day was determined for a lower rate of body weight gain during the 101-day premating period in F_1 females, and a NOAEL of 7 mg/kg/day, the lowest dose tested, was identified.

5. Developmental Effects

- Todd et al. (1972d) administered tebuthiuron (purity not stated) in the diet to groups of 25 adult Wistar-derived female rats (245 to 454 g) at levels of 0, 600, 1,200 or 1,800 ppm based on the active ingredient (0, 30, 60 or 90 mg/kg/day, based on Lehman, 1959) on days 6 to 15 of gestation. Fetal and uterine parameters were normal and the fetal defects that occurred were not attributed to the test compound. The NOAEL for developmental effects was greater than 1,800 ppm (90 mg/kg/day), the highest dose tested.

- Todd et al. (1975) administered tebuthiuron (purity not stated) by gavage to groups of 15 adult female Dutch Belted rabbits at levels of 10 or 25 mg/kg/day on days 6 to 18 of gestation. No developmental or toxic effects were observed.

6. Mutagenicity

- Hill (1984) reported that primary cultures of adult rat hepatocytes incubated with concentrations of tebuthiuron ranging from 0.5 to 1,000 μg/mL did not exhibit unscheduled DNA synthesis.

- Rexroat (1984) reported that tebuthiuron did not induce *Salmonella* revertants (strains TA1535, TA1537, TA1538, TA98 and TA100) when tested at concentrations ranging between 100 and 5,000 μg/plate, with or without metabolic activation. It was concluded that tebuthiuron was not mutagenic in the Ames *Salmonella*/mammalian microsome test for bacterial mutation.

- Neal (1984) reported that tebuthiuron did not induce sister chromatid exchange *in vivo* in bone marrow cells of Chinese hamsters administered oral doses of 200, 300, 400 or 500 mg/kg tebuthiuron.

- Cline et al. (1978) reported that histadine auxotrophs of *Salmonella typhimurium* (strains G46, TA1535, TA100, TA1537, TA1538, TA98, C3076 and D3052) and tryptophan auxotrophs of *Escherichia coli* were not reverted to the prototype by tebuthiuron at levels of 0.1 to 1,000 μg/mL, with or without metabolic activation.

7. *Carcinogenicity*

- Todd et al. (1976a) administered tebuthiuron (purity > 97%) in the diet to groups of Harlan rats (40/sex/dose) at levels of 0, 400, 800 or 1,600 ppm based on the active ingredient (0, 20, 40 or 80 mg/kg/day, based on Lehman, 1959) for 2 years. The authors reported no influence of the test compound on the incidence of neoplasms at any dose level.

- Todd et al. (1976b) administered tebuthiuron in the diet to groups of Harlan ICR mice (40/sex/dose) at levels of 0, 400, 800 or 1,600 ppm (0, 60, 120 or 240 mg/kg/day, based on Lehman, 1959) for 2 years. The authors reported no statistical evidence of increased incidence of tumors at any dose level.

IV. Quantification of Toxicological Effects

A. One-day Health Advisory

No information was found in the available literature that was suitable for the determination of the One-day HA value for tebuthiuron. It is therefore recommended that the Ten-day value for a 10-kg child, 2.5 mg/L (3,000 μg/L, calculated below), be used at this time as a conservative estimate of the One-day HA value.

B. Ten-day Health Advisory

The study by Todd et al. (1975) has been selected to serve as the basis for the Ten-day HA value for tebuthiuron because the NOAEL in the Dutch Belted rabbit was the lowest end point observed in a short-term developmental study. This study identified a NOAEL of 25 mg/kg/day (the highest dose tested) based on an absence of maternal toxicity. In another developmental study in rats by Todd et al. (1972d), a NOAEL of 90 mg/kg/day (the highest dose tested) was recorded. Since it is unknown whether the rabbit or the rat is more sensitive, the lower NOAEL was conservatively chosen in deriving the 10-day HA.

Using a NOAEL of 25 mg/kg/day, the Ten-day HA for a 10-kg child is calculated as follows:

$$\text{Ten-day HA} = \frac{(25 \text{ mg/kg/day}) (10 \text{ kg})}{(100) (1 \text{ L/day})} = 2.5 \text{ mg/L} (3{,}000 \text{ } \mu\text{g/L})$$

C. Longer-term Health Advisories

The two-generation reproduction study in rats (Hoyt et al., 1981) has been chosen to serve as the basis for the Longer-term HA for tebuthiuron. In this study, four groups of Wistar rats (25/sex) were fed tebuthiuron at 0, 7, 14 and 28 mg/kg/day in the diet for 101 days (F_0 rats) or 121 days (F_1 rats) and then for a further period sufficient to mate, deliver and rear two successive litters of young to 21 days of age. No adverse effects were reported in this study except for a lower body weight gain during the premating period in F_1 females at the dietary levels of 14 and 28 mg/kg. The NOAEL was identified as 7 mg/kg/day. The chronic study by Todd et al. (1976b) in mice was not selected because the weight loss and vacuolation of pancreatic acinar cells noted in rats was not observed in mice even at dose levels as high as 160 mg/kg/day, indicating that the mouse is less sensitive than the rat. The subchronic (90 day) feeding study in beagle dogs reported by Todd et al. (1972c) was not selected because the NOAEL (12.5 mg/kg/day) identified in that study was significantly higher than that of the rat study. In addition, the duration of the rat study was more appropriate for the derivation of a Longer-term HA.

Using the NOAEL of 7 mg/kg/day, the Longer-term HA for a 10-kg child is calculated as follows:

$$\text{Longer-term HA} = \frac{(7 \text{ mg/kg/day}) (10 \text{ kg})}{(100) (1 \text{ L/day})} = 0.7 \text{ mg/L} (700 \text{ } \mu\text{g/L})$$

The Longer-term HA for the 70-kg adult is calculated as follows:

$$\text{Longer-term HA} = \frac{(7 \text{ mg/kg/day}) (70 \text{ kg})}{(100) (2 \text{ L/day})} = 2.45 \text{ mg/L} (2{,}000 \text{ } \mu\text{g/L})$$

D. Lifetime Health Advisory

The two-generation reproduction study in rats (Hoyt et al., 1981) has been selected to serve as the basis for the Lifetime HA value for tebuthiuron. In this study, four groups of Wistar rats (25/sex) were fed tebuthiuron at 0, 7, 14 or 28 mg/kg/day in the diet for 101 days (F_0 rats) or 121 days (F_1 rats) and then for a further period sufficient to mate, deliver and rear two successive litters of young to 21 days of age (i.e., the test diet was fed throughout mating, gestation and lactation). The F_{1a} rats were parents of the F_2 offspring. No adverse effects were reported in this study except for a lower rate of body weight gain during the premating period in F_1 females at dietary levels of 14 and 28 mg/kg. The NOAEL was identified as 7 mg/kg/day. The chronic study by Todd et al. (1976b) in mice was not selected because the weight loss and vacuolation of pancreatic acinar cells noted in rats was not observed in mice even at dose levels as high as 160 mg/kg/day, indicating that the mouse is less sensitive than the rat.

Using the NOAEL of 7 mg/kg/day, the Lifetime HA is calculated as follows:

Step 1: Determination of the Reference Dose (RfD)

$$RfD = \frac{(7 \text{ mg/kg/day})}{(100)} = 0.07 \text{ mg/kg/day}$$

Step 2: Determination of the Drinking Water Equivalent Level (DWEL)

$$DWEL = \frac{(0.07 \text{ mg/kg/day}) (70 \text{ kg})}{(2 \text{ L/day})} = 2.45 \text{ mg/L } (2,000 \text{ } \mu g/L)$$

Step 3: Determination of the Lifetime Health Advisory

$$\text{Lifetime HA} = (2.45 \text{ mg/L}) (20\%) = 0.49 \text{ mg/L } (500 \text{ } \mu g/L)$$

E. Evaluation of Carcinogenic Potential

- The International Agency for Research on Cancer has not evaluated the carcinogenic potential of tebuthiuron.

- Applying the criteria described in EPA's guidelines for assessment of carcinogenic risk (U.S. EPA, 1986), tebuthiuron may be classified in Group D: not classifiable. This category is for substances with inadequate human and animal evidence of carcinogenicity.

V. Other Criteria, Guidance and Standards

- No other criteria, guidance or standards were found in the available literature.

VI. Analytical Methods

- Analysis of tebuthiuron is by a gas chromatographic (GC) method applicable to the determination of certain nitrogen-phosphorus-containing pesticides in water samples (U.S. EPA, 1988). In this method, approximately 1 liter of sample is extracted with methylene chloride. The extract is concentrated and the compounds are separated using capillary column GC. Measurement is made using a nitrogen phosphorus detector. This method has been validated in a single laboratory, and the estimated detection limit for the analytes, such as tebuthiuron, is 1.3 μg/L.

VII. Treatment Technologies

• No information on treatment technologies capable of effectively removing tebuthiuron from contaminated water was found in the available literature.

VIII. References

Adams, E., J. Magnussen, J. Emmerson et al.* 1982. Radiocarbon levels in the milk of lactating rats given ^{14}C-tebuthiuron (compound 75503) in the diet. Unpublished study. Greenfield, IN: Eli Lilly and Company. MRID 00106081.

Berard, D.F.* 1977. ^{14}C-Tebuthiuron degradation study in anaerobic soil. Prepared and submitted by Eli Lilly and Company, Greenfield, IN. MRID 00900098.

Cline, J.C., G.Z. Thompson and R.I. McMahon.* 1978. The effect of Lilly Compound 75503 (tebuthiuron) upon bacterial systems known to detect mutagenic events. Unpublished study. Greenfield, IN: Eli Lilly and Company. MRID 000416090.

Day, E.W.* 1976a. Laboratory soil leaching studies with tebuthiuron. Unpublished study. Received Feb. 18, 1977, under 1471–109. Submitted by ELANCO Products Co., Div. of Eli Lilly and Company, Indianapolis, IN. CDL:095854-I. MRID 00020782.

Day, E.W.* 1976b. Aged soil leaching study with herbicide tebuthiuron. Unpublished study. Received Feb. 18, 1977, under 1471–109. Submitted by ELANCO Products Co., Div. of Eli Lilly and Company, Indianapolis, IN. CDL:095854-J. MRID 00020783.

ELANCO Products Company.* 1972. Environmental safety studies with EL-103. Unpublished study. Received Mar. 13, 1973, under 1471–97. Prepared in cooperation with United States Testing Co., Inc. CDL:120339-I. MRID 00020730.

Hill, L.* 1984. The effect of tebuthiuron (Lilly Compound 75503) on the induction of DNA repair synthesis in primary cultures of adult rat hepatocytes. Unpublished study. Greenfield, IN: Eli Lilly and Company. MRID 00141692.

Holzer, F.J., R.F. Sieck, R.L. Large et al.* 1972. EL-103: Leaching study.

Unpublished study. Received Mar. 13, 1973, under 1471–97. Prepared in cooperation with Purdue Univ., Agronomy Dept., and United States Testing Co., Inc. Submitted by ELANCO Products Co., Division of Eli Lilly and Company, Indianapolis, IN. CDL:120339-K. MRID 00020732.

Hoyt, J.A., E.R. Adams and N.V. Owens.* 1981. A two-generation reproductive study with tebuthiuron in the Wistar rat. Unpublished study. Greenfield, IN: Eli Lilly and Company. MRID 00090108.

Lehman, A.J. 1959. Appraisal of the safety of chemicals in foods, drugs, and cosmetics. Assoc. Food Drug Off. U.S., Q. Bull.

Meister, R., ed. 1983. Farm chemicals handbook. Willoughby, OH: Meister Publishing Company.

Morton, D.M. and D.G. Hoffman. 1976. Metabolism of a new herbicide, tebuthiuron (1-(5-(1,1-dimethylethyl)-1,3,5-thiadiazol-2-yl)-1,3-dimethylurea), in mouse, rat, rabbit, dog, duck and fish. J. Toxicol. Environ. Health. 1:757–768.

Mosier, J.W. and D.G. Saunders.* 1976. A hydrolysis study on the herbicide tebuthiuron. Includes undated method. Unpublished study. Received Feb. 18, 1977 under 1471–109. Submitted by ELANCO Products Co., Div. of Eli Lilly and Company, Indianapolis, IN. CDL:095854-F. MRID 00020779.

Neal, S.B.* 1984. The effect of tebuthiuron (Lilly Compound 75503) on the *in vivo* induction of sister chromatid exchange in bone marrow of Chinese hamsters. Unpublished study. Greenfield, IN: Eli Lilly and Company. MRID 00141693.

Rainey, D.P. and J.D. Magnussen.* 1976a. Behavior of ^{14}C-tebuthiuron in soil. Unpublished study. Received Feb. 18, 1977 under 1471–109. Prepared in cooperation with A & L Agricultural Laboratories and United States Testing Co., Inc. Submitted by ELANCO Products Co., Div. of Eli Lilly and Company, Indianapolis, IN. CDL:095854-C. MRID 00020777.

Rainey, D.P. and J.D. Magnussen.* 1976b. Photochemical degradation studies with ^{14}C-tebuthiuron. Unpublished study. Received Feb. 18, 1977 under 1471–109. Submitted by ELANCO Products Co., Div. of Eli Lilly and Company, Indianapolis, IN. CDL:095854-D. MRID 00020778.

Rainey, D.P. and J.D. Magnussen.* 1978. Behavior of ^{14}C-tebuthiuron in soil: addendum report. Unpublished study. Received June 1, 1978 under 1471–109. Submitted by ELANCO Products Co., Div. of Eli Lilly and Company, Indianapolis, IN. CDL:097100-C. MRID 00020693.

Rexroat, M.* 1984. The effect of tebuthiuron (Lilly Compound 75503) on the induction of reverse mutations in *Salmonella typhimurium* using the Ames test. Unpublished study. Greenfield, IN: Eli Lilly and Company. MRID 00140691.

STORET. 1988. STORET Water Quality File. Office of Water. U.S. Environmental Protection Agency (data file search conducted in May, 1988).

Todd, G.E., W.J. Griffing, W.R. Gibson et al.* 1972a. Special subacute rat toxicity study. Unpublished study. Greenfield, IN: Eli Lilly and Company. MRID 00020798.

Todd, G.C., W.R. Gibson and G.F. Kiplinger.* 1972b. The toxicological evaluation of EL-103 in rats for 3 months. Unpublished study. MRID 00020662.

Todd, G.C., W.R. Gibson and G.F. Kiplinger.* 1972c. The toxicological evaluation of EL-103 in dogs for 3 months. Unpublished study. MRID 00020663.

Todd, G.C., W.R. Gibson and C.C. Kehr. 1974. Oral toxicity of tebuthiuron (1-(5-tert-butyl-1,3,4-thiadiazol-2-yl)-1,3-dimethylurea) in experimental animals. Food Cosmet. Toxicol. 12:461–470.

Todd, G.C., J.K. Markham, E.R. Adams et al.* 1972d. Rat teratology study with EL-103. Unpublished study. MRID 00020803.

Todd, G.C., J.K. Markham, E.R. Adams, N.V. Owens, F.O. Gossett and D.M. Morton.* 1975. A teratology study with EL-103 in the rabbit. Unpublished study. Greenfield, IN: Eli Lilly and Company. MRID 00020644.

Todd, G.C., W.R. Gibson, D.G. Hoffman, S.S. Young and D.M. Morton.* 1976a. The toxicological evaluation of tebuthiuron (EL-103) in rats for two years. Unpublished study. Greenfield, IN: Eli Lilly and Company. MRID 00020714.

Todd, G.C., W.R. Gibson, D.G. Hoffman, S.S. Young and D.M. Morton.* 1976b. The toxicological evaluation of tebuthiuron (EL-103) in mice for two years. Unpublished study. Greenfield, IN: Eli Lilly and Company. MRID 00020717.

U.S. EPA. 1986. U.S. Environmental Protection Agency. Guidelines for carcinogen risk assessment. Fed. Reg. 51(185):33992–34003. September 24.

U.S. EPA. 1988. U.S. Environmental Protection Agency. EPA Method 507 — Determination of nitrogen and phosphorus containing pesticides in water by GC/NPD. Draft. April 15. Available from U.S. EPA's Environmental Monitoring and Support Laboratory, Cincinnati, OH.

---

* Confidential Business Information submitted to the Office of Pesticide Programs.

Terbacil

I. General Information and Properties

A. CAS No. 5902–51–2

B. Structural Formula

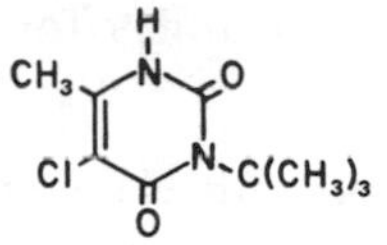

5-Chloro-3-(1,1-dimethylethyl)-6-methyl-2,4(1H,3H)-pyrimidinedione
or 3-tert-butyl-5-chloro-6-methyluracil

C. Synonyms

- Sinbar; Geonter (Meister, 1988).

D. Uses

- Terbacil is an herbicide used for the selective control of annual and perennial weeds in crops such as sugarcane, alfalfa, apples, peaches, blueberries, strawberries, citrus, pecans and mint (Meister, 1988).

E. Properties (Meister, 1988)

Chemical Formula	$C_9H_{13}O_2N_2Cl$
Molecular Weight	216.65
Physical State (at 25°C)	White crystals
Boiling Point (at 25 mm Hg)	--
Melting Point	175–177°C
Density	1.34 g/mL (25°C)
Vapor Pressure (29.5°C)	4.8×10^{-7} mm Hg
Specific Gravity	1.34 (25°C/25°C)
Water Solubility (25°C)	710 mg/L
Log Octanol/Water Partition Coefficient	–1.41

Taste Threshold   --
Odor Threshold   --
Conversion Factor   --

F. Occurrence

- Terbacil was not sampled at any water supply stations listed in the STORET database (STORET, 1988). No information was found in available literature on the occurrence of terbacil.

G. Environmental Fate

- ^{14}C-Terbacil at 5 ppm was stable (less than 2% degraded) in buffered aqueous solutions at pH 5, 7 and 9 for 6 weeks at 15°C in the dark (Davidson et al., 1978).

- After 4 weeks of irradiation with UV light (300 to 400 nm), about 16% of the applied ^{14}C-terbacil (5 ppm) was photodegraded in distilled water (pH 6.2) (Davidson et al., 1978).

- Soil metabolism studies indicate that terbacil is persistent in soil. At 100 ppm, terbacil was slowly degraded in an aerobic sandy loam soil (80% remained after 8 months) (Marsh and Davies, 1978). Terbacil at 8 ppm had a half-life of about 5 months in aerobic loam soil (Zimdahl et al., 1970). ^{14}C-Terbacil at 2 ppm had a half-life of 2 to 3 months in aerobic silt loam and sandy loam soils (Rhodes, 1975; Gardiner, 1964; Gardiner et al., 1969). The formation of carbon dioxide is slow; for example, 28% of the applied ^{14}C-terbacil at 2.88 ppm on sandy loam soil degraded to carbon dioxide in 600 days (Wolf, 1973; Wolf, 1974; Wolf and Martin, 1974).

- Degradation of terbacil in an anaerobic soil environment is also slow. In anaerobic silt loam and sandy soils, ^{14}C-terbacil at 2.1 ppm was slightly degraded (less than 5% after 60 days) in the dark (Rhodes, 1975). Only trace amounts of ^{14}C-terbacil, applied at 2.88 ppm, were degraded to ^{14}C-carbon dioxide after 145 days in an anaerobic environment when metabolized by microbes in the dark (Rhodes, 1975). At least 90% of the label remained as terbacil after 90 days of incubation in both sterile and nonsterile soils. Small amounts (0.8 to 1.5%) of the label of carbon dioxide were evolved from nonsterile soil, whereas 0.01% was evolved from sterile soil (Rhodes, 1975).

- Terbacil was mobile in soil columns of sandy loam and fine sandy soil (Rhodes, 1975; Mansell et al., 1972). However, in a silt loam soil column, only 0.4% of the applied ^{14}C-terbacil leached with 20 inches of water (Rhodes, 1975). In an aged soil column leaching study of the leaching characteristics of degradates, about 52% and 4% of the applied radioactivity in aged sandy loam and silt loam soils, respectively, leached (Rhodes, 1975). Terbacil phytotoxic residues were mobile to depths of 27.5 to 30 cm in a sandy soil column treated with terbacil at 5.6 kg/ha and eluted with 10 or 20 cm water (Marriage, 1977). Terbacil was negligibly adsorbed to soils ranging in texture from sand to clay (Davidson et al., 1978; Liu et al., 1971; Rao and Davison, 1979). Terbacil was adsorbed (54%) to a muck soil (36% organic matter) (Liu et al., 1971).

- Data from field dissipation studies showed that terbacil persistence in soil varied with application rate, soil type and rainfall. In the field, terbacil phytotoxic residues persisted in soil for up to 16 months following a single application of terbacil. Residues were found at the maximum depths sampled (3 to 43 inches) (Gardiner, no date a,b; Gardiner et al., 1969; Isom et al., 1969; Isom et al., 1970; Liu et al., no date; Mansell et al., 1977; Mansell et al., 1979; Morrow and McCarty, 1976; Rahman, 1977; Rhodes, 1975).

- Phytotoxic residues resulting from multiple applications of terbacil persisted for 1 to more than 2 years following the final application (Skroch et al., 1971; Tucker and Phillips, 1970; Benson, 1973; Doughty, 1978).

- Terbacil has not been found in ground water; however, its soil persistence and mobility indicate that it has the potential to get into ground water.

II.   Pharmacokinetics

A. Absorption

- No information was found in the available literature on the absorption of terbacil. However, evidence of systemic toxicity indicates that terbacil is absorbed following ingestion.

B. Distribution

- No information was found in the available literature on the distribution of terbacil.

C. Metabolism

- No information was found in the available literature on the metabolism of terbacil.

D. Excretion

- No information was found in the available literature on the excretion of terbacil.

III.   Health Effects

A. Humans

- No information was found in the available literature on the health effects of terbacil in humans.

B. Animals

1. *Short-term Exposure*

- It was not possible to determine a lethal dose of terbacil in dogs because repeated emesis prevented dosing with terbacil in amounts in excess of 5,000 mg/kg (Paynter, 1966). However, in a dog receiving one oral dose of terbacil at 250 mg/kg followed 5 days later by a dose of 100 mg/kg, emesis, diarrhea and mydriasis were noted.

- In rats (details not available), the LD_{50} was between 5,000 and 7,500 mg/kg (Sherman, 1965). At 2,250 mg/kg, inactivity, weight loss and incoordination were noted.

2. *Dermal/Ocular Effects*

- Hood (1966) reported that no compound-related clinical or pathological changes were observed when terbacil was applied to the clipped dorsal skin of rabbits (five males, five females) at a dose level of 5,000 mg/kg (as a 55% aqueous paste), for 5 hours/day, 5 days/week for 3 weeks (15 applications). The parameters observed included body weight, dermal reaction, organ weights and histopathology.

- Reinke (1965) reported that no dermal reactions were observed when terbacil was administered to the intact dorsal skin of 10 guinea pigs as a 15% solution in 1:1 acetone:dioxane containing 13% guinea pig fat.

- Reinke (1965) reported no observed sensitization in 10 albino guinea pigs when terbacil was administered nine times during a 3-week period, with half of the animals in each group receiving dermal applications on abraded dorsal skin and the others receiving intradermal injections. After 2 weeks, the animals were challenged by application of terbacil to intact and abraded skin. The challenge application was repeated 2 weeks later.

3. *Long-term Exposure*

- Wazeter et al. (1964) administered terbacil, 82.7% active ingredient (a.i.), in the diet to Charles River pathogen-free albino rats (20/sex/level) at levels of 0, 100, 500 or 5,000 ppm of a.i. for 90 days. This corresponds to doses of about 0, 5, 25 or 250 mg/kg/day based on the dietary assumptions of Lehman (1959). The parameters observed included body weight, food consumption, hematology, liver function tests, urinalyses, organ weights and gross and histologic pathology. No adverse effects with respect to behavior and appearance were noted. All rats survived to the end of the study. No effect on body weight gain was observed in either sex when terbacil was administered at 5 or 25 mg/kg/day. Females administered 250 mg/kg/day gained slightly less weight (15%) than controls. Males at this level showed no effect. No compound-related hematological or biochemical changes were found, and urinalyses were normal at all times. No gross or microscopic pathological changes were noted in animals administered terbacil at 5 or 25 mg/kg/day. Morphological changes in animals receiving the highest dose level were limited to the liver and consisted of statistically significant increases in liver weights. This change was accompanied by a moderate-to-marked hypertrophy of hepatic parenchymal cells associated with vacuolation of scattered hepatocytes. Similar microscopic changes, but with reduced severity, were found in one rat at the 25 mg/kg/day level. This study identified a Lowest-Observed-Adverse-Effect Level (LOAEL) of 25 mg/kg/day and a No-Observed-Adverse-Effect Level (NOAEL) of 5 mg/kg/day.

- Goldenthal et al. (1981) administered terbacil (97.8% a.i.) in the diet to CD-1 mice (80/sex/level) at levels of 0, 50, 1,250 or 5,000 to 7,500 ppm for 2 years. Based on the dietary assumptions of Lehman (1959), 1 ppm in the diet of mice is equivalent to 0.15 mg/kg/day; therefore, these levels correspond to doses of about 0, 7.5, 187 or 750 to 1,125 mg/kg/day. The 5,000-ppm dose level was increased slowly to 7,500 ppm by week 54 of the study. Mortality was significantly higher (p < 0.05) in mice at the high dosage levels throughout the study. No changes considered biologically important or compound-related occurred in the hematological parameters. An increased incidence of hepatocellular hypertrophy was seen microscopically in male and female mice administered 750 to 1,125 mg/kg/day and in male mice administered 187 mg/kg/day. An increased incidence of hyperplastic liver nodules also occurred in male mice administered 750 to 1,125 mg/kg/day. Female mice from the 187-mg/kg/day group and both male and female mice from the 7.5-mg/kg/day group were free of compound-related microscopic lesions. This study identified a LOAEL of 187 mg/kg/day and a NOAEL of 7.5 mg/kg/day.

- Wazeter et al. (1967b) administered terbacil (80% a.i.) in the diet to Charles River CD albino rats (36/sex/level) at levels of 0, 50, 250 or 2,500 ppm to 10,000 ppm of a.i. for 2 years. Based on the dietary assumptions of Lehman (1959), 1 ppm in the diet of a rat corresponds to 0.05 mg/kg/day; therefore, these dietary levels correspond to doses of about 0, 2.5, 12.5 or 125 to 500 mg/kg/day. The 2,500 ppm level was increased slowly to 10,000 ppm by week 46 of the study. No adverse compound-related alterations in behavior or appearance occurred in any test group. No significant differences in body weight gain in males and females administered 2.5 or 12.5 mg/kg/day were observed. Rats administered 125 to 500 mg/kg/day exhibited a significantly lower rate of body weight gain. This difference occurred early and became more pronounced with time in the female rats than in the male rats. Maximum differences were 14 to 17% in the male rats and 24 to 27% in the females when compared to the controls. No compound-related gross pathological lesions were seen at necropsy in rats from any groups. The only compound-related variation in organ weights was a slight increase in liver weights among rats from the 125- to 500-mg/kg/day dose level at final sacrifice. Histological changes were observed in the livers of rats fed terbacil at 12.5 mg/kg/day for 1 year and in the high-dose group fed 125 to 500 mg/kg/day for 1 and 2 years. These changes consisted of enlargement and occasional vacuolation of centrilobular hepatocytes. Due to an outbreak of respiratory congestion observed in all study groups at week 27, all animals were placed on antibiotic treatment (tetracycline hydrochloride), and the therapy was successful. This study identified a LOAEL of 125 to 500 mg/kg/day, based on irreversible histological changes in the liver, and a NOAEL of 12.5 mg/kg/day.

- Wazeter et al. (1966) administered terbacil (80% a.i.) in the diet to young purebred beagle dogs (4 to 6 months old, 4/sex/dose) at dose levels of 0, 50, 250 or 2,500 to 10,000 ppm of a.i. for 2 years. Based on the dietary assumptions of Lehman (1959), 1 ppm in the diet of a dog corresponds to 0.025 mg/kg/day; therefore, these dietary levels correspond to approximately 0, 1.25, 6.25 or 62.5 to 250 mg/kg/day. The 2,500-ppm level was gradually increased to 10,000 ppm from week 26 to week 46 of the study. All animals underwent periodic physical examinations; hematologic tests; and determinations of 24-hour alkaline phosphatase, prothrombin time, serum glutamate oxaloacetate transaminase (SGOT), serum glutamate pyruvate transaminase (SGPT) and cholesterol. No adverse compound-related alterations in behavior or appearance occurred among any of the control or treated dogs. No mortalities occurred during the 2-year course of treatment. Although there were some fluctuations in body weight throughout the study, these were not considered to be compound-related. No alterations in hematology, plasma biochemistry or urinalysis were observed. No compound-related gross or microscopic pathological changes were seen in any of the dogs sacrificed after 1 or 2 years of feeding. A slight increase in relative liver weights and elevated alkaline phosphatase occurred in dogs from the 62.5- to 250-mg/kg/day group and the 6.25-mg/kg/day group, which were sacrificed after 1 or 2 years. Also at 6.25 mg/kg/day, there was an increase in thyroid-to-body weight ratio. This study identified a LOAEL of 6.25 mg/kg (250 ppm) and a NOAEL of 1.25 mg/kg/day (50 ppm).

4. Reproductive Effects

- Wazeter et al. (1967a) administered terbacil (80% a.i.) in the diet to male and female Charles River CD rats of three generations (10 males and 10 females per level per generation) at dietary levels of 0, 50 or 250 ppm of a.i. Based on the dietary assumptions of Lehman (1959), 1 ppm in the diet of a rat is equivalent to 0.05 mg/kg/day; therefore, these dietary levels correspond to doses of about 2.5 or 12.5 mg/kg/day. Each parental generation was administered terbacil in the diet for 100 days prior to mating. No abnormalities in behavior, appearance or food consumption of the parental rats were observed in any of the three generations. Males at the 12.5 mg/kg/day level in all three generations exhibited reduced body weight gains. Females in all three generations were similar to controls in body weight gain. No abnormalities were observed in the breeding cycle of any of the three generations relative to the fertility of the parental male and female rats, development of the embryos and fetuses, abortions, deliveries, live births, sizes of the litters, viability of the newborn, survival of the pups until weaning or growth of the pups during the nursing period. Gross examination of pups surviving at weaning from both litters of all three generations did not reveal any evidence of abnormalities. No compound-related histopathological lesions were observed in any of the tissues examined from weanlings of the F_{3b} litter.

This study identified a LOAEL of 12.5 mg/kg/day and a NOAEL of 2.5 mg/kg/day.

5. *Developmental Effects*

- E.I. duPont (1984a) administered terbacil by gavage as a 0.5% suspension in methyl cellulose to groups of 18 female New Zealand White rabbits (5 months old) from days 7 to 19 of gestation at dose levels of 0, 30, 200 or 600 mg/kg/day. Maternal mortality was significantly increased (p $\leq$ 0.05) at the 600 mg/kg/day level. Additional indicators of maternal toxicity at 600-mg/kg/day were a significant increase (p $\leq$ 0.05) in adverse clinical signs (anorexia and liquid or semisolid yellow, orange or red discharges found below the cages) and a significant decrease (p $\leq$ 0.05) in body weight gain. Mean body weight gains and the incidence of adverse effects were similar in controls and in the 30- and 200-mg/kg/day groups. Fetal toxicity at doses of 600 mg/kg/day included a significant decrease (p $\leq$ 0.05) in fetal body weight and a significant increase (p $\leq$ 0.05) in the frequency of extra ribs and partially ossified and unossified phalanges and pubes. This increase was not due to a statistically significant increase in any specific malformation, and occurred only at a dosage level that was overtly toxic to the dams, suggesting to the authors that it may be the result of maternal toxicity. No increase in the incidence of adverse effects was noted among fetuses produced by animals administered 30 or 200 mg/kg/day terbacil. Based on maternal and fetal toxicity, this study identified a LOAEL of 600 mg/kg/day and a NOAEL of 200 mg/kg/day.

- Culik et al. (1980) administered terbacil (96.6% a.i.) in the feed to female Charles River CD rats from days 6 to 15 of gestation at levels of 0, 250, 1,250 or 5,000 ppm. Based on the measured food consumption, these dietary levels correspond to doses of about 0, 23, 103 or 391 mg/kg/day. Maternal parameters observed included clinical signs of toxicity and changes in behavior, body weight and food consumption. Statistically significant (p $\leq$ 0.05), compound-related reductions in mean body weight, weight gain and food consumption were seen in animals administered 103 or 391 mg/kg/day. No other clinical signs or gross pathological changes were observed in any animals. The mean number of live fetuses per litter and mean final maternal body weight were significantly lower (p $\leq$ 0.05) in the groups administered 103 or 391 mg/kg/day than in the control group; the mean number of implantations per litter was also significantly lower (p $\leq$ 0.05) than in control animals. Anomalies occurred in the renal pelvis, and ureter dilation was found in all the treatment groups. This study identified a LOAEL (lowest dose tested) of

23 mg/kg/day, based on anomalies of the renal pelvis and ureter dilation.

6. *Mutagenicity*

- E.I. duPont (1984b) reported that terbacil did not induce unscheduled DNA synthesis in primary cultures of rat hepatocytes (0.01 and 1.0 μM); did not exhibit mutagenic activity in the CHO/HGPRT assay (0 to 5.0 μM) with or without metabolic activation; and did not produce statistically significant differences between mean chromosome numbers, mean mitotic indices or significant increases in the frequency of chromosomal aberrations when tested by *in vivo* bone marrow chromosome studies in Sprague-Dawley CD rats (15/sex/level) administered a single dose of terbacil by gavage at 0, 20, 100 or 500 mg/kg.

- Murnik (1976) reported that terbacil significantly elevated the rates of apparent dominant lethals when tested in *Drosophila melanogaster*, but the authors concluded that the significant reductions in egg hatch were probably due to physiological toxicity of the treatment, since genetic assays did not indicate the induction of chromosomal breakage or loss.

7. *Carcinogenicity*

- Goldenthal et al. (1981) administered terbacil (97.8% a.i.) in the diet to CD-1 mice (80/sex/level) at levels of 0, 50, 1,250 or 5,000 to 7,500 ppm for 2 years. These levels correspond to doses of about 0, 7.5, 187 or 750 to 1,125 mg/kg/day (Lehman, 1959). The 5,000-ppm dose level was increased slowly to 7,500 ppm by week 54 of the study. No increased incidence of cancer in the treated animals was found.

- Wazeter et al. (1967b) administered terbacil (80% a.i.) in the diet to Charles River CD albino rats (36/sex/level) at levels of 0, 50, 250 or 2,500 to 10,000 ppm of a.i. for 2 years. These levels correspond to doses of about 0, 2.5, 12.5 or 125 to 500 mg/kg/day (Lehman, 1959). No evidence of compound-related carcinogenic effects was noted.

IV.   Quantification of Toxicological Effects

A. One-day Health Advisory

No information was found in the available literature that was suitable for determination of the One-day HA value for terbacil. It is, therefore, recommended that the Ten-day HA value for a 10-kg child, 0.25 mg/L (300 μg/L), calculated below, be used at this time as a conservative estimate of the One-day HA value.

B. Ten-day Health Advisory

The dietary reproductive study in rats by Wazeter et al. (1967a) has been selected to serve as the basis for the Ten-day HA value for terbacil. It identifies a LOAEL of 12.5 mg/kg/day, based on a reduced body weight gain in the males in all three generations, and a NOAEL of 2.5 mg/kg/day. The teratology study in rats by Culik et al. (1980) provides support for this conclusion. This teratology study identifies a LOAEL of 23 mg/kg/day (no doses lower than 23 mg/kg/day were tested), and essentially the same Ten-day HA value (0.23 mg/L) can be derived from this LOAEL by using an uncertainty factor of 1,000.

The Ten-day HA for a 10-kg child is calculated as follows:

$$\text{Ten-day HA} = \frac{(2.5 \text{ mg/kg/day}) (10 \text{ kg})}{(100) (1 \text{ L/day})} = 0.25 \text{ mg/L} (300 \text{ } \mu\text{g/L})$$

C. Longer-term Health Advisory

The dietary reproductive study in rats by Wazeter et al. (1967a) has been selected to serve as the basis for the Longer-term HA values for terbacil. A NOAEL of 2.5 mg/kg/day is identified in this study. A 90-day subchronic study in rats (Wazeter et al., 1964) identifying a NOAEL of 5 mg/kg/day supports this conclusion.

The Longer-term HA for a 10-kg child is calculated as follows:

$$\text{Longer-term HA} = \frac{(2.5 \text{ mg/kg/day}) (10 \text{ kg})}{(100) (1 \text{ L/day})} = 0.25 \text{ mg/L} (300 \text{ } \mu\text{g/L})$$

The Longer-term HA for a 70-kg adult is calculated as follows:

$$\text{Longer-term HA} = \frac{(2.5 \text{ mg/kg/day}) (70 \text{ kg})}{(100) (2 \text{ L/day})} = 0.875 \text{ mg/L} (900 \text{ } \mu\text{g/L})$$

D. Lifetime Health Advisory

The 2-year dog feeding study by Wazeter et al. (1966), selected to serve as the basis for the Lifetime HA value for terbacil, identifies a NOAEL of 1.25 mg/kg/day, based on relative liver weight increases and an increase in alkaline phosphatase. A number of other studies provide information that supports the conclusion that the overall NOAEL for lifetime exposure of rats, mice and dogs to terbacil is less than 25 mg/kg/day. These include a 2-year feeding study in mice that identifies a NOAEL of 7.5 mg/kg/day for liver changes (Goldenthal, 1981) and a 2-year feeding study in rats that identifies a NOAEL of 12.5 mg/kg/day for lower body weight gain and liver effects (Wazeter et al., 1967b).

Using a NOAEL of 1.25 mg/kg/day, the Lifetime HA is calculated as follows:

Step 1: Determination of the Reference Dose (RfD)

$$RfD = \frac{(1.25 \text{ mg/kg/day})}{(100)} = 0.013 \text{ mg/kg/day (rounded from 0.0125 mg/kg/day)}$$

Step 2: Determination of the Drinking Water Equivalent Level (DWEL)

$$DWEL = \frac{(0.0125 \text{ mg/kg/day}) (70 \text{ kg})}{(2 \text{ L/day})} = 0.44 \text{ mg/L (400 } \mu g/L)$$

Step 3: Determination of the Lifetime Health Advisory

$$\text{Lifetime HA} = (0.44 \text{ mg/L}) (20\%) = 0.09 \text{ mg/L (90 } \mu g/L)$$

E. Evaluation of Carcinogenic Potential

- The International Agency for Research on Cancer has not evaluated the carcinogenic potential of terbacil.

- Applying the criteria described in EPA's guidelines for assessment of carcinogenic risk (U.S. EPA, 1986), terbacil may be classified in Group E: evidence of noncarcinogenicity for humans. This category is used for substances that show no evidence of carcinogenicity in at least two adequate animal tests or in both epidemiologic and animal studies. Studies by Goldenthal et al. (1981) and Wazeter et al. (1967b) reported no induction of any carcinogenic effect in mice or rats, respectively, administered terbacil in the diet for 2 years.

V.   Other Criteria, Guidance and Standards

- Tolerances have been established for residues of terbacil in or on many agricultural commodities by the U.S. EPA Office of Pesticide Programs (U.S. EPA, 1985a).

VI.   Analytical Methods

- Analysis of terbacil is by a gas chromatographic (GC) method applicable to the determination of certain organonitrogen pesticides in water samples (U.S. EPA, 1987). This method requires a solvent extraction of approximately 1 L of sample with methylene chloride using a separatory funnel. The methylene chloride extract is dried and exchanged to acetone during concentration to a volume of 10 mL or less. The compounds in the extract are separated by gas chromatography, and the measurement is made with a thermionic bead detector. This method has been validated in a single laboratory, and the estimated detection limit for terbacil is 4.5 μg/L.

VII.   Treatment Technologies

- Treatment technologies currently available have not been tested for their effectiveness in removing terbacil from drinking water.

VIII.   References

Benson, N.R. 1973. Efficacy, leaching and persistence of herbicides in apple orchards. Bulletin No. 863. Washington State University, College of Agriculture Research Center.

Culik, R., C.K. Wood, A.M. Kaplan et al.* 1980. Teratogenicity study in rats with 3-tert-butyl-5-chloro-6-methyluracil. Haskell Laboratory Report No. 481-79. Unpublished study. MRID 00050467. Newark, DE: Haskell Laboratory for Toxicology and Industrial Medicine.

Davidson, J.M., L.T. Ou and P.S.C. Rao. 1978. Adsorption, movement, and biological degradation of high concentrations of selected herbicides in soils. In: Land disposal of hazardous wastes. EPA 600/9-78-016. U.S. Environmental Protection Agency, Office of Research and Development. pp. 233-244.

Doughty, C.C. 1978. Terbacil phytotoxicity and quackgrass (Agropyron repens) control in highbush blueberries (Vaccinium corymbosum). Weed Sci. 26:448-492.

E.I. duPont de Nemours and Company, Inc.* 1984a. Embryo-fetal toxicity and teratogenicity study of terbacil by gavage in the rabbit. Haskell Laboratory for Toxicology and Industrial Medicines, Newark, DE. Unpublished Study.

E.I. duPont de Nemours and Company, Inc.* 1984b. *In vitro* testing of terbacil. Unpublished Study. Newark, DE: Haskell Laboratory for Toxicology and Industrial Medicines.

Gardiner, J.A.* No date a. Examination of ^{14}C-terbacil-treated soil for the possible presence of 5-chlorouracil. Unpublished study. Submitted by E.I. du Pont de Nemours and Company, Inc., Wilmington, DE.

Gardiner, J.A.* No date b. Exposure of 2-^{14}C-labeled terbacil to field conditions, Supplement I. Unpublished study submitted by E.I. du Pont de Nemours and Company, Inc., Wilmington, DE.

Gardiner, J.A.* 1964. Laboratory exposure of 2-^{14}C-terbacil to moisture, light, and nonsterile soil. Unpublished study submitted by E.I. du Pont de Nemours and Company, Inc., Wilmington, DE.

Gardiner, J.A., R.C. Rhodes, J.B. Adams, Jr. and E.J. Soboezenski. 1969. Synthesis and studies with 2-C^{14}-labeled bromacil and terbacil. J. Agric. Food Chem. 17:980–986.

Goldenthal, E., S. Homan and W. Richter.* 1981. Two-year feeding study in mice (terbacil). IRDC No. 125–027. Unpublished study. MRID 00126770. International Research and Development Corporation.

Hood, D.* 1966. Fifteen exposure skin absorption studies with 3-tert-butyl-5-chloro-6-methyluracil. Report No. 33–66. Unpublished study. MRID 00125785.

Isom, W.H., H.P. Ford, M.P. Lavalleye and L.S. Jordan.* 1969. Persistence (sic) of herbicides in irrigated soils. Unpublished study prepared by Sandoz-Wander, Inc., submitted by American Carbonyl, Inc. Tenafly, NJ.

Isom, W.H., H.P. Ford, M.P. Lavalleye and L.S. Jordan. 1970. Persistence (sic) of herbicides in irrigated soils. Proc. Ann. Calif. Weed Conf. 22:58–63.

Lehman, A.J. 1959. Appraisal of the safety of chemicals in foods, drugs and cosmetics. Assoc. Food Drug Off. U.S., Q. Bull.

Liu, L.C., H.R. Cibes-Viade and J. Gonzalez-Ibanez. No date. Persistence of several herbicides in a soil cropped to sugarcane. J. Agric. Univ. Puerto Rico (volume no. not available):147–152.

Liu, L.C., H. Cibes-Viade and F.K.S. Koo. 1971. Adsorption of atrazine and terbacil by soils. J. Agric. Univ. Puerto Rico 55(4):451–460.

Mansell, R.S., D.V. Calvert, E.E. Stewart, W.B. Wheeler, J.S. Rogers, D.A. Graetz, L.E. Allen, A.F. Overman and E.B. Knipling. 1977. Fertilizer and pesticide movement from citrus groves in Florida flatwood soils. Report No. EPA 600/2-77-177. Athens, GA: U.S. Environmental Protection Agency, Environmental Research Laboratory. Also available from NTIS, Springfield, VA, PB-272 889.

Mansell, R.S., W.B. Wheeler, D.V. Calvert and E.E. Stewart. 1979. Terbacil movement in drainage waters from a citrus grove in a Florida flatwood soil. Proc. Soil Crop Sci. Soc. Fl. 37:176–179.

Mansell, R.S., W.B. Wheeler, L. Elliott and M. Shaurette. 1972. Movement of acarol and terbacil pesticides during displacement through columns of Wabasso fine sand. Proc. Soil Crop Sci. Soc. Fl. 31:239–243.

Marriage, P.B., S.U. Kahn and W.J. Saidak. 1977. Persistence and movement of terbacil in peach orchard soil after repeated annual applications. Weed Res. 17:219–225.

Marsh, J.A.P. and H.A. Davies. 1978. The effect of herbicides on respiration and transformation of nitrogen in two soils. III. Lenacil, terbacil, chlorthiamid and 2,4,5-T. Weed Res. 18:57–62.

Meister, R., ed. 1988. Farm chemicals handbook. Willoughby, OH: Meister Publishing Company.

Morrow, L.A. and M.K. McCarty. 1976. Selectivity and soil persistence of certain herbicides used on perennial forage grasses. J. Environ. Qual. 5:462–465.

Murnik, M.R. 1976. Mutagenicity of widely used herbicides. Genetics. 83(54):S54. Abstract.

Paynter, O.F.* 1966. Final report. Acute oral toxicity study in dogs. Unpublished study. MRID 00012206. Newark, DE: Haskell Laboratory for Toxicology and Industrial Medicines.

Rahman, A. 1977. Persistence of terbacil and trifluralin under different soil and climatic conditions. Weed Res. 17:145–152.

Rao, P.S.C. and J.M. Davidson. 1979. Adsorption and movement of selected pesticides at high concentrations in soils. Water Res. 13:375–380.

Reinke, R.E.* 1965. Primary irritation and sensitization skin tests. Haskell Laboratory Report No. 79–65. E.I. duPont deNemours and Company, Inc. Newark, DE: Haskell Laboratory for Toxicology and Industrial Medicine. Unpublished study. MRID 0006803.

Rhodes, R.C.* 1975. Biodegradation studies with 2-^{14}C-terbacil in water and soil. Unpublished study prepared in cooperation with University of Delaware, College of Agricultural Sciences. Submitted by E.I. duPont deNemours and Company, Inc., Wilmington, DE.

Sherman, H.* 1965. Oral LD_{50} test. Haskell Laboratory Report No. 160–65. E.I. duPont de Nemours and Company, Inc. Newark, DE: Haskell Laboratory for Toxicology and Industrial Medicine. Unpublished study. MRID 00012235.

Skroch, W.A., T.J. Sheets and J.W. Smith. 1971. Herbicides effectiveness, soil residues, and phytotoxicity to peach trees. Weed Sci. 19:257–260.

STORET. 1988. STORET Water Quality File. Office of Water. U.S. Environmental Protection Agency (data file search conducted in May, 1988).

Tucker, D.P. and R.L. Phillips. 1970. Movement and degradation of herbicides in Florida citrus soil. Citrus Ind. 51(3):11–13.

U.S. EPA. 1985a. U.S. Environmental Protection Agency. Terbacil; tolerances for residues. 40 CFR 180.209.

U.S. EPA. 1986. U.S. Environmental Protection Agency. Guidelines for carcinogen risk assessment. Fed. Reg. 51(185):33992–34003. September 24.

U.S. EPA. 1987. U.S. Environmental Protection Agency. Draft document. Method 507 — Determination of nitrogen and phosphorus-containing pesticides in water by gas chromatography with a nitrogen-phosphorus detector. Available from U.S. Environmental Protection Agency, Environmental Monitoring and Support Laboratory, Cincinnati, OH.

Wazeter, F.X., R.H. Buller and R.G. Geil.* 1964. Ninety-day feeding study in rats. IRDC No. 125–004. International Research and Development Corp. Unpublished study. MRID 00068035.

Wazeter, F.X., R.H. Buller and R.G. Geil.* 1966. Two-year feeding study in the dog. IRDC No. 125–011. International Research and Development Corp. Unpublished study. MRID 00060851.

Wazeter, F.X., R.H. Buller and R.G. Geil.* 1967a. Three-generation reproduction study in the rat. IRDC No. 125–012. International Research and Development Corp. Unpublished study. MRID 00060852.

Wazeter, F.X., R.H. Buller and R.G. Geil.* 1967b. Two year feeding study in the albino rat. IRDC No. 125–100. International Research and Development Corp. Unpublished study. MRID 00060850.

Wolf, D.C. 1973. Degradation of bromacil, terbacil, 2,4-D and atrazine in soil and pure culture and their effect on microbial activity. Ph.D. Dissertation. University of California, Riverside.

Wolf, D.C. 1974. Degradation of bromacil, terbacil, 2,4-D and atrazine in soil and pure culture and their effects on microbial activity. Dissertation Abstracts International B. 34(10):4783–4784.

Wolf, D.C. and J.P. Martin. 1974. Microbial degradation of 2-carbon-14-bromacil and terbacil. Proc. Soil Sci. Soc. Am. 38:921–925.

Zimdahl, R.L., V.H. Freed, M.L. Montgomery and W.R. Furtick. 1970. The degradation of triazine and uracil herbicides in soil. Weed Res. 10:18–26.

---

* Confidential Business Information submitted to the Office of Pesticide Programs.

Terbufos

I. General Information and Properties

A. CAS No. 13071-79-9

B. Structural Formula

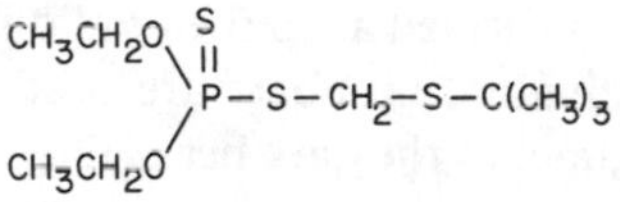

$$CH_3CH_2O \diagdown \underset{CH_3CH_2O}{\overset{\displaystyle S}{\underset{\diagup}{\overset{\|}{P}}}} - S - CH_2 - S - C(CH_3)_3$$

S-[[(1,1-Dimethylethyl)thio]methyl]O,O-diethyl phosphorodithioate

C. Synonyms

- Counter; Contraven (Meister, 1986).

D. Uses

- Terbufos is used in the control of corn rootworm and other soil insects and nematodes infesting corn, sugarbeet maggots in sugarbeets and green bug on grain sorghum (Meister, 1986).

E. Properties (Windholz et al., 1983; Meister, 1986)

Chemical Formula	$C_9H_{21}O_2PS_3$
Molecular Weight	288.41
Physical State (room temp.)	Clear, slightly brown liquid
Boiling Point	69°C/0.01 mm Hg
Melting Point	−29.2°C
Density (24°C)	1.105
Vapor Pressure (25°C)	34.6 mPa
Specific Gravity	1.1
Water Solubility (25°C)	15 mg/L
Log Octanol/Water Partition Coefficient	595
Taste Threshold	--
Odor Threshold	--

759

Conversion Factor   --
Technical Grade   87 to 97% pure

F. Occurrence

- Terbufos has been found in 134 of 2,016 surface water samples analyzed and in 0 of 283 ground-water samples (STORET, 1988). The 85th percentile of all nonzero samples was .10 μg/L in surface water, with a maximum concentration found of 2.25 μg/L. This information is provided to give a general impression of the occurrence of this chemical in ground and surface waters as reported in the STORET database. The individual data points retrieved were used as they came from STORET and have not been confirmed as to their validity. STORET data are often not valid when individual numbers are used out of the context of the entire sampling regime, as they are here. Therefore, this information can only be used to form an impression of the intensity and location of sampling for a particular chemical.

G. Environmental Fate

- ^{14}C-Terbufos hydrolyzes rapidly in buffer solutions (Miller and Jenney, 1973). At a concentration of 4.6 ppm, the hydrolysis half-lives were 4.5, 5.5 and 8.5 days at pH 5, 7 and 9, respectively. Principal hydrolysis products were identified as formaldehyde (accounting for 50 to 70% of the applied radioactivity), t-butyl mercaptan and O,O-diethylphosphorodithioic acid.

- ^{14}C-Terbufos photodegrades rapidly on silica gel glass surfaces (Miller and Jenney, 1973). Approximately 30% of the radioactivity was recovered as terbufos within 8 days.

- Terbufos degrades with a half-life of about 10 days in a silt loam soil under aerobic conditions (U.S. EPA, 1977). Terbufos residues dissipate fairly rapidly in the field. In Illinois, residues from application of terbufos (Counter 15-G) at 1 lb active ingredient (a.i.)/A dissipated to 0.31 ppm within 40 days; residues were nondetectable (<0.05 ppm) by day 135 (Steller et al., 1973a). In Colorado, residues from a similar application of terbufos declined from 5.9 ppm to 1.56 ppm 100 days later; residues of 0.15 ppm were present at 309 days after application.

- ^{14}C-Terbufos residues are immobile in four different soil types: sand and sandy loam from New Jersey, Wisconsin silt loam and a North Dakota silt clay (Hui, 1973). Terbufos residues were slightly more mobile when sandy loam soil was aged for 30 days before leaching it with 22.5 inches of water; 18% of the applied radioactivity was detected in the 3.5- to 7.0-inch layer. Formaldehyde was the only degradate identified (Steller et al., 1973b).

II. Pharmacokinetics

A. Absorption

- North (1973) reported that 83% of a single oral dose of technical ^{14}C-terbufos (0.8 mg/kg) was excreted in the urine of rats 168 hours after dosing. (The carbon atom of the thiomethyl portion of terbufos was radiolabeled.) An additional 3.5% was recovered in feces. This study indicates that terbufos was well absorbed (about 80 to 85%) from the gastrointestinal tract.

B. Distribution

- North (1973) reported that maximum residues of cholinesterase-inhibiting compounds (phosphorylated metabolites), resulting from a single oral dose of technical ^{14}C-terbufos (0.8 mg/kg) given to rats, were found in rat liver (0.08 ppm) 6 hours after dosing. In the same study, residues of hydrolysis (nonphosphorylated metabolites) products reached a maximum in rat kidney 12 hours after dosing (0.9 ppm). After 168 hours, each body tissue in the rat contained less than 0.1 ppm radiolabeled terbufos.

C. Metabolism

- North (1973) reported that terbufos was extensively metabolized in the rat. ^{14}C-Radiolabeled terbufos was administered in a single dose to 16 male Wistar rats at a dose level of 0.8 mg/kg via gavage. Examination of urine extracts by thin-layer chromatography (TLC) showed the presence of 10 radiometabolites in the rat urine. Approximately 96% of the radioactivity present in the urine was composed of an S-methylated series of metabolites, which result from the cleavage of the sulfur-phosphorus bond, methylation of the liberated thiol group and oxidation of the resulting sulfide to sulfoxides and sulfones. Of the remaining radioactivity, about 2% was composed of various oxidation products of the intact parent organophosphorus compound and 2% was an unknown metabolite.

D. Excretion

- North (1973) reported that technical terbufos and its metabolites were rapidly excreted in the urine of the rat. Radiolabeled terbufos was administered in a single dose to male Wistar rats at a dose level of 0.8 mg/kg by gavage. Of all the radioactivity recovered in the urine, 50% was excreted after 15 hours. After 168 hours, the termination of the test, 83% of the terbufos was excreted via the urine and 3.5% was recovered in the feces.

III.   Health Effects

A. Humans

- Peterson et al. (1984) reported the results of farm worker exposure to Counter 15-G (a 15% granular formulation of terbufos). Five farmers (one loader, one flagger and three scouts) were exposed for varying time periods (loader, 5 minutes; flagger, 15 minutes; scouts, twice each for 30 minutes) during a typical workday while Counter 15-G was applied aerially to a young corn crop. The mean exposure via inhalation was <0.25 µg/hour, the sensitivity of the monitoring method, for all samples collected. The exposure values for the five farm workers were 331 µg/hour for the loader, 0 µg/hour for the flagger, 381 µg/hour for the scouts (after 3 days) and 250 µg/hour for the scouts (after 7 days). All of the farm workers were men and weighed between 65.9 and 90.9 kg. Analysis of urinary metabolites showed no indication of significant absorption by any of the exposed workers. For example, all urinary alkyl phosphate analyses were negative (detection level, 0.1 ppm). Plasma and red blood cell cholinesterase values of the exposed workers showed no significant (95% confidence level) decrease in activity when compared to preexposed samples, indicating no adverse physiological effects from exposures.

- Devine et al. (1985) reported results similar to Peterson et al. (1984) for 11 farmers who were exposed to terbufos during a typical workday while planting corn and applying Counter 15-G. The average estimated dermal exposure was 72 µg/hour, and the estimated respiratory exposure was 11 µg/hour. The results of urinary alkyl phosphate analyses were all negative, showing no detectable absorption of terbufos. Plasma and red blood cell cholinesterase (ChE) values of the exposed farmers showed no significant difference in activity when compared to preexposure or control values, indicating no adverse physiological effects from the exposure. The report concluded that, based on the study results, the use of Counter 15-G does not present a significant hazard, in terms of acute toxicity, to farmers using this product for the control of corn insects.

B. Animals

1. *Short-term Exposure*

- Parke and Terrell (1976) reported that the acute oral LD_{50} value of technical-grade (86%) terbufos in Wistar rats was 1.73 mg/kg. Terbufos was administered in doses of 1.0 to 3.0 mg/kg via gavage in corn oil to a total of 50 rats (25 female/25 male; 10/dose). Average weight of the rats ranged from 200 to 300 g. The lowest dose (1.0 mg/kg) did not result in any mortality. Observed effects to the rats were respiratory depression, piloerection, clonic convulsions, exophthalmus, ptosis, lacrimation, hemorrhage and decreased motor activity.

- Consultox Laboratories (1975) reported that the acute oral LD_{50} value of technical-grade (86%) terbufos in male Wistar rats was 1.5 mg/kg. Terbufos was administered by gavage in doses of 0.50 to 2.5 mg/kg to a total of 50 male rats (10/dose) at an average weight of 200 ± 20 g. No mortality was reported at the low dose (0.50 mg/kg). Ten percent mortality was reported at the 0.75-mg/kg dose. Other effects reported were salivation, diuresis, diarrhea, disorientation, chromodocryorrhea, piloerection and body tremors.

- American Cyanamid Company (1972a) reported acute oral LD_{50} values (for 96.7% technical-grade terbufos) in dogs, mice and rats of 4.5 mg/kg (male)/6.3 mg/kg (female), 3.5 mg/kg (male)/9.2 mg/kg (female) and 4.5 mg/kg (male)/9.0 mg/kg (female), respectively. No details were given as to age or weight.

- American Cyanamid Company (1972b) reported additional acute oral LD_{50} values in male Wistar rats and female CFl mice of 1.6 mg/kg and 5.0 mg/kg, respectively. Other effects reported included cholinesterase inhibition in both sexes.

- Berger (1977) reported that plasma ChE was inhibited by as much as 79% in eight beagle dogs that were dosed via corn oil with 0.05 mg/kg/day (only dose tested) technical terbufos for 28 days. Red blood cell ChE was not inhibited at the dose tested.

- Tegeris Laboratories (1987) fed dogs (6/sex/dose) technical terbufos (purity 89.6%) in a corn oil vehicle (capsule) at doses of 0, 1.25, 2.5, 5.0 and 15 μg/kg/day for 28 days in order to define a plasma cholinesterase

NOAEL, which was not previously defined in a 1-year study conducted by American Cyanamid (1986). Based upon the evaluation of plasma, RBC and brain cholinesterase (ChE) activities at 1, 2 and 4 weeks during a 28-day dosing regime, a plasma ChE activity LOAEL (ranging from 10 to 20% depression) was determined to be 2.5 μg/kg/day in male and female dogs. A plasma ChE NOAEL was identified at 1.25 μg/kg/day.

2. Dermal/Ocular Effects

- Kruger and Feinman (1973) conducted a subacute dermal toxicity test in New Zealand White rabbits. Technical-grade terbufos was administered at doses varying from 0.004 to 0.1 mg/kg to the shaved, intact or abraded backs of male and female rabbits (2.5 to 3.5 kg). All animals survived the 30-day test and showed no adverse effects with regard to food and water intake, elimination, behavior, pharmacological effects and weight gain differences. There were no observed changes in hematological determinations (hematocrit, total erythrocyte and total leukocyte levels). Minor changes reported were increased numbers of eosinophils and basophils in all groups, occasional minimal edema that abated by day 21 and occasional mild erythema. All observed changes occurred on intact and abraded skin sites.

- American Cyanamid (1972a,b) conducted a series of tests with 96.7 and 85.8% terbufos using New Zealand White rabbits. Twenty male rabbits (2.56 to 2.73 kg) were administered doses of 0.4 to 3.5 mg/kg terbufos to their shaved backs. Dermal contact with terbufos was maintained for 24 hours. The dermal LD_{50} value was 1.0 mg/kg. An acute dermal test with 96.7% terbufos resulted in an LD_{50} of 1.1 mg/kg in male rabbits (no other details were given). In another test with 96.7% terbufos, 0.5 mL (500 mg) of terbufos was applied to the backs of rabbits; all of these animals died within 24 hours after dosing.

- American Cyanamid (1972a) reported the results of an application of 0.1 mg of technical-grade (96.7%) terbufos to the eyes of New Zealand albino rabbits. All animals died within 2 to 24 hours after dosing.

3. Long-term Exposure

- Daly et al. (1979) administered terbufos (90% a.i.) in the diet to groups of male and female Sprague-Dawley rats (20/sex/group, 24 to 39 days old, 95 to 150 g) at levels of 0, 0.125, 0.25, 0.5 or 1.0 ppm (estimated doses of 0, 0.01, 0.02, 0.046 or 0.09 mg/kg/day based on feed conversions given by the authors) for 90 days. Body weights and food consumption were measured weekly. Blood samples were obtained weekly and analyzed for plasma, and erythrocyte ChE. Brain ChE was analyzed at the study's termination. Body organs were weighed and analyzed for histopathology. The No-Observed-Adverse-Effect Level (NOAEL) was determined to be 0.02 mg/kg/day, based on the absence of effects on ChE. The statistically significant Lowest-Observed-Adverse-Effect Level (LOAEL) was determined to be 0.046 mg/kg based on the observed 17% decrease in plasma ChE in females. There were no depressions of erythrocyte or brain ChE at the highest dose tested (0.09 mg/kg/day). In addition, gross postmortem observations revealed no findings related to the test substance. Systemically, the LOAEL for increased liver weight in females and for a dose-related increase in liver extramedullary hematopoiesis was 0.046 mg/kg/day. No other histological lesions were found to be compound-related. The systemic NOAEL based on absence of liver effects was determined to be 0.02 mg/kg in this study.

- Biodynamics, Inc. (1987) administered terbufos (purity 89.6%) to Charles River CD rats (30/sex/dose) in a corn oil vehicle at doses of 0, 0.125, 0.5 and 1.0 ppm (estimated doses of 0.006, 0.025 and 0.05 mg/kg/day, based on Lehman, 1959) for 1 year. The systemic NOAEL was identified at 0.05 mg/kg/day, the highest dose tested. The cholinesterase NOAEL was defined at 0.025 mg/kg/day and the ChE LOAEL was observed at 0.05 mg/kg/day based on plasma and brain cholinesterase decreases (10 to 30%).

- Morgareidge et al. (1973) administered technical-grade terbufos in the diet to groups of male and female beagle dogs (4/sex/group, 10 to 13 months old, 9.0 to 13.8 kg) at levels of 0.0025, 0.01 and 0.04 mg/kg/day, 6 days a week for 26 weeks. Plasma, red blood cell and brain ChE levels; body weight and food; urinalysis; gross necropsy examination; and histopathology were evaluated. Observed effects included a decrease in ChE activity in plasma at all dose levels; however, decreased ChE activity was statistically significant only for doses of 0.01 mg/kg/day and above. At 0.01 mg/kg/day, plasma ChE was inhibited by 26% and red blood cell ChE was inhibited by 14%. No statistical analyses were performed on body weight changes, food consumption, hematology, clinical chemistry, urinalyses and organ weight data. The LOAEL (based on ChE effects) determined by the study was 0.01 mg/kg/day and the NOAEL was determined to be 0.0025 mg/kg/day.

- Rapp et al. (1974b) administered technical-grade terbufos in the diet to groups of Long-Evans rats (60/sex/dose, weanlings, 122 to 138.8 g) at levels of 0.25, 1.0, 2.0, 4.0 and 8.0 ppm for 2 years. These doses correspond to 0.0125, 0.05, 0.1, 0.2 and 0.4 mg/kg/day (Lehman, 1959). The original high doses (2.0 ppm) were increased to 4.0 and then to 8.0 ppm for males, and were increased from 2.0 to 4.0 to 8.0 and then reduced to 4.0 ppm for females. Body weight and food consumption were measured weekly. Hematology, clinical chemistry and urinalyses were also performed. Red blood cell ChE and brain ChE were significantly inhibited at 0.05 mg/kg/day (20% inhibition for brain ChE and 43% for red blood cell ChE in females) and above. Red blood cell ChE was also inhibited at 0.0125 mg/kg/day (12% in males and 15% in females). At the high dose (0.1 to 0.4 mg/kg/day), there was a noticeable decrease in mean body weight and mean food consumption. Mortality rates were 24 and 27% (males and females, respectively) at the high dose, 19% (males) at the mid-dose and 10% (males) at the low dose. The incidence of exophthalmia was in high-dose females (exophthalmia was also noted in low- and mid-dose control females). This study did not establish a NOAEL. The LOAEL was equivalent to the lowest dose tested (0.0125 mg/kg/day).

- McConnell (1983) conducted a followup pathology study of the results obtained from Rapp et al. (1974). Effects reported were gastric ulceration and/or erosion of glandular and nonglandular stomach mucosa in high-dose rats. No similar effect was seen in low- and mid-dose rats. Acute bronchopneumonia and granuloma of lungs occurred in high-dose rats more frequently than in low-dose, mid-dose or control rats. The authors reported that lung inflammation did not appear directly associated with the compound.

- Shellenberger (1986) administered technical-grade terbufos (89.6% a.i.) in capsule form to groups of beagle dogs (6/sex/dose, 6.8 to 7.5 kg, 5 to 6 months old) at doses of 0, 0.015, 0.120, 0.240 and 0.480 mg/kg/day for 1 year. The high doses were eventually reduced to 0.090 and 0.060 mg/kg/day after the eighth week of the study. Body weight and food consumption were measured, together with assessment of urinalyses, organ weights and cholinesterase levels. One male and one female at the high dose and one female at 0.240 mg/kg/day were found dead. At the two highest doses (0.240 and 0.480 mg/kg/day), decreased body weights and food consumption were observed. Mean erythrocytic parameters of high-dose males and females were significantly reduced at 3 months but not at 6 months or at termination of the study. Plasma ChE activity was significantly inhibited to 55% of controls at 0.015 mg/kg/day. Slight inhibition of erythrocyte ChE activity occurred at 0.120 mg/kg/day in females but not in males. No inhibition of erythrocyte ChE in males or females was observed at the lower doses. Brain ChE activities were similar for both sexes at all dose levels. Urinalyses and organ weight data revealed no significant differences. The report suggests that the NOAEL was 0.120 mg/kg/day in males and 0.090 mg/kg/day in females based on lack of erythrocyte ChE inhibition.

- American Cyanamid (1986) administered terbufos (purity 89.6%) in corn oil vehicle (capsule) to beagle dogs (6/sex/dose) at doses of 0, 0.015, 0.06, 0.09 and 0.12 mg/kg/day for 1 year. The major effect of terbufos was upon cholinesterase activity. There was a substantial depression in male and female dog plasma cholinesterase activity (30 to 60%) in all treatment groups compared to controls at weeks 13 through 52. RBC ChE activity in all dogs was moderately inhibited (20%) in both mid and high doses. Brain ChE inhibition was more variable but depression in ChE activity was apparent at mid and high doses. There was no evidence of compound-related effects upon mean organ weights or organ-to-body or brain weight ratios nor upon histopathology of non-neoplastic lesions. A plasma ChE NOAEL could not be established. The LOAEL for this study was identified as 0.015 mg/kg/day, the lowest dose tested.

4. *Reproductive Effects*

- Smith and Kasner (1972a) administered technical terbufos via the diet to Long-Evans and Blue Spruce rats (10 males/dose, weighing 276.3 g; 20 females/dose, weighing 185.6 g) for a period of 6 months at levels of 0, 0.25 and 1 ppm. These levels correspond to doses of 0, 0.0125 and 0.05 mg/kg/day, based on a conversion factor of 0.05 for rats (Lehman, 1959). The first parental generation (F_0) was dosed for 60 days. A reproductive LOAEL of 0.05 mg/kg was determined based on an increased percentage of litters with dead offspring in each of three generations as compared to controls. A NOAEL of 0.0125 mg/kg was identified.

5. *Developmental Effects*

- MacKenzie (1984) administered terbufos (87.8% a.i.) by gavage to groups of 18 female New Zealand White rabbits (3.5 kg) at levels of 0, 0.1, 0.2 and 0.4 mg/kg/day on days 7 to 19 of gestation. Reproductive indices monitored were female mortality, corpora lutea or implants, sex ratio, implantation efficiency, fetal body weight, fetal mortality and skeletal development. Cesarean sections were performed on day 29 of gestation. Survival of adult female rabbits was 100% in controls and in the 0.2-mg/kg/day dose group; 89% in the 0.1-mg/kg/day dose group; and 67% in the high-dose (0.4 mg/kg/day) group. There were no statistically significant dose-related differences in mean body weight, weight changes or gravid uterine weights, mean number of corpora lutea, implantation efficiency, sex ratio, fetal body weight or number of live or resorbing fetuses. The incidence of fetuses with an accessory left subclavian artery was significantly greater in the high-dose (0.4 mg/kg/day) group. The incidence of an extra unilateral rib and of chain fusion of sternebrae was significantly lower in the high-dose group than in the controls. According to the author, terbufos appears to be maternally toxic at 0.4 mg/kg/day, the highest dose tested.

- Rodwell (1985) administered terbufos (87.8% a.i.) via gavage to groups of 25 Charles River female rats (226 to 282 g, 71 days old) at doses of 0.05, 0.10 and 0.20 mg/kg/day on days 6 to 15 of gestation. Cesarean sections were performed on day 20; half of the fetuses were stained for skeletal evaluation. Parent survivability, body weight and embryonic and fetal development were all assessed. All parents survived the test. No changes in general appearance or behavior were observed. No doses led to teratogenic effects. Slightly decreased maternal mean body weights were observed during days 12 to 16 and following treatment in the 0.10- and 0.20-mg/kg/day dose groups. The study demonstrates that terbufos is slightly maternally toxic at dose levels of 0.10 and 0.20 mg/kg/day. A maternal NOAEL of 0.05 mg/kg/day, the lowest dose tested, was identified.

6. *Mutagenicity*

- Thilager et al. (1983) reported that Chinese hamster ovary cells tested with and without S-9 rat liver activation at concentrations of 100, 50, 25, 10, 5 and 2.5 nL/mL (ppm) terbufos (purity not specified) did not cause any significant increase in the frequencies of chromosomal aberrations. Only a concentration of 100 nL/mL proved to be cytotoxic.

- Allen et al. (1983) conducted mutagenicity tests with terbufos (87.8% a.i.) in the presence of S-9 metabolic activation and Chinese hamster ovary cells and in the absence of S-9 activation. Initial tests were conducted with doses of 100 to 10 µg/L, and then followed up with S-9 activation at doses of 50, 42, 33, 25, 10 and 5 µg/mL. Terbufos proved to be cytotoxic at 75 to 100 µg/mL with activation and at 50 to 70 mg/mL without activation. There were no increases in the frequency of chromosomal aberrations. The authors concluded that terbufos reflected a negative mutagenic potential.

- Godek et al. (1983) conducted a rat hepatocyte primary culture/DNA repair test with terbufos (87.8% a.i.) at doses ranging from 100 to 33 µg/well (a well contains 2 mL of media). Unscheduled DNA repair synthesis was quantified by a net nuclear increase of black silver grains for 50 cells/slide. This value was determined by taking a nuclear count and three adjacent cytoplasmic counts (100 µg/well was cytotoxic). The results for terbufos were negative in the rat hepatocyte primary culture/DNA repair test. These findings are based on the inability of terbufos to produce a mean grain count of 5, or greater than the vehicle-control mean grain count at any level. The authors concluded that terbufos reflected a negative mutagenic potential.

7. Carcinogenicity

- Rapp et al. (1974a) administered technical terbufos in the diet to groups of mice (75/sex/dose) at levels of 0, 0.5, 2.0 and 8.0 ppm for 18 months. These doses correspond to 0.075, 0.3 and 1.2 mg/kg/day (Lehman, 1959). The authors reported no signs of tumors or neoplasia. Effects noted include alopecia and signs of ataxia, exophthalmia in males, corneal cloudiness and opacity and eye rupture. Organ tissues examined were liver, kidney, heart and lung. No pathological changes attributable to terbufos were observed in these four organs.

- Rapp et al. (1974b) administered technical terbufos in the diet to groups of Long-Evans rats (60/sex/dose) at levels of 0, 0.25, 1.0, 2.0, 4.0 and 8.0 ppm for 2 years. These doses correspond to 0.0125, 0.05, 0.1, 0.2 and 0.4 mg/kg/day (Lehman, 1959). There were no indications of tumorigenic effects at any dose tested.

- McConnell (1983) conducted a followup pathology evaluation of the results obtained from Rapp et al. (1974b) and concluded that the compound had no effect on tumorigenesis.

- Tegeris Labs, Inc. (1986) administered terbufos (purity 89.6%) in the diet to groups of Charles River CD-1 mice (65/sex/dose) at levels of 0, 3, 6 and 12 ppm for 18 months. These doses correspond to 0.45, 0.90 and 1.8 mg/kg/day based on Lehman (1959). At these dose levels, there was no indication of oncogenic effect.

IV. Quantification of Toxicological Effects

A. One-day Health Advisory

No information was found in the available literature that was suitable for the determination of the One-day HA value for terbufos. It is, therefore, recommended that the Ten-day HA value for a 10-kg child (0.005 mg/L, calculated below) be used at this time as a conservative estimate of the One-day HA value.

B. Ten-day Health Advisory

The teratogenicity study in rats by Rodwell (1985) has been selected to serve as the basis for the Ten-day HA value for terbufos. Pregnant rats administered terbufos via gavage at a level of 0.05 mg/kg/day from days 6 to 15 of gestation showed no clinical signs of toxicity in the adult animals and no reproductive or teratogenic effects in the fetuses. The study identified a NOAEL of 0.05 mg/kg/day. These results are supported by the results of studies by MacKenzie (1984) with rabbits and by Smith and Kasner (1972a) with rats. In addition, the Rodwell (1985) study is of a most appropriate duration (10 days) for deriving a 10-day HA.

Using a NOAEL of 0.05 mg/kg/day, the Ten-day HA for a 10-kg child is calculated as follows:

$$\text{Ten-day HA} = \frac{(0.05 \text{ mg/kg/day}) (10 \text{ kg})}{(100) (1 \text{ L/day})} = 0.005 \text{ mg/L } (5 \text{ } \mu\text{g/L})$$

C. Longer-term Health Advisories

The 28-day feeding study in beagle dogs by Tegeris Labs (1987) has been selected to serve as the basis for the Longer-term HA value for terbufos. In this study, dogs were administered terbufos in a corn oil vehicle (capsule) at doses of 0, 1.25, 2.5, 5.0 and 15 μg/kg/day. A plasma cholinesterase NOAEL, which was not previously defined in a 1-year dog study conducted by American Cyanamid (1986), was defined at 1.25 μg/kg/day in the Tegeris Labs (1987) study. The plasma ChE LOAEL was determined to be 2.5 μg/kg/day based on decreased plasma ChE activity in male and female dogs. Other studies of longer duration were not selected for a number of reasons. First, the Tegeris Labs (1987) study was designed to define the plasma ChE NOAEL, which was not achieved in the 1-year study conducted by American Cyanamid (1986). It is a followup study employing the same parameters as the American Cyanamid (1986) study, only the doses are lower. The results of the two studies, evaluated together, adequately define the chronic end point of ChE activity depression. In other studies, either the ChE NOAEL is an order of magnitude larger than the NOAEL defined by Tegeris Labs (1987) or only a LOAEL has been defined. For example, Daly et al. (1979) identified a NOAEL an order of magnitude higher, as is also the case in Shellenberger (1986). Rapp et al. (1974b) and American Cyanamid (1986) were rejected since these studies identified LOAELs only. Morgareidge et al. (1973) was rejected because there was a high degree of uncertainty associated with the actual dose consumed by the test animals.

Using a NOAEL of 0.0013 mg/kg/day, the Longer-term HA is calculated as follows:

$$\text{Longer-term HA} = \frac{(0.0013 \text{ mg/kg/day})(10 \text{ kg})}{(10) (1 \text{ L/day})} = 0.0013 \text{ mg/L } (1.0 \text{ μg/L})$$

The Longer-term HA for a 70-kg adult is calculated as follows:

$$\text{Longer-term HA} = \frac{(0.0013 \text{ mg/kg/day})(70 \text{ kg})}{(10) (2 \text{ L/day})} = 0.0045 \text{ mg/L/day } (5 \text{ μg/L})$$

D. Lifetime Health Advisory

The 28-day feeding study in beagle dogs by Tegeris Labs (1987) has been selected to serve as the basis for the Longer-term HA value for terbufos. In this study, dogs were administered terbufos in a corn oil vehicle (capsule) at doses of 0, 1.25, 2.5, 5.0 and 15 μg/kg/day. A plasma cholinesterase NOAEL, which was not previously defined in a 1-year dog study conducted by American Cyanamid (1986), was defined at 1.25 μg/kg/day in the Tegeris Labs (1987) study. The plasma ChE LOAEL was determined to be 2.5 μg/kg/day based on decreased plasma ChE activity in male and female dogs. Other studies of longer duration were not selected for a number of reasons. First, the Tegeris Labs (1987) study was designed to define the plasma ChE NOAEL that was not achieved in the 1-year study conducted by American Cyanamid (1986). It is a followup study employing the same parameters as the American Cyanamid (1986) study, only the doses are lower. The results of the two studies, evaluated together, adequately define the chronic end point of ChE activity depression. In other studies, either the ChE NOAEL is an order of magnitude larger than the NOAEL defined by Tegeris Labs (1987) or only a LOAEL has been defined. For example, Daly et al. (1979) identified a NOAEL an order of magnitude higher, as is also the case in Shellenberger (1986). Rapp et al. (1974b) and American Cyanamid (1986) were rejected since these studies identified LOAELs only. Morgareidge et al. (1973) was rejected because there was a high degree of uncertainty associated with the actual dose consumed by the test animals.

Using this study, the Lifetime HA is calculated as follows:

Step 1: Determination of the Reference Dose (RfD)

$$\text{RfD} = \frac{(0.0013 \text{ mg/kg/day})}{(10)} = 0.00013 \text{ mg/kg/day}$$

Step 2: Determination of the Drinking Water Equivalent Level (DWEL)

$$DWEL = \frac{(0.00013 \text{ mg/kg/day}) (70 \text{ kg})}{(2 \text{ L/day})} = 0.0045 \text{ mg/L/day} (5.0 \text{ } \mu g/L)$$

Step 3: Determination of the Lifetime Health Advisory

Lifetime HA = (0.0045 mg/L) (20%) = 0.0009 mg/L (0.9 μg/L)

E. Evaluation of Carcinogenic Potential

- The International Agency for Research on Cancer has not evaluated the carcinogenic potential of terbufos.

- The U.S. EPA's Cancer Assessment Group (CAG) has assessed the carcinogenic potential of terbufos and has concluded that there are not enough data to determine whether terbufos is carcinogenic.

- Applying the criteria described in EPA's guidelines for assessment of carcinogenic risk (U.S. EPA, 1986a), terbufos may be classified in Group D: not classifiable. This category is for substances with inadequate human and animal evidence of carcinogenicity. Although two carcinogenicity studies are available for at least two animals, these studies have been considered flawed by the Agency.

V.   Other Criteria, Guidance and Standards

- No other criteria, guidance or standards were found in the available literature.

VI.   Analytical Methods

- Analysis of terbufos is by a gas chromatographic (GC) method applicable to the determination of certain nitrogen-phosphorus containing pesticides in water samples (U.S. EPA, 1988). In this method, approximately 1 liter of sample is extracted with methylene chloride. The extract is concentrated and the compounds are separated using capillary column GC. Measurement is made using a nitrogen-phosphorus detector. This method has been validated in a single laboratory and estimated detection limits for analytes such as terbufos are at 0.5 μg/L.

VII.   Treatment Technologies

- No data were found on the removal of terbufos from drinking water by conventional treatment.

- No data were found on the removal of terbufos from drinking water by activated carbon adsorption. However, due to its low solubility and high molecular weight, terbufos probably would be amenable to activated carbon adsorption.

- No data were found on the removal of terbufos from drinking water by ion exchange. However, the structure of this ester indicates that it is not ionic and, thus, probably would not be amenable to ion exchange.

- No data were found for the removal of terbufos from drinking water by aeration. However, the Henry's Coefficient can be estimated from available data on solubility (10 to 15 mg/L) and vapor pressure (0.01 mm Hg at 69°C). Terbufos probably would not be amenable to aeration or air stripping because its Henry's Coefficient is approximately 12 atm.

VIII.   References

Allen, J., E. Johnson and B. Fine.* 1983. Mutagenicity testing of AC 92,100 in the *in vitro* CHO/HGPRT mutation assay. Project No. 0402. Final report. Unpublished study. MRID 133297.

American Cyanamid Company.* 1972a. Summary of data: investigations made with respect to the safety of AC 92,100. Summary of studies 093580-A through 093580-D. Unpublished study. MRID 35960.

American Cyanamid Company.* 1972b. Toxicity data: O,O-diethyl-S(tert,butyl thiomethyl) phosphorodiothiolate technical 85.8% AC 2162–42. Report No. A-72–95. Unpublished study. MRID 37467.

American Cyanamid Company.* 1986. One-year toxicity study in purebred beagle dogs with AC 92,100. Unpublished study. EPA Accession No. 263678-263680.

Berger, H.* 1977. Toxicology report on experiment L-1680 and L-1680-A: cholinesterase activity of dogs receiving Counter soil insecticide for 28 days. Toxicology Report No. A A77–158. Unpublished study. MRID 63189.

Biodynamics, Inc.* 1987. A one-year dietary toxicity study with AC 92,100 in rats. Unpublished study. EPA Accession No. 400986.

Consultox Laboratories.* 1975. Acute oral and percutaneous toxicity evaluation. Unpublished study. MRID 29863.

Daly, I., W. Rinehart and A. Martin.* 1979. A three-month feeding study of

Counter terbufos insecticide in rats. Project No. 78-2343. Unpublished study. MRID 109446.

Devine, J.M., G.B. Kinoshita, R.P. Peterson and G.L. Picard. 1985. Farm worker exposure to terbufos during planting operations of corn. Arch. Environ. Contam. Toxicol. 15(1):113–120.

Godek, E., R. Naismith and R. Mathews.* 1983. Rat hepatocyte primary culture/DNA repair test: AC 92,100. PH 311-AC-001-83. Unpublished study. MRID 133298.

Hui, T.* 1973. Counter soil insecticide: soil leaching studies of 92,100. PD-M 10:455–483. Final report. Unpublished study. Received May 1, 1974, under 4F1496. Submitted by American Cyanamid Company, Princeton, NJ MRID 87693

Kruger, R. and H. Feinman.* 1973. 30-Day subacute dermal toxicity in rabbits of AC-92,100. Unpublished study. Prepared by Food and Drug Research Labs, Inc. July 17. Submitted to American Cyanamid Company, Princeton, NJ.

Lehman, A.J. 1959. Appraisal of the safety of chemicals in foods, drugs and cosmetics. Assoc. Food Drug Off. U.S., Q. Bull.

MacKenzie, K.* 1984. Teratology study with AC 92,100 in rabbits. Study No. 6123-116. Unpublished study. Prepared by Hazelton Laboratories America, Inc. MRID 147532.

McConnell, R.* 1983. Twenty-four month oral toxicity and carcinogenicity study in rats: AC 92,100. Pathology report. Unpublished study. Biodynamics, Inc. April 22. MRID 130845.

Meister, R.T., ed. 1986. Farm chemicals handbook. Willoughby, OH: Meister Publishing Company.

Miller, P. and K. Jenney.* 1973. CL 92, 100 Counter insecticide: metabolic studies of ^{14}C-labeled CL 92, 100 in hydrolytic and photolytic environments. PD-M 10:959–1007. Progress report, Apr. 5, 1973 – Oct. 15, 1973. Unpublished study. Received May 1, 1974, under 4F1496. Submitted by American Cyanamid Company, Princeton, NJ MRID 87694.

Morgareidge, K., S. Sistner, M. Daniels et al.* 1973. Final report: six-month feeding study in dogs on AC-92,100. Laboratory No. 1193. Unpublished study. Food and Drug Laboratories, Inc. February 14. MRID 41139.

North, N.H.* 1973. Counter® insecticide: rat metabolism of CL 92,100. PD-M 10:1008–1080. Progress report, March 1, 1973 - Sept. 28, 1973. Unpublished study. Submitted by American Cyanamid Company, Princeton, NJ. MRID 87695.

Parke, G.S.E. and Y. Terrell.* 1976. Acute oral toxicity in rats: compound: Enlist technical insecticide (terbufos). EPA file symbol 2749-VEL. Laboratory No. 6E-3164. Unpublished study. MRID 35121.

Peterson, R., G. Picard, J. Higham et al.* 1984. Farm worker study with aerial application of Counter 15-G. Report No. C-2370. Unpublished study. MRID 137760.

Rapp, R., A. Tebaldi, N. Wilson et al.* 1974a. An 18-month carcinogenicity study of AC 92,100 in mice. Project No. 71R-728. Unpublished study. Biodynamics, Inc. MRID 85170.

Rapp, W., N. Wilson, M. Mannion et al.* 1974b. A three- and 24-month oral toxicity and carcinogenicity study of AC-92,100 in rats. Project No. 71R-725. Unpublished study. Biodynamics, Inc. July 31. MRID 49236.

Rodwell, D.* 1985. A teratology study with AC 92,100 in rats. Project No. WIL-35014. Final report. Unpublished study. Prepared by WIL Research Laboratories, Inc. MRID 147533.

Shellenberger, T.* 1986. One-year oral toxicity study in purebred beagle dogs with AC 92,100. Final report. Report No. 8414. Unpublished study. Report No. 981-84-118. Prepared by Tegeris Laboratories, Inc. for American Cyanamid Company, Princeton, NJ. MRID 161572.

Smith, J.M. and J. Kasner.* 1972a. A three-generation reproduction study of AC-92,100 in rats. Status report for American Cyanamid Company, Nov. 28, 1972. Project No. 71R-727. Unpublished study. MRID 37473.

Steller, W., A. Schopbach, P. Ogg,et al.* 1973a. Counter 15-G: total residues of Counter (CL 92,100) and its metabolites in soil. Report No. C-480. Unpublished study. Received May 1, 1974, under 4F1496. Submitted by American Cyanamid Company, Princeton, NJ MRID 87706.

Steller, W., P. Ogg and W. Van Scoik.* 1973b. Counter 15-G: residues of Counter (CL92,100) and its individual toxic metabolites in soil. Report No. C-385. Unpublished study. Received May 1, 1974, under 4F1496. Submitted by American Cyanamid Company, Princeton, NJ MRID 87708.

STORET. 1988. STORET Water Quality File. Office of Water. U.S. Environmental Protection Agency (data file search conducted in May, 1988).

Tegeris Laboratories, Inc.* 1986. Chronic dietary toxicity and oncogenicity study with AC 92,100 (Terbufos) in mice. Unpublished study. EPA Accession No. 400986.

Tegeris Laboratories, Inc.* 1987. 28-Day oral toxicity study in the dog with AC 92,100. Unpublished study. EPA Accession Nos. 4037401–4037402.

Thilager, A., P. Kumaroo and S. Knott.* 1983. Chromosome aberration in Chinese hamster ovary cells (test article AC-92,100). Microbiological Associate Study No. T1906 337006. Sponsor Study No. 981-83-106. Unpublished study. MRID 133296.

U.S. EPA. 1977. U.S. Environmental Protection Agency. The degradation of selected pesticides in soil: a review of the published literature. EPA 600/9-77-022. Municipal Environmental Research Laboratory, Cincinnati, OH.

U.S. EPA. 1986a. U.S. Environmental Protection Agency. Guidelines for carcinogen risk assessment. Fed. Reg. 51(185):33992–34003. September 24.

U.S. EPA. 1986b. U.S. Environmental Protection Agency. Code of Federal Regulations. 40 CFR 180.352.

U.S. EPA. 1988. U.S. Environmental Protection Agency. Method 507 — Determination of nitrogen and phosphorous containing pesticides in water by GC/NPD. Draft. April 15. Available from U.S. EPA's Environmental Monitoring and Support Laboratory, Cincinnati, OH.

Windholz, M., S. Budavari, R.F. Blumetti and E.S. Otterbein. 1983. The Merck index, 10th ed. Rahway, NJ: Merck and Company.

---

* Confidential Business Information submitted to the Office of Pesticide Programs.

2,4,5-Trichlorophenoxyacetic Acid

I. General Information and Properties

A. CAS No. 93-76-5

B. Structural Formula

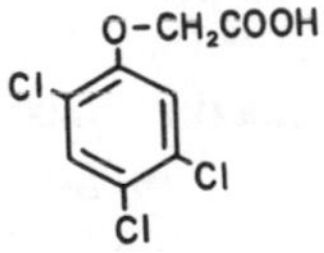

2,4,5-Trichlorophenoxyacetic acid

C. Synonyms

- 2,4,5-T; Dacamine; Ded-Weed; Fencerider; Forron; Inverton 245; Linerider; T-Nox; U-46 (Meister, 1988).

D. Uses

- The use of 2,4,5-T in the United States has been canceled since 1985. Some or all applications may be classified by the U.S. EPA as Restricted Use Pesticides (Meister, 1988).

E. Properties (BCPC, 1983; Meister, 1988; Windholz et al., 1983; Khan, 1985; CHEMLAB, 1985)

Chemical Formula	$C_8H_5O_3Cl_3$
Molecular Weight	255.49
Physical State (25°C)	Crystals
Boiling Point	--
Melting Point	153°C
Density	--
Vapor Pressure (25°C)	6.46×10^{-6} mm Hg
Specific Gravity	--
Water Solubility (25°C)	Solubility of acid is 150 g/L; amine salts are soluble at 189 g/L (20°C); esters are insoluble

777

Log Octanol/Water
   Partition Coefficient   3.00 (calculated)
Taste Threshold   --
Odor Threshold   --
Conversion Factor   --

F. Occurrence

- 2,4,5-T has been found in 4,021 of 21,616 surface water samples analyzed and in 99 of 2,905 ground-water samples (STORET, 1988). Samples were collected at 3,838 surface water locations and 1,853 ground-water locations, and 2,4,5-T was found in 44 states. The 85th percentile of all nonzero samples was 0.1 μg/L in surface water and 1 μg/L in ground-water sources. The maximum concentration found was 370 μg/L in surface water and 38 μg/L in ground water. This information is provided to give a general impression of the occurrence of this chemical in ground and surface waters as reported in the STORET database. The individual data points retrieved were used as they came from STORET and have not been confirmed as to their validity. STORET data are often not valid when individual numbers are used out of the context of the entire sampling regime, as they are here. Therefore, this information can only be used to form an impression of the intensity and location of sampling for a particular chemical.

G. Environmental Fate

- Salts dissociate in aqueous media, and esters are rapidly hydrolyzed to acids by enzymatic action in animals, plants and soil (Loos, 1975).

- The phenoxy ether link is stable in plants and animals, but is cleaved by bacterial action in soil and in the rumen of cattle and sheep (Leng, 1977). The rate of cleavage is inhibited by substitution of a third chlorine at the meta position on the ring and is sterically hindered by the angular methyl group in the 2-propionic acid side-chain.

- The first step in degradation of chlorophenoxy acids and chlorophenols occurs by hydroxylation at the 4-position with shift of the 4-chloro group to the 3- or 5-position (NIH shift). This reaction also is inhibited by the presence of an m-chloro substituent; thus, residues of 2,4,5-trichlorophenol appeared in the milk of cows and in the liver and kidney of cattle maintained on feed containing 100 or 300 ppm 2,4,5-T, respectively (Leng, 1977).

- The rate of microbial degradation of chlorophenoxyalkanoic acids in the environment depends on a number of factors, including sunlight, temperature, moisture and organic matter in soil, and whether the organisms were adapted by repeated treatment (Loos, 1975; NRCC, 1978; Bovey and Young, 1980). Under field conditions favorable for degradation, 2,4-D disappears within about 2 to 3 weeks, whereas other phenoxy herbicides disappear more slowly (Loos, 1975). The average half-lives in a laboratory study were 4, 10, 17 and 20 days for 2,4-D, dichloroprop, fenoprop, and 2,4,5-T, respectively, when added as dimethylamine salts at about 5 ppm in three soils containing partially decomposed litter obtained from two forest sites and one grassland site in Oklahoma (Altom and Stritzke, 1973).

- Photodegradation is rapid when phenoxy herbicides are exposed to sunlight in water containing dissolved organic matter, and the resultant chlorophenols are photolyzed even more rapidly than the parent acids (NRCC, 1978; Bovey and Young, 1980). The dioxin impurity TCDD is also degraded rapidly by sunlight, particularly in the presence of its herbicide carrier on the surface of leaves or in the presence of other hydrogen donors dissolved in water (NRCC, 1978).

- Plants treated with phenoxy herbicides rapidly hydrolyze esters and salts to the parent acids, which form ether-soluble conjugates with amino acids or water-soluble conjugates with sugars, depending on the relative susceptibility or resistance of the plants to the herbicides (Loos, 1975).

II.   Pharmacokinetics

A. Absorption

- In a study by Gehring et al. (1973), single oral doses of 5 mg/kg 2,4,5-T were ingested by five male volunteers. Essentially all the 2,4,5-T was excreted unchanged via the urine, indicating that gastrointestinal absorption was nearly complete.

- Fang et al. (1973) administered single doses of ^{14}C-labeled 2,4,5-T in corn oil by gavage to pregnant and nonpregnant female Wistar rats at dose levels of 0.17, 4.3 or 41 mg/kg. Expired air, urine, feces, internal organs and tissues were analyzed for radioactivity. During the first 24 hours, an average of 75 ± 7% of the radioactivity was excreted in the urine, indicating that at least 75% of the dose had been absorbed.

- Piper et al. (1973) administered single oral doses of ^{14}C-labeled 2,4,5-T in corn oil-acetone (9:1) to adult female Sprague-Dawley rats at dose levels of 5, 50, 100 or 200 mg/kg, and to adult female beagle dogs at 5 mg/kg. Fecal excretion was 3% at the lowest dose (5 mg/kg) and increased to 14% at the highest dose (200 mg/kg) in rats. In dogs given the 5 mg/kg dose, fecal excretion was 20%. These data indicated that absorption was somewhat dose dependent, but was 80% or higher at all doses.

B. Distribution

- Gehring et al. (1973) administered single oral doses of 5 mg/kg of 2,4,5-T to five male volunteers. Essentially all the 2,4,5-T was absorbed; 65% of the absorbed dose remained in the plasma where 98.7% was bound reversibly to protein. The volume of distribution was 0.097 L/kg. Utilizing the kinetic constants from the single-dose experiment, the expected concentrations of 2,4,5-T in the plasma of individuals receiving repeated doses of 2,4,5-T were calculated. From these calculations, it was determined that the plasma concentrations would essentially reach a plateau value after 3 days. If the daily dose ingested in mg/kg is A_0, the concentrations in the plasma after attaining plateau would range from 12.7 A_0 to 22.5 A_0 μg/mL (Gehring et al., 1973).

- Fang et al. (1973) administered single oral doses of ^{14}C-labeled 2,4,5-T to pregnant and nonpregnant female Wistar rats and internal organs and tissues were analyzed for radioactivity. Radioactivity was detected in all tissues, with the highest concentration found in the kidney. The maximum concentration in all tissues was generally reached between 6 and 12 hours after administration of the dose (0.17, 4.3 or 41 mg/kg) by gavage, and then declined rapidly. Radioactivity also was detected in the fetuses and in the milk of the pregnant rats. The average biological half-life of 2,4,5-T in the organs was 3.4 hours for the adult rats and 97 hours for the newborn.

- Piper et al. (1973) administered single oral doses of 5, 50, 100 or 200 mg/kg 2,4,5-T to Sprague-Dawley rats, and found that the apparent volume of distribution increased with dose, indicating that distribution of 2,4,5-T in the body was dose-dependent.

C. Metabolism

- Gehring et al. (1973) administered single oral doses of 5 mg/kg 2,4,5-T to human volunteers. Essentially all the chemical was excreted in the urine as parent compound, indicating that there is little metabolism of 2,4,5-T in humans.

- Grunow et al. (1971) investigated the metabolism of 2,4,5-T in male Wistar (AF/Han) rats that received single oral doses of 50 mg/kg. The 2,4,5-T was dissolved in peanut oil and administered by gavage. Urine was collected for 7 days after dosing and examined by gas chromatography for 2,4,5-T and its conjugates and metabolites. From 45 to 70% of the administered dose was recovered in urine. In general, about 10 to 30% of this was as acid-hydrolyzable conjugates, and the remainder was unchanged 2,4,5-T. Three animals were given doses of 75 mg/kg, and their urine pooled. A metabolite isolated from this pooled urine was identified as N-(2,4,5-trichlorophenoxy-acetyl)glycine.

- Piper et al. (1973) administered single oral doses of 2,4,5-T to female Sprague-Dawley rats at dose levels of 5, 50, 100 or 200 mg/kg. A small amount of an unidentified metabolite was detected in urine at the high doses, but not at the lower doses. In adult beagle dogs given oral doses of 5 mg/kg, three unidentified metabolites were detected in urine, suggesting a difference in metabolism between rats and dogs.

- In a study by Fang et al. (1973) in female Wistar rats, urinalysis revealed that 90 to 95% of the radioactivity was unchanged 2,4,5-T. The authors also found three unidentified minor metabolites, two of which were nonpolar, in the urine.

D. Excretion

- In a study by Gehring et al. (1973), single doses of 5 mg/kg 2,4,5-T were ingested by five male volunteers. The concentrations of 2,4,5-T in plasma and its excretion were measured at intervals after ingestion. The clearances from the plasma, as well as the body, occurred via apparent first-order rate processes with half-lives of 23.1 and 29.7 hours, respectively. Essentially all the 2,4,5-T was excreted unchanged via the urine.

- In a study by Fang et al. (1973), 2,4,5-T labeled with ^{14}C was orally administered to pregnant and nonpregnant female Wistar rats at various dosages, and expired air, urine and feces were analyzed for radioactivity. During the first 24 hours, 75 ± 7% of the radioactivity was excreted in the urine and 8.2% was excreted in the feces. No ^{14}C was found in the expired air. There was no significant difference in the rate of elimination between the pregnant and nonpregnant rats, or among the dosages used (0.17, 4.3 and 41 mg/kg). The average biological half-life of 2,4,5-T in the organs was 3.4 hours for the adult rats and 97 hours for the newborn.

- Grunow et al. (1971) investigated the excretion of 2,4,5-T in male Wistar (AF/Han) rats after single oral doses of 50 mg/kg. The 2,4,5-T was dissolved in peanut oil and administered by gavage. From 45 to 70% of the administered dose was recovered in urine within 7 days.

- Clearance of ^{14}C activity from the plasma and its elimination from the body of rats and dogs were determined after single oral doses of labeled 2,4,5-T (Piper et al., 1973). The half-life values for clearance from the plasma of Sprague-Dawley (Spartan strain) rats given doses of 5, 50, 100 or 200 mg/kg were 4.7, 4.2, 19.4 and 25.2 hours, respectively; half lives for elimination from the body were 13.6, 13.1, 19.3 and 28.9 hours, respectively. Urinary excretion of unchanged 2,4,5-T accounted for 68 to 93% of the radioactivity eliminated from the body of the rats. Fecal excretion was 3% at 5 mg/kg, and increased to 14% at 200 mg/kg. These results indicate that the excretion of 2,4,5-T is altered when large doses are administered. In adult beagle dogs given doses of 5 mg/kg, the half-life values for clearance from plasma and elimination from the body were 77.0 and 86.6 hours, respectively. By 9 days posttreatment, 11% of the dose had been recovered in urine and 20% had been recovered in feces.

III.   Health Effects

- Technical 2,4,5-T contains traces of the highly toxic compound 2,3,7,8-tetrachlorodibenzo-p-dioxin (TCDD) as an impurity (NAS, 1977). Preparations of 2,4,5-T formerly contained TCDD at levels of 1 to 80 ppm, a concentration sufficiently high to cause toxic effects that are characteristic of TCDD. It has not been feasible to completely eliminate TCDD from technical 2,4,5-T, but NAS (1977) reported it to be present in commercial 2,4,5-T at less than 0.1 ppm. In the following sections, the purity of 2,4,5-T or the level of TCDD impurity is given when known. When the generic term "dioxin" is used, no further information is provided, and the 2,4,5-T is presumed to contain a variety of dioxin species as well as other phenoxy compounds and assorted intermediates and breakdown products.

A. Humans

1. *Short-term Exposure*

- No clinical effects were observed in five volunteers who ingested single oral doses of 5 mg/kg of 2,4,5-T (Gehring et al., 1973).

- After an explosion in a chemical plant producing 2,4,5-T in 1949, symptoms in exposed workers included chloracne, nausea, headache, fatigue, and muscular aches and pains (Zack and Suskind, 1980).

2. *Long-term Exposure*

- The mortality experience in a cohort of 1,926 men who had sprayed 2,4,5-T acid during 1955 to 1971 was followed prospectively from 1972 to 1980. Exposure was generally rather low because the duration of work had mostly been less than 2 months. In the period 1972 to 1976, mortality from all natural causes in this group was only 54% of the expected value (based on age-specific rates for the general population), and in the next 4-year period, 81% of the expected value. In the assessment of cancer, mortality allowance was made for 10- and 15-year periods of latency between the first exposure and the start of the recording of vital status during the followup. No increase in cancer mortality was detected, and the distribution of cancer types was unremarkable. No cases of death from lymphomas or soft-tissue sarcomas were found. It was noted, however, that the study results should be interpreted with caution due to the small size of the cohort, the low past exposure, and the brief followup period (Riihimaki et al., 1982).

- An investigation of the rate of birth malformations in the Northland region of New Zealand was analyzed with reference to the exposure in the area to 2,4,5-T, which was applied as frequently as once a month from 1960 to 1977. The chosen area was divided into sectors rated as high, intermediate or low, based on the frequency of aerial spraying. During this period, there were 37,751 babies born in the hospitals in these sectors. It was estimated that well over 99% of all births in the Northland area occur in hospitals. The epidemiological analysis of the birth data gave no evidence that any malformation of the central nervous system, including spina bifida, was associated with the spraying of 2,4,5-T. Heart malformations, hypospadias and epispadias increased with spraying density, but the increases were not statistically significant ($p > 0.05$). The only anomaly that increased in a statistically significant ($p < 0.05$) manner with respect to the spraying was talipes (club foot) (Hanify et al., 1981).

- The relationship between the use of 2,4,5-T in Arkansas and the concurrent incidence of facial clefts in children was studied retrospectively. The estimated levels of exposure were determined by categorizing 75 counties into high, medium and low exposure groups on the basis of their rice acreage during 6- to 7-year intervals beginning in 1943. A total of 1,201 cases of cleft lip and/or cleft palate for the 32 years (until 1974) was detected by screening birth certificates and hospital records. Facial cleft rates, presented by sex, race, time period and exposure group, generally rose over time. No significant differences were found for any race or sex combination. The investigators concluded that the general increase seen in facial cleft incidence in the high- and low-exposure groups was attributable to better case finding rather than maternal exposure to 2,4,5-T (Nelson et al., 1979).

- Ott et al. (1980) reported no effects in a survey of 204 workers engaged in 2,4,5-T production at estimated airborne levels of 0.2 to 0.8 mg/m^3 for 1 month to 10 years.

- Numerous epidemiological studies on the relationship between exposure to chlorophenoxyacetic acids and cancer induction are reviewed in U.S. EPA (1985). The conclusion in this review is that there is "limited" evidence for the carcinogenicity of chlorinated phenoxyacetic herbicides and/or chlorophenols with chlorinated dibenzodioxin impurities, primarily based on Swedish case-control studies that associated induction of soft-tissue sarcomas with exposure to these agents.

A. Animals

1. *Short-term Exposure*

- The acute oral toxicity of 2,4,5-T was determined in mice, rats and guinea pigs by Rowe and Hymas (1954) over a 2-week period. The LD$_{50}$ values were 500 mg/kg for rats, 389 mg/kg for mice and 381 mg/kg for guinea pigs.

- Drill and Hiratzka (1953) investigated the acute oral toxicity of 2,4,5-T in adult mongrel dogs given single oral doses of 50, 100, 250 or 400 mg/kg by gelatin capsule. Animals were observed for 14 days, at which time survivors were necropsied. The number of deaths at the four dose levels were 0/4, 1/4, 1/1 and 1/1, respectively. The LD$_{50}$ value was estimated to be 100 mg/kg or higher. Marked changes were not observed in animals that died, effects being limited to weight loss, slight to moderate stiffness in the hind legs and ataxia (at the highest doses).

- Weanling male Wistar rats were fed diets containing 2,4,5-T for 3 weeks to investigate effects on the immune system (Vos et al., 1983). Doses of 2,4,5-T (>99% purity, TCDD content not specified) was fed at levels of 200, 1,000 or 2,500 ppm (approximately 20, 100 or 250 mg/kg/day, assuming 1 ppm equals 0.1 mg/kg/day in a younger rat, from Lehman, 1959). Following the 3-week feeding period, the animals were sacrificed and the organs of the immune system, as well as other parameters of general toxicity, were examined. Even at the lowest dose level of 200 ppm in the diet, 2,4,5-T caused a significant ($p < 0.05$) decrease in relative kidney weight and a significant ($p < 0.05$) increase in serum IgG level, the most sensitive indicators of its effects. In this study, based on general toxicologic and specific immunologic effects in the rat, the Lowest-Observed-Adverse-Effect Level (LOAEL) was 20 mg/kg/day.

2. Dermal/Ocular Effects

- Gehring and Betso (1978) summarized the effects of 2,4,5-T on the skin and the eye. The dry material is slightly irritating to the skin and the eye. Highly concentrated solutions may burn the skin with prolonged or repeated contact and can strongly irritate the eye and possibly cause corneal damage. Preparations of 2,4,5-T formerly contained 1 to 80 ppm 2,3,7,8-TCDD, a concentration high enough to cause chloracne in industrial workers (NAS, 1977).

3. Long-term Exposure

- Drill and Hiratzka (1953) investigated the subchronic toxicity of 2,4,5-T in adult mongrel dogs. One or two dogs of each sex per group were fed capsules in food containing 0, 2, 5, 10 or 20 mg/kg 2,4,5-T, 5 days per week for 13 weeks. Animals were weighed twice weekly, and blood was taken on days 0, 30 and 90. Upon death or completion of the study, animals were necropsied with histological examination of a number of tissues. No deaths occurred at doses of 10 mg/kg/day or less, but 4/4 animals receiving 20 mg/kg/day died. No effects on body weight, hematology and pathology were seen except in animals that died. The No-Observed-Adverse-Effect Level (NOAEL) was identified as 10 mg/kg/day.

- McCollister and Kociba (1970) examined the effects of 2,4,5-T administered in the diet for 90 days to male and female Sprague-Dawley rats (Spartan strain). The 2,4,5-T (99.5% pure, <0.5 ppm dioxin) was included in the diet at levels corresponding to doses of 0, 3, 10, 30 or 100 mg/kg/day. Five animals of each sex were used at each dose level. At the conclusion of the study, necropsy, urinalyses, blood counts and clinical chemistry assays were performed. There was no mortality in any group. At 100 mg/kg, animals of both sexes had depressed ($p < 0.05$) body weight gain, a slight but significant ($p < 0.05$) decrease in food intake and elevated ($p < 0.05$) serum alkaline phosphatase (AP) levels. Necropsy revealed paleness and an accentuated lobular pattern of the liver, with some inconsistent hepatocellular swelling. Males (but not females) had slightly elevated serum glutamic pyruvic transaminase (SGPT) levels, and slight decreases in red blood cell counts and in hemoglobin. Males given 100 mg/kg/day had increased ($p < 0.05$) kidney/body and liver/body weights. At the 30 mg/kg/day dose level, males exhibited increased ($p < 0.05$) liver/body and kidney/body weight ratios and kidney weights. Females given 30 mg/kg/day had slightly but significantly ($p < 0.05$) elevated AP and SGPT levels, but the authors felt that the clinical significance of these latter findings was doubtful. No treatment-related effects were observed at the 3 or 10 mg/kg dose level. From this study, a NOAEL of 10 mg/kg/day and a LOAEL of 30 mg/kg/day were identified.

- Groups of Sprague-Dawley rats (50/sex/level) were maintained on diets supplying 3, 10 or 30 mg/kg/day of 2,4,5-T for 2 years (Kociba et al., 1979). The 2,4,5-T was approximately 99% pure, containing 1.3% (w/w) other phenoxy acid impurities. Dioxins were not detected, the limit of detection for TCDD being 0.33 ppb. An interim sacrifice was performed on an additionally included group of 10 animals of each sex at 118 to 119 days. Control groups included 86 animals of each sex. The highest dose level was associated with some degree of toxicity, including a decrease in body weight gain ($p < 0.05$ in females) and an increase in relative kidney weight ($p < 0.05$ in males). Increases ($p < 0.05$) in the volume of urine excreted and in the urinary excretion of coproporphyrin and uroporphyrin were also observed at this dose level. Increased ($p < 0.05$) morphological changes were observed in the kidney, liver and lungs of animals administered 30 mg/kg/day. The kidney changes involved primarily the presence ($p < 0.05$) of mineralized deposits in the renal pelvis in females. Effects noted at the 10 mg/kg dose level were primarily an increased ($p < 0.05$) incidence of mineralized deposits in the renal pelvis in females. During the early phase of the study there was an increase ($p < 0.05$) in urinary excretion of coproporphyrin in males. At the lowest dose level (3 mg/kg), there were no changes that were considered to be related to treatment throughout the 2-year period. From this study in rats, a NOAEL of 3 mg/kg/day was identified.

4. *Reproductive Effects*

- Male and female Sprague-Dawley rats (F_0) were fed lab chow containing 2,4,5-T (<0.03 ppb TCDD) to provide dose levels of 0, 3, 10 or 30 mg/kg/day for 90 days and then were bred (Smith et al., 1981). At day 21 of lactation, pups were randomly selected for the following generation (F_1) and the rest were necropsied. Subsequent matings were conducted to produce F_2, F_{3a} and F_{3b} litters, successive generations being fed from weaning on the appropriate test or control diet. Fertility was decreased ($p < 0.05$) in the matings of the F_{3b} litters in the group given 10 mg/kg/day. Postnatal survival was significantly ($p < 0.05$) decreased in the F_2 litters of the 10 mg/kg group and in the F_1, F_2 and F_3 litters of the 30 mg/kg group. A significant decrease ($p < 0.05$) in relative thymus weight was seen only in the F_{3b} generation of the 30 mg/kg group, but the relative liver weights of weanlings was significantly ($p < 0.05$) increased in the F_2, F_{3a} and F_{3b} litters of this dosage group. Smith et al. (1981) concluded that dose levels of 2,4,5-T that were sufficiently high to cause signs of toxicity in neonates had no effect on the reproductive capacity of the rats, except for a tendency toward a reduction of postnatal survival at a dose of 30 mg/kg. Reproduction was not impaired at the lowest dose of 3 mg/kg. The apparent NOAEL with respect to reproductive capacity and fetotoxic effects in this study is 3 mg/kg/day, taking into consideration the report, on apparently the same study, by Smith et al. (1978), who noted a significant ($p < 0.05$) decrease in F_1 (10 and 30 mg/kg on days 14 and 21) and F_3 (3 mg/kg on day 14, and 10 and 30 mg/kg on day 21) litters. The study concluded that there was no effect of 2,4,5-T on rat reproduction except for a tendency toward a reduction in neonatal survival at 10 and 30 mg/kg.

5. *Developmental Effects*

- Sparschu et al. (1971) tested 2,4,5-T (commercial grade, 0.5 ppm TCDD) at levels of 50 or 100 mg/kg/day in pregnant rats (strain not specified) on either days 6 to 15 (50 mg/kg) or days 6 to 10 (100 mg/kg) of gestation. The 2,4,5-T was administered by oral intubation in a solution of Methocel, and controls were given an appropriate volume of Methocel. At the 50 mg/kg dose, there was a slightly higher incidence of delayed ossification of the skull bones, but this was not considered a teratogenic response. The 100 mg/kg dose (administered on days 6 to 10) was toxic to the dams and caused a high incidence of maternal deaths (only 4 of the 25 pregnant rats survived). Of these, 3 had complete early resorptions, and 1 had a litter of 13 viable fetuses that showed toxic effects (not further described) but no terata. From these data for maternal effects, a NOAEL of 50 mg/kg and a LOAEL of 100 mg/kg were identified. Also identified were a NOAEL of 100 mg/kg for teratogenicity and a LOAEL of 50 mg/kg for fetotoxicity.

- A sample of 2,4,5-T (technical grade) containing 0.5 ppm TCDD as well as other phenoxy compounds was administered to CD-l rats by oral intubation on days 6 through 15 of gestation at dose levels of 10, 21.5, 46.4 or 80 mg/kg/day (Courtney and Moore, 1971). Examination of offspring revealed that the sample was not teratogenic at these dose levels. There was a significant ($p < 0.05$) increase in fetal mortality at the 80 mg/kg/day dose levels (the maternal LD_{40}). In two 2,4,5-T-treated fetuses, mild gastrointestinal hemorrhages were observed as a fetotoxic effect. Kidney anomalies were also slightly increased with the effect most pronounced at the 80 mg/kg level, but the number of litters examined was too small to evaluate this observation. In a separate study, rats were administered 50 mg/kg/day in an identical protocol, but in this case they were allowed to litter, and the neonates were examined and weighed on day 1 and followed for 21 days. Postnatal growth and development were comparable to that of the control animals. A NOAEL of 46.4 mg/kg/day for both fetotoxicity and teratogenicity in the CD-l rat was identified from these data.

- Sprague-Dawley rats (50/group) and New Zealand White rabbits (20/group) were given oral doses (gavage for rats, capsules for rabbits) of 2,4,5-T (containing 0.5 ppm TCDD) during gestation (Emerson et al., 1971). The rats received daily doses of 1, 3, 6, 12 or 24 mg/kg on days 6 through 15, while the rabbits were administered 10, 20 or 40 mg/kg on days 6 through 18 of gestation. In both species, animals were observed daily, weighed periodically and subjected to Cesarean section prior to parturition. Rabbit pups were kept for observation for 24 hours and then sacrificed. There were no observable adverse effects in dams of either species treated with the 2,4,5-T. Litter size, number of fetal resorptions, birth weights and sex ratios all appeared to be unaffected in the treated groups. Detailed visceral and skeletal examinations were performed on the control and high-dose groups for each species, and no embryotoxic or teratogenic effects were revealed. A NOAEL for fetotoxic and maternal effects identified from this study was 24 mg/kg/day for the rat and 40 mg/kg/day for the rabbit.

- Several different samples of 2,4,5-T (containing <0.5 ppm TCDD) were tested in pregnant Wistar rats by daily oral administration on days 6 through 15 of gestation at dose levels between 25 and 150 mg/kg/day (Khera and McKinley, 1972). In some cases, fetuses were removed by

Cesarean section for examination; some animals were allowed to litter, and the offspring were observed for up to 12 weeks. At doses of 100 mg/kg, there was an increase (p < 0.05) in fetal mortality and an increase (p < 0.05) in skeletal variations; a visceral anomaly was noted (dilatation of the renal pelvis), which was slightly increased over the control level, but was not statistically significant (p > 0.05). The survival of the progeny was not affected up to doses of 100 mg/kg, and in only one trial was there a low average litter size and viability. This effect was not duplicated in a repeat test with the same sample. At the 25 and 50 mg/kg dose levels, significant (p < 0.05) differences from controls were not apparent. With respect to fetotoxicity, this study identified a NOAEL of 50 mg/kg/day in the rat.

- The teratogenic effects of 2,4,5-T were examined in golden Syrian hamsters after oral dosing (by gavage) on days 6 through 10 of gestation at dose levels of 20, 40, 80 or 100 mg/kg/day (Collins et al., 1971). Four samples of 2,4,5-T with dioxin levels of 45, 2.9, 0.5 or 0.1 ppm were administered. Three samples, which had no detectable dioxin (based on TCDD), were also tested. The 2,4,5-T samples induced fetal death and terata. The incidence of effects increased with increasing content of the TCDD impurity. The 2,4,5-T samples with no detectable dioxin produced no malformations below the 100 mg/kg dose level. Using the data from the 2,4,5-T samples with no detectable dioxin, a NOAEL of 80 mg/kg/day for the hamster was identified.

- Behavioral effects resulting from *in utero* exposure to 2,4,5-T were examined in Long-Evans rats after single oral doses were administered during gestation (Crampton and Rogers, 1983). The sample of 2,4,5-T contained <0.03 ppm TCDD. Novelty response (latency, ambulation, rearing, grooming, and defecation in open field testing) abnormalities in offspring were detected after single doses as low as 6 mg/kg were administered on day 8 of gestation. Examination of the brain in the affected offspring failed to reveal any changes of a qualitative or quantitative structural nature in various areas of the brain. With respect to behavioral effects, the LOAEL for this study is 6 mg/kg.

- The teratogenic effects of technical 2,4,5-T (TCDD content 0.1 ppm) were studied using large numbers of pregnant mice of C57BL/6, C3H/He, BALB/c and A/JAX inbred strains and CD-1 stock (Gaines et al., 1975). Dose-response curves were determined for the incidence of cleft palate, embryo lethality and fetal growth retardation. These determinations were replicated 6 to 10 times for each inbred strain and 35 times for the CD-1. The number of litters studied ranged from 236 for BALB/c mice to 1,485 for CD-1 mice. Treatment was by gavage on days 6 to 14 of pregnancy, and dose levels of 2,4,5-T ranged from 15 to 120 mg/kg/day. The lowest dose tested in the A/JAX was 15 mg/kg, and this dose was teratogenic. The other strains and CD-1 demonstrated teratogenicity at 30 mg/kg, the lowest dose tested. There were significant ($p < 0.05$) differences in sensitivities among the strains for the parameters measured. Based on this study in the mouse, the LOAEL for teratogenic effects is 15 mg/kg/day for the A/JAX strain and 30 mg/kg/day for the other strains.

- Neubert and Dillmann (1972) studied the effects of 2,4,5-T in pregnant NMRI mice. Three samples of 2,4,5-T were utilized: one had < 0.02 ppm dioxin, and was considered "dioxin-free"; a second sample had a dioxin content of 0.05 + 0.02 ppm; and the third sample had an undetermined dioxin content. The 2,4,5-T was administered by gavage on days 6 through 15 of gestation at dose levels from 8 to 120 mg/kg/day. Fetuses were removed on day 18 and examined. Cleft palate frequency exceeding ($p < 0.05$) that of the controls was observed with doses higher than 30 mg/kg with all samples. Reductions ($p < 0.05$) in fetal weight were observed with all samples tested at doses as low as 10 to 15 mg/kg. There was no clear increase in embryo lethality over that of controls at these lower doses. With the purest sample of 2,4,5-T, single oral doses of 150 to 300 mg/kg were capable of producing significant ($p < 0.05$) incidences of cleft palate. The maximal teratogenic effect was seen when the 2,4,5-T was administered on days 12 to 13 of gestation. Based on the data obtained with the purest sample of 2,4,5-T, the teratogenic NOAEL is 15 mg/kg/day and the fetotoxic NOAEL is 8 mg/kg/day.

- Roll (1971) examined the teratogenic effects of 2,4,5-T in NMRI-Han mice after oral administration on days 6 to 15 of gestation at dose levels of 0, 20, 35, 60, 90 or 130 mg/kg/day. The 2,4,5-T sample had a purity of 99.6%, with a dioxin content of < 0.01 ppm (measured by the DOW method), or 0.05 ± 0.02 ppm [measured by the U.S. Food and Drug Administration (FDA) method]. Peanut oil was used as the vehicle. Animals were sacrificed on day 18 and examined for defects. Fetal weight was significantly ($p < 0.05$) lower than control at all doses. Resorptions were significantly ($p < 0.05$) increased at 60 mg/kg and above. The incidence of cleft palates was significantly ($p < 0.05$) higher at 35 mg/kg and higher, but there was no effect at 20 mg/kg. There were also dose-dependent increases in ossification defects of sternum and various other bones. The authors concluded that 2,4,5-T alone (independent of TCDD contamination) was teratogenic in mice, and that the teratogenic NOAEL in this strain was 20 mg/kg/day. In view of the significantly (p < 0.05) lower fetal weight at 20 mg/kg/day, this level may also be considered the LOAEL for fetotoxicity.

- No teratogenic effects were observed in the offspring of female Rhesus monkeys that were given oral doses of 0.05, 1.0 or 10.0 mg 2,4,5-T (containing 0.05 ppm TCDD)/kg/day in capsules during gestation days 22 through 38. Nor was toxicity evident in the mothers (Dougherty et al., 1976).

6. *Mutagenicity*

- At 250 and 1,000 ppm 2,4,5-T (with no detectable TCDD), mutation rate was significantly (p < 0.05) increased at the higher dose in the sex-linked recessive lethal test in *Drosophila* as carried out by Majumdar and Golia (1974). The sex-linked test was not affected by 920 or 1,804 ppm of the sodium salt of 2,4,5-T at pH 6.8 in a study carried out by Vogel and Chandler (1974). Although they found no cytogenetic effects in *Drosophila*, Magnusson et al. (1977) concluded that 1,000 ppm 2,4,5-T (<0.1 ppm TCDD) did cause an increase (p < 0.05) in the number of recessive lethals compared to the controls. Rasmusson and Svahlin (1978) treated *Drosophila* larvae with food containing 100 and 200 ppm 2,4,5-T; survival was low at 200 ppm, but 2,4,5-T had no observable effect on somatic mutational activity.

- Anderson et al. (1972) found that neither 2,4,5-T nor its butyric acid form showed any mutagenic action when tested on histidine-requiring mutants of *Salmonella typhimurium*.

- Buselmaier et al. (1972) found that intraperitoneal injection of 2,4,5-T (dioxin levels not given) had no effect in the host-mediated assay (500 mg/kg) or in the dominant lethal test (100 mg/kg) with NMRI mice. Styles (1973), likewise, found no increase in back mutation rates with the serum of rats treated orally with 2,4,5-T in the host-mediated assay with *Salmonella typhimurium* (dosages and purity of the samples not given).

- Shirasu et al. (1976) found that 2,4,5-T did not induce mitotic gene conversion in a diploid strain of *Saccharomyces cerevisiae*. When the pH of the treatment solution was less than 4.5, Zetterberg (1978) found that 2,4,5-T was mutagenic in haploid, DNA-repair-defective *S. cerevisiae*.

- Jenssen and Renberg (1976) investigated the cytogenetic effects of 2,4,5-T in mice by examining the ability of the herbicide to induce micronuclei formation in the erythrocytes of mouse bone marrow. CBA mice were treated at 8 to 10 weeks of age (20 to 30 g) with a single intraperitoneal injection of 100 mg/kg of 2,4,5-T (<1 ppm TCDD) dissolved in Tween 80 and physiological saline. Cytogenetic examination at 24 hours and 7 days after treatment showed no detectable increase in micronuclei in the erythrocytes compared to controls. A weak toxic effect on the mitotic activity was indicated, as judged by a decrease in the percentage of polychromatic erythrocytes.

7. *Carcinogenicity*

- Innes et al. (1969) investigated the potential carcinogenic effects of 2,4,5-T in two hybrid strains of mice derived by breeding SPF C57BL/6 female mice to either C3H/Anf or AKR males. Beginning at 6 days of age, 2,4,5-T was administered by gavage in 0.5% gelatin to a group of 72 mice at a dose level of 21.5 mg/kg/day. This was reported to be the maximum tolerated dose. At 28 days of age, the 2,4,5-T was added to the diet at a level of 60 ppm, corresponding to a dose of about 9 mg/kg/day (assuming that 1 ppm equals 0.15 mg/kg/day in the diet, from Lehman, 1959). This dose was fed for 18 months, at which time the study was terminated. All animals were necropsied and the tissues were examined both grossly and microscopically. There were no significant (p > 0.05) increases in tumors in either strain of treated mice.

- A lifetime study using oral administration of 2,4,5-T in both sexes of two strains of mice, C3Hf and XVII/G, was performed by Muranyi-Kovacs et al. (1976). The 2,4,5-T, which contained less than 0.05 ppm of dioxins, was administered in the water (1,000 mg/L) for 2 months beginning at 6 weeks of age, and thereafter in the diet at 80 ppm (12 mg/kg/day) until death or when the mice were sacrificed *in extremis*. In the treated C3Hf mice there was a significant (p < 0.03) increase in the incidence of total tumors found in female mice and a significant (p < 0.001) increase in total nonincidental tumors in each sex, which the authors interpreted as life-threatening. No significant (p > 0.05) difference was found in the XVII/G strain between the treated and control mice. The authors felt that although 2,4,5-T demonstrated carcinogenic potential in the C3Hf strain, additional studies in other strains and in other species of animals needed to be performed before a reliable conclusion with respect to carcinogenicity could be made.

- Groups of Sprague-Dawley rats (50 males and 50 females) were maintained on diets supplying 3, 10 or 30 mg/kg/day of 2,4,5-T for 2 years (Kociba et al., 1979). The 2,4,5-T was approximately 99% pure, containing 1.3% (w/w) other phenoxy acid impurities. Dioxins were not detected, the limit of detection for TCDD being 0.33 ppb. An interim sacrifice was performed on an additionally included group of 10 animals of each sex at 118 to 119 days. Control groups included 86 animals of each sex. At the end of the 2-year period, there was no significant (p > 0.05) increase in tumor incidence in any treated group compared to the control for either male or female animals.

IV.   Quantification of Toxicological Effects

A. One-day Health Advisory

No information was found in the available literature that was suitable for determination of the One-day HA value for 2,4,5-T. The study in humans by Gehring et al. (1973) was not selected because observations of the subjects were reported simply as clinical effects without further details. The behavioral study in rats by Crampton and Rogers (1983) was not selected because the interpretation of altered novelty response behavior in the absence of other toxic signs needs further investigation before definitive conclusions can be made. It is therefore recommended that the Ten-day HA value for a 10-kg child (0.8 mg/L, calculated below) be used at this time as a conservative estimate of the One-day HA value.

B. Ten-day Health Advisory

The study by Neubert and Dillman (1972) has been selected to serve as the basis for determination of the Ten-day HA value for 2,4,5-T. This developmental study in rats identified a NOAEL of 8 mg/kg/day and a LOAEL of 15 mg/kg/day, based on reduced body weights in pups from dams exposed on days 6 to 15 of gestation. This LOAEL is supported by a number of other developmental studies in rodents that identified LOAELs ranging from 15 to 100 mg/kg/day (Roll, 1971; Sparschu et al., 1971; Khera and McKinley, 1972; Gaines et al., 1975). In the 21-day feeding study in rats by Vos et al. (1983), a LOAEL of 20 mg/kg/day was identified based on effects on kidney weight and the immune system. The 8 mg/kg/day NOAEL for fetal effects selected from the Neubert and Dillman (1972) study may not be applicable to a 10-kg child; however, the assumptions for a 10-kg child are used with this NOAEL in this case since, although a NOAEL was not found in the 21-day study by Vos et al. (1983) where the observed effects are applicable to a 10-kg child, the LOAEL of 20 mg/kg/day is 2.5 times higher than the NOAEL used for the Ten-day HA.

Using a NOAEL of 8 mg/kg/day, the Ten-day HA for a 10-kg child is calculated as follows:

$$\text{Ten-day HA} = \frac{(8 \text{ mg/kg/day}) (10 \text{ kg})}{(100) (1 \text{ L/day})} = 0.8 \text{ mg/L } (800 \text{ } \mu\text{g/L})$$

C. Longer-term Health Advisory

The reproduction study by Smith et al. (1978, 1981) has been selected to serve as the basis for the Longer-term HA value for 2,4,5-T because the reduction in neonatal survival over multiple generations is concluded to be relevant to the Longer-term HA for a 10-kg child. The NOAEL identified was 3 mg/kg/day, and the LOAEL was 10 mg/kg/day. Other possible selections have a higher NOAEL [10 mg/kg/day in the 90-day feeding study in rats by McCollister and Kociba (1970) and the 90-day oral treatment study in dogs by Drill and Hiratzka (1953)].

Using a NOAEL of 3 mg/kg/day, the Longer-term HA for a 10-kg child is calculated as follows:

$$\text{Longer-term HA} = \frac{(3 \text{ mg/kg/day}) (10 \text{ kg})}{(100) (1 \text{ L/day})} = 0.3 \text{ mg/L } (300 \text{ } \mu\text{g/L})$$

The Longer-term HA for a 70-kg adult is calculated as follows:

$$\text{Longer-term HA} = \frac{(3 \text{ mg/kg/day}) (70 \text{ kg})}{(100) (2 \text{ L/day})} = 1.05 \text{ mg/L } (1{,}000 \text{ } \mu\text{g/L})$$

D. Lifetime Health Advisory

The study by Kociba et al. (1979) has been selected to serve as the basis for the Lifetime HA value for 2,4,5-T. In this study, rats were fed 2,4,5-T in the diet for 2 years. Based on observations of effects of 2,4,5-T on various biochemical parameters in addition to gross and microscopic observations related to general toxicity in the rats, this study identified a NOAEL of 3 mg/kg/day and a LOAEL of 10 mg/kg/day. This study is supported by the three-generation rat study (Smith et al., 1978, 1981) that identified a NOAEL of 3 mg/kg/day.

Using this study, the Lifetime HA is calculated as follows:

Step 1: Determination of the Reference Dose (RfD)

$$RfD = \frac{(3.0 \text{ mg/kg/day})}{(100)\,(3)} = 0.01 \text{ mg/kg/day}$$

where: 3 = modifying factor used by U.S. EPA Office of Pesticide Programs
to account for data gaps (chronic feeding study in dogs), which do not
make it possible to establish the most sensitive end point for 2,4,5-T.

Step 2: Determination of the Drinking Water Equivalent Level (DWEL)

$$DWEL = \frac{(0.01 \text{ mg/kg/day})\,(70 \text{ kg})}{(2 \text{ L/day})} = 0.35 \text{ mg/L } (400 \text{ } \mu g/L)$$

Step 3: Determination of the Lifetime Health Advisory

$$\text{Lifetime HA} = (0.35 \text{ mg/L})\,(20\%) = 0.07 \text{ mg/L } (70 \text{ } \mu g/L)$$

E. Evaluation of Carcinogenic Potential

- Chronic feeding studies with 2,4,5-T in Sprague-Dawley rats (Kociba et al., 1979) and C57BL/6×C3H/Anf, C57BL/6×AKR and XVII/G strains of mice (Innes et al., 1969; Muranyi-Kovacs et al., 1976) were negative for carcinogenic effects. A chronic feeding study with 2,4,5-T in C3Hf mice was inconclusive (Muranyi-Kovacs et al., 1976).

- IARC (1982) concluded that the carcinogenicity of 2,4,5-T is indeterminant (Group 3, inadequate evidence in animals and humans).

- Applying the criteria described in EPA's guidelines for assessment of carcinogenic risk (U.S. EPA, 1986), 2,4,5-T may be classified in Group D: not classifiable. This category is for agents with inadequate animal evidence of carcinogenicity.

- The Carcinogen Assessment Group (CAG) of the U.S. EPA classified chlorophenoxyacetic acids and/or chlorophenols containing 2,3,7,8-TCDD in IARC category 2A (probably carcinogenic in humans on the basis of limited evidence in humans), but only a quantitative cancer risk estimate for 2,3,7,8-TCDD itself was made. The CAG considered the human evidence for the carcinogenicity of 2,3,7,8-TCDD alone to be "inadequate" because of the difficulty in attributing observed effects solely to the presence of 2,3,7,8-TCDD, which occurs as an impurity in the phenoxyacetic acids and chlorophenols (U.S. EPA, 1985).

V. Other Criteria, Guidance and Standards

- The U.S. EPA Office of Pesticide Programs has calculated a Provisional Acceptable Daily Intake (PADI) value of 0.003 mg/kg/day, based on the results of a rat chronic oral NOAEL of 3 mg/kg/day with an uncertainty factor of 300 (used because of a data gap).

- The National Academy of Sciences (NAS, 1977) has calculated an ADI of 0.1 mg/kg/day, using a NOAEL of 10 mg/kg/day (identified in a 90-day feeding study in dogs) and an uncertainty factor of 100. A chronic Suggested-No-Adverse-Effect Level (SNARL) of 0.7 mg/L was calculated based on the ADI of 0.1 mg/kg/day.

- The American Conference of Governmental Industrial Hygienists (ACGIH, 1981) has recommended a Threshold Limit Value-Time-weighted Average (TLV-TWA) of 10 mg/m^3 and a Threshold Limit Value-Short-term Exposure Limit (TLV-STEL) of 20 mg/m^3.

- The ADI recommended by the World Health Organization is 0 to 0.03 mg/kg (Vettorazzi and van den Hurk, 1983).

VI. Analytical Methods

- Determination of 2,4,5-T is by a liquid-liquid extraction gas chromatographic procedure (U.S. EPA, 1978; Standard Methods, 1985). Specifically, the procedure involves the extraction of chlorophenoxy acids and their esters from an acidified water sample with ethyl ether. The esters are hydrolyzed to acids and extraneous organic material is removed by a solvent wash. The acids are converted to methyl esters, which are extracted from the aqueous phase. Separation and identification of the esters are performed by gas chromatography. Detection and measurement are accomplished by an electron capture, microcoulometric or electrolytic conductivity detector. Identification may be corroborated through the use of two unlike columns. The detection limit is dependent on the sample size and instrumentation used. Typically, using a 1 L sample and a gas chromatograph with an electron capture detector results in an approximate detection limit of 10 ng/L for 2,4,5-T.

VII.   Treatment Technologies

- Available data indicate that granular-activated carbon (GAC) and powdered-activated carton (PAC) adsorption will effectively remove 2,4,5-T from water.

- Robeck et al. (1965) experimentally determined adsorption isotherms for the butoxy ethanol ester of 2,4,5-T on PAC. Based on these results, it was calculated that 14 mg/L PAC would be required to remove 90% of 2,4,5-T, while 44 mg/L PAC would be required to remove 99% of 2,4,5-T (Pershe and Goss, 1979; Robeck et al., 1965).

- Robeck et al. (1965) reported the results of a GAC column operating under pilot plant conditions. At a flow rate of 0.5 gpm/ft^3, 99 + % of 2,4,5-T was removed. By comparison, treatment with 5 to 20 mg/L PAC removed 80 to 95% of the same concentration of 2,4,5-T.

- In a laboratory study conducted with an exchange resin, Rees and Au (1979) reported 89 ± 2% removal efficiency of 2,4,5-T from contaminated water by adsorption onto synthetic resins.

- Conventional water treatment techniques of coagulation with alum, sedimentation and sand filtration removed 63% of the 2,4,5-T ester present in spiked river water (Robeck et al., 1965).

- Treatment technologies for the removal of 2,4,5-T from water are available and have been reported to be effective. However, selection of individual or combinations of technologies to attempt 2,4,5-T removal from water must be based on a case-by-case technical evaluation, and an assessment of the economics involved.

VIII.   References

ACGIH. 1981. American Conference of Governmental Industrial Hygienists. Threshold limit values for chemical substances and physical agents in the workroom environment. Cincinnati, OH: ACGIH. p. 27.

Altom, J.T. and J.F. Stritzke. 1973. Degradation of dicamba, picloram, and four phenoxyherbicides in soil. Weed Sci. 21:556–560.

Anderson, K.J., E.G. Leighty and M.T. Takahashi. 1972. Evaluation of herbicides for possible mutagenic properties. J. Agric. Food Chem. 20:649–656.

BCPC. 1983. British Crop Protection Council. 2,4,5-T. In: Worthing, C.R., ed. The pesticide manual: a world compendium, 7th ed. p. 11120.

Bovey, R.W. and A.L. Young. 1980. The science of 2,4,5-T and associated phenoxy herbicides. New York, NY: Wiley-Interscience. p. 462.

Buselmaier, W., G. Roehrborn and P. Propping. 1972. Mutagenicity investigations with pesticides in the host-mediated assay and the dominant lethal test in mice. Biol. Zentralbl. 91:311–325.

CHEMLAB. 1985. The chemical information system. CIS, Inc., Baltimore, MD.

Collins, T.F.X., G.H. Williams and G.C. Gray. 1971. Teratogenic studies with 2,4,5-T and 2,4-D in the hamster. Bull. Environ. Contam. Toxicol. 6(6):559–67.

Courtney, K.D. and J.A. Moore. 1971. Teratology studies with 2,4,5-tri-chlorophenoxyacetic acid and 2,3,7,8-tetrachlorodibenzo-p-dioxin. Toxicol. Appl. Pharmacol. 20:396–403.

Crampton, M.A. and L.J. Rogers. 1983. Low doses of 2,4,5-trichlorophenoxyacetic acid are behaviorally teratogenic in rats. Experientia. 39:891–2.

Dougherty, W.J., F. Coulston and L. Goldberg. 1976. The evaluation of the teratogenic effects of 2,4,5-trichlorophenoxyacetic acid in the Rhesus monkey. Environ. Qual. Saf. 5:89–96.

Drill, V.A. and T. Hiratzka. 1953. Toxicity of 2,4-dichlorophenoxyacetic acid and 2,4,5-trichlorophenoxyacetic acid: a report on their acute and chronic toxicity in dogs. Arch. Ind. Hyg. Occup. Med. 7:61–67.

Emerson, J.L., D.J. Thompson, R.J. Strebing, C.G. Gerbig and V.B. Robinson. 1971. Teratogenic studies on 2,4,5-trichlorophenoxyacetic acid in the rat and rabbit. Food Cosmet. Toxicol. 9:395–404.

Fang, S.C., E. Fallin, M.L. Montgomery and V.H. Freed. 1973. Metabolism and distribution of 2,4,5-trichlorophenoxyacetic acid in female rats. Toxicol. Appl. Pharmacol. 24(4):555–563.

Gaines, T.B., J.F. Holson, C.J. Nelson and H.J. Schumacher. 1975. Analysis of strain differences in sensitivity and reproducibility of results in assessing 2,4,5-T teratogenicity in mice. Abstract No. 30. Toxicol. Appl. Pharmacol. 33:174–175.

Gehring, P.J. and J.E. Betso. 1978. Phenoxy acids: effects and fate in mammals. Ecol. Bull. 27:122–133.

Gehring, P.J., C.G. Krammer, B.A. Schwetz, J.Q. Rose, V.K. Rowe and J.S. Zimmer. 1973. The fate of 2,4,5-trichlorophenoxyacetic acid (2,4,5-T) following oral administration to man. Toxicol. Appl. Pharmacol. 25(3):441.

Grunow, W., C. Bohme and B. Budczies. 1971. Renal excretion of 2,4,5-T by rats. Food Cosmet. Toxicol. 9:667–670.

Hanify, J.A., P. Metcalf, C.L. Nobbs and K.J. Worsley. 1981. Aerial spraying of 2,4,5-T and human birth malformation: an epidemiological investigation. Science. 212(4492):349–351.

IARC. 1982. International Agency for Research on Cancer. IARC monographs on the evaluation of carcinogenic risk of chemicals to man. Suppl. 4. Lyon, France: IARC.

Innes, J.R.M., B.M. Ulland, M.G. Valerio, L. Petrucelli, L. Fishbein, E.R. Hart, A.J. Pallotta, R.R. Bates, H.L. Falk, J.J. Gart, M. Klein, I. Mitchell and J. Peters. 1969. Bioassay of pesticides and industrial chemicals for tumorigenicity in mice: a preliminary note. J. Natl. Cancer Inst. 42(6):1101–1114.

Jenssen, D. and L. Renberg. 1976. Distribution and cytogenetic test of 2,4,-D and 2,4,5-T phenoxyacetic acids in mouse blood tissues. Chem. Biol. Interact. 14(3–4):291–299.

Khan, M.A.Q. 1985. Personal communication to Environmental Criteria and Assessment Office, U.S. Environmental Protection Agency, Cincinnati, OH. January.

Khera, K.S. and W.P. McKinley. 1972. Pre- and postnatal studies on 2,4,5-trichlorophenoxyacetic acid, 2,4,-dichlorophenoxyacetic acid and their derivatives in rats. Toxicol. Appl. Pharmacol. 22:14–28.

Kociba, R.J., D.J. Keyes, R.W. Lisowe, R.P. Kalnins, D.D. Dittenber, C.E. Wade, S.J. Gorzinski, N.H. Mahle and B.A. Schwetz. 1979. Results of a two-year chronic toxicity and oncogenic study of rats ingesting diets containing 2,4,5-trichlorophenoxyacetic acid (2,4,5-T). Food Cosmet. Toxicol. 17:205–221.

Lehman, A.J. 1959. Appraisal of the safety of chemicals in foods, drugs and cosmetics. Assoc. Food Drug Off. U.S., Q. Bull.

Leng, M.L. 1977. Comparative metabolism of phenoxy herbicides in animals. In: Ivie, G.W. and H.W. Dorough, eds., Fate of Pesticides in Large Animals. New York, NY: Academic Press pp. 53–76.

Loos, M.A. 1975. Phenoxyalkanoic acids. In: Kearney, P.C. and D.D. Kaufman, eds., Herbicides chemistry, degradation and mode of action, 2nd ed., Vol. 1. New York, NY: Marcel Dekker. pp. 1–128.

Magnusson, J., C. Ramel and A. Eriksson. 1977. Mutagenic effects of chlorinated phenoxyacetic acids in *Drosophila melanogaster*. Hereditas. 87(1):121–123.

Majumdar, S.K. and J.K. Golia. 1974. Mutation test of 2,4,5-trichlorophenoxyacetic acid on *Drosophila melanogaster*. Can. J. Genet. Cytol. 16(2):465–466.

McCollister, S.B. and R.J. Kociba. 1970. Results of 90-day dietary feeding study on 2,4,5-trichlorophenoxyacetic acid. Unpublished study by Dow Chemical. MRID 00092151.

Meister, R., ed. 1988. Farm chemicals handbook. Willoughby, OH: Meister Publishing Company.

Muranyi-Kovacs, I., G. Rudali and J. Imbert. 1976. Bioassay of 2,4,5-trichlorophenoxyacetic acid for carcinogenicity in mice. Br. J. Cancer. 33:626–633.

NAS. 1977. National Academy of Sciences. Drinking water and health. Vol. 1. Washington, DC: National Academy Press.

Nelson, C.J., J.F. Holson, H.G. Green and D.W. Gaylor. 1979. Retrospective study of the relationship between agricultural use of 2,4,5-T and cleft palate occurrence in Arkansas. Teratology. 19:(3)377–384.

Neubert, D. and I. Dillmann. 1972. Embryotoxic effects in mice treated with 2,4,5-trichlorophenoxyacetic acid and 2,3,7,8-tetrachlorodibenzo-p-dioxin. Naunyn-Schmiedeberg's Arch. Pharmacol. 272:243–264.

NRCC. 1978. National Research Council of Canada. Phenoxy herbicides — their effects on environmental quality. NRC No. 16075. Ottawa, Canada: NRCC CNRC Publications. p. 440.

Ott, M.G., B.B. Holder and R.D. Olson. 1980. A mortality analysis of employees engaged in the manufacture of 2,4,5-trichlorophenoxyacetic acid. J. Occup. Med. 22(1):47–50.

Pershe, E.R. and J. Goss. 1979. Uses of powdered and granular activated carbon in water treatment. J. New Eng. Water Works Assoc. (9):254–286.

Piper, W.N., J.Q. Rose, M.L. Leng and P.J. Gehring. 1973. The fate of 2,4,5-trichlorophenoxyacetic acid (2,4,5-T) following oral administration to rats and dogs. Toxicol. Appl. Pharmacol. 26:339–351.

Rasmusson, B. and H. Svahlin. 1978. Mutagenicity tests of 2,4-dichlorophenoxyacetic acid and 2,4,5-trichlorophenoxyacetic acid in genetically stable and unstable strains of *Drosophila melanogaster*. Ecol. Bull. 27:190–192.

Rees, G.A.V. and L. Au. 1979. Use of XAD-2 macroreticular resin for the recovery of ambient trace levels of pesticides and industrial organic pollutants from water. Bull. Environ. Contam. Toxicol. 22(4/5):561–566.

Riihimaki, V., S. Asp and S. Hernberg. 1982. Mortality of 2,4-dichlorophenoxyacetic acid and 2,4,5-trichlorophenoxyacetic acid herbicide applicators in Finland: first report of an ongoing prospective cohort study. Scand. J. Work Environ. Health. 8(1):37–42.

Robeck, G.G., K.A. Dostal, J.M. Cohen and J.F. Kreissl. 1965. Effectiveness of water treatment processes in pesticide removal. J. Am. Water Works Assoc. (2):181–199.

Roll, R. 1971. Studies of the teratogenic effect of 2,4,5-T in mice. Food Cosmet. Toxicol. 9(5):671–676.

Rowe, V.K. and T.A. Hymas. 1954. Summary of toxicological information on 2,4-D and 2,4,5-T type herbicides and an evaluation of the hazards to livestock associated with their use. Am. J. Vet. Res. 15:622–629.

Shirasu, Y., M. Moriya, K. Kato, A. Furuhashi and T. Kada. 1976. Mutagenicity screening of pesticides in the microbial system. Mutat. Res. 40:19–30.

Smith, F.A., B.A. Schwetz, F.J. Murray, A.A. Crawford, J.A. John, R.J. Kociba and C.J. Humiston. 1978. Three-generation study of rats ingesting 2,4,5-trichlorophenoxyacetic acid in the diet. Abstract. Toxicol. Appl. Pharmacol. 45:293.

Smith, F.A., F.J. Murray, J.A. John, K.D. Nitschke, R.J. Kociba and B.A. Schwetz. 1981. Three-generation reproduction study of rats ingesting 2,4,5-trichlorophenoxyacetic acid in the diet. Food Cosmet. Toxicol. 19:41–45.

Sparschu, G.L., F.L. Dunn, R.W. Lisowe and V.K. Rowe. 1971. Study of the effects of high levels of 2,4,5-trichlorophenoxyacetic acid on fetal development in the rat. Food Cosmet. Toxicol. 9:527–530.

Standard Methods. 1985. Method 509B, chlorinated phenoxy acid herbicides. Standard Methods for the Examination of Water and Wastewater, 16th ed. American Public Health Association, American Water Works Association, Water Pollution Control Federation.

STORET. 1988. STORET Water Quality File. Office of Water. U.S. Environmental Protection Agency (data file search conducted in May, 1988).

Styles, J.A. 1973. Cytotoxic effects of various pesticides *in vivo* and *in vitro*. Mutat. Res. 21(1):50–51.

U.S. EPA. 1978. U.S. Environmental Protection Agency. Method for chlorophenoxy acid herbicides in drinking water. Methods for organochlorine pesticides and chlorophenoxy acid herbicides in drinking water and raw source water. Interim draft. Washington, DC: Office of Drinking Water. July.

U.S. EPA. 1985. U.S. Environmental Protection Agency. Health assessment document for polychlorinated dibenzo-p-dioxins. EPA 600/8-84/014F. Cincinnati, OH: Office of Health and Environmental Assessment.

U.S. EPA. 1986. U.S. Environmental Protection Agency. Guidelines for carcinogen risk assessment. Fed. Reg. 51(185):33992–34003. September 24.

Vettorazzi, G. and G.W. van den Hurk, eds. 1985. Pesticides reference index. J.M.P.R. p. 41.

Vogel, E. and J.L.R. Chandler. 1974. Mutagenicity testing of cyclamate and some pesticides in *Drosophila melanogaster*. Experientia. 30(6):621–623.

Vos, J.G., E.I. Krajnc, P.K. Beekhof and M.J. van Logten. 1983. Methods for testing immune effects of toxic chemicals: evaluation of the immunotoxicity of various pesticides in the rat. Pestic. Chem. Hum. Welfare Environ., Proc. 5th Int. Congr. Pestic. Chem. 3:497–504.

Windholz, M., S. Budavari, R.F. Blumetti and E.S. Otterbein, eds. 1983. The Merck index—an encyclopedia of chemicals and drugs, 10th ed. Rahway, NJ: Merck and Company, Inc.

Zack, J.A. and R.R. Suskind. 1980. The mortality experience of workers exposed to tetrachlorodibenzodioxin in a trichlorophenol process accident. J. Occup. Med. 22(1):11–14.

Zetterberg, G. 1978. Genetic effects of phenoxy acids on microorganisms. Ecol. Bull. 27:193–204.

Trifluralin

I. General Information and Properties

A. CAS No. 1582-09-8

B. Structural Formula

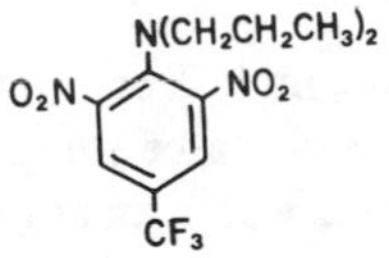

$$N(CH_2CH_2CH_3)_2$$

$$O_2N \quad\quad NO_2$$

$$CF_3$$

alpha, alpha, alpha-Trifluoro-2,6-dinitro-N,N-dipropyl-p-toluidine

C. Synonyms

- 2,6-Dinitro-N, N-dipropyl-4-trifluoromethylaniline; Agreflan; Crisalin; Treflan; L-36352; Trifluralin (U.S. EPA, 1985b,c).

D. Uses

- Trifluralin is a selective herbicide (preemergent) for control of annual grasses and broad-leafed weeds. It is applied to soybean, cotton and vegetable crops; fruit and nut trees, shrubs; and roses and other flowers. It is also used on golf courses, rights-of-way and domestic outdoor and industrial sites (U.S. EPA, 1985c).

E. Properties (Meister, 1983; U.S. EPA, 1985c)

Chemical Formula	$C_{13}H_{16}F_3N_3O_4$
Molecular Weight	335.2
Physical State (25°C)	Orange, crystalline solid
Boiling Point	139 to 140°C
Melting Point	46 to 49°C
Density	--
Vapor Pressure (25°C)	1.1×10^{-4} mm Hg
Specific Gravity	--
Water Solubility (25°C)	>1 mg/L

Log Octanol/Water
    Partition Coefficient   4.69
Taste Threshold   --
Odor Threshold   --
Conversion Factor   --

F. Occurrence

- Trifluralin is not a strong potential ground-water contaminant due to its strong adsorption to soil and negligible leaching (U.S. EPA, 1985c).

- Trifluralin has been detected in finished drinking water supplies (NAS, 1977).

- Trifluralin has been found in 172 of 2,047 surface water samples analyzed and in 1 of 507 ground-water samples (STORET, 1988). Samples were collected at 249 surface water locations and 386 ground-water locations, and trifluralin was found in 7 states. The 85th percentile of all nonzero samples was 0.54 μg/L in surface water. This information is provided to give a general impression of the occurrence of this chemical in ground and surface waters as reported in the STORET database. The individual data points retrieved were used as they came from STORET and have not been confirmed as to their validity. STORET data are often not valid when individual numbers are used out of the context of the entire sampling regime, as they are here. Therefore, this information can only be used to form an impression of the intensity and location of sampling for a particular chemical.

G. Environmental Fate

- Trifluralin at 5 ppm degraded with 15% of the applied trifluralin lost after 20 days in a silt loam soil (aerobic metabolism) study (Parr and Smith, 1973). The samples were incubated in the dark at 25°C and 0.33 bar moisture.

- Trifluralin, applied alone or in combination with chlorpropham or chlorpropham plus PPG-124, dissipated with a half-life of 42 to 84 days in sandy loam or silt loam soil incubated at 72 to 75°F and 18% moisture content under laboratory conditions (Maliani, 1976).

- In an anaerobic soil metabolism study, trifluralin at 5 ppm degraded in nonsterile silt loam soil, with less than 1% of applied trifluralin detected after 20 days of incubation (0.33 bar moisture in the dark at 25°C; anaerobicity was maintained with nitrogen gas). Autoclaving and flooding the soil decreased the degradation rate of the compound (Parr and Smith, 1973).

- ^{14}C-Trifluralin at 1.1 kg/ha was relatively immobile in sand, sandy loam, silt, loam and clay loam soil columns (30-cm height) eluted with 60 cm of water, with more than 90% of the applied radioactivity remaining in the top 0- to 10-cm segment (Gray et al., 1982).

- Trifluralin concentrations in runoff (water/sediment suspensions) were less than 0.04% of the applied amount for 3 consecutive years following treatment at 1.4 kg/ha and 13 to 27 cm of rainfall (Willis et al., 1975). The field plots (silty clay loam soil, 0.2% slope) were planted with cotton or soybeans.

- In the field, ^{14}C-trifluralin (99% pure) at 0.84 to 6.72 kg/ha dissipated in the top 0- to 0.5-cm layer of a silt loam soil, with 14, 4 and 1.5% of the applied amount remaining 1, 2 and 3 years, respectively, after application (Golab et al., 1978). Approximately 30 minor degradates were identified and quantified; none represented more than 2.8% of the applied amount. Trifluralin (4 lb/gal EC) at 0.75 and 1.5 lb/A dissipated in a medium loam soil, with 20 and 32%, respectively, of the applied remaining 120 days after treatment (Helmer et al., 1969; Johnson, 1977).

- Trifluralin (4 lb/gal EC) dissipated from a sandy loam soil treated at 1.0 lb active ingredient (a.i.)/A, with a half-life of 2 to 4 months (Miller, 1973).

- Trifluralin was detected in 107 soil samples taken nationwide at less than 0.01 to 0.98 ppm in fields treated with trifluralin at various rates for 1, 2, 3 or 4 consecutive years (Parka and Tepe, 1969).

- Trifluralin was detected in 12% of the soil samples taken from 80 sites in 15 states in areas considered to be regular pesticide-use areas based on available pesticide-use records (Stevens et al., 1970). Concentrations detected in soils ranged from less than 0.01 to 0.48 ppm. Trifluralin residues were detected in only 3.5% of the 1,729 agricultural soils sampled in 1969 (Wiersma et al., 1972).

- Trifluralin was detected at a maximum concentration of 0.25 ppm. Residues of volatile nitrosamines (dimethylnitrosoamine, N-nitrosodipropylamine, or N-butyl-N-ethyl-N-nitrosoamine) were not detected in water samples taken from ponds and wells located in or near fields that had been treated with trifluralin at various rates (Day et al., 1977).

II. Pharmacokinetics

A. Absorption

- Emmerson and Anderson (1966) indicated that trifluralin is not readily absorbed from the gastrointestinal (GI) tract and that the fraction that is absorbed is completely metabolized. Of an orally administered dose (100 mg/kg), only 11 to 14% was excreted in the bile after 24 hours, indicating low enterohepatic circulation.

B. Distribution

- No information was found in the available literature on the distribution of trifluralin.

C. Metabolism

- Four metabolites of trifluralin were identified in rats. Twelve rats were given 100 mg/kg $^{14}CF_3$-trifluralin in corn oil by gavage for 2 weeks. The metabolites, identified by thin-layer chromatography, were produced by removal of both propyl groups or dealkylation and reduction of a nitro group to an amine (Emmerson and Anderson, 1966).

- An *in vitro* study using rat hepatic microsomes indicated that trifluralin undergoes aliphatic hydroxylation of the N-alkyl substituents, N-dealkylation and reduction of a nitro group (Nelson et al., 1976).

- There are insufficient data to characterize the general metabolism of trifluralin in animals (U.S. EPA, 1986a).

D. Excretion

- Rats given an oral dose (100 mg/kg) of $^{14}CF_3$-trifluralin excreted virtually all of the dose within 3 days. The radioactivity was excreted during the first 24 hours. Approximately 78% of the dose was eliminated in the feces and 22% in the urine (Emmerson and Anderson, 1966).

III.   Health Effects

A. Humans

1. *Short-term Exposure*

- The Pesticide Incident Monitoring System database indicated 105 incident reports involving trifluralin from 1966 to April of 1981. Of the 105 reports, 49 cases involved humans exposed to trifluralin alone. Twenty-seven cases involved human exposure to mixtures containing trifluralin. The remaining incidents involved nonhuman exposures (U.S. EPA, 1981a).

- Among reports of human exposure to trifluralin alone, one fatality was reported. A 9-year-old girl suffered cardiac arrest following the ingestion of an unknown amount of trifluralin (U.S. EPA, 1981a).

- Verhalst (1974, as cited in U.S. EPA, 1985b) reported that the symptoms observed in trifluralin poisonings appeared to be related to the solvent used (e.g., acetone or xylene) rather than trifluralin itself.

2. *Long-term Exposure*

- The majority of reported trifluralin exposure cases were occupational in nature. Trifluralin exposure has resulted in dermal and ocular irritation in humans. Other reported symptoms include respiratory involvement, abdominal cramps, nausea, diarrhea, headache, lethargy and paraesthesia following dermal or inhalation exposure. Specific exposure levels or durations were not reported (U.S. EPA, 1981a).

B. Animals

1. *Short-term Exposure*

- The acute oral toxicity of trifluralin is low. The following oral LD_{50} values have been reported: mice >5 g/kg; rats >10 g/kg; dogs, rabbits and chickens >2 g/kg (Meister, 1983; RTECS, 1985).

- An inhalation LC_{50} value of a liquid formulation containing 41% trifluralin (species not stated) of >2.44 mg/L/hour was reported (U.S. EPA, 1985d). No other information was available.

2. *Dermal/Ocular Effects*

- The results of a primary dermal-irritation study in the rabbit were negative at 72 hours following application of a 41.2% trifluralin solution (U.S. EPA, 1985d).

- Treflan, containing 10% trifluralin, was tested for sensitization in female guinea pigs. A dose of 50 mg was applied to the skin of 12 animals, three times a week for 2 weeks. No dermal irritation or contact sensitization developed during this time (Negelski et al., 1984).

- In a similar study, a 95% technical trifluralin solution was shown to be a potential skin sensitizer in guinea pigs using the Buehler topical-patch method (U.S. EPA, 1985d).

- A 14-day study in which rabbits were exposed to 2 mL/kg trifluralin topically produced diarrhea and slight dermal erythema in exposed animals. No other effects were reported (ELANCO, 1979).

- Technical-grade trifluralin applied as a powder to rabbit eyes was reported as nonirritating. Slight conjunctivitis developed but cleared within a week (U.S. EPA, 1985d).

- When applied as a 41.2% liquid to rabbit eyes, trifluralin produced corneal opacity that cleared in 7 days (U.S. EPA, 1985d).

3. *Long-term Exposure*

- In a modified subacute study, female Harlan Wistar-derived rats were given 0, 0.05, 0.1 or 0.2% (0, 500, 1,000 or 2,000 ppm) trifluralin in their diet for 3 months. Assuming that 1 ppm in the diet of rats equals 0.05 mg/kg/day (Lehman, 1959), these levels correspond to doses of 0, 25, 50 and 100 mg/kg/day. Physical appearance, behavior, body and organ weights, mortality and clinical chemistries were monitored in progeny from 10 females. No significant effects were observed in survival or appearance. Liver weights in progeny continuously fed diets of 0.1% and 0.2% trifluralin were increased over those of control animals. The study identified a No-Observed-Adverse-Effect Level (NOAEL) in progeny of 0.05% (25 mg/kg) trifluralin (Worth et al., 1977)

- In a 90-day study, male Fischer 344 rats were fed trifluralin at dietary levels of 0 (n = 60), 0.005% (n = 60), 0.02% (n = 45), 0.08% (n = 45), 0.32% (n = 45) and 0.64% (n = 45). These concentrations are equivalent to dietary levels of 0, 50, 200, 800, 3,200 and 6,400 ppm trifluralin, respectively (Emmerson et al., 1985). Assuming that 1 ppm in the diet of a rat equals 0.05 mg/kg/day (Lehman, 1959), these levels correspond to doses of 0, 2.5, 10, 40, 160 and 320 mg/kg/day. After 90 days, alpha-1, alpha-2 and beta-globulin levels were significantly increased in treatment groups receiving 10 mg/kg/day or greater. Other effects included increased aspartate transaminase, urinary calcium, inorganic phosphorus and magnesium at levels ≥ 160 mg/kg/day. A NOAEL of 2.5

mg/kg/day and a Lowest-Observed-Adverse-Effect Level (LOAEL) of 10 mg/kg/day can be identified from this study.

- Sixty weanling Harlan rats were fed 0, 20, 200, 2,000 or 20,000 ppm trifluralin in the diet for 729 days (24 months). Assuming that 1 ppm in the diet of a rat equals 0.05 mg/kg (Lehman, 1959), these concentrations correspond to doses of 0, 1, 10, 100 or 1,000 mg/kg/day. No significant effects were observed in growth rate, mortality or food consumption of treated animals at the three lower dose levels. Animals in the highest dose group (1,000 mg/kg/day) were significantly smaller than controls and ranked lower in food consumption. No effects on hematology were noted. Animals in the high-dose group displayed a slight proliferation of the bile ducts. No other histopathological effects were observed. A NOAEL of 2,000 ppm (100 mg/kg/day) was reported (Worth et al., 1966c).

- In a 2-year chronic carcinogenicity study with Fischer 344 rats, doses greater than 128 mg/kg/day in males and 154 mg/kg/day in females were reported to produce overt toxicity. Groups of 60 animals/sex/dose were fed dietary levels of 0.08, 0.3 or 0.65% (30, 128 or 272 mg/kg/day for males, and 37, 154 or 336 mg/kg/day for females) trifluralin. Body weights of the high-dose groups were significantly decreased in both sexes. This may be related to the decreased food consumption observed in those groups. Increased blood urea nitrogen (BUN) levels and increased liver and testes weights were noted in the two high-dose groups. Kidney and heart weights were significantly decreased in females in the 0.3%- and 0.65%-trifluralin groups. Other noncarcinogenic effects included decreased hemoglobin values and erythrocyte counts in both sexes of the high-dose group (Emmerson and Pierce, 1980). This study appears to identify a NOAEL of 0.08% trifluralin (30 to 37 mg/kg/day).

- B6C3F$_1$ mice (40/sex/group) were exposed to dose levels of 40, 180 or 420 mg/kg/day trifluralin in the diet for 2 years. Animals exposed to the two higher levels exhibited decreased body weight and renal toxicity. Other noncarcinogenic effects included decreased erythrocytic and leukocytic values in the high-dose group, increased BUN and alkaline phosphatase levels in the 180- and 420-mg/kg/day group, decreased kidney weights in the two higher-dose groups and decreased spleen and uterine weights with increased liver weights in the high-dose group (Emmerson and Owen, 1980). No effects were noted at the low-dose level (40 mg/kg/day).

- Occasional emesis and increased liver-to-body weight ratios were observed in dogs (3/sex/dose) fed 25 mg/kg/day trifluralin for 3 years. No adverse effects were observed in animals fed 10 mg/kg/day (Worth, 1970). An intermediate dose was not tested.

4. *Reproductive Effects*

- In a four-generation reproduction study (Worth et al., 1966b), rats were given 0, 200 or 2,000 ppm trifluralin in the diet (0, 10 or 100 mg/kg/day). A reproductive NOAEL of 200 ppm (10 mg/kg/day) was identified. The number of animals used in the study was not reported. However, a review of this study (U.S. EPA, 1985d) indicated that an insufficient number of animals were used and that several other deficiencies in the study may have compromised the integrity of the results.

- In a 3-year feeding study in dogs, a NOAEL of 10 mg/kg/day was identified in adults (Worth et al., 1966b). Dogs (2/sex/dose at 400 ppm and 3/sex/dose at 1,000 ppm) were given 10 or 25 mg/kg/day trifluralin in the diet. When bred after 2 years of exposure, no differences in litter size, survival or growth of the pups were reported. An occasional emesis and increased liver weights were reported in adults in the 25-mg/kg/day group.

5. *Developmental Effects*

- Female rabbits (number not specified) were fed 0, 100, 225, 500 or 800 mg/kg/day by gavage during pregnancy (ELANCO, 1984). No adverse reproductive effects were observed at the two lower dose levels. The 500 and 800 mg/kg/day levels resulted in anorexia, aborted litters and decreased live births. The NOAEL for maternal effects was identified as 225 mg/kg/day.

- Rabbits (number not specified) exposed to 225 or 500 mg/kg/day trifluralin during pregnancy exhibited anorexia and cachexia and aborted litters (U.S. EPA, 1985d). Fetotoxicity, as evidenced by decreased fetal weight and size, was observed at the high-dose level. These effects were not observed at the 100-mg/kg/day level.

- In a rabbit teratology study, a total of 32 mated females were given 225, 450 or 1,000 mg/kg/day trifluralin by gavage (Worth et al., 1966a). Animals were dosed until the 25th day of gestation and then sacrificed. Does in the 1,000 mg/kg/day group weighed slightly less than controls. Two fetuses were found to be underdeveloped in the high-dose group; however, this was not considered by the investigators to be treatment related. Average litter size and weight were not significantly affected.

The authors reported that their results identified a safe level of 1,000 mg/kg/day.

- Rabbit does (number per group not specified) were given 100, 225, 500 or 800 mg/kg/day trifluralin by gavage during pregnancy (ELANCO, 1984). The 500 and 800 mg/kg/day levels resulted in decreased live births, cardiomegaly and wavy ribs in the progeny. No effects on progeny were observed at 225 mg/kg/day or less (ELANCO), 1984).

6. *Mutagenicity*

- Anderson et al. (1972) reported that trifluralin did not induce point mutations in any of the three microbial systems tested. No further details were provided in the review.

- Trifluralin was tested for genotoxicity in several *in vivo* and *in vitro* systems (ELANCO, 1983). No reverse mutations were observed in *Salmonella typhimurium* or *Escherichia coli* when incubated with 25 to 400 mg trifluralin/plate without activation; trifluralin was also negative when tested at levels of 50 to 800 mg/plate with activation. Negative results were obtained in mouse lymphoma L5178Y TK$^+$ cells incubated with 0.5 to 20 μg/mL trifluralin with and without activation. An *in vivo* sister-chromatid exchange study in Chinese hamster ovary (CHO) cells following exposure to 500 mg/kg trifluralin was also negative.

7. *Carcinogenicity*

- NCI (1978) conducted bioassays on B6C3F$_1$ mice and Osborne-Mendel rats using technical-grade trifluralin (which contained 84 to 88 ppm of the contaminant dipropylnitrosamine). Two dietary levels were used in each bioassay. Mice (50/sex/group) were exposed to trifluralin at dose levels of 2,000 or 3,744 ppm (males) or 2,740 or 5,192 ppm (females) for 78 weeks and observed for an additional 13 weeks after exposure. A significant dose-related increase in hepatocellular carcinomas was observed in female mice (0/20 control, 12/47 low dose, 21/44 high dose). An increased incidence of alveolar/bronchiolar adenomas was also observed (0/19 control, 6/43 low dose, 3/30 high dose) in female mice. Squamous cell carcinomas in the forestomach of a few treated female mice were also observed. Although the incidence of squamous cell carcinomas in the forestomach was not statistically significant when compared to pooled and matched controls, NCI deemed this finding to be treatment related, since it was an unusual type of lesion. Male mice were not significantly affected by trifluralin exposure.

- Rats (50/sex/group) were exposed to two levels of trifluralin in the feed (4,125 or 8,000 ppm for males; 4,125 or 7,917 ppm for females) for 78 weeks followed by a 33-week observation period (NCI, 1978). Assuming 1 ppm in the diet of rats equals 0.05 mg/kg/day (Lehman, 1959), these doses correspond to approximately 206 or 400 mg/kg/day. Several neoplasms were observed and compared to pooled and matched controls. These neoplasm types were reported to occur spontaneously in the Osborne-Mendel strain and were not considered treatment related by NCI.

- In a 2-year feeding study, B6C3F$_1$ mice were given 0, 563, 2,250 or 4,500 ppm trifluralin. These doses correspond to 0, 40, 180 or 420 mg/kg/day (Lehman, 1959) in the diet (Emmerson and Owen, 1980). Levels of a nitrosamine contaminant of trifluralin, NDPA, were below the 0.01-ppm analytical detection limit. A total of 40 animals/sex/treatment group and 60/sex for controls was used. At the lowest dose level, 40 mg/kg/day, no adverse effects were observed in either sex. Decreased body weight and renal effects were noted in mice in the mid- and high-dose groups. Pathology revealed progressive glomerulonephritis in females of the high-dose group. Hepatocellular hyperplasia and hypertrophy were also observed in the treated mice. The specific dose level was not reported. No evidence of increased incidence or decreased latency for any type of neoplasm was found in any of the mice.

- Trifluralin was administered to Fischer 344 rats (60/sex/group) at dose levels of 813, 3,250 or 6,500 ppm (corresponding to 30, 128 or 272 mg/kg/day for males and 37, 154 or 336 mg/kg/day for females) in the diet for 2 years (Emmerson and Pierce, 1980). A significant increase in malignant renal neoplasms and thyroid tumors in male rats and in neoplasms of the bladder in both sexes was reported. A high incidence (20/30) of renal calculi was also observed in animals in the high-dose groups.

IV.   Quantification of Toxicological Effects

A. One-day Health Advisory

No information was found in the available literature that was suitable for determination of the One-day HA value for trifluralin. Therefore, it is recommended that a modified DWEL (0.03 mg/L, calculated below) for a 10-kg child be used as a conservative estimate for the One-day HA value.

For a 10-kg child, the adjusted DWEL is calculated as follows:

$$\text{DWEL} = \frac{(0.0025 \text{ mg/kg/day}) (10 \text{ kg})}{1 \text{ L/day}} = 0.025 \text{ mg/L } (30 \text{ }\mu\text{g/L})$$

where: 0.0025 mg/kg/day = Rfd (see Lifetime Health Advisory Section).

B. Ten-day Health Advisory

No information was found in the available literature that was suitable for determination of the Ten-day HA value for trifluralin. It is therefore recommended that a modified DWEL (0.03 mg/L) for a 10-kg child be used as a conservative estimate for the Ten-day HA value.

C. Longer-term Health Advisory

No information was found in the available literature that was suitable for determination of the Longer-term HA value for trifluralin. It is, therefore, recommended that a modified DWEL (0.03 mg/L) for a 10-kg child be used as a conservative estimate for a Longer-term exposure for a child and the DWEL (0.1 mg/L, calculated below) be used for the adult.

D. Lifetime Health Advisory

The Emmerson et al. (1985) study has been selected to serve as the basis for the Lifetime HA value for trifluralin. Fischer 344 rats were fed diets containing 0.005, 0.02, 0.08, 0.32 or 0.64% trifluralin (2.5, 10, 40, 160 or 320 mg/kg/day) for 90 days. Significant increases in urinary alpha-1, alpha-2 and betaglobulins were observed in animals receiving 10 mg/kg/day or greater. A NOAEL of 2.5 mg/kg/day was identified. Other longer-term studies report NOAELs at higher doses.

Using a NOAEL of 2.5 mg/kg/day, the Lifetime HA is calculated as follows:

Step 1: Determination of the Reference Dose (RfD)

$$\text{RfD} = \frac{(2.5 \text{ mg/kg/day})}{(1,000)} = 0.0025 \text{ mg/kg/day } (0.003 \text{ mg/kg/day})$$

Step 2: Determination of the Drinking Water Equivalent Level (DWEL)

$$\text{DWEL} = \frac{(0.003 \text{ mg/kg/day}) (70 \text{ kg})}{(2 \text{ L/day})} = 0.105 \text{ mg/L } (100 \text{ }\mu\text{g/L})$$

Step 3: Determination of the Lifetime Health Advisory

$$\text{Lifetime HA} = \frac{(0.105 \text{ mg/L}) (20\%)}{10} = 0.0021 \text{ mg/L} (2 \text{ } \mu\text{g/L})$$

E. Evaluation of Carcinogenic Potential

- Applying the criteria described in EPA's guidelines for assessment of carcinogenic risk (U.S. EPA, 1986b), trifluralin may be classified in Group C: possible human carcinogen. This category is used for substances that show limited evidence of carcinogenicity in animals and inadequate evidence in humans.

- In an NCI (1978) study of female B6C3F$_1$ mice, significant dose-related increases in hepatocellular carcinomas and alveolar adenomas were observed when the animals were exposed to 33 or 62 mg/kg/day trifluralin in the diet for 78 weeks. The trifluralin used in this study contained 84 to 88 ppm dipropylnitrosamine. Male rats, when exposed to 30, 128 or 272 mg/kg/day trifluralin in the diet for 2 years, exhibited significant increases in the incidences of kidney, urinary bladder and thyroid tumors (Emmerson and Pierce, 1980).

- The evidence from the Emmerson and Pierce (1980) and NCI (1978) studies indicates that trifluralin has carcinogenic potential. Based on the results of the Emmerson and Pierce (1980) study, the U.S. EPA Carcinogen Assessment Group (CAG) has prepared a quantitative risk estimate of trifluralin exposure (U.S. EPA, 1981b). The CAG estimated a potency factor (q_1*) of 7.66×10^{-3} mg/kg/day based on the combined incidence of tumors in male rats. Assuming that a 70-kg human adult consumes 2 liters of water per day over a 70-year lifespan, the estimated cancer risk would be 10^{-4}, 10^{-5} and 10^{-6} at concentrations of 500, 50 and 5 μg/L, respectively.

V.   Other Criteria, Guidance and Standards

- Residue tolerances ranging from 0.05 to 2.0 ppm trifluralin have been established for a variety of agricultural commodities (U.S. EPA, 1985a).

- NAS (1977) has calculated an Acceptable Daily Intake (ADI) of 0.1 mg/kg bw/day with a Suggested-No-Adverse-Response-Level (SNARL) of 700 μg/L.

VI. Analytical Methods

- Determination of trifluralin is by Method 508 (U.S. EPA, 1987). In this procedure, a measured volume of sample of approximately 1 liter is solvent-extracted with methylene chloride by mechanical shaking in a separatory funnel or mechanical tumbling in a bottle. The methylene chloride extract is isolated, dried and concentrated to a volume of 5 mL after solvent substitution with methyl tert-butyl ether (MTBE). Chromatographic conditions are described that permit the separation and measurement of the analytes in the extract by GC with an electron capture detector (ECD). An alternative manual liquid-liquid extraction method using separatory funnels is also described. This method has been validated in a single laboratory, and the estimated detection limit determined for the analytes in this method, including trifluralin, is 0.025 μg/L.

VII. Treatment Technologies

- Available data indicate that reverse osmosis (RO), granular-activated carbon (GAC) adsorption, conventional treatment and possibly air stripping will remove trifluralin from water.

- U.S. EPA investigated the amenability of a number of compounds, including trifluralin, to removal by GAC. No system performance data were given.

- Conventional water treatment techniques of coagulation with alum, sedimentation and filtration proved to be 100% effective in removing trifluralin from contaminated water (Nye, 1984).

- Sanders and Seibert (1983) determined experimentally water solubility, vapor pressure, Henry's Law constant and volatilization rates for trifluralin; 100% of the compound volatilized under laboratory conditions.

- Treatment technologies for the removal of trifluralin from water are available and have been reported to be effective. However, selection of an individual technology or combination of technologies to attempt trifluralin removal from water must be based on a case-by-case technical evaluation, and an assessment of the economics involved.

VIII.   References

Anderson, K.J., E.G. Leighty and M.T. Takahashi. 1972. Evaluation of herbicides for possible mutagenic properties. J. Agric. Food Chem. 20:649–656. Cited in U.S. EPA, 1985b.

Day, E., S. West and M. Amundson. 1977. Residues of volatile nitrosamines in water samples from fields treated with Treflan: pre-RPAR Review submission #8. Unpublished study. Submitted by ELANCO Products Company, Division of Eli Lilly and Company, Indianapolis, IN, to the Office of Pesticide Programs.

ELANCO. 1979.* Acute hazard evaluation of Treflan 4EC, Lot X-27572, including oral, dermal, ocular and inhalation studies. Eli Lilly and Company.

ELANCO. 1983.* Genetic toxicology studies with trifluralin (compound 36352). MRID 00126659. Eli Lilly and Company.

ELANCO. 1984.* Teratology study in rabbits. Cited in U.S. EPA, 1985b. Eli Lilly and Company.

Emmerson, J.L. and R.C. Anderson. 1966. Metabolism of trifluralin in the rat and dog. Toxicol. Appl. Pharmacol. 9:84–97.

Emmerson, J.L. and N.V. Owen. 1980.* The chronic toxicity of compound 36352 (trifluralin) given as a component of the diet to the B6C3F$_1$ mouse for 24 months. MRID 00044338.

Emmerson, J.L. and E.C. Pierce. 1980.* The chronic toxicity of compound 36352 (trifluralin) given as a component of the diet to Fischer 344 rats for two years. MRID 00044337.

Emmerson, J.L., W.H. Jordan and R.W. Usher. 1985.* Special urinalysis study in Fischer 344 rats maintained on diets containing trifluralin (compound 36352) for three months. Eli Lilly and Company. EPA Accession No. 261912.

Golab, T., W. Althaus and H. Wooten. 1978. Fate of ^{14}C-trifluralin in soil. Unpublished study. Submitted by ELANCO Products Company, Division of Eli Lilly and Company, Indianapolis, IN.

Gray, J.E., A. Loh, R.F. Sieck et al. 1982. Laboratory leaching of ethylfluralin. Unpublished study. Submitted by ELANCO Products Company, Division of Eli Lilly and Company, Indianapolis, IN.

Helmer, J.D., W.S. Johnson and T.W. Waldrep. 1969. Experiment No. WB(F)9-132: Soil persistence data. Unpublished study. Submitted by ELANCO Products Company, Division of Eli Lilly and Company, Indianapolis, IN.

Johnson, W. 1977. Determination of trifluralin in agricultural crops and soil. Procedure No. 5801616. Unpublished study. Submitted by ELANCO Products Company, Division of Eli Lilly and Company, Indianapolis, IN.

Lehman, A.J. 1959. Appraisal of the safety of chemicals in foods, drugs and cosmetics. Assoc. Food Drug Off. U.S., Q. Bull.

Maliani, N. 1976. CIPC and CIPC + PPG-124 interaction study (Exhibit E). Laboratory No. 97021. Unpublished study. Prepared by Morse Laboratories, Inc. Submitted by PPG Industries, Barberton, OH.

Meister, R., ed. 1983. Farm chemicals handbook. Willoughby, OH: Meister Publishing Company.

Miller, J.H. 1973. Residue Report AGA 2527−2nd Report Project No. 120002.

Mosier, J. and D. Saunders. 1978. A hydrolysis study on the herbicide trifluralin. Unpublished study. Submitted by ELANCO Products Company, Division of Eli Lilly and Company, Indianapolis, IN.

NAS. 1977. National Academy of Sciences. Drinking water and health, Vol. I. Washington, DC: National Academy Press.

NCI. 1978. National Cancer Institute. Bioassay of trifluralin for possible carcinogenicity. NCI-CG-TR-34.

Negelski, D., J.L. Emmerson, B. Arthur et al. 1984.* Guinea pig sensitization study of treflan 10G, a granular formulation (FN-1199) containing 10% trifluralin. MRID 00137468.

Nelson, J.O., P.C. Kearney, J.R. Plimmer and P.E. Menzer. 1976. Metabolism of trifluralin, profluralin and fluchloralin by rat liver microsomes. Pest. Biochem. Phys. 7:73–82. Cited in U.S. EPA, 1985b.

Nye, J.C. 1984. Treating pesticide-contaminated wastewater: development and evaluation of a system. American Chemical Society.

Parka, S. and J. Tepe. 1969. The disappearance of trifluralin from field soils. Weed Sci. 17(1):119–122.

Parr, J.F. and S. Smith. 1973. Degradation of trifluralin under laboratory conditions and soil anaerobiosis. Soil Sci. 115(1):55–63.

RTECS. 1985. Registry of Toxic Effects of Chemical Substances. Washington, DC: National Institute of Occupational Safety and Health.

Sanders, P.F. and J.N. Seibert. 1983. A chamber for measuring volatilization of pesticides from model soil and water disposal systems. Chemosphere. 12(7/8):999–1012.

Stevens, L., C. Collier and D. Woodham. 1970. Monitoring pesticides in soils from areas of regular, limited, and no pesticide use. Pestic. Monit. J. 4(3):145–166.

STORET. 1988. STORET Water Quality File. Office of Water. U.S. Environmental Protection Agency (data file search conducted in May, 1988).

U.S. EPA. 1981a. U.S. Environmental Protection Agency. Summary of reported incidents involving trifluralin. Pesticide Incident Monitoring System. Report No. 441. Washington, DC: Office of Pesticide Programs.

U.S. EPA. 1981b. U.S. Environmental Protection Agency. Carcinogenic potency of trifluralin, including N-nitroso-di-n-propylamine (NDPA) and diethylitrosamine (DENA). Memo from Chao Chen and Bernard Haberman to Marcia Williams. July 29.

U.S. EPA. 1985a. U.S. Environmental Protection Agency. Code of Federal Regulations. 40 CFR 180.201.

U.S. EPA. 1985b. U.S. Environmental Protection Agency. Pesticide survey chemical profile. Final report. Contract No. 68-01-6750. Washington, DC: Office of Drinking Water.

U.S. EPA. 1985c. U.S. Environmental Protection Agency. Post phase II registration standard support team meeting for trifluralin. I. Regulatory position and rationale. Memo from Robert Ikeda to Registration Standard Support Team. September 4.

U.S. EPA. 1985d. U.S. Environmental Protection Agency. Trifluralin registration standard: toxicology chapter. Memo from Roland Gessert to Richard Mountfort. June 25.

U.S. EPA. 1986a. U.S. Environmental Protection Agency. Draft guidance for the registration of pesticide products containing trifluralin.

U.S. EPA. 1986b. U.S. Environmental Protection Agency. Guidelines for carcinogen risk assessment. 51 FR 33992. September 24.

U.S. EPA. 1987. U.S. Environmental Protection Agency. Method 508 — Determination of chlorinated pesticides in water by gas chromatography with an electron capture detector. Available from EPA's Environmental Monitoring and Support Laboratory, Cincinnati, OH.

Verhalst, H. 1974. Personal communication to Eli Lilly and Company. Cited in U.S. EPA, 1985b.

Whittaker, K.F. et al. 1982. Collection and treatment of wastewater generated by pesticide applicators. EPA 600/2-82-028.

Wiersma, G.B., H. Tai and P.F. Sand. 1972. Pesticide residue levels in soils, FY 1969 — National Soils Monitoring Program. Pestic. Monit. J. 6(3):194-201.

Willis, G.H., R.L. Rogers and L.M. Southwick. 1975. Losses of diuron, linuron, fenac, and trifluralin in surface drainage water. J. Environ. Qual. 4(3):399-402.

Worth, H.M. 1970. The toxicological evaluation of benomyl and trifluralin. Pesticide Symposia 6th Conference, August, 1966. Miami, FL: Halos and Assoc. pp. 263-267.

Worth, H.M., E.R. Adams, J.K. Markham, N.V. Owen, S.S. Young and D.M. Morton. 1977.* A modified subacute toxicity study with trifluralin. MRID 00134326.

Worth, H.M., R.M. Small, W.R. Gibson and E.C. Pierce. 1966a.* Teratology studies with trifluralin. MRID 00083647.

Worth, H.M., R.M. Small, W.R. Gibson, W.J. Griffing, E.C. Pierce and

P.N.Harris. 1966b.* Effects of trifluralin treatment on reproduction in rats and dogs. MRID 00083646.

Worth, H.M., R.M. Small, P.N. Harris and R.C. Anderson. 1966c.* Chronic toxicity studies with trifluralin. MRID 00076447.

---

* Confidential Business Information submitted to the Office of Pesticide Programs.